Covalent Organic Frameworks

Covalent organic frameworks-based nanomaterials have emerged as promising candidates for energy applications owing to their superior electrochemical properties, surface area, nano-device integration, multifunctionality, printability, and mechanical flexibility. This book provides fundamentals, various synthesis approaches, and applications of covalent organic frameworks-based nanomaterials and their composites for generating energy.

The main objective of this book is to provide current, state-of-the-art knowledge about covalent organic frameworks-based nanomaterials and their composites for supercapacitors, batteries, photovoltaics, and fuel cells, covering almost the entire spectrum in the energy field under one title. Aimed at widening fundamental understanding about covalent organic frameworks and mechanisms for realization and advancement in devices with improved energy efficiency and high storage capacity, this book will provide new directions for scientists, researchers, and students to better understand the principles, technologies, and applications of covalent organic frameworks.

Covalent Organic Frameworks
Chemistry, Properties, and Energy Applications for a Sustainable Future

Edited by
Tuan Anh Nguyen and Ram K. Gupta

CRC Press
Taylor & Francis Group
Boca Raton London New York

CRC Press is an imprint of the
Taylor & Francis Group, an **informa** business

CRC Press
Boca Raton and London
First edition published 2023
by CRC Press
6000 Broken Sound Parkway NW, Suite 300, Boca Raton, FL 33487-2742
and by CRC Press

4 Park Square, Milton Park, Abingdon, Oxon, OX14 4RN

CRC Press is an imprint of Taylor & Francis Group, LLC

© 2023 selection and editorial matter, Tuan Anh Nguyen and Ram K. Gupta; individual chapters, the contributors

ISBN: 9781032069883 (hbk)
ISBN: 9781032073514 (pbk)
ISBN: 9781003206507 (ebk)

DOI: 10.1201/9781003206507

Typeset in Palatino
by KnowledgeWorks Global Ltd.

Contents

Contributors

Hani Nasser Abdelhamid
Advanced Multifunctional Materials
  Laboratory
Department of Chemistry, Faculty of
  Science
Assiut University
Assiut, Egypt

Proteomics Laboratory for Clinical
  Research and Materials Science
Department of Chemistry
Assiut University
Assiut, Egypt

Kayode Adesina Adegoke
Department of Chemical Sciences
University of Johannesburg
Doornfontein, South Africa

Mohaddeseh Afshari
Department of Chemistry
Isfahan University of Technology
Isfahan, Islamic Republic of Iran

Syed Shoaib Ahmad Shah
Hefei National Laboratory for Physical
  Sciences at the Microscale
CAS Key Laboratory of Soft Matter
  Chemistry
Department of Chemistry
University of Science and Technology of
  China
Hefei, Anhui, China

Hassan Arafat
Center for Membranes and Advanced
  Water Technology (CMAT)
Department of Chemical
  Engineering
Khalifa University
Abu Dhabi, United Arab Emirates

Christos Argirusis
Laboratory of Inorganic Materials
  Technology
School of Chemical Engineering
National Technical University of Athens
Athens, Greece

TU Clausthal
Clausthaler Zentrum für Materialtechnologie
Clausthal-Zellerfeld, Germany

Nikolaos Argirusis
mat4nrg GmbH
Clausthal-Zellerfeld, Germany

Humaira Bashir
Department of Botany
University of the Punjab
Lahore, Pakistan

Stockbridge School of Agriculture
University of Massachusetts
Amherst, USA

Muhammad Sohail Bashir
Hefei National Laboratory for Physical
  Sciences at the Microscale
CAS Key Laboratory of Soft Matter
  Chemistry
Department of Polymer Science and
  Engineering
University of Science and Technology of
  China
Hefei, Anhui, China

Karan Basve
Material Application Research Laboratory
Department of Nano Sciences and
  Materials
Central University of Jammu
Jammu, India

Şahika Sena Bayazit
Department of Chemical Engineering
Beykent University
Istanbul, Turkey

Olugbenga Solomon Bello
Department of Pure and Applied
   Chemistry
Ladoke Akintola University of Technology
Ogbomoso, Oyo State, Nigeria

Shiva Bhardwaj
Department of Physics
Kansas Polymer Research Center
Pittsburg State University
Pittsburg, Kansas, USA

Yu Cao
Key Laboratory for Green Chemical
   Technology of Ministry of Education
School of Chemical Engineering and
   Technology
Tianjin University
Tianjin, China

Pratim Kumar Chattaraj
Department of Chemistry
Indian Institute of Technology
Kharagpur, India

Karan Chaudhary
Department of Chemistry
University of Delhi
Delhi, India

Ferda Civan Çavuşoğlu
Department of Chemical Engineering
Beykent University
Istanbul, Turkey

Moein Darabi Goudarzi
Department of Chemistry
Iran University of Science and Technology
Tehran, Iran

Gopal Das
Department of Chemistry
Indian Institute of Technology Guwahati
Guwahati, Assam

Prasenjit Das
Department of Chemistry
Indian Institute of Technology
Kharagpur, India

Felipe M. de Souza
Kansas Polymer Research Center
Pittsburg State University
Pittsburg, Kansas, USA

Mohammad Dinari
Department of Chemistry
Isfahan University of Technology
Isfahan, Islamic Republic of Iran

Ludovic F. Dumée
Center for Membranes and
   Advanced Water Technology
   (CMAT)
Research and Innovation Center on CO2
   and H2 (RICH)
Department of Chemical
   Engineering
Khalifa University
Abu Dhabi, United Arab Emirates

Sana Eid
Center for Membranes and Advanced
   Water Technology (CMAT)
Department of Chemical
   Engineering
Khalifa University
Abu Dhabi, United Arab Emirates

El-Sayed M. El-Sayed
State Key Laboratory of Structural
   Chemistry
Fujian Institute of Research on the
   Structure of Matter
Chinese Academy of Sciences
Fujian, Fuzhou, P. R. China
University of the Chinese Academy of
   Sciences
Beijing, P.R. China

Chemical Refining Laboratory
Refining Department
Egyptian Petroleum Research Institute
Cairo, Egypt

Ahmed Gulzar
Research and Innovation Center on CO2
    and H2 (RICH)
Department of Chemical Engineering
Khalifa University
Abu Dhabi, United Arab Emirates

Ram K. Gupta
Department of Chemistry
Kansas Polymer Research Center
Pittsburg State University
Pittsburg, Kansas, USA

Georgios N. Karanikolos
Research and Innovation Center on CO2
    and H2 (RICH)
Center for Membranes and Advanced
    Water Technology (CMAT); Center for
    Catalysis and Separation (CeCaS)
Department of Chemical
    Engineering
Khalifa University
Abu Dhabi, United Arab Emirates

Gagandeep Kaur
Material Application Research
    Laboratory
Department of Nano Sciences and
    Materials
Central University of Jammu
Jammu, India

Varinder Kaur
Department of Chemistry
Panjab University
Chandigarh, India

Sandeep Kaushal
Department of Chemistry
Sri Guru Granth Sahib World University
    Fatehgarh Sahib
Punjab, India

Navid Keshmiri
Surface Coating and Corrosion
    Department
Institute for Color Science and
    Technology
Tehran, Iran

Negin Khosroshahi
Department of Chemistry
Iran University of Science and
    Technology
Tehran, Iran

Turkan Kopac
Department of Chemistry
Zonguldak Bülent Ecevit
    University
Zonguldak, Turkey

Pawan Kumar
Material Application Research Laboratory
Department of Nano Sciences and
    Materials
Central University of Jammu
Jammu, India

Yong Li
Department of Applied Biology and
    Chemical Technology
The Hong Kong Polytechnic
    University
Hung Hom, Kowloon, Hong Kong SAR,
    China

Encarnación Lorenzo
Departamento de Química Analítica
    y Análisis Instrumental & Institute
    for Advanced Research in Chemical
    Sciences (IAdChem)
Universidad Autónoma de Madrid
Madrid, Spain

IMDEA-Nanociencia
Ciudad Universitaria de Cantoblanco
Madrid, Spain

Thobeka Makhunga
Department of Chemical Sciences
University of Johannesburg
Doornfontein, South Africa

Marcos Martínez-Fernández
Departamento de Química Orgánica I
Universidad Complutense
    de Madrid
Madrid, Spain

Rafael Martínez-Palou
Dirección de Investigación en
  Transformación de Hidrocarburos
Instituto Mexicano del Petróleo
San Bartolo Atepehuacan
CDMX, México

Emiliano Martínez-Periñán
Departamento de Química Analítica y
  Análisis Instrumental
Universidad Autónoma de Madrid
Madrid, Spain

Tebogo Mashola
Department of Chemical Sciences
University of Johannesburg
Doornfontein, South Africa

Dhanraj T. Masram
Department of Chemistry
University of Delhi
Delhi, India

Thabo Matthews
Department of Chemical Sciences
University of Johannesburg
Doornfontein, South Africa

Nobanathi Wendy Maxakato
Department of Chemical Sciences
University of Johannesburg
Doornfontein, South Africa

Siyabonga Mbokazi
Department of Chemical Sciences
University of Johannesburg
Doornfontein, South Africa

Maria Mechili
Laboratory of Inorganic Materials
  Technology
School of Chemical Engineering
National Technical University of Athens
Athens, Greece

Parisa Miri
Department of Chemistry
Iran University of Science and Technology
Tehran, Iran

Nazanin Mokhtari
Department of Chemistry
Isfahan University of Technology
Isfahan, Islamic Republic of Iran

Sukanta Mondal
Department of Education
A. M. School of Educational Sciences
Assam University
Silchar, Assam

Kudzai Mugadza
Department of Chemical Sciences
University of Johannesburg
Doornfontein, South Africa

Institute of Materials Science
Processing and Engineering Technology
Chinhoyi University of Technology
Chinhoyi, Zimbabwe

Tayyaba Najam
Institute for Advanced Study and Institute
  of Microscale Optoelectronics
Shenzhen University
Shenzhen, China

Ali Al Najjar
Center for Membranes and Advanced
  Water Technology (CMAT)
Department of Chemical Engineering
Khalifa University
Abu Dhabi, United Arab Emirates

Parisa Najmi
Surface Coating and Corrosion
  Department
Institute for Color Science and Technology
Tehran, Iran

Biswajit Nayak
Department of Chemistry
Indian Institute of Technology Guwahati
Guwahati, Assam

Gülsüm Özçelik
Department of Chemical Engineering
Beykent University
Istanbul, Turkey

Fusheng Pan
Key Laboratory for Green Chemical
Technology of Ministry of Education
School of Chemical Engineering and
Technology
Tianjin University, Tianjin, China

Pavlos K. Pandis
Laboratory of Inorganic Materials
Technology
School of Chemical Engineering
National Technical University of Athens
Athens, Greece

Twinkle Paul
Research and Innovation Center on CO2
and H2 (RICH)
Department of Chemical Engineering
Khalifa University
Abu Dhabi, United Arab Emirates

Bahram Ramezanzadeh
Surface Coating and Corrosion
Department
Institute for Color Science and Technology
Tehran, Iran

Mohammad Ramezanzadeh
Surface Coating and Corrosion
Department
Institute for Color Science and Technology
Tehran, Iran

Federico Roncaroli
Departamento de Física de la Materia
Condensada
Instituto de Nanociencia y Nanotecnología,
Centro Atómico Constituyentes
Comisión Nacional de Energía Atómica
San Martín, Buenos Aires, Argentina

Departamento de Química Inorgánica
Analítica y Química Física
Universidad de Buenos Aires, Ciudad
Universitaria, Pabellón II
Ciudad de Buenos Aires, Argentina
Consejo Nacional de Investigaciones
Científicas y Técnicas-CONICET
Ciudad de Buenos Aires, Argentina

Vahid Safarifard
Department of Chemistry
Iran University of Science and
Technology
Tehran, Iran

Aqsa Safdar
School Education Department
Punjab, Pakistan
Department of Chemistry
University of the Punjab
Lahore, Pakistan

Gabriele Scandura
Research and Innovation Center on CO2
and H2 (RICH)
Department of Chemical
Engineering
Khalifa University
Abu Dhabi, United Arab Emirates

José L. Segura
Departamento de Química Orgánica I
Universidad Complutense
de Madrid
Madrid, Spain

Cyril Selepe
Department of Chemical Sciences
University of Johannesburg
Doornfontein, South Africa

Dinesh Shetty
Center for Catalysis and Separation
(CeCaS)
Department of Chemistry
Khalifa University
Abu Dhabi, United Arab Emirates

Raghubir Singh
Department of Chemistry
DAV College
Chandigarh, India

Georgia Sourkouni
TU Clausthal
Clausthaler Zentrum für
Materialtechnologie
Clausthal-Zellerfeld, Germany

Jie Sun
Key Laboratory for Green Chemical
    Technology of Ministry of Education
School of Chemical Engineering and
    Technology
Tianjin University
Tianjin, China

Pooja Upadhyay
Material Application Research Laboratory
Department of Nano Sciences and
    Materials
Central University of Jammu
Jammu, India

Christos Vaitsis
Laboratory of Inorganic Materials
    Technology
School of Chemical Engineering
National Technical University of Athens
Athens, Greece

Heriberto Díaz Velázquez
Dirección de Investigación en
    Transformación de Hidrocarburos
Instituto Mexicano del Petróleo
San Bartolo Atepehuacan
CDMX, México

Lawrence Yoon Suk Lee
Department of Applied Biology and
    Chemical Technology
The Hong Kong Polytechnic University
Hung Hom, Kowloon, Hong Kong SAR,
    China

Daqiang Yuan
State Key Laboratory of Structural
    Chemistry
Fujian Institute of Research on the
    Structure of Matter
Chinese Academy of Sciences
Fujian, Fuzhou, P. R. China

University of the Chinese Academy of
    Sciences
Beijing, P.R. China

Weiran Zheng
Department of Chemistry
Guangdong Technion - Israel Institute of
    Technology
Shantou, China

Research Institute for Smart Energy
The Hong Kong Polytechnic University
Hung Hom, Kowloon, Hong Kong SAR,
    China

Memory Zikhali
Department of Chemical Sciences
University of Johannesburg
Doornfontein, South Africa

Editor Biographies

Dr. Tuan Anh Nguyen has completed his BSc in Physics from Hanoi University in 1992, and his PhD in Chemistry from Paris Diderot University (France) in 2003. He was a Visiting Scientist at Seoul National University (South Korea, 2004) and the University of Wollongong (Australia, 2005). He then worked as a Postdoctoral Research Associate & Research Scientist at Montana State University (USA), 2006–2009. In 2012, he was appointed as Head of the Microanalysis Department at the Institute for Tropical Technology (Vietnam Academy of Science and Technology). He has managed four PhD theses as thesis director and three are currently in progress. He is Editor-In-Chief of *Kenkyu Journal of Nanotechnology & Nanoscience* and Founding Co-Editor-In-Chief of *Current Nanotoxicity & Prevention*. He is the author of four Vietnamese books and Editor of 32 Elsevier books in the Micro & Nano Technologies Series.

Dr. Ram K. Gupta is an associate professor at Pittsburg State University. Dr. Gupta's research focuses on conducting polymers and composites, green energy production and storage using biowastes and nanomaterials, optoelectronics and photovoltaics devices, organic-inorganic hetero-junctions for sensors, bio-based polymers, flame-retardant polymers, bio-compatible nanofibers for tissue regeneration, scaffold and antibacterial applications, corrosion inhibiting coatings, and bio-degradable metallic implants. Dr. Gupta has published over 245 peer-reviewed articles, written over 50 book chapters, made over 300 national, international, and regional presentations, chaired many sessions at national/international meetings, and edited many books. He has received several million dollars for research and educational activities from many funding agencies. He serves as Editor-in-Chief, Associate Editor, and editorial board member of numerous journals.

1

Covalent Organic Framework: An Introduction

Gagandeep Kaur, Pooja Upadhyay, Karan Basve, and Pawan Kumar
Material Application Research Laboratory, Department of Nano Sciences and Materials,
Central University of Jammu, Jammu, India

CONTENTS

1.1 Introduction

Porous organic material (POM) is another class of unique materials with a profoundly ordered structure and tunable porosity. Among all known POMs, metal-organic frameworks (MOFs) and covalent organic frameworks (COFs) have acquired an enormous interest of specialists due to their excellent properties. MOFs being the first category of porous materials are a sort of coordination polymers that ties metal ions and organic linkers by organizing coordinate bonds. Whereas, COFs are the second advanced category of porous materials in which metal ions are replaced by light metals. Basically, COFs are composed of strong covalent holding between organic linker and light metals (B, N, O, H, C, etc.) that form an association between natural science and polymer science (Figure 1.1). COFs act as alternative materials to MOFs with improved crystalline and excellent thermal stability [1]. Apart from this, they have properties like huge surface territory, high porosity, regular pore size, low density, high functionality, high-temperature tolerance, etc. Importantly, these advanced materials were reported for the utilization in catalysts, sensors, drug

FIGURE 1.1
Different properties and applications of COFs.

delivery, structural modifications, water purification, magnetic separations, gas stockpiling, and adsorptions [2].

Aside from well-known applications of COFs, the primary objective of scripting this section is to appraise the energy applications. With the rapid increase in pollution, there is a huge demand for energy which is a matter of anguishing. Energy storing applications of COFs are of incredible interest as it incorporates supercapacitors, metal ion/sulfur/air-based batteries, solar cells, etc. [3]. COF-based energy applications, such as photocatalytic processes, i.e., hydrogen-evolved responses (HER), hydrogen reduced responses (HRR), oxygen evolved responses (OER), and oxygen reduced responses (ORR), might be researched further in the near future. Importantly, COFs-based energy applications, i.e., photocatalytic processes including hydrogen-evolved responses (HER), hydrogen reduced responses (HRR), oxygen evolved responses (OER), and oxygen reduced responses (ORR) possible more explored soon. In this chapter, a fundamental overview of COFs was presented. Importantly, this chapter covers the linkage, synthetic methods, properties, and applications based on the present literature available on COFs. In addition, the future perspective of COFs was explored for diverse applications.

1.2 Linkage, Synthesis, and Properties of COFs

1.2.1 Linkage

Till now, different type of monomers and chemical reactions between them has offered different type of chemical linages. For example, COF-1 comprising boroxine (B_3O_3) linkage synthesized by self-condensation of 1, 4 benzene diboronic acid (BDBA) involving -B-O-linkage. In boronate ester linkage (-B-O-), a planar linkage is formed via the condensation of catechol derivatives and boronic acids. Furthermore, COF-5 synthesized by microwave irradiation method and conventional solvothermal method contains boronate ester linkage. Additionally, COF with boronate linkages has been actively explored for 2D COFs synthesis. Imine-based COFs containing -C-N- linkage are acquired via Schiff-based condensation reactions between polyaldehydes and polyamines in presence of an organic solvent. In comparison to COF-bearing boronate linkages, imine-type COFs are most stable due to being more resistant to hydrolytic decomposition [4].

Similarly, synthesis of COFs with azine and hydrazone linkages formed through the reversible reaction between symmetrical hydrazides or hydrazines and aldehyde, respectively. COF-type hydrazone linkage shows enhanced hydrolysis and oxidation stability which is maintained by partially prolonged hydrogen bonds of hydrazine in the pore spaces. COFs with other kinds of linkages and linkers (alkenes, ethers, triazine) were also reported and summarized in Table 1.1 with synthesis and properties [4–6]. A variety of monomers have been studied to develop new COFs with divergent properties and applications. Till now, various organic building units have been established but it is no easy task to build a COF with high crystallinity and porosity [5–9]. Usually, COF is synthesized by simple imine and aldehyde Schiff-based condensation reaction using dynamic covalent chemistry through the reversible formation of strong covalent bonds [7]. These strong covalent bonds can be formed, split, or rebuilt under certain conditions. The thermodynamically stable reversible covalent bonds help the products to form self-restoring and error-correction structures. To achieve high-crystalline COFs, linkage formation must be reversible and rate of reaction must be fast enough to autocorrect the structural flaws. Hence, the linkages of reversible covalent bonds are recognized to be an essential condition for the manufacturing of crystalline COFs. In 2D COFs, this error-correction range in the process of lattice formation will give high crystallinity, where the covalent properties give high thermal stability.

1.2.2 Synthesis Methods

The different synthesis methods including solvothermal, microwave-assisted, mechanochemical, sonochemical, and ionothermal have been reported for COFs (Figure 1.2). These methods are briefly discussed below.

TABLE 1.1

Represents COFs with Different Linkage, Linkage-Type, Synthesis Methods, Properties

Sr. No.	Linkage	Linkage Type	Method/Process	Properties/Characterization	Examples	Ref
1.	-B-O-	Boroxine	Sonochemical, Solvothermal method	• High thermal stability up to 500°C • High porosity and high surface area • High reversibility • Unstable in acid, base, water, and nucleophile	COF-1, COF-5	[4]
2.	-B-O-	Boronate Ester	Condensation reaction	• High thermal stability up to 600°C • High surface area and low density • High reversibility • Unstable in acid, base, water, alcohol, and nucleophile	COF-108	[4]
3.	-C-N-	Imine	Condensation reaction	• High thermal stability up to 500°C • High porosity and high surface area • Requires catalyst, high low crystallinity compared to boroxine • Chemically stable in water, acid, and alcohol than boronate ester and boroxine	COF-300, COF-5	[4], [5]
4.	-C-N-	Hydrazone	Condensation	• Good crystallinity • Chemical stability (due to hydrogen bonding interactions) • Thermal stability up to 280°C and high porosity	COF-42, COF-43	[4], [5]
5.	-C-N-	β-Ketoenamine	Condensation	• Chemical stability in boiling water, acid, and base • Structural versatility and large surface area • High thermal stability up to 350°C	BF-COF-2, EB-COF	[6]
6.	-C-N-	Imide	Condensation	• High stability and high porosity • Good crystallinity • High thermal stability up to 530°C • Better than B-O linked COFs	PI-COF, PI-COF-2, PI-COF-5	[6]
7.	-C-N-	Triazine	Ionothermal, Poly-condensation	• High chemical stability • Porosity and Poor crystallinity • High thermal stability up to 400°C	CTF-1, CTF-2	[4], [6]
8.	-C=C-	Alkenes	Aldol-Condensation hydrothermal vapor-assisted	• Crystalline nature • Permanent micro-porosity and high surface area • High thermal stability up to 350°C	COF-701	[4]
9.	-C-O-	Ethers	Condensation	• Moderate crystallinity • High porosity chemical stability	PAE-COF, JUC-505, JUC-506	[4]

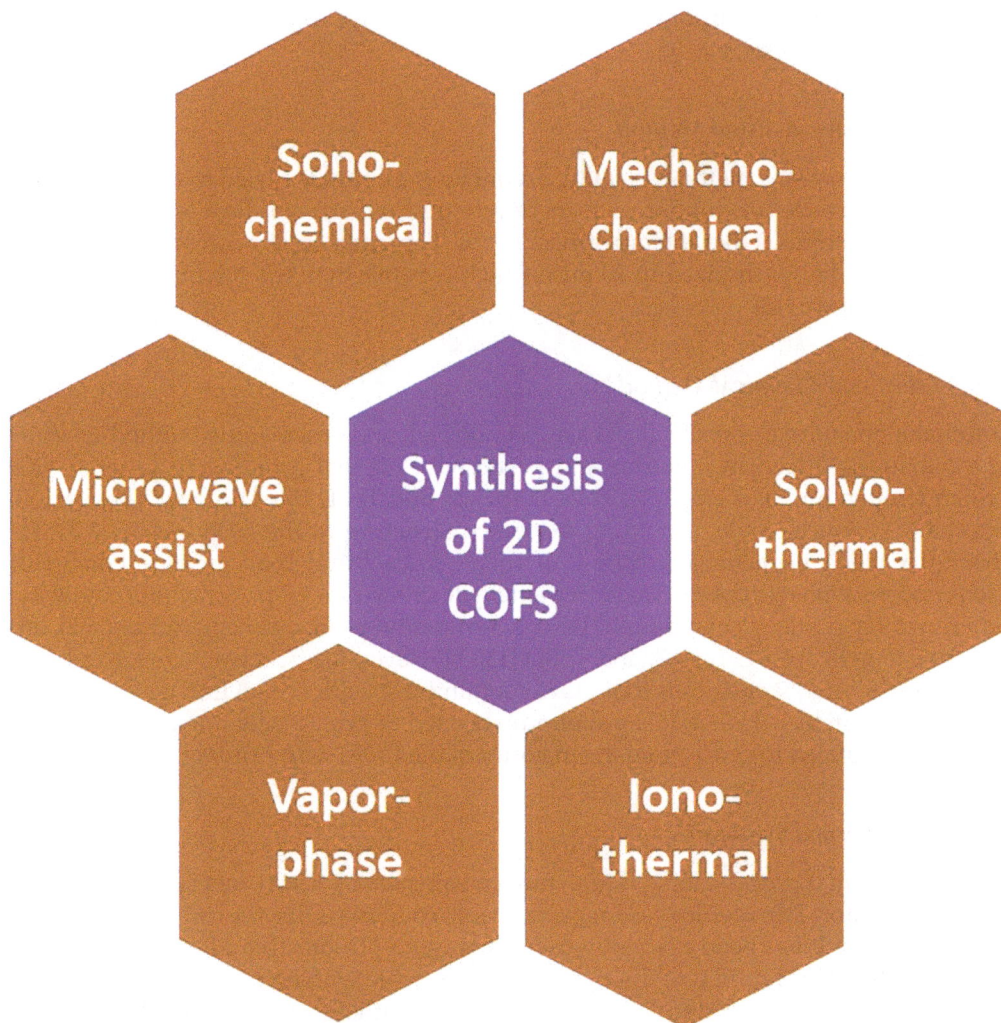

FIGURE 1.2
Different routes for the synthesis of COFs.

1.2.2.1 Solvo-Thermal Method

Among various synthetic approaches known till now, the solvothermal method is the most commonly used method where the reaction conditions are highly dependent on the reactivity of building units, solubility, and reversibility of the reactions. However, this conventional method suffers some limitations like a long time, selected solvent mixture (mesitylene,1-4 dioxane acetone), high temperature, high pressure, inert environment, pyrex tube, degassing apparatus, etc. For instance, synthesis of 3D COF-5 via solvothermal method often takes multiple days and even several months which creates obstacles to the explorations and applications of functional COFs [10, 11]. To mitigate such issues, new techniques were utilized to integrate distinctive sorts of COFs. Different synthesis techniques incorporate ionothermal, microwave assist, sonochemical, vapor phase assisted synthesis, etc. [4].

1.2.2.2 Microwave-Assisted Method

In the microwave assist method, usually, a monomer mixture is sealed in appropriate solvents in a microwave tube under vacuum or nitrogen and stir and heat for 60 minutes at the specified conditions like at temperature 100° C. For instance, in microwave conditions, a 2D COF-5 can be synthesized in 20 min, which is better than the solvothermal method as mentioned above [12].

1.2.2.3 Mechano-Chemical Method

A mechano-chemical approach is an easy, economical, and eco-friendly method in which the monomer is placed in a mortar and pestle with manual grinding to generate COF at room temperature. For example, TpPa-1, TpPa-2, and TpBD have been synthesized via a mechano-chemical route, providing exfoliated structure to the COF (Figure 1.3). The solvent-free mechanochemical synthesis of TpPa-1, TpPa-2, and TpBA are based on Schiff-based condensation reaction between free aldehyde group of 1,3,5-triformylphloroglucinol and imines say p-phenylenediamine (Pa-1), 2,5-dimethyl-p-phenylenediamine (Pa-2), and benzidine to give TpPa-1, TpPa-2, and TpBD [13]. However, the disadvantage looked at by these manufactured strategies is that COFs containing specific sort of linkage are blended by specific techniques. Hence, it is profoundly needed to grow a new manufactured technique that is utilized to orchestrate the majority of the COFs with brilliant crystallization.

1.2.2.4 Ionothermal Synthesis

In the ionothermal process, ionic liquid/molten salt such as $ZnCl_2$ acts as both a solvent and an accelerator. This method also requires harsh conditions like the conventional solvothermal method. It has been efficiently used to produce 3D ionic-liquid-containing COFs. In addition, the ionic liquid was discovered to be reusable with no noticeable activity loss. This research not only introduces a novel method for synthesizing COFs but also paves the door for green large-scale COF production in the industry [12].

1.2.2.5 Sonochemical Synthesis

The sonochemical method is also an alternative method for conventional solvothermal synthesis. The organic linkers dissolved in a particular solvent in a closed vessel under ultrasonic waves. For instance, COF-1 and COF-5 were synthesized by the sonochemical method. Even though this approach has still not been applied to a broad variety of connection types, it indicates the technology's promise for the synthesis of extremely porous materials [14]. The properties of 2D-COFs are tunable and therefore can be adjusted by changing the functionality of the linker or secondary building unit (SBU) [4]. For instance, nanoparticle-covalent organic framework (NP-COF) acts as a cathode and can be utilized to directly convert solar energy into electrochemical energy [15]. Furthermore, COF framed by the synergistic impact of 1,3,5-triazine2,4,6-triamine (Tt) and 2,4,6-triformyphlorogucinol (Tp) aldehyde utilizing triazine functional group is utilized for the photocatalytic response for E-Z isomerism of olefins or alkenes (C=C) [16]. Additionally, COFs bearing diacetylene and pyrene functionality were successfully utilized for HER [17, 18].

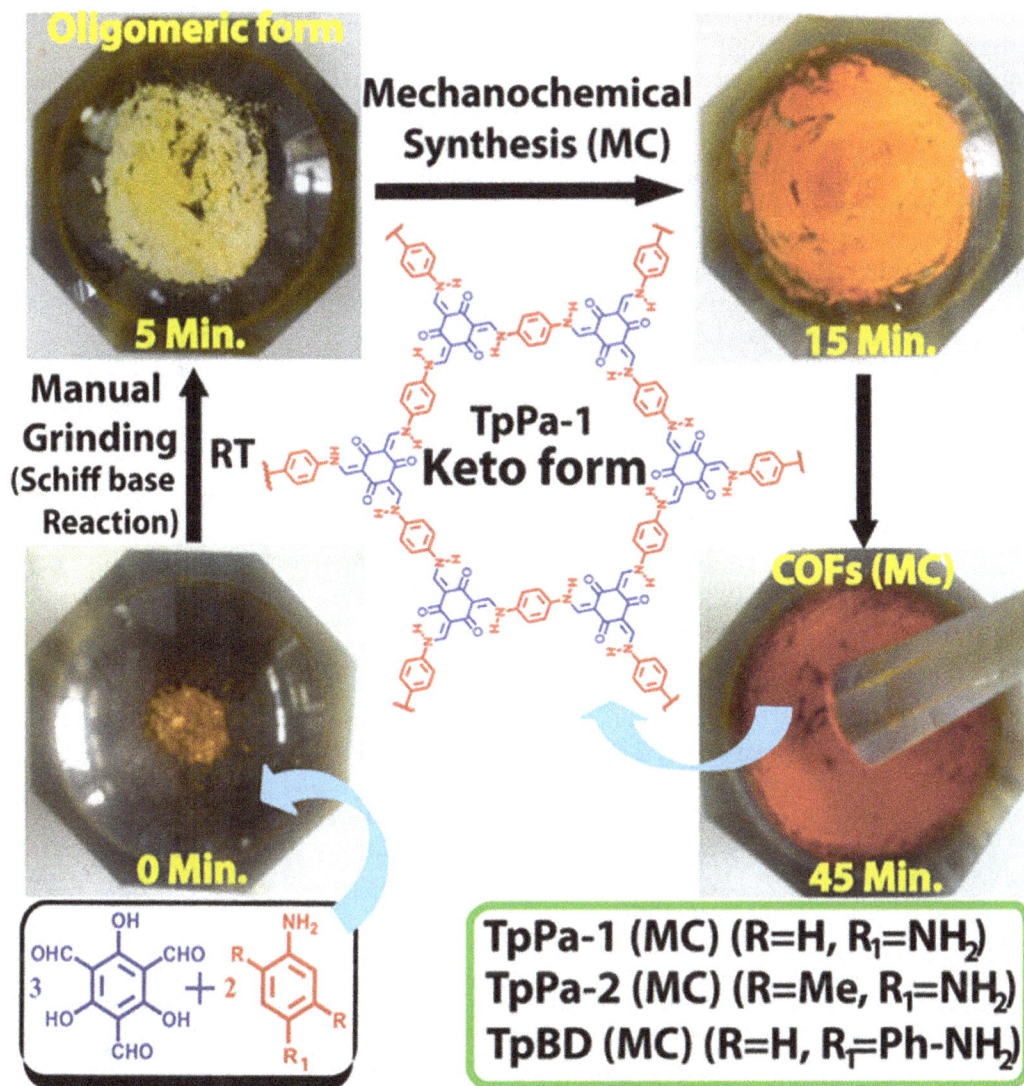

FIGURE 1.3
Synthesis of TpPa-1, TpPa-2, and TpBD using a mechanochemical method.

(Adapted with permission [13]. Copyright (2013) American Chemical Society.)

1.3 Applications of COFs

1.3.1 Photocatalysts

With the expanding demand for energy applications and the sun being the inexhaustible and never-ending source of energy, COFs can be utilized for photocatalyzed reactions. COFs commanding their adaptable properties and tunable nature can be used as photocatalysts by directly converting solar energy into electrochemical energy. The first photocatalytic reaction was reported in 2010. As referenced above, diacetylene and pyrene functionalized COF can be broadly used to produce hydrogen (H_2) from water. The basic principle behind the photocatalytic interaction of COFs is to retain a sufficient amount of solar energy to produce electrons and holes, which are then subsequently passed to catalysts for desirable photocatalyzed reactions. The photocatalytic effectiveness of COFs can effectively be increased by adding electron donor or acceptor functionality to it [19].

Hydrogen being a sustainable and continual source of energy is considered as quite possibly the most reliable source of energy during undersupply of fossil fuels. Generating hydrogen and oxygen from sunlight and water is simple and cost-efficient. Based on this enormous number of COFs reported till now showed phenomenal HER. For example, COF-bearing hydrazone linkage (TFPT-COF) utilizing platinum (Pt) as a catalyst can develop 1970 $\mu molg^{-1}h^{-1}$, whereas COFs with azine linkage (N_x-COF where $x = 0, 1, 2, 3$) also show photocatalyzed HER. It was reported that on varying the quantity of nitrogen atoms, the amount of hydrogen-producing capacity also varies. Greater the quantity of nitrogen atoms, more will be the hydrogen evolving tendency, and as a result, it follows order (N_3-COF > N_2-COF > N_1-COF > N_0-COF) [19, 20].

The photocatalytic HER also depend on the functionality of the group. It can be better perceived by taking illustration of β-ketoenamine-based COF, TpPa-COF-X (where X can be -CH_3, -NO_2, -H, and -$(CH_3)_2$) with four distinct functional groups. It was observed that all the four distinctive COFs evolves different amount of hydrogen [21]. Apart from hydrogen-evolved and hydrogen-reduced reactions, oxygen-evolved and oxygen-reduced reactions are also of great importance. For instance, metal oxides like iridium oxide and rubidium oxide can be utilized for oxygen-evolved and oxygen reduced reactions [22, 23]. Yet, because of enormous expense, it was fundamental for its alternative. COFs are not just utilized for photocatalyzed HER but at the same time give cost-friendly alternatives to OER and ORR. Furthermore, Ni- and Fe-based COFs (Ni/Fe-COF@CNT_{900}, $Ni_{0.5}Fe_{0.5}$@COF-SO3, and Co_3O_4/NPC-2) also show a phenomenal outcome for oxygen evolved photocatalyzed reactions [24, 25]. Apart from this cobalt-based COFs, Co-COF-900 and Co-TpBpy-800 also show radiant execution for oxygen reduction reactions (ORR) in the basic medium [26, 27].

In another study, covalent triazine frameworks were effectively used for ORR and HER reactions. BINOL-CTFs with different functionality say pyridinic-N, quaternary-N, pyrrolic-N, and pyridine-N-oxide were designed. However, these materials promisingly showed enhanced oxygen reduction and HER. It was found that the material showed an onset potential of 0.793V and first-order kinetics in the four-electron pathway. Furthermore, a half-wave potential of 0.737 V was also observed for ORR reactions. For the same material, the HER was observed at 0.31 V. Therefore, ORR performance varies with varying the number of N-atoms in the HER (Figure 1.4) [28].

1.3.2 Gas Storage

Hydrogen being a noncontaminant and financially cordial wellspring of energy should be safeguarded or stored. Both 2D and 3D-COFS are utilized for hydrogen stockpiling. The gas stockpiling capacity of COFs relies upon its pore size and enormous BET surface territory. More prominent the pore size, large will be the hydrogen absorbance capacity. 3D COFs generally possess more notable pore size and BET surface territory shows more hydrogen stockpiling capacity in contrast with 2D COFs. For instance, 2D (COF-10) with BET surface space of 1760 $m^2\,g^{-1}$ has hydrogen engrossing capacity of 39.2 mg g^{-1} at 77 K while at same temperature conditions 3D (COF-102) with superior BET surface territory showed practically two-fold increased hydrogen stockpiling limit of 72.4 mg g^{-1}. Doping

(a)

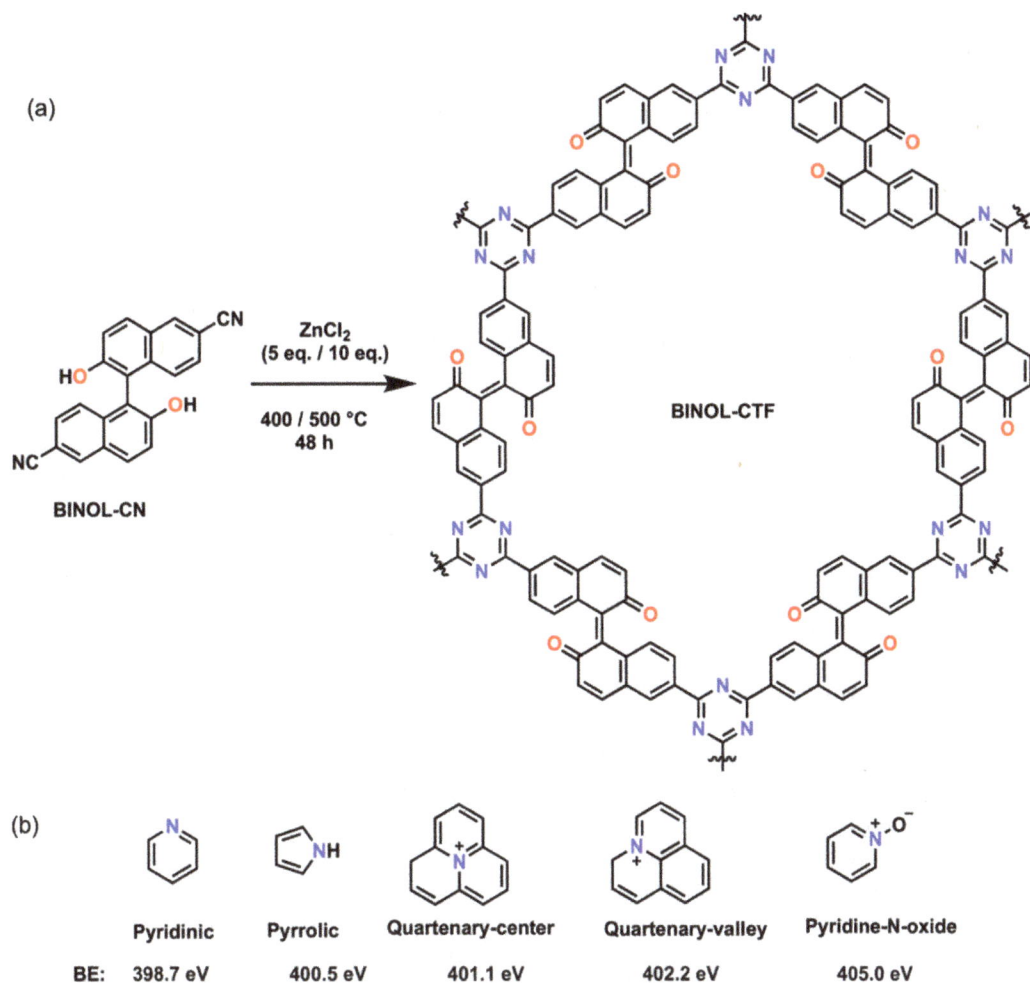

(b)

Pyridinic	Pyrrolic	Quartenary-center	Quartenary-valley	Pyridine-N-oxide
BE: 398.7 eV	400.5 eV	401.1 eV	402.2 eV	405.0 eV

FIGURE 1.4

BINOL-CTFs synthesis with five different functionalities uses ionothermal synthesis method for effective ORR and HER.

(Adapted with permission [28]. Copyright (2020) American Chemical Society.)

the 2D COFs can lift or upgrade their hydrogen stockpiling capacity. For example, 2D COF-1 doped with pyridine atoms between the level layers revealed more hydrogen absorbance limit in contrast with straightforward or undoped COF-1. Another model with an expanded hydrogen consumption limit was accounted for by adding lithium (Li) atoms on Pc-PBBA-COF additionally showed improved hydrogen take-up capacity [30]. Additionally, COFs like COF-103, COF-5, COF-10, COF-8, COF-16, COF-102, COF-1, and so on show hydrogen take-up limit of 70.5, 35.8, 39.2, 35.0, 22.6, 72.4, and 14.8 mg g^{-1}, respectively [30–33].

Like hydrogen, methane additionally gives a spotless and noncontaminated source of energy. Methane with the general formula CH_4 contains a carbon-hydrogen ratio (1:4) which produces less carbon dioxide (CO_2) in the climate. COFs possessing huge pore size and BET surface territory have large gas uptake limits. Methane engrossing limit of two-dimensional COFs: COF-10, C0F-8, COF-6, and COF-5 was accounted for 80, 87, 65, and 89 mg g^{-1}, respectively, at 35 bar and 77 K. Three-dimensional COF with more BET surface territory and enormous porosity can allow even more amount of methane into it. Both COF-103 and COF-102 under the same conditions consumed 175 and 187 mg g^{-1}, respectively [29, 30].

In another study, substituting or modifying COF-102 with dihalogen atoms additionally shows better outcomes for methane take-up. For instance, subbing two H molecules by two halogens (X = Cl, Br, and I) in COF-120 created tunable or functionalized COFs (COF-102-1,3-2X and COF-102-1,4-2X) without changing or contorting its structural orientation. Curiously, the best methane take-up outcomes were accounted for by iodine-subbed COF and follows methane take-up pattern as (Cl < Br < I) [29–31].

1.3.3 Batteries

Another interesting energy application of COFs was reported for advanced batteries. With persistent usage of energy, there is a serious need for energy delivering gadgets. Batteries can go about as the best gadget which can create a ceaseless stock of energy. Metals going through a simple oxidation process have more electrons giving tendency [34]. Metals like lithium (Li) and sodium (Na) are easily accessible and can go about as an easy source of metal-ion batteries with great electric potential. These are cost cordial moreover. Other heavier metals like Co, Ni, and Zn require significant expense and hence can't be utilized for enormous scope. COFs because of their large surface territory, low thickness, and high porosity offer additional promising outcomes for batteries [35]. COF when attached to carbon nanotubes (CNTs) creates COF-CNT and shows good cyclic stability.

A metal-ion battery also serves as an energy-generating gadget because of its easy and simple accessibility, long cycles, great thermal limits, specific energy density, etc. Regardless of these highlights, it experiences a few drawbacks. One of the drawbacks is the presence of intermediate redox ions (lithium polysulfides – LiPSs) which interact with electrolytes and are delocalized between the electrodes. This free delocalization of LiPSs particles between anode and cathode leads to a process known as the shuttle process. The presence of a shuttle in a battery impedes the efficiency of the battery by lowering cyclization and easy discharge. Hence, this process should be decreased to deliver profoundly productive batteries. It was observed that COF-1 CNTs absorb PSs in Li-S batteries. Later it was found doping of tripositive boron and dinegative oxygen show synchronous absorbance of S^{2-} and Li^{+}. This leads to proficient catching of LiPSs [36]. The other method of lowering the shuttle process is the introduction of hybrid crystal phase titanium oxide TiO$_2$ (HCPT) into the layers of COF (HCPT@COF) that showed superior performance in lithium-sulfur-based batteries. It shows high cyclic stability, low capacity degrade rate, high reversibility, and so on [37]. Apart from these, another revealed technique to increase the efficiency of batteries was observed on nitrogen-doped graphitic carbon attached to COF (N-GC-COF). The initially reported imine-linked covalent organic framework (TpTt-COF) and nitrogen-doped graphitic carbon (N-GC) was reported as (N-GC-TpTt-COF). This nitrogen-doped COF showed great cycling capacity and high discharge capacity of 869 mAhg^{-1}. Even after 200 cycles, the discharge limit was found to be 670 mAhg^{-1}. This demonstrates that (N-GC-TpTt-COF) showed great conformity for the high productivity of lithium-sulfur batteries [38].

Metal-air electrochemical cell is also a remarkable electrochemical cell that utilizes pure metal as an anode, air as cathode, and aqueous electrolyte. Like metal-ion and lithium-sulfur batteries, metal-air batteries additionally fill in as an energy storage gadget for a sustainable future. Zinc (Zn)–air batteries are well-known metal-air batteries because of their low cost, easy availability, natural abundance, less toxicity, etc. Despite owning all these properties, Zn-air batteries suffer slow kinetics for OER and ORR which affect the reactivity and efficiency of the battery. However, it was noted that pyridine-based COF was utilized as a cathode in electrochemical Zn-air batteries. Later, the pyridine-based COF was linked with CNTs and was reported as CNT@POF. Zinc-air batteries with CNT@POF as cathode showed high energy efficiency, low energy gap, high power capacity, and served as saving energy storage and converter device [39].

In contrast to inflexible batteries, the present-day world depends on changed, tunable batteries which are named adaptable or flexible batteries. Adaptable batteries have acquired a colossal interest in analysts because of their following characteristics: adaptability, shape, ease to handle, lightweight, and so on. The essential condition for designing these batteries is a polymer-restricting terminal which upgrades the conductivity of batteries. The redox dynamic nature of COF can trigger the electrochemical and energy-giving limit of batteries. For instance, COF containing β-ketoenamine linkage was synthesized by copolymerization of 1, 3, 5-triformylphloroglucinol (TFP) and 2, 6-diaminoanthraquinone (DAAQ) monomers were reported as DAAQ-TFP COF. DAAQ-TFP COF when linked to CNT was created as c-CNT@COF which improved flexibility, high capacitance, and high cycle stability (Figure 1.5) [40]. Another illustration of COF linked to carbon nanofibers (CNF) which were manufactured as COF-CNF hybrid thin sheet additionally revealed high adaptability and compactness, improved capacitance (42 mF cm^{-2}), energy-storing capacity (464 mF cm^{-2}), and electrical conductivity of 0.25×10^{-3} S cm^{-1} [41].

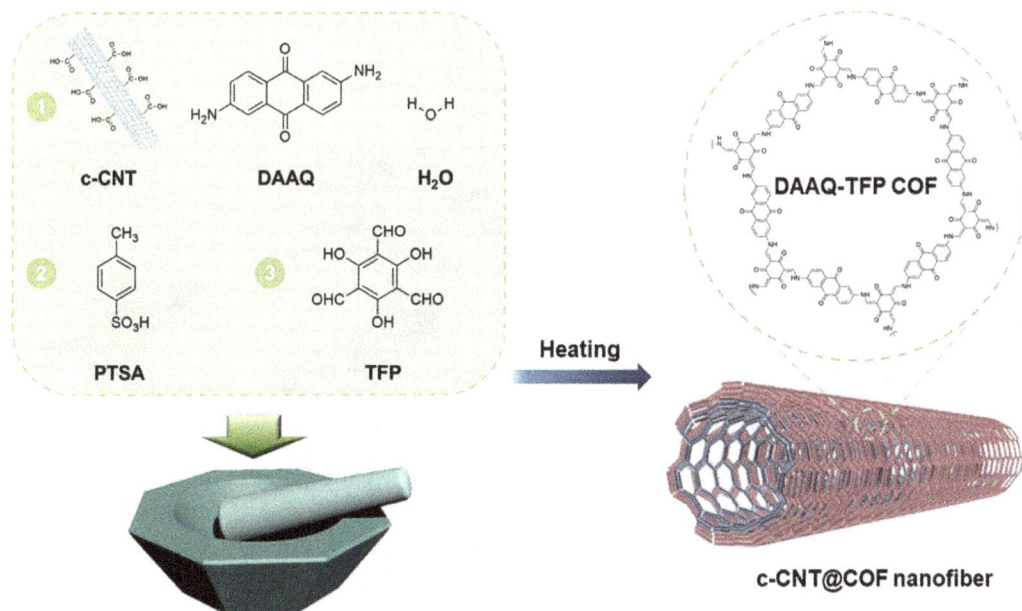

FIGURE 1.5
Schematic of the synthesis of c-CNT@COF. (Adapted with permission [40]. Copyright (2020) The Authors. Published by Elsevier Ltd. This is an open-access article under the CC BY license, http://creativecommons.org/licenses/by/4.0/.)

1.3.4 Supercapacitors

Supercapacitors offer several advantages such as short charging and discharging time, high power densities, a wide range of working temperatures, and a long-life cycle due to which they are attracting a lot of interest [42, 43]. In supercapacitors, charge storage occurs through a very fast absorption/desorption reaction. The efficiency of supercapacitors can be increased by doping nitrogen and carbon materials and as it offers a pseudo-capacitance effect [43]. Numerous examinations have been accounted for on COFs as a supercapacitor. The primary benefit of COFs presents redox-active functionalities because of which they are appropriate for pseudo-capacitive energy storage. The electrode fabrication is troublesome because of its insoluble nature. Because of their non-conductive nature and poor electrochemical stabilities, they display poor capacitive performance and low productivity which restricts their use in supercapacitors. In comparison to electrical double-layer capacitors (EDLCs), it has limited cycling stability and low capacitive performance. In another study, TpOMe-DAQ COF was built into thin sheets with a width of 200 μm (Figure 1.6). This COF showed crystalline structure, excellent chemical stability, and long-range ordered pores. It was observed that this COF showed excellent stability up to 100,000 cycles with a capacitance of 1280 mF cm^{-2} and 135 F g^{-1} [44].

Additionally, aza-π conjugated COFs, a particular class of permeable organic solid, show great conductivity because of broadened 2D aromatized π-conjugated linkage with π-π overlapping contributing a consistent track for the electrons bringing about the leading material [45]. A study on 2D COF synthesized by solvothermal method showed capacitance of 163 F/g at 0.5 A/g at a voltage of 0–2.5 V [46, 47]. Additionally, supercapacitor

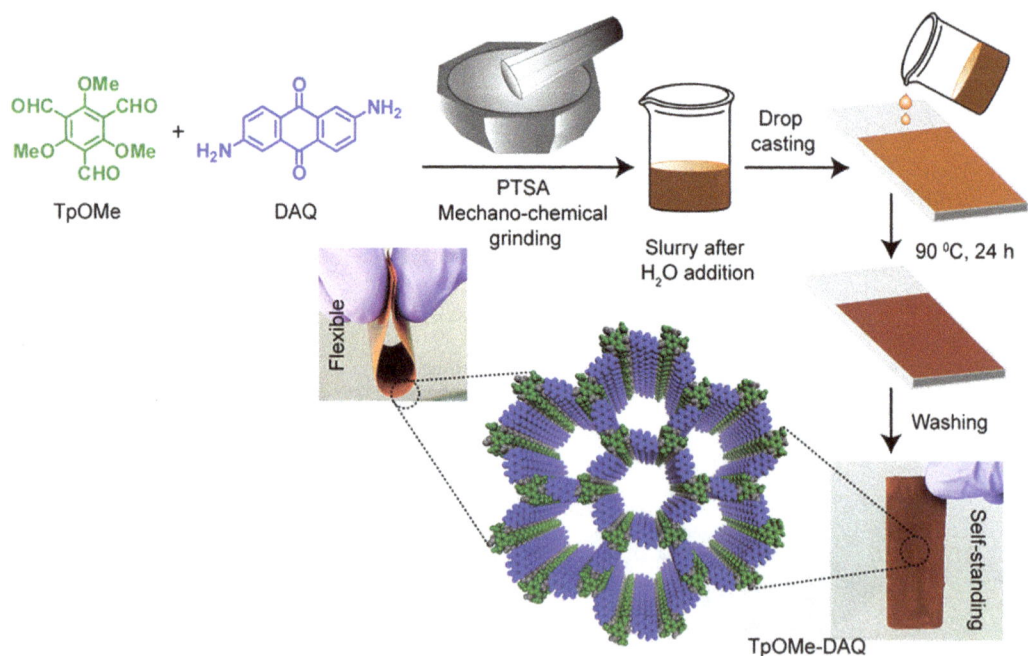

FIGURE 1.6

Schematic representation of TpOMe-DAQ COF nanosheets using a mechanochemical grinding method for improved capacitance. (Adapted with permission [44]. Copyright (2018) American Chemical Society.)

TABLE 1.2
Some Examples of COFs, Synthesis/Methods, and Supercapacitor Capacity

Sr. No.	COF Examples	Synthesis Method/Reaction	Capacitance	Reference
1	TFP-COFs	Polycondensation	291.1 F g^{-1}	[42]
2	TpPa-COF@PANI	One-pot solvothermal	95 F g^{-1}	[43]
3	Hex-Aza-COFs	Solvothermal condensation reaction	585 F g^{-1}	[45]
4	PI-COF	Solvothermal method	163 F g^{-1}	[46]
5	PDC-MA-COF	Aldehyde-amine condensation reaction	335 F g^{-1}	[47]

PDC-MA-COF produced by aldehyde amine condensation reaction displayed an enhanced energy density of 29.2 W h/kg and force thickness of 750 W/kg. Additionally, it showed brilliant cyclic stability and could hold 88% capacitance after 20000 charge-discharge cycles. Table 1.2 summarizes some examples of COFs, synthesis/methods, and supercapacitor capacity.

1.3.5 Solar Cells

Because of the worldwide utilization of energy, the request for environmentally clean energy sources has constantly raised. Solar energy can be effectively trapped by a tool popularly known as a solar cell. A solar cell is a device that uses the sun's energy and converts it into electric energy. COFs with crystalline and conjugated characteristics and high-dimensional structure have attracted a lot of interest from researchers who want to use them in photovoltaic applications including dye-sensitized solar cells (DSSCs), perovskite solar cells (PVSCs), and so on. COFs possessing crystalline and conjugated properties with high-dimensional geometry have gained incredible attention of researchers and implementing that in photovoltaic applications including dye-sensitized solar cells (DSSCs), perovskite solar cells (PVSCs), etc. Both two-dimensional and three-dimensional COFs can be used as photovoltaic cells. For instance, 2D COF with the core building unit of tetraphenylethylene and linkage of bicarbazole and pyrene showed a remarkable increase in power conversion energy (PCE). Furthermore, Pyrene and Bicarbazole when attached to 4, 4', 4", 4'''-(ethane-1, 1, 2, 2-tetrayl) tetranilino (ETTA) using the solvothermal process produces TFPPy-ETTA-COF and Car-ETTA-COF. Both COFs showed high crystallinity, good thermal stability, good π-π interactions, and well-conjugated properties. The PCE for Car-ETTA-COF was 19.79% whereas 19.72% was reported for TFPPy-ETTA-COF [48]. Additionally, to increase the capability of solar cells, 2D covalent organic nanosheets can also be used. For example, both CON-16 and CON-10 synthesized by the Stille coupling reaction showed a remarkable increase in PCE. It was reported that CON-16 showed PCE of 6.10% whereas CON-10 showed a PCE of 10.17% which is almost 4% greater than CON-16 [49].

Three-dimensional COFs consisting of large surface area and more defined porosity can also be used to enhance perovskite solar cells or photovoltaic cells. Three-dimensional COFs are limited; therefore, work on them is less reported to date. A 3D COF consists of spirobifluorene as a building unit with imine bonds. SP-3D-COFs when doped between the layers of perovskite solar cells showed enhancement in photoconduction, thermal stability, crystallinity, permanent porosity, etc. The reported enhancement was 18.0% for SP-3D-COF2 and 15.9% for SP-3D-COF1 while which was high compared to its reference undoped solar cells [50]. Table 1.3 gives the PCE of different COFs.

TABLE 1.3

Some Device Parameters for COF-Based Solar Cells

No.	Dimension (2D/3D)	COFs Used	Fill Factor (FF) (%)	PCE (%)	References
1.	2D-COF	TFPPy-ETTA-COF	78.34	19.72	[48]
2.	2D-COF	Car-ETTA-COF	77.84	19.79	[48]
3.	2D-COF	CON-10	0.65	10.17	[49]
4.	2D-COF	CON-16	0.45	6.10	[49]
5.	3D-COF	SP-3D-COF1	79.8	15.9	[50]
6.	3D-COF	SP-3D-COF2	78.3	18.0	[50]

1.4 Conclusion and Future Perspective

COFs being inventive categories of porous materials bearing high surface territory, controlled porosity, geometry, and tunability are capable of confronting environment and energy-related issues. From the past few decades, the structural frameworks, functionality, and synthesis of 2D COFs have been actively explored, but several important issues regarding COFs are still to be resolved. Firstly, synthesis, properties, kinetics, thermodynamics as well its crystallization is still not well clear about their future utility. Different types of COFs are synthesis methods. No general synthesis method is reported till now which can synthesize the maximum number of COFs. Secondly, COFs synthesis methods require selected solvent systems say mesitylene, 1–4 dioxane, acetone in particular ratios. Furthermore, COFs with high crystallinity requires harsh conditions and sealed vessel. For instance, COF-5 synthesized by solvothermal method requires a long time (72 hours) under harsh conditions whereas the same, when synthesized by microwave method, requires only a few hours required. However, the structural properties and crystallinity of the COFs also change on changing the synthetic method. To avoid these hindrances, room temperature synthesis methods for COF synthesis are required. Therefore, synthesis of COFs at moderate conditions is still challenging. Furthermore, COFs possess poor moisture tolerance capacity which hinders their gas storage applications. The photocatalytic and photovoltaic activity of COFs depends on light absorbance to produce more electron-hole pairs. Thus, constructing new COFs with high solar absorbance to improve the photocatalyzed reactions and stable crystalline porous framework is required which provides the best screening for covalent organic framework energy application in the future.

References

1. Mancheño, María José, Sergio Royuela, A. de la Peña, Mar Ramos, Félix Zamora, and José L. Segura. "Introduction to covalent organic frameworks: an advanced organic chemistry experiment." *Journal of Chemical Education* 96, no. 8 (2019): 1745–1751.
2. Zhao, Wei, Lieyin Xia, and Xikui Liu. "Covalent organic frameworks (COFs): perspectives of industrialization." *CrystEngComm* 20, no. 12 (2018): 1613–1634.
3. Wang, De-Gao, Tianjie Qiu, Wenhan Guo, Zibin Liang, Hassina Tabassum, Dingguo Xia, and Ruqiang Zou. "Covalent organic framework-based materials for energy applications." *Energy & Environmental Science* 14, no. 2 (2021): 688–728.

4. Bhunia, Sukanya, Kaivalya A. Deo, and Akhilesh K. Gaharwar. "2D covalent organic frameworks for biomedical applications." *Advanced Functional Materials* 30, no. 27 (2020): 2002046.
5. Waller, Peter J., Felipe Gándara, and Omar M. Yaghi. "Chemistry of covalent organic frameworks." *Accounts of Chemical Research* 48, no. 12 (2015): 3053–3063.
6. Lohse, Maria S., and Thomas Bein. "Covalent organic frameworks: structures, synthesis, and applications." *Advanced Functional Materials* 28, no. 33 (2018):1705553.
7. Lyle, Steven J., Peter J. Waller, and Omar M. Yaghi. "Covalent organic frameworks: organic chemistry extended into two and three dimensions." *Trends in Chemistry* 1, no. 2 (2019): 172–184.
8. Fan, Hongwei, Alexander Mundstock, Jiahui Gu, Hong Meng, and Jürgen Caro. "An azine-linked covalent organic framework ACOF-1 membrane for highly selective CO 2/CH 4 separation." *Journal of Materials Chemistry A* 6, no. 35 (2018): 16849–16853.
9. Song, Yanpei, Qi Sun, Briana Aguila, and Shengqian Ma. "Opportunities of covalent organic frameworks for advanced applications." *Advanced Science* 6, no. 2 (2019): 1801410.
10. Kandambeth, Sharath, Kaushik Dey, and Rahul Banerjee. "Covalent organic frameworks: chemistry beyond the structure." *Journal of the American Chemical Society* 141, no. 5 (2018): 1807–1822.
11. Ma, Tianqiong, Eugene A. Kapustin, Shawn X. Yin, Lin Liang, Zhengyang Zhou, Jing Niu, Li-Hua Li et al. "Single-crystal x-ray diffraction structures of covalent organic frameworks." *Science* 361, no. 6397 (2018): 48–52.
12. Ding, San-Yuan, and Wei Wang. "Covalent organic frameworks (COFs): from design to applications." *Chemical Society Reviews* 42, no. 2 (2013): 548–568.
13. Biswal, B. P., Chandra, S., Kandambeth, S., Lukose, B., Heine, T., and Banerjee, R. "Mechanochemical synthesis of chemically stable isoreticular covalent organic frameworks." *Journal of the American Chemical Society, 135,* no. 14 (2013): 5328–5331.
14. Lohse, M. S., and Bein, T. "Covalent organic frameworks: structures, synthesis, and applications." *Advanced Functional Materials, 28,* no. 33 (2018): 1705553.
15. Lv, Jiangquan, Yan-Xi Tan, Jiafang Xie, Rui Yang, Muxin Yu, Shanshan Sun, Ming-De Li, Daqiang Yuan, and Yaobing Wang. "Direct solar-to-electrochemical energy storage in a functionalized covalent organic framework." *Angewandte Chemie* 130, no. 39 (2018): 12898–12902.
16. Bhadra, Mohitosh, Sharath Kandambeth, Manoj K. Sahoo, Matthew Addicoat, Ekambaram Balaraman, and Rahul Banerjee. "Triazine functionalized porous covalent organic framework for photo-organocatalytic E–Z isomerization of olefins." *Journal of the American Chemical Society* 141, no. 15 (2019): 6152–6156.
17. Pachfule, Pradip, Amitava Acharjya, Jérôme Roeser, Thomas Langenhahn, Michael Schwarze, Reinhard Schomäcker, Arne Thomas, and Johannes Schmidt. "Diacetylene functionalized covalent organic framework (COF) for photocatalytic hydrogen generation." *Journal of the American Chemical Society* 140, no. 4 (2018): 1423–1427.
18. Stegbauer, Linus, Sebastian Zech, Gökcen Savasci, Tanmay Banerjee, Filip Podjaski, Katharina Schwinghammer, Christian Ochsenfeld, and Bettina V. Lotsch. "Tailor-made photoconductive pyrene-based covalent organic frameworks for visible-light driven hydrogen generation." *Advanced Energy Materials* 8, no. 24 (2018): 1703278.
19. Wang, Guang-Bo, Sha Li, Cai-Xin Yan, Fu-Cheng Zhu, Qian-Qian Lin, Ke-Hui Xie, Yan Geng, and Yu-Bin Dong. "Covalent organic frameworks: emerging high-performance platforms for efficient photocatalytic applications." *Journal of Materials Chemistry A* 8, no. 15 (2020): 6957–6983.
20. Yao, Yu-Hao, Jing Li, Hao Zhang, Hong-Liang Tang, Liang Fang, Gu-Dan Niu, Xiao-Jun Sun, and Feng-Ming Zhang. "Facile synthesis of a covalently connected RGO–COF hybrid material by in situ reaction for enhanced visible-light induced photocatalytic H$_2$ evolution." *Journal of Materials Chemistry A* 8, no. 18 (2020): 8949–8956.
21. Luo, Maolan, Qing Yang, Kewei Liu, Hongmei Cao, and Hongjian Yan. "Boosting photocatalytic H$_2$ evolution on GC 3 N 4 by modifying covalent organic frameworks (COFs)." *Chemical Communications* 55, no. 41 (2019): 5829–5832.
22. Chakraborty, Debanjan, Shyamapada Nandi, Rajith Illathvalappil, Dinesh Mullangi, Rahul Maity, Santosh K. Singh, Sattwick Haldar, Chathakudath P. Vinod, Sreekumar Kurungot,

and Ramanathan Vaidhyanathan. "Carbon derived from soft pyrolysis of a covalent organic framework as a support for small-sized RuO2 showing exceptionally low overpotential for oxygen evolution reaction." *ACS Omega* 4, no. 8 (2019): 13465–13473.

23. Zhuang, Gui-lin, Yi-fen Gao, Xiang Zhou, Xin-yong Tao, Jian-min Luo, Yi-jing Gao, Yi-long Yan, Pei-yuan Gao, Xing Zhong, and Jian-guo Wang. "ZIF-67/COF-derived highly dispersed Co3O4/N-doped porous carbon with excellent performance for oxygen evolution reaction and Li-ion batteries." *Chemical Engineering Journal* 330 (2017): 1255–1264.

24. Xu, Qing, Jing Qian, Dan Luo, Guojuan Liu, Yu Guo, and Gaofeng Zeng. "Ni/Fe clusters and nanoparticles confined by covalent organic framework derived carbon as highly active catalysts toward oxygen reduction reaction and oxygen evolution reaction." *Advanced Sustainable Systems* 4, no. 9 (2020): 2000115.

25. Gao, Zhi, Le Le Gong, Xiang Qing He, Xue Min Su, Long Hui Xiao, and Feng Luo. "General strategy to fabricate metal-incorporated pyrolysis-free covalent organic framework for efficient oxygen evolution reaction." *Inorganic Chemistry* 59, no. 7 (2020): 4995–5003.

26. Ma, Wenjie, Ping Yu, Takeo Ohsaka, and Lanqun Mao. "An efficient electrocatalyst for oxygen reduction reaction derived from a Co-porphyrin-based covalent organic framework." *Electrochemistry Communications* 52 (2015): 53–57.

27. Chen, Hao, Qiao-Hong Li, Wensheng Yan, Zhi-Gang Gu, and Jian Zhang. "Templated synthesis of cobalt subnanoclusters dispersed N/C nanocages from COFs for highly-efficient oxygen reduction reaction." *Chemical Engineering Journal* 401 (2020): 126149.

28. Jena, Himanshu Sekhar, Chidharth Krishnaraj, Shaikh Parwaiz, Florence Lecoeuvre, Johannes Schmidt, Debabrata Pradhan, and Pascal Van der Voort. "Illustrating the role of quaternary-N of BINOL covalent triazine-based frameworks in oxygen reduction and hydrogen evolution reactions." *ACS Applied Materials & Interfaces* 12, no. 40 (2020): 44689–44699.

29. Zhan, Xuejun, Zhong Chen, and Qichun Zhang. "Recent progress in two-dimensional COFs for energy-related applications." *Journal of Materials Chemistry A* 5, no. 28 (2017): 14463–14479.

30. Furukawa, Hiroyasu, and Omar M. Yaghi. "Storage of hydrogen, methane, and carbon dioxide in highly porous covalent organic frameworks for clean energy applications." *Journal of the American Chemical Society* 131, no. 25 (2009): 8875–8883.

31. Hu, Jinghao, Jianfei Zhao, and Tianying Yan. "Methane uptakes in covalent organic frameworks with double halogen substitution." *The Journal of Physical Chemistry C* 119, no. 4 (2015): 2010–2014.

32. Zhao, Hui, Yurou Guan, Hailong Guo, Renjun Du, and Cuixia Yan. "Hydrogen storage capacity on Li-decorated covalent organic framework-1: A first-principles study." *Materials Research Express* 7, no. 3 (2020): 035506.

33. Zhao, Jianfei, and Tianying Yan. "Effects of substituent groups on methane adsorption in covalent organic frameworks." *RSC Advances* 4, no. 30 (2014): 15542–15551.

34. Miner, Elise M., and Mircea Dincă. "Metal-and covalent-organic frameworks as solid-state electrolytes for metal-ion batteries." *Philosophical Transactions of the Royal Society A* 377, no. 2149 (2019): 20180225.

35. Xie, Jian, Peiyang Gu, and Qichun Zhang. "Nanostructured conjugated polymers: toward high-performance organic electrodes for rechargeable batteries." *ACS Energy Letters* 2, no. 9 (2017): 1985–1996.

36. Ghazi, Zahid Ali, Lingyun Zhu, Han Wang, Abdul Naeem, Abdul Muqsit Khattak, Bin Liang, Niaz Ali Khan, Zhixiang Wei, Lianshan Li, and Zhiyong Tang. "Efficient polysulfide chemisorption in covalent organic frameworks for high-performance lithium-sulfur batteries." *Advanced Energy Materials* 6, no. 24 (2016): 1601250.

37. Yang, Ziyi, Chengxin Peng, Ruijin Meng, Lianhai Zu, Yutong Feng, Bingjie Chen, Yongli Mi, Chi Zhang, and Jinhu Yang. "Hybrid anatase/rutile nanodots-embedded covalent organic frameworks with complementary polysulfide adsorption for high-performance lithium-sulfur batteries." *ACS Central Science* 5, no. 11 (2019): 1876–1883.

38. Zhang, Xue, Lu Yao, Shuai Liu, Qin Zhang, Yiyong Mai, Nantao Hu, and Hao Wei. "High-performance lithium sulfur batteries based on nitrogen-doped graphitic carbon derived from covalent organic frameworks." *Materials Today Energy* 7 (2018): 141–148.

39. Li, Bo-Quan, Shu-Yuan Zhang, Bin Wang, Zi-Jing Xia, Cheng Tang, and Qiang Zhang. "A porphyrin covalent organic framework cathode for flexible Zn-air batteries." *Energy & Environmental Science* 11, no. 7 (2018): 1723–1729.

40. Kong, Xueying, Shengyang Zhou, Maria Strømme, and Chao Xu. "Redox active covalent organic framework-based conductive nanofibers for flexible energy storage device." *Carbon* 171 (2021): 248–256.

41. Mohammed, Abdul Khayum, Vidyanand Vijayakumar, Arjun Halder, Meena Ghosh, Matthew Addicoat, Umesh Bansode, Sreekumar Kurungot, and Rahul Banerjee. "Weak intermolecular interactions in covalent organic framework-carbon nanofiber based crystalline yet flexible devices." *ACS Applied Materials & Interfaces* 11, no. 34 (2019): 30828–30837.

42. EL-Mahdy, Ahmed FM, Ying-Hui Hung, Tharwat Hassan Mansoure, Hsiao-Hua Yu, Yu-Shen Hsu, Kevin CW Wu, and Shiao-Wei Kuo. "Synthesis of [3+ 3] β-ketoenamine-tethered covalent organic frameworks (COFs) for high-performance supercapacitance and CO_2 storage." *Journal of the Taiwan Institute of Chemical Engineers* 103 (2019): 199–208.

43. Liu, Shuai, Lu Yao, Yi Lu, Xiaolin Hua, Jiaqiang Liu, Zhi Yang, Hao Wei, and Yiyong Mai. "All-organic covalent organic framework/polyaniline composites as stable electrode for high-performance supercapacitors." *Materials Letters* 236 (2019): 354–357.

44. Halder, Arjun, Meena Ghosh, Abdul Khayum M, Saibal Bera, Matthew Addicoat, Himadri Sekhar Sasmal, Suvendu Karak, Sreekumar Kurungot, and Rahul Banerjee. "Interlayer hydrogen-bonded covalent organic frameworks as high-performance supercapacitors." *Journal of the American Chemical Society* 140, no. 35 (2018): 10941–10945.

45. Kandambeth, Sharath, Jiangtao Jia, Hao Wu, Vinayak S. Kale, Prakash T. Parvatkar, Justyna Czaban Jóźwiak and Sheng Zhou "Covalent organic frameworks as negative electrodes for high performance asymmetric supercapacitors." *Advanced Energy Materials* 10, no. 38 (2020): 2001673.

46. Iqbal, Rashid, Amir Badshah, Ying-Jie Ma, and Lin-Jie Zhi. "An electrochemically stable 2D covalent organic framework for high-performance organic supercapacitors." *Chinese Journal of Polymer Science* 38, no. 5 (2020): 558–564.

47. Li, Li, Feng Lu, Rui Xue, Baolong Ma, Qi Li, Ning Wu, Hui Liu, Wenqin Yao, Hao Guo, and Wu Yang. "Ultrastable triazine-based covalent organic framework with an interlayer hydrogen bonding for supercapacitor applications." *ACS Applied Materials & Interfaces* 11, no. 29 (2019): 26355–26363.

48. Mohamed, Mohamed Gamal, Chia-Chen Lee, Ahmed FM EL-Mahdy, Johann Lüder, Ming-Hsuan Yu, Zhen Li, Zonglong Zhu, Chu-Chen Chueh, and Shiao-Wei Kuo. "Exploitation of two-dimensional conjugated covalent organic frameworks based on tetraphenylethylene with bicarbazole and pyrene units and applications in perovskite solar cells." *Journal of Materials Chemistry A* 8, no. 22 (2020): 11448–11459.

49. Park, Soyun, Min-Sung Kim, Woongsik Jang, Jin Kuen Park, and Dong Hwan Wang. "Covalent organic nanosheets for effective charge transport layers in planar-type perovskite solar cells." *Nanoscale* 10, no. 10 (2018): 4708–4717.

50. Wu, Chenyu, Yamei Liu, Hui Liu, Chenghao Duan, Qingyan Pan, Jian Zhu and Fan Hu. "Highly conjugated three-dimensional covalent organic frameworks based on spirobifluorene for perovskite solar cell enhancement." *Journal of the American Chemical Society* 140, no. 31 (2018): 10016–10024.

2

Materials, Chemistry, and Synthesis of Covalent Organic Frameworks

Mohaddeseh Afshari and Mohammad Dinari

Department of Chemistry, Isfahan University of Technology, Isfahan, Islamic Republic of Iran

CONTENTS

2.1   Introduction

Covalent bonds have a special place in the synthesis of organic materials due to their strength and diversity. The various atoms are joined together through covalent bonds to form molecules. Covalent bonding of molecules results in giant lattice solids with long-range order such as silicon carbide or diamond, which are among the hardest natural materials on the earth [1].

DOI: 10.1201/9781003206507-2

It is necessary for organic researchers to know the formation chemistry of these bonds in the zero dimension and to have complete control over them to synthesize new structures [2]. Covalent chemistry has been extended to one dimension by polymer chemists. Although extended covalently networked solids are widely present; there was no synthetic analog until the last two decades. In 2005, Yaghi's research group opened new doors for scientists to synthesize organic materials in higher dimensions [3]. They synthesized porous crystalline covalent organic frameworks (COFs) by linking symmetrical organic building blocks.

COFs are a new class of porous polymers in which small organic building blocks are precisely integrated into extended structures with periodic skeletons and regular pores [4]. During the formation of networked covalent structures, unlike zero-dimensional and one-dimensional organic structures, several frameworks with different free energies may be created; as a result, it seems complicated to achieve a crystalline and perfectly regular structure. The chemical reaction process is always like a swing between the kinetic and thermodynamic equilibria. Over recent years, a new concept called dynamic covalent chemistry (DCC) has been introduced in which reversible chemical reactions are performed under thermodynamic control [5, 6]. The reversible nature of the reactions allows error checking and proof-reading in the generated covalent structure to obtain the thermodynamic product with the lowest free energy and convenient crystallinity (Figure 2.1) [7]. The point to note is the high energy of covalent bonds (50–110 kcal mol^{-1}), so to achieve a thermodynamic product, the reactions must be performed at high temperature and pressure conditions.

As only a few monomers can tolerate such harsh conditions, the variety of usable building blocks to prepare physicochemical stable COFs significantly reduces. Given the above, thermal reversibility could not be a proper solution for making crystalline COFs. In this regard, alternative conditions to provide reversible reactions to accede a crystalline thermodynamic product have been investigated. In DCC, to maintain reversible conditions, all influencing factors are fully monitored. For instance, in the chemically induced DCC, a special chemical agent such as water molecules is used to maintain the reversibility of the reaction during the formation of covalent bonds in two or three dimensions frameworks [8]. Under closed reaction conditions, water molecules, as a by-product, can reverse the equilibrium toward the reaction, eventually leading to the high crystalline regular COF structures.

One of the outstanding features of COFs that distinguishes them from other porous polymers is their designability structures. Since the geometry of the building blocks does not change during the covalent bond formation, the geometry and dimensions of building blocks can determine the final synthesized structure topology [9]. Topology diagrams help us design the shape and size of cavities in the final skeletons based on the intended application before synthesis [10].

In COFs synthesis, similar to many other polymers, various methods are used to institute covalent bonds between organic building units such as solvothermal, ionothermal, microwave, and so forth [11]. Developing new strategies for accurately designing structures to achieve targeted COFs with high crystallinity and porosity under mild conditions remains a significant challenge for researchers. In recent years, various strategies have been reported for synthesizing COFs that have not already been possible and upgrading the crystallinity and porosity of previous synthesized structures.

So far, various COFs have been made using a wide range of organic linkers by different types of symmetric combinations. The introduction of highly regular crystalline COFs opened a new research area for scientists. The ability to pre-design covalent frameworks allows the systematic arrangement of building units next to each other, leading to the development of extended networked solids with the desired physicochemical properties. These robust crystalline lattice solids can be suitable candidates for use in a wide range of

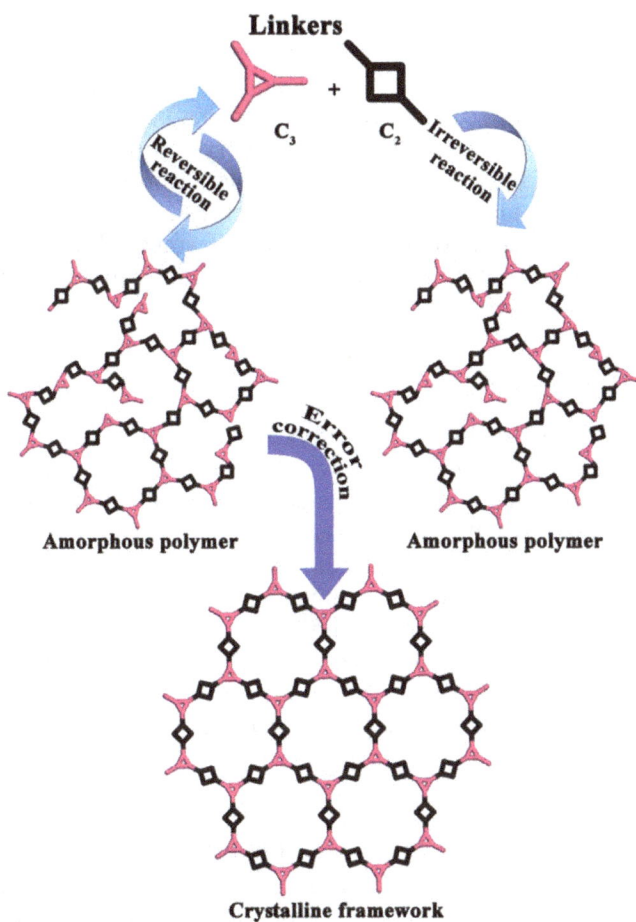

FIGURE 2.1
The schematic illustration of amorphous and crystalline structures formation through irreversible and reversible reactions.

applications such as the adsorption of guest molecules (dyes, drugs, toxins, etc.), optoelectronics, heterogeneous catalysts, and even ion transfer membranes due to heaving very regular π-stacked layered structures with designable shape and size pores

2.2 Affecting Parameters on the COFs Structure

In general, three key factors, monomer diversity, monomer geometry, and reaction chemistry, can determine the final structure of a covalent framework.

2.2.1 Chemical Diversity of Monomers

Monomers or units of construction used to prepare the extended crystalline covalent frameworks cover a wide range of π structures with different sizes and functions, including simple arenas, heterocycles, macrocycles, coordinated filaments, and extended π systems. Each of the monomers has special functional groups that determine the physicochemical properties of the final COF structure. What is important in choosing monomers to achieve directional connections and regular growth of polygons is their rigidity. Polycyclic aromatic structures can be a good option for preparing crystalline covalent frameworks. The more rigid monomers reduce the possibility of rotation around the bonds and irregularity, thus facilitating access to regular periodic structures. The wide variety of available monomers has resulted in a myriad of unique organic frameworks with well-defined wall environments for different applications. However, this alone cannot be enough to meet all expectations and build new structures.

2.2.2 Geometry of Monomer

Orientation of covalent bonds is an accurate tool for controlling how building units come together. As mentioned earlier, monomers have a definite geometry and symmetry. Since the geometry of the building units used in COF assembling is maintained during the reaction and the organic units directionally react with each other, it can be inferred that matching the symmetry of these units with each other and the geometry of these units control the development of polygonal skeletons in two or three dimensions, ultimately leading to the production of extended COF networks in predetermined directions. Polygons are usually composed of knot and linker units, the symmetry and size of which determine the shape of the polygon and the size of the cavities, respectively. So far, COFs with trigonal, tetragonal, hexagonal, kagome, and rhombic polygons have been prepared using C_2, C_3, C_4, and C_6- symmetric units.

2.2.3 Chemistry of COFs

It can be said that reaction chemistry is the most important factor in preparing a regular covalent network structure. The "COFs chemistry" or "reaction chemistry" refers to DCC, which has become more common in recent articles around network structures such as MOFs and COFs. According to this principle, conditions must be created to produce a crystalline structure so strong covalent bonds can be broken and re-formed. DCC allows the structure to overcome potential defects and produce the product with the lowest energy level and maximum stability. Since COFs are usually synthesized from organic units with a π system, in two-dimensional COFs, the interlayer forces resulting from the π-stacking between the aromatic rings also help achieve a crystalline structure with low free energy. In this way, to provide the reversible reaction, various conditions such as selecting the type of monomer, using condensation reactions with a poorly soluble by-product in the reaction medium, changing the polarity of the solvent, adjusting the space of the reaction vessel in an isolated system, and so on are used. Each of these issues directs the reaction process toward thermodynamic equilibrium and the production the more desirable products.

2.3 Design Principles

The integration of different synthetic reactions with topology diagrams allows the synthesis of many COF with pre-designed skeletons and pores. A topology design diagram can accurately determine the direction of covalent bonds formation and the growth of the polymer backbone. As mentioned, each monomer has a specific geometry that indicates the relative position of the reactive sites. Due to the retention of the monomers' geometry during the reaction, the created covalent bonds determine the relative position of the subsequent building units. This process continues at each connection so that the growth of the chains and sheets follows the designed topology diagram.

2.3.1 2D-COF Design

The combination of planer monomers limits the growth of COF structure in a 2D manner along the X and Y axes. The constituted 2D layers stack in such a way that similar monomer units are overlapped. This alignment of 2D layers along the z-direction and establishing noncovalent π–π stacking interactions between adjacent layers create high-order 2D COF structures. This feature of two-dimensional COFs not only increases their stability due to the π-arrays but also the well-defined one-dimensional channels formed along the z-axis, providing accessible sites for a variety of applications. So far, many 2D COFs have been synthesized using both regular and irregular polygonal skeletons. The diversity of available monomers alone cannot provide a wide variety of COFs. Using topological diagrams, it is possible to create different combinations of monomers geometries to design 2D and 3D COFs (Figures 2.2a and 2.2b, respectively) with different skeletons and pores. For instance, the combinations of $[C_3 + C_2]$ and $[C_3 + C_3]$ lead to the formation of two-dimensional COFs with hexagonal pores, but the resulting structures will have different pore sizes and π-density. Tetragonal structures can also be produced using $[C_4 + C_4]$ and $[C_4 + C_2]$ combinations [10].

A remarkable point in designing structures using these topology diagrams is the type and size of the pores. In addition to designing the structure of the polymer network, topology diagrams can also determine the size of pores. One attractive topology diagram is the $[C_2 + C_2]$ combination that yields two different kagome and rhombic structures (Figure 2.2a). The rhombic skeleton has one kind of pore. In contrast, the kagome skeleton has two different pore types, six periphery micro trigonal pores and one hexagonal mesopore in the middle of them. Even depending on the type of C_2-symmetric knots, the structure of the periphery pores may be pretty different. As another example, the generated

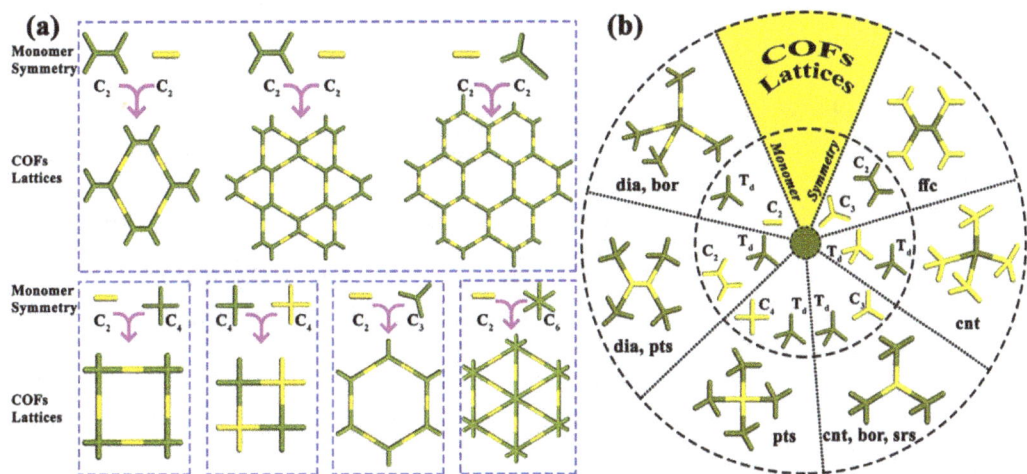

FIGURE 2.2
The fundamental topology diagrams for designing the (a) two-dimensional and (b) three-dimensional COFs.

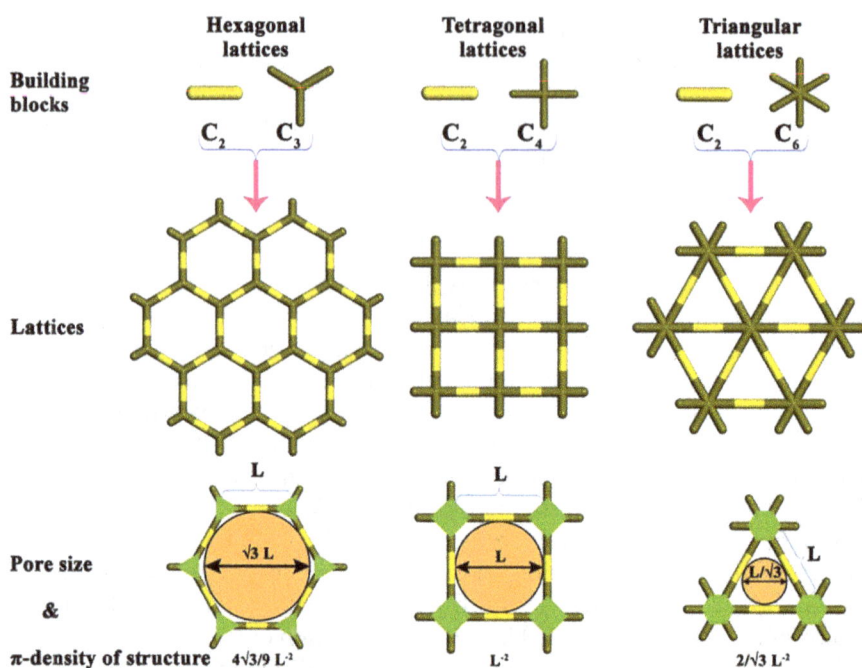

FIGURE 2.3
Determination of pore size and π-density of COF structures using designed topology diagrams.

COF structures from $[C_4 + C_2]$, $[C_3 + C_3]$, and $[C_3 + C_2]$ combinations often have meso-pores, but the combinations such as the $[C_4 + C_4]$, usually construct microporous COFs. Topology diagrams also provide information about the π-density of two-dimensional polygonal structures. As shown in Figure 2.3, the lattice structure based on the $[C_6 + C_2]$ combination has the smallest pore size $(L/\sqrt{3})$ and the highest π-density $(2/\sqrt{3}\ L^{-2})$ than $[C_4 + C_2]$ and $[C_3 + C_2]$ topology diagrams, which are more commonly used for the COFs synthesis [12].

2.3.2 3D-COF Design

COF structures are frequently made up of [1 + 1] systems consisting of one linker and one knot. To produce 3D COFs, at least one of these units must have a non-planar orthogonal or T_d geometry. Hitherto, various combinations of integrating orthogonal or T_d knots with planer linkers such as C_2, C_3, and C_4 and non-planer T_d linkers have been proposed. Unlike two-dimensional COFs, in which the backbone grows only in the X and Y direction, the 3D COF skeleton grows in the direction of all three axes, yielding an extended 3D COF net-work. The 3D COFs structures are classified into the bor, ctn, srs, dia, pts, and rra networks (Figure 2.2b) [13]. Some of these networks are accessible from different topology diagrams. For example, pts network is obtained from two $[T_d + C_4]$ and $[T_d + C_2]$ combinations, in which C_2- and C_4-symmetric units must contain four active sites. The 3D COFs, because of their interpenetrated structures, are hardly pre-designating and synthetically control-lable. These structures are mostly microporous and show less porosity than expected due to their interpenetrate networks.

2.3.3 Multiple-Component Topologies

So far, only the topology diagrams of two-component systems, including a type of linker and a type of knot, have been discussed that are alternately connected based on the geometry of the knot unit to form the final structure of COF. This combination of building units greatly limits the variety of COFs that can be produced. Consider that you have one monomer as a knot and 10 different monomers as linkers. Based on two-component topology diagrams, you can synthesize only 10 different COFs. To overcome this limitation, the multiple-component (MC) concept was introduced in the design of topology diagrams [14]. The multiple-component strategy essence is to use two or three different linkers simultaneously for the COF structures design [10]. For example, the [1 + 2] three components combination of one C_3-symmetric knot and two C_2-symmetric linkers based on the ratio of linkers leads to the production of two different types of COFs with altered hexagonal pores (Figure 2.4). A noteworthy point in the prepared structures using the MC strategy compared to the [1 + 1] two-component strategy is the existence of polygonal with unusual pore shapes, which results from the existing linkers with different lengths. Using this strategy, it is possible to have several types of pores with different dimensions and shapes in one COF structure, significantly increasing COF structures' applicability in particular applications.

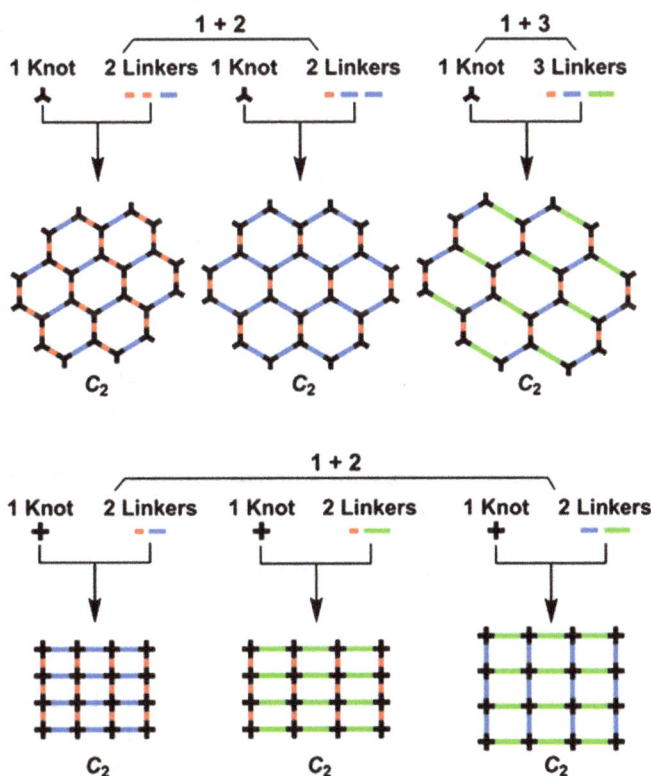

FIGURE 2.4
[1 + 2] or [1 + 3] multicomponent strategy for design the (a) hexagonal and (b) tetragonal COFs. Reproduced with permission from reference [14]. Copyright (2016) Springer Nature.

2.4 Synthesis of COFs

2.4.1 Synthetic Strategies

Up to now, various synthetic strategies have been studied and used to control the growth and regulate the properties of COFs to achieve structures with the highest crystallinity and order. Among the most important of them can be modulation strategy, to elevate the crystallinity; framework isomerism and post-synthetic linker exchange after synthesis strategies, to achieve structures that are not ordinarily synthesizable; finally, mixed linker/linkage strategy engineer the composition and function of COFs.

2.4.1.1 Modulation Strategy

The main focus of this strategy is on the reaction chemistry of the COFs and exploring ways to reinforce the dynamic nature of these reactions. As mentioned, the reversibility of the reaction can be modulated through factors such as the low-soluble by-product formation during condensation reactions (e. g., Shiff base reaction, formation of boronate ester, self-condensation of boronic acid), sealed reaction tube headspace, and the polarity of solvents. But it does not end there. The crystallization is the most challenging issue in manufacturing these structures and can only be acquired by designing an immense and time-consuming set of experiments. Therefore, more sophisticated methods have been used to modulate the reactions of COFs. Using monofunctional terminating ligands, which are also employed in the synthesis of MOF structures, is one of the most efficient methods in this regard [15–17]. These modulators are continuously attached to the end of the sheet and separated from it again, thereby reducing the rate of structure formation (Figure 2.5a). As a result, the time required is provided to check and eliminate the defects, then construct a high crystalline structure. The first studies on the growth mechanism of COFs were performed on COF-5 which was a 2D-COF based on boronic ester linkages [18–21]. Dichtel's research group used 4-tert-butylcatechol (TCAT) as a monofunctional ligand to synthesize COF-5 [18]. The obtained results indicated that the size of the crystals increased from 23 nm to 32 nm. They were able to increase the size of the crystals to 450 nm by increasing the amount of TCAT ($\geq$ 10 eq. to 1,4-benzenediboronic acid).

Bein et al. Used monoboronic acid derivatives instead of TCAT [22]. The results disclosed that monoboronic acid could be a more convenient modulator in synthesizing crystals up to several hundred nanometers in size from COF-5. The observations revealed that if 10% of the 4-mercaptophenylboronic acid is used (while the whole boric acid groups remain constant) as a monofunctional ligand, the surface area will be m^2 g^{-1} and a pore volume of 1.14 cm^3 g^{-1} (the values in the unmodulated structure were 1200 m^2 g^{-1} and 0.64 cm^3 g^{-1}, respectively). Unlike the TCAT, monoboronic acid modulators remain in the final structure; therefore, structural post-modification using various monoboronic acid derivatives is possible. For example, as shown in Figure 2.5b, methoxy-PEG-Maleimide has been successfully coated on the surface of COF-5 via Michael's addition reaction. Also, when 4-carboxyphenyl boronic acid was used as a modulator, the ATTO 633 dye was coated on the surface through the amide condensation reaction. The ability of the modulator ligands to form and break the bond with the structure is the most meaningful factor in selecting a ligand as the monofunctional terminating ligand.

Many modulator ligands have been studied in the imine condensation reaction. However, the results illustrated that only aniline could modulate the reaction to achieve

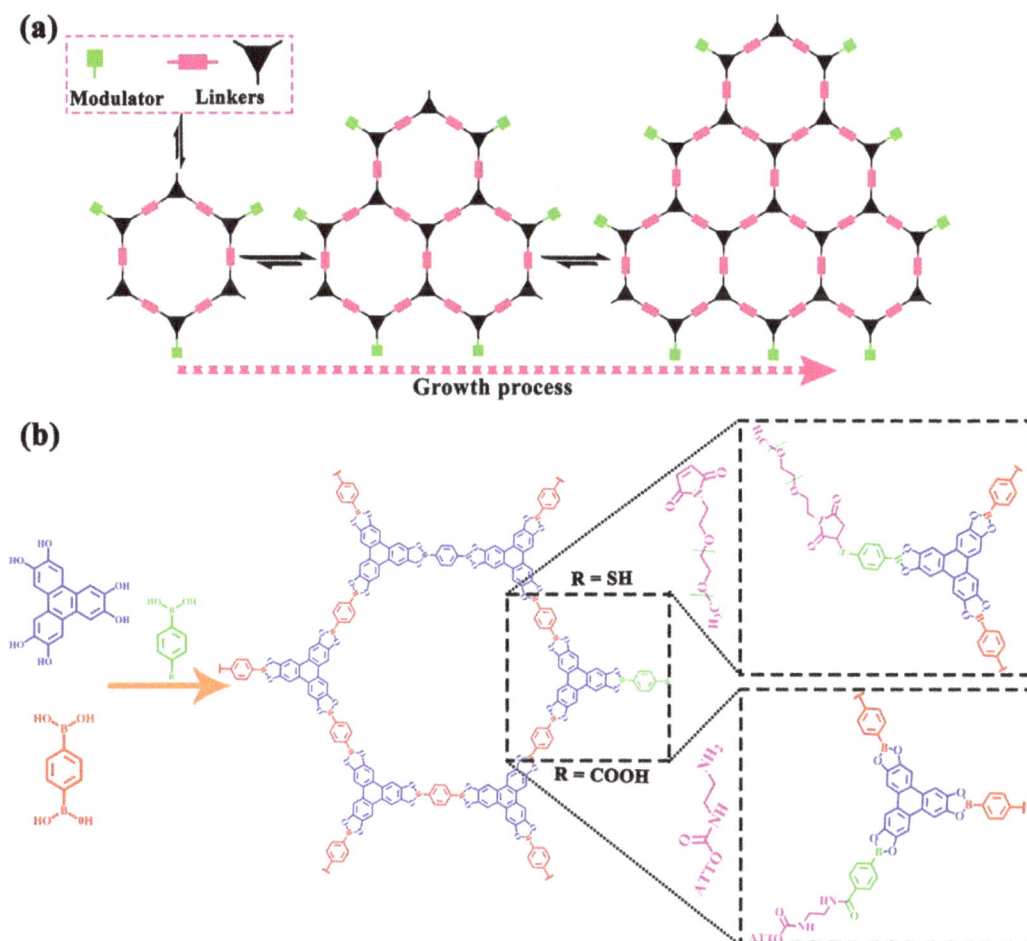

FIGURE 2.5
(a) Schematic representation of modulator's effect on the synthesis of polymers, (b) post-synthetically modification of COF-5 in the presence of MPBA modulator.

regular crystalline COFs. For instance, single crystals of COF-300 crystallized out with a size of about 60 μm within 40 days at ambient temperature in the presence of 15 eq. of aniline. If given enough time (about 80 days), the size of the crystal can even reach 100 μm [23, 24]. The observations of Liu, Wang, and colleagues showed that when two aniline and benzaldehyde ligands were used as modulators simultaneously, depending on the composition of the solvents and the monomer concentrations, highly regular COFs with controlled morphologies including spherical, short tubes, or hollow fibroids were obtained [25]. Here, Sc(OTf)$_3$ catalysts instead of acetic acid have paved the way for controlling morphology and achieving the selective structure. In conventional methods of COF synthesis, first swiftly deposited amorphous solid. Then, over time, the structure metamorphoses and crystallizes. This phenomenon demonstrates an uncontrolled nucleation and growth process. Theoretical and experimental studies portend that nucleation and growth are strongly dependent on the concentration of the primary monomers; as a result, a threshold limit can be considered for the concentration of the monomers. At below the threshold concentration, the growth process prevails over nucleation. For the nucleation rate-controlling, the concentration of monomers can be controlled in two ways: (I) the gradual increase of monomers (II) in situ release of protected monomers in the reaction medium. This strategy has been frequently used to synthesize imine-linked COFs by protonation of amine functional groups or protection of these with tertbutyloxycarbonyl and N-aryl benzophenone also using protected aldehyde monomers as imine or acetal structures. Reaction solvent can be mentioned as the last factor modulating to achieve the crystal structure. The mixture of the reaction solvents based on their solubility and polarity will significantly affect the final morphology. Dichtel and colleagues found that adding MeCN to the mixture of mesitylene and 1,4-dioxane solvents for the synthesis of COF-5 prevents the accumulation and stabilizes the formed crystal grains, which will lead to the growth of larger crystals [26].

2.4.1.2 Mixed Linker/Linkage Strategy

COFs are often produced by self-condensation or two-component condensation reactions. As a result, generating COFs with new and diverse topologies requires unique monomers obtained through complex organic reactions. As discussed earlier, using several types of linkers in one pot allows for frameworks with developed functions and exceptional structures. This strategy, which has also been used to synthesize MOFs, is denoted as the mixed linker/linkage approach. Depending on the type of used linker, this strategy is divided into two groups: the isostructural mixed linker (IML) strategy, in which the linkers provide a structure with the same topology but different composition than when only one type of linker is used, and the heterostructural mixed linker (HML) approach in which the obtained network from the linkers has a different structure than the case with one linker (Figure 2.6a). The used linkers in the IML strategy have similar dimensions and connection geometries; however, some secondary features, such as hanging functional groups, are different. Using a three-component reaction between 1,4-benzenediboronic acid (BDBA), 2,3,6,7,10,11 hexahydroxytriphenylene (HHTP), and azide-appended BDBA (N_3-BDBA), Jiang and colleagues gave a set of COFs with controlled amounts of azide groups [27]. A great point in the produced azide structures was post-synthetic click reaction with various alkynes, which has produced acetyl, ester, aromatic unit, chromophoric unit, and alkyl chain decorated COF walls (Figure 2.6b). Observations disclosed that increasing the percentage of azide-containing monomers in the structure decreases the surface areas, pore size, and pore volume.

In addition, linkers with a reduced number of reaction sites, called truncated monomers, also play an important role in IML strategy. A truncated monomer can participate in parts of the network and create a so-called defective structure. It can also operate as a capping agent at the grain boundary [28–30]. Of course, this should not be confused with the modulator monomers aforesaid above because truncated monomers have limited reversibility and cannot improve crystal growth. Unlike the modulation strategy, which focused on increasing the crystallization of frameworks, the main focus is on increasing the functional groups and modifying the structures after synthesis to produce specific structures for different applications. In the HML approach, ligands with different dimensions are used. Several COF structures with different topologies and cavities can be created by a constant combination utilizing the HML strategy. Zhao and colleagues synthesized two-dimensional imine-based COFs with three different types of pores using a condensation reaction between two C_2-symmetric dialdehydes (1:1 molar ratio) and D_{2h}-symmetric

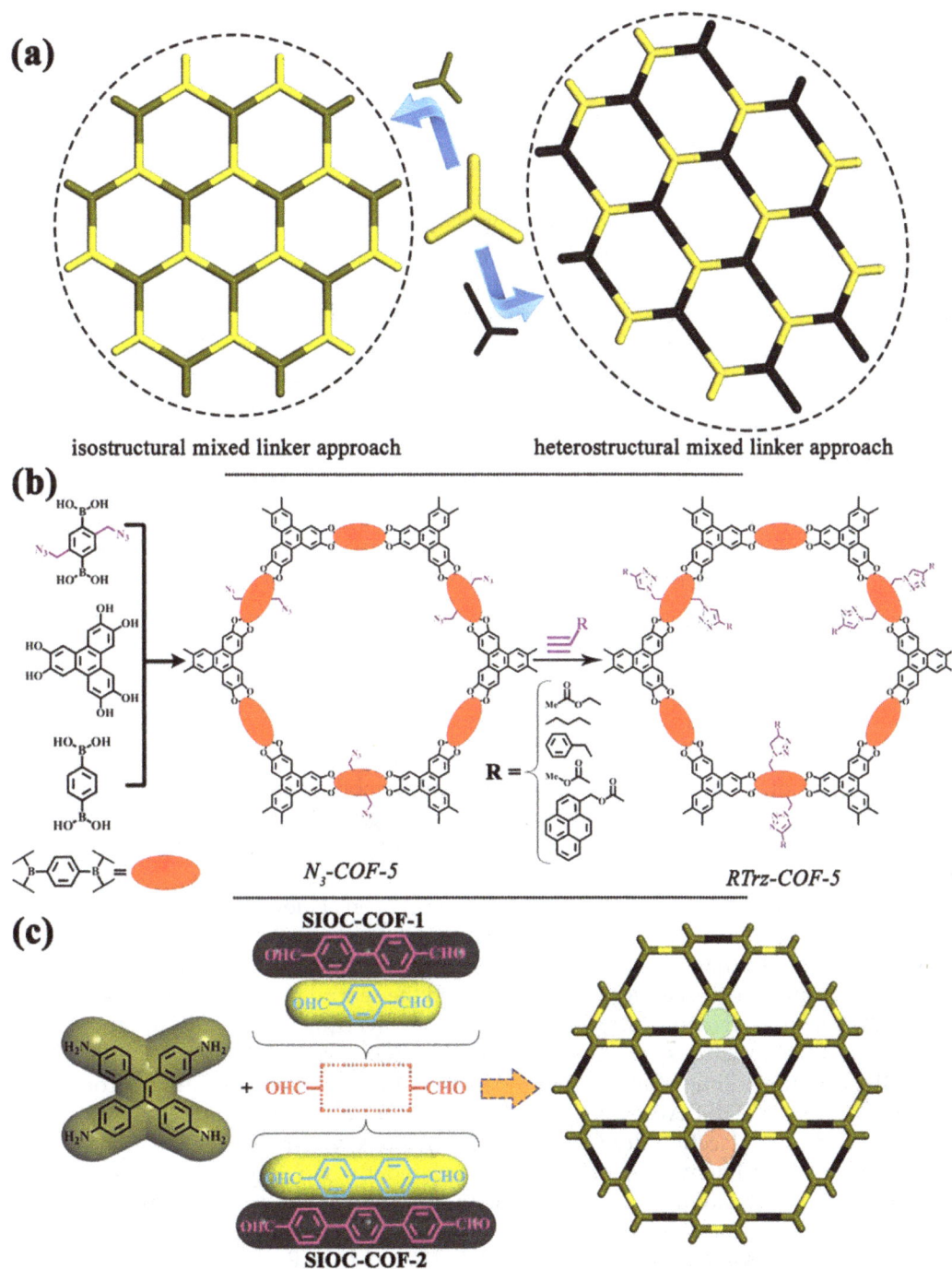

FIGURE 2.6

(a) Regular and stretched hexagonal pores formation from IML and HML approaches, respectively, schematic illustration of (b) N_3-COF-5 synthesis from HHTP, BDBA, and N_3-BDBA linkers and conversion of prepared COF to RTrz-COF-5 structure in the presence of various alkynes, and (c) triple-pore COFs synthesis of two different C_2-symmetric linkers and D_{2h}-symmetric knot.

4,4′,4″,4‴-(Ethene-1,1,2,2-tetrayl) tetraaniline monomers (Figure 2.6c) [31]. The existence of three different types of pores in the synthesized COFs was confirmed using BET analysis.

2.4.1.3 Framework Isomerism Strategy

In the chemistry of small organic molecules, compounds with different chemical structures but the same molecular formula are so-called structural isomers. The "framework isomerism" phrase was first defined for MOFs by Zhou et al. [32]. According to this definition, MOFs with different lattice structures composed of identical ligands and central metal nuclei are structural isomers relative to each other. Similarly, COF isomers are COFs with varying lattice structures constructed from the same linkers and knots. Unfortunately, there are limited examples in this area, so this issue has not been studied comprehensively. Recently, due to the need for broad structural diversity, this strategy has received more attention.

For example, TPE-COF-I and TPE-COF-II COFs are framework isomers. As seen in Figure 2.7a, TPE-COF-I has rhombic pores and lacks free functional groups, while in TPE-COF-II, the dimensions of the pores are larger (nearly 4 times), and half of the unreacted aldehyde functional groups are present in the network structure [33]. Using the imine condensation reaction between p-Phenylenediamine (PPD) and 4,4′,4″,4‴-(pyrene-1,3,6,8-tetrayl)tetrabenzaldehyde (PyTTBA), Wang and colleagues obtained three two-dimensional COFs with different morphologies based on the regulation of initial concentrations of monomers [34]. STM images of the COFs well confirm the distribution of species with different morphologies (Figure 2.7b).

2.4.1.4 Post-Synthetic Linker Exchange Strategy

The DCC concept, as the main factor in creating a thermodynamically reversible reaction, provides an opportunity to regulate COFs during the growth and crystallization process and allows post-synthesis restructuring through linker exchange. In MOFs chemistry, using solvent-assisted linker exchange approach, so-called SALE, has synthesized various MOF structures with high crystallinity that have not been achieved through conventional methods. It has not been long since the SALE was introduced to synthesize COF structures [35]. This may be due to the greater strength of covalent bonds in COF frameworks than the coordination interactions between the ligands and the central metal in the MOF structure. Using the linker exchange strategy, in addition to converting some COFs to other COFs, amorphous COP structures can also be turned into COF crystals.

For the first time, Zhao et al., in 2017, using the linker exchange technique, converted TP-COF-BZ to TP-COF-DAB by means of 10 eq. of p-Phenylenediamine in a mixture of mesitylene/dimethylacetamide/3 M AcOH (21/1/2.2, v/v) for 72 h at 120°C. The high concentration of p-Phenylenediamine is considered the driving force of the reaction (Figure 2.8a) [36]. As an interesting example in this regard, Yan and colleagues reported the synthesis of amine-functionalized imine-based COFs through the linker exchange strategy (Figure 2.8b). They could convert COF PTPA to COF PTBD-NH$_2$ and COF PTBD to COF PTPA-NH$_2$ in the presence of 10 eq. of 3,3′-diaminobenzidine (BD-NH$_2$) and 1 eq. of 1,2,4-benzenetriamine (PA-NH$_2$), respectively. The higher activity of PA-NH$_2$ rather than BD-NH$_2$ has led to the need for much smaller amounts of this linker [37]. Most importantly, the COFs obtained using this strategy usually are inaccessible and only obtained as amorphous materials. Using the linker exchange approach can prevail over many challenges associated with input monomer in the direct synthesis method, such

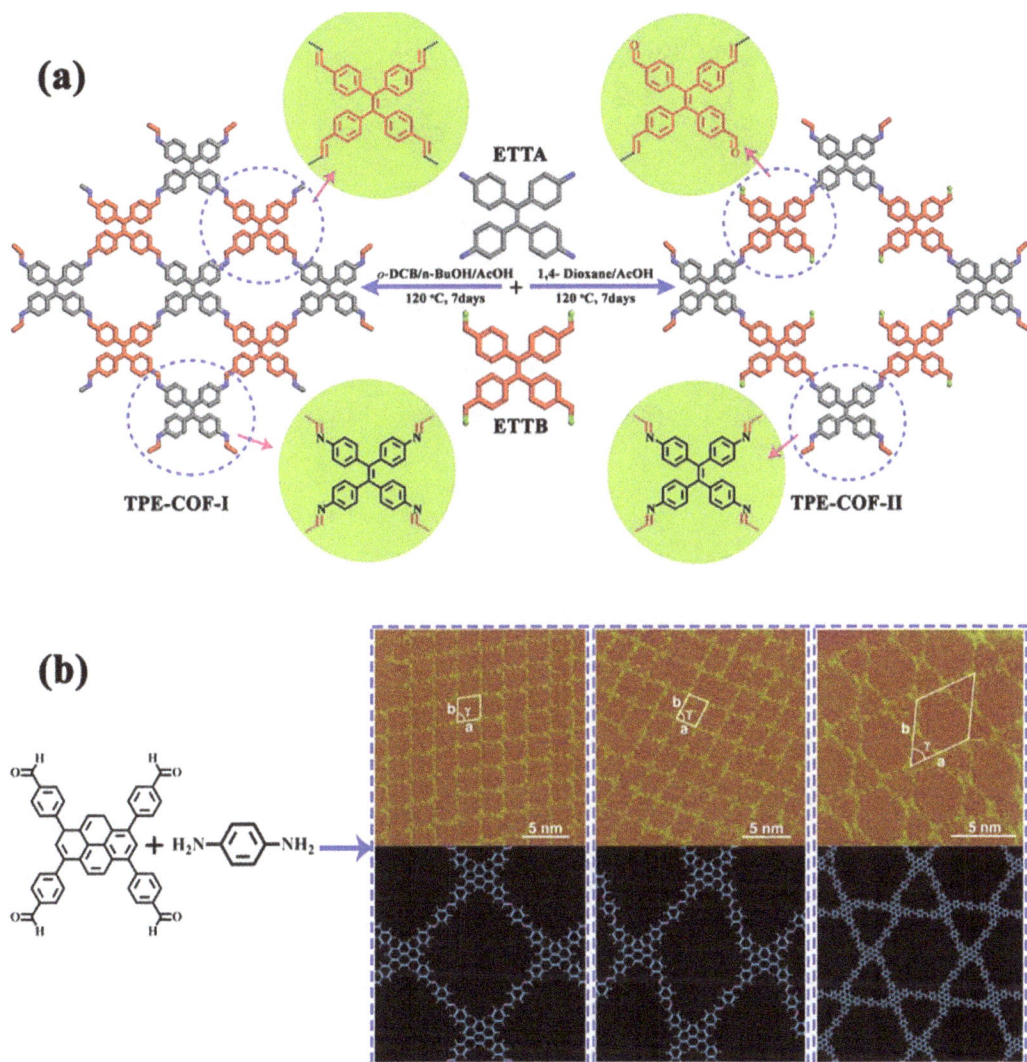

FIGURE 2.7

(a) Synthesis of TPE-COF-I and TPE-COF-II framework isomers through full condensation and partial condensation of tetraldehyde monomer, (b) and HR-STM images of the parallelogram, rhombus, and kagome morphological networks formed from TFPPy and 1,4-phenylenediamine condensation. Reproduced with permission from Ref. [34]. Copyright (2016) American Chemical Society.

as low solubility, insufficient thermal and chemical stability, low reactivity, and even undesirable topology.

2.4.2 Synthetic Methods

To use COFs in real-world applications, it is crucial to search and find suitable and low-cost solutions that do not require complicated and time-consuming processes. So far, many methods for synthesizing COFs have been developed in addition to the usual solvothermal methods, that here we briefly review and compare some of these methods.

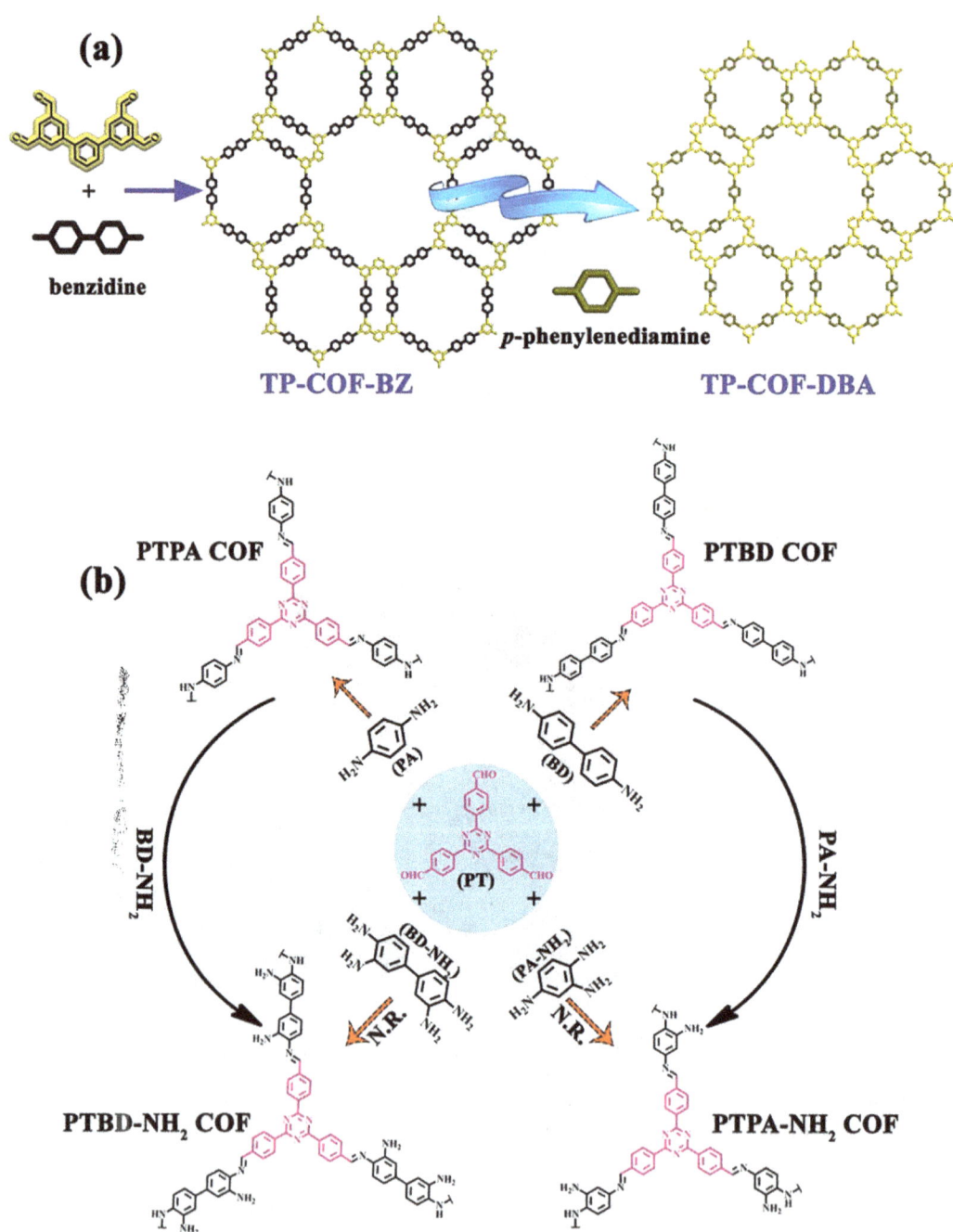

FIGURE 2.8
(a) Conversion of TP-COF-BZ to TP-COF-DAB by using post-synthetic linker exchange strategy and (b) amino-functionalized COFs preparation by indirect synthesis.

2.4.2.1 Solvothermal Synthesis

COFs are often synthesized by solvothermal methods. The solvothermal reaction conditions are greatly dependent on the solubility of the monomers, the reversibility of the linkages, and the activity of building units. Other parameters affecting this synthesis method include temperature, reaction time, and catalyst concentration. To enhance the solubility of the monomers, great care must be taken in the choice of the solvent. In most cases, a mixture of solvents (e.g., mesitylene and 1,4-dioxane) is used to mediate the release of monomers and increase the rate of nucleation.

As a general procedure, a mixture of linkers and knots in a combination of solvents is placed in a Pyrex tube with suitable volume and sonicated for a short time. Then the desired catalyst is added to it. After degassing the mixture using freeze-pump-thaw cycles, the reaction vessel is sealed by a burner. The sealed Pyrex tube is placed at 80 to 120°C for 3 to 7 days, motionless. Then, after cooling to room temperature, the product is collected and washed with a suitable solvent at ambient temperature or by using Soxhlet extraction to replace high boiling point solvents and remove trapped monomers or oligomers in the structure [38]. Some COFs can be produced on a large scale using this method.

2.4.2.2 Ionothermal Synthesis

In the ionothermal synthesis method, molten salt or ionic liquid is used as a solvent and catalyst for the cyclotrimerization reaction. As a general protocol, the monomer and $ZnCl_2$ are transferred to the reaction vessel of Pyrex or quartz, and after sealing, it is heated to 400–450°C. After cooling to ambient temperature, the resulting solid is washed well with water to remove the $ZnCl_2$. The COF powder is stirred in a dilute solution of hydrochloric acid for several hours to ensure the complete removal of $ZnCl_2$ [39]. What limits the use of ionothermal method is the temperature tolerance of monomers and undesirable by-products. Introducing a secondary energy source to perform the reaction under mild conditions increases the use of ionic liquid as a green solvent.

2.4.2.3 Microwave Synthesis

The microwave method can be considered the same as the solvothermal method with a complementary energy source. Compared to the solvothermal method, the microwave synthesis method yields cleaner products in a shorter time. On top of that, it is possible to control pressure and temperature simultaneously during the synthesis process. Cooper and colleagues synthesized boronate-ester-linked COF-5 using stirring under microwave irradiation (200 watts) at 100°C for 20 min [40, 41]. This synthesis approach was 200 times faster than the solvothermal synthesis of COF-5. To purify the product, the obtained COF was mixed with dry acetone, and after transfer to the microwave vessel, it was irradiated for another 20 min at 55°C. The purple color arising from oxidized 2,3,6,7,10,11-hexahydroxytriphenylene was wholly eliminated by repeating the microwave-assisted extraction process with acetone solvent. The microwave-assisted extraction can more effectively remove the trapped materials (e.g., monomers, oligomers, and solvent molecules); thus, the obtained product provides better porosity. In the Cooper Group research project, the BET surface area was also higher than that of the corresponding material prepared using the solvothermal method (2019 m^2 g^{-1} vs. 1590 m^2 g^{-1} for solvothermal method).

2.4.2.4 Mechanochemical Synthesis

Each of the aforementioned methods has unique and complex conditions such as using sealed Pyrex tube, accurate solvent selection, inert atmosphere, high temperature, etc. It is crucial to introduce a more straightforward synthesis method. Mechanochemical synthesis has attracted widespread attention due to its simplicity, relatively high speed, synthesis at ambient temperature, and environmental compatibility. In this method, the monomers are placed in a mortar and physically ground by a pestle. The grinding of the COFs during the mechanochemical synthesis causes the COFs to form graphene-like layered structures, which may be different from their solvothermal prepared structures. The formation of the structure can be followed by a solid color change over time. To develop this method, liquid-assisted grinding is used to increase the contact area between the monomers [42]. In this method, the monomers are ground with a small amount of catalyst solution, which leads to forming a more uniform phase and thus increases the crystallinity of the product.

2.4.3 Advanced Synthesis Techniques for COFs

Up to now, many COFs have been synthesized using the methods described above. However, these synthetic approaches often require a great deal of time and energy. On the other hand, in most cases, in addition to low efficiency, the scalability of the reaction is not possible. The COFs obtained using these methods are insoluble and make their use somewhat problematic. Due to the high sensitivity of the COFs quality to reaction conditions, there may be significant changes in the physicochemical properties of samples obtained from batch-to-batch synthesis. Therefore, we should always strive to find new ways for repeatable synthesis and without any complications.

2.4.3.1 Interfacial Synthesis

The interfacial synthesis approach is a new effective method for producing COFs in the form of thin films. Using this method, the limitations of COFs in bulk can be overcome and made processable films with controlled thickness. Many COFs, including Tp-Azo, Tp-Tts, Tp-Ttba, and Tp-Bpy have been fabricated using this technique [43]. According to the general method, the monomers are separately dissolved in two aqueous and organic phases. After adding the mixtures to each other, a thin layer of COF is formed between the two phases. Depending on the reaction time and monomer concentrations, films with adjustable thickness from 4 to 150 nm are obtained.

2.4.3.2 Room Temperature Vapor-Assisted Synthesis

The thin films of various linked-COFs can also be prepared at ambient temperature using this technique. Synthesis at ambient temperature is especially effective for cases where the building blocks are sensitive to high temperatures. In this method, a mixture of monomers in water and acetone solvents (~ 200 µl of low boiling point solvents) is prepared and poured on a piece of glass. The glass is then transferred to a desiccator and placed next to a small container containing a mixture of reaction solvents (1,4-dioxane and mesitylene mixture) at room temperature for approximately three days. After this time, a thin layer of COF will form on the glass piece. Selecting the suitable solvent combination is one of the most critical parameters in this technique [44]. Metal triflates can be very effective in accelerating the formation of the COFs layer. In the presence of an effective catalyst, this method leads to the product in a very short time (about a few minutes).

2.4.3.3 Sonochemical Synthesis

Sonochemical synthesis is a very inexpensive and fast method for organic material synthesis. Therefore, this method can be introduced as a suitable alternative to conventional COFs synthesis methods. In the sonochemical method, ultrasonic waves induce a phenomenon called "cavitation". Cavitation is a phenomenon in which the static pressure of a liquid reduces to below the liquid's vapor pressure, leading to the formation of small vapor-filled cavities in the liquid. The growth and collapse of the created cavities lead to an extraordinary increase in local temperature and pressure. Using sonochemical synthesis approach, Ahn et al. prepared COF-1 400 times faster than the solvothermal method [45]. Because each created cavity is considered a separate reaction vessel, this method is convenient for constructing micro-structured COFs from small building blocks.

2.5 Conclusion

COFs have been considerably studied as an attractive class of crystalline porous polymers. The COFs' advantages, in terms of their structural diversity and predictable design, over other porous materials are irrefutable. In this chapter, we discussed many structural aspects of COFs: how reaction chemistry will lead to the acquisition of porous crystalline polymers, how topological diagrams are used to design the shape and size of pores before synthesis proses, and modification of structures using the modulator agents. The advantages and limitations of general synthesis methods of COFs were pointed out, and emerging techniques for improving the synthesis conditions and structural crystallinity were investigated. However, many challenges remain, the rapid progress of COFs in the past few years promises a hopeful future for these crystalline structures.

References

1. R. Wentorf, R.C. DeVries, F. Bundy, Sintered superhard materials, Science, 208 (1980) 873–880.
2. E. Corey, Robert Robinson lecture. Retrosynthetic thinking – essentials and examples, Chemical Society Reviews, 17 (1988) 111–133.
3. A.P. Cote, A.I. Benin, N.W. Ockwig, M. O'Keeffe, A.J. Matzger, O.M. Yaghi, Porous, crystalline, covalent organic frameworks, Science, 310 (2005) 1166–1170.
4. N. Huang, P. Wang, D. Jiang, Covalent organic frameworks: a materials platform for structural and functional designs, Nature Reviews Materials, 1 (2016) 1–19.
5. Y. Jin, Q. Wang, P. Taynton, W. Zhang, Dynamic covalent chemistry approaches toward macrocycles, molecular cages, and polymers, Accounts of Chemical Research, 47 (2014) 1575–1586.
6. Y. Jin, C. Yu, R.J. Denman, W. Zhang, Recent advances in dynamic covalent chemistry, Chemical Society Reviews, 42 (2013) 6634–6654.
7. J. Hu, S.K. Gupta, J. Ozdemir, M.H. Beyzavi, Applications of dynamic covalent chemistry concept toward tailored covalent organic framework nanomaterials: A review, ACS Applied Nano Materials, 3 (2020) 6239–6269.

8. P.J. Waller, F. Gándara, O.M. Yaghi, Chemistry of covalent organic frameworks, Accounts of Chemical Research, 48 (2015) 3053–3063.

9. A.F. El-Mahdy, Y.H. Hung, T.H. Mansoure, H.H. Yu, T. Chen, S.W. Kuo, A hollow microtubular triazine-and benzobisoxazole-based covalent organic framework presenting sponge-like shells that functions as a high-performance supercapacitor, Chemistry – An Asian Journal, 14 (2019) 1429–1435.

10. K. Geng, T. He, R. Liu, S. Dalapati, K.T. Tan, Z. Li, S. Tao, Y. Gong, Q. Jiang, D. Jiang, Covalent organic frameworks: design, synthesis, and functions, Chemical Reviews, 120 (2020) 8814–8933.

11. S. Kim, H.C. Choi, Recent advances in covalent organic frameworks for molecule-based two-dimensional materials, ACS Omega, 5 (2019) 948–958.

12. S.Q. Xu, T.G. Zhan, Q. Wen, Z.F. Pang, X. Zhao, Diversity of covalent organic frameworks (COFs): a 2D COF containing two kinds of triangular micropores of different sizes, ACS Macro Letters, 5 (2016) 99–102.

13. H.R. Abuzeid, A.F. EL-Mahdy, S.W. Kuo, Covalent organic frameworks: design principles, synthetic strategies, and diverse applications, Giant, 6 (2021) 100054.

14. N. Huang, L. Zhai, D.E. Coupry, M.A. Addicoat, K. Okushita, K. Nishimura, T. Heine, D. Jiang, Multiple-component covalent organic frameworks, Nature Communications, 7 (2016) 1–12.

15. T. Tsuruoka, S. Furukawa, Y. Takashima, K. Yoshida, S. Isoda, S. Kitagawa, Nanoporous nanorods fabricated by coordination modulation and oriented attachment growth, Angewandte Chemie, 121 (2009) 4833–4837.

16. S. Hermes, T. Witte, T. Hikov, D. Zacher, S. Bahnmüller, G. Langstein, K. Huber, R.A. Fischer, Trapping metal-organic framework nanocrystals: an in-situ time-resolved light scattering study on the crystal growth of MOF-5 in solution, Journal of the American Chemical Society, 129 (2007) 5324–5325.

17. N. Stock, S. Biswas, Synthesis of metal-organic frameworks (MOFs): routes to various MOF topologies, morphologies, and composites, Chemical Reviews, 112 (2012) 933–969.

18. B.J. Smith, W.R. Dichtel, Mechanistic studies of two-dimensional covalent organic frameworks rapidly polymerized from initially homogenous conditions, Journal of the American Chemical Society, 136 (2014) 8783–8789.

19. H. Li, A.D. Chavez, H. Li, H. Li, W.R. Dichtel, J.L. Bredas, Nucleation and growth of covalent organic frameworks from solution: the example of COF-5, Journal of the American Chemical Society, 139 (2017) 16310–16318.

20. B.J. Smith, N. Hwang, A.D. Chavez, J.L. Novotney, W.R. Dichtel, Growth rates and water stability of 2D boronate ester covalent organic frameworks, Chemical Communications, 51 (2015) 7532–7535.

21. H. Li, A.M. Evans, I. Castano, M.J. Strauss, W.R. Dichtel, J.L. Bredas, Nucleation-elongation dynamics of two-dimensional covalent organic frameworks, Journal of the American Chemical Society, 142 (2019) 1367–1374.

22. M. Calik, T. Sick, M. Dogru, M. Döblinger, S. Datz, H. Budde, A. Hartschuh, F. Auras, T. Bein, From highly crystalline to outer surface-functionalized covalent organic frameworks – a modulation approach, Journal of the American Chemical Society, 138 (2016) 1234–1239.

23. T. Sun, W. Lei, Y. Ma, Y.B. Zhang, Unravelling crystal structures of covalent organic frameworks by electron diffraction tomography, Chinese Journal of Chemistry, 38 (2020) 1153–1166.

24. T. Sun, L. Wei, Y. Chen, Y. Ma, Y.-B. Zhang, Atomic-level characterization of dynamics of a 3D covalent organic framework by cryo-electron diffraction tomography, Journal of the American Chemical Society, 141 (2019) 10962–10966.

25. S. Wang, Z. Zhang, H. Zhang, A.G. Rajan, N. Xu, Y. Yang, Y. Zeng, P. Liu, X. Zhang, Q. Mao, Reversible polycondensation-termination growth of covalent-organic-framework spheres, fibers, and films, Matter, 1 (2019) 1592–1605.

26. A.M. Evans, L.R. Parent, N.C. Flanders, R.P. Bisbey, E. Vitaku, M.S. Kirschner, R.D. Schaller, L.X. Chen, N.C. Gianneschi, W.R. Dichtel, Seeded growth of single-crystal two-dimensional covalent organic frameworks, Science, 361 (2018) 52–57.

27. A. Nagai, Z. Guo, X. Feng, S. Jin, X. Chen, X. Ding, D. Jiang, Pore surface engineering in covalent organic frameworks, Nature Communications, 2 (2011) 1–8.
28. S.D. Brucks, D.N. Bunck, W.R. Dichtel, Functionalization of 3D covalent organic frameworks using monofunctional boronic acids, Polymer, 55 (2014) 330–334.
29. D.N. Bunck, W.R. Dichtel, Internal functionalization of three-dimensional covalent organic frameworks, Angewandte Chemie, 124 (2012) 1921–1925.
30. D.N. Bunck, W.R. Dichtel, Postsynthetic functionalization of 3D covalent organic frameworks, Chemical Communications, 49 (2013) 2457–2459.
31. Z.F. Pang, S.Q. Xu, T.Y. Zhou, R.R. Liang, T.G. Zhan, X. Zhao, Construction of covalent organic frameworks bearing three different kinds of pores through the heterostructural mixed linker strategy, Journal of the American Chemical Society, 138 (2016) 4710–4713.
32. M. Afshari, M. Dinari, M.M. Momeni, Ultrasonic irradiation preparation of graphitic-C3N4/polyaniline nanocomposites as counter electrodes for dye-sensitized solar cells, Ultrasonics Sonochemistry, 42 (2018) 631–639.
33. Q. Gao, X. Li, G.H. Ning, H.S. Xu, C. Liu, B. Tian, W. Tang, K.P. Loh, Covalent organic framework with frustrated bonding network for enhanced carbon dioxide storage, Chemistry of Materials, 30 (2018) 1762–1768.
34. Y.P. Mo, X.H. Liu, D. Wang, Concentration-directed polymorphic surface covalent organic frameworks: Rhombus, parallelogram, and kagome, ACS Nano, 11 (2017) 11694–11700.
35. O. Karagiaridi, W. Bury, J.E. Mondloch, J.T. Hupp, O.K. Farha, Solvent-assisted linker exchange: an alternative to the de novo synthesis of unattainable metal–organic frameworks, Angewandte Chemie International Edition, 53 (2014) 4530–4540.
36. C. Qian, Q.Y. Qi, G.F. Jiang, F.Z. Cui, Y. Tian, X. Zhao, Toward covalent organic frameworks bearing three different kinds of pores: the strategy for construction and COF-to-COF transformation via heterogeneous linker exchange, Journal of the American Chemical Society, 139 (2017) 6736–6743.
37. H.L. Qian, Y. Li, X.P. Yan, A building block exchange strategy for the rational fabrication of de novo unreachable amino-functionalized imine-linked covalent organic frameworks, Journal of Materials Chemistry A, 6 (2018) 17307–17311.
38. H. Lu, C. Wang, J. Chen, R. Ge, W. Leng, B. Dong, J. Huang, Y. Gao, A novel 3D covalent organic framework membrane grown on a porous α-Al 2 O 3 substrate under solvothermal conditions, Chemical Communications, 51 (2015) 15562–15565.
39. P. Kuhn, M. Antonietti, A. Thomas, Porous, covalent triazine-based frameworks prepared by ionothermal synthesis, Angewandte Chemie International Edition, 47 (2008) 3450–3453.
40. L.K. Ritchie, A. Trewin, A. Reguera-Galan, T. Hasell, A.I. Cooper, Synthesis of COF-5 using microwave irradiation and conventional solvothermal routes, Microporous and Mesoporous Materials, 132 (2010) 132–136.
41. N.L. Campbell, R. Clowes, L.K. Ritchie, A.I. Cooper, Rapid microwave synthesis and purification of porous covalent organic frameworks, Chemistry of Materials, 21 (2009) 204–206.
42. Y. Peng, G. Xu, Z. Hu, Y. Cheng, C. Chi, D. Yuan, H. Cheng, D. Zhao, Mechanoassisted synthesis of sulfonated covalent organic frameworks with high intrinsic proton conductivity, ACS Applied Materials & Interfaces, 8 (2016) 18505–18512.
43. K. Dey, M. Pal, K.C. Rout, S. Kunjattu H, A. Das, R. Mukherjee, U.K. Kharul, R. Banerjee, Selective molecular separation by interfacially crystallized covalent organic framework thin films, Journal of the American Chemical Society, 139 (2017) 13083–13091.
44. D.D. Medina, J.M. Rotter, Y. Hu, M. Dogru, V. Werner, F. Auras, J.T. Markiewicz, P. Knochel, T. Bein, Room temperature synthesis of covalent – organic framework films through vapor-assisted conversion, Journal of the American Chemical Society, 137 (2015) 1016–1019.
45. S.T. Yang, J. Kim, H.Y. Cho, S. Kim, W.S. Ahn, Facile synthesis of covalent organic frameworks COF-1 and COF-5 by sonochemical method, RSC Advances, 2 (2012) 10179–10181.

3

Recent Development in Synthesis of Covalent Organic Frameworks

Ferda Civan Çavuşoğlu, Gülsüm Özçelik, and Şahika Sena Bayazit
Department of Chemical Engineering, Engineering and Architecture Faculty, Beykent University, Istanbul, Turkey

CONTENTS

3.1 Introduction

The porous materials are known for their important properties such as porosities, channels, and voids. These key features make porous materials preferred in a wide variety of application areas. In recent years, the modular crystal structures composed of molecular components bonded to subunits stand out. MOFs and COFs are the leading structures of this porous material class. MOFs are formed by coordination bonds and COFs consist of covalently bonded carbon and other light elements. Yaghi and co-workers firstly reported COFs in 2005 [1, 2]. After they opened the door, numerous kinds of COFs have been prepared worldwide. COFs are very attractive materials because of their excellent properties; such as low densities and large surface areas. Their tunable surfaces are also important for fields such as catalysis, adsorption, sensing, and energy storage [3]. The organic linkers are used for the formation of COFs. The shape, connectivity, size, and

DOI: 10.1201/9781003206507-3

symmetry of COFs are defined according to organic linkers [4]. Different kinds of linkers have been used for COFs synthesis; ditopic, tritopic, and tetratopic linkers can be given as examples. Also, mixed linkers are developed for the construction of COFs. The building blocks materials used for the preparation of COFs are light elements such as C, H, O, Si, and B. COF crystals are obtained by combining these elements and organic linkers with various methods. Typical organic crystals generally cannot be prepared as 2D and 3D materials. The innovation of COFs is also emerging in this area. COF crystals have been obtained as 2D and 3D crystals [5]. Different kinds of COFs have been prepared due to their linkers. Boroxine, boronate, imine, hydrazine, azine, triazine, imide, and phenazine are some examples of linkages [6].

Different methods have been developed for the preparation of COFs. Solvothermal, ionothermal, microwave, room temperature, sonochemical, mechanochemical, and light promoted methods are utilized, generally. Microwave-assisted systems are efficient and the reaction rates are very high. These reasons make the microwave method attractive for COFs synthesis. The mechanochemical method is another attractive system because this method is solvent-free and fast. The solvothermal method is mostly the preferred one but the reaction conditions are very harsh and industrialization of this process is very difficult [3]. In this chapter, types of linkers and forming reactions of COFs are discussed, and some examples are given. The synthesis methods are explained clearly and produced COFs are given as examples.

3.2 Materials and Chemistry of COFs

COFs are composed of light elements (C, O, H, N, and Si) and linkers. The linkers are the most important part of COFs because they provide the shapes, sizes, connectivity, and symmetry of the crystals. Yaghi et al. gave some examples of the linkers [7], these linkers are presented in Figure 3.1, citing the research of Yaghi et al. [7]. The linkers are classified as ditopic, tritopic, and tetratopic linkers. The reaction between the linkers and light elements is a slightly reversible condensation reaction. Reversible coupling ensures that the basic units of the crystal structure are placed in the minimum thermodynamic state [8]. During the condensation reaction, covalent bonds occur between the units. The covalent bonding provides obtaining COFs, which have high thermal stability [4]. The crystallization problem is very important while synthesizing such materials. The coupling reaction has a reversible nature. This ability makes the products crystalline rather than an amorphous structure, and C-N, B-O, B-O-Si, etc. bonds are formed.

3.2.1 Forming Reactions of Covalent Bonds in COFs

3.2.1.1 Boron-Based COFs

Boroxines and Boronate Esters: Boronic acids transform into the boroxine ring by self-condensation reactions. Boroxine formation depends on the reaction between the boron and hydroxyl group of boronic acid. After the removal of the water molecule, B-O bonding occurs [9]. The first obtained COF structure was prepared according to this reaction. This crystal was COF-1 and boroxine rings were produced by self-condensation reaction using 1,4-benzene diboronic acid [4]. Some linkage types of COFs are given in Figure 3.2.

FIGURE 3.1

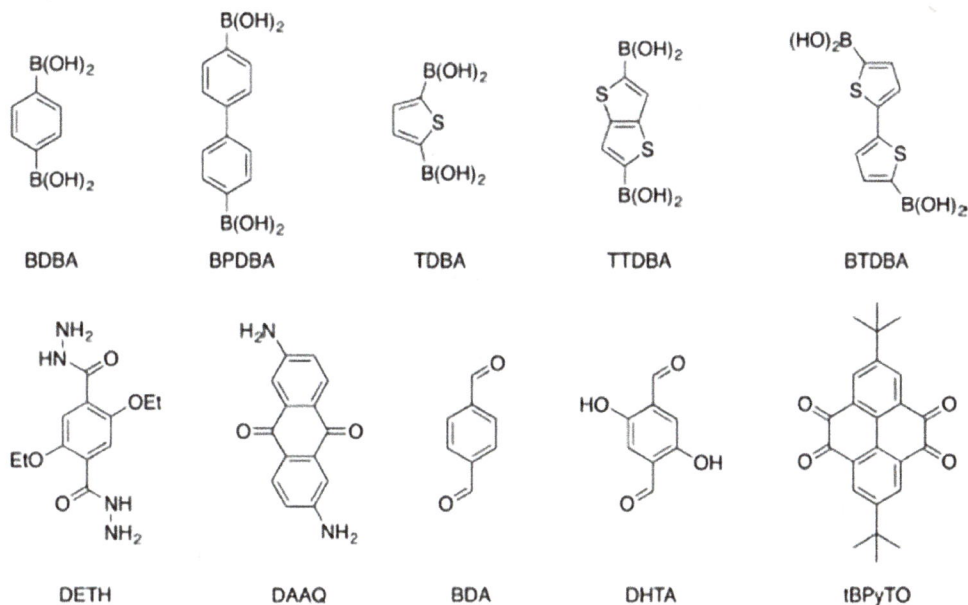

Some examples of ditopic, tritopic, and tetratopic linkers for the production of COFs. (Adapted with permission [7]. Copyright (2015) American Chemical Society.)

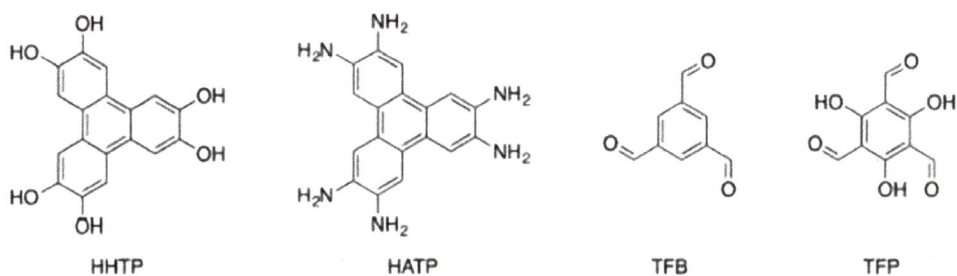

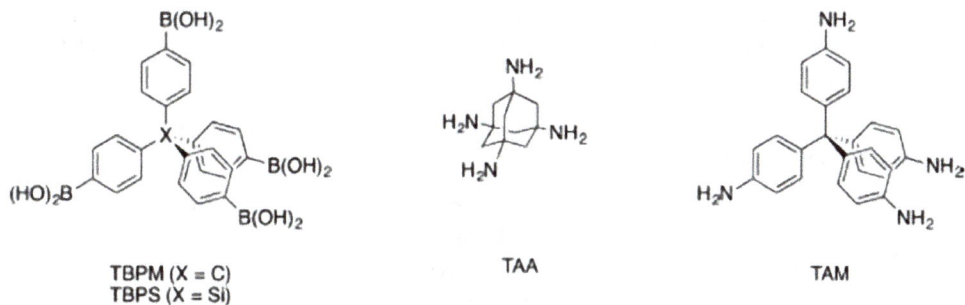

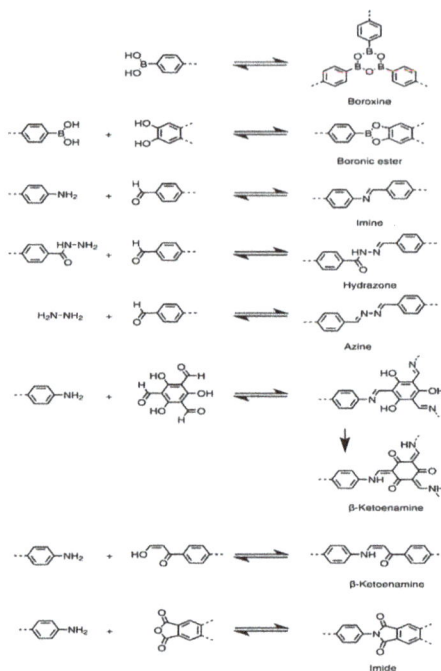

FIGURE 3.2
Linkage types of COFs. (Adapted with permission [4]. Copyright (2018) Wiley-VCH.)

The formation of boronate esters is similarly based on the condensation reaction of boronic acid. The difference in this reaction is hydroxyl group of the alcohol reacts with the boron so, after the reaction of boronic acid-alcohol, boronate esters form [9]. Some important COF examples of formed by boroxine and boronate esters; COF-1, COF-5, COF-6, COF-8, and COF-10. The surface areas of these COFs are 711, 1590, 980, 1400, and 2080 m^2/g, respectively [10]. The researchers also prepare borosilicate and spiroborate COFs. Borosilicate COFs form by cross-condensation of boronic acids with alcohols. COF-202 is a well-known borosilicate-based COF with 2690 m^2/g of the surface area [11]. Spiroborate-based COFs can be obtained by trimethyl borate and macrocycles functionalized by diol [4].

3.2.1.2 Imine-Based COFs

Imine-based COFs have been developed for increasing the chemical stability of the crystal structure. The imine-based COFs can be classified into two groups. One of them is the Schiff base reaction group, the other one is the hydrazine type group. Schiff base reaction depends on the co-condensation of aldehydes and amines. Schiff base reactions are known as a part of dynamic covalent chemistry [9]. The positive way of dynamic reaction chemistry is using an acid catalyst enables reversible covalent bonding. COF-300 was synthesized as the first imine-based COF by Yaghi et al. in 2009 [3]. They used Schiff base reaction for the preparation of COF-300. The surface area of COF-300 is 1360 m^2/g [12]. Hydrazine type of preparation method is based on the co-condensation reaction of aldehydes and hydrazides. The imine-based COFs are stable in acidic and basic media.

3.2.1.3 Hydrazone- and Triazine-Based COFs

Hydrazone-based COFs are another type of crystal formation reaction. This is a type of imine-based COFs but it cannot be limited to imines. Hydrazone linkages can be obtained by organic hydrazides and aldehydes. After the reaction C-N bonds form. The first examples of hydrazine-based COFs are COF-42 and COF-43 [13]. Triazine-based linkages provide COFs that have different geometry and sizes. Triazine-based linkages depend on the cyclomerization reaction of nitriles in the presence of a metal catalyst. The preparation of triazine COFs has some disadvantages. The reaction temperature is high and COFs must be purified in an acid solution. These effects can destruct the crystal structures. So, triazine-based COFs generally have lower crystallinity and porosity [3]. New researches show that triazine-based COFs can be prepared under better conditions, such as lower temperature. Tan et al., prepared triazine-based COFs making polycondensation reaction of aldehydes and amidines [14].

3.2.1.4 Imide-Based COFs

The condensation of amine and acetic acid anhydride at high-temperature forms imide linkages. Imide linkage reaction is sustained under reversible conditions. 2D and 3D imide-based COFs have been also produced. Polyimide COFs have a large surface area and great ability to adsorb dye molecules and drug delivery applications. Also, the thermal stabilities of polyimide COFs are very high [4].

3.3 Synthesis Methods of COFs

COFs are highly porous and crystalline polymers synthesized by reversibly holding together organic structural units by covalent bonds [15]. The structure and morphology of synthesized COFs vary significantly depending on the types of building monomers and the synthesis methods chosen. For instance, the temperature can be changed using different parameters that affect the synthesis conditions such as solvent and catalyst. Therefore, different preparation methods have been applied for the synthesis of COFs [16]. The solvothermal method is a general method mostly used in COFs synthesis. In time, new methods have been developed for synthesizing of COFs. Ionothermal, microwave, room temperature, sonochemical, mechanochemical, solvothermal, and light-assisted synthesis methods can be given as examples for this purpose. These newly developed methods can help shorten the reaction time while continuing to further improve the reaction efficiency and the crystallinity and porosity of COF products [16, 17].

3.3.1 Solvothermal Synthesis

Most prepared COFs are produced using the solvothermal synthesis method. Solvothermal synthesis of COF materials is carried out by mixing the monomers in a closed vessel for 2–9 days and under heating. The chosen temperatures are generally between 80 and 120°C [18]. The reaction conditions depend on the solubility and reactivity of monomer formation and the reversibility of the reactions, in this method [16]. The morphology and structure of the synthesized COFs depend on different parameters such as the time, solvent, and concentration of the applied catalyst, and the reaction media's temperature and pressure [16, 18]. The selection of an appropriate solvent system is a critical parameter for producing crystalline COFs. Solvent mixtures are often used to prepare COFs, especially binary mixtures are preferred and also an acidic catalyst can be used for the synthesis [19]. Binary solvent mixtures such as THF-methanol, dioxane-mesitylene, and DMAc-o-dichlorobenzene are used for the preparation of boronate ester and boroxine-based COFs. The dioxane-toluene solvent mixture is used to obtain borosilicate COFs. For imine-linked COFs, dioxane-aqueous acetic acid is used as the solvent. The mesitylene-dioxane-aqueous acetic acid solvent combination is efficient for hydrazone-linked COFs. In any event, proper concentration ratios of the solvent system are very important. The solvent ratio provides controlling the reaction under thermodynamic conditions [20]. Imine-based COFs are synthesized by condensation of arylamines and aldehydes under acid-catalyzed solvothermal conditions. It is known that it is chemically stable in water and organic solvents. Furthermore, imine-based COFs are predominant over other COFs in terms of crystallinity and structural morphology, allowing them to be used in various applications [21]. In addition to the morphology and structure of the synthesized COF, an appropriate temperature is important to provide the reversibility of the synthesis reaction. COFs are prepared at different temperatures according to the chemical reactivity of the building blocks. It is important to occur the reaction in a closed environment, so the water molecules can be present in reaction media that can initiate the reverse reaction [20].

COFs are usually obtained by a solvothermal reaction using organic solvents under vacuum conditions in a sealed Pyrex tube. If the starting material dissolves in the solvent system, it provides the controlling the diffusion of the basic units of COFs into the solution properly. And this enables the facilitation of crystal nucleation. Using a closed reaction system maintains the availability of H_2O to maintain favorable reversible conditions for crystallite growth. With this method, crystallization occurs gradually over days [22]. The preparation of COFs by this method can be briefly explained as follows. The monomers and solvents charged into the Pyrex tube are degassed by several freeze-pump-thaw cycles. The tube is closed and set to the desired temperature for a certain reaction time. The precipitate formed is then collected, after the washing and drying processes, solid powder COF is obtained [20]. Although high product quality is generally achieved in the most widely used solvothermal synthesis method, there are some obstacles. One of the biggest hurdles is that it is difficult to scale up to industrial applications. The reaction rates in many COFs syntheses need acceleration. Researching greener, more efficient, and safer synthetic techniques that can be an alternative to this method is one of the current issues [19].

The first COFs (COF-1: [$(C_3H_2BO)_6 \cdot (C_9H_{12})_1$], COF-5: $C_9H_4BO_2$) were synthesized using 1,4-benzenediboronic acid (BDBA) by the solvothermal self-condensation method in closed pyrex tubes. The synthesized COF-1 was determined to be a staggering hexagonal layered material bonded with planar boroxine rings (S_{BET} = 711 m^2/g, P6$_3$/mmc). The other synthesized COF-5 was determined to be a mesoporous material with a shaded boron nitride arrangement (S_{BET} = 1590 m^2/g, P6/mmm). The synthesized thermally stable materials have pore sizes ranging from 7 to 27 Å [2].

3.3.2 Ionothermal Synthesis

Another method used in COFs synthesis is the ionothermal method. The reaction required for synthesis is carried out at high temperatures. Ionic liquids or molten metal salts are used, which act as solvents or catalysts, in this method. Ionic liquids affect the reaction positively because of their high periodicity and stability. Their reusability is also a very important property. They have ion-exchange abilities and environmental friendliness. On the other hand, the need for high reaction temperatures limits the range of building blocks and the application of the ionothermal method. Also, the most commonly used monomers in this method are amorphous materials [16, 19]. In short, while the ionothermal method is considered to be a simple and green method, it has disadvantages such as requiring high temperature and long reaction time and being mostly applicable to amorphous materials [16]. The first COF synthesized using the ionothermal method was reported to be an amorphous but porous solid with a surface area of 2255 m^2/g at high temperatures [16, 23].

In a study with the ionothermal synthesis method, bicrystalline and porous imide-bonded COFs (TAPB-PTCDA- and TAPB-PMDACOF) were prepared in pure $ZnCl_2$. This synthesis is shown in Figure 3.3. The results showed that COF formation is simple and does not require the use of toxic solvents and additional base catalysts. It is also potentially scalable and quite fast, while in solvothermal method, it takes place between 3 and 7 days, it carries out within 10 hours under these conditions [24].

3.3.3 Microwave Synthesis

The microwave synthesis method provided advantages over solvothermal and ionothermal methods [16, 20]. This method enables faster and cleaner COFs synthesis to be completed, opening new possibilities for further applications at large scales [18, 19]. Furthermore, a closed vessel is not required for microwave synthesis in this method. The microwave solvent extraction process more efficiently removes the residues and impurities formed in the COFs, and the synthesized COFs have better porosity. Another advantage of this method is that the synthesis progress can be monitored visually and simultaneously control the reaction temperature and pressure [16, 20]. Therefore, the application of fast microwave-assisted methods is often preferred for crystal COFs synthesis.

The microwave method was first studied in 2009 to synthesize crystalline boron-based two-dimensional COF-5 and three-dimensional COF-102. The obtained COFs have very high surface areas (COF-5: 2019 m^2/g, COF-102: 2926 m^2/g) and have a shorter reaction time than the traditional solvothermal method. The reaction time of 20 minutes was obtained by the microwave heating method. This time is as short as 1/200th of the solvothermal reaction time [25, 26]. Two dimensional TpPa-COF (MW) was synthesized for the first time in a much shorter time and with a very high yield using the Schiff base reaction and the fast microwave-assisted (MW) solvothermal method. This synthesis is shown in Figure 3.4. TpPa-COF (MW) (724.6 m^2/g) has great porosity with higher BET surface area and high crystallinity than TpPa-COF (CS) (535 m^2/g) synthesized by the conventional solvothermal (CS) method [27].

3.3.4 Room-Temperature Synthesis

The room temperature method has many advantages over other methods in terms of being a simple method that does not require temperature and in terms of ease of use. Room temperature synthesis does not require the use of closed containers and allows mass production of COF materials, eliminating the hassle of controlling different synthetic parameters [18]. Most of the building units used are unstable at high temperatures, making the room temperature method for synthesizing COFs prominent. Also, as the nature of some materials used may change with temperature, the morphology of the prepared COFs may change and then produce side materials that can reduce its stability. For this reason, it

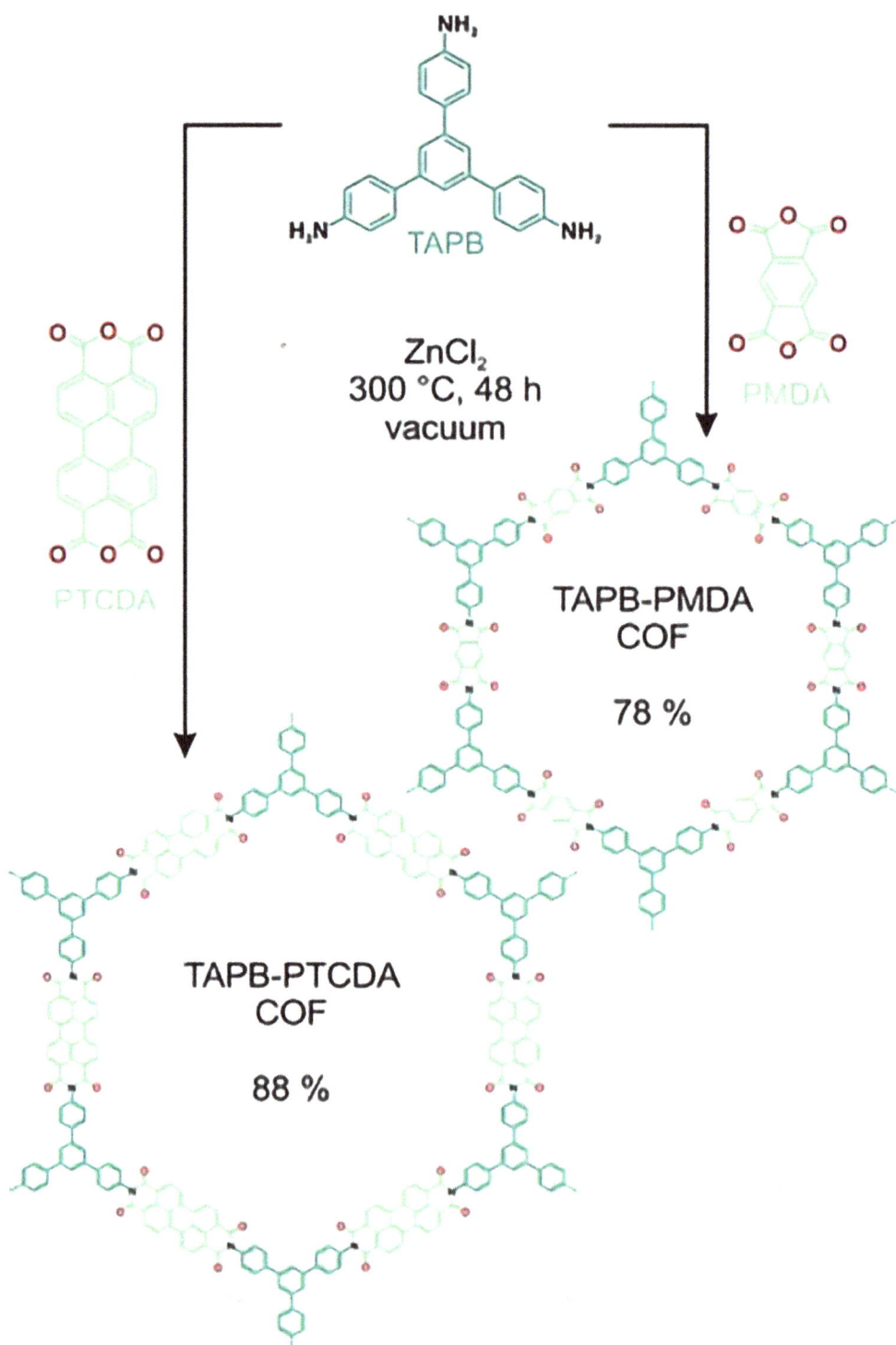

FIGURE 3.3
Synthesis of TAPB-PTCDA COF and TAPB-PMDA COF in ZnCl₂ by ionothermal method. (Adapted with permission [24]. Copyright (2020) Wiley-VCH.)

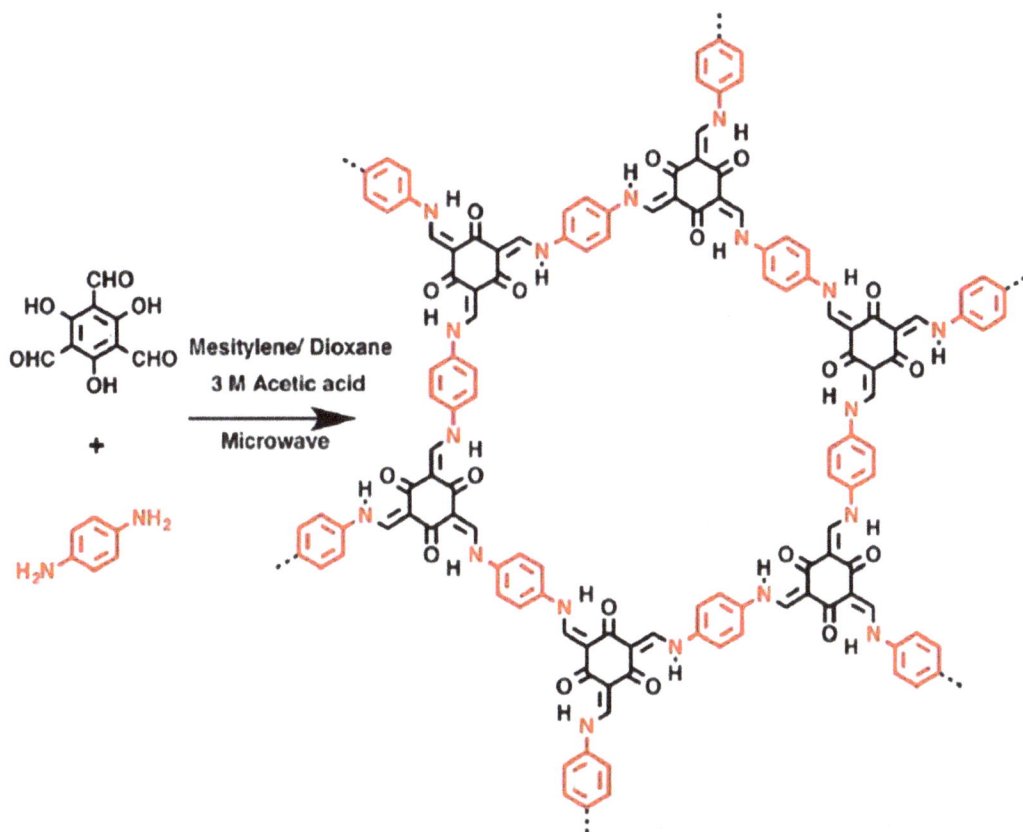

FIGURE 3.4
Synthesis of TpPa-COF (MW) by microwave method. (Adapted with permission [27]. Copyright (2015) Royal Society of Chemistry.)

is an undesirable situation, especially in the application of high-temperature synthesis approaches. In the synthesis of COFs, it is desirable not only to provide high crystallinity and porosity but also to be produced using high-speed methods with high stability. COFs prepared by room-temperature methods show superior stability in harsh environments and have excellent surface area and adsorption capacity. Other benefits of the room temperature synthesis method are its usability in large-scale applications as an easy and effective synthetic method [16]. Especially when sensitive monomers or substrates are involved, COFs synthesis at room temperature is favorable. Contrary to these advantages, large-scale synthesis is difficult to expand the applications of COFs. Because the reagents and ionic liquids are expensive [19]. Since COFs synthesis at room temperature applies to certain building monomers, its use remains limited [16].

The steam-assisted conversion (SAC) method is one of the applied room temperature methods and enables the synthesis of different types of COFs with excellent properties. In this context, a study has been carried out for large-scale commercial production under ambient conditions using batch solution-suspension approaches (SSA) and continuous flow systems. Three classical two-dimensional (2D) COFs (COF-LZU1, TpPa-1, and N3-COF) and two new imine-linked COFs (NUS-14 and NUS-15) were synthesized [17]. This work demonstrated the first example of synthesizing COFs in continuous flow systems.

The surface areas of the obtained COFs were calculated as 412 m^2/g for COF-LZU1, 834 m^2/g for TpPa-1, 1722 m^2/g for N3-COF, 1164 m^2/g for NUS-14, and 322 m^2/g for NUS-15. The continuous flow synthesis method, besides being fast and effective, has also been a pioneering method for the practical synthesis and application of COFs [16, 17].

Another method applied at room temperature is the vapor-assisted conversion (VAC) method. VAC is a variation of dry gel transformation (DGC) and vapor-assisted transformation (SAC) [16]. In a study conducted with this method, high-quality benzodithiophene films containing BDT-COF and COF-5 and adjustable in thickness were synthesized on different substrates. It has been determined that vapor-assisted transformation synthesis is a suitable method for the production of COF-based fine porous films, and it is thought to be a remarkable method for especially fragile building blocks that cannot survive the harsh conditions used in the solvothermal synthesis method [28].

3.3.5 Sonochemical Synthesis

The sonochemical synthesis method can accelerate the rate of crystallization in solutions that generate extremely high local temperatures and pressures (>5000 K and >1000 bar) to achieve extremely fast heating and cooling rates [29]. Since the sonochemical synthesis method does not require an induction period, the power consumption is very low and the synthesis time is very short. For these reasons, it is a very economical method [22]. With this method, the effects of the effect of temperature and sonication waves on the structure and morphology of COFs can be studied simultaneously. One of its best advantages is that it is considered a green method [16]. Crystals of COF-1 and COF-5 with superior properties were obtained for the first time in a study conducted in 2012, within 1–2 hours by a sonochemical method. This synthesis is shown in Figure 3.5. BET surface areas are calculated as 719 m^2/g for COF-1 and 2122 m^2/g for COF-5. It was significantly higher than the value reported for the solvothermal prepared COF-5 sample (1590 m^2/g) and closer to the value reported using microwave heating (2019 m^2/g) [22].

3.3.6 Mechanochemical Synthesis

The mechanochemical (MC) synthesis which is a simple and cost-effective method, has been successfully applied to a variety of reactions, including the MOFs and COFs synthesis. Synthesis by mechanochemistry is so efficient to achieve metastable solid phases. In addition, it is a superior technique for scanning and exploring which is the phase landscape of organic and metal-organic solids or new polymorphs [25, 30]. In a study with this method, chemically and thermally stable COFs were quickly prepared by solvent-free mechanical operation in 40 minutes at room temperature. It was synthesized using modified Schiff base reactions, which gained the framework extraordinary stability for water, base, and acid. This synthesis is shown in Figure 3.6. The progress of the reaction caused by the mechanical force can be facilely recognized by the color change of the powder during grinding. Though mechanochemically synthesized COFs show a low level of crystallinity and porosity, this method is very suitable in terms of environmental friendliness, fast reaction, high yield, and scale-up synthesis [25, 31].

3.3.7 Light Promoted Synthesis

Another approach used in COFs synthesis is the method light promoted, which synthesizes using simulated sunlight irradiation [16]. Light energy, which could initiate a reversible

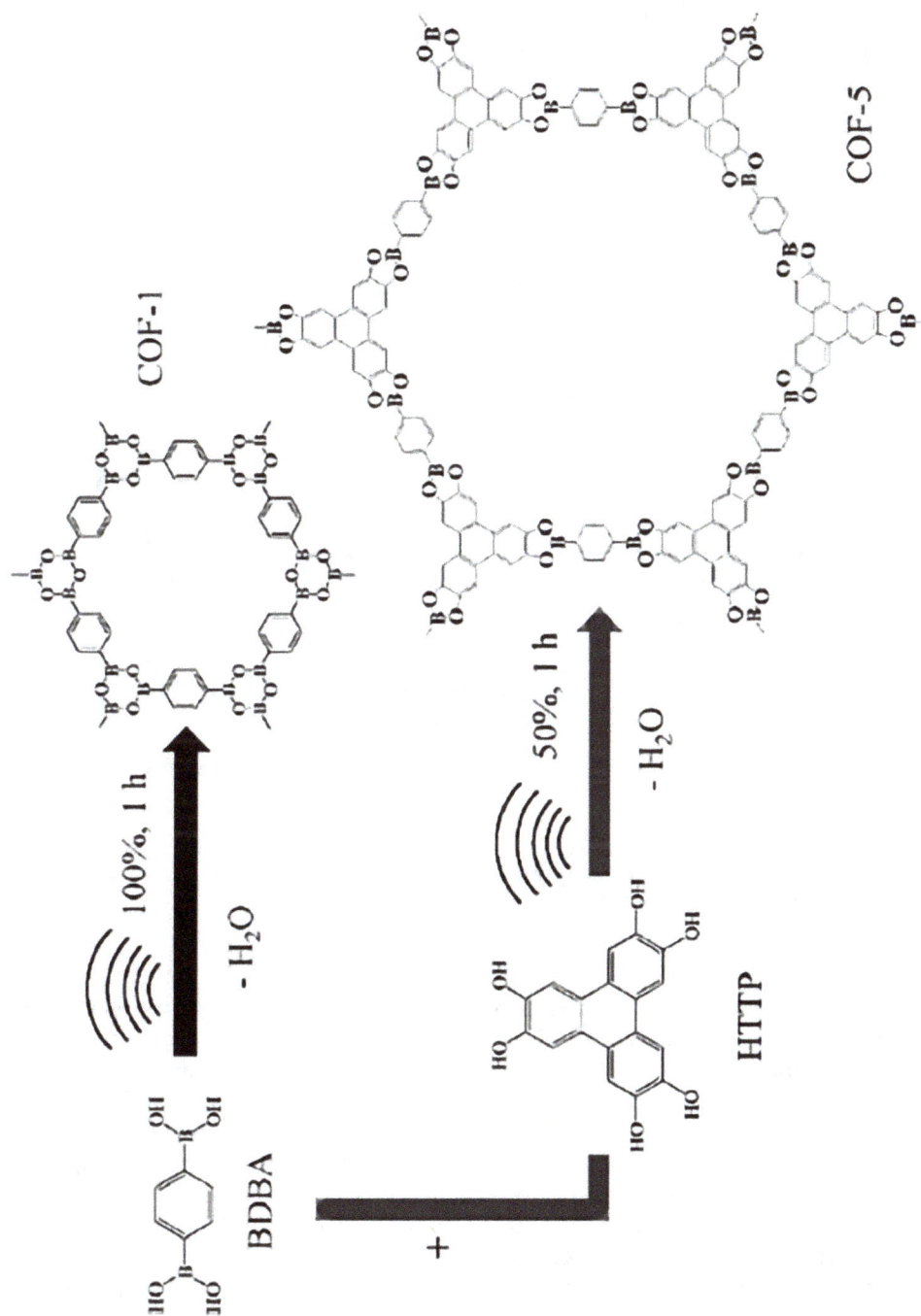

FIGURE 3.5
Synthesis of COF-1 and COF-5 by sonochemical method. (Adapted with permission [22]. Copyright (2012) Royal Society of Chemistry.)

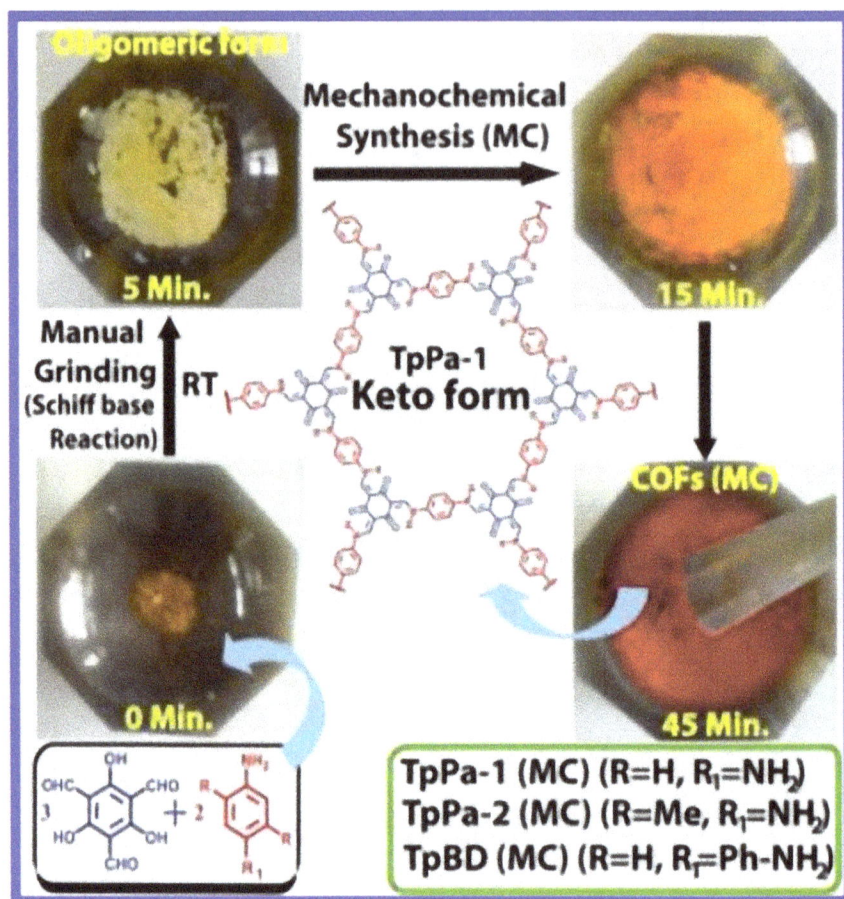

FIGURE 3.6
Experimental scheme and molecular structure of TpPa-1 (MC), TpPa-2 (MC), and TpBD (MC) obtained by MC synthesis method of simple Schiff base reactions performed by MC grinding. (Adapted with permission [31]. Copyright (2013) American Chemical Society.)

imine condensation reaction, is used to accelerate the imine condensation reaction to increase efficiency in the synthesis of COF. In one study, Lp-pi-COF material was synthesized by imine condensation with photon-assisted at the water interface and dehydration between precursors, which are PDA and TAPB. It was observed that in light-assisted poly-imine-based COF (pi-COF), photon energy accelerates the imine condensation reaction. In addition, it was seen to facilitate the formation of conversion from precipitates which are amorphous imine-containing to crystalline COF. This synthesis is shown in Figure 3.7 [32]. In another study using the light promoted technique, the imine condensation reaction, which yields extremely crystalline and conjugated COF (hcc-COF) within 3 hours at room temperature, was adapted and the surface area of the obtained COF was calculated as 598 m²/g [33]. This synthesis is shown in Figure 3.8.

There are many studies on COFs synthesis in the literature. The reaction conditions and properties of some COFs synthesized according to different synthesis methods from these studies are presented in Table 3.1.

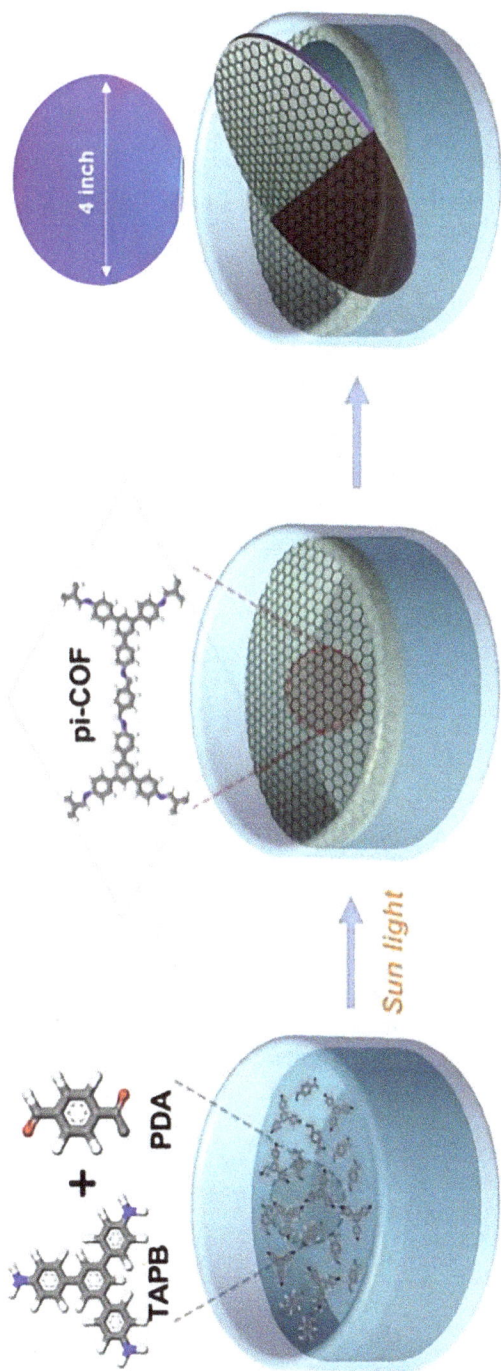

Experimental representation and molecular structure of Lp-pi-COF obtained by the light-assisted method. (Adapted with permission [32]. Copyright (2018) American Chemical Society.)

FIGURE 3.7

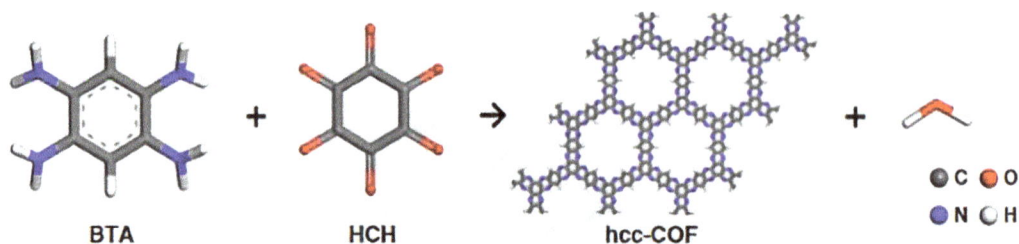

FIGURE 3.8
Molecular structure of hcc-COF obtained by condensation reaction between BTA and HCH precursors. (Adapted with permission [33]. Copyright (2019) Springer Nature.)

TABLE 3.1

Reaction Conditions and Properties of COFs Synthesized According to Different Synthesis Methods

Methods	COFs	Linkage	Reaction Time	Surface Area	References
Solvothermal	TPA-COF	Triphenylamine	72 h	398.59 m²/g	[34]
Solvothermal	TAPBB-COF	Porphyrin	72 h	843.8 m²/g	[35]
Solvothermal	COF1	Imine	72 h	735.9 m²/g	[36]
Solvothermal	3D-TPB-COF-HQ	Imine	7 d	842 m²/g	[37]
Ionothermal	TAPB-PTCDA COF	Imide	48 h	460 m²/g	[24]
	TAPB-PMDA COF	Imide	48 h	1250 m²/g	
Ionothermal	TFP-PA-COF	Keto-enamine	72 h	446/427/ 474 m²/g	[38]
	TFP-MPA-COF	Keto-enamine	72 h	604 m²/g	
	TFP-Azo-COF	Keto-enamine	72 h	809 m²/g	
	TFP-AHAn-COF	Keto-enamine	72 h	363/312/ 294 m²/g	
Ionothermal	3D-IL-COF-1	Imine	3 min	517 m²/g	[39]
	3D-IL-COF-2	Imine	3 min	653 m²/g	
	3D-IL-COF-3	Imine	3 min	870 m²/g	
	3D-IL-COF-1a	Imine	3 min	596 m²/g	
	3D-IL-COF-1b	Imine	3 min	537 m²/g	
Microwave	TAPB-PDA-COF	Imine	1 h	170 m²/g	[40]
	TAPB-TFA-COF	Imine	1 h	96 m²/g	
Microwave	COF-5	Boronate ester	20 min	2019 m²/g	[26]
Microwave	BTD-COF	Boronate ester	40 min	1000 m²/g	[41]
Microwave	SNW-1	Melamine	6 h	476 m²/g	[42]
Microwave	Fe_3O_4/M-COFs (2:100)	Melamine	4 h	600 m²/g	[43]
	Fe_3O_4/M-COFs (5:100)	Melamine	4 h	495 m²/g	
	Fe_3O_4/M-COFs (10:100)	Melamine	4 h	463 m²/g	
	Fe_3O_4/M-COFs (15:100)	Melamine	4 h	344 m²/g	
Room-temperature	COF-LZU1	Imine	72 h	412 m²/g	[17]
	TpPa-1	Enamine	72 h	834 m²/g	
	N_3-COF	Azine	72 h	1722 m²/g	
Room-temperature	Fe_3O_4@COFs	Aldehyde-amine	-	178.87 m²/g	[44]

(Continued)

TABLE 3.1 *(Continued)*

Reaction Conditions and Properties of COFs Synthesized According to Different Synthesis Methods

Methods	COFs	Linkage	Reaction Time	Surface Area	References
Room-temperature	Fe_3O_4@$[NH_2]$-COFs	Azide	-	50.05 m²/g	[45]
Room-temperature	COF-S	Imine	72 h	118 m²/g	[46]
	COF-L	Imine	72 h	398 m²/g	
Sonochemical	COF-5	Boronate ester	-	8.17 m²/g	[47]
	CNT@COF-5	Boronate ester	-	57.6 m²/g	
	Graphene@COF-5	Boronate ester	-	9.83 m²/g	
Sonochemical	COF-1	Imine	1 h	732 m²/g	[22]
	COF-5	Boronate ester	1 h	2122 m²/g	
Sonochemical	RIO-1	Imine	72 h	-	[48]
Mechanochemical	TpPa-1	Aldehyde-amine	1 h	61 m²/g	[31]
	TpPa-2	Aldehyde-amine	1 h	56 m²/g	
	TpBD	Aldehyde-amine	1 h	35 m²/g	
Solvothermal	TpBD	Aldehyde-amine	72 h	537 m²/g	
Mechanochemical	COF-LZU1	Imine	25 min	1097 m²/g	[30]
	TbBd-COF	Imine	25 min	799 m²/g	
Solvothermal	TpBpy ST	Bipyridine	72 h	1746 m²/g	[49]
Mechanochemical	TpBpy MC	Bipyridine	90 min	293 m²/g	
Mechanochemical	MCNTs@TpPa-1	Amine	30 min	218.05 m²/g	[50]
Light-promoted	hcc-COF	Imine	3 h	598 m²/g	[33]
Photon-assisted	2D-Lp-pi-COF film	Imine	1 h	-	[32]

3.4 Conclusion

Omar Yaghi and co-workers firstly reported COFs in 2005. Light elements such as boron, carbon, hydrogen, silicon, and oxygen are used for the preparation of COFs. These crystal materials are very effective in catalysis, separation, sensing, energy storage, and organic electronics fields. The surface area and porosities of COFs are very high and their thermal, mechanical, and chemical stabilities are very promising. The geometry, size, and shape of COFs depend on their reaction types and using materials. The linker type and production method are very important for COFs synthesis. Ditopic, tritopic, and tetratopic linkers have been used for the preparation of MOFs. Different reactions have been developed for obtaining better results. Generally, condensation and Schiff-base reactions are used for this aim. These reactions occur in different systems such as solvothermal, ionothermal, mechanochemical methods. Solvothermal methods are mostly used to carry out the condensation reaction, but more environmentally friendly methods have been started to be developed due to the reaction conditions. Mechanochemical methods can be given as examples. Solvent-free and environment-friendly methods should be developed for the industrialization of COFs production.

References

1. Song, Y., Sun, Q., Aguila, B., Ma, S.: Opportunities of Covalent Organic Frameworks for Advanced Applications. Adv. Sci. 6, 1801410 (2019).
2. Côté, A.P., Benin, A.I., Ockwig, N.W., O'Keeffe, M., Matzger, A.J., Yaghi, O.M.: Chemistry: Porous, Crystalline, Covalent Organic Frameworks. Science (80) 310, 1166–1170 (2005).
3. Zhao, W., Xia, L., Liu, X.: Covalent Organic Frameworks (COFs): Perspectives of Industrialization. CrystEngComm. 20, 1613–1634 (2018).
4. Lohse, M.S., Bein, T.: Covalent Organic Frameworks: Structures, Synthesis, and Applications. Adv. Funct. Mater. 28, 1705553 (2018).
5. Yaghi, O.M., Kalmutzki, M.J., Diercks, C.S.: Historical Perspective on the Discovery of Covalent Organic Frameworks. Introd. Reticular Chem. 177–195 (2019).
6. Wang, Z., Zhang, S., Chen, Y., Zhang, Z., Ma, S.: Covalent Organic Frameworks for Separation Applications. Chem. Soc. Rev. 49, 708–735 (2020).
7. Waller, P.J., Gándara, F., Yaghi, O.M.: Chemistry of Covalent Organic Frameworks. Acc. Chem. Res. 48, 3053–3063 (2015).
8. Haase, F., Lotsch, B. V.: Solving the COF Trilemma: Towards Crystalline, Stable and Functional Covalent Organic Frameworks. Chem. Soc. Rev. 49, 8469–8500 (2020).
9. Yaghi, O.M., Kalmutzki, M.J., Diercks, C.S.: Linkages in Covalent Organic Frameworks. Introd. Reticular Chem. 197–223 (2019).
10. Adrien P. Côté, Hani M. El-Kaderi, Hiroyasu Furukawa, Joseph R. Hunt, and, Yaghi, O.M.: Reticular Synthesis of Microporous and Mesoporous 2D Covalent Organic Frameworks. J. Am. Chem. Soc. 129, 12914–12915 (2007).
11. Hunt, J.R., Doonan, C.J., LeVangie, J.D., Côté, A.P., Yaghi, O.M.: Reticular Synthesis of Covalent Organic Borosilicate Frameworks. J. Am. Chem. Soc. 130, 11872–11873 (2008).
12. Uribe-Romo, F.J., Hunt, J.R., Furukawa, H., Klöck, C., O'Keeffe, M., Yaghi, O.M.: A Crystalline Imine-Linked 3-D Porous Covalent Organic Framework. J. Am. Chem. Soc. 131, 4570–4571 (2009).
13. Uribe-Romo, F.J., Doonan, C.J., Furukawa, H., Oisaki, K., Yaghi, O.M.: Crystalline Covalent Organic Frameworks with Hydrazone Linkages. J. Am. Chem. Soc. 133, 11478–11481 (2011).
14. Wang, K., Yang, L.M., Wang, X., Guo, L., Cheng, G., Zhang, C., Jin, S., Tan, B., Cooper, A.: Covalent Triazine Frameworks via a Low-Temperature Polycondensation Approach. Angew. Chemie Int. Ed. 56, 14149–14153 (2017).
15. Rogge, S.M.J., Bavykina, A., Hajek, J., Garcia, H., Olivos-Suarez, A.I., Sepúlveda-Escribano, A., Vimont, A., Clet, G., Bazin, P., Kapteijn, F., Daturi, M., Ramos-Fernandez, E.V., Llabrés Xamena, F.X.I., Van Speybroeck, V., Gascon, J.: Metal-Organic and Covalent Organic Frameworks as Single-Site Catalysts. Chem. Soc. Rev. 46, 3134–3184 (2017).
16. Bagheri, A.R., Aramesh, N.: Towards the Room-Temperature Synthesis of Covalent Organic Frameworks: A Mini-Review. J. Mater. Sci. 56, 1116–1132 (2020).
17. Peng, Y., Wong, W.K., Hu, Z., Cheng, Y., Yuan, D., Khan, S.A., Zhao, D.: Room Temperature Batch and Continuous Flow Synthesis of Water-Stable Covalent Organic Frameworks (COFs). Chem. Mater. 28, 5095–5101 (2016).
18. Ding, S.Y., Wang, W.: Covalent Organic Frameworks (COFs): From Design to Applications. Chem. Soc. Rev. 42, 548–568 (2013).
19. You, J., Zhao, Y., Wang, L., Bao, W.: Recent Developments in the Photocatalytic Applications of Covalent Organic Frameworks: A Review. J. Clean. Prod. 291, 125822 (2021).
20. Feng, X., Ding, X., Jiang, D.: Covalent Organic Frameworks. Chem. Soc. Rev. 41, 6010–6022 (2012).
21. Fan, H., Gu, J., Meng, H., Knebel, A., Caro, J.: High-Flux Membranes Based on the Covalent Organic Framework COF-LZU1 for Selective Dye Separation by Nanofiltration. Angew. Chemie Int. Ed. 57, 4083–4087 (2018).
22. Yang, S.T., Kim, J., Cho, H.Y., Kim, S., Ahn, W.S.: Facile Synthesis of Covalent Organic Frameworks COF-1 and COF-5 by Sonochemical Method. RSC Adv. 2, 10179–10181 (2012).

23. Bojdys, M.J., Jeromenok, J., Thomas, A., Antonietti, M.: Rational Extension of the Family of Layered, Covalent, Triazine-Based Frameworks with Regular Porosity. Adv. Mater. 22, 2202–2205 (2010).

24. Maschita, J., Banerjee, T., Savasci, G., Haase, F., Ochsenfeld, C., Lotsch, B. V.: Ionothermal Synthesis of Imide-Linked Covalent Organic Frameworks. Angew. Chemie Int. Ed. 59, 15750–15758 (2020).

25. Li, Y., Chen, W., Xing, G., Jiang, D., Chen, L.: New Synthetic Strategies toward Covalent Organic Frameworks. Chem. Soc. Rev. 49, 2852–2868 (2020).

26. Campbell, N.L., Clowes, R., Ritchie, L.K., Cooper, A.I.: Rapid Microwave Synthesis and Purification of Porous Covalent Organic Frameworks. Chem. Mater. 21, 204–206 (2009).

27. Wei, H., Chai, S., Hu, N., Yang, Z., Wei, L., Wang, L.: The Microwave-Assisted Solvothermal Synthesis of a Crystalline Two-Dimensional Covalent Organic Framework with High CO_2 Capacity. Chem. Commun. 51, 12178–12181 (2015).

28. Medina, D.D., Rotter, J.M., Hu, Y., Dogru, M., Werner, V., Auras, F., Markiewicz, J.T., Knochel, P., Bein, T.: Room Temperature Synthesis of Covalent-Organic Framework Films through Vapor-Assisted Conversion. J. Am. Chem. Soc. 137, 1016–1019 (2015).

29. Li, X., Yang, C., Sun, B., Cai, S., Chen, Z., Lv, Y., Zhang, J., Liu, Y.: Expeditious Synthesis of Covalent Organic Frameworks: A Review. J. Mater. Chem. A. 8, 16045–16060 (2020).

30. Emmerling, S.T., Luzia, S., Julien, P.A., Lotsch, B. V, Emmerling, S.T., Germann, L.S., Julien, P.A., Moudrakovski, I., Etter, M., Fri, T.: In Situ Monitoring of Mechanochemical Covalent Organic Framework Formation Reveals Templating Effect of Liquid Additive. Chem. 7, 1639–1652 (2021).

31. Biswal, B.P., Biswal, B.P., Chandra, S., Chandra, S., Kandambeth, S., Kandambeth, S., Lukose, B., Lukose, B., Heine, T., Heine, T., Banerjee, R., Banerjee, R.: Mechanochemical Synthesis of Chemically Stable Isoreticular Covalent Organic Frameworks. J. Am. Chem. Soc. 135, 5328–5331 (2013).

32. Kim, S., Lim, H., Lee, J., Choi, H.C.: Synthesis of a Scalable Two-Dimensional Covalent Organic Framework by the Photon-Assisted Imine Condensation Reaction on the Water Surface. Langmuir. 34, 8731–8738 (2018).

33. Kim, S., Choi, H.C.: Light-Promoted Synthesis of Highly-Conjugated Crystalline Covalent Organic Framework. Commun. Chem. 2, 1–8 (2019).

34. Xiong, S., Liu, J., Wang, Y., Wang, X., Chu, J., Zhang, R., Gong, M., Wu, B.: Solvothermal Synthesis of Triphenylamine-Based Covalent Organic Framework Nanofibers with Excellent Cycle Stability for Supercapacitor Electrodes. J. Appl. Polym. Sci. 139, 51510 (2022).

35. Wang, L., Wang, R., Zhang, X., Mu, J., Zhou, Z., Su, Z.: Improved Photoreduction of CO_2 with Water by Tuning the Valence Band of Covalent Organic Frameworks. ChemSusChem. 13, 2973–2980 (2020).

36. Dhankhar, S.S., Nagaraja, C.M.: Porous Nitrogen-Rich Covalent Organic Framework for Capture and Conversion of CO_2 at Atmospheric Pressure Conditions. Microporous Mesoporous Mater. 308, 110314 (2020).

37. Gao, C., Li, J., Yin, S., Sun, J., Wang, C.: Redox-Triggered Switching in Three-Dimensional Covalent Organic Frameworks. Nat. Commun. 11, 1–8 (2020).

38. Dong, B., Wang, W.J., Pan, W., Kang, G.J.: Ionic Liquid as a Green Solvent for Ionothermal Synthesis of 2D Keto-Enamine-Linked Covalent Organic Frameworks. Mater. Chem. Phys. 226, 244–249 (2019).

39. Guan, X., Ma, Y., Li, H., Yusran, Y., Xue, M., Fang, Q., Yan, Y., Valtchev, V., Qiu, S.: Fast, Ambient Temperature and Pressure Ionothermal Synthesis of Three-Dimensional Covalent Organic Frameworks. J. Am. Chem. Soc. 140, 4494–4498 (2018).

40. Chen, L., Du, J., Zhou, W., Shen, H., Tan, L., Zhou, C., Dong, L.: Microwave-Assisted Solvothermal Synthesis of Covalent Organic Frameworks (COFs) with Stable Superhydrophobicity for Oil/Water Separation. Chem. An Asian J. 15, 3421–3427 (2020).

41. Dogru, M., Sonnauer, A., Zimdars, S., Döblinger, M., Knochel, P., Bein, T.: Facile Synthesis of a Mesoporous Benzothiadiazole-COF Based on a Transesterification Process. CrystEngComm. 15, 1500–1502 (2013).

42. Zhang, W., Qiu, L.G., Yuan, Y.P., Xie, A.J., Shen, Y.H., Zhu, J.F.: Microwave-Assisted Synthesis of Highly Fluorescent Nanoparticles of a Melamine-Based Porous Covalent Organic Framework for Trace-Level Detection of Nitroaromatic Explosives. J. Hazard. Mater. 221–222, 147–154 (2012).
43. Ge, J., Xiao, J., Liu, L., Qiu, L., Jiang, X.: Facile Microwave-Assisted Production of Fe_3O_4 Decorated Porous Melamine-Based Covalent Organic Framework for Highly Selective Removal of Hg 2+. J. Porous Mater. 23, 791–800 (2016).
44. Lin, G., Gao, C., Zheng, Q., Lei, Z., Geng, H., Lin, Z., Yang, H., Cai, Z.: Room-Temperature Synthesis of Core-Shell Structured Magnetic Covalent Organic Frameworks for Efficient Enrichment of Peptides and Simultaneous Exclusion of Proteins. Chem. Commun. 53, 3649–3652 (2017).
45. Lu, Y.Y., Wang, X.L., Wang, L.L., Zhang, W., Wei, J., Lin, J.M., Zhao, R.S.: Room-Temperature Synthesis of Amino-Functionalized Magnetic Covalent Organic Frameworks for Efficient Extraction of Perfluoroalkyl Acids in Environmental Water Samples. J. Hazard. Mater. 407, 124782 (2021).
46. He, J., Luo, B., Zhang, H., Li, Z., Zhu, N., Lan, F., Wu, Y.: Surfactant-Free Synthesis of Covalent Organic Framework Nanospheres in Water at Room Temperature. J. Colloid Interface Sci. 606, 1333–1339 (2022).
47. Yoo, J., Lee, S., Hirata, S., Kim, C., Lee, C.K., Shiraki, T., Nakashima, N., Shim, J.K.: In Situ Synthesis of Covalent Organic Frameworks (Cofs) on Carbon Nanotubes and Graphenes by Sonochemical Reaction for CO2 Adsorbents. Chem. Lett. 44, 560–562 (2015).
48. Oliveira, C.J.F., Freitas, S.K.S., De Sousa, I.G.P.P., Esteves, P.M., Simao, R.A.: Solvent Role on Covalent Organic Framework Thin Film Formation Promoted by Ultrasound. Colloids Surfaces A Physicochem. Eng. Asp. 585, 124086 (2020).
49. Shinde, D.B., Aiyappa, H.B., Bhadra, M., Biswal, B.P., Wadge, P., Kandambeth, S., Garai, B., Kundu, T., Kurungot, S., Banerjee, R.: A Mechanochemically Synthesized Covalent Organic Framework as Proton-Conducting Solid Electrolyte. J. Mater. Chem. A. 4, 2682–2690 (2016).
50. Liu, G., Chen, H., Zhang, W., Ding, Q., Wang, J., Zhang, L.: Facile Mechanochemistry Synthesis of Magnetic Covalent Organic Framework Composites for Efficient Extraction of Microcystins in Lake Water Samples. Anal. Chim. Acta. 1166, 338539 (2021).

4

Architectural Aspects of Covalent Organic Frameworks for Energy Applications

Biswajit Nayak and Gopal Das

Department of Chemistry, Indian Institute of Technology Guwahati, Guwahati, Assam, India

CONTENTS

4.1 Introduction

Materials science has at present had a significant impact on our daily lives in the twenty-first century. Chemistry is the fundamental discipline that allows for the strategy and synthesis of novel molecules as well as the elucidation of molecular backgrounds of several physicochemical material goods and functions at many essential stages and times. One of chemistry's main goals is to integrate building pieces into organized structures [1]. This entails looking into two types of forces: covalent bonds and non-covalent interactions. The creation of diverse self-assembled and/or precisely planned fragments, polymers along arrangements is possible thanks to the extensive usage of supramolecular interactions. The introduction of covalent links for fundamental structural development, particularly through polymerization, has created an essential approach to joining organic molecules into chain arrangements, in which monomeric units are united in a repeated fashion [2]. Individual tiny molecule integration into organized structures, in particular,

DOI: 10.1201/9781003206507-4

is a dependable chemical illustration that was not up to simply build different polymers as well as organic constituents but to exhibit different qualities and functionalities. Covalently bonded organic moieties to construct artificial macromolecules using distinct and exact primary and also high-order assemblies, such as protein, and enzyme systems, is difficult [3]. Artificial polymers are similar to regular polymers for examples proteins, polysaccharides, DNA, and RNA in terms of chain architectures. Covalent bonds along with non-covalent interactions are used in biological polymer systems like proteins, DNA, and RNA to form well-defined arrangements where the covalent links control the direction of the main order chain configuration in addition to the non-covalent links accurately outline the high-order frameworks [4].

High-order structures are hardly ever controlled in artificial polymers, even though primary-order properties such as composition, links, and end groups can be conceivably synthetic polymers and chain length can be set in biological polymerization [5]. Polymers are conceivably made via either chain development or step development polymerization, depending on the method and the nature of the connections and monomer arrangement [6]. Chain-growth polymerization is useful for vinyl or heterocyclic subunits even though it is centered on the linking of C–C linkages to form the polymer framework or the ring expansion of heterocycles to form heterochains [7]. Step-growth polymerization, on the other hand, is designed for non-vinyl monomers with corresponding functional elements that can respond to form a covalent link [8]. By expanding covalent bonds along with non-covalent interactions in polymerization schemes, COFs implementing this area have evolved in the last decade [9]. The basic concept is to limit chain evolution to a two-dimensional plane in which the polymer framework is coordinated and organized [10]. The geometry-oriented topology map is necessary for directing individual monomers toward particular sites on the two-dimensional plane, and monomers possessing suitable designs and many reactive groups are required for assuring 2D polymer development [11]. As a result, the 2D polymers that arise produce expanded polygons having particular entities there at the joints and edges [12]. Systematic two-dimensional polymers can direct non-covalent interactions between planes utilizing polygon topology to produce layered as well as prolonged two-dimensional polymer structures having a distinct large structure called covalent organic frameworks (COFs) [13–16]. Similarly, 3D COFs with extended architectures will be achievable assuming the development of the polymer chain is accurately regulated in three-dimensional manner.

COFs are a type of crystalline polymer that allows organic units to be integrated into distinct primary as well as high-order arrangements [17]. COFs provide a foundation for creating polymeric materials by designing at times ordered organic structures. COFs use step-growth polymerization to propagate chains in 2 or 3 dimensions, ensuring that their inflexible supports are structurally guided and fundamental arrangements are maintained. To shape the definite yet prolonged crystalline arrangements, the polymerization process incorporates both covalent bonds along non-covalent interactions. COFs are totally predesignated and synthetically controlled polymers since the framework building is solely defined by monomers. Modern experimental technologies have recently been created, including that of microfluidic devices and gelation-mediated 3D printings, allowing researchers to examine easily interpretable COFs and systems [18]. Researchers review recent progress in fundamental aspects, conceptual design, and synthetic reactions; illustrate essential characteristics and diversities; consider the advancement and potential of different functionalities by elucidating structure-function correlations; and continue to be major; and predict future directions from chemistry.

4.2 Design Principle

The path of covalent bond construction and the polymer backbone progress is elaborated by the principle of the topology design plan. To make sure a distinct way of every single covalent bond formation, the monomers having relatively rigid backbones are required where the reactive sites are dispersed in a distinctive geometry. As shown in Figure 4.1, the basic concept of design can be streamlined and exemplified through the block model. Each monomer is representing a detailed geometry that reveals the reactive sites having relative positions. The spatial orientation is guided by the covalent bonds, which also regulate positions of the subsequent monomer units and in every connection, repeating this rule restricts the chain progression ways in an approach that firmly trails the topology diagram which is predesigned. The chemical foundation for the regulation and development of the primary-order arrangement is constructed by the comparable geometry of monomers in the topology diagram.

The arrangement of monomer units confines the development of backbones of the polymer in a two-dimensional fashion, which prominently leads in the direction of the creation of 2D atomic layers having particular arrangements, as shown in Figure 4.1 [12–16, 18–21]. At the same time, well-arranged free polygon regions are left by the development of the polymer chain mostly on the 2D x-y axis. The comprehensive primary-order structure having polygon-based along with in-built distinct nanopores constitute these 2D covalent polymers [10, 13]. Furthermore, in the 2D covalent polymer, the rigid skeleton gives the structure crystallinity and spatial lattice orientation. Certainly, by way of rigid backbones, the organic monomers which come across these requirements are broadly offered also they generally grasp the π-systems [18]. All these fundamental features allow additional liberty in discovering non-covalent contacts to monitor the construction of precise greater arrangements. The rigid two-dimensional polymers, where the comparative

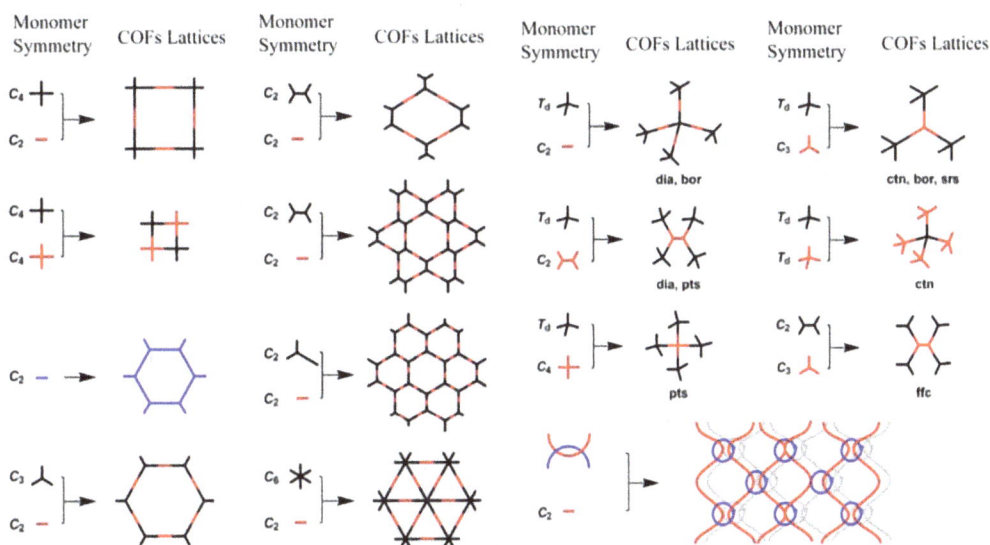

FIGURE 4.1

An elementary representation of topological pathways is used in the design of 2D and 3D COFs. (Reproduced with permission from reference [17]. Copyright (2020) American Chemical Society.)

positions of adjacent sheets are directed and commonly well-ordered by the interactions in the interlayer, are favored by the development of layered structures because the units are positioned at specific sites with desired orientations through two-dimensional covalent polymers. Every single monomer entity is overlapped through the stacking of the dimensional covalent polymers, to make most of the energy in the layered arrangements. Along the z-direction, this alignment of 2D covalent polymers creates the 2D COFs with a high-order arrangement. So, a distinctive characteristic of 2D COFs is to produce completely well-organized π-arrays along with 1D open channel.

4.2.1 Two-Dimensional COFs

Using a variability of diverse topologies; individually standard isotropic layouts and uneven anisotropic geometries polygon frameworks were settled, supporting the strategy of two-dimensional COFs. The lattice structure in every single instance is very much well-ordered, also the pore is discrete. There are enormous members of COFs behind every single topology with altered linkers and building blocks. The multiplicity of structural configurations in 2D COFs is accountable for the diversity.

4.2.1.1 Symmetric Topologies

Designing COFs with diverse pores and skeletons, the structural path agrees for dissimilar arrangements of a range of monomer geometries, as shown in (Figure 4.1a), leading to the different structural arrangements [13, 14, 16, 22]. For instance, hexagonal 2D COFs have been produced via the combination of $[C_3 + C_2]$, [9, 23, 24] while with diverse π-orderings and pore sizes, $[C_2 + C_2 + C_2]$ and $[C_3 + C_3]$ likewise both make hexagonal COFs [9, 23, 25]. The combination of $[C_4 + C_4]$ and $[C_4 + C_2]$ generates Tetragonal COFs [26–31]. The pore size is determined by the topology diagram. In case of $[C_4 + C_4]$ and $[C_2 + C_2 + C_2]$ arrangements, the resultant COFs make sure a probability of having micropores, however the $[C_3 + C_2]$, $[C_3 + C_3]$ along with $[C_4 + C_2]$ arrangements generally produce mesoporous frameworks. The $[C_6 + C_2]$ arrangements have developed the microporous COFs using maximum π-cloud as well as lowest pore [13, 32]. Furthermore, the structural design likewise regulates the π-cloud of the skeletons. As displayed in Figure 4.2, by a pore size of only $L/\sqrt{3}$, the resultant COFs ought to have triangular pores, which is $1/\sqrt{3}$ of tetragonal and 1/3 times of hexagonal COFs, and L is the space between the two vertices. Additionally, the π-cloud density for tetragonal and hexagonal COFs is $2/\sqrt{3}$ and 2/3 correspondingly, compared to $2/\sqrt{3}$ for L^{-2}.

The trigonal COFs with small micropores and great π-cloud provide to make molecular separation and semiconducting functionalities. Moreover, by dual pore configurations the construction of triangular lattices was assisted by the hexa-substituted C_3-symmetric vertices [32]. In (Figure 4.1a), by means of six peripheries triangular and one central dodecagonal pore, a kagome COF has been validated. The $[C_2 + C_2]$ combination is an interesting topology diagram that produces two structures having dissimilar skeletons and pores. The rhombic framework has only one type of pore, and with six peripheries trigonal micropores and one central hexagonal mesopore, kagome skeleton is the other type. The resulting topology is determined by the knot and linker arrangements. The type of topology is directed by the strength and the weight of the knot unit of interlayer interactions [33, 34]. With great π-systems, the C_2-symmetric knot has a general tendency to support solid π-π interactions which guide the way to form the rhombic polygons. The docking effect compensated for the kagome COF's dull interfacial interaction (π-π contacts) [35].

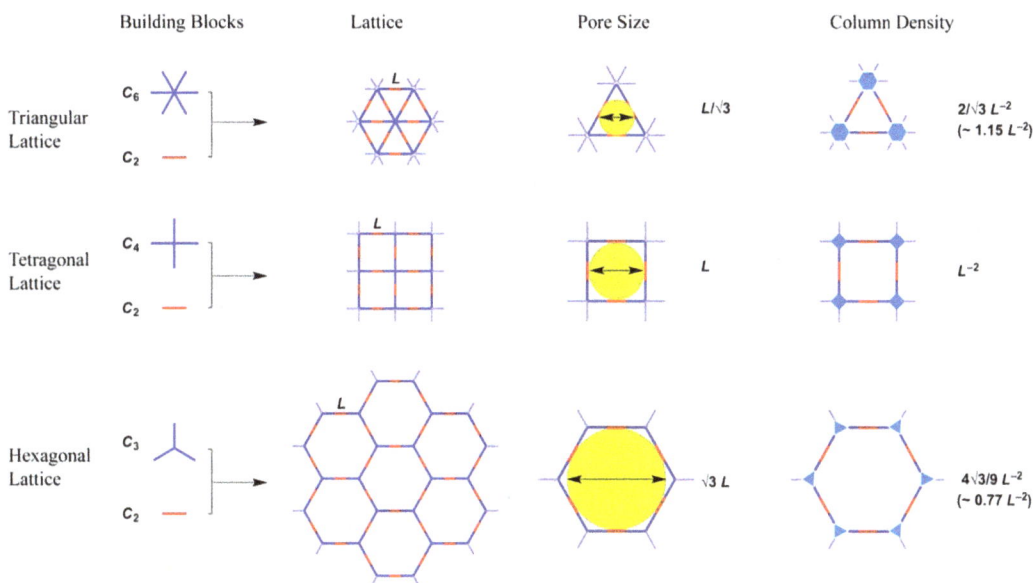

FIGURE 4.2
On structural diagrams, the pore size, as well as π-cloud of two-dimensional COFs, is depicted. (Reproduced with permission [17]. Copyright (2021) Royal Society of Chemistry.)

4.2.1.2 Asymmetric Topologies

The [1 + 1] two-element structures, which consist of one linker as well as one knot, some of which are in turn connected to generate standard polygons along with isotropic assemblies, are used in the topology diagrams above. The traditional [1 + 1] two-component arrangement requirement for COFs, on the other hand, severely confines the variety of COFs. In the event of a single knot as well as ten linkers, for example, only ten alternative COFs are conceivably built and created. The [1 + 1] arrangement-based variation limitation is eliminated by using a two or more components notion in COF design. As shown in Figure 4.3, the purpose of the multiple component technique is to create COF skeletons with 2 or 3 dissimilar linkers [11, 36]. Single C_3-symmetric component and ten linkers of C_2-symmetry, for example, might produce 210 dissimilar hexagonal frameworks. The [1 + 2] three-component approaches yield two unique hexagonal COFs based on the molar concentration of the two linkers (1/2 or 2/1). The four-component technique [1 + 3] produces hexagonal COFs with three dissimilar sets of edges. The structural complexity of the skeletons is substantially increased due to the presence of multicomponent. Following this, the lattice modifications change between isotopic constructions in [1+1] 2-component structures through anisotropic tiling, in addition, the pore changes from a standard pore to such an unevenly formed hexagonal pore in the multiple-component method. Certainly, ten distinct C_2-symmetric linkers having various π-arrangements, lengths, and redox actions were synthesized. 48 3-component COFs were made using 24 different permutations of the [1 + 2] method, while eight four-component frameworks were made using the [1 + 3] strategy [11]. The phase reliability of multicomponent COFs is suggested by the distinct lattice arrangement and single kind of pores found completely in all of these COFs. This technique is accepting of the multiple-component donor in addition to acceptor π-units being combined through a single framework. The unique displays of the donor as

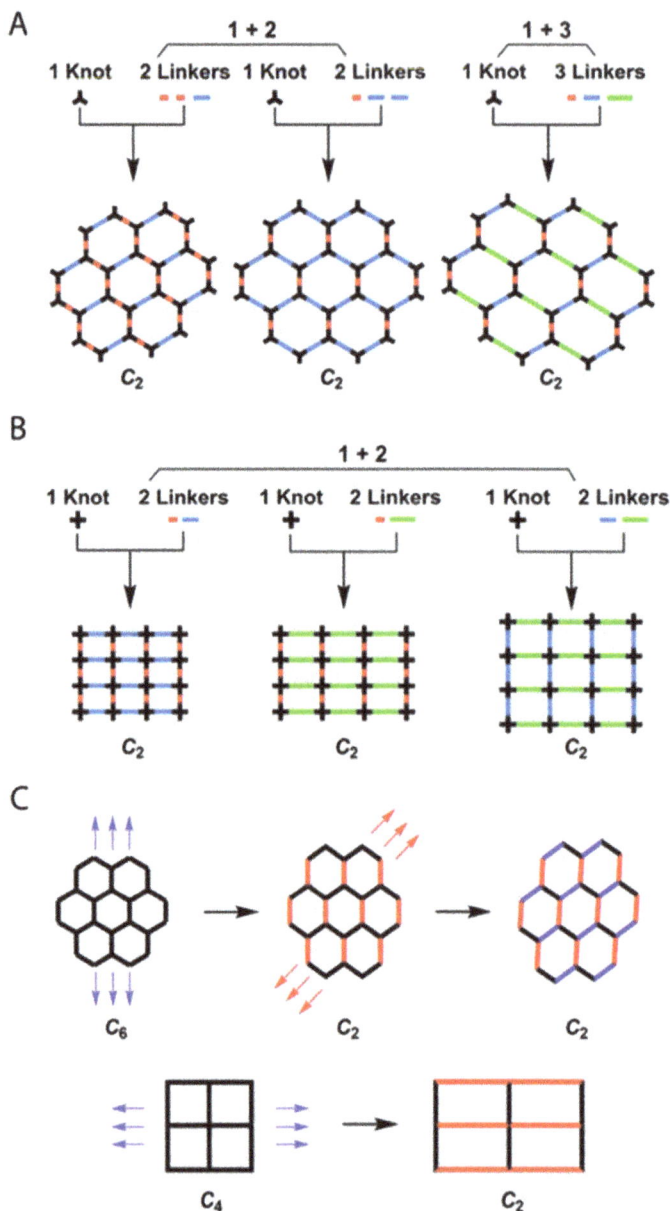

FIGURE 4.3

Multicomponent design technique aimed at (A) hexagonal, (B) tetragonal COFs, as well as (C) stable polygon extension. (Reproduced with permission [17]. Copyright (2020) American Chemical Society.)

well as acceptor π-supports in these frameworks may cause intercolumnar interactions, resulting in extraordinary conductivity that is not simply the totality of the [1 + 1] COF equivalents. As shown in Figure 4.3, this multiple-component method is not only relevant to hexagonal COFs but also tetragonal skeletons.

Figure 4.3 shows how combining one knot and two linkers with the [1 + 2] three-component approaches can expand the amount of tetragonal COFs and produce anisotropic tiling and distinctively shaped pores. Using phthalocyanine entities as a single knot, coupling them through ten distinct linkers, tetragonal asymmetric COFs can be generated, each one with two arrangements of edges that are asymmetrically tiled [11]. Alternatively, using the [2 + 1] multiple systems with two C_3-symmetric entities of varied sizes as knots and one C_2-symmetric link, an asymmetric hexagonal framework can be created [37]. The use of a desymmetrized unit has shown to be an operational method for constructing multiple-pore frameworks [38]. The desymmetry units 5-(4-formylphenyl) and 5-((4-formylphenyl) ethylene)-isophthalaldehyde have supports of varying lengths. As shown in Figure 4.1A, when all these vertices condense having C_2- symmetric ends, they create HP-COF-1, 2 that have double forms of hexagonal systems of varying forms and dimensions [39]. The multiple-component technique has been used also to a specific kagome topology which provides TPE constructed frameworks having an innate double pore arrangement as a potent device designed for creating heterogonous assemblies in single skeleton. For this case, when TPE knots (4,4′,4″,4‴-(ethene-1,1,2,2- tetrayl)tetraaniline (ETTA)) are combined using 2 dialdehydes of differing sizes, SIOC-COF-1, 2 and two triple pores COFs are formed [36].

4.2.2 Three-Dimensional COFs

The design of 3D COFs needs at least one building unit in comparison to 2D COFs where planar building units have frequently been involved. As shown in Figure 4.4, the building units' Td or orthogonal structure guides the extension lead of the backbone of the polymer into a covalently connected 3D system. To form 3D COFs, there are unlike groupings used for integration tetrahedral, or else an orthogonal unit with C_1, C_2, C_3, and C4 has been proposed [12, 16, 40]. 3D COFs are conceivably classified into dissimilar networks with respect to the topology figures, containing bor, ctn, srs, dia, pts, and rra. The $[T_d + C_4]$ and $[T_d + C_2]$ figures have been synthesized by pts network in which four reactive sites are required of the C_2- or C_4-symmetric unit, resulting in the production of two-folded three-dimensional COFs [41]. The $[T_d + C_3]$ diagram have been synthesized by the bor or ctn network, which provides a skeleton that is impervious to interpenetration, leading areas which have huge surface [42]. The $[T_d + T_d]$ can similarly generate the ctn network [42]. The largest family of 3D COFs is produced by the dia structure which is achieved by the $[T_d + C_2]$ arrangement [43]. In all these cases, a 1D channel is formed by the backbones of the polymer and these polygon networks are generally microporous having pore dimensions vary amid 0.7 to 1.5 nm. A C_2-symmetric linker is a linker that is symmetric in both directions. An intertwined 3D helix COF-505 is formed when an orthogonal complex of Cu(II) of phenanthroline is condensed [44].

Because of the interpenetrated structures, the morphology diagram of 3-dimensional COFs can only roughly estimate the number of times folds a particular COF will have. This is due to the chemical factor that controls the foldaway phenomenon is quite uncertain. On this basis, 3D COFs are barely prearranged and synthetically manageable, distinct from 2D COFs.

4.3 Structural Diversity and Building Blocks

COF diversity begins with the construction of COFs utilizing a variety of skeletons and pores, which is aided by the topology illustration. In the topology diagram depicted in Figure 4.4, the shape of monomers is crucial for the fundamental approach of COFs.

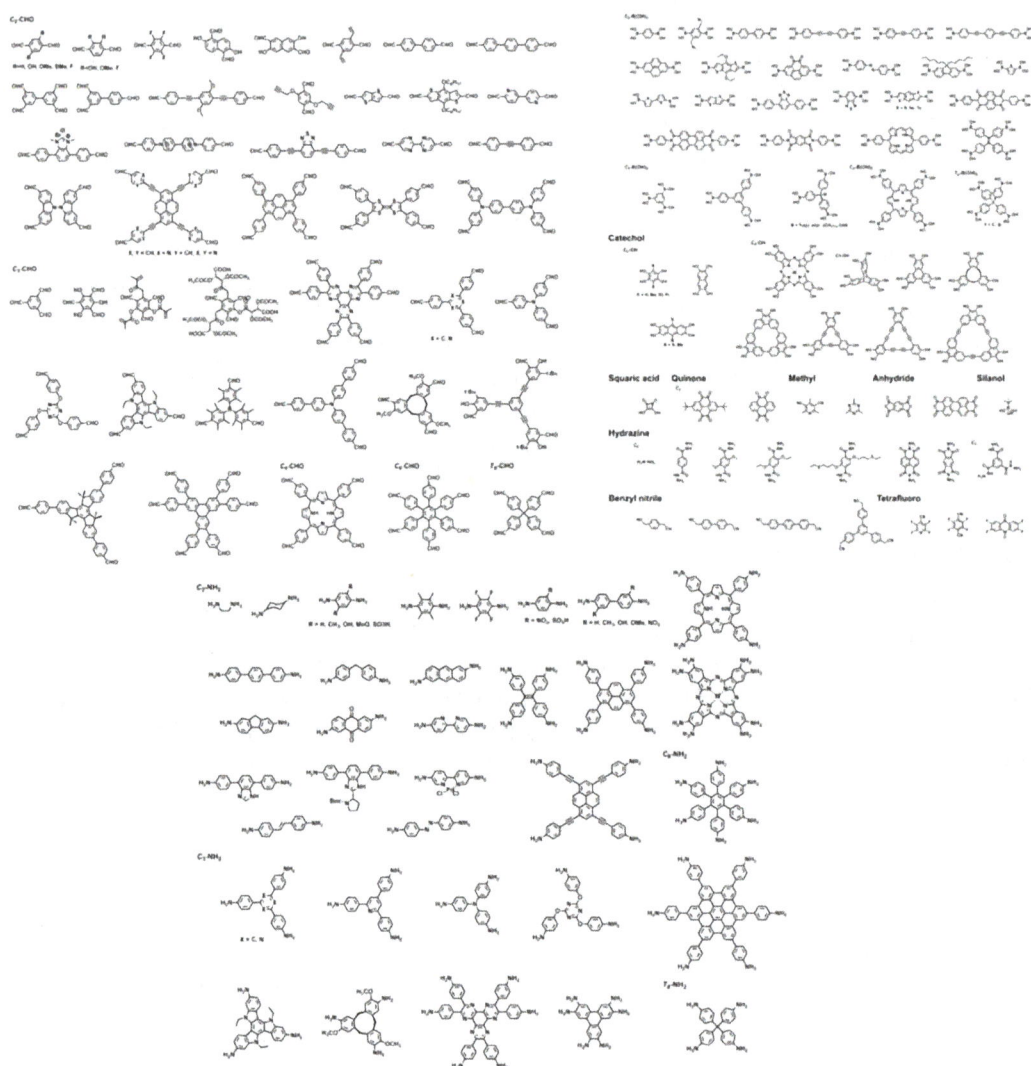

FIGURE 4.4
Symmetric monomers of C_2, C_3, C_4, C_6, and T_d with diverse well-designed units.

The building components often have backbones with a stiff conformation to adopt topology-focused bond development and keep the two-dimensional planarity of the protracted polygons. Apart from geometry, monomers were studied for their structural variety, which includes varied sizes, reactive groups, docking sites, and redox-active, photoactive groups, catalytic, and chiral centers. As a result of these differences, various COFs with diverse arrangements and functions emerge. 3D COFs have a smaller amount of structural variation, which is restricted through the accessibility of orthogonal or T_d-symmetric units because structural multiplicity is heavily reliant on the supply of monomer units. The [$C_4 + C_3$] method, which was recently developed, increases the fundamental variety of three-dimensional frameworks. Nonetheless, 3D frameworks have a lot of room to grow their repertoire by using several π-moieties as linkers intended for summarizing the orthogonal or T_d-symmetric monomers, or by looking into the use of GSUs to build 3D structures.

The monomer architectures with characteristic backbone along with reactive sites are shown in Figure 4.4. The backbone could be built to ensure a variety of C_2, C_3, C_4, C_6, and T_d topologies, ranging from benzene through basic arenes, heterocycles, and macrocycles. As a result, hundreds of monomers have been investigated.

4.3.1 Two-Dimensional COFs

In arrears to the wide range of building elements, two-dimensional COFs are the most common class of COFs. 2D COFs are conceivably planned and produced into diverse skeletons, topologies, and pores. This structural variety opens up new possibilities for the multipurpose resource platform.

4.3.1.1 Hexagonal COFs

Hexagonal, kagome, tetragonal, and trigonal COFs have been developed and constructed based on the varied topologies of monomers. These synthesized COFs have diverse yet distinct π-skeletons along with pore sizes and shapes. The building parts of hexagonal COFs on the root of the knot arrangement, are conceivably classified into dissimilar classes containing benzene, triazine, benzene having triphenyl moiety, and also triphenylene products as shown in Figures 4.4. The construction of various COFs supported by the smallest benzene knot, comprising COF-1, which is boroxine connected, the boronate ester accompanying COF-6,11,14,16, in addition to Ph-An-COF (pore dimension = 0.64–2.9 nm) [23], the imine connected COF-LZU1, NUS-9,10(R), TpBD-COFs, TpPa-COFs, TpPa- SO_3H-COF, TpPa-(SO_3H-Py)-COF, and DAAQ-TFP, PI-2-COF, having pore dimension = 0.84–2.5 nm Tp-Azo, Tp-Stb COFs, the hydrazine related COF-42 and COF-JLU4 (pore dimension = 2.2–2.3 nm) [45] with pore size = 0.94–0.96 nm the azine connected COF-JLU2 and ACOF-1, then the triazine allied CTF-1 (pore dimension = 1.2 nm) in addition to the C=C interconnected g-$C_{40}N_3$-COF, g-$C_3 1N_3$-COF, and g-$C_{37}N_3$-COF (pore dimension = 1.1–3.2 nm). Triphenylbenzene generates only mesoporous COFs, because of its large size, as well as the boronate ester connected BTP-COF and COF-8 (pore size = 1.64–4.0 nm) [23], the imine connected 2,3- PI-3-COF, DhaTab, COF-DhaTab, 2,3-DhaTta, LZU-20, TAPB-PDA COF, TPB-DMTP-COF, TPB- and TP-COFs (pore dimension = 3.26–3.9 nm) the azine associated Nx-COF (x = 0, 1, 2, 3), COF-JLU3, and LZU-22 (pore dimension = 2.4–2.6 nm), with pore size = 3.5–3.8 nm the hydrazine connected LZU 21, TFPT-COF and COF-43. The triphenylene units make the way to the produce many of ester (boronate) allied hexagonal frameworks, for example, COF-5,6,8,10, DTP-ANDI-COF, HHTPDBB COF, COF-316 (JUC-505), JUC-506, and COF-318 (pore dimension = 1.2–5.3 nm) [9, 11, 23, 24]. Hence, the combination [$C_3 + C_2$] is dominant in planning hexagonal mesoporous frameworks. The single pore assembly is illustrated in Figure 4.5.

4.3.1.2 Tetragonal COFs

For structural construction, tetragonal arrangements support the usage of C_2- as well as C_4-symmetric monomer units by way of knots and linkers. A large number of backbones are covered by the C_2-symmetric monomers such as pyridine, bipyridine, phenyl, biphenyl, thiophene, biphenylazo, stilbene, thiadiazole, tetraphenylethene (TPE), and porphyrins. In contrast, porphyrin and phthalocyanine knots are representing the C_4-symmetric monomers. The permutation of C_2-symmetric edges through C_4-symmetric vertices by modifying the knots and linkers adopts the produce a huge sum of tetragonal designs

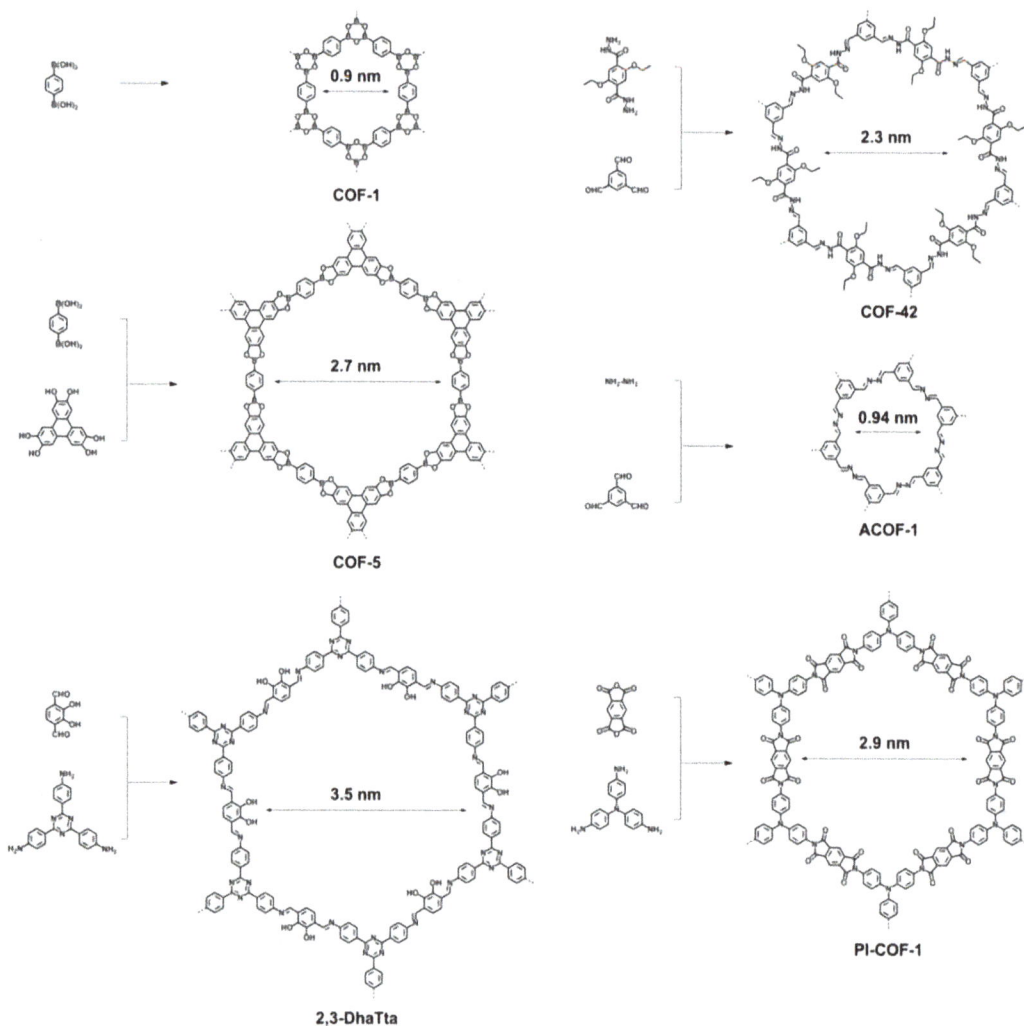

FIGURE 4.5
The basic structure of hexagonal morphologies is represented by the [C$_3$ + C$_2$] arrangement of building blocks.

with pore sizes ranging from 1.8 to 4.4 nm [12, 16, 27, 29, 30]. By using imine, boronate ester, double-stage linkages, and C=C double bond, tetragonal structures can be produced. The tetragonal frameworks are based on the structure of the knot, conceivably classified as phthalocyanine and porphyrin frameworks. The boronate-ester-linked frameworks for porphyrin-based COFs, have a wide range of unlike pore size and skeletons comprising MPCOFs (M = Cu, Co, Zn, H$_2$) (pore dimension = 2.3–2.5 nm) and COF-66 [28]. Correspondingly, with different linkers the imine connected frameworks were produced, where synthesized COFs have different pore sizes, for example Mp-DHPh-COFs (M = H$_2$, Cu, Ni), DhaTph-COF, DmaTph-COF, COF-367-M (M = Co, Co/Cu), Mp-DHPhx-COFs (x = 25, 50, 75%), COF-366-M (M = H, Co), Mp-2,3-DHPh-COF (M = Cu, Ni, Zn), Mp-PyTTPh-COF, Cup-DHNAPh-COF, CuP-BPy-Ph-COF (pore size = 1.8–2.9 nm) [30]. Phthalocyanine COFs having ester (boronate) link have planar layer assemblies as well as produce dissimilar structures, containing NiPc-BTDA NiPc-COF, MPc-COFs (M = Co, Cu, Zn), Pc-PBBA COF, MC-COFs-NiPc, M_1DPP-M2Pc COFs (M_1 = H$_2$, Zn, Cu; M_2 = Ni, Cu), ZnPc-DPB-COF, ZnPc-Py-COF, ZnPc-NDI-COF, D_{ZnPc}-A_{PDI}-COF, D_{CuPc}-A_{PyrDI}-COF, D_{NiPc}-A_{NDI} COF, D_{CuPc}-A_{PDI}-COF, and D_{NiPc}-A_{PyrDI}-COF, (pore size = 2.0-4.4nm) [11, 26, 27, 29]. The arrangement of ester (boronate), imine, hydrazone, as well as boroxine links composed organized through a range of knots offers a double-stage strategy, together with phthalocyanine and porphyrin derivatives. The presence of anisotropic lattice points makes the skeletons very heterogeneous in all these cases. These COFs contain mesoporous (pore dimension = 2.1–3.7 nm) CuPc-FPBA-TABPy, PyTTA ZnP, TMBDA, DETHz COFs and microporous (pore dimension = 1.8 nm) CuPc-FPBA-ETTA.

4.3.1.3 Rhombic COFs

The [C_2 + C_2] arrangements by tetraphenyl pyrene units have synthesized Rhombic shaped COFs bringing about the production of azine associated, imine linked, and C=C double bond allied COFs, as well as Py-Azine COF, ILCOF-1, Py-DHPh COF, Py-1P, 2P, 2PE, 3PE COF, Py-2,21-BPyPh COF, Py-3PE$_{BTD}$ COF, and sp^{2}c-COF, (pore dimension = 1.7–2.79 nm) [31, 34]. Figure 4.6 elucidates the distinctive cases of rhombic lattices.

4.3.1.4 Kagome COFs

By means of the [C_3 + C_2] or [C_2 + C_2] figure Kagome sort frameworks were designed. TPE knot and linker having the C_2-symmetry typically used by the [C_2 + C_2] figure. The [C_2 + C_2] figure produces the imine allied COF-TPDA, SIOC-COF-1, 2, (4PE-3P COF), and

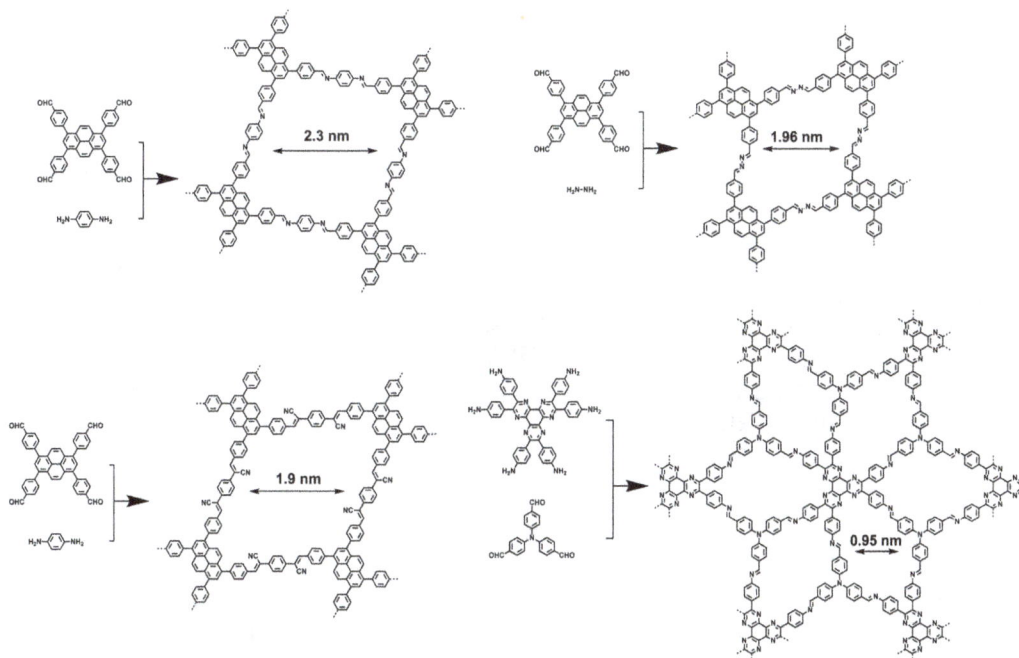

FIGURE 4.6
The [C_2 + C_2] and [C_2 + C_3] diagrams are used to represent the basic design of Rhombic COFs.

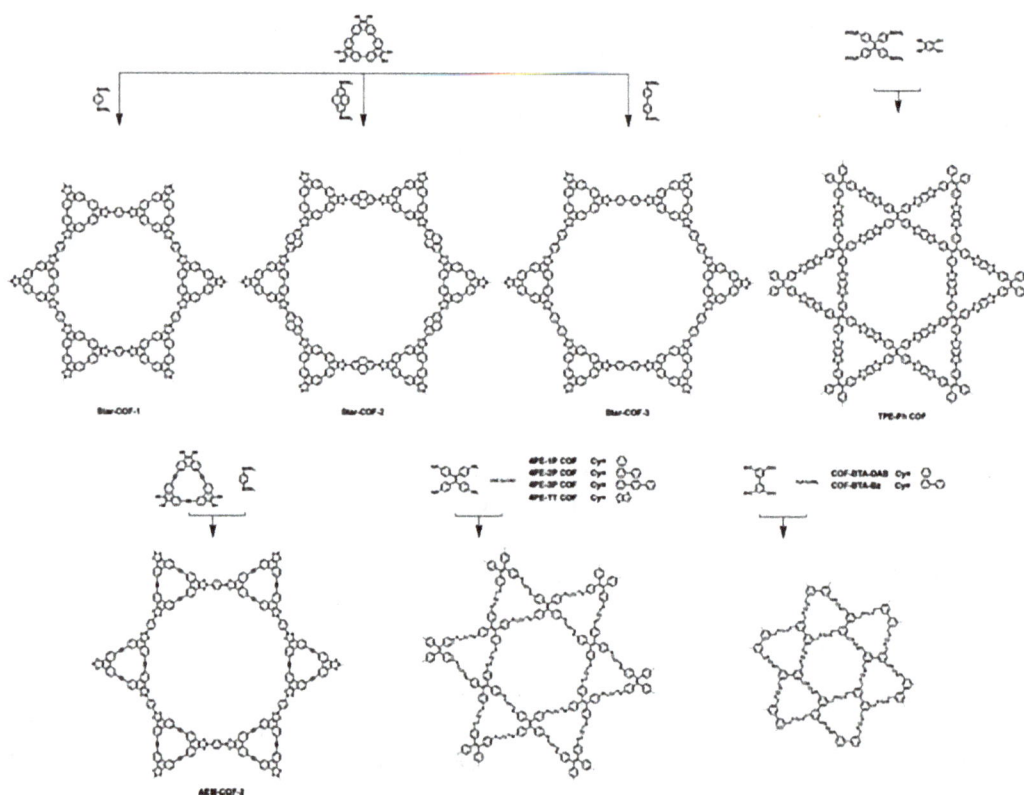

FIGURE 4.7
Representation of the $[C_2 + C_2]$ building components as a scheme of kagome frameworks (dual-pore).

boronate-ester connected TPE-Ph COF [33, 35, 36]. By means of (BTA) [1, 1'-biphenyl]-3,3',5,5'- Tetracarbaldehyde units COF-BTA-BA and COF-BTA-DAB correspondingly attain kagome topology. Using C_3-symmetric macrocycles, the $[C_3 + C_2]$ topology diagram can similarly produce kagome frameworks, comprising the boronic ester connected Star-COF-1, 2, 3, Py-MV-DBA-COF, AEM-COF-2, DBA-COF 1 (=AEM-COF-1), and Py-DBACOF-1,2 [22, 37]. Pore size varying from 0.71 to 1.8 nm for triangular micropore and with pore widths 2.6–4.5 nm the dodecagonal or hexagonal mesopore together represents in the kagome lattice. Bring together substituents into the linker of C_2-symmetry; the $[C_2 + C_2]$ diagram could potentially produce both single-pore rhombic frameworks and dual-pore kagome frameworks and that is extremely hooked on the bulk of the substituents. Distinctive cases of kagome lattices are illustrated in Figure 4.7.

4.3.1.5 Triangular COFs

By means of the C_6-symmetric units, the triangular topology diagram leads to the creation of microporous COFs having pore size varies between 1.1 and 1.8 nm and the examples are HEX-COF-1, HBC-COF, and HPBCOF [12]. The condensation reaction of the hexaazat-riphenylene analog or triphenylene moiety, the 1,1'-biphenyl 4,4'-diamine, or C_2-symmetric BDA produces triangular COFs [32]. HAT-COF has pore diameters of 1.13 and 1.52 nm, while

HFPTPBPDA has pore sizes of 1.27 plus 1.55 nm. Recently, researchers looked at the carbon (sp^{2}c) coupled CCP-HATN two-dimensional COF with pore widths of 0.68 and 1.28 nm [46].

4.3.1.6 Heteropore COFs

COFs having three unlike classes of openings established on a several connecting site approaches can be designed over and done with the combination of vertex truncation. By means of the tetratopic linker used for pyrene tetra-aniline and benzene tripicolinaldehyde as the heteropore or tritopic linker used for triazine tribenzaldehyde, PY$_2$B- and PT$_2$B-COFs with pore size varying from 1.83 to 2.37 nm conceivably made [47]. [1,1':3',1''-terphenyl] feasibly used to create triple-pore frameworks TP-COF-BZ and TP-COF-DAB. The knot is -3,3'',5,5'' tetracarbaldehyde (TPTCA). The pore sizes of TP-COF-BA, TP-COF-DAB are 1.61 plus 3.18 nm, correspondingly; because these rectangular shape pores are too small to be detected, the sizes of their pores are unlikely to be determined. Heteropore frameworks can be formed by relating C$_2$-symmetrical building units with linear linkers. COFBABD-BZ with pore dimension = 1.8–2.5 nm and COF-BABD-DB with pore dimension = 0.98–1.72 nm synthesized by the polycondensation reaction of benzidine or 1,4-diaminobenzene with 4'-(bis(4-formylphenyl)amino)-[1,1'-biphenyl]-3,5-dicarbaldehyde in n-butyl alcohol/o-dichlorobenzene/aqueous acetic acid [48]. Likewise, COF-DA-DB-TB along with COF-DA-DB by a diformylphenylboronic unit adopts heteropore architecture via orthogonal processes [49]. COF-DA-DB has a 1.84 nm pore size, but COF-DA-TB contains two separate pore diameters of 1.28 and 1.52 nm.

4.3.2 Three-Dimensional COFs

The [T$_d$ + T$_d$], [T$_d$ + C$_2$], [Td + C$_3$], [Td + C$_4$], also [C$_2$ + C$_3$] illustrations have been used to create 3D COFs as shown in Figure 4.4 and Figure 4.8. The above topology diagrams yield bor otherwise dia, ffc, pts, also ctn frames [40]. In order to make 3-dimensional COFs, diverse kinds of the T$_d$-symmetric nodes were created together with Cu(PDB)$_2$PO$_2$Ph$_{e,}$ complex salt of Cu(I)-bis[4,4'-(1,10-phenanthroline-2,9-diyl)dibenzaldehyde]-tetrafluoroborate, 1,3,5,7 tetraaminoadamantane (TAA), tetra(4-dihydroxyborylphenyl) silane, tetra(4-dihydroxyborylphenyl) methane, tert-butylsilane triol, and tetra-(4-aminophenyl) methane [12, 16, 42, 43]. ETTA has recently been formed as a C$_4$-symmetric unit on the arrangement. 3D COFs have been synthesized using imine, imide, boroxine, boronate ester, spiroborate, and borosilicate links [12, 16, 42, 43]. The boroxine allied 3-dimensional COF-102, 103 have been produced using the ctn structure having pore sizes lie between 1.15 and 1.25 nm [42]. The DBA-3D-COF-1 (pore size = 2.96 nm, bor), COF-105,108 (1.83 nm, ctn; 2.6 nm, bor), also boronate-ester connected frameworks have been arranged with various topologies, whilst borosilicate allied COF-202 has pore size = 1.1 nm with ctn structure [42]. A Imine related 3D frameworks were synthesized using dia topology, particularly LZU-301 (particle sizes ranging 0.58 × 1.04 nm^2 from 0.96 × 1.04 nm^2), 3D-ionic-COF-1 (pore size = 0.86 nm), SP-3D-COF-1, 2 (pore size = 1.2 nm, 1.5 nm), 3D-Salphen- COFs (JUC-508, pore size = 1.3 nm; JUC-509, pore size = 1.2 nm), 3D-CCOF-5 (pore size = 0.62 and 0.74 nm), and COF-300, 320, 505; the ctn topology is included in BF-COF-1, 2 (pore size = 0.83 nm, 0.81 nm); and COF-500 (pore size = 1.23 nm), 3D-Por-COF (pore size = 0.60 and 1.07 nm), 3D-CuPor-COF (pore size = 0.63 and 1.18 nm), 3D-Py-COF (pore size = 0.59 nm), and 3D-TPE-COF (pore size = 0.57 nm) implements the pts diagram [41, 43]. The usual examples of 3D COFs are shown in Figure 4.8.

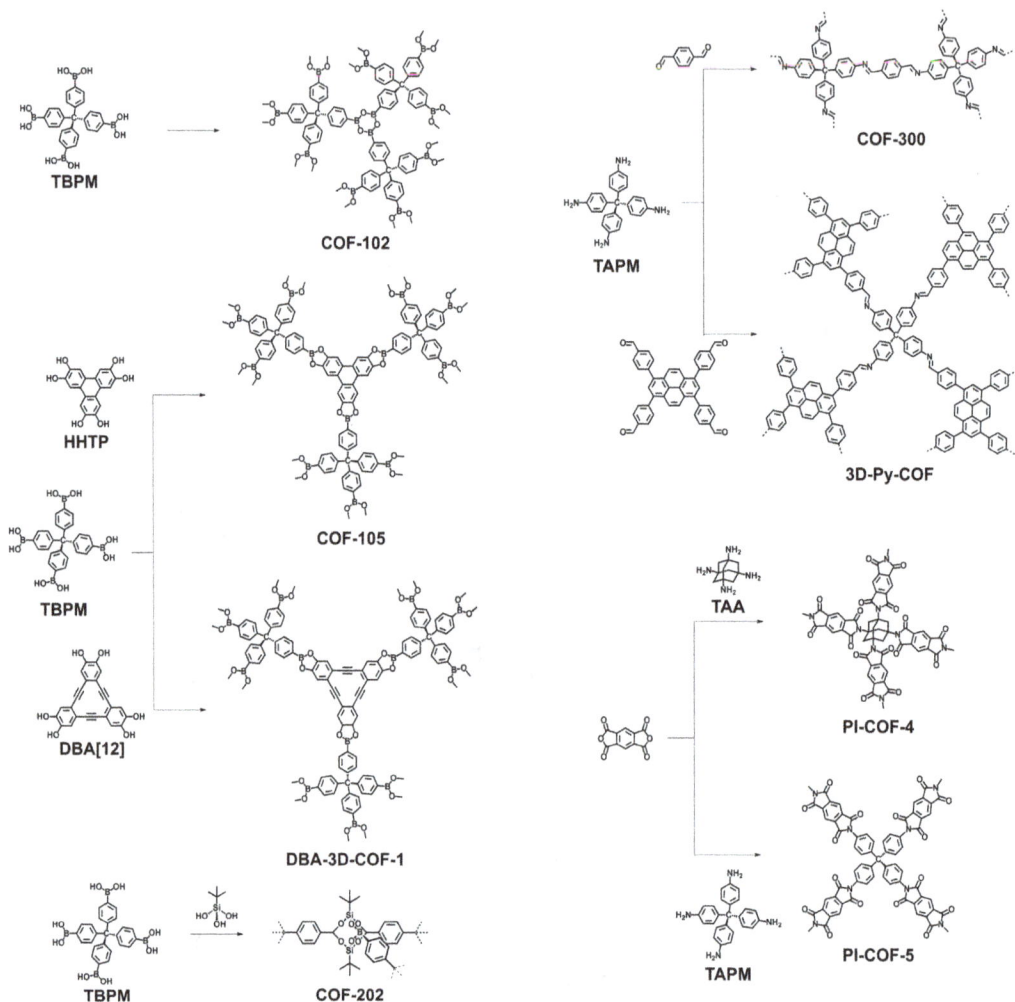

FIGURE 4.8
T$_d$-symmetric topology is used to represent the structural design of 3D frameworks.

4.4 Conclusion

The field of COFs has been shaped by rapid advances in the strategy and synthesis of organic supplies over the last decade. The field's foundation has been established by the scheme principle established on geometry equivalent and topology structures, whereas artificial reactions have formed a great total of arrangements and resources, representing that frameworks provide an extremely rare platform for the planned synthesis of appropriate organic materials and frameworks with long-term structural organizations. Investigation of topology illustrations is one of the significant subjects that will be tackled in the future from the viewpoint of chemistry. The existing structural illustration is based mostly on the use of monomers that have symmetric geometry. For certain circumstances, the use of asymmetric components as monomers in place of polycondensation has been established; additional research on knots as well as linkers having asymmetric structure would allow for the investigation of detailed topologies while maintaining extraordinary crystallinity plus porosity. For connecting organic components into well-arranged frameworks to generate anisotropic patterning besides unique pores that are unreachable toward new porous materials, the exploration of the multi-component technique including both knots plus linkers is of significant importance. To construct chiral COFs with chiral components, monomers, as well as skeleton diversities, must be expanded further. COFs have shown promise in energy storage applications such as photocatalysis, electrocatalysis, supercapacitors, metal-ion/sulfur batteries, and others due to their flexible supramolecular and synthetic techniques, high conjugated and changeable structures, specific surface area, and porosity. COFs must demonstrate the power of chemistry in producing well-arranged organic materials which are superior to standard polymers as well as new porous materials; such accomplishment can be considered a significant advancement in polymer science during the last century.

References

1. C. Reinhardt, Chemical Sciences in the 20th Century: Bridging Boundaries, John Wiley & Sons, 2001.
2. R. J. Young, Introduction to Polymers, Chapman & Hall, 1987.
3. Gardel, M. L. Synthetic Polymers with Biological Rigidity. Nature 2013, 493, 619.
4. Fukui, T.; Kawai, S.; Fujinuma, S.; Matsushita, Y.; Yasuda, T.; Sakurai, T.; Seki, S.; Takeuchi, M.; Sugiyasu, K. Control over Differentiation of a Metastable Supramolecular Assembly in One and Two Dimensions. Nat. Chem. 2017, 9, 493–499.
5. J. R. Fried, Polymer Science and Technology, Prentice Hall, 2nd edn, 2003.
6. H. R. Allcock, F. W. Lampe and J. F. Mark, Contemporary Polymer Chemistry, Prentice Hall, 3rd edn, 2003.
7. P. J. Flory, Principles of Polymer Chemistry, Cornell University Press, 1953.
8. G. Odian, Principles of Polymerization, John Wiley & Sons, 1991.
9. Côté, A. P.; Benin, A. I.; Ockwig, N. W.; O'Keeffe, M.; Matzger, A. J.; Yaghi, O. M. Porous, Crystalline, Covalent Organic Frameworks. Science 2005, 310, 1166–1170.
10. Colson, J. W.; Dichtel, W. R. Rationally Synthesized Two-Dimensional Polymers. Nat. Chem. 2013, 5, 453–465.
11. Huang, N.; Zhai, L.; Coupry, D. E.; Addicoat, M. A.; Okushita, K.; Nishimura, K.; Heine, T.; Jiang, D. Multiple-Component Covalent Organic Frameworks. Nat. Commun. 2016, 7, 12325.
12. Dalapati, S.; Addicoat, M.; Jin, S.; Sakurai, T.; Gao, J.; Xu, H.; Irle, S.; Seki, S.; Jiang, D. Rational Design of Crystalline Supermicroporous Covalent Organic Frameworks with Triangular Topologies. Nat. Commun. 2015, 6, 7786.
13. Keyu Geng, Ting He, Ruoyang Liu, Sasanka Dalapati, Ke Tian Tan, Zhongping Li, Shanshan Tao, Yifan Gong, Qiuhong Jiang, and Donglin Jiang. Covalent Organic Frameworks: Design, Synthesis, and Functions. Chem. Rev. 2020, 120, 8814–8933.
14. Ruoyang Liu, a Ke Tian Tan, a Yifan Gong, a Yongzhi Chen, a Zhuoer Li, a Shuailei Xie, a Ting He, a Zhen Lu, a Hao Yanga and Donglin Jiang. Covalent Organic Frameworks: An Ideal Platform for Designing Ordered Materials and Advanced Applications. Chem. Soc. Rev., 2021, 50, 120
15. Huang, N.; Wang, P.; Jiang, D. Covalent Organic Frameworks: A Materials Platform for Structural and Functional Designs. Nat. Rev. Mater. 2016, 1, 16068.
16. Chen, X.; Geng, K.; Liu, R.; Tan, K. T.; Gong, Y.; Li, Z.; Tao, S.; Jiang, Q.; Jiang, D. Covalent Organic Frameworks: Chemical Approaches to Designer Structures and Built-in Functions. Angew. Chem., Int. Ed. 2019, DOI: 10.1002/anie.201904291.

17. K. Geng, T. He, R. Liu, S. Dalapati, K. T. Tan, Z. Li, S. Tao, Y. Gong, Q. Jiang and D. Jiang, Chem. Rev., 2020, 120, 8814–8933.
18. Kandambeth, S.; Dey, K.; Banerjee, R. Covalent Organic Frameworks: Chemistry beyond the Structure. J. Am. Chem. Soc. 2019, 141, 1807–1822.
19. Yang, L.; Wei, D.-C. Semiconducting Covalent Organic Frameworks: A Type of Two-Dimensional Conducting Polymers. Chin. Chem. Lett. 2016, 27, 1395–1404.
20. Xu, L. Q.; Ding, S. Y.; Liu, J. M.; Sun, J. L.; Wang, W.; Zheng, Q.Y. Highly Crystalline Covalent Organic Frameworks from Flexible Building Blocks. Chem. Commun. 2016, 52, 4706–4709.
21. Kouwer, P. H.; Koepf, M.; Le Sage, V. A.; Jaspers, M.; van Buul, A. M.; Eksteen-Akeroyd, Z. H.; Woltinge, T.; Schwartz, E.; Kitto, H. J.; Hoogenboom, R.; et al. Responsive Biomimetic Networks from Polyisocyanopeptide Hydrogels. Nature 2013, 493, 651.
22. Feng, X.; Dong, Y.; Jiang, D. Star-Shaped Two-Dimensional Covalent Organic Frameworks. CrystEngComm 2013, 15, 1508–1511.
23. Côté, A. P.; El-Kaderi, H. M.; Furukawa, H.; Hunt, J. R.; Yaghi, O. M. Reticular Synthesis of Microporous and Mesoporous 2D Covalent Organic Frameworks. J. Am. Chem. Soc. 2007, 129, 12914–12915.
24. Wan, S.; Guo, J.; Kim, J.; Ihee, H.; Jiang, D. A Belt-Shaped, Blue Luminescent, and Semiconducting Covalent Organic Framework. Angew. Chem., Int. Ed. 2008, 47, 8826–8830.
25. Xie, Y. F.; Ding, S. Y.; Liu, J. M.; Wang, W.; Zheng, Q. Y. Triazatruxene Based Covalent Organic Framework and Its Quick-Response Fluorescence-on Nature towards Electron Rich Arenes. J. Mater. Chem. C 2015, 3, 10066–10069.
26. Spitler, E. L.; Dichtel, W. R. Lewis Acid-Catalysed Formation of Two-Dimensional Phthalocyanine Covalent Organic Frameworks. Nat. Chem. 2010, 2, 672.
27. Ding, X.; Guo, J.; Feng, X.; Honsho, Y.; Guo, J.; Seki, S.; Maitarad, P.; Saeki, A.; Nagase, S.; Jiang, D. Synthesis of Metallophthalocyanine Covalent Organic Frameworks that Exhibit High Carrier Mobility and Photoconductivity. Angew. Chem., Int. Ed. 2011, 50, 1289–93.
28. Feng, X.; Chen, L.; Dong, Y.; Jiang, D. Porphyrin-Based Two-Dimensional Covalent Organic Frameworks: Synchronized Synthetic Control of Macroscopic Structures and Pore Parameters. Chem. Commun. 2011, 47, 1979–81.
29. Ding, X.; Feng, X.; Saeki, A.; Seki, S.; Nagai, A.; Jiang, D. Conducting Metallophthalocyanine 2D Covalent Organic Frameworks: the Role of Central Metals in Controlling π-Electronic Functions. Chem. Commun. 2012, 48, 8952–4.
30. Lin, S.; Diercks, C. S.; Zhang, Y. B.; Kornienko, N.; Nichols, E. M.; Zhao, Y. B.; Paris, A. R.; Kim, D.; Yang, P.; Yaghi, O. M.; Chang, C. J. Covalent Organic Frameworks Comprising Cobalt Porphyrins for Catalytic CO_2 Reduction in Water. Science 2015, 349, 1208–1213.
31. Chen, R.; Shi, J.-L.; Ma, Y.; Lin, G.; Lang, X.; Wang, C. Designed Synthesis of a 2D Porphyrin-Based sp^2 Carbon-Conjugated Covalent Organic Framework for Heterogeneous Photocatalysis. Angew. Chem., Int. Ed. 2019, 58, 6430–6434.
32. Xu, S.-Q.; Zhan, T.-G.; Wen, Q.; Pang, Z.-F.; Zhao, X. Diversity of Covalent Organic Frameworks (COFs): A 2DCOF Containing Two Kinds of Triangular Micropores of Different Sizes. ACS Macro Lett. 2016, 5, 99–102.
33. Zhou, T.-Y.; Xu, S.-Q.; Wen, Q.; Pang, Z.-F.; Zhao, X. One-Step Construction of Two Different Kinds of Pores in a 2D Covalent Organic Framework. J. Am. Chem. Soc. 2014, 136, 15885–15888.
34. Chen, X.; Huang, N.; Gao, J.; Xu, H.; Xu, F.; Jiang, D. Towards Covalent Organic Frameworks with Predesignable and Aligned Open Docking Sites. Chem. Commun. 2014, 50, 6161–3.
35. Ascherl, L.; Sick, T.; Margraf, J. T.; Lapidus, S. H.; Calik, M.; Hettstedt, C.; Karaghiosoff, K.; Döblinger, M.; Clark, T.; Chapman, K. W.; Auras, F.; Bein, T. Molecular Docking Sites Designed for the Generation of Highly Crystalline Covalent Organic Frameworks. Nat. Chem. 2016, 8, 310–316.
36. Pang, Z. F.; Xu, S. Q.; Zhou, T. Y.; Liang, R. R.; Zhan, T. G.; Zhao, X. Construction of Covalent Organic Frameworks Bearing Three Different Kinds of Pores through the Heterostructural Mixed Linker Strategy. J. Am. Chem. Soc. 2016, 138, 4710–3.

37. Crowe, J. W.; Baldwin, L. A.; McGrier, P. L. Luminescent Covalent Organic Frameworks Containing a Homogeneous and Heterogeneous Distribution of Dehydrobenzoannulene Vertex Units. J. Am. Chem. Soc. 2016, 138, 10120–1013.

38. Jin, Y.; Hu, Y.; Zhang, W. Tessellated Multiporous Two-Dimensional Covalent Organic Frameworks. Nat. Rev. Chem. 2017, 1, 0056.

39. Zhu, Y.; Wan, S.; Jin, Y.; Zhang, W. Desymmetrized Vertex Design for the Synthesis of Covalent Organic Frameworks with Periodically Heterogeneous Pore Structures. J. Am. Chem. Soc. 2015, 137, 13772–5.

40. Trewin, A.; Cooper, A. I. Predicting Microporous Crystalline Polyimides. CrystEngComm 2009, 11, 1819.

41. Lin, G.; Ding, H.; Chen, R.; Peng, Z.; Wang, B.; Wang, C. 3D Porphyrin-Based Covalent Organic Frameworks. J. Am. Chem. Soc. 2017, 139, 8705–8709.

42. El-Kaderi, H. M.; Hunt, J. R.; Mendoza-Cortés, J. L.; Côté, A. P.; Taylor, R. E.; O'Keeffe, M.; Yaghi, O. M. Designed Synthesis of 3D Covalent Organic Frameworks. Science 2007, 316, 268–272.

43. Uribe-Romo, F. J.; Hunt, J. R.; Furukawa, H.; Klöck, C.; O'Keeffe, M.; Yaghi, O. M. A Crystalline Imine-Linked 3-D Porous Covalent Organic Framework. J. Am. Chem. Soc. 2009, 131, 4570–4571.

44. Liu, Y. Z.; Ma, Y. H.; Zhao, Y. B.; Sun, X. X.; Gandara, F.; Furukawa, H.; Liu, Z.; Zhu, H. Y.; Zhu, C. H.; Suenaga, K.; Oleynikov, P.; Alshammari, A. S.; Zhang, X.; Terasaki, O.; Yaghi, O. M. Weaving of Organic Threads into a Crystalline Covalent Organic Framework. Science 2016, 351, 365–369.

45. Zhang, Y.; Shen, X.; Feng, X.; Xia, H.; Mu, Y.; Liu, X. Covalent Organic Frameworks as pH Responsive Signaling Scaffolds. Chem. Commun. 2016, 52, 11088–11091.

46. Xu, S.; Wang, G.; Biswal, B. P.; Addicoat, M.; Paasch, S.; Sheng, W.; Zhuang, X.; Brunner, E.; Heine, T.; Berger, R.; Feng, X. A Nitrogen- Rich 2D sp2-Carbon-Linked Conjugated Polymer Framework as a High-Performance Cathode for Lithium-Ion Batteries. Angew. Chem. Int. Ed. 2019, 58, 849–853.

47. Banerjee, T.; Haase, F.; Trenker, S.; Biswal, B. P.; Savasci, G.; Duppel, V.; Moudrakovski, I.; Ochsenfeld, C.; Lotsch, B. V. Sub- Stoichiometric 2D Covalent Organic Frameworks from Tri- and Tetratopic Linkers. Nat. Commun. 2019, 10, 2689.

48. Zhu, M.-W.; Xu, S.-Q.; Wang, X.-Z.; Chen, Y.; Dai, L.; Zhao, X. The Construction of Fluorescent Heteropore Covalent Organic Frameworks and Their Applications in Spectroscopic and Visual Detection of Trinitrophenol with High Selectivity and Sensitivity. Chem. Commun. 2018, 54, 2308–2311.

49. Liang, R.-R.; Xu, S.-Q.; Pang, Z.-F.; Qi, Q.-Y.; Zhao, X. Self- Sorted Pore-Formation in the Construction of Heteropore Covalent Organic Frameworks Based on Orthogonal Reactions. Chem. Commun. 2018, 54, 880–883.

5

Functionalized Covalent Organic Frameworks for Improved Energy Applications

Yong Li[1], Weiran Zheng[2], and Lawrence Yoon Suk Lee[1,3]

[1]*Department of Applied Biology and Chemical Technology, The Hong Kong Polytechnic University, Hung Hom, Kowloon, Hong Kong SAR, China*

[2]*Department of Chemistry, Guangdong Technion – Israel Institute of Technology, Shantou, China*

[3]*Research Institute for Smart Energy, The Hong Kong Polytechnic University, Hung Hom, Kowloon, Hong Kong SAR, China*

CONTENTS

5.1 Introduction

Recently, the synthetic methods of covalent organic frameworks (COFs) have rapidly progressed, greatly expanding their library with new members. This has brought many new opportunities for COFs to be engaged in several applications with high performances. The most notable applications for the COFs are from the energy-related fields, such as electrocatalysis, photocatalysis, battery, and capacitors. Compared with their inorganic counterparts (metal-organic frameworks), COFs are solely made up of organic molecules, and their periodic skeleton and well-defined pores are easier to be modified and/or functionalized

DOI: 10.1201/9781003206507-5

by chemical means, and thereby endow a wider diversity of structures [1]. In applying COFs toward different applications of specific requirements, the modification and functionalization of COF's structure become crucial to regulate their intrinsic properties.

A typical COF consists of two distinct components, namely linkers (*i.e.*, building blocks) and linkages (*i.e.*, the chemical structures connecting individual building blocks). The wide selection of linkers for COFs allows the decoration of their backbone with functional moieties, offering greater design flexibility. Also, the COFs are a suitable scaffold for embedding novel functional groups because their tunable pore size and chemical environment can serve as a reactor for catalytic reactions and molecule transport. In this regard, two strategies are commonly referred to for the functionalization of COFs. The first one is known as the bottom-up approach (Figure 5.1A), which employs pre-designed functionalized linkers and/or linkages for COF synthesis. The other method is the post-synthetic modification (PSM) approach (Figure 5.1B), which uses COFs as the reactants and introduces functional groups to their surface and/or pores.

The bottom-up approach allows the uniform distribution of functional groups throughout the bulk COF structure. The functional groups attached to the linkers need to be stable during the synthesis of COF. Otherwise, a protection–deprotection process is required to ensure the structural integrity of the functional groups. Moreover, the existence of functional groups may disturb the reaction path and kinetics of COF formation, leading to the

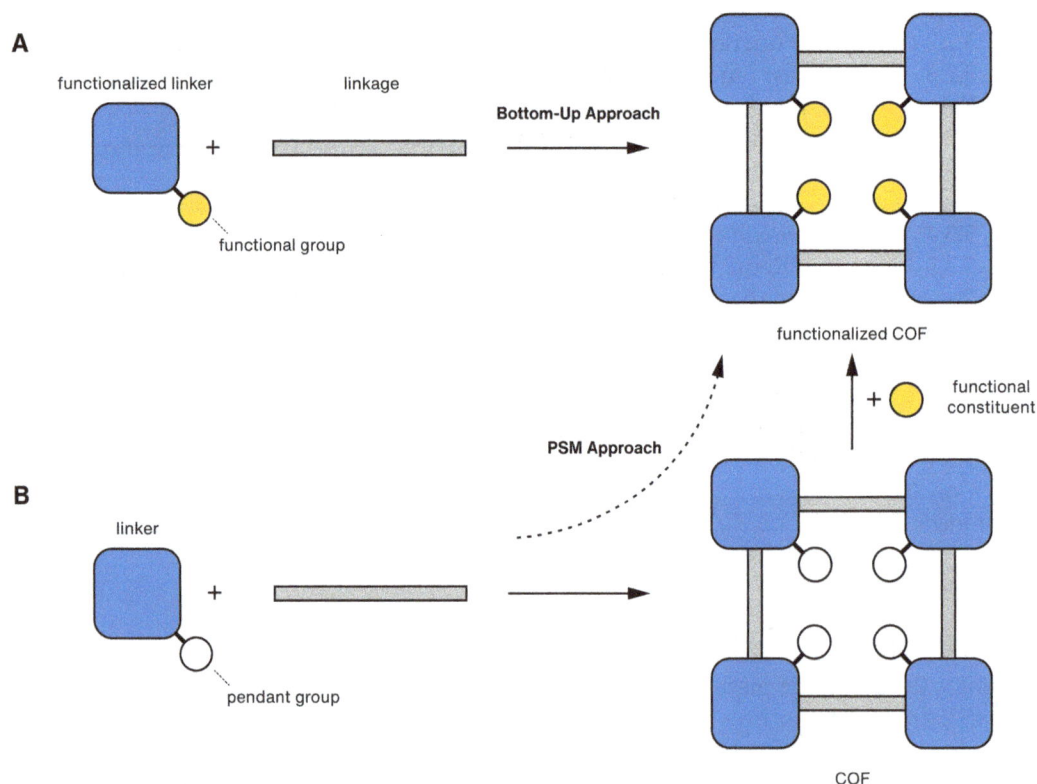

FIGURE 5.1
Schematic illustration of the functionalization approaches for COFs. (A) Bottom-up approach; and (B) Post-synthetic modification (PSM) approach.

altered pore structure (size and chemical environment) or poor crystallinity. On the other hand, the PSM approach provides a flexible way to insert a greater range of functional moieties into COFs while preventing undesired changes in the skeleton. More importantly, it prevents or reduces the occurrence of unfavorable side effects during reticulation and functional integration. Therefore, the PSM method is often favored over the bottom-up method. In the following contents, we will mainly focus on the PSM approach for further functionalization of COFs and their improved performance in several energy-related applications.

5.2 Post-Synthetic Functionalization of COFs

Depending on the nature of the functionalization requirements, the PSM of COFs usually involves bond construction, chemical reaction, and host–guest interaction between the functional ingredients and the COF structure. In general, three categories of functionalization are proposed based on the position of the functional groups: pendant group and skeleton functionalization, linkage functionalization, and host–guest functionalization [2].

5.2.1 Pendant Group and Skeleton Functionalization

A range of structural functionalization can be achieved by reacting specific pendant groups and/or skeleton of the COFs with the desired reactants. Depending on the reaction type, one can introduce covalent/ionic/coordination bonds to the pores of COFs. Some commonly used pendant groups for the construction of covalent bonds are listed in Table 5.1. Usually, the pendant groups are pre-installed on the linkers to facilitate the synthetic process of COFs. The pendant groups are designed to be located at the pore channels once the COFs are constructed such that the pendant groups are still accessible by the functional constituents during the PSM process. The functional groups, such as triazole, ester, ether, thiocarbamate, amide, amidoxime, and thioether, can be incorporated into COFs following this principle. Figure 5.2A shows one example where a vinyl-functionalized COF (COF-V) is functionalized with thioether groups (COF-S-SH) using

TABLE 5.1

Typical Reactive Pendant Groups Appended on COF Backbone Useful for Further Functionalization

Pendant groups	Coupling agents	Functionalization
$-N_3$ (azide)	$-C{\equiv}C-$ (alkyne)	Triazole
$-OH$ (hydroxyl)	$-C(O)-O-C(O)-$ (anhydride)	Ester
	$-NCS$ (isothiocyanate)	Thiocarbamate
	$-Br$ (halide)	Ether
$-NH_2$ (amine)	$-C(O)-O-C(O)-$ (anhydride)	Amide
$-{\equiv}N$ (nitride)	$-NH_2$ (amine)	Amidoxime
$-SH$ (thiol)	$-C{=}C-$ (alkene)	Thioether

Source: Adapted with permission from reference [2]. Copyright (2019) The author(s). The article was printed under a CC-BY license.

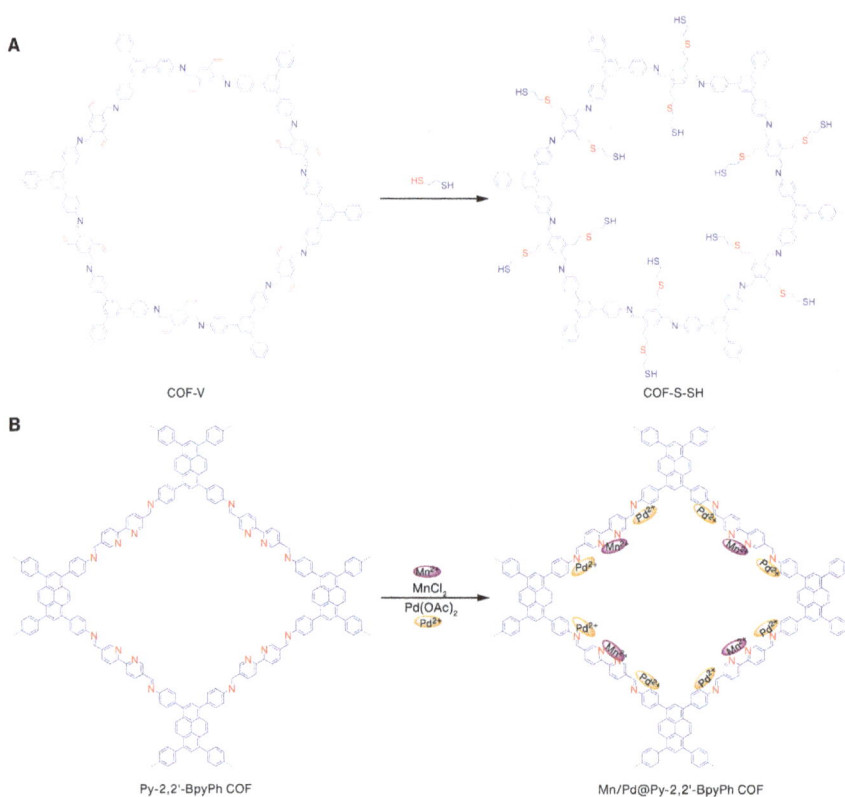

FIGURE 5.2
(A) Construction of thioether-functionalized COF-S-SH from COF-V. (B) Construction of Mn/Pd-coordinated Py-2,2'-BPyPh COF. (Adapted with permission [2]. Copyright (2019) The authors, some rights reserved; exclusive licensee Oxford University Press. Distributed under a Creative Commons Attribution License 4.0 (CC BY).)

1,2-ethanedithiol to introduce sulfur species (mass loading = 20.9 wt.%) to the channels of COF-V [3].

In addition to the covalent bonds, the ionic/coordination bonds are also commonly introduced to the COFs to establish a polar environment in the pores, which can benefit further incorporation of metal/molecular ions. Figure 5.2B shows an example of a metal ion/COF hybrid that utilizes the coordination environment of N atoms to stabilize metal centers [4]. Depending on the specific atomic arrangements of coordination environments, it is possible to anchor various metal/molecular species onto the channels of COFs.

5.2.2 Linkage Functionalization

The linkages are also widely used to introduce functional groups to the COF structure. Typically, linkages are stable structures constructed by covalent bonds. Yet, a certain group (*e.g.*, –SH, –NH) present on the linkages can be reduced/oxidized and acts as the reaction sites to construct larger functional groups. For instance, the imine linkages can be oxidized to amide linkages to form amide-COFs, and they can be further reduced to amine linkages. However, the functionalization of linkage is not as popular as the linker modification, due to the limited availability of linkages containing redox functional groups.

5.2.3 Host–Guest Functionalization

Embedding molecules/clusters/particles into the channels of COFs is another way of functionalizing the COFs and has been gaining wide interest recently. The porous trait of COFs makes them a perfect material to structurally confine and accommodate other components. By adjusting the pendant groups on the COF, the interaction of the COF with the hosted components can be regulated.

The formed COF–guest hybrids normally inherit the advantages of both components, especially the highly ordered porous structure of COFs. Metal nanoparticles (NPs) are the most popular guest species due to their vast applications in catalysis. Figure 5.3A illustrates a typical method of doping Pt NPs into the channels of COFs: first, the pendant groups with S atoms act as the ligands to host $PtCl_4^{2-}$ species; second, the Pt ions are chemically reduced by reducing agents ($NaBH_4$ in this example) to form Pt NPs inside the channels [5]. One significant advantage of this host–guest functionalization method is that the size of the Pt NPs can be well-determined by the size of the pore, and the NPs are uniformly distributed. Figure 5.3B shows another example where a C60 molecule is incorporated into the channel of CS-COF (conjugated and stable COF) [6]. Such effort is accomplished by a thermal sublime diffusion method and the interaction between conjugated groups from both C60 and COF.

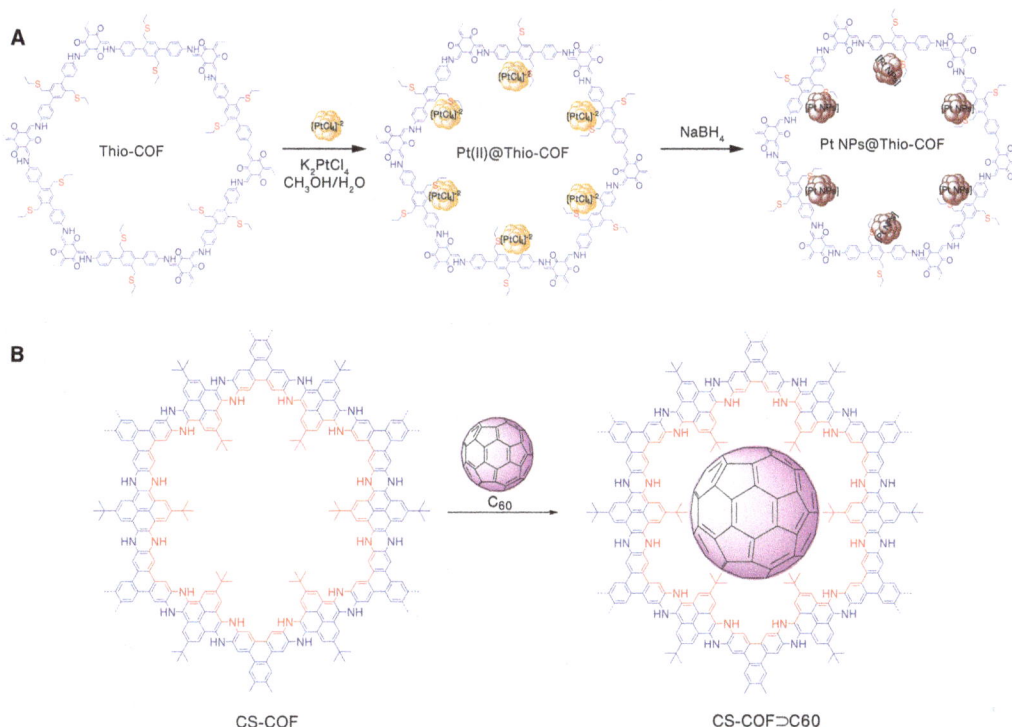

FIGURE 5.3

(A) Construction of Pt NPs deposited COF *via* the coordination between –SH and Pt ions. (B) Construction of C60-embedded CS-COF. (Adapted with permission [2]. Copyright (2019) The authors, some rights reserved; exclusive licensee Oxford University Press. Distributed under a Creative Commons Attribution License 4.0 (CC BY).)

5.3 Rational Functionalization for Improved Electrocatalytic Performance

Recently, electrocatalysis has been widely engaged in the water, carbon, and nitrogen cycles for energy conversion and storage applications. Specifically, the water cycle contains four hydrogen- and oxygen-involving electrochemical reactions: hydrogen evolution reaction (HER), hydrogen oxidation reaction (HOR), oxygen evolution reaction (OER), and oxygen reduction reaction (ORR). The combination of HER and OER is the well-known water splitting reaction that produces renewable and clean H_2 fuel. The ORR can be applied to the hydrogen–oxygen fuel cells and the rechargeable metal–air batteries. The carbon and nitrogen cycles are also important to address several problems in environmental and energy crises, which include the electrocatalytic CO_2 reduction reaction (CO_2RR) and N_2 reduction reaction (NRR).

Recently, the applications of COFs in electrocatalysis have become an appealing route due to their excellent chemical and thermal stability, high porosity, tunable pore size, large surface area, and designable skeleton. Compared with pristine COFs, the functionalized groups-, metal/non-metal atoms-, or metal complex-modified COFs have received huge attention and showed great potential to improve the efficiency of the aforementioned electrocatalysis.

5.3.1 Hydrogen Evolution Reaction/Hydrogen Oxidation Reaction

One of the first examples of COF-catalyzed HER (*i.e.*, pyrene–porphyrin–linked COF) was reported by Bhunia *et al.* in 2017 [7]. Although it showed a low onset potential of 50 mV and good electrochemical stability with 500 cycles of cyclic voltammetry, the overpotential of hydrogen evolution was still *ca.* 200 mV to reach a current density of 1 mA cm^{-2}. Such a large overpotential and slow reaction kinetics (Tafel slope of 116 mV dec^{-1}) indicate that the unmodified COFs are far from the practical application of HER. Recently, metal centers have been incorporated into COFs for highly active HER. As one of the most active HER catalysts, ultrasmall Pt NPs (size = 3 nm) were modified on bpyTPP-COFs (bpy = 2,2′-bipyridine, TPP = *meso*-tetraphenylporphyrin) by Park *et al.* [8]. The HER efficiency of the as-prepared PtNPs@COF was 13 times higher than that of the commercial Pt/C, and a faradaic efficiency of *ca.* 100% was reported. However, a survey of recent works indicates that the application of COFs in HER is not the primary trend due to the semiconducting nature of COFs. Furthermore, the HOR, the reverse reaction of HER, has been rarely reported to date.

5.3.2 Oxygen Evolution Reaction

The investigation of OER using COFs attracts more attention than HER owing to the importance of OER in energy conversion and storage technologies. According to the literature, the OER catalysts based on COFs are normally designed by incorporating metal complex or anchoring metal ions into nitrogen-enriched COF skeletons (*e.g.*, bipyridine and porphyrin structures) [9]. These newly added metal sites are believed to play the role of active centers. For example, Vaidhyanathan's group synthesized $Co_xNi_y(OH)_2$ NPs on an sp^3 N-rich IISERP–COF2 and Ni_3N NPs on IISERP–COF3 [10, 11]. Both COFs showed outstanding OER activities and exceptional kinetics with comparable activity to that of the reported NiFe LDH. The overpotentials of $Co_xNi_y(OH)_2$–COF and Ni_3N–COF were as low as 258 and 230 mV at 10 mA cm^{-2}, respectively. The superb OER activity was attributed to

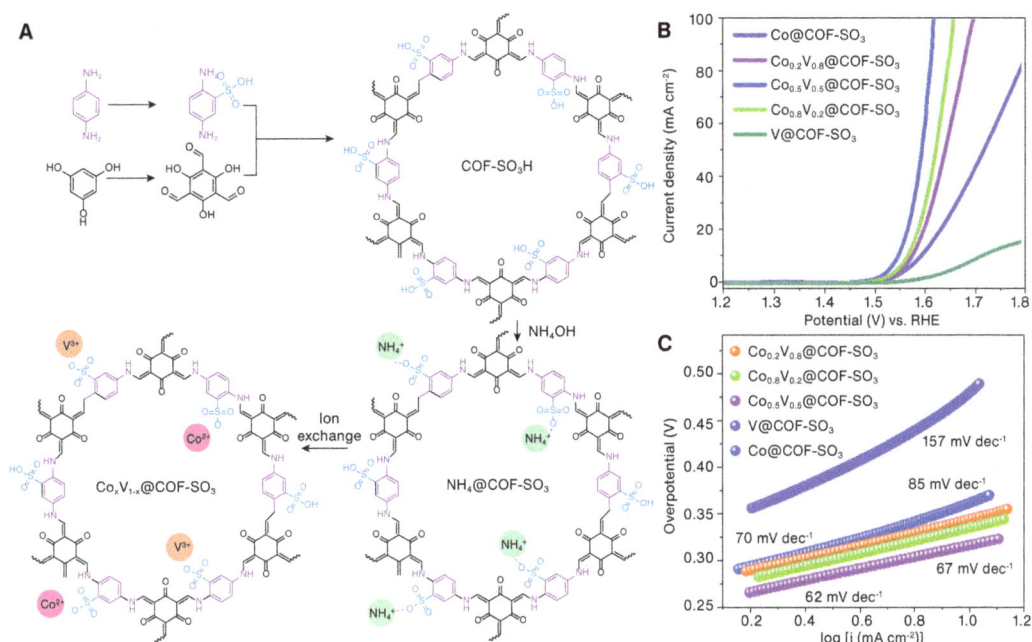

FIGURE 5.4

(A) Schematic illustration of the synthetic route to prepare Co_xV_{1-x}@COF–SO_3^-. (B) LSV curves and (C) Tafel plots of Co_xV_{1-x}@COF–SO_3^- (x = 0, 0.2, 0.5, 0.8, 1) in the OER. (Adapted with permission [13]. Copyright (2020) The Royal Society of Chemistry.)

the strong electronic interaction between the ultrasmall NPs and COF support, and the COF skeletons also provided the pathways for effective electron transfer.

Coordinating metal ions into the COFs with bipyridine or porphyrin units to form stable M–N_4 sites has been suggested as a novel and feasible approach to further improve OER activity and maximize the utilization of active metal centers. Recently, Lin *et al.* used crystal field stabilization energy (CFSE) to predict the OER activity of 3*d* transition metals, incorporated the porphyrin-containing COFs [12]. A volcano plot was identified, and the predicted OER trend of metal–COFs followed the order: Fe > Co > Ni > Cr > V > Ti > Sc > Mn. As learned from the free energy diagrams of metal–COFs, the formation of O* is considered as the rate-determining step (RDS) for Co, Cu, and Zn. In contrast, the RDS of Ti, Cr, and Fe is believed to be the formation of HOO*. In addition to the M–N_4 anchoring strategy, a cation-exchange strategy was recently reported by Gao *et al.* who prepared Co/V-incorporated bimetallic COF–SO_3^- for OER (Figure 5.4A) [13]. The COF was firstly modified by –SO_3^- group followed by an ammoniation reaction in concentrated ammonia water to obtain NH_4@COF–SO_3^-. Then, Co^{2+} and V^{3+} ions were anchored on COF–SO_3^- to replace NH_4^+ species. In the OER catalysis, the as-prepared $Co_{0.5}V_{0.5}$@ COF–SO_3^- required 300 mV to deliver 10 mA cm^{-2}, and the Tafel slope was 67 mV dec^{-1} (Figures 5.4B–C), which is much better than most Co-based OER catalysts.

5.3.3 Oxygen Reduction Reaction

The ORR is the reverse reaction of OER and is also an important electrocatalytic reaction. Generally, the ORR activity of catalysts can be estimated on a rotating disk electrode

(RDE) or a flow cell. The former is to study the catalytic mechanism of ORR, and the latter can simulate the practical application in a high current density. It is well-known that the ORR can occur *via* two pathways: $2e^-$ transfer to form H_2O_2 ($O_2 + 2H^+ + 2e^- \rightarrow H_2O_2$ in the acidic electrolyte) and $4e^-$ transfer to form OH^- or H_2O ($O_2 + 4H^+ + 4e^- \rightarrow 2H_2O$ in the acidic electrolyte), which can be calculated by the Koutecky–Levich (K–L) equation based on the results obtained from RDE. Similar to the case of OER, the functionalized COFs, especially some transition metal (TM: *e.g.,* Sc, Ti, V, Cr, Mn, Fe, Co, Ni, Cu, and Zn)-anchored COFs, show great potentials to improve ORR efficiency (Figure 5.5A). Using the free energy diagrams, CFSE, and configuration energy (CE) descriptors (Figures 5.5B–D), Lin *et al.* predicted that Fe- and Co-modified COFs would be active in the $4e^-$ transfer pathway due to the spontaneously exothermic reactions [12], while Cu- and Zn-functionalized COFs would prefer the $2e^-$ transfer pathway, thanks to the smaller theoretical overpotentials (1.15 V for Cu and 0.97 V for Zn) for the formation of H_2O_2 compared with that for the H_2O formation (1.456 V for Cu and 1.785 V for Zn). Additionally, the alkaline-earth metals (*e.g.,* Be, Mg, Ca, Sr, and Ba) were also evaluated in the metal–COF structures. The

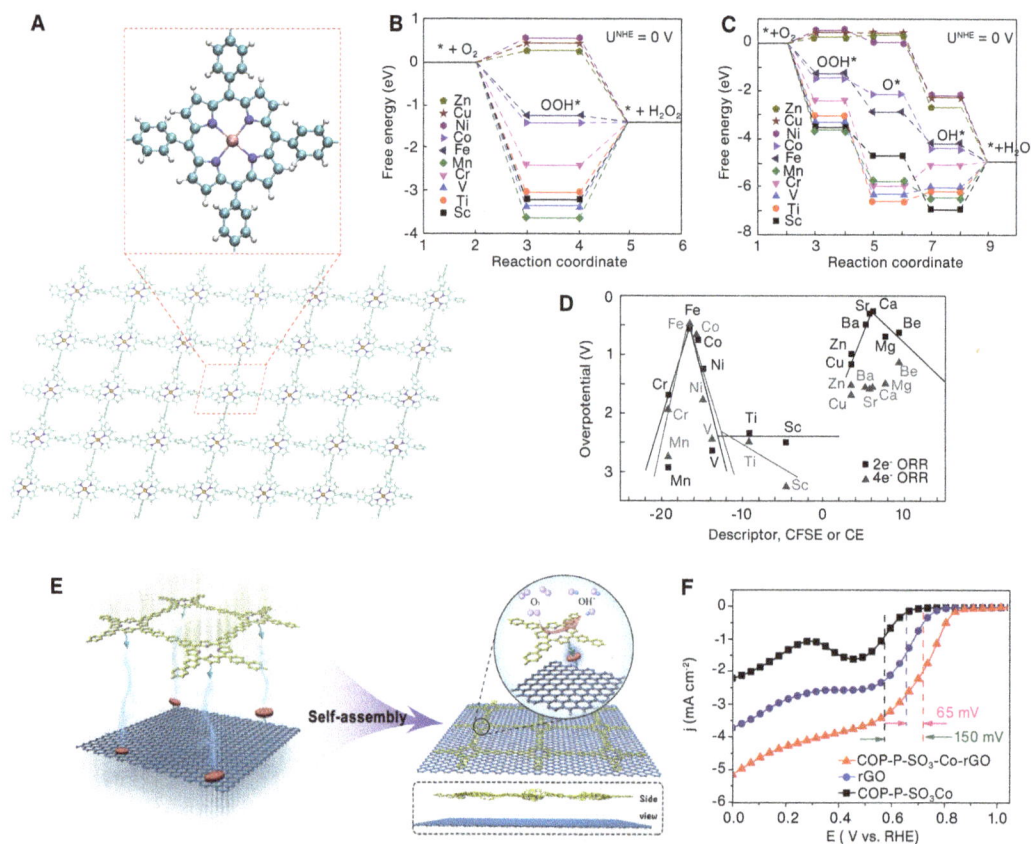

FIGURE 5.5

(A) Schematic illustration of metal–COF structures. (B) Two-electron and (C) four-electron transfer ORR of transition metal–COFs, in which Zn, Cu, Ni, Co, Fe, Mn, Cr, V, Ti, and Sc are anchored on the COFs. (D) Overpotentials as a function of CFSE and configuration energy (CE) for transition metal–COFs and alkaline metal–COFs in $2e^-/4e^-$ electron pathways. Adapted with permission [12]. Copyright (2017) WILEY–VCH. (E) Schematic illustration of self-assembled COP–P–SO$_3$–Co–rGO. (F) LSV curves of COP–P–SO$_3$–Co–rGO, COP–P–SO$_3$–Co, and rGO in the ORR region in 0.1 M KOH at 1,600 rpm. (Adapted with permission [14]. Copyright (2018) WILEY–VCH.)

alkaline-earth metal–COFs were predicted to have relatively higher theoretical overpotentials (> 1.4 V) in 4e⁻ transfer ORR, and thus it was expected to generate H_2O_2 following the 2e⁻ transfer pathway (Figure 5.5D).

Considering the semiconducting nature of COFs, the hybridization of COFs with some conductive carbon materials (*e.g.*, graphene and carbon nanotube) is also an efficient way to improve the conductivity. For example, Guo *et al.* synthesized a well-defined porphyrin macrocycle-based covalent organic polymers (COPs–P) modified with sulfonic acid side groups (–SO_3H) as shown in Figure 5.5E [14]. To a mixed solution containing the COP–P–SO_3H and reduced graphene oxide (rGO), Co^{2+} ions were added to form a self-assembled structure COP–P–SO_3–Co–rGO by electrostatic adsorption between Co^{2+} and both of sulfonic acid group and electronegative graphene oxide. The COP–P–SO_3–Co–rGO showed a greatly improved electrical conductivity of $2.56 \times 10^{-1}\,S\,m^{-1}$, which is seven orders of magnitude larger than that of pristine COP ($3.06 \times 10^{-9}\,S\,m^{-1}$). The synergetic effect in the COP–P–SO_3–Co–rGO resulted in a higher ORR activity than the pure COP–P–SO_3Co (Figure 5.5F).

5.3.4 Carbon Dioxide CO_2 Reduction Reaction

Electrochemical CO_2RR is a meaningful route to reduce the greenhouse gas into value-added chemicals, such as CO, formic acid, methane, methanol, ethanol, and ethylene. Generally, the CO_2RR proceeds in three steps: 1) the adsorption of CO_2 molecules on the active sites; 2) the breaking of C=O bonds and the formation of C–H bonds, and 3) the desorption of the new products. The second step is the most important and complicated process that determines the types of products based on the number of the involved electrons and protons. Owing to the high energy barrier to break C=O bonds and the strong competitive reaction of HER, electrocatalysts with high selectivity and activity are required for CO_2RR. Some metal–COF structures were studied for CO_2RR by Yaghi and co-workers in 2015 [15]. They compared two Co-based COFs (COF–366–Co and COF–367–Co). The COF–366–Co (pore size = 2.35 nm) was prepared by imine condensation of Co(TAP) (TAP = 5,10,15,20-tetrakis(4-aminophenyl)porphinato]cobalt) and 1,4-benzenedicarboxaldehyde (BDA). It showed a high selectivity of CO with a faradaic efficiency of 90% and yielded 36 mL of CO in 24 h per milligram catalyst. In comparison, the COF–367–Co was synthesized with a larger pore size of 2.65 nm by replacing the BDA with biphenyl-4,4′-dicarboxaldehyde (BPDA). The catalytic activity was further enhanced, and 100 mL of CO was generated with an equivalent catalyst at the same period without the loss of faradic efficiency (91%). The high activities of both Co–based COFs indicate the great potential of metal–COF structures for CO_2RR. Although the recently reported COFs have already shown high activity and selectivity in the production of CO, further reduction of CO remains a challenge because of the high energy barrier, and more investigation on new types of COFs is needed to obtain a wide range of products as well as high selectivity.

5.3.5 Nitrogen Reduction Reaction

The electrochemical NRR is one of the most promising methods to produce ammonia (NH_3). Different from the traditional Haber–Bosch process in which the NH_3 is obtained by the reaction between N_2 and H_2 under high temperature and high pressure, the electrochemical NRR normally works under mild conditions, and the adsorption of N_2, breaking of N≡N bonds, protonation, and the subsequent desorption of products take place on active sites under a suitable voltage at ambient conditions (Figure 5.6A). It is believed that

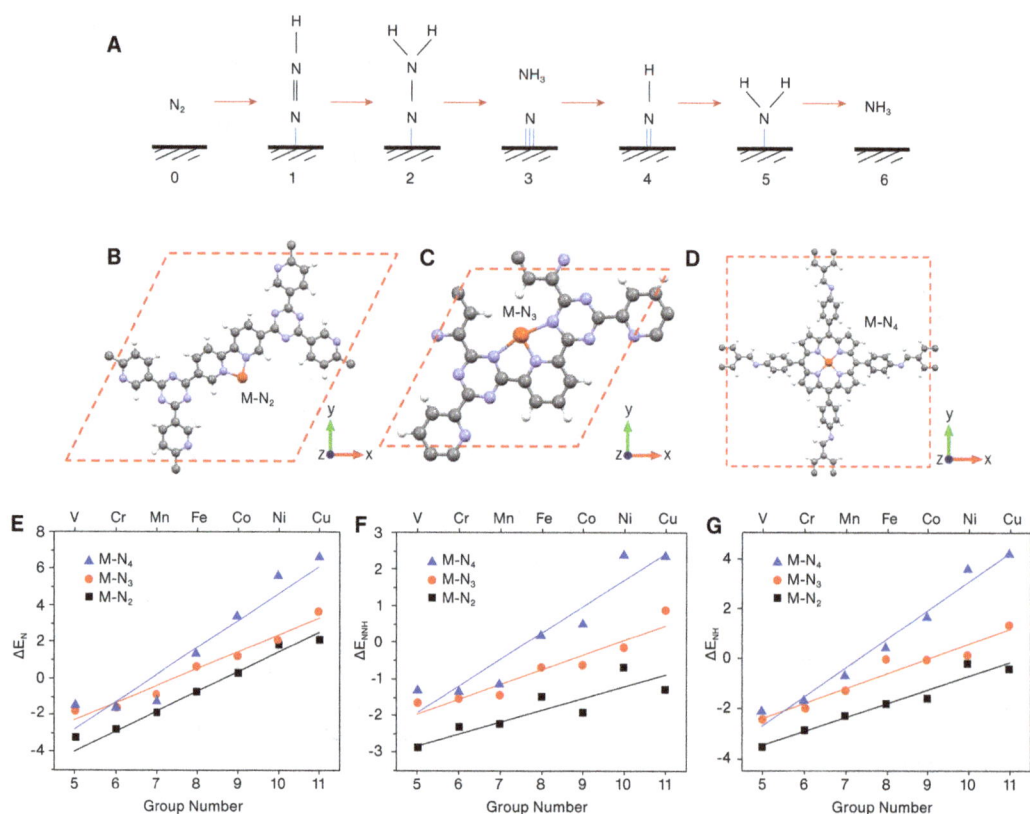

FIGURE 5.6

(A) The proposed reaction pathway for NRR on the catalyst surface. (B) Bipyridine-type (M–N$_2$), (C) terpyridine-type (M–N$_3$), and (D) porphyrin-type (M–N$_4$) metal–nitrogen bonds in metal–COFs. The theoretical adsorption energies of intermediates (N*, NNH*, and NH*) as a function of the group number of the metal species (V, Cr, Mn, Fe, Co, Ni, and Cu) with the above three coordination numbers: (E) ΔE_N, (F) ΔE_{NNH}, and (G) ΔE_{NH}. (Adapted with permission [17]. Copyright (2021) American Chemical Society.)

the electrochemical approach will dominate industrial-scale NH$_3$ production in the future. Yan *et al.* reported a boron-rich COF that was incorporated with conductive nitrogen-doped carbon nanosheets (B-COF/NC) [16]. The B-COF/NC showed a good NRR performance with a high NH$_3$ production rate of 12.5 µg h^{-1} mg^{-1} and a faradaic efficiency of 45.4%. They concluded that boron sites were the active sites, and the crystalline boron–COF would convert to an amorphous phase during the catalytic process, which could facilitate the adsorption of N$_2$ and decrease the energy barrier of N$_2$ dissociation.

The metal–COF structures are also considered as the promising candidates for NRR. Using density functional theory, Kamiya's group studied the relationships between the coordination numbers/numbers of d–electrons and the adsorption strength of reaction intermediates in various 3d-metal single atoms (*e.g.*, V, Cr, Mn, Fe, Co, Ni, and Co) anchored on COFs [17]. The adsorption energies of the intermediates, including N*, NNH*, and NH*, on various metal sites with different coordination environments, such as M–N$_2$, M–N$_3$, and M–N$_4$, were compared (Figures 5.6B–G). All the intermediates showed a similar linear trend; the adsorption energies increased with coordination numbers or d–electron numbers, indicating the existence of a scaling relationship for NRR on metal–COFs. The theoretical onset potential as the function of the *d*–electron number for M–N$_2$, M–N$_3$, and M–N$_4$ predicted that the optimal metal for NRR would shift from Co to Fe and Mn when the coordination number increased from 2 to 3 and 4. Compared with single-atom catalysts, dual-metal-site catalysts tended high catalytic activity, which has been already proved in ORR. Very recently, Zhang *et al.* also studied the NRR activity of the COFs embedded with dual-atomic active sites using the first-principles calculations [18]. Various dual-atomic site-functionalized COFs were constructed with the combination of Mo, W, Fe, Co, and Ru elements, and the theoretical results predicted that MoRu–, MoMo–, MoCo–, and WRu–COFs would have great potentials for high activity of NRR owing to the low limiting potentials (–0.29 to –0.57 V) and high faradic efficiency (76–100%).

5.4 Rational Functionalization for Improved Photocatalytic Performance

The applications of COFs in photocatalytic reactions are also widely investigated in recent years due to their thermal and chemical stability, light-harvesting capability, charge separation, and tunable bandgaps. In 2014, Lotsch's group first reported a hydrazine-based COF, a kind of intrinsic covalent bonding framework, for photocatalytic water splitting [19]. The hydrogen evolution rate of this COF could reach 230 μmol h^{-1} g^{-1} when sodium ascorbate was used as the sacrificial electron donor. Furthermore, the activity went up to 1,970 μmol h^{-1} g^{-1} once the sacrificial donor was changed to triethanolamine (TEOA). Thereafter, a series of new type COFs (azine-linked COFs) were reported by the same group with an impressive photocatalytic hydrogen evolution rate (N$_3$–COF, 1.7 mmol h^{-1} g^{-1}), which is known as the highest activity among the unmodified COFs [20]. To further improve the photocatalytic water splitting efficiency, the pristine COFs were functionalized with some functional groups (fused sulfone and triazine) [21, 22], semiconducting metal complexes (Ni(OH)$_2$, CdS, Mo$_3$S$_{13}$, C$_3$N$_4$, and TiO$_2$) [23–26], and metal/non-metal ions (cobalt and fluorine) [27, 28]. According to the reports, the functionalized COFs possess a good wettability that ensures the full contact and interactions with water and sacrificial donor. Besides, the increased external quantum efficiency, decreased optical gap, and longer weighted lifetime, compared with unmodified COFs, indicate that the functionalized COFs have a huge potential for highly efficient photocatalytic hydrogen evolution.

Photocatalytic CO$_2$RR has become another major application of COFs since 2016, thanks to their high porosity, large surface area, and stable crystalline structure. Different from normal COFs, a new type of donor–acceptor (D–A)-bridged COFs were constructed by mixing two building blocks and studied for photocatalytic CO$_2$RR in recent years, such as porphyrin–tetrathiafulvalene COF and carbazole–triazine COF [29, 30]. This kind of COFs exhibited effective electron–hole pairs separation and transformation properties owing to the inner electron transfer networks from the donor moieties to accepter ones, which was responsible for high photocatalytic CO$_2$RR rates. More importantly, water can act as the electron donor for CO$_2$ reduction without extra photosensitizers or co-catalysts during the process. Inspired by this, some water-oxidation semiconducting complexes (*e.g.*, TiO$_2$, Bi$_2$WO$_6$, and α–Fe$_2$O$_3$) were also modified on COFs for photocatalytic CO$_2$RR [31]. They showed a similar mechanism that can be described as follows: the photogenerated electrons separated from the electron–hole pairs in the semiconductor are first trans-ferred to the active sites on the COF to reduce CO_2, and the positively charged holes in the semiconductor lead to the H_2O oxidation. Additionally, metal (Ni, Co, Cu, and Zn)-functionalized COFs were also employed for CO_2 photoreduction [32–34]. Metal cen-ters were proposed as the active sites for CO_2 adsorption and activation, which exhib-ited superb selectivity and activity. Nevertheless, the building blocks surrounding the metal centers are equally important to affect the CO_2 photoreduction by facilitating the intermediates in the catalytic reaction. For example, Zhong *et al.* reported that Ni–TpBpy COFs (2,2'-bipyridine-based COFs with single Ni sites) could produce CO at an impres-sive rate of 4,057 µmol g^{-1} with a CO selectivity of 96%, whereas Ni–TbBpy COFs, a simi-lar Ni binding environment to Ni–TpBpy COFs, showed a poor CO selectivity of 58% and a low CO production rate [32].

5.5 Rational Functionalization for Improved Energy Storage Performance

Energy storage is an important technique to store extra electricity for use on demand. The large surface area, high porosity, and high electrons mobility of COFs make them a promising material class for the applications of energy storage, such as supercapacitors, metal-ion batteries, and metal sulfur batteries.

5.5.1 Supercapacitors

A supercapacitor is one of the most efficient energy storage devices with a high power density and long cycling life. Based on the energy storage mechanism, two categories are identified: electrochemical double-layer capacitors (EDLCs) and pseudocapacitors. Specifically, the EDLCs store electricity by constructing electrode–electrolyte interfaces without electrons transfer. In contrast, pseudocapacitors are derived from reversible redox reactions or intercalation on the surface of the electrode. The pseudocapacitors normally show a higher performance: more than 10 times that of EDLCs. The COFs with designable skeletons and high porosity, especially functionalized COFs, have been widely applied in pseudocapacitors. In 2013, Dichtel and co-workers first applied functionalized COFs for the fabrication of supercapacitors [35]. As shown in Figure 5.7A, the β-ketoenamine-linked COFs were functionalized with two redox-active molecules, 2,6-diaminoanthraquinone (DAAQ) and inactive *p*-diaminobenzene (DAB), to compare the charge storage property. According to the cyclic voltammograms (CVs) presented in Figure 5.7B, the strong revers-ible redox peaks of DAAQ–TFP–COF (TFP = 1,3,5-triformylphloroglucinol) and DAAQ indicated that the protonation and deprotonation reactions took place due to the transfor-mation between redox-active C=O and C–OH groups. The DAB–TFP–COF and pure DAB only exhibited small double-layer capacitances. The overall capacity of DAAQ–TFP–COF was 40 ± 9 F g^{-1} after 10 charge/discharge cycles, which was almost 3 times larger than that of DAB–TFP–COF (15 ± 6 F g^{-1}). The DAAQ–TFP–COF also showed outstanding stability even after 5,000 cycles (Figures 5.7C–D). To further improve the capacitance, an oriented thin film of DAAQ–TFP–COF was grown on Au electrode by the same group. The oriented DAAQ–TFP–COF possessed a greatly improved capacitance of 3 mF cm^{-2} compared with the random DAAQ–TFP–COFs (0.4 mF cm^{-2}) [36]. Furthermore, Dichtel's group further

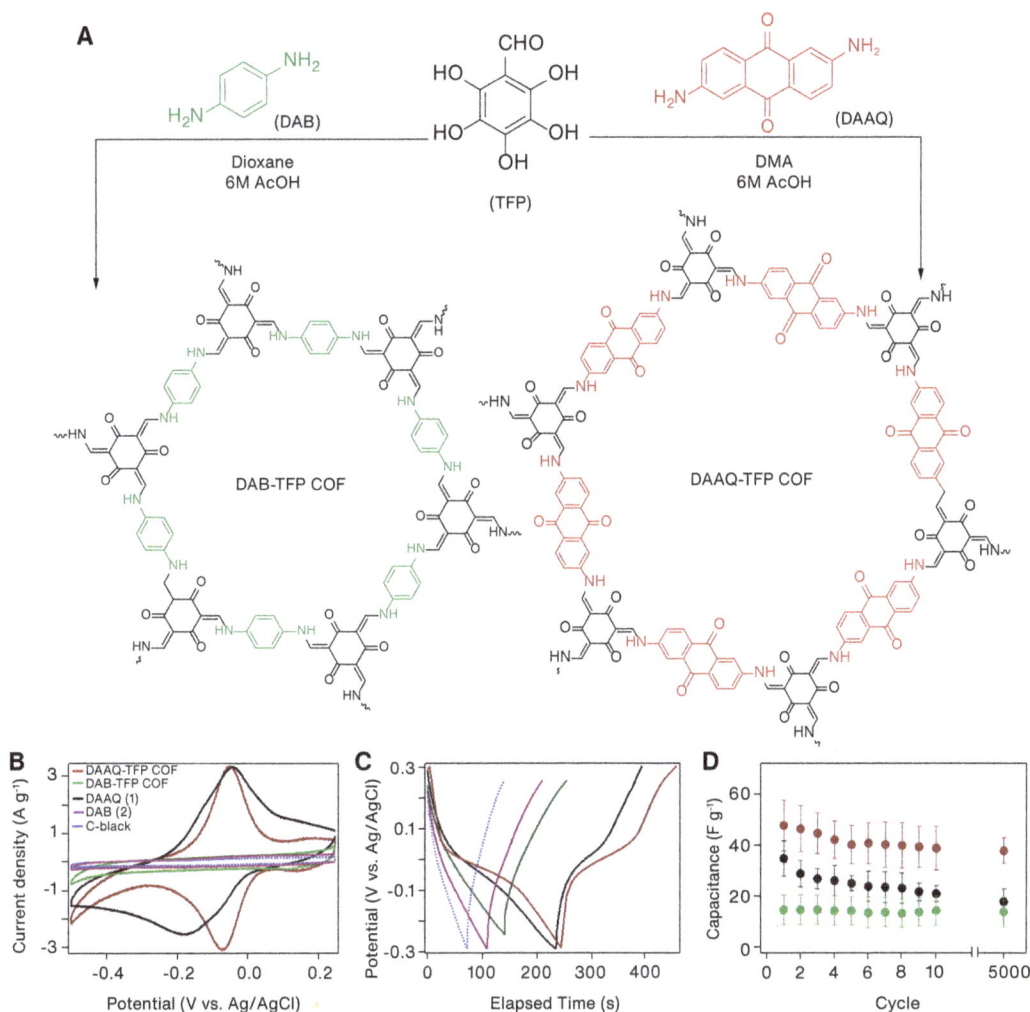

FIGURE 5.7
(A) Synthetic route of redox-active DAAQ–TFP–COF and inactive DAB–TFP–COF. (B) Cyclic voltammograms and (C) Galvanostatic charge–discharge curves of DAAQ–TFP–COF, DAB–TFP–COF, DAAQ, DAB, and C-black. (D) Capacitance as a function of the cycles from 1 to 5,000 cycles. (Adapted with permission [35]. Copyright (2013) American Chemical Society.)

introduced a conductive polymer, 3,4-ethylenedioxythiophene (EDOT), into the pore space of DAAQ–TFP–COFs to improve the electrical conductivity and found that the capacitance can reach 350 F cm^{-3} at a charging rate of 10 C [37].

Hybridization of metal complex or carbon-based materials with COFs is another effective approach to improve the capacities of pseudocapacitors and EDLCs. For example, Thomas *et al.* introduced Fe$_3$O$_4$ NPs into imine-linked COFs, and the capacitance was significantly enhanced to 112 F g^{-1} at a current density of 0.5 A g^{-1} due to the strong redox activity [38]. Very recently, the same group reported an ultralight COF/rGO aerogel (a low density of ~7.0 mg cm^{-3}) that showed a specific capacity of 269 F g^{-1} at the same current density. The high capacity was attributed to the synergistic effect of EDLCs and pseudocapacitors due to the incorporation of conductive rGO. Besides, a high specific surface area (246 m^2 g^{-1}) of COF/rGO was believed to provide abundant redox-active sites [39]. Even after 5,000 cycles, the capacity of COF/rGO could remain 96%.

5.5.2 Metal-Ion Batteries

Metal-ion batteries (MIBs) are one of the most widely used energy storage devices, which play a vital role in the transition from fossil fuels to renewable energies with zero-carbon emission. Various metal ions have been explored for use in MIBs, such as lithium, sodium, and potassium. Undoubtedly, lithium-ion batteries (LIBs) are the most successful ones in the global market of MIBs due to their high energy density and large cycling stability. To further improve the energy capacity of LIBs to satisfy the requirements in electric vehicles, novel anode materials are required to replace traditional carbon materials or metal/metal oxides. Recently, COF-based materials have been studied as anode materials in LIBs owing to the desirable properties of high porosity with controllable pore size. The internal channels in COFs can facilitate the charge carrier (Li$^+$) transport and offer the potential to increase the storage capacity. In 2016, Zhao *et al.* presented two conjugated COFs for LIBs anodes, N2– and N3–COF, which showed high charge capacities of 689 and 707 mA h g^{-1} for the first cycle, respectively [40]. The corresponding capacities decreased to 600 and 593 mA h g^{-1} after 500 cycles, respectively, both of which were ~81% capacity retention.

Considering that the poor conductivity of pristine COFs may suffer the potential drops and undesirable energy loss, the functionalized COFs, such as those integrated with redox-active sites and carbon-based materials have been explored. Sun *et al.* designed a dual redox-active site-modified COF (Tp–Azo–COF: Tp = 1,3,5-triformylphloroglucinol, Azo = 4,4′-azodianiline) that showed a large specific surface area of 632 m^2 g^{-1} and a small pore size of ~2.6 nm [41]. The introduced C=O and N=N sites largely improved the performance; the capacity reached 305.97 mA h g^{-1} and almost 100% capacity retention was recorded even after 3,000 cycles at a high current density of 1,000 mA g^{-1}. Wang's group reported a few-layer 2D COFs (COF@CNTs) grown around the surface of carbon nanotube (CNT) by utilizing the π–π interaction [42]. The COF@CNTs exhibited an extremely high capacity of 1,536 mA h g^{-1} after 500 cycles, thanks to the C=N functional groups and the efficient utilization of carbon atoms in the benzene ring.

In addition to the application in LIBs, the functionalized COFs also show great potentials as the anode in other MIBs, such as sodium-ion batteries (SIBs), potassium-ion batteries (PIBs), and zinc-ion batteries (ZIBs). For example, Lu's group synthesized β-ketoenamine-linked DAAQ–COF with the enriched C=O groups. When used as the anode in SIB, a high capacity of 420 mA h g^{-1} at 100 mA g^{-1} and 99% capacity retention after 10,000 cycles at 5 mA g^{-1} were achieved [43]. The oxidation of C=O to C–O and α–C radical intermediates facilitated the coordination of sodium ions with four steps in the discharge process, which resulted in a high specific capacity.

5.5.3 Metal Sulfur Batteries

Metal sulfur batteries, especially lithium–sulfur batteries (LSBs), have attracted great attention due to their higher theoretical capacity and energy density that largely exceed those of LIBs. Generally, the LSBs are assembled by a sulfur cathode and a Li metal anode. During the discharging process, sulfur is first reduced to soluble intermediates (Li_2S_x, $3 \leq x \leq 8$) and then to semiconductive polysulfides (PS: Li_2S_2/Li_2S). Thus, some common drawbacks exist in LSBs, such as poor conductivity, shuttle effect, and volume changes. Compared with the traditional cathode where carbon materials are applied to anchor sulfur, the COFs are considered as a better sulfur host material due to the designable skeleton, large space for sulfur loading, and chemical stability. In 2014, Wang's group is one of the firsts who employed COFs in LSBs by impregnating sulfur into the pores of COFs [44]. The as-prepared CTF–1/S showed a high sulfur loading mass of ~34 wt.%, demonstrating a discharge capacity of 1,497 mA h g^{-1} at the first cycle. However, the capacity dropped to 762 mA h g^{-1} after 50 cycles owing to the formation of soluble polysulfide intermediates and the shuttle effect. To improve the long-term stability, Tang *et al.* synthesized boronate ester COF–1 that fully contained B–O groups (Figure 5.8A). The strong adsorption of S_x^{2-} (*e.g.*, S_8^{2-}, S_6^{2-}, and S_4^{2-}) and Li$^+$ on positively charged B and negatively charged O (Figure 5.8B), respectively, greatly reduced the dissolution of polysulfides and the shuttle effect [45]. The COF–1/S cathode showed a high initial capacity of 1,628 mA h g^{-1} at 0.2 C, and 929 mA h g^{-1} was retained after 100 cycles. Recently, Zhang *et al.* prepared fluorine-functionalized COFs as the sulfur host material for LSBs and found that the initial discharge capacity could reach 1,120 mA h g^{-1} at 0.1 C, but rapidly dropped to 962 mA h g^{-1} after 3 cycles and faded to 645 mA h g^{-1} after 100 cycles [46]. The substantial reduction of capacity indicated the existence of dissolution and shuttling of polysulfides. To inhibit the shuttling effect, Zang's group synthesized a cationic COF (EB–COF–Br: EB = ethidium bromide) that is functionalized with Br$^-$ ions (Figure 5.8C) [47]. Subsequently, an anion exchange process was employed to replace Br with polysulfides to form EB–COF–PS. Compared with the electrically neutral COF (TpPa–COF) prepared by Tp and *p*-phenylenediamine, the specific capacities of EB–COF–PS were larger at various current densities (Figure 5.8D), especially at the high regime (4.0 and 5.0 C). Owing to the strong trapping behavior of polysulfides, the EB–COF–PS showed an initial capacity of 787 mA h g^{-1} at 0.5 C and a retained capacity of 555 mA h g^{-1} after 500 cycles, which was ~70% capacity retention (Figure 5.8E).

5.5.4 Solar Cells

Solar cells are photovoltaic devices that can directly convert light energy into electricity. Currently, the most promising perovskite solar cells (PSCs) show the highest power conversion efficiency (PCE) of up to ~22%. The COFs are also considered as a potential candidate to assemble highly efficient solar cells as they can provide high carrier mobilities, semiconductive nature, and uniform 2D structures. In 2013, Bein *et al.* studied thieno[2,3-b] thiophene-based COFs (TT–COF) as the electron donor that was incorporated with an electron acceptor [6]–phenly–C$_{61}$–butyric acid methyl ester (PCBM) in an organic photovoltaic device [48]. However, the poor FCE of 0.053% in indium tin oxide (ITO)/TT–COF:PCBM/Al device indicated that the individual application of COFs as photoactive components was not suitable. Inspired by this, Chen's group synthesized two functional groups (formyl and amino groups)-modified 2D pyrene-based COFs (Py–COF) on ITO and employed them as the hole transporting layer for PSCs [49]. The PCE of Py–COF-modified PSCs went up to 6.36%. Later, Zhao and co-workers reported the simple bulk doping of 3D COFs, spirobifluorene core-linked COFs (SP–3D–COFs), into perovskite layers (CH$_3$NH$_3$PbI$_3$) in perovskite solar cells [50]. The SP–3D–COFs-doped PSCs showed a higher PCE of 18% compared with the unmodified ones.

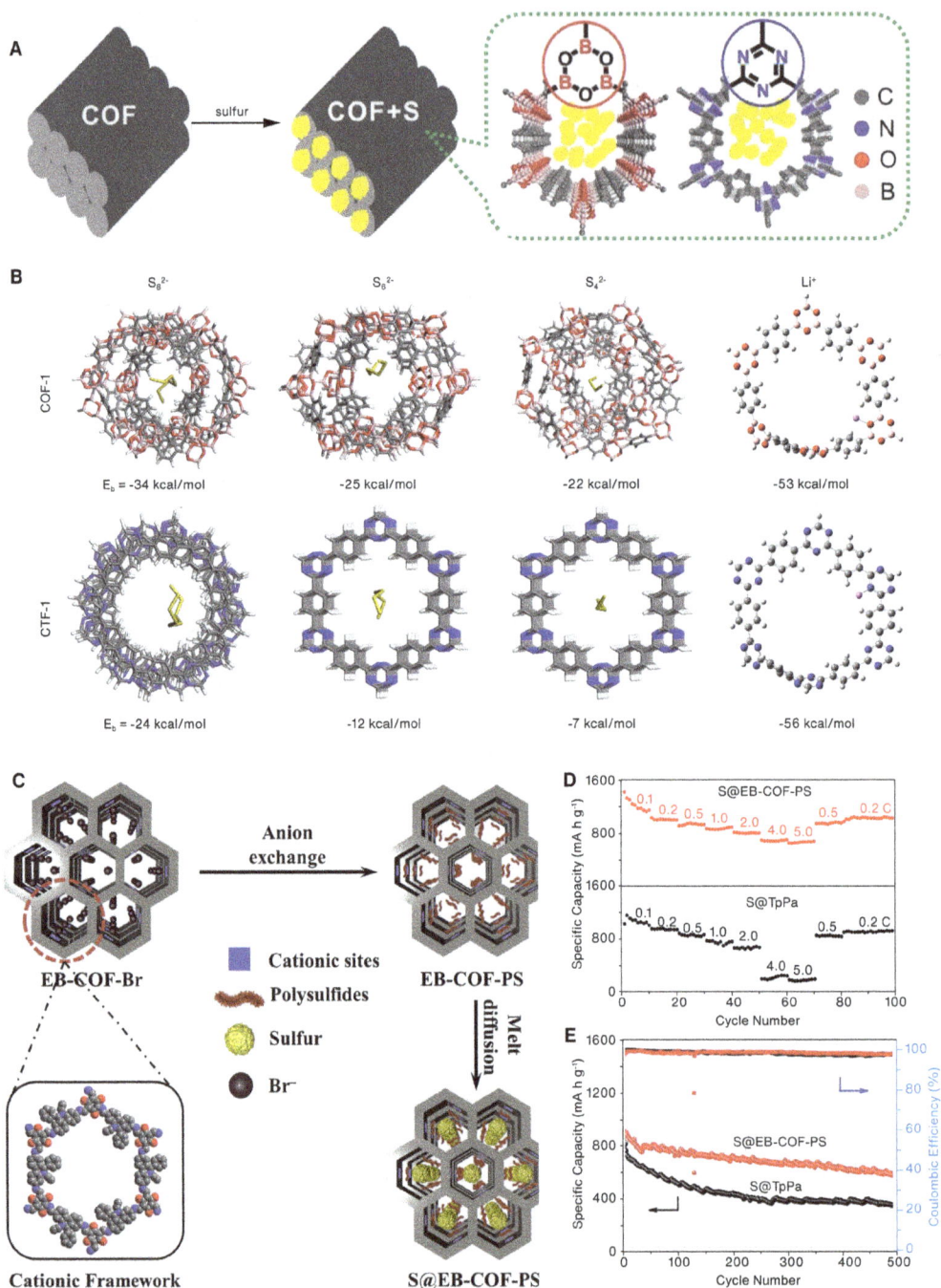

FIGURE 5.8

(A) Diagrams of sulfur loading on nitrogen-doped COF (CTF–1) and boron-doped COF (COF–1). (B) Adsorption energies of intermediate species (S_8^{2-}, S_6^{2-}, S_4^{2-}, and Li^+) on COF–1 and CTF–1. Adapted with permission from reference [45]. Copyright (2016) WILEY–VCH. (C) Schematic illustration of S@EB–COF–PS synthesis. (D) Specific capacities of S@EB–COF–PS and S@TpPa at various current densities (0.1, 0.2, 0.5, 1.0, 2.0, 4.0, 5.0 C). (E) Stability tests of S@EB–COF–PS and S@TpPa at a current density of 0.5 C for 500 cycles with the corresponding coulombic efficiency. (Adapted with permission [47]. Copyright (2020) WILEY–VCH.)

5.6 Conclusion

To conclude, we have provided a brief introduction of current methods for COFs' functionalization (pendant group and skeleton functionalization, linkage functionalization, and host–guest functionalization) with some important examples of their performance improvement in energy applications, from energy conversion to storage. These applications have different structural requirements for the COFs, including the design of pore size, functional groups, conductivity, and stability. As a result, COFs need to be functionalized specifically to fulfill the needs, either with non-metallic species such as S- and N-containing groups or metal species such as metal ions or even clusters. COFs can be constructed by the nearly infinite combination of linkers and linkages, and their functionalization would further increase the number of COFs that are possible to be constructed. In search of the optimal candidates for the application in energy-related applications, we expect that the involvement of theoretical prediction together with the development of organic synthetic methods boost their success.

References

1. Zheng, W.; Tsang, C.-S.; Lee, L. Y. S.; Wong, K.-Y., Two-Dimensional Metal-Organic Framework and Covalent-Organic Framework: Synthesis and their Energy-Related Applications. Mater. Today Chem. 2019, *12*, 34–60.
2. Yusran, Y.; Guan, X.; Li, H.; Fang, Q.; Qiu, S., Postsynthetic Functionalization of Covalent Organic Frameworks. Natl. Sci. Rev. 2020, *7*(1), 170–190.
3. Sun, Q.; Aguila, B.; Perman, J.; Earl, L. D.; Abney, C. W.; Cheng, Y.; Wei, H.; Nguyen, N.; Wojtas, L.; Ma, S., Postsynthetically Modified Covalent Organic Frameworks for Efficient and Effective Mercury Removal. J. Am. Chem. Soc. 2017, *139*(7), 2786–2793.
4. Leng, W. G.; Ge, R. L.; Dong, B.; Wang, C.; Gao, Y., Bimetallic Docked Covalent Organic Frameworks with High Catalytic Performance Towards Tandem Reactions. RSC Adv. 2016, *6* (44), 37403–37406.
5. Lu, S.; Hu, Y.; Wan, S.; McCaffrey, R.; Jin, Y.; Gu, H.; Zhang, W., Synthesis of Ultrafine and Highly Dispersed Metal Nanoparticles Confined in a Thioether-Containing Covalent Organic Framework and Their Catalytic Applications. J. Am. Chem. Soc. 2017, *139* (47), 17082–17088.
6. Guo, J.; Xu, Y.; Jin, S.; Chen, L.; Kaji, T.; Honsho, Y.; Addicoat, M. A.; Kim, J.; Saeki, A.; Ihee, H.; Seki, S.; Irle, S.; Hiramoto, M.; Gao, J.; Jiang, D., Conjugated Organic Framework with Three-Dimensionally Ordered Stable Structure and Delocalized π Clouds. Nat. Commun. 2013, *4*, 2736.
7. Bhunia, S.; Das, S. K.; Jana, R.; Peter, S. C.; Bhattacharya, S.; Addicoat, M.; Bhaumik, A.; Pradhan, A., Electrochemical Stimuli-Driven Facile Metal-Free Hydrogen Evolution from Pyrene-Porphyrin-Based Crystalline Covalent Organic Framework. ACS Appl. Mater. Interfaces 2017, *9* (28), 23843–23851.
8. Park, E.; Jack, J.; Hu, Y.; Wan, S.; Huang, S.; Jin, Y.; Maness, P. C.; Yazdi, S.; Ren, Z.; Zhang, W., Covalent Organic Framework-Supported Platinum Nanoparticles as Efficient Electrocatalysts for Water Reduction. Nanoscale 2020, *12* (4), 2596–2602.
9. Zhao, X.; Pachfule, P.; Thomas, A., Covalent Organic Frameworks (COFs) for Electrochemical Applications. Chem. Soc. Rev. 2021, *50* (12), 6871–6913.
10. Nandi, S.; Singh, S. K.; Mullangi, D.; Illathvalappil, R.; George, L.; Vinod, C. P.; Kurungot, S.; Vaidhyanathan, R., Low Band Gap Benzimidazole COF Supported Ni3N as Highly Active OER Catalyst. Adv. Energy Mater. 2016, *6* (24), 1601189.

11. Mullangi, D.; Dhavale, V.; Shalini, S.; Nandi, S.; Collins, S.; Woo, T.; Kurungot, S.; Vaidhyanathan, R., Low-Overpotential Electrocatalytic Water Splitting with Noble-Metal-Free Nanoparticles Supported in a sp³N-Rich Flexible COF. Adv. Energy Mater. 2016, 6 (13), 1600110.

12. Lin, C. Y.; Zhang, L.; Zhao, Z.; Xia, Z., Design Principles for Covalent Organic Frameworks as Efficient Electrocatalysts in Clean Energy Conversion and Green Oxidizer Production. Adv. Mater. 2017, 29 (17), 1606635.

13. Gao, Z.; Yu, Z. W.; Huang, Y. X.; He, X. Q.; Su, X. M.; Xiao, L. H.; Yu, Y.; Huang, X. H.; Luo, F., Flexible and Robust Bimetallic Covalent Organic Frameworks for the Reversible Switching of Electrocatalytic Oxygen Evolution Activity. J. Mater. Chem. A 2020, 8 (12), 5907–5912.

14. Guo, J.; Lin, C. Y.; Xia, Z.; Xiang, Z., A Pyrolysis-Free Covalent Organic Polymer for Oxygen Reduction. Angew. Chem. Int. Ed. 2018, 57 (38), 12567–12572.

15. Lin, S.; Diercks, C. S.; Zhang, Y. B.; Kornienko, N.; Nichols, E. M.; Zhao, Y.; Paris, A. R.; Kim, D.; Yang, P.; Yaghi, O. M.; Chang, C. J., Covalent Organic Frameworks Comprising Cobalt Porphyrins for Catalytic CO_2 Reduction In Water. Science 2015, 349 (6253), 1208–1213.

16. Liu, S.; Wang, M.; Qian, T.; Ji, H.; Liu, J.; Yan, C., Facilitating Nitrogen Accessibility to Boron-Rich Covalent Organic Frameworks via Electrochemical Excitation for Efficient Nitrogen Fixation. Nat. Commun. 2019, 10 (1), 3898.

17. Ohashi, K.; Iwase, K.; Harada, T.; Nakanishi, S.; Kamiya, K., Rational Design of Electrocatalysts Comprising Single-Atom-Modified Covalent Organic Frameworks for the N_2 Reduction Reaction: A First-Principles Study. J. Phys. Chem. C 2021, 125 (20), 10983–10990.

18. Zhang, Z. H.; Qi, S. Y.; Wang, J.; Zhao, M. W., Bi-Atom Active Sites Embedded in a Two-Dimensional Covalent Organic Framework for Efficient Nitrogen Reduction Reaction. Appl. Surf. Sci. 2021, 563, 150352.

19. Stegbauer, L.; Schwinghammer, K.; Lotsch, B. V., A Hydrazone-Based Covalent Organic Framework for Photocatalytic Hydrogen Production. Chem. Sci. 2014, 5(7), 2789–2793.

20. Vyas, V. S.; Haase, F.; Stegbauer, L.; Savasci, G.; Podjaski, F.; Ochsenfeld, C.; Lotsch, B. V., A Tunable Azine Covalent Organic Framework Platform for Visible Light-Induced Hydrogen Generation. Nat. Commun. 2015, 6, 8508.

21. Wang, X.; Chen, L.; Chong, S. Y.; Little, M. A.; Wu, Y.; Zhu, W. H.; Clowes, R.; Yan, Y.; Zwijnenburg, M. A.; Sprick, R. S.; Cooper, A. I., Sulfone-Containing Covalent Organic Frameworks For Photocatalytic Hydrogen Evolution From Water. Nat. Chem. 2018, 10 (12), 1180–1189.

22. Yang, Y. L.; Niu, H. Y.; Xu, L.; Zhang, H.; Cai, Y. Q., Triazine Functionalized Fully Conjugated Covalent Organic Framework for Efficient Photocatalysis. Appl. Catal., B 2020, 269, 118799.

23. Dong, H.; Meng, X. B.; Zhang, X.; Tang, H. L.; Liu, J. W.; Wang, J. H.; Wei, J. Z.; Zhang, F. M.; Bai, L. L.; Sun, X. J., Boosting Visible-Light Hydrogen Evolution of Covalent-Organic Frameworks by Introducing Ni-Based Noble Metal-Free Co-Catalyst. Chem. Eng. J. 2020, 379, 122342.

24. Thote, J.; Aiyappa, H. B.; Deshpande, A.; Diaz, D.; Kurungot, S.; Banerjee, R., A Covalent Organic Framework-Cadmium Sulfide Hybrid as a Prototype Photocatalyst for Visible-Light-Driven Hydrogen Production. Chemistry 2014, 20 (48), 15961–15965.

25. Cheng, Y. J.; Wang, R.; Wang, S.; Xi, X. J.; Ma, L. F.; Zang, S. Q., Encapsulating $[Mo_3S_{13}]^{2-}$ Clusters in Cationic Covalent Organic Frameworks: Enhancing Stability and Recyclability by Converting a Homogeneous Photocatalyst to a Heterogeneous Photocatalyst. Chem. Commun. 2018, 54 (96), 13563–13566.

26. Li, C. C.; Gao, M. Y.; Sun, X. J.; Tang, H. L.; Dong, H.; Zhang, F. M., Rational Combination of Covalent-Organic Framework and Nano TiO_2 by Covalent Bonds to Realize Dramatically Enhanced Photocatalytic Activity. Appl. Catal., B 2020, 266, 118586.

27. Liu, C.; Xiao, Y. M.; Yang, Q.; Wang, Y. C.; Lu, R. W.; Chen, Y. X.; Wang, C. J.; Yan, H. J., A Highly Fluorine-Functionalized 2D Covalent Organic Framework for Promoting Photocatalytic Hydrogen Evolution. Appl. Surf. Sci. 2021, 537, 148082.

28. Aiyappa, H. B.; Thote, J.; Shinde, D. B.; Banerjee, R.; Kurungot, S., Cobalt-Modified Covalent Organic Framework as a Robust Water Oxidation Electrocatalyst. Chem. Mater. 2016, 28 (12), 4375–4379.

29. Lu, M.; Liu, J.; Li, Q.; Zhang, M.; Liu, M.; Wang, J. L.; Yuan, D. Q.; Lan, Y. Q., Rational Design of Crystalline Covalent Organic Frameworks for Efficient CO_2 Photoreduction with H_2O. Angew. Chem. Int. Ed. 2019, *58* (36), 12392–12397.

30. Lei, K.; Wang, D.; Ye, L.; Kou, M.; Deng, Y.; Ma, Z.; Wang, L.; Kong, Y., A Metal-Free Donor-Acceptor Covalent Organic Framework Photocatalyst for Visible-Light-Driven Reduction of CO_2 with H_2O. ChemSusChem 2020, *13* (7), 1725–1729.

31. Zhang, M.; Lu, M.; Lang, Z. L.; Liu, J.; Liu, M.; Chang, J. N.; Li, L. Y.; Shang, L. J.; Wang, M.; Li, S. L.; Lan, Y. Q., Semiconductor/Covalent-Organic-Framework Z-Scheme Heterojunctions for Artificial Photosynthesis. Angew. Chem. Int. Ed. 2020, *59* (16), 6500–6506.

32. Zhong, W.; Sa, R.; Li, L.; He, Y.; Li, L.; Bi, J.; Zhuang, Z.; Yu, Y.; Zou, Z., A Covalent Organic Framework Bearing Single Ni Sites as a Synergistic Photocatalyst for Selective Photoreduction of CO2 to CO. J. Am. Chem. Soc. 2019, *141* (18), 7615–7621.

33. Lu, M.; Li, Q.; Liu, J.; Zhang, F. M.; Zhang, L.; Wang, J. L.; Kang, Z. H.; Lan, Y. Q., Installing Earth-Abundant Metal Active Centers to Covalent Organic Frameworks for Efficient Heterogeneous Photocatalytic CO_2 Reduction. Appl. Catal., B 2019, *254*, 624–633.

34. Wang, X. K.; Liu, J.; Zhang, L.; Dong, L. Z.; Li, S. L.; Kan, Y. H.; Li, D. S.; Lan, Y. Q., Monometallic Catalytic Models Hosted in Stable Metal-Organic Frameworks for Tunable CO_2 Photoreduction. ACS Catal. 2019, *9* (3), 1726–1732.

35. DeBlase, C. R.; Silberstein, K. E.; Truong, T. T.; Abruna, H. D.; Dichtel, W. R., β-Ketoenamine-Linked Covalent Organic Frameworks Capable of Pseudocapacitive Energy Storage. J. Am. Chem. Soc. 2013, *135* (45), 16821–16824.

36. DeBlase, C. R.; Hernandez-Burgos, K.; Silberstein, K. E.; Rodriguez-Calero, G. G.; Bisbey, R. P.; Abruna, H. D.; Dichtel, W. R., Rapid and Efficient Redox Processes within 2D Covalent Organic Framework Thin Films. ACS Nano 2015, *9* (3), 3178–3183.

37. Mulzer, C. R.; Shen, L.; Bisbey, R. P.; McKone, J. R.; Zhang, N.; Abruna, H. D.; Dichtel, W. R., Superior Charge Storage and Power Density of a Conducting Polymer-Modified Covalent Organic Framework. ACS Cent. Sci. 2016, *2* (9), 667–673.

38. Liao, Y. Z.; Li, J. H.; Thomas, A., General Route to High Surface Area Covalent Organic Frameworks and Their Metal Oxide Composites as Magnetically Recoverable Adsorbents and for Energy Storage. ACS Macro Lett. 2017, *6* (12), 1444–1450.

39. Li, C.; Yang, J.; Pachfule, P.; Li, S.; Ye, M. Y.; Schmidt, J.; Thomas, A., Ultralight Covalent Organic Framework/Graphene Aerogels with Hierarchical Porosity. Nat. Commun. 2020, *11* (1), 4712.

40. Bai, L. Y.; Gao, Q.; Zhao, Y. L., Two Fully Conjugated Covalent Organic Frameworks as Anode Materials for Lithium Ion Batteries. J. Mater. Chem. A 2016, *4* (37), 14106–14110.

41. Zhao, G. F.; Zhang, Y. Y.; Gao, Z. H.; Li, H. N.; Liu, S. M.; Cai, S.; Yang, X. F.; Guo, H.; Sun, X. L., Dual Active Site of the Azo and Carbonyl-Modified Covalent Organic Framework for High-Performance Li Storage. ACS Energy Lett. 2020, *5* (4), 1022–1031.

42. Lei, Z.; Yang, Q.; Xu, Y.; Guo, S.; Sun, W.; Liu, H.; Lv, L. P.; Zhang, Y.; Wang, Y., Boosting Lithium Storage in Covalent Organic Framework via Activation of 14-Electron Redox Chemistry. Nat. Commun. 2018, *9* (1), 576.

43. Gu, S.; Wu, S.; Cao, L.; Li, M.; Qin, N.; Zhu, J.; Wang, Z.; Li, Y.; Li, Z.; Chen, J.; Lu, Z., Tunable Redox Chemistry and Stability of Radical Intermediates in 2D Covalent Organic Frameworks for High Performance Sodium Ion Batteries. J. Am. Chem. Soc. 2019, *141* (24), 9623–9628.

44. Liao, H. P.; Ding, H. M.; Li, B. J.; Ai, X. P.; Wang, C., Covalent-Organic Frameworks: Potential Host Materials for Sulfur Impregnation in Lithium-Sulfur Batteries. J. Mater. Chem. A 2014, *2* (23), 8854–8858.

45. Ghazi, Z. A.; Zhu, L. Y.; Wang, H.; Naeem, A.; Khattak, A. M.; Liang, B.; Khan, N. A.; Wei, Z. X.; Li, L. S.; Tang, Z. Y., Efficient Polysulfide Chemisorption in Covalent Organic Frameworks for High-Performance Lithium-Sulfur Batteries. Adv. Energy Mater. 2016, *6* (24), 1601250.

46. Wang, D. G.; Li, N.; Hu, Y.; Wan, S.; Song, M.; Yu, G.; Jin, Y.; Wei, W.; Han, K.; Kuang, G. C.; Zhang, W., Highly Fluoro-Substituted Covalent Organic Framework and Its Application in Lithium-Sulfur Batteries. ACS Appl. Mater. Interfaces 2018, *10* (49), 42233–42240.

47. Liu, X. F.; Chen, H.; Wang, R.; Zang, S. Q.; Mak, T. C. W., Cationic Covalent-Organic Framework as Efficient Redox Motor for High-Performance Lithium-Sulfur Batteries. Small 2020, *16* (34), e2002932.

48. Dogru, M.; Handloser, M.; Auras, F.; Kunz, T.; Medina, D.; Hartschuh, A.; Knochel, P.; Bein, T., A Photoconductive Thienothiophene-Based Covalent Organic Framework Showing Charge Transfer towards Included Fullerene. Angew. Chem. Int. Ed. 2013, *52* (10), 2920–2924.

49. Li, Y.; Chen, Q.; Xu, T.; Xie, Z.; Liu, J.; Yu, X.; Ma, S.; Qin, T.; Chen, L., De Novo Design and Facile Synthesis of 2D Covalent Organic Frameworks: A Two-in-One Strategy. J. Am. Chem. Soc. 2019, *141* (35), 13822–13828.

50. Wu, C.; Liu, Y.; Liu, H.; Duan, C.; Pan, Q.; Zhu, J.; Hu, F.; Ma, X.; Jiu, T.; Li, Z.; Zhao, Y., Highly Conjugated Three-Dimensional Covalent Organic Frameworks Based on Spirobifluorene for Perovskite Solar Cell Enhancement. J. Am. Chem. Soc. 2018, *140* (31), 10016–10024.

6

Covalent Organic Frameworks: Fundamentals to Advanced Energy Applications

Shiva Bhardwaj[1,2], Felipe M. de Souza[2], and Ram K. Gupta[2,3]

[1]*Department of Physics, Pittsburg State University, Pittsburg, Kansas, USA*

[2]*Kansas Polymer Research Center, Pittsburg State University, Pittsburg, Kansas, USA*

[3]*Department of Chemistry, Pittsburg State University, Pittsburg, Kansas, USA*

CONTENTS

6.1 Introduction

The development of society impeccably depends on energy, primarily on renewable energy sources. Due to the continuous depletion of fossil fuels, growing energy demand, and the adverse greenhouse effect, the effective use of renewable energies (such as geothermal, tidal, wind, and solar) has become very important [1]. These clean energy options can be used to power electric vehicles, electronics in general, and compose the energy grid. For that, there is a constant need to find novel materials that can serve as an alternative to generate and store energy. In that regard, the class of materials known as covalent organic frameworks (COFs) is emerging as multifunctional materials with great potential in energy production and storage (Figure 6.1) [2]. They can present two- or three-dimensional (2-D, 3-D) structures, at which 2-D refers to building blocks that repeat and restrict themselves covalently among the two directions [1]. Also, they can have triangular, tetragonal, and hexagonal polymeric layers that are further stacked to generate different types of structures according to the geometries of monomers. 3-D refers to geometrically and chemically extended space units that use tetrahedral linkages in an extra dimension to form solid organic structures through reactions. They include extended systems with building blocks attached via stable chemical bonds. What makes them unique is their molecular design that shows the individuality in their formation from different types of polymers. Some are linear, hyper-branched, and cross-linked which may allow available photons, excitons, electrons, holes, molecules, or ions to interplay with created platforms for structural design and functional development. They tend to have two

DOI: 10.1201/9781003206507-6

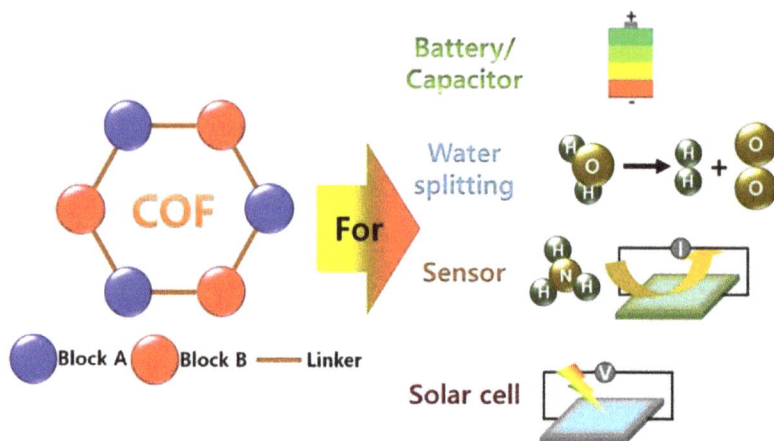

FIGURE 6.1

Energy application of COFs. (Adapted with permission [2]. Copyright (2020), American Chemical Society.)

3-D geometries, i.e., the structure that reflex continuity and extends the porosity, which generally looks like the structure of metal-organic framework (MOF), termed as eclipsed. The other ones are usually non-continuous with short porous arrangements which are less favorable, termed as staggered. Different mixed and single types of organic molecules function as linkers and nodes. These choices determine the dimensionality of the resultant structure.

The advantage of COFs is that they are highly porous, allowing them to have a higher surface area compared to other available organic frameworks and considerably more than zeolites. They can act as robust hosts and self-active materials, which mainly have available active sites for the action of electrons and other subatomic particulates on the COFs tunnel wall. This active areal property makes them beneficial for energy applications like water splitting, solar cells, metal-ion batteries, supercapacitors, and many more. High surface area facilitates the electrolyte ions to be adsorbed on the electrode's surface to maximize the double layer capacitances, and their porous structure allows the fast transportation of ions, which helps increase the rate capability of electrodes. Some of these synthesized materials also show high electrochemical performances which indicate their chances of having high chemical and thermal stability. When used as cathode material in lithium-ion batteries, cation filtration, polysulfide adsorption, and zinc-air batteries, they can overcome several issues of capacity fading over repeated use and allow the increase of both the lifespan and capacity of energy storage devices especially, batteries.

COFs can be used for gas storage, catalysis, and drug delivery systems. Materials having a 2-D structure with few-layer nanosheets accelerate ion diffusion transportation into the crystalline structure or onto the system's surface. Crystalline nature allows a subatomic particle to pass smoothly throughout the structure's layers, although most of the COFs offer somewhat crystallinity up to a certain level. After completing the condensation of 3-D COFs, the reversibility during the formation of covalent bonds is a must-required property. Continuous research is going on for the development of crystalline COFs materials. A way more important look toward these materials is that they are organic, abundant, reliable, and sustainable [3].

6.2 Synthesis of COFs

COFs are organic-based compounds which during synthesis, usually deploy the kinetically controlled reactions for the fabrication of structure into comprehensive 2-D and 3-D networks. During the synthesis of COFs, there must be reversibility in the formation of covalent bonds, which leads to some mismatched covalent linkages that break up during the reaction after the successful condensation of the self-healing structure. This structural order is possible due to the dynamic covalent chemistry (DCC) that allows the repairing and formation of highly ordered frameworks. Several reversible organic reactions allow the fabrication of crystalline COFs as follows Schiff base reaction (reaction between aldehydes and amines), trimerization of nitriles (formation of 1,3,5-triazine by heating the nitrile in the presence of copper carbonate catalysts), and boronate ester formation (formed by the reaction between boronic acid and an alcohol).

Apart from reaction mechanism development, this is a time-consuming process to obtain COFs with high porosity and crystallinity. Therefore, there is a requirement to find suitable temperature and pressure conditions for different synthetic methods proposed for developing COFs by researchers worldwide, including solvothermal, microwave-assisted, ionothermal, mechanothermal interfacial, post-treatment, salt-mediated, high-temperature calcination, and many other techniques [3]. The generalized, high temperature (80–120°C) combined with low reaction pressure and optimal synthesis parameters is crucial for preparing COFs. These conditions can only be obtained in the solvothermal synthesis route, one of the most common and repeated methods for developing COFs.

The building block's solubility, reactivity, and reaction reversibility determine the conditions for the COF's synthesis. Generally, a suitable monomer is selected based on the process to multiply and grow the vertices and edges. Then, catalyst, and solvents or a mixture of solvents are transferred to the Pyrex tube. After removing the excess amount of gas via freeze-pump-thaw cycles, this system is sealed with a gas burner and kept at a suitable temperature and pressure. The filtrate is transferred and dried under vacuum at 80–120°C through the continuous presence of N_2 or Ar in the dark. In the case of the synthesis of boronate-ester linkage, some solvents such as dioxane-mesitylene, DMAc-o-dichlorobenzene, and tetrahydrofuran (THF) have been used for the construction of boronate ester-linked COFs. For the case of imine-linked COFs, a mixture of dioxane-ethanol can be used as a solvent. A mixture of N-methylpyrrolidone (NMP), mesitylene, and isoquinoline yields polyarylimide (PI)-COF-4 and PI-COF-5 after the completion of five days of reaction. TPT-COF-1 was obtained from 2,4,6-tris (4-amino phenoxy)-1,3,5-triazine (TPT-NH$_2$) and 2,4,6-tris(4-formylphenoxy)-1,3,5-triazine (TPT-CHO), which exhibited a high crystalline, porosity, and surface area of 1589 m²/g. The material's property can be tuned based on the building blocks available, enabling several possibilities that can be pre-designed [1, 3].

A suitable alternative to the solvothermal synthesis route is the microwave-assisted method. It provides a way to quickly synthesize COFs under the same condition as solvothermal. It saves time, can increase the yield as well as crystallinity with a higher Brunauer-Emmet-Teller (BET) surface area. Boronate-ester-linked COF-5, COF-102, and imine-linked triphenyl-tricarbaldehyde (TpTa-COF) have been successfully synthesized using this method. The microwave-assisted synthesis route was 200 times faster and improved the structure's porosity [4, 5]. In addition, the surface area enhanced from 901 to 2019 m²/g, which was considerably higher than 1590 m²/g obtained from the solvothermal method. The synthesis of COF using a microwave reactor is shown in Figure 6.2. It was concluded that after microwave synthesis and extraction with acetone, the particle size increased, and

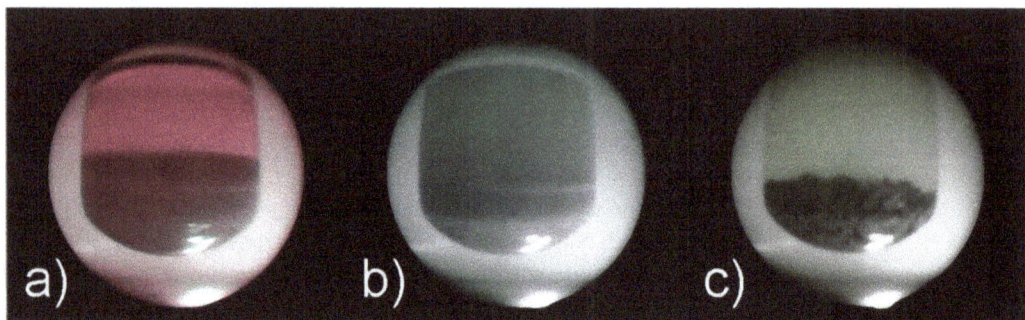

FIGURE 6.2
Camera images were taken as an observation part of COF-5 reaction and purification in a microwave reactor: (a) purple is obtained after initial steps, (b) removal of impurities using acetone, (c) gray powdered COF-5 with S_{BET} = 2019 m^2/g. (Adapted with permission [5]. Copyright (2009), American Chemical Society.)

the COF's morphology changed. This variation affected the surface area, which resulted in higher values when compared to conventional methods. The above obtained 2-D COFs are typical examples. However, the 3-D COFs are the one that possesses the promising properties derived from extended linkages in 3-D extra space with high porosity and low density. However, with limited choices of tetrahedral building blocks, for example, a COF-102 prepared using boronate-ester linkages with carbon nitride or boracite topology from self-condensation of tetra (4-dihydroxy boryl phenyl) methane (TBPM) and its silane analog (TBPS) through the irradiation at 100°C for 20 minutes. An obtained sample had a BET surface area of 2926 m^2/g, comparable to the surface area of 3472 m^2/g obtained by the conventional method. Based on that, it is notable that microwaves for developing COFs have several benefits like faster reaction, higher crystallinity, and higher reproducibility.

The ionothermal method is a convenient approach as it avoids the usage of environmentally harmful solvents and catalysts in large quantities while reducing the reaction time compared to conventional synthesis methods. It is a process in which ionic liquids/molten salts, i.e., organic salts with a low melting point (below 100°C) are used as a liquid phase to act as solvent as well as catalysts. Also, they can potentially be used as a template or structure-directing agent during the growing of single crystals. In general, the monomer and ZnCl$_2$ in a Pyrex sealed glass tube containing liquid are evacuated and heated up to 400°C for 40 h. The obtained mixture is then cooled to room temperature and washed thoroughly with water followed by diluted HCl solution for 15 h to remove ZnCl$_2$. Finally, the collected amount of material is cleaned with water and THF after filtration and dried under vacuum to produce (CTF-1), which follows the aromatic nitrile building units using trimerization reaction [6]. Figure 6.3 shows the synthesis route of ionothermal method.

A series of 3-D ionic liquid containing COFs was successfully synthesized by using a solvent 1-butyl-3-methylimidazolium bis-[(trifluoromethyl)sulfonyl] imide [BMIm], yielding a 3D-IL-COF-1 as a light yellow crystalline solid [7]. The obtained COF had high crystallinity, surface area, and adsorption selectivity for gases like CO$_2$/N$_2$ and CO$_2$/CH$_4$. The pursue of less complicated approaches led to a mechanochemical synthesis route for the development of porous COFs because it includes simple steps that comprise monomers grounded at room temperature. Liquid-assisted grinding can also be performed to improve the system's homogeneity. As an example, solvent-free mechanochemical synthesis of 1,3,5-Triformylphloroglucinol (Tp), aromatic diamine [p-phenylenediamine (TpPa-1), 2,5-dimethyl-p-phenylenediamine (TpPa-2) and benzidine (TpBD)] under the aldehyde-amine Schiff base condensation reaction

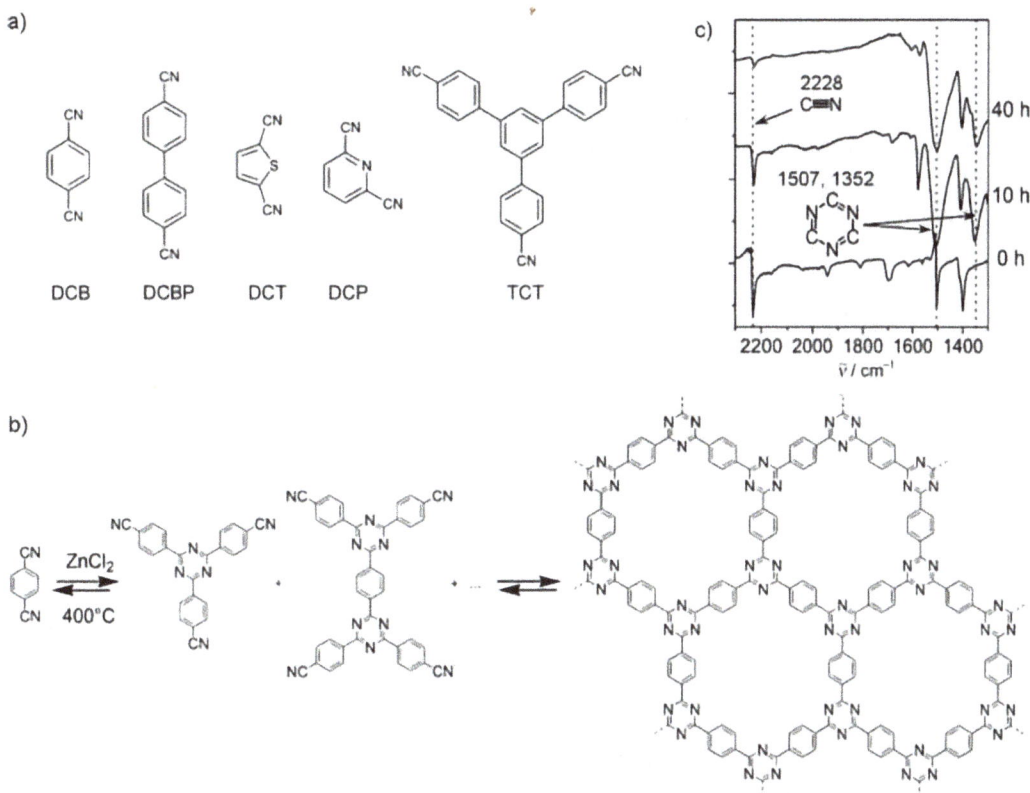

FIGURE 6.3
(a) Monomers used for ionothermal method. (b) Trimerization of dicyanobenzene in $ZnCl_2$ to form CTF-1. (c) FTIR spectra from 0 to 40 h. (Adapted with permission [6]. Copyright (2008), John Wiley and Sons.)

along which 1-2 drop of mesitylene:dioxane (1:1) was placed in a mortar and grounded by using pestle at room temperature which after 5 min was converted to a light yellow powder and again ground for 15 min. The powder color changed to orange and later after 40 min to a dark red color which was mopped with anhydrous acetone and dichloromethane to remove the oligomeric impurities. Then it was dried at 180°C under vacuum for 24 h. COFs obtained through this method presented similar properties to those synthesized through solvothermal [8, 9].

6.3 COFs for Supercapacitors

Capacitors are devices that store energy in electrical form. Two electrical conductors separated by a certain distance embedded with a dielectric insulator between its plates form the capacitor. Supercapacitors are an upgraded version of capacitors, which can store more energy due to extensive surface area and parallel plate's porosity along with a smaller distance between them. One of its energy storage mechanisms is based on the formation of an electric double layer of opposite charges at the electrode's walls, thus defining them as electric double-layer capacitors (EDLCs). Supercapacitors outperform batteries when compared in power density, allowing them to store more energy for a short period. As discussed, COFs have several properties that could directly benefit from developing super-capacitors electrodes. The presence of chemically rigid and highly π conjugated bonding allows the design of a controllable pore structure that can provide room for the electrolytes. Its semiconducting skeleton prompts EDLCs mechanism for charge storage. That aspect is also attributed to the lighter elements that compose COFs, leading to high gravi-metric performance for energy storage devices.

A triazine-based COF is synthesized using a Schiff-base condensation reaction between 1,4-piperazinedicarboxaldehyde (PDC) and melamine (MA) as structural units [10]. A three-electrode system was used to determine the electrochemical capacitive perfor-mances of PDC-MA-COF in a 6M KOH aqueous solution. The assembled PDC-MA-COF//AC asymmetric supercapacitor (ASC) is shown in Figure 6.4. The cyclic voltammetry (CV) curves at different scan rates ranging from 10 to 100 mV/s within the potential window range 0 to 0.5 V are shown in Figure 6.5a. Symmetrical redox peaks suggested the quasi-reversible redox process, which indicated pseudocapacitive characteristics. CV profile area increased in direct relation with the scan rate while it maintained asymmetry profile at higher scan rate. Galvanostatic charge-discharge (GCD) curves at different current densi-ties also showed asymmetric profile with a maximum specific capacitance of 335 F/g at a density of 1 A/g and 248 F/g at a density of 10 A/g. It retained 74% of specific capacitance as represented in Figure 6.5b–c. Electrochemical impedance spectroscopy (EIS) was per-formed within the frequency range of 0.01 Hz to 100 kHz, and the Nyquist plot of the PDC-MA-COF electrode is shown in Figure 6.5d. Their small semicircle in the high-frequency region indicated lower charge transfer resistance, higher conductivity, and an almost ver-tical straight line in the low-frequency region suggesting a slow ionic diffusion rate. The ASC exhibited a high energy density of 29.2 Wh/kg with a power density of 750 W/kg and excellent cyclic stability retention of 88% after 20000 GCD cycles.

Regarding future research work, scientists are expecting low-cost, highly conductive, redox-active COF materials to be developed, for which Khattak et al. [11] synthesized a series of pyridine-based COFs and demonstrated the redox activity of the pyridyl

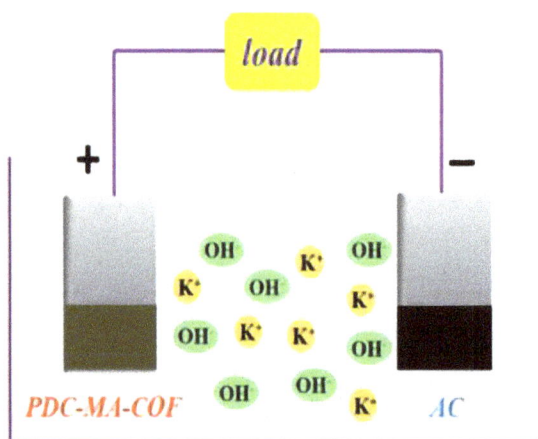

FIGURE 6.4

Schematic of PDC-MA-COF//AC ASC. (Adapted with permission [10]. Copyright (2019), American Chemical Society.)

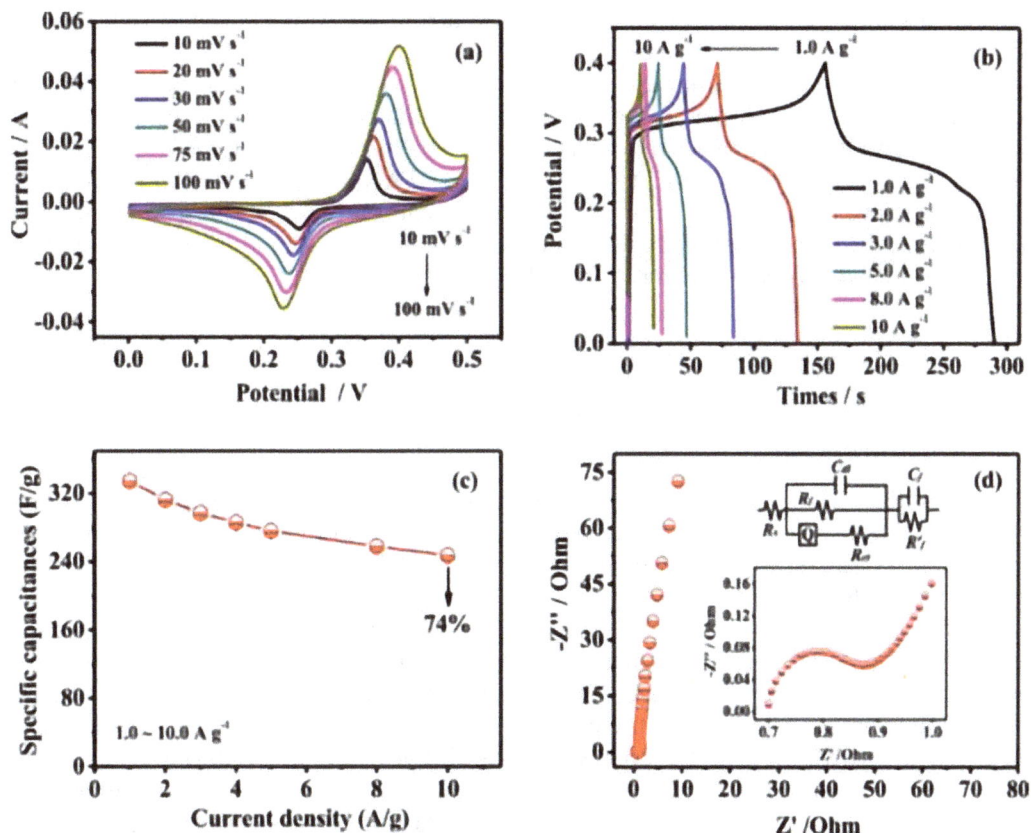

FIGURE 6.5

(a) CV curves at different scan rates. (b) GCD curves at various current densities. (c) Specific capacitances at various current densities. (d) The Nyquist plot of PDC–MA–COF electrode; the insets are the magnification of the high-frequency region and the fitting equivalent circuit diagram. (Adapted with permission [10]. Copyright (2019), American Chemical Society.)

functional group in H_2SO_4 solution through a controlled method. A long-term lifetime of up to 6000 charge/discharge cycles has been demonstrated for the pyridine-containing COF, and a 2D–2D electrode material with an edge-ion (vertical) orientation has also been developed. This innovative structure was achieved by vertically coordinating the diboronic acid onto graphene sheets before the growth process. After carbonization treatment, the electrochemical performance was evaluated. It was found that electrochemical activity can also be improved by enriching the π-electron communication throughout the framework.

Structural engineering plays a significant role in high-performance capacitors [12, 13]. Following that line, graphene appears as an important component for capacitor construction [2]. Based on that, Shi et al. [14] demonstrated the use of sodium anthraquinone-2-sulfonate (AQS) as an organic redox-active compound along with graphene nanosheets that were incorporated to enhance the electronic conductivity. The -SO_3^- functional group of AQS offered excellent hydrophilicity, promoting the molecular level binding of AQS with reduced graphene oxide (rGO) and leading to a 3D interconnected aerogel (AQS@rGO). The composite exhibited a high specific capacitance of 567.1 F/g at 1 A/g with a stable capacity retention of 89.1% over 10000 cycles at 10 A/g. More importantly, the optimized composite maintains a high capacitance of 315.1 F/g even at 30 A/g due to the high pseudocapacitance of AQS and the capacitive contribution of rGO. A novel all-organic composite was successfully prepared, by in-situ polycondensation of 1,3,5-triformylphloroglucinol and p-phenylenediamine (TpPa-COF) on polyaniline (PANI) under a one-pot solvothermal condition, TpPa-COF covered the well-distributed intricate network structure of TpPa-COF@PANI and PANI formed a core-shell structure. Combining the highly porous and redox-active TpPa-COF and conductive PANI, TpPa-COF@PANI composite electrodes showed significantly improved supercapacitor application performance. After 30000 cycles, the battery's capacity retention rate was as high as 83%, which indicated that the electrode had excellent cycling stability [15].

6.4 COFs for Batteries

The demand for energy storage components is increasing with the development of a society that strongly relies on electronic devices. A key to improving the electrochemical performance of batteries is to develop advanced electrode materials. Currently, electrode studies mainly focus on the development of materials with a high energy density and high power density by regulating the morphological structure, but most of the electrodes are of inorganic materials, which implies recycling costs, metal mining, and environmental pollution. Transition metal-based electrode materials work on the concept of change in the oxidation state, which affects the energy storage capacity inside the battery that gets limited by the lattice structure, reducing the versatility of inorganic materials. Along with transition metal-based electrode materials, scientists also search for environment-friendly, relatively low-cost, and diversely structured organic compounds as potential substitutes to these materials.

The main advantage for the organic compound is that they have many redox-active sites that can store energy with good structural design ability, and electrochemical performance due to the presence of different functional groups. Still, these organic compounds have some issues like small organic molecules that are soluble in electrolytes, resulting in poor cyclic stability and low efficiency. Apart from all, organic materials provide the possibility of designing materials with different structures and morphologies. Many COFs have been designed due to the diversity of organic compounds along with their insolubility in organic electrolytes. COFs that are light element-based and possess adjustable structures with designable periodic skeleton and ordered nanopores have become a promising organic electrode material for rechargeable advanced batteries.

Chen et al. [16] synthesized an N-rich COF for high performance on batteries based on the triple condensation reaction under mild acid media and inert atmosphere between triquinoxalinylene and benzoquinone (TQBQ) with tetraminophenone (TABQ) or cyclohexanehexaone (CHHO). These TQBQ-COF materials consist of multiple carbonyls and pyrazine groups, where carbonyl are used as active redox sites and pyrazine act as linkages block to form the 2D conjugated framework for which theoretical capacity found to be 515 mAh/g and structure is shown in Figure 6.6a. Powder X-ray diffraction (PXRD), high-resolution transmission electron microscopy (HR-TEM), and selected area electron microscopy (SAED) pattern are shown in Figure 6.6b–d confirming the formation of the obtained organic framework. The pristine TQBQ-COF had a surface area of 46.95 m^2/g and after annealing, it increased to 94.36 m^2/g. The GCD profile of the TQBQ-COF electrode in the sodium battery delivered an initial discharge capacity of 452 mAh/g and a high

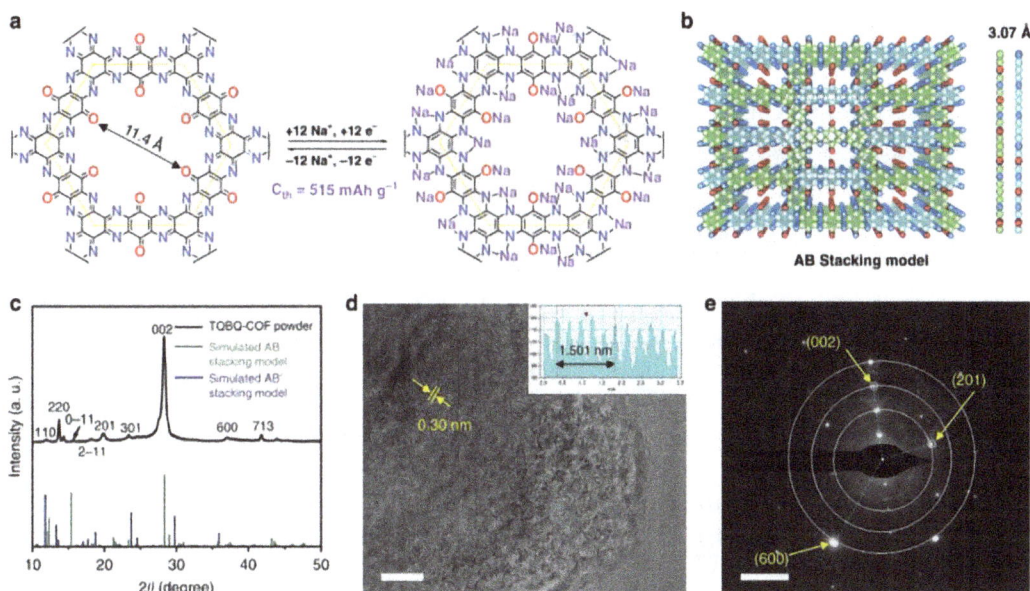

FIGURE 6.6

(a) The chemical structure and possible electrochemical redox mechanism of TQBQ-COF. (b) Top and side views of the schematic AB stacking model of TQBQ-COF layers. (c) PXRD pattern and the simulated AB/AB' stacking models of TQBQ-COF powder. (d) HR-TEM image of TQBQ-COF. (e) SAED pattern of TQBQ-COF. (Adapted with permission [16]. Copyright (2020), The Authors (Published by Springer Nature). The article is licensed under a Creative Commons Attribution 4.0 International License.)

reversible capacity of up to ~ 400 mAh/g after 10 cycles and a capacity of 352.3 mAh/g after 100 cycles. CV measurement was employed to estimate the diffusion kinetics of Na^+ at different scan rates ranging from 0.2 to 2 mV/s using Equation (6.1).

$$i = a\gamma^b \tag{6.1}$$

where i is the peak current, γ is the scan rate, and a and b are the constants where the value of b is close to 0.5, it shows a Na^+ diffusion-controlled process. The excellent cycling stability of the TQBQ-COF electrode could be ascribed to its inherent stable feature. Additionally, the weak peak in the UV-vis spectra of TQBQ-COF electrodes in $NaPF_6$/ DEGDME (Di-ethylene glycol-Di- methyl ether) indicated the poor solubility of the discharged and recharged products. Moreover, the TQBQ-COF-based pouch type battery shows a capacity of 81 mAh/g and a voltage plateau of 1.86 V, corresponding to the gravimetric and volumetric energy density of 101.1 Wh/kg and 78.5 Wh/L based on the whole pouch-type cell, respectively. This work shows the design and application of COFs with multiple redox sites for high energy and power densities for Na batteries.

Covalent triazine frameworks (CTFs) are another kind of COF. CTFs are usually N-rich, which are believed to be lithiophilic. In 2019, Coskun and co-workers [17] fabricated a CTF (CTF-1) through a cyclotrimerization reaction between 1,4-dicyanobenzene and Li bis (trifluoromethylsulphonyl) imide (LiTFSI). On the one hand, LiTFSI could act as a catalyst. On the other hand, it could result in the formation of homogeneous LiF particles among the CTF framework. They coated the formed CTF/LiF on a 3D air laid-paper (AP) scaffold by an immersion method to generate AP-CTF-LiF, which was able to serve as a host for Li

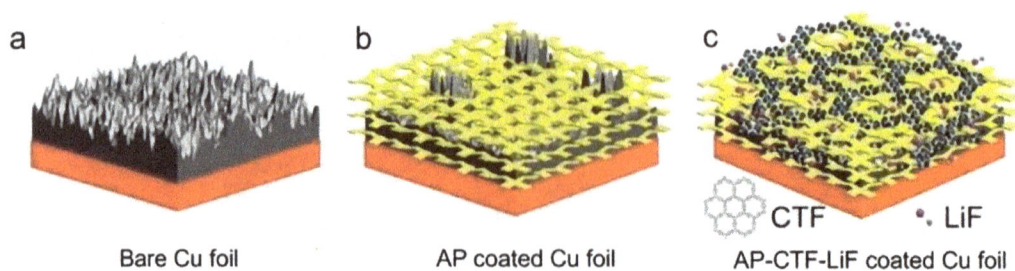

FIGURE 6.7
Schematics showing the Li deposition behaviors on bare for (a) Cu foil, (b) AP coated Cu foil, and (c) AP-CTF-LiF coated Cu foil, respectively. (Adapted with permission [18]. Copyright (2019), John Wiley and Sons.)

metal anode. In this host, a lithiophilic CTF was utilized as nucleation seeds to regulate the uniform Li deposition. The LiF could improve the interfacial stability of the Li metal anode. The 3D AP could alleviate the volume fluctuation of the electrode during the plating/stripping process. As a result, Li dendrite growth was inhibited because of the synergistic effect of the AP-CTF-LiF host which can be depicted in Figure 6.7.

The AP-CTF-LiF could achieve a stable coulombic efficiency of 96.2% for 220 cycles at (10 -1 mAh/cm^2), which was better than bare Cu foil (92.4% for 60 cycles). Symmetric cells with AP-CTF-LiF could remain stable for 700 cycles (5 -1 mAh/cm^2). While symmetric cells with bare Cu foil failed after no more than 10 h. Designing polymer electrolytes with well-defined porous structures enabled extra ion channels that addressed this issue. Jiang and co-workers prepared solid-state polyelectrolyte COFs through integrating flexible oligo (ethylene oxide) chains on the pore walls of COFs [19]. By using 2,5-bis[(2-methoxy ethoxy) methoxy]terephthalaldehyde (BMTP) and 1,3,5-tri(4-aminophenyl)benzene (TPB) as the precursors, TPB-BMTP-COF was obtained. The surface area, pore size, and pore volume of TPB-BMTP-COF were 1746 m^2/g, 3.02 nm, and 0.96 cm^3/g, respectively. The TPB-BMTPCOF could remain stable at 300°C, which was better than PEO solid-state electrolyte that suffered from shrinkage above 120°C. To evaluate the Li$^+$ conductivity, TPB-BMTPCOF powders were loaded with LiClO$_4$ by a solution diffusion method in LiClO$_4$/CH$_3$OH solution and then dried to generate Li$^+$@TPB-BMTP-COF. The Li$^+$ conductivity of Li$^+$@TPB-BMTP-COF was 6.04×10^{-6}, 2.85×10^{-5}, 1.66×10^{-4}, and 5.49×10^{-4} S/cm at 40, 60, 80, and 90°C, respectively. On the other hand, the Li$^+$ conductivity of PEO-Li$^+$ was only 8.0×10^{-8} S/cm at 40°C. The superior Li$^+$ conductivity of Li$^+$@TPBBMTP-COF was owed to the existence of dense oligo ethylene oxide chains, which could be complex with Li$^+$ to generate a polyelectrolyte interface in the COF channels. The built-in polyelectrolyte interface was able to promote the dissociation of the ionic bond and provide a pathway for Li$^+$ transport.

6.5 COFs for Fuel Cells

A promising solution to the energy-related challenges is to introduce clean and sustainable energy sources such as electrochemical water-splitting to produce H$_2$ and O$_2$ for clean combustion. These emerging energy technologies are under intensive research owning to their sustainability and zero carbon emission [20]. There are many types of fuel cells depending on anode, cathode, electrolyte, and temperature operating range. At the anode, the catalysis

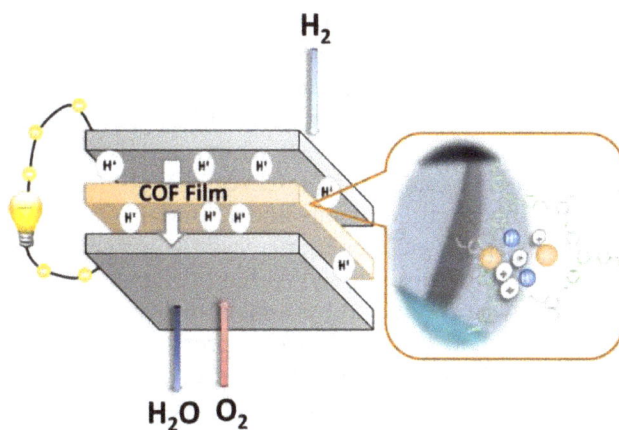

FIGURE 6.8
General representation of COF film as the separator with its structure inside the fuel cell. (Adapted with permission [20]. Copyright (2017), American Chemical Society.)

causes the fuel to undergo an oxidation reaction that generates ions and electrons. The ions move from anode to cathode through electrolyte as electrons flow from anode to cathode through external circuit producing electricity (Figure 6.8). The anode reaction in fuel cells is either due to direct oxidation of hydrogen, or methanol, or indirect oxidation through reforming step for hydrocarbon fuels [21]. The cathode reaction is oxygen reduction from the air in most fuel cells. For hydrogen/oxygen (air) fuel cells, the overall reaction is

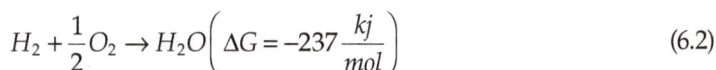

$$H_2 + \frac{1}{2}O_2 \rightarrow H_2O \left(\Delta G = -237 \frac{kj}{mol} \right) \tag{6.2}$$

Where ΔG is the change in Gibbs free energy of formation. Water is released during the reaction at cathode or anode depending on the type of fuel cell. For an ideal fuel cell, the theoretical voltage at standard conditions of 25°C and 1 atm pressure is 1.23 V under the typical operating voltage range of 0.6–0.7 V for the high-performance fuel cells.

Extensive efforts and great progress have been made toward the fabrication of electrocatalysts with earth-abundant materials, including metal-derived, polymers, and carbon nanomaterials that can be used in fuel cells. In that sense, the known properties of COFs such as well-defined pore structures, allow them to be employed as proton exchange membrane fuel cells (PEMFC) [22]. Nafion, which is a membrane composed of a perfluoro sulfonic acid with polytetrafluorethylene may decrease its performance in fuel cells with the increase in temperature due to the decrease in its conductibility. COF materials are relatively lightweight, provide a variety of functionalities, and have membrane processability like polymers through chemical or physical modifications. The grafting of acid groups on COFs and doping with proton carriers is an efficient way to introduce proton conductivity and layout the platform for the study of intrinsic and extrinsic protonic transference in porous materials. Yang et al. [23] developed a series of COFs with a phenolic hydroxyl group (proton donors) and azo groups (proton acceptors) integrated inside the COFs skeleton. H_3PO_4 is then loaded into the series of COF materials (H_3PO_4@COFs) and converted into the solid electrolyte membrane through membrane electrode assembly (MEA). The fuel cell's power density was 81 mW/cm^2 under H_2/O_2 with 100% relative humidity (RH) at 60°C, which surpassed previously reported COF-based MEAs, but still far inferior from the conventional Nafion-based MEAs for practical use. The possible adsorption effect of phosphate ions causes the catalyst activity which can affect the proton conductivity. To rescue this effect leaching of guest molecules (e.g., H_3PO_4) should be considered during the formation of COF for long-term PEMFC operations. It has been noted that H_3PO_4 loaded with Schiff base networks (SNW) type COFs [24], which is well compatible with the Nafion ionomer, can be employed in the catalyst layer (CL) of PEMFC to serve as the hygroscopic agent due to its intrinsic hydrophilic group and abundant cavities. The immobilized H_3PO_4 could hold the water in the cavities to increase the water retention of the CL with low humidity conditions. The synthesis of SNW-1 has been discussed in the literature and the impregnation of H_3PO_4 was conducted by a vacuum-assisted method [25]. A catalyst-coated membrane (CCM) method was employed to prepare MEA [26]. The performances of these MEAs might be attributable to the unique properties of the H_3PO_4 loaded COF network, i.e., the proton transferability and the abundant cavities containing H_3PO_4, which made it a proper proton conductor and water reservoir.

A self-humidifying MEA with an H_3PO_4-loaded COF network as anode additive was developed to improve the PEMFC performance under low humidity conditions. The results from the single-cell tests and EIS measurements revealed that an optimal content of 10 wt.% COF in the anode CL delivered the highest performance at various humidity conditions due to the minimum cell ohmic resistance and charge transfer resistance. At 60°C and 38% RH, the maximum power density of the optimized MEA was up to 582 mW/cm², which was almost seven times higher than that for the routine MEA (85 mW/cm²). Aside from the doping of porous materials with acids, heterocyclic bases can also be used as proton conducting materials. This may affect the strength of hydrogen bonding interactions but increases the chances of rotation of acid and bases molecules. To achieve a high level of proton conductivity in COFs, Jiang et al. [27] reported a mesoporous COF (TPB–DMTP-COF) through the reaction between 2,5-dimethoxy-terephthalaldehyde (DMTP) and TPB via the solvothermal method. The proton conduction across mesoporous channels was studied, which could be applied as a host material for loading proton carriers. In this work, triazole and imidazole were chosen as N-heterocyclic proton carriers, providing trz@TPB–DMTP-COF and im@TPB–DMTP-COF with high loading capacities [27]. The frameworks achieved satisfying proton conductivities (4.37×10^{-3} S/cm at 130°C) that were 2–4 orders of magnitude higher than those of microporous and non-porous polymers, implying the importance of doping N-heterocyclic proton carriers for enhanced proton conductivity. Another work about proton conductive COF was reported by Zhu's group [28], who introduced a cationic COF that was synthesized by combining a cationic monomer, ethidium bromide (EB) (3,8-diamino-5-ethyl-6-phenylphenanthridinium bromide), with 1,3,5-triformylphloroglucinol (TFP) via Schiff base reactions. By immobilizing $PW_{12}O_{40}{}^{3-}$ into this porous cationic framework, the proton conductivity enhanced to 3.32×10^{-3} S/cm at 97% RH. These contributions have shown the promising applications of COF materials in fuel cells and will probably encourage researchers to further design functional COF-based proton conducting materials for fuel cells.

6.6 COFs for Flexible Devices

Electronic and energy storage flexible devices have attracted a significant amount of attention in recent years for their potential applications in modern society. The redox-active, porous structural, and nanocrystalline nature of obtained materials include several grain boundaries which could act as a roadblock for flexible devices like supercapacitors, micro-supercapacitors (MSC) among others. Hence, the inherent properties of COFs may provide a feasible approach for the development of optimized flexible energy storage devices. One example of the use of COFs in this manner can be performed by compositing them with carbon nanotubes to enhance their conductivity. Based on this concept, Mohammed et al. [29] developed a COF and carbon nanofiber matrix (CNF@COF) using the in-situ solid-state mechanomixing method. The load of 10 to 20 wt.% of CNF into the COF's structure led to an increase in conductivity. The structure of COF and a short depiction of the process is shown in Figure 6.9.

Electrochemical measurements for CNF@COF showed the areal capacitance of 464 mF/cm^2 calculated from GCD curves at a current density of 0.25 mA/cm^2. To support the industrial standard for commercial electrodes, the testing was performed on an aluminum substrate in which the volumetric capacitance was 92.8 F/cm^3. The symmetric supercapacitor device was fabricated by taking 1 cm^2 geometrical area of CNF@COF as electrodes. These COF electrodes were placed on the grafoil sheets, which served as current collectors of the supercapacitor. A thin layer of PVA-H$_2$SO$_4$ electrolyte gel was uniformly coated on the COF@CNF thin sheet and promoted wetting over the electrode. Two electrodes were made by the procedure, which was sandwiched in between a polypropylene separator. Through that, the cell presented an areal capacitance of 167 mF/cm^2.

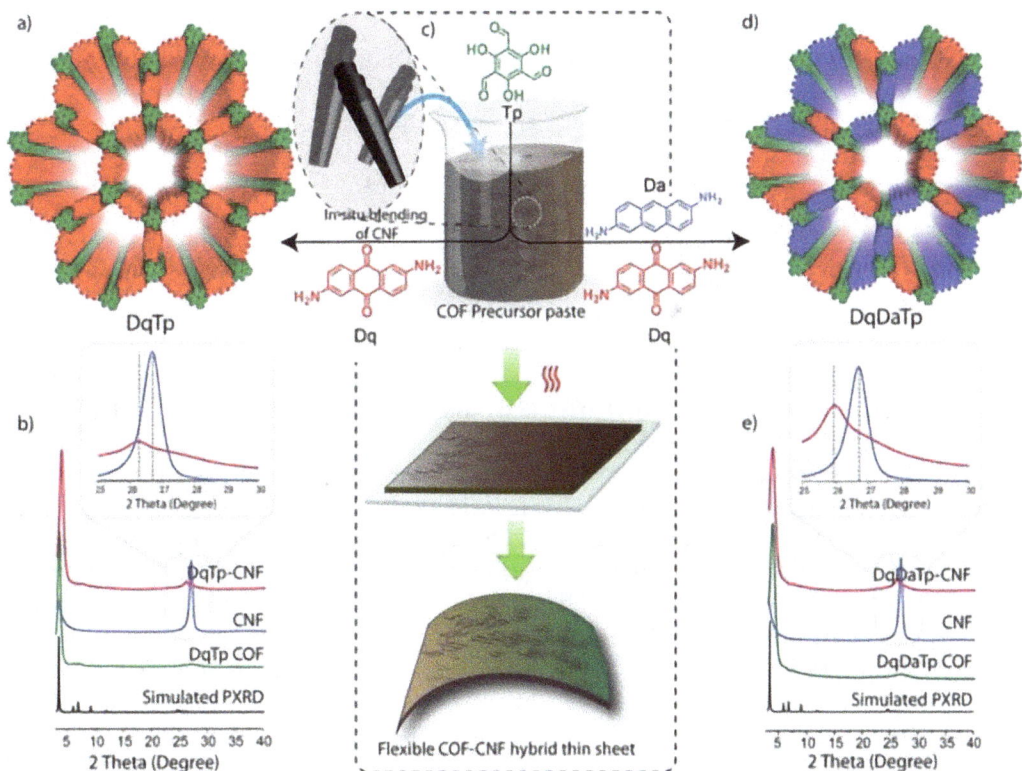

FIGURE 6.9

(a, d) Eclipsed models of DqTp and DqDaTp COFs; (b, e) PXRD patterns of DqTp-CNF and DqDaTp-CNF hybrids; (c) schematic representation of the synthesis of crystalline yet flexible DqTp-CNF and DqDaTp-CNF COFs. (Adapted with permission [29]. Copyright (2019), American Chemical Society.)

Zhang et al. [30] reported the development of fully conjugated olefin-linked COF (g-$C_{34}N_6$-COF) with unsubstituted C=C bond linkages. This allowed efficient electron delocalization which endowed such kinds of COFs with significantly improved electron conductivity over the 2D frameworks. The BET surface area of the observed COF was 1003 m^2/g. Structural morphology was detected by scanning electron microscopy and transmission electron microscopy, which showed the fibrous morphology with a uniform diameter around ~100 nm. The top-view of SEM images of thin-film revealed the uniform dispersion of COF fibers with entangled single-wall carbon nanotubes (SWCNTs) providing abundant active sites and intrinsic regular porous structure. Based on that, an electrode was built using a flexible polyethylene terephthalate (PET) film as a substrate to adhere to the g-$C_{34}N_6$-COF. A quasi-rectangular shape at different scan rates in CV was obtained indicating its high and stable capacitive performance. Based on the CV curve measurements the specific areal capacitances were 15.2, 9.4, and 5.1 mF/cm^2 at a scan rate of 2, 100, and 500 mV/s, respectively. These values were higher than graphene/CNT (5.1 mF/cm^2), thiophene nanosheets/graphene (3.9 mF/cm^2), V_2O_5/graphene (3.9 mF/cm^2), and phosphorene/graphene (9.8 mF/cm^2) from other reports [31]. GCD curves presented a triangular shape, which suggested supercapacitive behavior along with good rate capability. The capacitance retention was 45% when the current density went from 0.05 to 5 mA/cm^2. However, it was only 20% of the CNT-based MSC's. The Ragone plot depicted the overall performance of the fabricated device (thickness ~ 3 μm), which showed the energy densities of 7.3, 4.9, and 2.3 mWh/cm^3 at power densities of 0.05, 1.04, and 10.4 W/cm^3.

Khayum et al. [32] stated that the cutting-edge wearable supercapacitor requires flexible and free-standing electrodes to be applicable and take the challenges to develop it. In that sense, the granular form of COF is not sufficient to stand alone for flexible devices. Hence, unique strategies are required to convert the granular form of COFs into porous, crystalline, flexible, and free-standing electrodes. In that sense, β-ketoenamine linked COF was inserted by redox-active 2,6-diaminoanthraquinone (Dq) and 2,6-diamino anthracene (Da) for the construction of heterolinked COF with 1,3,5-triformylphloroglucinol (Tp). A layer of a thin sheet of DqTp-COF, DaTp-COF, DqDaTp-COF (with different ratios) was prepared using a solvothermal synthesis route and for the formation of paste, p-toluene-sulfonic acid (PTSA) was mixed thoroughly. The as-synthesized material was crystalline which is shown by the obtained XRD pattern. Different peaks of DqTp (3.6 and 6.2), DaTp (3.6 and 6.2), Dq_1Da_2Tp (3.7 and 6.0), Dp_1Da_1Tp (3.5 and 6.1), and Dq_2Da_1Tp (3.5 and 5.9) corresponds to the 100 and 110 planes. Excellent stability was reported after the thermal stability test by all COF thin sheets at 430°C via thermogravimetric analysis (TGA). Also, the N_2 adsorption isotherm confirmed the crystalline COF thin sheets with microporous nature. DqTp and DaTp COFs exhibit BET surface areas of 940 and 577 m^2/g, respectively. Dq_1Da_2Tp, Dq_1Da_1Tp, and Dq_2Da_1Tp COFs maintained an ordered porous nature with a surface area as high as 1400, 804, and 1004 m^2/g.

Apart from having two distinct linkers, the regular arrangement of building blocks maintained the high crystallinity into the same framework. The free-standing COF thin-sheet electrode with the thickness ~25 μm (thickness was measured by the electronic screw gauge) was prepared for the electrochemical characterization in a three-electrode configuration. CV curves were recorded at various scan rates (10–500 mV/s) in a potential window of -0.7 to 0.3 V. GCD study was performed at various current densities (1.56, 3.12, and 6.25 mA/cm^2).The devices preparation, CV and GCD curves are shown in Figure 6.10.

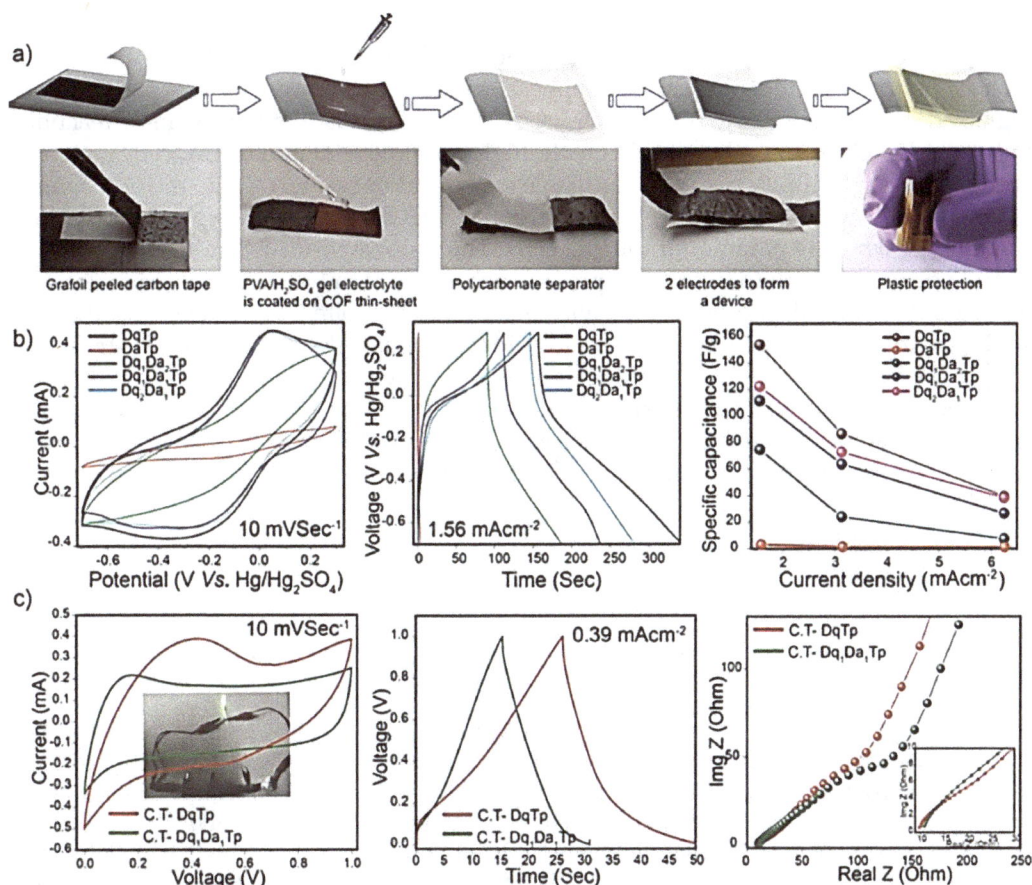

FIGURE 6.10
(a) Fabrication of the CT-COF supercapacitor device. (b) CV, GCD, and current density vs specific capacitance plot. (c) Device characterization: CV (inset: 3.5 V LED lighted up by the series connection of four flexible devices), GCD, and impedance analysis of CT-DqTp and CT-Dq$_1$Da$_1$Tp COF supercapacitor devices. (Adapted with permission [32]. Copyright (2018), American Chemical Society.)

6.7 Conclusion

Throughout the discussion in this chapter, it was noticeable that COF materials are continuously developing toward the organic nature of clean energy and helping to step toward sustainable energy development. The main challenge faced by COF materials is their weak interaction between the bonds which decreases the ionic or electronic conductivity making them less conducting. Also, temperature allows them to distort their shape beyond the operating region, which tends to increase the microporous size of particles, decreasing their specific surface area. At present, only limited COFs with few linkages have been developed and applied for mass transfer. Improving the processability while maintaining high proton or ionic conductivity is the core idea to design effective mass transfer via COFs. To address these scientists are looking for suitable monomers, chemical linkages, functional groups, dimensions of COF among other aspects. In that sense, the current aim lies in the development of a clear structure-performance relationship that can provide COFs with high-temperature operating regions with low-cost production and high conductivity. Further, efforts to widen the range of applications in addition to fuel cells, sensors, or biological applications are still in the infant phase, which can be explored by further developing novel synthetic routes to obtain COFs. One critical issue is that the chemical reactions to synthesize COFs must be able to create high crystalline materials. Indeed, materials with limited crystallinity may not be suitable to name as or categorize into the COF family. Owing to their well-defined 1-D channel structures, COFs have been explored for proton conduction, which is paramount for fuel cells. The innovative structural design of COFs would open a new path to a high-rate proton-conducting membrane, aiming to replace the high-cost Nafion with proton conductivity of 0.1 S/cm. Despite these drawbacks, the possibility of manufacturing electronic devices, ideally in any shape of 2D or 3D structures that can be bent or twisted to some degree is still a challenge. Also, obtaining smaller size particles whereas delivering the same performance as their bulkier and rigid counterparts is the next step on technological advance for these devices. Through that, is likely that a considerable part of the research in the field of energy storage applications will be devoted to COFs.

References

1. Geng K, He T, Liu R, Dalapati S, Tian Tan K, Li Z, Tao S, Gong Y, Jiang Q, Jiang D (2020) Covalent Organic Frameworks: Design, Synthesis, and Functions. Chem Rev 120:8814–8933.
2. Zhang K, Kirlikovali KO, Varma RS, Jin Z, Jang HW, Farha OK, Shokouhimehr M (2020) Covalent Organic Frameworks: Emerging Organic Solid Materials for Energy and Electrochemical Applications. ACS Appl Mater Interfaces 12:27821–27852.
3. Xu L, Ding S-Y, Liu J, Sun J, Wang W, Zheng Q-Y (2016) Highly crystalline covalent organic frameworks from flexible building blocks. Chem Commun 52:4706–4709.
4. Díaz De Greñu B, Torres J, García-González J, Muñoz-Pina S, De Los Reyes R, Costero AM, Amorós P, Ros-Lis JV, Reviews C (2021) Microwave-Assisted Synthesis of Covalent Organic Frameworks: A Review. ChemSusChem 14:208–233.
5. Campbell NL, Clowes R, Ritchie LK, Cooper AI (2009) Rapid microwave synthesis and purification of porous covalent organic frameworks. Chem Mater 21:204–206.
6. Kuhn P, Antonietti M, Thomas A (2008) Porous, Covalent Triazine-Based Frameworks Prepared by Ionothermal Synthesis. Angew Chemie Int Ed 47:3450–3453.
7. Guan X, Ma Y, Li H, Yusran Y, Xue M, Fang Q, Yan Y, Valtchev V, Qiu S (2018) Fast, Ambient Temperature and Pressure Ionothermal Synthesis of Three-Dimensional Covalent Organic Frameworks. J Am Chem Soc 140:4494–4498.
8. Biswal BP, Chandra S, Kandambeth S, Lukose B, Heine T, Banerjee R (2013) Mechanochemical synthesis of chemically stable isoreticular covalent organic frameworks. J Am Chem Soc 135:5328–5331.
9. Emmerling ST, Germann LS, Julien PA, Moudrakovski I, Etter M, Friščić T, Dinnebier RE, Lotsch BV. (2021) In situ monitoring of mechanochemical covalent organic framework formation reveals templating effect of liquid additive. Chem 7:1639–1652.
10. Li L, Lu F, Xue R, Ma B, Li Q, Wu N, Liu H, Yao W, Guo H, Yang W (2019) Ultrastable Triazine-Based Covalent Organic Framework with an Interlayer Hydrogen Bonding for Supercapacitor Applications. ACS Appl Mater Interfaces 11:26355–26363.
11. Khattak AM, Ghazi ZA, Liang B, Khan NA, Iqbal A, Li L, Tang Z (2016) A redox-active 2D covalent organic framework with pyridine moieties capable of faradaic energy storage. J Mater Chem A 4:16312–16317.

12. Sun J, Klechikov A, Moise C, Prodana M, Enachescu M, Talyzin AV (2018) A Molecular Pillar Approach To Grow Vertical Covalent Organic Framework Nanosheets on Graphene: Hybrid Materials for Energy Storage. Angew Chemie Int Ed 57:1034–1038.

13. Beidaghi M, Wang C (2012) Micro-Supercapacitors Based on Interdigital Electrodes of Reduced Graphene Oxide and Carbon Nanotube Composites with Ultrahigh Power Handling Performance. Adv Funct Mater 22:4501–4510.

14. Shi R, Han C, Duan H, Xu L, Zhou D, Li H, Li J, Kang F, Li B, Wang G (2018) Redox-Active Organic Sodium Anthraquinone-2-Sulfonate (AQS) Anchored on Reduced Graphene Oxide for High-Performance Supercapacitors. Adv Energy Mater 8:1802088.

15. Sajjad M, Lu W (2021) Covalent organic frameworks based nanomaterials: Design, synthesis, and current status for supercapacitor applications: A review. J Energy Storage 39:102618.

16. Shi R, Liu L, Lu Y, Wang C, Li Y, Li L, Yan Z, Chen J (2020) Nitrogen-rich covalent organic frameworks with multiple carbonyls for high-performance sodium batteries. Nat Commun 11:178.

17. Wei C, Tan L, Zhang Y, Zhang K, Xi B, Xiong S, Feng J, Qian Y (2021) Covalent Organic Frameworks and Their Derivatives for Better Metal Anodes in Rechargeable Batteries. ACS Nano 15:12741–12767.

18. Zhou T, Zhao Y, Choi JW, Coskun A (2019) Lithium-Salt Mediated Synthesis of a Covalent Triazine Framework for Highly Stable Lithium Metal Batteries. Angew Chemie Int Ed 58:16795–16799.

19. Xu Q, Tao S, Jiang Q, Jiang D (2018) Ion Conduction in Polyelectrolyte Covalent Organic Frameworks. J Am Chem Soc 140:7429–7432.

20. Montoro C, Rodríguez-San-Miguel D, Polo E, Escudero-Cid R, Ruiz-González ML, Navarro JAR, Ocón P, Zamora F (2017) Ionic Conductivity and Potential Application for Fuel Cell of a Modified Imine-Based Covalent Organic Framework. J Am Chem Soc 139:10079–10086.

21. Lin CY, Zhang L, Zhao Z, Xia Z (2017) Design Principles for Covalent Organic Frameworks as Efficient Electrocatalysts in Clean Energy Conversion and Green Oxidizer Production. Adv Mater 29:1606635.

22. Xie Z, Tian L, Zhang W, Ma Q, Xing L, Xu Q, Khotseng L, Su H (2021) Enhanced low-humidity performance of proton exchange membrane fuel cell by incorporating phosphoric acid-loaded covalent organic framework in anode catalyst layer. Int J Hydrogen Energy 46:10903–10912.

23. Yang Y, He X, Zhang P, Andaloussi YH, Zhang H, Jiang Z, Chen Y, Ma S, Cheng P, Zhang Z (2020) Combined Intrinsic and Extrinsic Proton Conduction in Robust Covalent Organic Frameworks for Hydrogen Fuel Cell Applications. Angew Chemie Int Ed 59:3678–3684.

24. Georg Schwab M, Fassbender B, Wolfgang Spiess H, Thomas A, Feng X, Müllen K (2009) Catalyst-free Preparation of Melamine-Based Microporous Polymer Networks through Schiff Base Chemistry. J Am Chem Soc 131:7216–7217.

25. Yin Y, Li Z, Yang X, Cao L, Wang C, Zhang B, Wu H, Jiang Z (2016) Enhanced proton conductivity of Nafion composite membrane by incorporating phosphoric acid-loaded covalent organic framework. J Power Sources 332:265–273.

26. Kim E-Y, Yim S-D, Bae B, Yang T-H, Park S-H, Choi H-S (2016) Study of a highly durable low-humidification membrane electrode assembly using crosslinked polyvinyl alcohol for polymer electrolyte membrane fuel cells. J Solid State Electrochem 20:1723–1730.

27. Xu H, Tao S, Jiang D (2016) Proton conduction in crystalline and porous covalent organic frameworks. Nat Mater 15:722–726.

28. Zhao X, Pachfule P, Thomas A (2021) Covalent organic frameworks (COFs) for electrochemical applications. Chem Soc Rev 50:6871–6913.

29. Mohammed AK, Vijayakumar V, Halder A, Ghosh M, Addicoat M, Bansode U, Kurungot S, Banerjee R (2019) Weak Intermolecular Interactions in Covalent Organic Framework-Carbon Nanofiber Based Crystalline yet Flexible Devices. ACS Appl Mater Interfaces 11:30828–30837.

30. Xu J, He Y, Bi S, Wang M, Yang P, Wu D, Wang J, Zhang F (2019) An Olefin-Linked Covalent Organic Framework as a Flexible Thin-Film Electrode for a High-Performance Micro-Supercapacitor. Angew Chemie Int Ed 58:12065–12069.
31. Hu R, Zhao J, Wang Y, Li Z, Zheng J (2019) A highly stretchable, self-healing, recyclable and interfacial adhesion gel: Preparation, characterization and applications. Chem Eng J 360:334–341.
32. Khayum MA, Vijayakumar V, Karak S, Kandambeth S, Bhadra M, Suresh K, Acharambath N, Kurungot S, Banerjee R (2018) Convergent Covalent Organic Framework Thin Sheets as Flexible Supercapacitor Electrodes. ACS Appl Mater Interfaces 10:28139–28146.

7

Covalent Organic Frameworks for Fuel Cell Applications

Federico Roncaroli[1,2,3]

[1]*Departamento de Física de la Materia Condensada, Instituto de Nanociencia y Nanotecnología, Centro Atómico Constituyentes, Comisión Nacional de Energía Atómica, San Martín, Buenos Aires, Argentina*

[2]*Departamento de Química Inorgánica, Analítica y Química Física, Facultad de Ciencias Exactas y Naturales, Universidad de Buenos Aires, Ciudad Universitaria, Pabellón II, Ciudad de Buenos Aires, Argentina*

[3]*Consejo Nacional de Investigaciones Científicas y Técnicas-CONICET, Ciudad de Buenos Aires, Argentina*

CONTENTS

7.1 Introduction

Fuel cells (FCs) are electrochemical devices that convert the free energy of the reaction into electricity. Only hydrogen-feed FCs will be discussed within this chapter. Hydrogen provides an alternative to the employ of fossil fuels and a possible solution to the accumulation of greenhouse gases in the atmosphere. Proton exchange membrane fuel cells (PEM-FCs) and alkaline anion exchange membrane fuel cells (AAEM-FCs) are composed of two electrodes: the anode (negative) where the hydrogen oxidation reaction (HOR) takes place and the cathode (positive) where the oxygen reduction reaction (ORR) takes place. The ion exchange membrane allows ions migration selectively, prevents gas diffusion, and has low

DOI: 10.1201/9781003206507-7

electron conductivity. The operation temperature of the PEM-FCs is 80–90°C, and they are composed of two gas diffusion layers (GDL), two catalyst layers, and PEM (Figure 7.1). These three components, in the form of thin layers, are attached forming the membrane electrode assembly (MEA). The two GDL are sheets of carbon paper that allow the gases to diffuse to the catalyst layer and extract the generated water. The surface of each GDL is coated by the corresponding catalyst. Nanoparticulated Pt supported on carbon is the typical catalyst for both the anode and cathode. The catalyst particles are usually adhered to each other with the same polymer as the membrane (ionomer/binder) to allow ions to reach the catalyst active sites. In this way protons generated upon H_2 oxidation on the anode, migrate through the PEM to the cathode where O_2 is reduced to produce water, and electrons flow through the external circuit (electrical current) (Figure 7.1) [1]. AAEM-FCs share several similarities in their design to PEM-FCs. The major difference is that the membrane is an AEM, i.e., a hydroxide ion conductive membrane. Carbonate ion formation from atmospheric CO_2 or gas impurities can severely reduce cell performance and durability. AEMs exhibit lower conductivity and durability than PEMs. On the other hand, AAEM-FCs could allow the possibility to employ non-platinum group metals (non-PGM) on both electrodes as catalysts [1].

Covalent-organic frameworks (COFs), together with metal-organic frameworks (MOFs), hydrogen-bonded organic frameworks (HOFs), and metal hydrogen-bonded organic frameworks (M-HOFs), form a big and growing family of open-framework materials [2]. COFs are crystalline 2D and 3D porous organic polymers, whose molecular building units are linked through covalent bonds. Due to their permanent porosity, high specific surface area, and the possible incorporation of heteroatoms, metal coordination, or functionalization, they have gained attention for multiple applications like gas absorption and separation, membrane fabrication, energy conversion, and electro-catalysis. Typical reactions employed for the COF synthesis are boronic ether formation, aldehyde/amine condensation, and Schiff-base reactions. The reversibility of these reactions allows the formation of crystalline materials [3, 4].

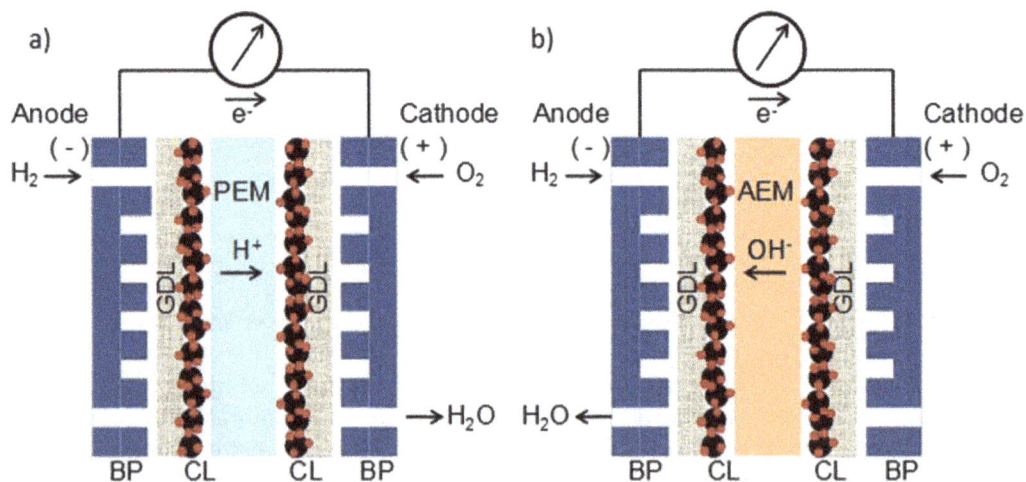

FIGURE 7.1

Schemes of the components of PEM-FC (a), and AAEM-FC (b). GDL = Gas Diffusion Layer, BP = Bipolar Plates, CL = Catalyst Layer.

7.2 COFs and Their Derived Materials as Cathodic Catalysts in Fuel Cells

The ORR takes place on the cathode. Particularly in the PEM-FCs, its rate is considered to be slower than the anodic reaction, since the complex mechanism involves the consumption of 4 electrons and several elementary steps. A competitive reaction pathway is the formation of H_2O_2, which occurs through a 2-electron reduction of O_2. This reaction not only reduces the power of the cell, but also H_2O_2 exerts a corrosive effect on different parts of the cell. Typically, noble metals are employed as catalysts, being Pt nanoparticles supported on carbon (20% w/w) the standard catalyst. Pt and noble metals are scarce and expensive, which would prevent, for example, the massive utilization of PEM-FCs in electrical vehicles. Extensive effort has been done to obtain noble metal-free-catalysts. In this direction, Co and Fe based catalysts, obtained from thermal treatment or pyrolysis of complexes, coordination polymers, and MOFs exhibit higher stability and catalytic activity than their precursor compounds, comparable in many cases to those of Pt catalysts in acidic medium and even better, in some cases, in alkaline medium. Another approach is the development of metal-free-catalysts which have shown competitive activity in alkaline media. In this way, COFs can serve as ORR catalysts in four ways: (1) pristine metal-free COFs, (2) metal coordinated COFs, (3) COF derived X-doped-metal-free carbons, or (4) COF derived Fe/N or Co/N-doped carbons [5]. Quantitative electrochemical parameters from some recent results are summarized in Table 7.1 and the corresponding synthetic strategies are discussed in the following sections.

7.2.1 Pristine Metal-Free COFs

A naphthalene diimide-based covalent organic framework (NDI-COF) has been synthesized through condensation reaction between 1,4,5,8-naphthalenetetracarboxylic dianhydride (NTCDA) and 1,3,5-tris(4-aminophenyl) benzene (TAPB) under solvothermal conditions by *Segura et al*. It was exfoliated into nanosheets and drop-casted onto a glassy carbon electrode. Electrochemical measurements showed that the naphthalene diimide groups acted as efficient metal-free electrocatalysts for the ORR in alkaline media. The redox activity of the naphthalene diimide groups and the semiconducting nature of the COF may be responsible for the ORR catalytic activity [6].

Nejati *et al*. electro-polymerized (5,10,15,20-tetrakis(4-aminophenyl)porphyrin) on glassy carbon in the presence of pyridine (Py-POR-COF). It exhibited an ORR process at *ca*. 0.6 V vs RHE (pH 13). The presence of pyridine during electro-polymerization increased the POR-COF crystallinity and the ORR catalytic performance [7].

A π-conjugated 2D covalent organic radical framework (PTM-CORF) based on the stable polychlorotriphenylmethyl (PTM) radical was obtained by Wu *et al*. from liquid/liquid interfacial acetylenic homo-coupling of a triethynylpolychlorotriphenylmethane monomer followed by deprotonation and oxidation of the polychlorotriphenylmethane units. The material exhibited antiferromganetic coupling of neighboring PTM radicals, an interesting ORR activity after mixing with carbon black (20% wt), and good durability (40000 s) [8].

Metal-free thiophene-sulfur COFs were obtained by Yao *et al*., from condensation of 2,4,6-tris(4-aminophenyl)benzene (TAPB) with 2,5-thiophenedicarboxaldehyde (TDC) or 2,2′-bithiophenyl-5,5′-dicarbaldehyde (bTDC), producing JUC-527 or JUC-528, respectively. The experimental evidence and DFT calculations showed that the thiophene-sulfur building blocks acted as active sites for the ORR. The JUC-528 COF showed the best electrochemical performance, 25 h durability, and was successfully used to construct a Zn/air battery [9].

TABLE 7.1

Electrokinetic Parameters Toward the ORR in Alkaline Medium of Some COF-Derived Catalysts

	Catalyst	E_o (V vs RHE)	$E_{1/2}$ (V vs RHE)	Electron transfer number	Tafel slope (mV dec⁻¹)	H_2O_2 yield (%)	Ref.
Pristine Metal-Free COFs	NDI-COF	0.75	*ca* 0.70	3.6			[6]
	Py-POR-COF		*ca*0.6	3.97			[7]
	PTM-CORF		0.70	3.9			[8]
	JUC-528		0.70	3.81	65.9		[9]
Pristine metal coordinated COFs	Pt@ TM-TPT-COF	1.04	0.89	3.8–3.9		4.5–8	[10]
	Cu-CTF/CPs	0.91	*ca* 0.80	3.75–3.95			[11]
	pfSAC-Fe-0.2		0.910	ca 4.0	31.7	< 7	[12]
	Fe-COF-BTC	0.965	0.900				[13]
	Fe0.5Co0.5Pc-CP NS@G	1.006	0.927	3.9		< 5	[14]
COF derived X-doped- metal free- carbons	4"-NP from 2D [4 + 4] COFs	0.88	0.81	*ca* 4.0	70		[16]
	N,P carbon (TAPB-DMTA)	0.87	0.81	*ca* 4.0	72		[17]
	B-N-C capsules		0.85	3.7	69	13–18	[18]
	PA@TAPT–DHTA–COF1000NH3	0.97	0.857	3.77–3.98	110	1.5–11	[19]
	CPN-NS		0.868	*ca* 4	47	*ca* 20	[20]
COF derived metal-X-doped carbons	Co-TpBpy-800 nanocages	0.91	0.831	3.9	72		[21]
	(Co)GC@COF-NC0.04	0.923	0.841	3.90	78.4	6	[22]
	Ni/Fe-COF@CNT900	1.01	0.87	3.9		5.5	[23]
	Cu₆Sn₅@S-N-C-900		0.86	3.9–4.0	75		[24]
	Co@COF₉₀₀	0.93	0.86	3.75–3.91	89	2–6	[25]
	mesoC-TpBpy-Fe	0.920	0.845	ca 4.0			[26]
	FeS/Fe₃C@N-S-C-800	1.02	0.87	3.9	90		[27]
	LTHT-FeP aerogel	0.92	0.83	*ca* 4.0		0–6	[28]
Pt /C		0.91	0.85	*ca* 4.0	113		[26]

E_o = Onset potential, $E_{1/2}$ = half wave potential.

7.2.2 Pyrolysis-Free Metal Coordinated or Metal Supported on COFs

Pt nanoparticles were supported on a conjugated nitrogen-rich COF, TM-TPT-COF, by Jiang *et al*. The COF was obtained from 2,4,6-trimethyl-1,3,5-trizaine (TM) and 1,3,5-tris-(4-formylphenyl)-triazine (TPT) condensation. Pyridinic N atoms acted as nucleation sites for Pt nanoparticles on the COF surface and within the pore channels, resulting in uniform Pt distribution and more accessible Pt active sites. This resulted in ultrahigh ORR activity. The strong Pt and N interactions and the chemical stability of COF support were responsible for its high stability and long durability (50 h) [10].

Cu ions in low coordination geometry were immobilized on a covalent triazine framework (CTF) by Kamiya *et al*. The CTF was obtained by polymerization of 2,6-dicyanopyridine in molten $ZnCl_2$ containing carbon nanoparticles (CPs). The coordination number of Cu was 3.4 from EXAFS measurements. The unsaturated coordination structure of Cu sites would allow strong binding to O_2 and high ORR activity. An alternative hypothesis is that CTF stabilizes the Cu(I) state at a relatively high redox potential, as seen in redox proteins. The higher stability of this catalyst (Cu-CTF/CPs) than other Cu-based catalysts is probably due to the rigid cross-linked covalent bonds within the COF [11].

Xiang *et al.* obtained a Fe-single atom catalyst through a pyrolysis-free synthetic approach using a fully-conjugated Fe-phthalocyanine (FePc)–rich COF. The alternative pyrolysis method would generate FeN_x moieties randomly located on a carbon matrix (Fe-N-C). Benzene-1,2,4,5-tetracarbonitrile was polymerized in the presence of Fe producing a fully closed -conjugated COF, which was subsequently deposited on a graphene matrix (pfSAC-Fe-x, where x is the mass ratio). This catalyst was tested in a Zn-air battery delivering a similar discharge behavior but a higher power density compared to a battery using Pt/C [12].

Another quasi-phthalocyanine Fe-COF was prepared by Xiang *et al.* through the assembly of benzene-1,2,4,5-tetracarbonitrile (BTC) with $FeCl_3$, and 1,8-Diazabicyclo(5,4,0)undec-7-ene (DBU) as the catalysts, under mild microwave heating (see Figure 7.2). The closed conjugated systems provided rigid structures to the Fe-COF-BTC and also ensured a uniform 2D structure, homogeneous elemental distribution, and long-range regular coordinated single Fe atoms. In alkaline medium, OH^- adsorption on positively charged centers, allowed the Fe-COF-BTC to be exfoliated and suspended. The Fe-COF-BTC dissolved in water was used as ORR catalyst with carbon paper as working electrode. It was also employed in a Zn/air flow battery, showing better cycling stability than a cell with a Pt/C-coated air electrode. The high activity of the catalysts was attributed to the low work function and the fully closed conjugated structure [13].

Ultrathin 2D coordination polymer nanosheets (MPc-CP NSs, M = $Fe_{0.5}Co_{0.5}$, Fe, and Co) with fully conjugated electronic structure, were exfoliated from the ethynyl-linked phthalocyanine (Pc) polymers by Yao *et al.* They were prepared from the phthalocyanine monomers, $M[Pc(I)_4]$ and $M[Pc(ethynyl)_4]$, through Sonagashira–Hagihara coupling reaction. The monomers corresponded to four regio-isomers, what led to highly disordered polymers. This allowed easy exfoliation through a simple sonication procedure and deposition on graphene nanosheets ($Fe_{0.5}Co_{0.5}Pc$-CP NS@G). This composite exhibited better ORR (0.1 M KOH) than the buk MPc-CP and even better than Pt 20% supported on carbon, good stability and methanol tolerance. These materials were tested in Zn/air batteries [14].

7.2.3 COF-Derived X-Doped-Metal-Free Carbons

COFs like MOFs, in general, suffer from low electrical conductivity, in this way, pyrolysis under an inert atmosphere (N_2 or Ar) produces a carbonaceous matrix (graphitic in part) with higher conductivity and stability. The employment of COFs and also MOFs as precursors for X-doped (X = O, N, S, P) materials and also to produce MN_xC_y moieties (M = Fe, Co) within the carbonaceous matrix is probably their most spread application within this topic and has been extensively reviewed [15]. The incorporation of metallic centers and nanoparticles will be discussed in the next section.

2D [4 + 4] COFs were obtained using tetratopic building blocks (amines and aldehydes) by Maenosono *et al.* Metal-free N, P-doped carbon electrocatalysts were obtained upon pyrolysis under Ar in the presence of Na_2HPO_2 (upstream side) of this nanoporous 2D [4 + 4] COFs. The N, P doped carbons exhibited very good electrochemical parameters and high durability in 0.1 M KOH. These catalysts behaved as bifunctional catalysts toward the ORR and Hydrogen Evolution Reaction (HER) [16], being this last reaction relevant in electrolyzers.

Duan *et al.* obtained a metal-free N, P-codoped carbon from TAPB-DMTA COF carbonization and subsequent phosphorization. (TAPB = 1,3,5-Tris(4-aminophenyl)benzene, DMTA = dimethoxyterephthaldehyde). The electrochemical parameters for the ORR were comparable to those of Pt 20% in 0.1 M KOH, with high durability and methanol tolerance.

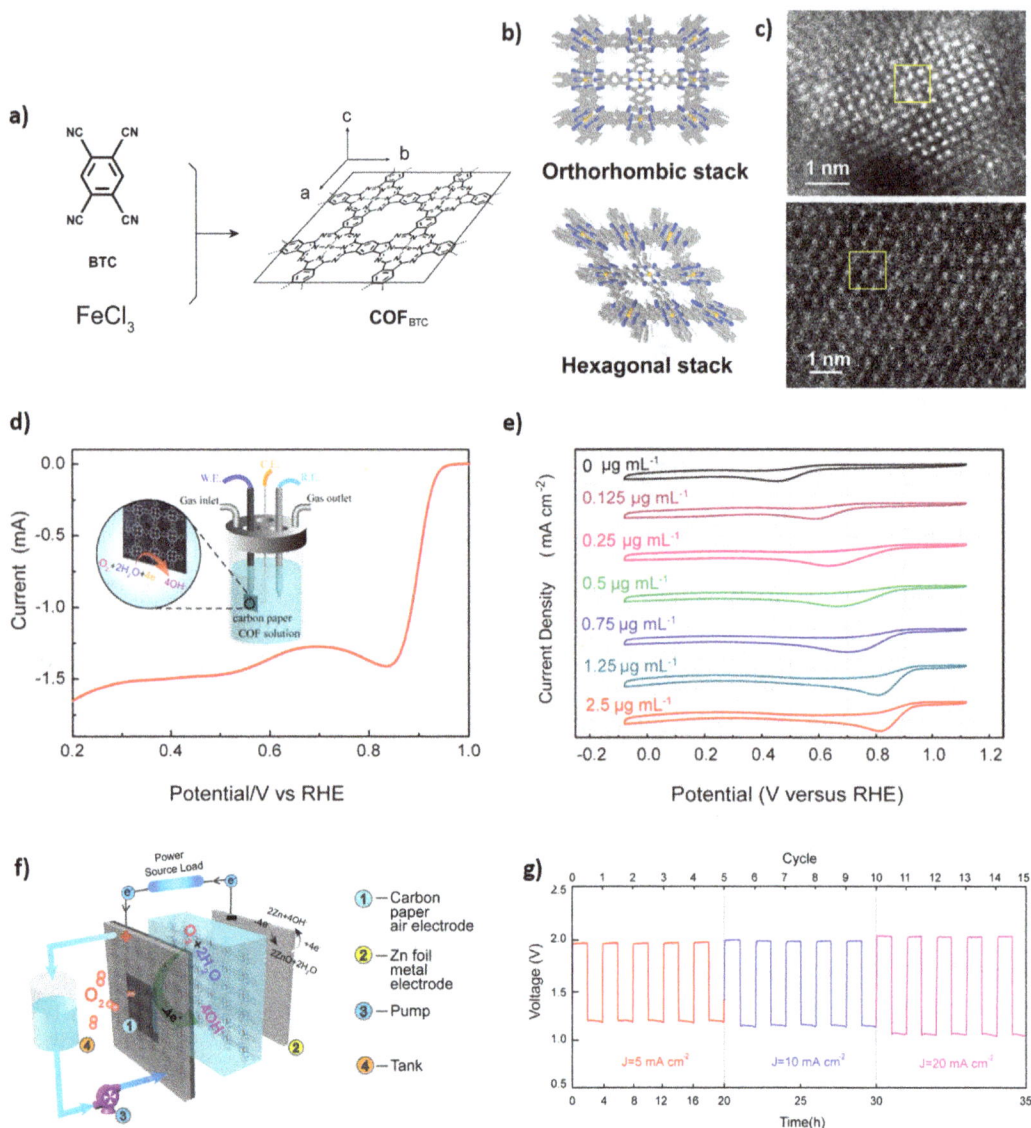

FIGURE 7.2

Synthesis, characterization, and electrochemical studies toward the ORR of a pyrolysis-free metal coordinated COF: fully-conjugated iron phthalocyanine Fe-COF-BTC. (a) Scheme of the synthesis of Fe-COF-BTC. (b) Reconstructed crystal structure of Fe-COF-BTC for the orthorhombic stack and hexagonal stack, respectively. (c) High-Angle Annular Dark-Field Scanning Transmission Electron Microscopy (HAADF-STEM) of Fe-COF-BTC with orthorhombic stack (upper) and hexagonal stacks (lower) directly observed. (d) Linear Sweep Voltammetry (LSV) measurements of the Fe-COF-BTC solution with carbon paper as electrodes. (e) Cyclic Voltammetry (CV) curves of a Fe-COF-BTC solution at different concentrations in O_2-saturated 0.1 M KOH at a sweep rate of 100 mV s^{-1}. (f) Scheme of the Zn–air flow battery by coupling the Zn electrode with two air electrodes to separate ORR and OER: a Fe-COF-BTC solution was used for ORR, and IrO_2 was precoated for the OER. (g) Charging/discharging cycling of Zn–air flow batteries at various current densities. (Adapted with permission [13]. Copyright 2019 American Chemical Society.)

It has been proposed that doping with a second or third heteroatom induces charge redistribution, decreases the charge transfer resistance, and modifies the wettability [17].

Shan *et al.* obtained a metal-free B, N-codoped mesoporous carbon (B-N-C) through copper-assisted thermal conversion of a melamine-boroxine COF. The hollow capsule morphology together with the dominated mesoporosity, the suitable graphitization degree, and the simultaneous doping of electron-rich nitrogen and electron-deficient boron would be responsible for the high ORR activity. This mesoporous carbon also served as electrode material for supercapacitors with a specific capacitance of 230 F g^{-1} [18].

To obtain small-sized N, P-doped carbon sheets, phytic (PA) acid was used as a template during the TAPT-DHTA COF pyrolysis by Jiang *et al.* (1000°C, TAPT = 4,4′,4″-(1,3,5-triazine-2,4,6-triyl) trianiline, DHTA = 2,5-dihydroxyterephthalaldehyde). Phytic acid was proposed to help to exfoliate the 2D COF layers and to guide the conversion of 2D COF into a 2D carbon. The decomposition of the phytic acid generated pores within the carbon structure and was the P source. A thermal treatment with NH$_3$ (900°C) produced a catalyst (PA@TAPT–DHTA–COF1000NH3) with lower O content, higher N content, higher specific BET surface area and presence of mesopores. This material showed competitive electrokinetic parameters (0.1 M KOH), methanol tolerance, and high durability for the ORR. It additionally served as catalysts for the Oxygen Evolution Reaction (OER), relevant for electrolyzers [19].

Naphthalene was polymerized through a Friedel–Crafts reaction in the presence of AlCl$_3$, CCl$_4$, and SBA-15 (mesoporous silica nanoparticles, template) by Liao *et al.* (Figure 7.3). The reaction product was mixed with sulfur and pyrolyzed under an N$_2$/NH$_3$ atmosphere (900°C). HF leaching removed the template and generated the CPN-NS catalyst with rod-like morphology, and micro and mesoporosity. The presence of S in the catalyst reduced the oxidized-N content and increased the total N content compared with a sample without S. In this way, the CPN-NS catalyst exhibited excellent ORR activity, methanol tolerance, and stability (20000 s) [20].

7.2.4 COF-Derived Metal-X-Doped Carbons

Core–shell SiO$_2$@Co-TpBpy nano-spheres were obtained by Zhang *et al.*, upon the reaction of 1,3,5-Trifomylphloroglucinol (Tp) and 2,2′-bipyridine-5,5′-diamine (Bpy) in the presence of NH$_2$-functionalized SiO$_2$ nanoparticles, and further coordination to Co^{2+} ions. After pyrolysis and silica template removal, Co subnanoclusters dispersed in N-doped carbon nanocages were obtained. The Co-TpBpy-800 nanocages thus obtained exhibited high activity toward ORR and high stability in alkaline medium [21].

Yamauchi *et al.* used pyrolyzed ZIF-67 (Co^{2+} zeolitic imidazolate framework) as growth support for 2,6-diaminoanthraquinone (DAAQ)- 1,3,5-triformylphloroglucinol (TFP)-COF nanofibers. This composite was further pyrolyzed (800°C) to obtain the GC@COF-NC$_{0.04}$catalyst, which exhibited ORR activity in 0.1 M KOH very similar to that of Pt 20% supported on carbon and high durability [22].

2,4,6-trihydroxybenzene-1,3,5-tricarbaldehyde (TP) and [2,2′-bipyridine]-5,5′-diamine (BPY) reacted under solvothermal conditions in the presence of carbon nanotubes (CNTs) to produce COF@CNT. This was subsequently coordinated to Ni and Fe ions, and pyrolyzed by Xu *et al.* to obtain Ni/Fe-COF@CNT900. The catalyst exhibited bifunctional ORR and OER activity, which made it suitable for the O$_2$ electrode in Zn–air batteries, showing a good discharging/charging capability (over 200 cycles) [23].

Ultrafine nanoalloyed Cu$_6$Sn$_5$ particles (2–10 nm) supported on the surface of porous N, S-doped carbons (Cu$_6$Sn$_5$@N–S–Cs) were obtained upon pyrolysis (800–1000°C) of

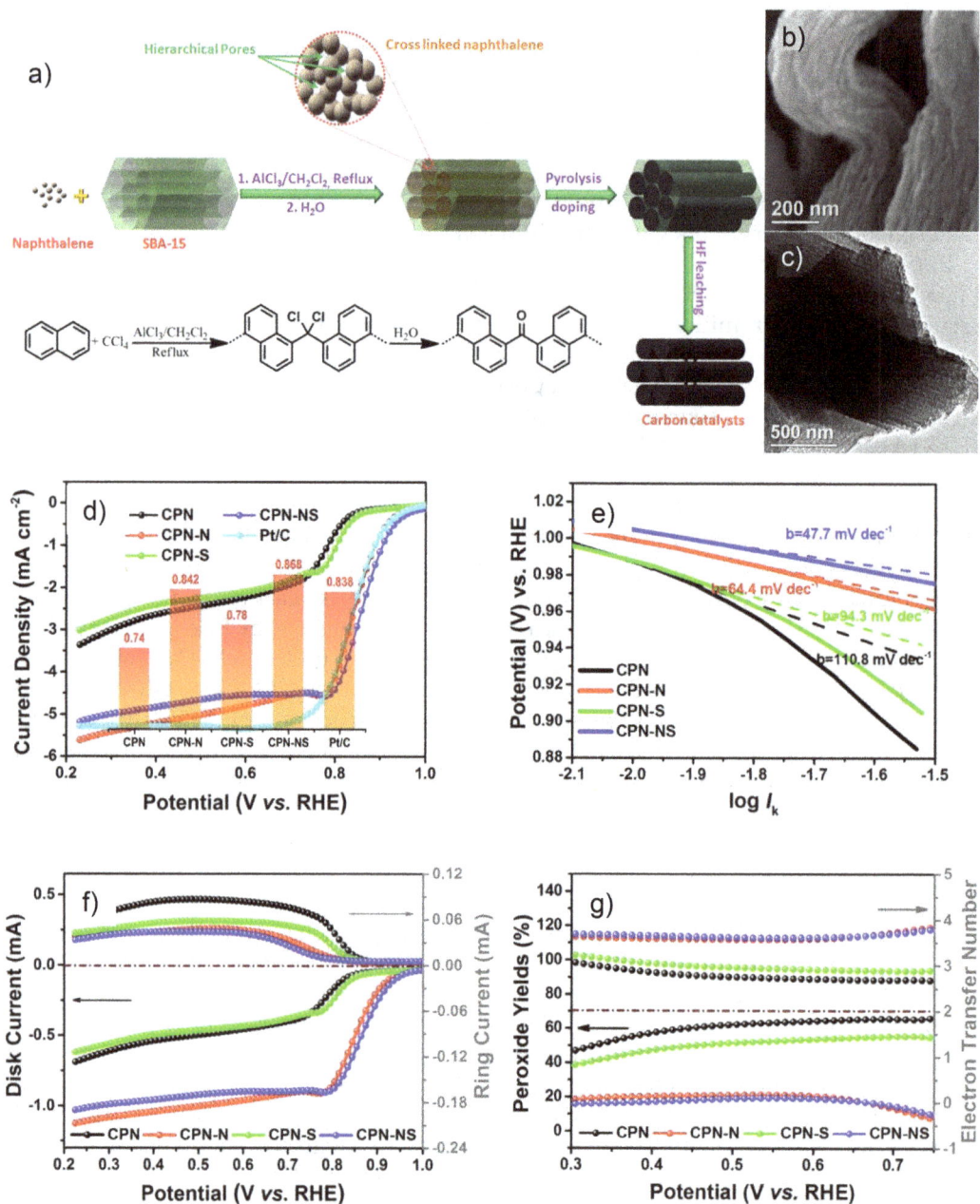

FIGURE 7.3

Synthesis, characterization, and electrochemical studies of a metal-free N, S-doped carbon (CPN-NS) as a catalyst for the ORR. (a) Scheme of the synthesis. (b) SEM image. (c) TEM image. (d) LSV curves in O_2-saturated solution and $E_{1/2}$ of different related catalysts. (e) Corresponding Tafel plots. (f) Results of rotating ring-disk electrode (RRDE) measurements. (g) H_2O_2 yields and electron transfer numbers based on the RRDE measurements. (Reproduced with permission [20]. Copyright 2017 American Chemical Society.)

a Cu-Sn-COF precursor by Shan *et al.* The precursor was obtained as an amorphous material from the reaction of 2,6-pyridinedicarboxaldehyde and 5,10,15,20-tetrakis-(4-aminophenyl)-porphyrin (TAPP) dissolved in dimethylsulfoxide (S source), in the presence of $CuCl_2$ and $Sn(OH)_x$. Spectroscopic results were interpreted in terms of the presence of $Sn(OH)_x$ and $Cu-N_x$ coordinated species. The presence of Cu and Sn might facilitate the O, N, and S doping and the graphitization of the COF during pyrolysis. Sn promoted the formation of small Cu_6Sn_5 particles instead of $Cu_{1.92}S$ and allowed the interaction of Cu with the carbon matrix. Electrochemical studies of Cu_6Sn_5@S-N-C-900 in alkaline medium showed very competitive parameters for the ORR, high durability, and methanol tolerance. Cu_6Sn_5@SnO_x could activate the tightly contacted N-S-doped carbon layers, which would be the actual active sites for ORR. Finally, the lithium-ion storage capacity was evaluated, showing a promising performance as anode material for lithium-ion batteries [24].

Aiming to obtain a single-atom catalyst, Sun *et al.* loaded Co^{2+} ions onto COF-300. The COF was obtained upon condensation of 4, 4', 4'', 4'''-methanetetrayltetraaniline (TAM) and terephthalaldehyde (TP) under solvothermal conditions. The Co^{2+}@COF-300 was subsequently pyrolyzed at 900°C, to produce Co@COF$_{900}$. (Final Co content was 0.17 wt%). The electrochemical parameters for the ORR were much better than the ones obtained without Co loading, evidencing the contribution of $Co-N_x$ sites to the ORR [25].

Thomas *et al.* obtained a silica templated COF, which was obtained from a mechano-chemical reaction between 2,2'-bipyridine-5,5'-diamine and 1,3,5-triformylphloroglucinol. The COF was converted into a Fe-N-doped mesoporous carbon upon Fe coordination, pyrolysis (900°C), and silica removal (mesoC-TpBpy-Fe). This material exhibited a large pore volume and specific surface area, which significantly enhanced the mass transfer efficiency and increased the accessibility to the active sites, yielding a high ORR activity. The presence of bipyridine groups in the precursor COF facilitated the formation of FeN_x active sites during pyrolysis. The catalyst was tested in a Zn/air battery showing better discharge stability than one with Pt/C as cathode [26].

A porous covalent phenanthroline framework (Fe-Phen-COFs) coordinated to Fe-DMSO (dimethylsulfoxide) complex was synthesized by Shan *et al.* The COF was obtained using 3,8-dibromophenanthroline and 1,3,5-benzenetriboronic acid trivalent alcohol ester as a rigid building block via Suzuki coupling reaction. The Fe-Phen-COFs were pyrolyzed (800°C) to produce N-S-doped carbons with embedded core-shell Fe_3C and FeS composite nanostructures (FeS/Fe_3C@N-S-C). This material exhibited trifunctional catalytic activity toward ORR, OER and HER, and was employed in the construction of zinc-air batteries showing good charge-discharge performance and stability [27].

Fe-porphyrin aerogels were synthesized by Elbaz *et al.*, upon condensation of tetra-4-aminophenyl porphyrin with terephthalaldehyde in the presence of $FeCl_2$, using dimethylsulfoxide as solvent. The gel thus formed was dried using CO_2 under supercritical conditions. Finally, it was pyrolyzed at 600°C (low-temperature heat-treatment, LTHT), aiming to increase the electronic conductivity and catalytic activity, while preserving its macro-structure and keeping a high concentration of atomically dispersed catalytic Fe sites (9.0 wt.%) [28].

As it could be seen, much work has been done employing rotating disk electrodes, however few reports on single-cell tests (FCs) constructing MEAs are available. Unfortunately, phenomena like ion transport, mass transfer, and water/heat management within the FC can only be adequately studied in a MEA.

7.3 COFs and Their Derived Materials as Anodic Catalysts in Fuel Cells

In the previous sections, it was shown that non-platinum group metal (non-PGM) catalysts have reached satisfactory activities for the ORR, comparable to those of the Pt group metal (PGM) containing catalysts, particularly in alkaline medium. However, the activity and durability of non-PGM electrocatalysts for the HOR are still much lower than PGM ones, in acidic and alkaline media. The development of non-PGM catalysts for HOR opens up the possibility to build PEM-FCs and AAEM-FCs completely free of PGMs, which would lead to cost reduction and massive distribution of this technology. The HOR occurs, in most reported cases, on metallic surfaces. The acid conditions of PEM-FCs, practically exclude the use of non-PGM catalysts on the anode. This leaves AAEM-FCs as the alternative to employ earth-abundant, non-PGM metals like Ni, Co, Fe, and Mo. Ni is the most widely employed non-PGM metal in HOR catalysts [29, 30].

A covalent triazine framework (CTF) was obtained from 2,6-dicyanopyridine polymerization on conductive carbon particles by Nakanishi *et al.* After impregnation with a $K_2[PtCl_4]$ solution, the N atoms with electron lone pairs were coordinated to Pt atoms, producing a single atom Pt catalyst (Figure 7.4). The absence of Pt clusters was probed by Extended X-Ray Absorption Fine Structure (EXAFS) and by HAADF-STEM. LSV experiments showed that 2.8 wt% Pt-CTF and 20 wt% Pt nanoparticles supported on carbon (Pt/C) had similar HOR activities (0.1 M $HClO_4$), and were much higher than 3.0 wt% Pt/C. While 20% Pt/C is a benchmark as ORR catalyst, the 2.8 wt% Pt-CTF catalyst exhibited a very poor ORR activity. The presence of single Pt atoms is responsible for the selectivity of this last catalyst toward HOR, even in the presence of O_2, which is very relevant for the PEM-FCs construction, since O_2 can cross over to the anode during the start-up/shut-down of the device, producing degradation and corrosion of the anode. A PEM-FC was constructed employing 2.8 wt% Pt-CTF (0.020 mg-Pt cm^{-2}) as anode catalyst and 47 wt% Pt/C (0.50 mg-Pt cm^{-2}) as cathode catalysts. The maximum power density was (487 mW cm^{-2} at 1.2 A cm^{-2},) and the open-circuit voltage was *ca.* 1 V, which were very similar parameters to those obtained employing 20 wt% Pt/C (0.10 mg-Pt cm^{-2}) as the anode catalyst, i.e., 5 times higher Pt load [31].

In general, significant effort has been made to tailor or engineer the atomic structure of the metallic nanoparticles of the HOR catalysts. This allows the modulation of the electron distribution and the band structures, which control the activation, adsorption, and desorption of reaction intermediates during the HOR [32]. Many strategies have been adopted to increase the HOR activity, some include heteroatom doping, alloying, and heterostructure engineering. The hydrogen binding energy (HBE) of Ni can be lowered by the interstitial N doping, for example as Ni_3N, and the free energy for hydrogen oxidation is, in this way, decreased. A strategy employed to prevent Ni surface passivation, oxidation, and/or corrosion is to encapsulate it within a thin carbon layer, BN, or graphene oxide [29, 30]. Although COFs have been rarely employed to obtain HOR catalysts, they are very suitable materials for the above-described strategies, which could be used as supports for metallic particles or single atoms. On the other hand, COFs can be used as catalyst precursors, contributing to heteroatom doping and producing an electron conductive carbonaceous matrix during thermal treatment. Pyrolysis has been proved to be a simple and effective technique to obtain single-atom catalysts [32]. An accurate temperature and atmosphere control, and appropriate COF precursor could lead to the formation of either metal single atoms or metallic nanoparticles with different sizes.

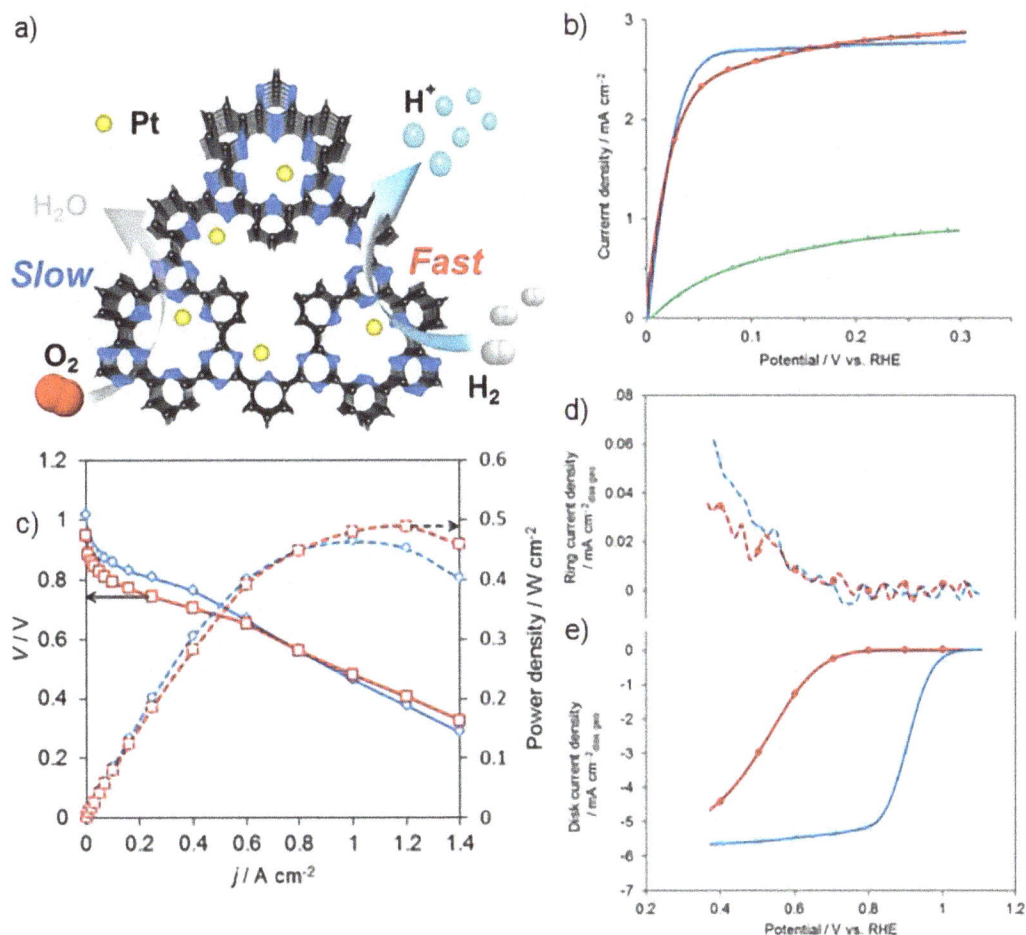

FIGURE 7.4
CTF with atomically dispersed Pt atoms as a catalyst for the HOR. (a) Scheme of the Pt-CTF structure. (b) LSV of 2.8 wt% Pt-CTF (red), 20 wt% Pt/C (blue), and 3.0 wt% Pt/C (green) for the HOR (0.1 M $HClO_4$). (c) Performances of MEAs: 2.8 wt% Pt-CTF (0.02 mg-Pt cm^{-2}) or 20 wt% Pt/C (0.10 mg-Pt cm^{-2}) as the anode catalyst. (d) H_2O_2 oxidation currents recorded at 1.2 V vs. RHE on a ring electrode during the ORR on disc electrodes with the catalysts 20 wt% Pt/C and 2.8 wt% Pt-CTF. (e) Corresponding polarization curves for the ORR. (Reproduced with permission [31]. Copyright 2016, Wiley-VCH.)

7.4 Ion Conducting Membranes Made from COFs

Extensive work has been done on the development of ion-conducting membranes that could be used for FCs. Most reports deal with proton conducting membranes; however, there are reports on OH⁻ conducting membranes as well. Particularly in 2D COFs, vertical columnar arrays formed by the stacking of aromatic building units as well as conjugated polymeric layers can provide continuous 1D channels which are very suitable for ion diffusion [4]. Some recent results are summarized in Table 7.2 and discussed in the following sections.

TABLE 7.2

Some Recent Results on Proton and OH⁻ Ion Conduction in COFs

	Compound/s	Conductivity (S cm^{-1})	Condition	E_a (eV)	IEC (mmol g^{-1})	Tensile strength (MPa)	Ref.
Extrinsic proton conducting COFs	H$_3$PO$_4$@NKCOF-10	0.0904	80°C 90% RH	0.06			[34]
	SPEEK/H$_3$PO$_4$@CTFp15	0.313	80°C	0.2507	1.47	47	[35]
	HPW@TAPT-DHTA	0.53	80°C 100% RH				[36]
	COF(TfBD)@Nafion/GO	0.30	80°C 95% RH	0.27			[37]
	H$_3$PO$_4$@TPB-DMeTP-COF	0.191	160°C (anhydrous)	0.34			[38]
	COF-F6-H	0.042	140°C (anhydrous)	0.09–0.54			[39]
Intrinsic proton conducting COFs	SPEEK/TpPa–SO$_3$H–5%	0.346	80°C 95% RH	0.2868	2.34	74.5	[41]
	TpPa-SO$_3$H (self-standing)	0.54	80°C, water	0.19			[42]
	IPC-COF (NUS-9)	0.38	80°C (98% RH)		3.2	91.2	[43]
NAFION ™		0.1–0.01		0.38		43	[33, 37]
OH⁻ conducting COFs	HPSf@QCOF 20%	0.0143	30°C		0.38		[45]
	TPB-BPTA-COF	0.0152	80°C, water	0.20			[46]
	QA@COF-LZU1/PPO-5	0.168	80°C		2.91	18.8	[47]
	COF-QA-TFB(P)	0.200	80°C, 100% RH			*ca* 50	[48]
	COF-QA-2	0.212	80°C	0.12	2.24	49–53	[49]
Heterocyclic benzyl quaternary ammonium groups radiation-grafted to ETFE		0.159	80°C, 95 % RH			29	[50]

IEC = ion exchange capacity, E_a = activation energy, RH = relative humidity, ETFE = poly(ethylene-co-tetrafluoroethylene)

7.4.1 Proton Conductive or Exchange Membranes (PEMs)

Nafion™ is the commercial name of the polymer developed by DuPont, which is the standard ionomer and PEM material in acid FCs. It is generated from copolymerization of a perfluorinated vinyl ether (with sulfonyl acid fluoride) co-monomer with tetrafluoroethylene. The hydrophilic and strongly acidic sulfonic groups are solvated in the presence of water, producing interconnected channels (phase separation), which allow proton conduction by hoping mechanism. The hydrophobic polytetrafluoroethylene backbone does not conduct anions nor electrons, but provides structural support. The high proton conductivity (10^{-1}–10^{-2} S cm^{-1}), the mechanical properties (what allows films extrusion), and the recyclability without loss of performance, are reasons for its success in PEM-FCs. There are three major disadvantages concerning Nafion employment: the high relative humidity level (RH) required for operation, the low operating temperature, and the high cost. Dehydration of Nafion above 100°C and its consequent conductivity loss impedes using it at a higher temperature. The high temperature would increase reaction kinetics and reduce poisoning on the electrodes. This promotes developing and testing new materials for its replacement [33].

There are two main strategies to convert COFs into highly proton conductive materials: non-volatile acid doping or adsorption within the pores (H_3PO_4, phytic acid, phosphotungstic acid, etc.), and sulfonate, phosphonate, imidazole, and phenolic hydroxyl group functionalization. Examples of both techniques are given below. Membrane conductivity measurements are usually made through electrochemical impedance spectroscopy, at different temperatures and relative humidity values (RH). The activation energy obtained from the temperature dependence of the conductivity allows distinguishing between the two mechanisms currently employed. These are the Grotthuss mechanism, with activation energy lower than 0.4 eV, and the vehicle mechanism for which the activation energy is higher than 0.4 eV. In the Grotthuss mechanism of proton hopping, the proton is transferred through the hydrogen-bonded network via the formation and subsequent breaking of O-H bonds. Adsorbed water molecules together with the available proton sources play a key role in forming a continuous H-bonded network within the pores and channels. Alternatively, in the vehicle or diffusion mechanism, protons are directly bound to carrier molecules (H_2O, NH_3), which diffuse through the pores, channels, or pathways [33]. COFs are usually crystalline materials, with high proton conductivity in some cases, however, their high rigidity usually makes them not suitable to be used as a PEM or AEM. To overcome these limitations, obtaining composite materials or blending COFs with polymers, are some of the strategies [33].

7.4.1.1 Extrinsic Proton Conduction or Doped COFs

An olefin-linked COF (NKCOF-10) was synthesized by Zhang *et al.*, through the aldol reaction between 2,5-dimethylpyrazina and 1,3,5-triformylbenzene (Figure 7.5). The COF exhibited a layered honeycomb-like crystalline framework with high surface area and stability toward harsh conditions, particularly in strong acid/base, what made it suitable for FC applications. The proton conductivity depended on water adsorption. The 1.8 nm pores could be loaded with H_3PO_4 (H_3PO_4@NKCOF-10), which bound to the pyrazine groups of the COF and increased the proton conductivity to 6.97×10^{-2} S cm^{-1} at 298 K and 9.04×10^{-2} S cm^{-1} at 353 K (RH 90%). The adsorbed water molecules are proposed to allow a hydrogen-bond mediated proton transport mechanism. The H_3PO_4@NKCOF-10 was used to construct membranes for PEM-FCs. The electrochemical parameters of the prototype were: 0.87 V (open circuit voltage), a maximum power density of 135 mW cm^{-2}, a maximum current density of 676 mA cm^{-2} at 323 K [34].

Ding *et al.* impregnated H_3PO_4 into a CTF, which was further blended with sulfonated poly(ether-ether ketone) to form a composite membrane (SPEEK/H_3PO_4@CTFp). The abundant basic N atoms of triazine provided binding sites for H_3PO_4. The proton conductivity of the composite membrane was 0.313 S cm^{-1} (80°C, filler content 15%). The activation energy was 24.19 kJ/mol (0.2507 eV). Single-cell test of this membrane showed the following results: 325.8 mW cm^{-2} maximum power density, 0.762 V open-circuit voltage and 1899.8 mA cm^{-2}, maximum current density (65°C, 30% RH) [35].

A disorder-to-order transformation from amorphous polymeric membrane to crystalline COF membrane was achieved through monomer exchange by Jiang *et al.* The polymeric membrane was synthesized through the reaction of 1,3,5-tris-(4-amidophenyl)triazine and (TAPT) and phthalaldehyde (PA). Reversible imine bonds allowed the new monomers to replace the pristine ones within the amorphous membrane, driving the transformation from disordered network to ordered framework. The amorphous TAPT-PA membrane supported on ITO glass reacted with 2,5-dihydroxyterephthaldeyde (DHTA) under solvothermal conditions and was subsequently doped with phosphotungstic acid (HPW). This

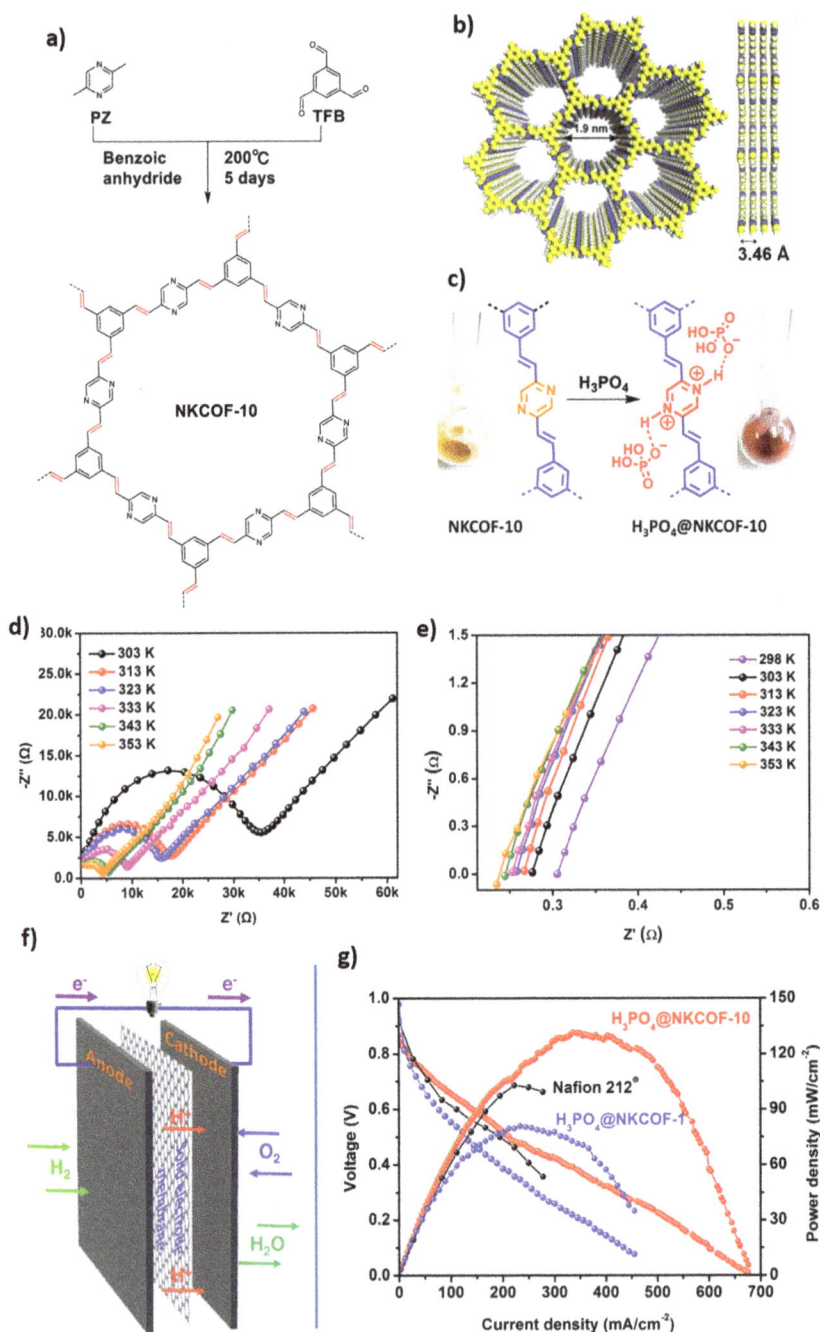

FIGURE 7.5

Extrinsic proton conduction in an H_3PO_4-doped COF (NKCOF-10). (a) Scheme of the synthesis of NKCOF-10. (b) Top and side views of the eclipsed AA-stacking model. (c) Scheme of the protonation of the pyrazine groups with H_3PO_4. (d,e) Nyquist plots of NKCOF-10 (d) and H_3PO_4@NKCOF-10 (e). (f) Scheme of the PEM-FC using H_3PO_4@NKCOF-10, H_3PO_4@NKCOF-1, and Nafion 212 as solid electrolyte membranes of MEAs. (g) Polarization curves and power density curves. (Adapted with permission [34]. Copyright (2021) The Authors, some rights reserved; exclusive licensee Springer Nature. Licensed under a Creative Commons Attribution 4.0 International License. http://creativecommons.org/licenses/by/4.0/.)

last reagent also dissolved the ITO and released the membrane from the substrate. The proton conductivity of the HPW@TAPT-DHTA membrane reached 0.53 S cm^{-1} (80 °C, 100% RH). This membrane was blended with Nafion and employed in a PEM-FC. The open-circuit voltage was 0.55 V and the peak power density was *ca.* 120 mW cm^{-2} [36].

A similar strategy was employed by Guo *et al.* to transform polyazomethine (PAM)-based amorphous fibers into crystalline COF fibers keeping pristine fibrous morphology and size but gaining periodic and oriented micropore channels. The transformation took place during a reaction of the PAM fibers with dichlorobenzene under solvothermal conditions. These imine-linked COF nanofibers provided oriented pore channels suitable for proton carrier immobilization (i.e., Nafion). They also worked as a binder for graphene oxide nanosheets producing a membrane with water-proofing behavior and ductility (COF(TfBD)@Nafion/GO). Proton conductivity of 0.30 S cm^{-1} (80°C, 95% RH) was reached, which is higher than a Nafion/graphene oxide membrane. The activation energy of the fibrous COF(TfBD)@Nafion/GO membrane was 0.27 eV (40°C to 80°C, 40% RH), lower than that of Nafion (0.38 eV), evidencing that the COF@Nafion fibers present long-range proton conduction channels [37].

Anhydrous proton "superflow" was achieved in H$_3$PO$_4$@TPB-DMeTP-COF by Jiang *et al.* Topology guided synthesis from 1,3,5-tri(4-aminophenyl)benzene (TPB) and 2,5-dimethyl-terephthalaldehyde (DMeTP) was employed to construct dense and aligned 1D-dimension nanochannels (3.36 nm). DMeTP induced hyperconjugation and inductive effects to stabilize the pore structure. The nitrogen sites on pore walls confined and stabilized the H$_3$PO$_4$ network in the channels via hydrogen-bonding interactions. The proton conductivity of H$_3$PO$_4$@TPB-DMeTP-COF was 1.91×10^{-1} S cm^{-1} 160°C, under anhydrous conditions, which was higher than pure H$_3$PO$_4$ (*ca* 1×10^{-1} S cm^{-1}). Three factors are proposed to determine good anhydrous proton conduction: (i) the stability of the framework; (ii) the stability of proton network within the pores; and (iii) the porosity of material that determines the density of proton networks in the material [38].

To obtain anhydrous proton conducting COFs to be used in high-temperature PEM-FCs (> 100°C), a series of perfluoroalkyl functionalized hydrazone-linked 2D COFs were synthesized through a bottom-up self-assembly strategy by Horike *et al.* The COFs were obtained by solvothermal reaction of 1,3,5-triformylbenzene with three hydrazide monomers with fluorinated side chains of different length. The enhanced hydrophobicity conferred high stability toward strong acids. The superhydrophobic 1D channels had plenty of NH groups which offered sites for H$_3$PO$_4$ binding. The highest conductivity, 4.2×10^{-2} S cm^{-1}, was obtained employing 62% wt H$_3$PO$_4$, 140°C with 6 carbon atoms in the fluoroalkyl chain (COF-F6-H) [39].

Other organic polymers have been synthesized and doped with H$_3$PO$_4$, like poly[2,5-benzimidazole] (ABPBI), as an alternative to Nafion usage. H$_3$PO$_4$ doped ABPBI membranes have reached proton conductivities above 200 mS cm^{-1} at 180°C, even without humidification (Table 7.2) [1, 40]. In this view, COFs are very promising and competitive PEM materials for PEM-FCs.

7.4.1.2 *Intrinsic Proton Conduction or Functionalized COFs*

Sulfonated COF nanosheets (TpPa-SO$_3$H) were synthesized via interfacial polymerization from 1,3,5-triformylphloroglucinol (Tp) and 2,5-diaminobenzenesulfonic acid (Pa-SO$_3$H) by Jiang *et al.* This COF was incorporated into a sulfonated poly (ether-ether ketone) (SPEEK) matrix. The composite thus obtained was cast to form PEMs. The SPEEK/TpPa–SO$_3$H–5% membrane reached a proton conductivity of 0.346 S cm^{-1} at 80°C. A single fuel cell test with the SPEEK/TpPa–SO$_3$H–5% membrane exhibited higher open voltage (1.01 V) and power density (86.54 mW cm^{-2}) than the pristine SPEEK membrane, indicating an improved fuel cell performance after incorporation of TpPa-SO$_3$H nanosheets [41].

An ultrathin (10–100 nm) free-standing TpPa-SO$_3$H COF membrane was obtained by Zhu *et al.*, on an amine functionalized silicon surface, which initiated the aldimine condensation polymerization of Tp and Pa-SO$_3$H. The surface anchored COF was subsequently protected by a thin layer of poly(methylmethacrylate). The membrane was removed from the silicon substrate using 2% HF solution and the poly(methylmethacrylate) was dissolved by acetone, without any decomposition of the COF structure. The proton conductivity of the membrane in pure water resulted in 0.54 S cm^{-1} at 80°C [42].

Intrinsic proton-conducting COF (IPC-COF) nanosheets (NUS-9) were synthesized in aqueous solutions via diffusion and solvent co-mediated modulation of (1,3,5-Triformylphloroglucinol (TFP) in octanoic acid and 2,5-diaminobenzenesulfonic acid (DABA) in water) (see Figure 7.6) by Jiang *et al.* This technique permitted controlled nucleation and in-plane-dominated growth of COF nanosheets (slow monomer diffusion and increased interaction between IPC-COF and solvent lead to slow nucleation and in-plane growth). Particle aggregation, typically observed in the solvothermal technique (fast nucleation and random growth), was avoided. These nanosheets facilitated membrane fabrication through vacuum-assisted self-assembly. Freestanding IPC-COF membranes with crystalline, rigid ion nanochannels exhibited a weakly humidity-dependent conductivity over a wide range (30–98% RH), and proton conductivity of 0.38 S cm^{-1} at 80°C (98% RH), due to ordered 1D proton transport pathways and high concentration of proton carriers. In contrast, the proton conductivity of Nafion 212 declines from 0.04 S cm^{-1} (98% RH) to *ca* 10^{-4} S cm^{-1} (30% RH) at 40°C. Usually, high proton conductivity is reached through a high concentration of proton carriers. However, this usually leads to high water uptake and a high swelling ratio, which reduces the mechanical stability of the membrane. In the present case, the IPC-COF membrane exhibited a negligible swelling ratio (0–98% RH). Finally, a prominent fuel cell performance of 0.93 Wcm^{-2} at 35% RH and 80°C was measured, a consequence of its high water retention capacity [43].

7.4.2 Hydroxide Ion Conduction and Anion Exchange Membranes (AEMs)

Although porous frameworks (MOFs and COFs) with well-organized channels, which act as ion conduction pathways, can achieve excellent conductivity and high structural stability, OH$^-$ conduction on porous frameworks has been reported with much less extent than proton conduction in COFs [44].

A quaternized COF (QCOF) was synthesized by a mechanochemical method from 1,3,5-triformylphloroglucinol, and ethidium bromide by Zhang *et al.* The QCOF was incorporated into a hydroxyl functionalized polysulfone matrix (HPSf) to make a mixed matrix AEM (or MMAEM). The resultant HPSf@QCOF contained abundant hydrogen-bond networks and micropores, which synergistically enhanced hydroxide conduction. At a QCOF content of 20%, it showed a very low ion exchange capacity (IEC) of 0.38 mmol g^{-1} but a good conductivity of 14.3 mS cm^{-1} and a very low swelling ratio of 2.5% at 30°C. The membranes showed good alkali stability (64% anion conductivity retention after 192 h at 60°C, 1 M KOH) [45].

Jiang *et al.* anchored imidazolium groups on the surface of the 1D-channels of TPB-BPTA-COF (TAPB = 1,3,5-tris(4-aminophenyl)benzene, BPTA = 2,5-bis (2-propynyloxy) terephthalaldehyde). The structure hosted hexagonal channels with cationic chains

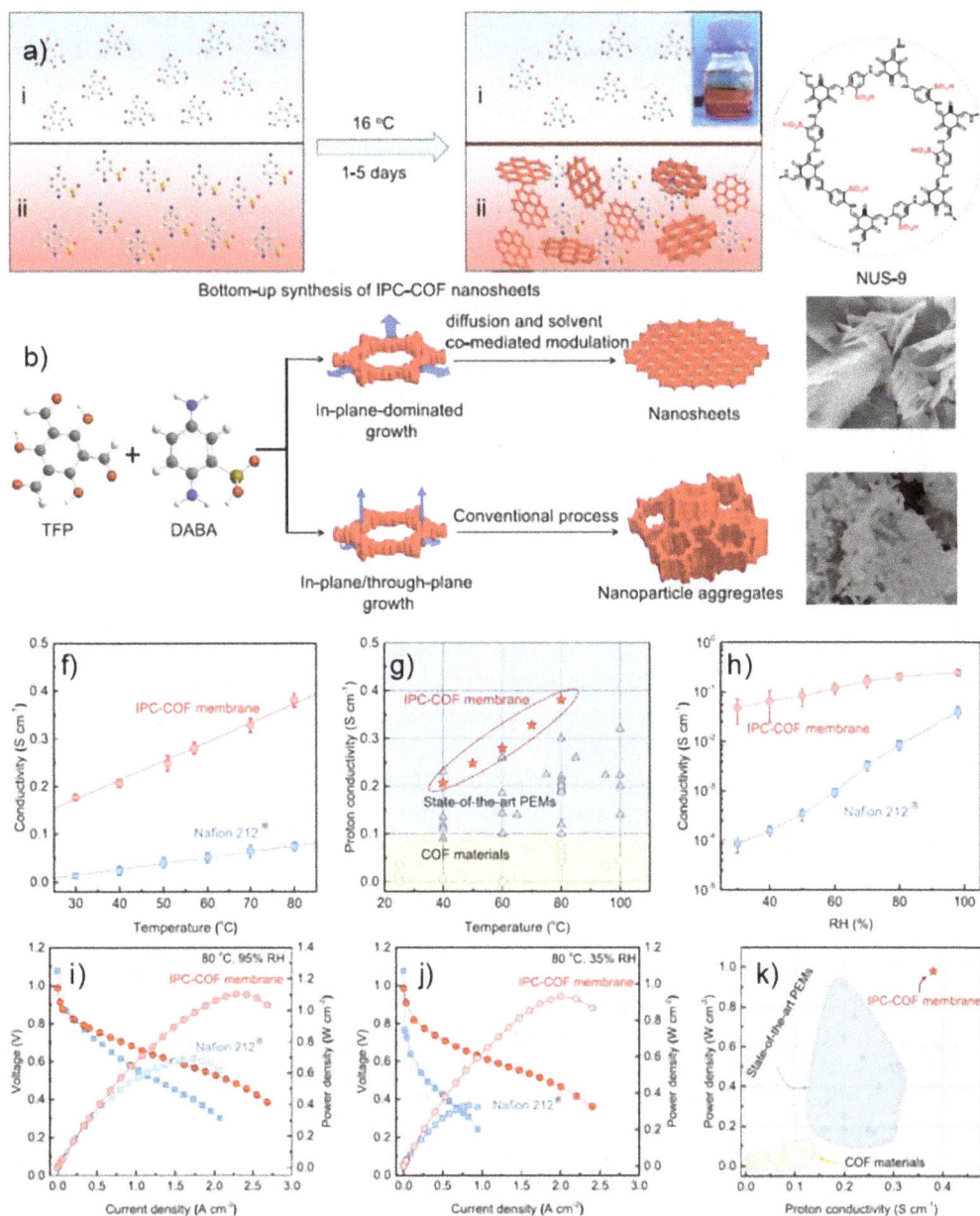

FIGURE 7.6

Intrinsic proton conducting (sulfonic acid functionalized) COF (NUS-9) nanosheets (IPC-COF). (a) Scheme of the bottom-up synthesis, right inset: chemical structure of NUS-9. "i" correspond to TFP organic solution; "ii" corresponds to DABA aqueous solution. (b) Scheme of the controlled IPC-COF growth via both: diffusion and solvent co-mediated modulation process. (f) Temperature dependence of the proton conductivity of IPC-COF membrane and Nafion 212 at 98% RH. (g) Comparison of proton conductivity data for IPC-COF membrane (solid stars), COF materials (open diamonds), and state-of-the-art PEMs (open triangles). (h) Proton conductivity of IPC-COF membrane and Nafion 212 versus RH at 40°C. (i, j) j-V polarization curves and power densities of MEAs made from IPC-COF membrane and Nafion 212 measured at varied RH and temperatures. (k) Single fuel cell power density at low RH (<35%) versus proton conductivity for IPC-COF membrane (solid stars), COF materials (open squares), and state-of-the-art PEMs (open circles). (Reproduced with permission [43]. Copyright 2020 Wiley-VCH.)

stacked along the z-direction, which constituted a continuous anionic phase. This structure promoted OH⁻ transport reaching a conductivity of 1.53 x 10⁻² S cm⁻¹at 80°C in ultrapure water. This value could be further improved in water vapor reaching a conductivity of 5.34 x 10⁻² S cm⁻¹ at 80°C. Impedance spectroscopy measurements at different temperatures and studies on deuterated samples evidenced a proton-exchange hopping mechanism for hydroxide anions transport [46].

Another hybrid anion exchange membrane was prepared by Tang *et al.*, incorporating quaternary ammonium (QA)-functionalized COF-LZU1 into brominated poly(2,6-dimethyl-1,4-phenylene oxide) (BPPO). (COF-LZU1 was prepared from 1,3,5-triformylbenzene and 1,4-diaminobenzene). The hydroxide ion conductivity of QA@COF-LZU1/PPO-5 hybrid membrane was 39 mS cm⁻¹ at 30°C and 168 mS cm⁻¹ at 80°C. Quaternary ammonium groups uniformly distributed in the highly ordered 1D channels of COF-LZU1 produced a well-developed hydrogen-bonding network and they were responsible for the high OH⁻ conductivity. Confinement of quaternary ammonium groups within the high alkaline stable COF-LZU1 resulted in the high stability of the hybrid membrane (5 days in 2 M NaOH solution at 60°C) [47].

Tight COF membranes were synthesized in an organic-aqueous reaction system by assembling functional hydrazides and different aldehyde precursors, producing six different quaternary ammonium-functionalized COFs. The membranes were obtained in a phase-transfer process. Higher hydrophilicity of the aldehyde precursors moved the reaction zone from the organic-aqueous interface region to the aqueous phase of the reaction system. In this way, higher anion conductivity was observed compared with the loose membranes obtained through an interfacial polymerization process. The highest hydroxide conductivity was 200 mS cm⁻¹ [48].

Inspired by the efficient anion transport found in living organisms and trying to obtain dense and ordered alignment of quaternary ammonium-functionalized side chains along the 1D channels within the COFs, a phase-transfer polymerization process (instead of interfacial polymerization) was developed by Jiang *et al.*, to obtain self-standing COFs membranes instead of powder samples (Figure 7.7). Four different hydrazide building units with ethyl, butyl, hexyl, and diethyl ether were prepared and dissolved in a mixture of water and acetic acid, and reacted with 1,3,5-triformylbenzene in mesitylene through phase-transfer polymerization. This technique decoupled the formation of COF layers and the membrane assembly, producing self-standing, densely packed membranes with very low roughness (1.28 nm) and high mechanical strength (49–53 MPa). This *de novo* design allowed the incorporation of more quaternary ammonium groups than a post-functionalization technique. Relatively high water uptake and relatively low area swelling were observed for these membranes (ca 80% and 18%, respectively at 30–80°C for COF-QA-2), as a consequence of the rigid carbon backbone and the porous structure with adequate space to host water. It was found that shorter, more hydrophilic side chains improved the anion conduction. In this way, the membrane with two carbon side-chain (COF-QA-2) reached the highest conductivity of the series, i.e., 212 mS cm⁻¹ (80°C), due to simultaneously increased ion mobility and ion concentration within the 1D channels. Finally, the COF-QA-2 membrane was employed to construct a fuel cell (Pt 0.5 mg cm⁻² as anode and cathode, 60°C, 100% RH), reaching an open circuit voltage of 0.989 V and a peak power density of 163.7 mW cm⁻² [49].

In comparison, other materials like organic polymers have reached hydroxide ion conductivities around 0.16 S cm⁻¹ (Table 7.2), which makes COFs very competitive materials for AEMs and AAEM-FCs in general [50].

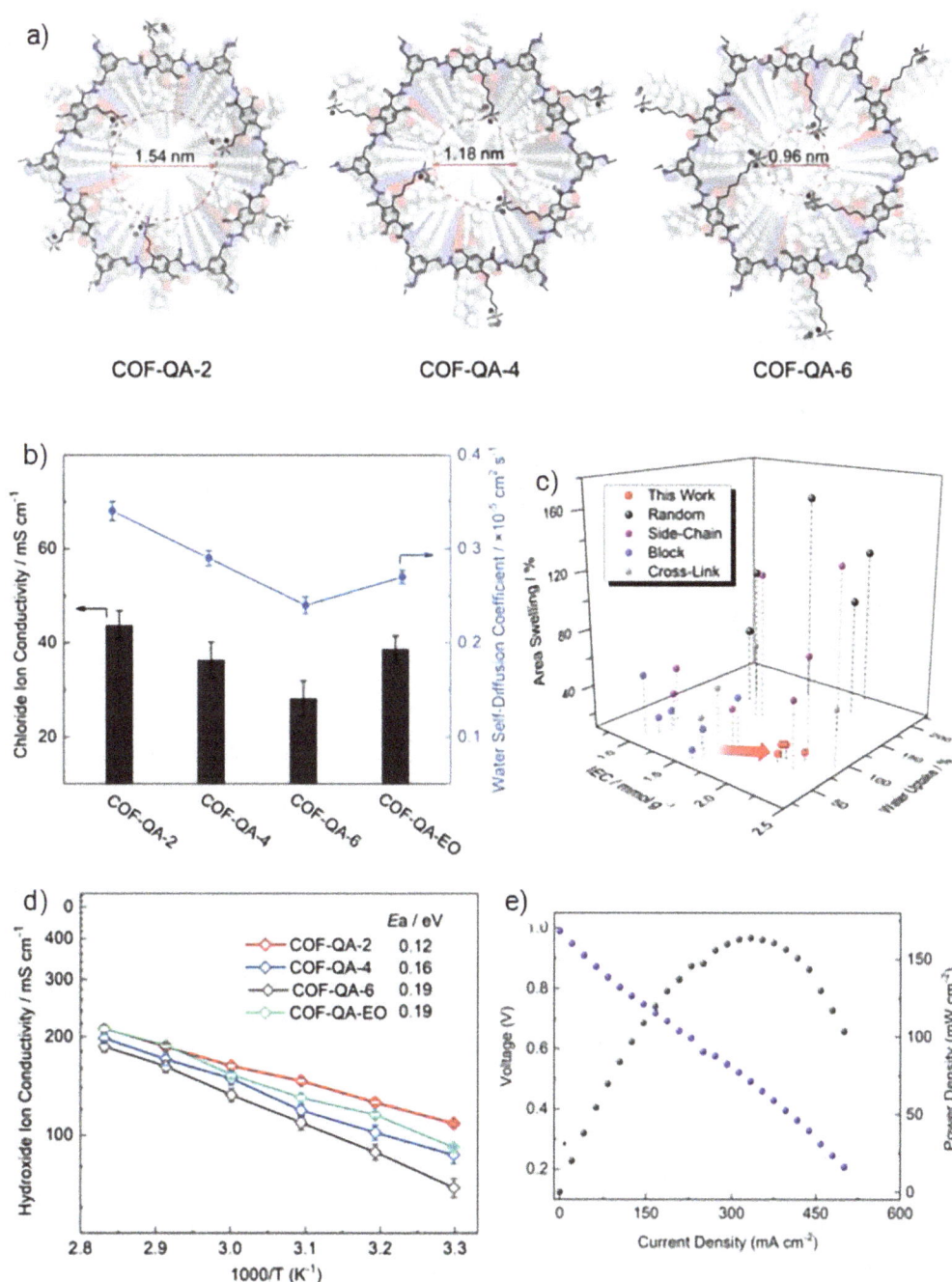

FIGURE 7.7

Synthesis and hydroxide ion conduction of a quaternary ammonium-functionalized COF. (a) Scheme of the pore channels and theoretical effective pore sizes of the COF-QAs with alkyl spacers, based on structural models. (b) Chloride ion conductivity of the COF-QAs membranes. Water self-diffusion coefficients of the membranes in chloride form. (c) Comparison of area swelling, water uptake, and IEC of the COF-QAs membranes and other typical AEMs in the literature. (d) Temperature-dependence of the hydroxide ion conductivity of the COF-QAs membranes. (e) Polarization and power density curves of the COF-QA-2 membrane in H_2/O_2 single-cell AEMFC test. (Reproduced with permission [49]. Copyright 2020 Wiley-VCH.)

7.5 Summary, Conclusions, and Perspectives

This chapter summarizes some of the most recent or relevant results concerning the application of COFs to fuel cells. In this regard, two topics have experienced significant advance: the development of PEMs and ORR catalysts, particularly non-PGM catalysts. However, massive deployment of fuel cells would require the reduction or, ideally, substitution of Pt and PGMs, which are scarce and expensive. In this way, two strategies could be followed. One would be the development of non-PGM HOR catalysts for PEM-FCs with high activity and, especially, durability in an acidic medium. This has been proved to be a very challenging task. In turn, the second strategy would require also the development of non-PGM HOR catalysts in alkaline medium, and also AEMs. Both of them are currently under development as has been shown in this chapter. However, non-PGM HOR catalysts with higher activity and durability are still required, and AEMs with higher hydroxide ion conductivity and durability are also needed to obtain AEM-FCs with high power density, suitable for automotive and other massive uses. COFs are very suitable materials to accomplish all these challenges, and some of them are presently in progress. There is no doubt that COF will contribute to a significant advance of fuel cell technologies soon.

Acknowledgments

This work was supported by the Agencia Nacional de Promoción Científica y Tecnológica (project PICT-2016-3017), the National Atomic Energy Commission of Argentina (CNEA), and the Research Council of Argentina (CONICET). FR is a researcher of CONICET. The author wishes to thank Natalia Guazzone (CIES-CNEA) for providing part of the literature and also to Prof. Horacio R. Corti for valuable discussions.

References

1. H. R. Corti, E. R. Gonzalez (Eds). *Direct Alcohol Fuel Cells: Materials, Performance, Durability and Applications*; Springer: Netherlands, 2014.
2. Li, W.; Mukerjee, S.; Ren, B.; Cao, R.; Fischer, R. A. Open Framework Material Based Thin Films: Electrochemical Catalysis and State-of-the-Art Technologies. *Adv. Energy Mater.*, 2021, *n/a*, 2003499.
3. Zhao, X.; Pachfule, P.; Thomas, A. Covalent Organic Frameworks (COFs) for Electrochemical Applications. *Chem. Soc. Rev.*, 2021, *50*(12), 6871–6913.
4. Li, J.; Jing, X.; Li, Q.; Li, S.; Gao, X.; Feng, X.; Wang, B. Bulk COFs and COF Nanosheets for Electrochemical Energy Storage and Conversion. *Chem. Soc. Rev.*, 2020, *49*(11), 3565–3604.
5. Cui, X.; Lei, S.; Wang, A. C.; Gao, L.; Zhang, Q.; Yang, Y.; Lin, Z. Emerging Covalent Organic Frameworks Tailored Materials for Electrocatalysis. *Nano Energy*, 2020, *70*, 104525.
6. Royuela, S.; Martínez-Periñán, E.; Arrieta, M. P.; Martínez, J. I.; Ramos, M. M.; Zamora, F.; Lorenzo, E.; Segura, J. L. Oxygen Reduction Using a Metal-Free Naphthalene Diimide-Based Covalent Organic Framework Electrocatalyst. *Chem. Commun.*, 2020, *56*(8), 1267–1270.

7. Tavakoli, E.; Kakekhani, A.; Kaviani, S.; Tan, P.; Ghaleni, M. M.; Zaeem, M. A.; Rappe, A. M.; Nejati, S. In Situ Bottom-up Synthesis of Porphyrin-Based Covalent Organic Frameworks. *J. Am. Chem. Soc.*, 2019, *141*(50), 19560–19564.

8. Wu, S.; Li, M.; Phan, H.; Wang, D.; Herng, T. S.; Ding, J.; Lu, Z.; Wu, J. Toward Two-Dimensional π-Conjugated Covalent Organic Radical Frameworks. *Angew. Chemie Int. Ed.*, 2018, *57*(27), 8007–8011.

9. Li, D.; Li, C.; Zhang, L.; Li, H.; Zhu, L.; Yang, D.; Fang, Q.; Qiu, S.; Yao, X. Metal-Free Thiophene-Sulfur Covalent Organic Frameworks: Precise and Controllable Synthesis of Catalytic Active Sites for Oxygen Reduction. *J. Am. Chem. Soc.*, 2020, *142*(18), 8104–8108.

10. Zhai, L.; Yang, S.; Yang, X.; Ye, W.; Wang, J.; Chen, W.; Guo, Y.; Mi, L.; Wu, Z.; Soutis, C.; Xu, Q.; Jiang, Z. Conjugated Covalent Organic Frameworks as Platinum Nanoparticle Supports for Catalyzing the Oxygen Reduction Reaction. *Chem. Mater.*, 2020, *32*(22), 9747–9752.

11. Iwase, K.; Yoshioka, T.; Nakanishi, S.; Hashimoto, K.; Kamiya, K. Copper-Modified Covalent Triazine Frameworks as Non-Noble-Metal Electrocatalysts for Oxygen Reduction. *Angew. Chemie Int. Ed.*, 2015, *54*(38), 11068–11072.

12. Peng, P.; Shi, L.; Huo, F.; Mi, C.; Wu, X.; Zhang, S.; Xiang, Z. A Pyrolysis-Free Path toward Superiorly Catalytic Nitrogen-Coordinated Single Atom. *Sci. Adv.*, 2019, *5*(8), eaaw2322.

13. Peng, P.; Shi, L.; Huo, F.; Zhang, S.; Mi, C.; Cheng, Y.; Xiang, Z. In Situ Charge Exfoliated Soluble Covalent Organic Framework Directly Used for Zn–Air Flow Battery. *ACS Nano*, 2019, *13*(1), 878–884.

14. Liu, W.; Wang, C.; Zhang, L.; Pan, H.; Liu, W.; Chen, J.; Yang, D.; Xiang, Y.; Wang, K.; Jiang, J.; Yao, X. Exfoliation of Amorphous Phthalocyanine Conjugated Polymers into Ultrathin Nanosheets for Highly Efficient Oxygen Reduction. *J. Mater. Chem. A*, 2019, *7*(7), 3112–3119.

15. Zhao, H.; Sheng, L.; Wang, L.; Xu, H.; He, X. The Opportunity of Metal Organic Frameworks and Covalent Organic Frameworks in Lithium (Ion) Batteries and Fuel Cells. *Energy Storage Mater.*, 2020, *33*, 360–381.

16. Yang, C.; Tao, S.; Huang, N.; Zhang, X.; Duan, J.; Makiura, R.; Maenosono, S. Heteroatom-Doped Carbon Electrocatalysts Derived from Nanoporous Two-Dimensional Covalent Organic Frameworks for Oxygen Reduction and Hydrogen Evolution. *ACS Appl. Nano Mater.*, 2020, *3*(6), 5481–5488.

17. Yang, C.; Maenosono, S.; Duan, J.; Zhang, X. COF-Derived N,P Co-Doped Carbon as a Metal-Free Catalyst for Highly Efficient Oxygen Reduction Reaction. *ChemNanoMat*, 2019, *5*(7), 957–963.

18. Zhou, Z.; Zhang, X.; Xing, L.; Liu, J.; Kong, A.; Shan, Y. Copper-Assisted Thermal Conversion of Microporous Covalent Melamine-Boroxine Frameworks to Hollow B, N-Codoped Carbon Capsules as Bifunctional Metal-Free Electrode Materials. *Electrochim. Acta*, 2019, *298*, 210–218.

19. Xu, Q.; Tang, Y.; Zhang, X.; Oshima, Y.; Chen, Q.; Jiang, D. Template Conversion of Covalent Organic Frameworks into 2D Conducting Nanocarbons for Catalyzing Oxygen Reduction Reaction. *Adv. Mater.*, 2018, *30*(15), 1706330.

20. You, C.; Jiang, X.; Wang, X.; Hua, Y.; Wang, C.; Lin, Q.; Liao, S. Nitrogen, Sulfur Co-Doped Carbon Derived from Naphthalene-Based Covalent Organic Framework as an Efficient Catalyst for Oxygen Reduction. *ACS Appl. Energy Mater.*, 2018, *1*(1), 161–166.

21. Chen, H.; Li, Q.-H.; Yan, W.; Gu, Z.-G.; Zhang, J. Templated Synthesis of Cobalt Subnanoclusters Dispersed N/C Nanocages from COFs for Highly-Efficient Oxygen Reduction Reaction. *Chem. Eng. J.*, 2020, *401*, 126149.

22. Zhang, S.; Xia, W.; Yang, Q.; Valentino Kaneti, Y.; Xu, X.; Alshehri, S. M.; Ahamad, T.; Hossain, M. S. A.; Na, J.; Tang, J.; Yamauchi, Y. Core-Shell Motif Construction: Highly Graphitic Nitrogen-Doped Porous Carbon Electrocatalysts Using MOF-Derived Carbon@COF Heterostructures as Sacrificial Templates. *Chem. Eng. J.*, 2020, *396*, 125154.

23. Xu, Q.; Qian, J.; Luo, D.; Liu, G.; Guo, Y.; Zeng, G. Ni/Fe Clusters and Nanoparticles Confined by Covalent Organic Framework Derived Carbon as Highly Active Catalysts toward Oxygen Reduction Reaction and Oxygen Evolution Reaction. *Adv. Sustain. Syst.*, 2020, *4*(9), 2000115.

24. Zhang, X.; Liu, L.; Liu, J.; Cheng, T.; Kong, A.; Qiao, Y.; Shan, Y. Ultrafine Cu6Sn5 Nanoalloys Supported on Nitrogen and Sulfur-Doped Carbons as Robust Electrode Materials for Oxygen Reduction and Li-Ion Battery. *J. Alloys Compd.*, 2020, *824*, 153958.

25. Xu, Q.; Zhang, H.; Guo, Y.; Qian, J.; Yang, S.; Luo, D.; Gao, P.; Wu, D.; Li, X.; Jiang, Z.; Sun, Y. Standing Carbon-Supported Trace Levels of Metal Derived from Covalent Organic Framework for Electrocatalysis. *Small*, 2019, *15*(50), 1905363.

26. Zhao, X.; Pachfule, P.; Li, S.; Langenhahn, T.; Ye, M.; Tian, G.; Schmidt, J.; Thomas, A. Silica-Templated Covalent Organic Framework-Derived Fe–N-Doped Mesoporous Carbon as Oxygen Reduction Electrocatalyst. *Chem. Mater.*, 2019, *31*(9), 3274–3280.

27. Kong, F.; Fan, X.; Kong, A.; Zhou, Z.; Zhang, X.; Shan, Y. Covalent Phenanthroline Framework Derived FeS@Fe3C Composite Nanoparticles Embedding in N-S-Codoped Carbons as Highly Efficient Trifunctional Electrocatalysts. *Adv. Funct. Mater.*, 2018, *28*(51), 1803973.

28. Zion, N.; Cullen, D. A.; Zelenay, P.; Elbaz, L. Heat-Treated Aerogel as a Catalyst for the Oxygen Reduction Reaction. *Angew. Chemie Int. Ed.*, 2020, *59*(6), 2483–2489.

29. Zhao, G.; Chen, J.; Sun, W.; Pan, H. Non-Platinum Group Metal Electrocatalysts toward Efficient Hydrogen Oxidation Reaction. *Adv. Funct. Mater.*, 2021, *31*(20), 2010633.

30. Zhao, R.; Yue, X.; Li, Q.; Fu, G.; Lee, J.-M.; Huang, S. Recent Advances in Electrocatalysts for Alkaline Hydrogen Oxidation Reaction. *Small*, 2021, *n/a*, 2100391.

31. Kamai, R.; Kamiya, K.; Hashimoto, K.; Nakanishi, S. Oxygen-Tolerant Electrodes with Platinum-Loaded Covalent Triazine Frameworks for the Hydrogen Oxidation Reaction. *Angew. Chemie Int. Ed.*, 2016, *55*(42), 13184–13188.

32. An, L.; Zhao, X.; Zhao, T.; Wang, D. Atomic-Level Insight into Reasonable Design of Metal-Based Catalysts for Hydrogen Oxidation in Alkaline Electrolytes. *Energy Environ. Sci.*, 2021, *14*(5), 2620–2638.

33. Pal, S. C.; Das, M. C. Superprotonic Conductivity of MOFs and Other Crystalline Platforms Beyond 10–1 S cm–1. *Adv. Funct. Mater.*, 2021, *31*(31), 2101584.

34. Wang, Z.; Yang, Y.; Zhao, Z.; Zhang, P.; Zhang, Y.; Liu, J.; Ma, S.; Cheng, P.; Chen, Y.; Zhang, Z. Green Synthesis of Olefin-Linked Covalent Organic Frameworks for Hydrogen Fuel Cell Applications. *Nat. Commun.*, 2021, *12*(1), 1982.

35. Sun, X.; Song, J.-H.; Ren, H.; Liu, X.; Qu, X.; Feng, Y.; Jiang, Z.-Q.; Ding, H. Phosphoric Acid-Loaded Covalent Triazine Framework for Enhanced the Proton Conductivity of the Proton Exchange Membrane. *Electrochim. Acta*, 2020, *331*, 135235.

36. Fan, C.; Wu, H.; Guan, J.; You, X.; Yang, C.; Wang, X.; Cao, L.; Shi, B.; Peng, Q.; Kong, Y.; Wu, Y.; Khan, N. A.; Jiang, Z. Scalable Fabrication of Crystalline COF Membranes from Amorphous Polymeric Membranes. *Angew. Chemie Int. Ed.*, 2021, *60*(33), 18051–18058.

37. Kong, W.; Jia, W.; Wang, R.; Gong, Y.; Wang, C.; Wu, P.; Guo, J. Amorphous-to-Crystalline Transformation toward Controllable Synthesis of Fibrous Covalent Organic Frameworks Enabling Promotion of Proton Transport. *Chem. Commun.*, 2019, *55*(1), 75–78.

38. Tao, S.; Zhai, L.; Dinga Wonanke, A. D.; Addicoat, M. A.; Jiang, Q.; Jiang, D. Confining H3PO4 Network in Covalent Organic Frameworks Enables Proton Super Flow. *Nat. Commun.*, 2020, *11*(1), 1981.

39. Wu, X.; Hong, Y.; Xu, B.; Nishiyama, Y.; Jiang, W.; Zhu, J.; Zhang, G.; Kitagawa, S.; Horike, S. Perfluoroalkyl-Functionalized Covalent Organic Frameworks with Superhydrophobicity for Anhydrous Proton Conduction. *J. Am. Chem. Soc.*, 2020, *142*(33), 14357–14364.

40. Diaz, L. A.; Abuin, G. C.; Corti, H. R. Acid-Doped ABPBI Membranes Prepared by Low-Temperature Casting: Proton Conductivity and Water Uptake Properties Compared with Other Polybenzimidazole-Based Membranes. *J. Electrochem. Soc.*, 2016, *163*(6), F485–F491.

41. Yin, Z.; Geng, H.; Yang, P.; Shi, B.; Fan, C.; Peng, Q.; Wu, H.; Jiang, Z. Improved Proton Conduction of Sulfonated Poly (Ether Ether Ketone) Membrane by Sulfonated Covalent Organic Framework Nanosheets. *Int. J. Hydrogen Energy*, 2021, *46*(52), 26550–26559.

42. Liu, L.; Yin, L.; Cheng, D.; Zhao, S.; Zang, H.-Y.; Zhang, N.; Zhu, G. Surface-Mediated Construction of an Ultrathin Free-Standing Covalent Organic Framework Membrane for Efficient Proton Conduction. *Angew. Chemie Int. Ed.*, 2021, *60*(27), 14875–14880.

43. Cao, L.; Wu, H.; Cao, Y.; Fan, C.; Zhao, R.; He, X.; Yang, P.; Shi, B.; You, X.; Jiang, Z. Weakly Humidity-Dependent Proton-Conducting COF Membranes. *Adv. Mater.*, 2020, 32(52), 2005565.

44. Kang, D. W.; Kang, M.; Yun, H.; Park, H.; Hong, C. S. Emerging Porous Solid Electrolytes for Hydroxide Ion Transport. *Adv. Funct. Mater.*, 2021, 31(19), 2100083.

45. Li, L.; Akhtar Qaisrani, N.; Ma, L.; Bai, L.; Zhang, A.; He, G.; Zhang, F. Mixed Matrix Anion Exchange Membrane Containing Covalent Organic Frameworks: Ultra-Low IEC but Medium Conductivity. *Appl. Surf. Sci.*, 2021, 560, 149909.

46. Tao, S.; Xu, H.; Xu, Q.; Hijikata, Y.; Jiang, Q.; Irle, S.; Jiang, D. Hydroxide Anion Transport in Covalent Organic Frameworks. *J. Am. Chem. Soc.*, 2021, 143(24), 8970–8975.

47. Chen, J.; Guan, M.; Li, K.; Tang, S. Novel Quaternary Ammonium-Functionalized Covalent Organic Frameworks/Poly(2,6-Dimethyl-1,4-Phenylene Oxide) Hybrid Anion Exchange Membranes with Enhanced Ion Conductivity and Stability. *ACS Appl. Mater. Interfaces*, 2020, 12(13), 15138–15144.

48. Kong, Y.; He, X.; Wu, H.; Yang, Y.; Cao, L.; Li, R.; Shi, B.; He, G.; Liu, Y.; Peng, Q.; Fan, C.; Zhang, Z.; Jiang, Z. Tight Covalent Organic Framework Membranes for Efficient Anion Transport via Molecular Precursor Engineering. *Angew. Chemie Int. Ed.*, 2021, 60(32), 17638–17646.

49. He, X.; Yang, Y.; Wu, H.; He, G.; Xu, Z.; Kong, Y.; Cao, L.; Shi, B.; Zhang, Z.; Tongsh, C.; Jiao, K.; Zhu, K.; Jiang, Z. De Novo Design of Covalent Organic Framework Membranes toward Ultrafast Anion Transport. *Adv. Mater.*, 2020, 32(36), 2001284.

50. Arges, C. G.; Zhang, L. Anion Exchange Membranes' Evolution toward High Hydroxide Ion Conductivity and Alkaline Resiliency. *ACS Appl. Energy Mater.*, 2018, 1(7), 2991–3012.

8

Covalent Organic Frameworks-Based Nanomaterials for Hydrogen Reduction Reactions

Sandeep Kaushal

Department of Chemistry, Sri Guru Granth Sahib World University, Fatehgarh Sahib, Punjab, India

CONTENTS

8.1 Introduction

Nature motivates researchers to develop novel materials with intricate structures capable of performing complex functions. Scientists have investigated a variety of chemical designs by utilizing a variety of techniques to link diverse building components, emulating natural structures [1]. These attempts have emerged in various types of porous materials, like conjugated microporous polymers, silica, mesoporous zeolites, phenolic resins and carbon, organosilica, metal-organic frameworks (MOFs), etc. [2, 3]. The kinetically regulated irreversible coupling reaction, which can induce structural irregularity and build oligomers that reduce the efficiency of the work, is mostly responsible for the formation of these structures [4]. Subsequently, novel everlasting porous materials with comprehensive structures and reasonable crystallinity are still required. COFs are a novel type of crystalline porous organic polymer constructed by dynamic condensations of light-component key components connected by strong covalent bonds, according to Yaghi et al. [5]. Long-range-ordered crystalline forms emerge from COFs' capacity to self-heal and their thermodynamically regulated covalent bonding. COFs also have exceptional chemical stability in organic solvents and can maintain their ordered structures and crystallinity under challenging circumstances (e.g., acidic and basic environments). The material's completely covalently bonded nature resulted in such incredible stability [6]. Furthermore, in COFs, hydrogen bonding and π-stacking interactions can reinforce their porosity framework and defend them against solvation and hydrolysis.

Moreover, compared to similar inorganic zeolites and other porous silicas, COFs can have greater porosities and bigger pores, constructing them appropriate for use in

DOI: 10.1201/9781003206507-8

catalysis, where larger pores accelerate reactant diffusion and product desorption, enhancing the yield of the product and selectivity (Figure 8.1). In the meantime, COFs have small densities, numerous open sites, large surface areas, and easy molecular functionalization, thus they have expanded considerably and are suited for prospective practice like gas storage, separation, catalysis, and optoelectronics [7]. COFs are promising candidate materials, particularly in catalytic applications. Their distinct channel structure and variable porosity allow for quick access to active areas as well as rapid bulk transport for catalytic reactions [8]. Moreover, the vast surface area of COFs permits the formation of catalytic sites that are widely distributed and react competently with substrates. Furthermore, the polymeric frame materials are constructed of simple organic building blocks interconnected by covalent bonding and have excellent thermal and chemical stability. At the same time, COFs permit the automatic incorporation of organic building blocks into periodic and organized structures with atomic precision. To precisely manufacture the functional crystalline porous frameworks catalyst, which is made up of organic small molecules, a flexible and acceptable design may be utilized. Various kinds of fundamental active sites, comprising metal catalytic sites and chiral compounds, can be incorporated into the frameworks, which can significantly increase turnover numbers and catalytic activity. Being a solid catalyst, it is simple to recover products and separate COFs material after reactions, making it possible to reuse the catalysts. Because of these particular advantages, the research and search for novel COFs materials and their potential use in catalysis [9].

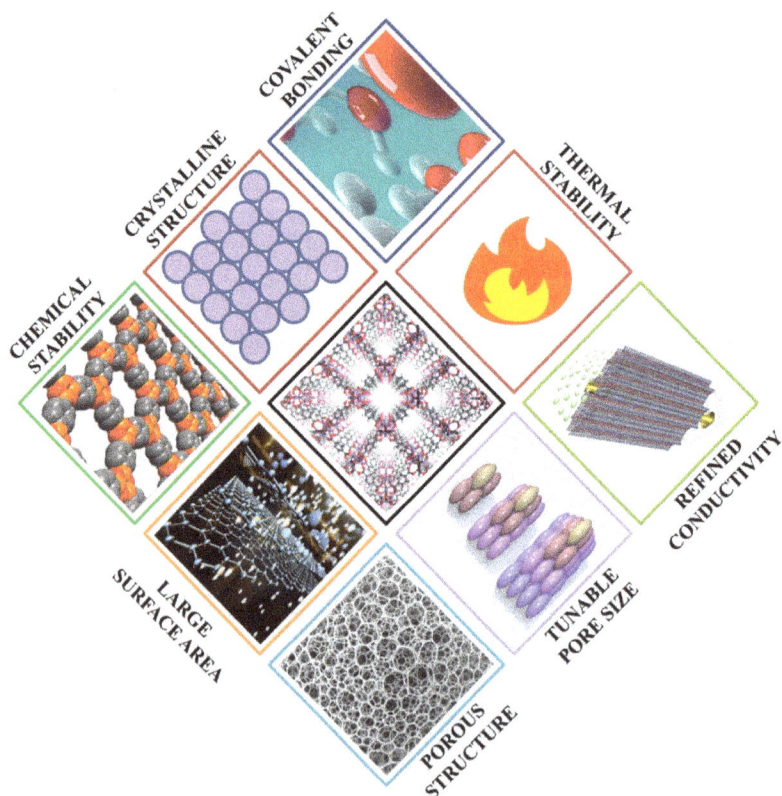

FIGURE 8.1
Various properties of COFs.

8.2 Morphology of COFs

COFs have an adaptable structure thanks to ample building blocks and effective covalent bonds. Several efforts have been made to explore the properties of COFs with unusual structures, such as 0D, 1D, 2D, and 3D structures [10]. Understanding the photocatalytic activity of COFs requires morphological and structural research. The next section discusses the fabrication and photocatalytic properties of COFs on various morphologies.

8.2.1 0-Dimensional Structures

Because of their high surface area, 0D-structural materials are regarded to be potential photocatalysts. However, their photocatalytic activity is still limited because of the large agglomeration. COF monomers are usually only moderately soluble in reaction solvents, resulting in a heterogeneous growth environment that makes the crystallization process difficult to comprehend. Presently, the majority of COFs are synthesized with a morphology that is poorly regulated, resulting in insoluble and unprocessable aggregates that are formed. A homogeneous polymerization technique was recently developed to diminish irreversible crystallite aggregation and precipitation, resulting in stable COF nanoparticle colloidal suspensions [11]. A translucent solution containing nanoparticles was achieved by incorporating a specified quantity of CH_3CN to a standard solvothermal mixture of COF-5. The CH_3CN has been proven to stabilize discrete crystallites in solution and prevent them from aggregating (Figure 8.2a–b). The coupling of COF and nitrile functional group was found to be necessary for the creation of nanoparticles. The real-time development of

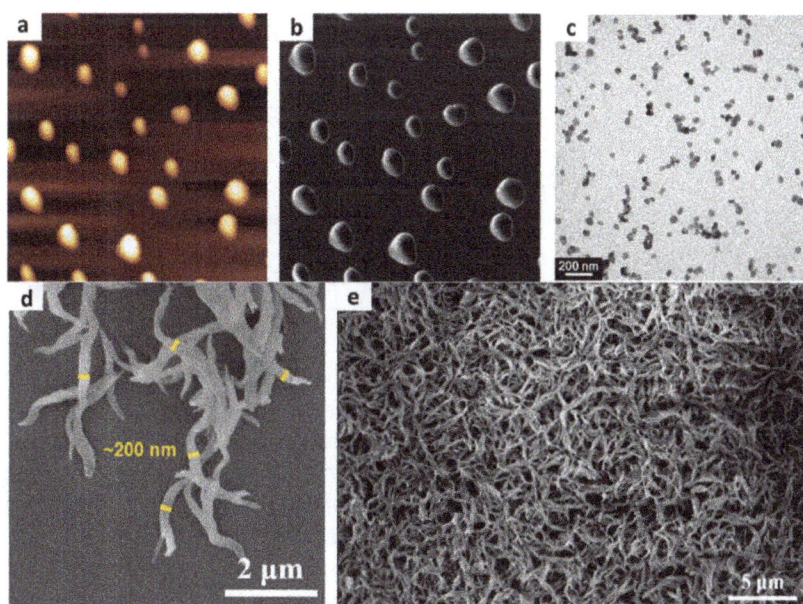

FIGURE 8.2

(a and b) AFM of COF colloids; (c) Characteristic VT-LCTEM image of COF-5 nanoparticles; (d and e) FESEM micrographs of g-C18N3-COF. (Adapted with permission [11]. Copyright 2017, American Chemical Society.)

nanoparticles was observed using variable-temperature liquid cell transmission electron microscopy (VT-LCTEM) imaging (Figure 8.2c).

These kinds of stable porous nanoparticles with a functional inner surface allowed for site-isolated catalysis. A two-step technique was also used to regulate the creation of 2D COFs, resulting in single-crystalline, micrometer-sized particles. The 0D structures with high surface speciation are considered to have better photocatalytic activity when compared to their conventional counterparts. Furthermore, when particle size is decreased, discrete energy levels appear at the band-edges of both the CB and the VB owing to the quantum confinement effect, increasing the redox potential of photogenerated electrons and holes. Nevertheless, investigating COF photocatalysts with a 0D structure is problematic.

8.2.2 1-Dimensional Structures

Because of their high surface-to-volume ratio, 1D structures including nanofibers, nanoribbons, and nanowires have seen a surge in research over the years. The investigation of COF morphology in relation to 1D structures is extremely valuable. To date, crystalline COF fibers have been made via solvent-assisted solid-state synthesis, vapor-assisted solid-state synthesis, and bottom-up microfluidic synthesis [12]. Novel crystalline COF nanofibers, for instance, were made using a solvothermal approach based on the co-polymerization of 2,4,6-tris(4-aminophenyl)-pyridine (TAPP) with 2,6-dihydroxynaphthalene-1,5-dicarbaldehyde (DHNDA) at 180°C. The COF was fabricated as homogenous nanofibers with a length of up to tens of micrometers. The morphology of the nanoparticles altered significantly, from irregular nanoparticles to homogenous nanofibers with greater crystallinity, which may be explained by the dissolution–recrystallization process. COF nanohybrids with outstanding optical and electrical properties were made possible because of this change. Furthermore, vapor-assisted solid-state fabrication might be used to make nanofibers. Even a minimal amount of solvent vapor was required in this approach, and the nanofibrous morphology was shown to change with reaction time and solvent vapor composition. When compared to the thread-shaped COF, the optical absorption edge of the ribbon-like COF displayed a red-shift, improving solar application competency and therefore resulting in a higher photocatalytic degradation rate of phenol. Such study indicates that morphology influenced COF-based materials' photocatalytic activity, which might be explained by agglomeration behavior, dispersity, and incident light-harvesting capabilities in an aqueous system.

8.2.3 2-Dimensional Structures

If the morphology and structure of materials are modified, their optical, photochemical, and photoelectrical properties may be influenced directly or indirectly. Aside from 0D and 1D structures, 2D structures such as thin films and few-layered nanosheets have also been extensively researched in photocatalytic activity. The thin films' strong photocatalytic performance is due to their high homogeneity and aspect ratio, as well as the small travel distance of photoexcited messengers [13]. Mechanical delamination, solvent-assisted exfoliation, solvothermal interfacial synthesis have all been used to produce COF thin films as free-standing forms or coated on specified substrates in recent years. Solvothermal synthesis is one of the most popular since it is easy and simple. For instance, 200 nm thick TT-COF thin films were created on a washed glass substrate by just dipping the substrate in a bulk TT-COF solution. COF exfoliation with axial ligand inclusion again resulted in ultrathin 2D porphyrin nano-disks with improved photocatalytic activity. COF DhaTph (Dha: 2,5-dihydroxyterephthalaldehyde, Tph: 5,10,15,20-tetrakis (4-aminophenyl)-21H,23Hporphyrin) was removed by introducing 4-ethylpyridine and copper (Cu) ion ligands into the porphyrin core at the same time, yielding e-CON(Cu, epy) [14]. The findings show that 2-D COF thin films and nanosheets with broad light absorption, an optical band gap, and excellent charge separation have a lot of ability to improve photocatalytic activity.

8.2.4 3-Dimensional Structures

Heterogeneous nucleation and development, template-directed technique, self-assembly strategy, and multiple-linking-site approaches are used to make COFs with 3D structures. The hollow TpPa COF, for example, was developed and synthesized using the ZnO-nanorod template. Optimizing the synthetic circumstances of TpPa-1 and TpPa-2 COFs resulted in flower-shaped morphology in addition to hollow morphology [15].

8.3 COF as Photocatalyst in Hydrogen Evolution Reactions

Non-renewable energy sources have been the main thrust for monetary development in our reality since the beginning of the mechanical transformation. Hydrogen, as a practical and sustainable power source, is widely regarded as the best alternative to petroleum products, and producing hydrogen through photocatalytic water splitting by combining water and sunlight, two readily available and limitless resources on the planet, is widely regarded as the substance and of critical importance in addressing global issues. Photocatalytic hydrogen production using COFs as photocatalysts has received increasing study attention in recent years, and it has emerged as a potential research subject. In 2014, Stegbauer et al. synthesized COF having layered structure by linkage of triazine and phenyl units. This visible light-activated COF was further doped with Pt metal which was capable of producing H_2 from water splitting reaction with a constant HER (hydrogen evolution rate) value of 1970 μmol g^{-1}h^{-1} [16]. This group of research scholars further studied the effect of dopants in COF materials. Vyas et al. outlined various azine-bonded COFs having a variable number of nitrogen atoms in their aromatic ring structure. Pt metal-doped azine-linked COFs showed an increment in HER value as the number of nitrogen atoms was added to the lattice. The highest HER value (1703 μmol g^{-1}h^{-1}) was obtained from N_3-COF attributed to the large surface area and better charge transfer capacity provided by extra nitrogen atoms in the lattice. It has been displayed by SEM images that the morphology of N_0-COF & N_0-COF has changed from ball-like agglomerations to rod-like in N_2 and N_3–COF [17]. Further, TEM and SAED patterns confirmed the formation of hexagonal pores and a nanocrystalline in-plane structure with sizes ~50–100 nm.

To get a better vision, Sun and his team analyzed the H_2 evolution rate of different functional groups linked to one COF molecule. Four different COFs were synthesized using ketoenamine linkage named TpPa-COF-X, where X can be –$(CH_3)_2$, -NO_2, -H, -CH_3. From experimental data, the highest HER value was exhibited by TpPa-COF-$(CH_3)_2$ i.e 8.3 μmol g^{-1}h^{-1}, and no decrease was shown in 30 h. This high catalytic activity was due to the high electron donor power of the alkyl group, which enhanced charge separation capacity and appeared to hike photocatalytic performance [18]. Thomas et al. synthesized β-ketoenamine COF (TP-EDDA-COF based on acetylene and TP-BDDA-COF based on diacetylene) and Pt metal acts as co-catalyst during hydrogen evolution photocatalytic reaction. Among these two COFs diacetylene-based composite showed HER value of 32 µmol g^{-1} h^{-1} whereas acetylene-based moiety showed HER value of 30 µmol g^{-1} h^{-1} only, provided similar reaction conditions [19].

Rivers, oceans, and sea covers approx. 97% of earth's water and they are a huge source of fuel energy. Splitting of seawater using thioether-linked TTR-COF was reported by Yu et al. and a significant HER value was obtained. Due to the high affinity of thioether to adsorb Au ions, active charged species were successfully separated and transferred during the photocatalytic reaction. This research work has gained huge attention as seawater was splitted for the first time to produce hydrogen energy [20]. Cooper et al. made a comparative study of two sulfone-based COFs and named S-COF and FS-COF and Pt metal was considered as co-catalyst. Upon photocatalytic reduction of water, hydrogen energy was produced and obtained HER value for S-COF was 4.44 µmol $g^{-1}h^{-1}$ and for FS-COF was 10.1 µmol g^{-1} h^{-1}. Here benzothiophene sulfone group enhanced the light absorption capacity of COFs by expanding their planarity and lowering bandgap energy. As compared to bare COF, sulfone-linked COFs have high crystallinity and large surface area and hence improve photocatalysis efficiency [21].

To optimize the bandgap energy of COFs and increase their crystallinity, stability, and photocatalytic activity their combination with a semiconductor photocatalyst leads to the formation of modified COFs having enhanced chemical and physical properties. Banerjee and his co-workers reported TpPa-2-COF doped with CdS semiconductor and employed this photocatalyst for hydrogen evolution reaction. This resulted in HER value being much higher than undoped COF hence intermixing of organic and inorganic materials generate upgraded photocatalyst [22]. Using the same concept, Zhang et al. reported MoS_2/TpPa-1-COF nanocomposite and utilized for photocatalytic hydrogen evolution [23]. Zang and co-workers further modified the method of selecting two organic and inorganic units to synthesize a hybrid cluster. A novel material Mo_3S_{13}/EB-COF was fabricated by putting together anionic $[Mo_3S_{13}]^{2-}$ unit and cationic EB-COF. Under the presence of visible light irradiation, photocatalytic water splitting was performed, and obtained HER value was 13125 µmol g^{-1} h^{-1} for 18h [24]. Numerous photocatalytic composites are available in the literature that is based upon graphitic carbon nitride (g-C_3N_4). Yan et al. fabricated g-C_3N_4/COF where building blocks were linked with imine linkage. Photocatalytic results of water splitting reaction produced a higher HER value of 10.1 mmol g^{-1} h^{-1} as compared to COF. Observed modifications resulted from lower bandgap energy and visible light absorbance capability of photocatalyst [25]. Recently, Lan and co-workers studied bandgap energy interaction possibilities between NH_2-UiO-66 with TpPa-1-COF and synthesized MOF-COF hybrid composite. As synthesized covalently bonded composite comprises almost 20 times higher HER value of 23.41 mmol g^{-1} h^{-1} as compared to TpPa-1-COF [26].

Most of the research work in the field of COFs is based on azine linkage, imine linkage, or hydrazone linkage, which is proven to be stable for a limited period due to the presence of partial conjugation between polymer units. To overcome this limitation, Jiang and co-workers synthesized completely conjugated COF which showed high stability in acidic, basic, and neutral solvents as well as synthesized sp^2 carbon-COF showed HER OF 1360 µmol g^{-1} h^{-1} under the presence of visible light radiations [27]. Zhang et al. reported two semiconductors-based triazine-linked sp^2 hybridized carbon-COFs named g-$C_{18}N_3$-COF and g-$C_{33}N_3$-COF having band gap energy values of 2.42 and 2.54 eV, respectively. After their photocatalytic activity, the calculated average HER of g-$C_{18}N_3$-COF was higher (292 µmol g^{-1} h^{-1}) as compared to g-$C_{33}N_3$-COF (74 µmol g^{-1} h^{-1}) during irradiation of visible light [28]. To further enhance hydrogen production, a condensation reaction of aromatic methyl groups was performed to synthesize sp^2 carbon COFs and showed higher photocatalytic activity [29].

As of recently, most photocatalytic frameworks have consolidated the costly honorable platinum nanoparticles as the co-impetus for hydrogen advancement, which appears to restrict the turn of events of ecologically kind-hearted photocatalytic hydrogen advancement frameworks. Accordingly, the improvement of novel respectable without metal photocatalysts is of principal significance for supportable efficient photocatalysis. In this regard, Lotsch and co-workers fabricated N$_2$-COF having azine linkage and joined with a co-catalyst chloro-pyridine cobaloxime. This metal-free photocatalyst acts as a good photo absorber with HER of 782 µmol g^{-1} h^{-1} [30]. The same group of research scholars synthesized COF having thiazolo-thiazole linkage to utilize it for hydrogen production and NiME acts as co-catalyst during photocatalytic activity. As synthesized TpDTz-COF was capable of showing higher HER of 941 µmol g^{-1} h^{-1} [31] (Figure 8.3).

Another noble metal-free photocatalyst was synthesized by Sun et al. to improve hydrogen production activity. For this, Ni(OH)$_2$ was incorporated into TpPa-2-COF system. Obtained results showed a higher value of HER 1896 µmol g^{-1} h^{-1} which is approximately 27 times higher than TpPa-2-COF [32]. Wang et al. worked on different methods of incorporation of Co metal ions (earth-abundant) into two-dimensional COF NUS-55. To synthesize NUS-55(Co), post-synthetic metalation of vacant sites present in COF was filled whereas NUS-55/[Co(bpy)$_3$]Cl$_2$ was fabricated via molecular co-catalyst incorporation and also showed seven times higher value of HER 2480 µmol g^{-1} h^{-1} in the presence of visible light. This research work proved that the method of metal ion incorporation into COF lattice plays an important role in their photocatalytic efficiency [33].

Covalent triazine framework (CTFs) falling in the category of COFs comprises high chemical and thermal stability. Li et al. synthesized CTF-BT/Th-x molecular heterojunction to improve bandgap properties via polymerization technique. Observed results confirmed successful separation of active charged species and showed HER of 6.6 mmol g^{-1} h^{-1}

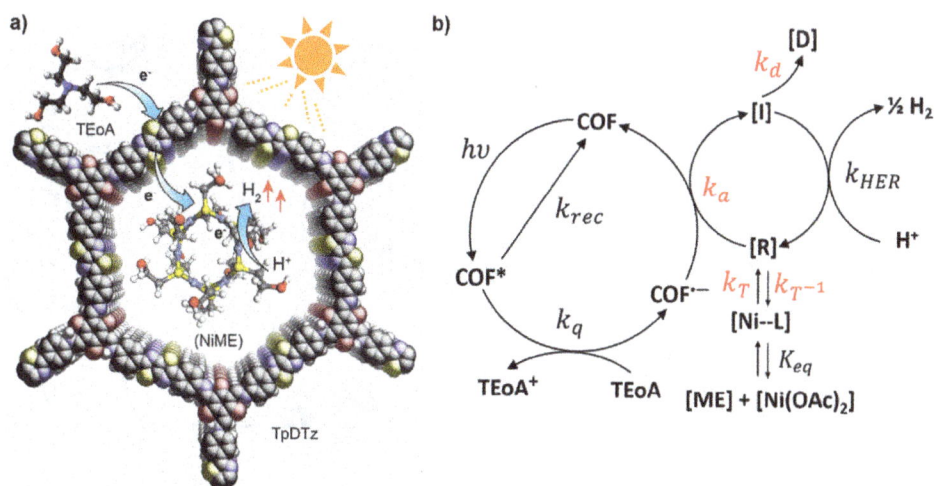

FIGURE 8.3

(a) Graphical representation of the planned pathway for H$_2$ evolution and (b) Projected important phases of the photocatalytic H$_2$ evolution reaction with TpDTz COF. (Adapted with permission [31]. Copyright 2019, American Chemical Society.)

under the presence of visible light radiations. With intermixing of these two components approximately 6 times higher photocatalytic efficiency was achieved as compared to a single lattice structure [34]. Substitution of CTFs with heteroatoms (N or O) was reported by Tan and co-workers and to examine their photocatalytic efficiency hydrogen evolution rate was monitored under favorable conditions and found HER of CTF-O 72 μmol g^{-1} h^{-1} and for CTF-N hydrogen evolution rate was 538 μmol g^{-1} h^{-1} [35]. Su et al. synthesized sulfur-doped CTFs by following the annealing method. Different lattice structures were synthesized based on different sulfur content and utilized for hydrogen evolution reactions. The highest HER value 2000 μmol g^{-1} h^{-1} was exhibited by CTF10 formed with 10% sulfur content. Comparative results confirmed that doped CTFs showed higher HER value as compared to pristine covalent triazine frameworks [36]. Cheng et al. synthesized phosphorous atoms doped CTFs by using red phosphorous via the hydrothermal method. Obtained HER value was 4.5 times higher as compared to un-doped covalent triazine frameworks [37] (Figure 8.4).

In addition to these methods and structures, Jin and co-workers studied the effect of intermixing of two monomeric units out of which one acts as electron-donor and the other as

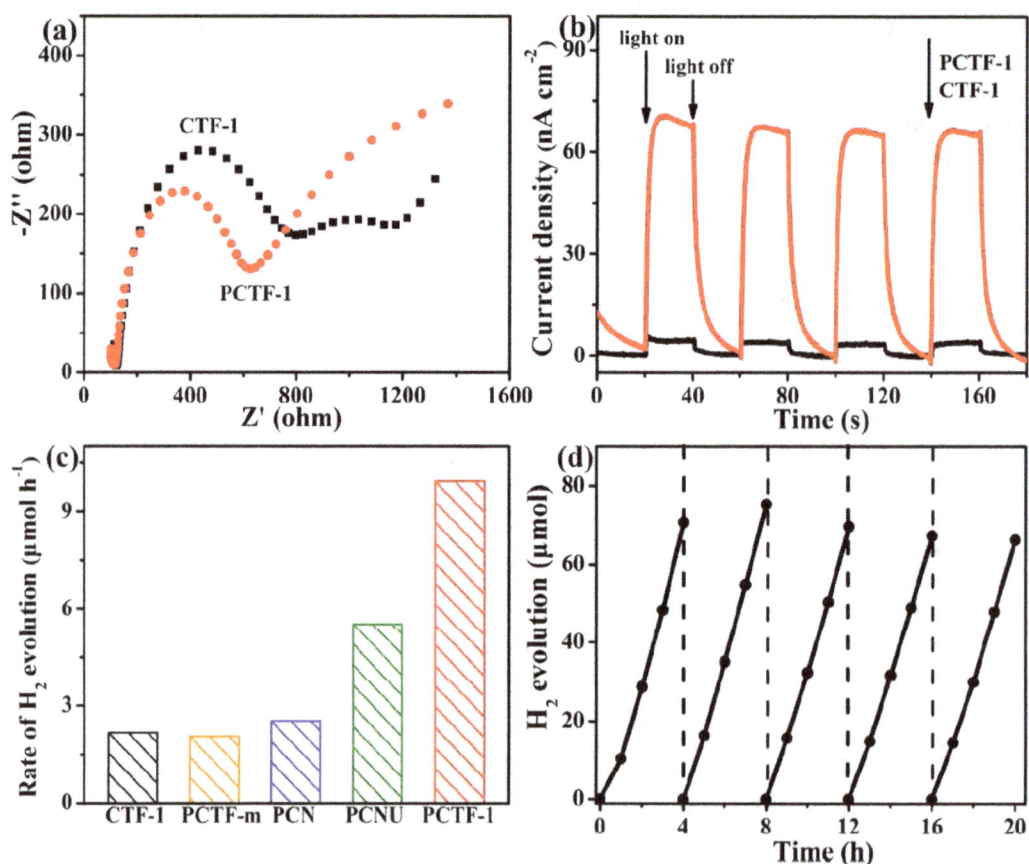

FIGURE 8.4
(a) Electrochemical impedance spectroscopy plots; (b) photocurrent responses; (c) H$_2$ evolution rates; and (d) reusability up to 5 runs of CTF-1 and PCTF-1. (Adapted with permission [37]. Copyright 2018, American Chemical Society.)

electron deficient. Based on this approach, a series of CTFs were synthesized named D-A1-A2 where A1 and A2 are electron acceptors (triazine units and benzothiazole respectively) and carbazole (D) is the electron donor unit. With these polymeric structures, obtained HER of 19.3 mmol g^{-1} h^{-1} under visible light [38]. Tan et al. synthesized CTFs having highly crystalline structure and arranged them in a series of CTF-HUST-1 to HUST-4 by following condensation reaction of aldehydes with amidines at very low-temperature conditions. Being a good photocatalyst, under the presence of visible light maximum HER was shown by CTF-HUST-2 sample (2647 μmol g^{-1} h^{-1}) [39]. Optimization of the morphology of CTFs can provide variable photocatalytic efficiencies. During research work synthesized hollow and well as bulk CTFs via polycondensation technique. Upon water-splitting reaction, a huge increase in HER of the hollow structure was observed (6040 μmol g^{-1} h^{-1}) as compared to CTF-HUST-1 bulky structure (1540 μmol g^{-1} h^{-1}) provided similar reaction conditions [40].

Recently, Zhou et al. synthesized a highly stabilized 2D β–ketoamine linked covalent organic framework by incorporating polyethylene glycol in vacant pores of the lattice structure. This unique structure is much better than 2D stacked structure as PEG restricts dislocation of π–bonded layers and showed higher photocatalytic activity as compared to 2D-COFs [41]. For better understanding, Lotsch et al. studied the effect of lattice structure on the physical and chemical properties of COFs. The previously reported method for optimization of nitrogen atoms in COF was further improved by introducing pyridine-linked COF which enhances photocatalytic activity by changing the polarity of COF lattice. In addition to this, they also conclude that by changing the molecular dimensions, several electronic and crystalline properties can be improved successfully [42].

Schwinghammer et al. synthesized hydrazone-based COF named TEPT-COF by condensation reaction. To further enhance reducing capability of COF incorporation of Pt metal leads to the formation of Pt/TEPT-COF. With this photocatalyst, 1970 μmol g^{-1} h^{-1} HER was gained, which was approximately 3 times higher than other hydrazone-based crystalline COFs. Retention of photocatalytic activity was observed for 92 h with a quantum efficiency of 2.2%. Residual catalysts can be regenerated after photocatalytic activity by passing the amorphous sample into initial experimental conditions [43]. As examined previously, perhaps the most charming characters of COFs is primary tunability, which takes into consideration structure-to-work plan at a nuclear level. Without a doubt, numerous sorts of examinations considered on COF-based photocatalysts for water parting have been finished by fitting the structure squares and linkages. Extension of alkynes by adding variable hetero-aromatic monomeric units leads to the formation of pyrene-linked COFs (A-TEXPY). Pt metal was used as a co-catalyst which enhances HER under visible light irradiation, upon photocatalytic activity, as synthesized COFs exhibited HER of 98 μmol g^{-1} h^{-1} [44]. Zou and team workers synthesized novel MOF/COFs composite (NH_2-MIL-125/B-CTF-1) by using covalent triazine linked units and obtained HER of 360 μmol g^{-1} h^{-1} in the presence of visible light. Photocatalytic studies revealed high stability of the covalently bonded hybrid structure as compared to frameworks having Van der Waals interactions. The high photocatalytic activity could be attributed to the formation of amide linkage between MOF and COF materials which strengthen high charge separation and reduced recombination of active charged species [45]. L. Wu et al. synthesized a series of covalent triazine-based units by incorporating halogen atoms by providing high temperature to CTFs and addition of a high amount of ammonium halide. Obtained COFs were utilized for H_2 production reaction in the presence of visible light energy. Among all the halogen-doped COFs, the highest HER was acquired by Cl-CTFs as compared to the triazine framework. The enhanced optical and photochemical activity of hybrid materials was due to the electronegativity and the atomic size of halogen units [46].

Bhaumik and co-workers reviewed the biological and chemical activity of porphyrin and metal-based porphyrins [47]. Based on these studies, they synthesized porphyrin-linked porous organic frameworks photocatalyst for water splitting reaction [48]. In order to hike the photocatalytic activity of COFs, several metal atoms were doped into the porphyrin ring. Villagran and his co-workers reported cobalt-doped porphyrin ring-based COFs and utilized them for hydrogen production reaction under provided acidic medium. Due to the presence of a large porphyrin ring, exposed surface area, porous platform, and a large number of active sites were delivered for HER activity [49]. Furthermore, Bhunia and his team workers synthesized a quasi 2D covalent organic framework by solvothermal method. Obtained 3D triclinic COF comprised porphyrin ring and pyrene units on the edges located in an alternative manner due to which a large surface area is generated by electro-catalyst. The hydrogen evolution rate produced by SB-POR-PY@COF showed a low overpotential of 50 mV and high strength in the provided acidic conditions and retained efficiency after 500 catalytic cycles [50]. Another metal-free covalent organic framework was designed by condensation between triformyl phloroglucinol and the substituted porphyrin ring. As synthesized TpPAM exhibited an overpotential of 250 mV at provided current density of 10 mA cm^{-2} with high durability and high retention rate up to 1000 testing cycles [51].

Covalent organic framework worked as a pillar for small-sized nanoparticles utilized for water splitting reactions by stabilizing and immobilizing nano units. Among COFs, the covalent triazine (C_3N_4) framework was proved to be the most stable porous photocatalyst [52, 53]. Based on this conclusion, Hu et al. fabricated CTF@MoS$_2$ hybrid nanocomposite. MoS$_2$ nanoparticles were grown on the surface of the covalent triazine framework by in situ pathway. During water-splitting reaction, an overpotential of 93 mV was exhibited by CTF@MoS$_2$ composite, attributed to the presence of MoS$_2$ active sites, fast electron transmission between a ground state and excited state [53]. Yao et al. reported rGO@COF hybrid composite connected via covalent bonds for water splitting reaction. By following the one-pot method, graphene oxide was added into TpPa-1-COF and treated under reaction conditions. During the photocatalytic activity, HER of 11.9 mmol g^{-1}h^{-1} was exhibited as-synthesized building blocks under the presence of visible light energy [54]. Recently, Shan at. al. fabricated Fe$_2$P/Fe$_4$N nanocomposite and incorporated it into COF material for utilizing it as water splitting electro-catalyst. Synthesized nanoparticles upon incorporation in nitrogen-doped building blocks lead to the formation of Fe$_2$P/Fe$_4$N@N-Carbon having a rough sheet-like morphology with the stacking of small particles microporous structure and large pore volume (Figure 8.5) [55].

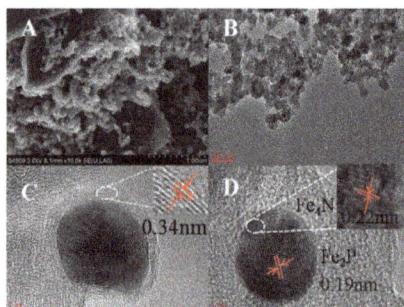

FIGURE 8.5
(A, B) SEM and (C, D) HRTEM micrographs of Fe$_2$P/Fe$_4$N@C-800. (Adapted with permission [55]. Copyright 2017, American Chemical Society.)

Nakanishi et al. fabricated a covalent triazine framework encapsulated onto a covalent organic framework by forming a coordinate bond through N atoms and exhibited high electrocatalytic efficiency during water splitting reactions [56]. Wang et al. reported CdS@ CTF-1 synthesized by a photo deposition method. High photocatalytic activity during the water splitting mechanism was shown by hybrid material due to the presence of large surface area and more active sites [57]. While these are promising models for Covalent Organic Framework-based HER photocatalyst, the so-far introduced electrocatalytic exhibitions are still lower than those for Pt-based frameworks. Up until now, the helpless conductivity of COFs and their lower security, particularly under a solid electrolyte climate appear to be the significant bottlenecks to accomplishing viable application for huge scope H_2 creation. Future work ought to accordingly additionally focus on the improvement of COFs with improved dependability and new ways to deal which guarantee a cozy contact with leading backings to advance the HER execution of COF-based electrocatalyst and photocatalyst.

8.4 Summary and Outlook

Since Yaghi and colleagues discovered covalent organic frameworks in 2005, many COFs have been created and constructed for a variety of purposes. As photocatalysts, COFs, a new form of crystalline porous material, has a promising future. In addition, COFs are interesting materials for energy conversion and storage applications because of their regulated and well-defined structure, high porosity, and surface area. COFs do have some features that set them apart from other commonly utilized materials in this industry. In comparison to carbon- and graphene-based materials, there is excellent structural control initially. Metal coordinating moieties in COFs, for instance, can serve as a comprehensible binding site for electrocatalysis or redox-active sites in supercapacitors, allowing for improved pseudocapacitive efficiency or reversible reactions in rechargeable batteries. Furthermore, COFs also have improved light-absorbing capacity and accelerated charge carrier mobility due to the expanded in-plane conjugation and well-defined interlayer p-stacking structures. The latest achievements and innovations in COF development and photocatalytic application were highlighted in this chapter. Controllable morphologies such as 0D, 1D, 2D, and 3D structures were characterized, as well as a rising number of covalent bonds responsive to shape and function design. It was also addressed how COFs may be used as photocatalysts for photocatalytic H_2 evolution. While there have been some significant advancements and achievements, the research of COFs and COF-based photocatalysts is still in its early phases, and some difficulties must be overcome for future development to proceed.

COFs' structures, morphologies, and characteristics are likely to fluctuate as a result of varied biosynthetic pathways and reaction conditions, resulting in varying photocatalytic performance. Simple, low-cost synthetic methods with benign reaction conditions are eagerly sought for the synthesis of COFs with enhanced photocatalytic activity. It is necessary to develop new COFs that are both stable and efficient. It is critical to consider how to regulate the bandgap configuration of COFs with ease. As a result, the development of COF photocatalysts with high molar absorption coefficients and enhanced light absorption is promoted. The underlying mechanism of the photocatalytic system based on COF is still unknown. Theoretical calculations are capable of determining structures and characteristics along with recreating the photocatalytic procedure, making them a very useful tool. Furthermore, advanced characterization approaches should be investigated to reveal the mechanism behind most photocatalytic activities, which might lead to the development of more productive COF-based photocatalysts.

References

1. Nath I, Chakraborty J, Verpoort F (2016) Metal organic frameworks mimicking natural enzymes: a structural and functional analogy. Chem Soc Rev 45:4127–4170.
2. Manzano M, Regi MV (2020) Mesoporous silica nanoparticles for drug delivery, Adv Funct Mater 30:1902634.
3. Lakhi KS, Park DH, Bahily KA, Cha W, Viswanathan B, Choy JH, Vinu A (2017) Mesoporous carbon nitrides: synthesis, functionalization, and applications, Chem Soc Rev 46:72–101.
4. Lohse MS, Bein T (2018) Covalent organic frameworks: structures, synthesis, and applications. Adv Funct Mater 28:1705553.
5. Cote AP, Benin AI, Ockwig NW, O'Keeffe M, Matzger AJ, Yaghi OM (2005) Chemistry: porous, crystalline, covalent organic frameworks. Science 310:1166–1170.
6. Diercks CS, Yaghi OM (2017) The atom, the molecule, and the covalent organic framework. Science 355:1585.
7. Kaderi HM, Hunt JR, Mendoza-Cortes JL, Cote AP, Taylor RE, OKeeffe M, Yaghi OM (2007) Designed synthesis of 3D covalent organic frameworks. Science. 316:268–272.
8. Segura JL, Mancheno MJ, Zamora F (2016) Covalent organic frameworks based on Schiff-base chemistry: synthesis, properties and potential applications. Chem Soc Rev 45:5635–5671.
9. Jin EQ, Asada M, Xu Q, Dalapati S, Addicoat MA, Brady MA, Xu H, Nakamura T, Heine T, Chen QH, Jiang DL (2017) Two-dimensional sp^2 carbon–conjugated covalent organic frameworks. Science 357:673–676.
10. He S, Zeng T, Wang SH, Niu HY, Cai YQ (2017) Facile synthesis of magnetic covalent organic framework with three-dimensional bouquet-like structure for enhanced extraction of organic targets, ACS Appl. Mater. Interfaces, 9, 2959–2965.
11. Smith BJ, Parent LR, Overholts AC, Beaucage PA, Bisbey RP, Chavez AD, Hwang N, Park C, Evans AM, Gianneschi NC, Dichtel WR (2017) Colloidal covalent organic frameworks. ACS Cent Sci 3:58–65.
12. San-Miguel DR, Abrishamkar A, Navarro JAR, Trujillo RR, Amabilino DB, Balleste RM, Zamora F, Luis JP (2016) Crystalline fibres of a covalent organic framework through bottom-up microfluidic synthesis. Chem. Commun. 52:9212–9215.
13. Colson JW, Woll AR, Mukherjee A, Levendorf MP, Spitler EL, Shields VB, Spencer MG, Park J, Dichtel WR (2011) Oriented 2D covalent organic framework thin films on single-layer graphene. Science 332:228–231.
14. Fan ZY, Nomura K, Zhu MS, Li XX, Xue JW, Majima T, Osakada Y (2019) Synthesis and photocatalytic activity of ultrathin two-dimensional porphyrin nanodisks via covalent organic framework exfoliation. Commun. Chem. 55:4363–4366.
15. Pachfule P, Kandmabeth S, Mallick A, Banerjee R (2015) Hollow tubular porous covalent organic framework (COF) nanostructures. Chem. Commun. 51:11717–11720.
16. Stegbauer L, Schwinghammer K, Lotsch BV (2014) A hydrazone-based covalent organic framework for photocatalytic hydrogen production. Chem Sci 5:2789–2793.
17. Vyas VS, Haase F, Stegbauer L, Savasci G, Podjaski F, Ochsenfeld C, Lotsch BV (2015) A tunable azine covalent organic framework platform for visible light-induced hydrogen generation. Nat Commun. 6:8508.

18. Sheng JL, Dong H, Meng XB, Tang HL, Yao YH, Liu DQ, Bai LL, Zhang FM, Wei JZ, Sun XJ (2019) Effect of different functional groups on photocatalytic hydrogen evolution in covalent-organic frameworks. Chem Cat Chem. 11:2313–2319.

19. Pachfule P, Acharjya A, Roeser J, Langenhahn T, Schwarze M, Schomacker R, Thomas A, Schmidt J (2018) Diacetylene functionalized covalent organic framework (COF) for photocatalytic hydrogen generation. J Am Chem Soc 140:1423–1427.

20. Li LY, Zhou ZM, Li LY, Zhuang ZY, Bi JH, Chen JH, Yu Y, Yu JG (2019) Thioether-functionalized 2D covalent organic framework featuring specific affinity to Au for photocatalytic hydrogen production from seawater. ACS Sustainable Chem Eng 7:18574–18581.

21. Wang X, Chen L, Chong SY, Little MA, Wu Y, Zhu WH, Clowes R, Yan Y, Zwijnenburg MA, Sprick RS, Cooper AI (2018) Sulfone-containing covalent organic frameworks for photocatalytic hydrogen evolution from water. Nat Chem 10:1180– 1189.

22. Thote J, Aiyappa HB, Deshpande A, Diaz Diaz D, Kurungot S, Banerjee R (2014) A Covalent organic framework–cadmium sulfide hybrid as a prototype photocatalyst for visible-light-driven hydrogen production. Chem Eur J 20:15961–15965.

23. Gao MY, Li CC, Tang HL, Sun XJ, Dong H, Zhang FM (2019) Boosting visible-light-driven hydrogen evolution of covalent organic frameworks through compositing with MoS_2: a promising candidate for noble-metal-free photocatalysts. J Mater Chem A 7:20193–20200.

24. Cheng YJ, Wang R, Wang S, Xi XJ, Ma LF, Zang SQ (2018) Encapsulating $[Mo_3S_{13}]^{2-}$ clusters in cationic covalent organic frameworks: enhancing stability and recyclability by converting a homogeneous photocatalyst to a heterogeneous photocatalyst. Chem. Commun. 54:13563–13566.

25. Luo M, Yang Q, Liu K, Cao H, Yan H (2019) Boosting photocatalytic H_2 evolution on g-C_3N_4 by modifying covalent organic frameworks (COFs). Chem Commun 55:5829–5832.

26. Zhang FM, Sheng JL, Yang SD, Sun XJ, Tang HL, Lu M, Dong H, Shen FC, Liu J, Lan YQ, Rational design of MOF/COF hybrid materials for photocatalytic H_2 evolution in the presence of sacrificial electron donors. Angew Chem Int Ed 57:12106–12110.

27. Jin E, Lan Z, Jiang Q, Geng K, Li G, Wang X, Jiang D, 2D sp^2 carbon-conjugated covalent organic frameworks for photocatalytic hydrogen production from water. Mater Chem Front 5:1632–1647.

28. Wei S, Zhang F, Zhang W, Qiang P, Yu K, Fu X, Wu D, Bi S, Zhang F (2019) Semiconducting 2D triazine-cored covalent organic frameworks with unsubstituted olefin linkages. J Am Chem Soc 141:14272–14279.

29. Bi S, Yang C, Zhang W, Xu J, Liu L, Wu D, Wang X, Han Y, Liang Q, Zhang F (2019) Two-dimensional semiconducting covalent organic frameworks via condensation at arylmethyl carbon atoms. Nat Commun 10:2467.

30. Banerjee T, Haase F, Savasci G, Gottschling K, Ochsenfeld C, Lotsch BV (2017) Single-site photocatalytic H_2 evolution from covalent organic frameworks with molecular cobaloxime Co-catalysts. J Am Chem Soc 139:16228–16234.

31. Biswal BP, Vignolo-Gonzalez HA, Banerjee T, Grunenberg L, Savasci G, Gottschling K, Nuss J, Ochsenfeld C, Lotsch BV (2019) Sustained solar H_2 evolution from a thiazolo[5,4- d]thiazole-bridged covalent organic framework and nickel-thiolate cluster in water. J Am Chem Soc 141:11082–11092.

32. Dong H, Meng XB, Zhang X, Tang HL, Liu JW, Wang JH, Wei JZ, Zhang FM, Bai LL, Sun XJ (2020) Boosting visible-light hydrogen evolution of covalent-organic frameworks by introducing Ni-based noble metal-free co-catalyst. Chem Eng J 379:122342.

33. Wang J, Zhang J, Peh SB, Liu G, Kundu T, Dong J, Ying Y, Qian Y, Zhao D (2019) Cobalt-containing covalent organic frameworks for visible light-driven. hydrogen evolution Sci China:Chem 63:192–197.

34. Huang W, He Q, Hu Y, Li Y (2019) Molecular heterostructures of covalent triazine frameworks for enhanced photocatalytic hydrogen production. Angew Chem Int Ed 58:8676–8680.

35. Guo L, Niu Y, Xu H, Li Q, Razzaque S, Huang Q, Jin S, Tan B (2018) Engineering heteroatoms with atomic precision in donor–acceptor covalent triazine frameworks to boost photocatalytic hydrogen production. J Mater Chem A 6:19775–19781.

36. Li L, Fang W, Zhang P, Bi J, He Y, Wang J, Su W (2016) Sulfur-doped covalent triazine-based frameworks for enhanced photocatalytic hydrogen evolution from water under visible light. J Mater Chem A 4:12402–12406.
37. Cheng Z, Fang W, Zhao T, Fang S, Bi J, Liang S, Li L, Yu Y, Wu L (2018) Efficient visible-light-driven photocatalytic hydrogen evolution on phosphorus-doped covalent triazine-based frameworks. ACS Appl Mater Interfaces 10:41415–41421.
38. Guo L, Niu Y, Razzaque S, Tan B, Jin S (2019) Design of D–A1–A2 covalent triazine frameworks via copolymerization for photocatalytic hydrogen evolution. ACS Catal 9:9438–9445.
39. Wang K, Yang LM, Wang X, Guo L, Cheng G, Zhang C, Jin S, Tan B, Cooper A (2017) Covalent triazine frameworks via a low-temperature polycondensation approach. Angew Chem Int Ed 6:14149–14153.
40. Wang N, Cheng G, Guo L, Tan B, Jin S (2019) Hollow covalent triazine frameworks with variable shell thickness and morphology. Adv Funct Mater 1904781.
41. Zhou T, Wang L, Huang X et al. (2021) PEG-stabilized coaxial stacking of two-dimensional covalent organic frameworks for enhanced photocatalytic hydrogen evolution Nat Commun 12:3934.
42. Haase F, Banerjee T, Savasci G, Ochsenfeld C, Lotsch BV (2017) Structure–property–activity relationships in a pyridine containing azine-linked covalent organic framework for photocatalytic hydrogen evolution. Faraday Discuss 201:247–264.
43. Senker J, Lotsch BV, Schwinghammer AK, Tuffy B, Mesch MB, Wirnhier E, Martineau C, Taulelle F, Schnick W (2013) Triazine-based carbon nitrides for visible-light-driven hydrogen evolution, Chem Int Ed 52:2435–2439.
44. Stegbauer L, Zech S, Savasci G, Banerjee T, Podjaski F, Schwinghammer K, Ochsenfeld C, Lotsch BV, (2018) Tailor-made photoconductive pyrene-based covalent organic frameworks for visible-light driven hydrogen generation, Adv Energy Mater 8:1703278.
45. Li F, Wang DK, Xing QJ, Zhou G, Liu SS, Li Y, Zheng LL, Ye P, Zou JP (2019) Design and syntheses of MOF/COF hybrid materials via postsynthetic covalent modification: An efficient strategy to boost the visible-light-driven photocatalytic performance, Appl Catal, B, 243:621–628.
46. Cheng Z, Zheng K, Lin G, Fang S, Li L, Bi J, Shen J, Wu L (2019) Constructing a novel family of halogen-doped covalent triazine-based frameworks as efficient metal-free photocatalysts for hydrogen production, Nanoscale Adv, 1:2674–2680.
47. Gao WY, Chrzanowski M, Ma SQ (2014) Metalmetalloporphyrin frameworks: a resurging class of functional materials, Chem Soc Rev, 43:5841–5866.
48. Patra BC, Khilari S, Manna RN, Mondal S, Pradhan D, Pradhan A, Bhaumik A (2017) A metal-free covalent organic polymer for electrocatalytic hydrogen evolution, ACS Catal, 7:6120–6127.
49. Wu YY, Veleta JM, Tang DY, Price AD, Botez CE, Villagran D (2018) Efficient electrocatalytic hydrogen gas evolution by a cobalt-porphyrin-based crystalline polymer, Dalton Trans, 47:8801–8806.
50. Bhunia S, Das SK, Jana R, Peter SC, Bhattacharya S, Addicoat M, Bhaumik A, Pradhan A (2017) Electrochemical stimuli-driven facile metal-free hydrogen evolution from pyrene-porphyrin-based crystalline covalent organic framework, ACS Appl Mater Interfaces, 9:23843–23851.
51. Patra BC, Khilari S, Manna RN, Mondal S, Pradhan D, Pradhan A, Bhaumik A, (2017) A metal-free covalent organic polymer for electrocatalytic hydrogen evolution, ACS Catal, 7:6120–6127.
52. Kuhn P, Antonietti M, Thomas A (2008) Porous covalent triazine-based frameworks prepared by ionothermal synthesis, Angew Chem Int Ed, 47:3450–3453.
53. Qiao SL, Zhang BY, Li Q, Li Z, Wang WB, Zhao J, Zhang XJ, Hu YQ (2019) Pore surface engineering of covalent triazine frameworks@MoS$_2$ electrocatalyst for the hydrogen evolution reaction, ChemSusChem, 12:5032–5040.
54. Yao YH, Li J, Zhang H et al. (2020) Facile synthesis of a covalently connected rGO–COF hybrid material by *in situ* reaction for enhanced visible-light induced photocatalytic H$_2$ evolution, J Mater Chem A 8:8949–8956.

55. Fan X, Kong F, Kong A, Chen A, Zhou Z, Shan Y (2017) Covalent porphyrin framework-derived Fe$_2$P@Fe$_4$N-coupled nanoparticles embedded in N-doped carbons as efficient trifunctional electrocatalysts, ACS Appl Mater Interfaces 9:32840−32850.

56. Kamiya K, Kamai R, Hashimoto K, Nakanishi S (2014) Platinum-modified covalent triazine frameworks hybridized with carbon nanoparticles as methanol-tolerant oxygen reduction electrocatalysts, Nat Commun 5:1–6.

57. Wang D, Zeng H, Xiong X, Wu MF, Xia M, Xie M, Zou JP, Luo SL (2020) Highly efficient charge transfer in CdS-covalent organic framework nanocomposites for stable photocatalytic hydrogen evolution under visible light, Sci Bull, 65:113–122.

9

Covalent Organic Frameworks-Based Nanomaterials for Hydrogen Evolution Reactions

Felipe M. de Souza[1] and Ram K. Gupta[1,2]

[1]*Kansas Polymer Research Center, Pittsburg State University, Pittsburg, Kansas, USA*

[2]*Department of Chemistry, Pittsburg State University, Pittsburg, Kansas, USA*

CONTENTS

9.1 Introduction

The scientific community has explored a myriad of materials that can be used in several applications for the generation of energy in devices such as, supercapacitors, batteries, fuel cells, among others. One of the technologies that have been gaining ground due to its environmental credential and sustainability is the water-splitting process, which consists of the chemical disruption of water into H_2 and O_2. Within that line, the researchers have been working on the development of catalysts to perform this process either through electrical or photocatalytic routes in a way that the goal is to make it commercially feasible. In that sense, there is a myriad of materials that can perform this water-splitting process which include mostly metal-based nanomaterials such as transition metal nitrides, sulfides, carbides, and oxides that can be further functionalized with carbon-based nanomaterials such as graphene, graphene oxide (GO), reduced graphene oxide (rGO). Another class of materials that has been gaining attention is the covalent-organic frameworks (COFs) which are crystalline porous structures composed of organic multifunctional segments that can arrange themselves in relatively organized patterns enabling the design of a macromolecular skeleton with a predictable number of nanopores and geometry. These highly organized organic materials have quickly drawn attention due to their atomically precise topology along with a plethora of different structures with high surface area. Based on that, their design consists of three factors which are retention of geometry, use of diverse building blocks, and the presence of reversible dynamic covalent bonds. Because of their molecular symmetry, COF's structure can form precise and uniform pores, which grants them tailored surface area. During the synthetic process, transition metals can be incorporated into their structure through the doping process. Hence, the controllable surface area

DOI: 10.1201/9781003206507-9

along with the uniformly distributed metallic centers can provide high selectivity and efficiency toward electrocatalysis. This effect can be achieved by properly selecting the organic building blocks for the design of a COF as this is another key aspect to enhance its catalytic activity. Also, this property can be tuned according to the ligand's degree of conjugation [1, 2]. In that sense, macrocyclic structures such as porphyrins, tetraazannulenes, and phthalocyanines, among many others can be incorporated. These types of ligands carry two aspects that influence the COF's properties. One is that the discrete nanosized pores of these cyclic ligands allow them to differentiate from other materials as transition metals can be homogeneously distributed within the cyclic ligand's nanosized pores. Because of that, the agglomeration or cluster of nanoparticles can be avoided as metallic atoms can be atomically distributed over the COF's structure.

The other aspect is that their morphology can be precisely controlled due to the rigid structure regularity of the aromatic cyclic ligands used as monomers. Through that, nanochannels can be created which facilitate the ionic transport along with nano reactive sites that can be better exposed to catalyze the reactions or perform electron transfer steps. Within that line, COFs differentiate from molecular organic frameworks (MOFs) since the former is based on strong covalent bonds between light elements, e.g., C, B, O, H, and N, whereas the latter is based on coordination bonds which include transition metals [3]. The combination of that factor and the tailored structure of COFs that allows their pores to be atomically designed leads to materials with a surface area that can reach values higher than 4,000 m^2/g along with low densities of around 0.17 g/cm^3. On top of that, the high crystallinity of COFs, despite their organic nature, ease charge transfer over a relatively longer range which can also be attributed to the large degree of conjugation of COFs due to the aromatic building blocks that are usually employed in its synthesis.

Another aspect of COFs is related to their high stability due to the strong covalent bonds which grant them high chemical stability in harsh environments such as the presence of highly reductive or oxidative species or high pH, making them more robust them other nanomaterials. Since COFs rely on metal-free molecular design to perform catalytic processes, they can potentially serve as more viable materials for this process. COFs can also perform electro or photocatalysis for both hydrogen evolution reaction (HER) and oxygen evolution reaction (OER), as their symmetrical structures allow the implementation of different building blocks that can act as catalytic sites to perform HER and OER simultaneously, which is a feature often challenging to achieve in other nanomaterials suitable for water splitting. There is a variety of precise 2D and 3D COF structures that can be designed during synthesis (Figure 9.1). Hence, COF's tailored porosity, highly versatile and organized structure, high surface area, and low weight to volume ratio demonstrate promising applications for these materials for gas storage, photo, and electrocatalysis with high selectivity.

9.2 Types of COFs and Approaches to Tuning Their Properties

The monomers utilized for the design of COFs must present a rigid chemical structure with its reactive sites distributed in specific regions to enable a guided growth of the corresponding COF structure. Hence, defining the proper geometry and active sites of the monomers is a major step for COF synthesis. In that sense, the design of 2D COFs can be

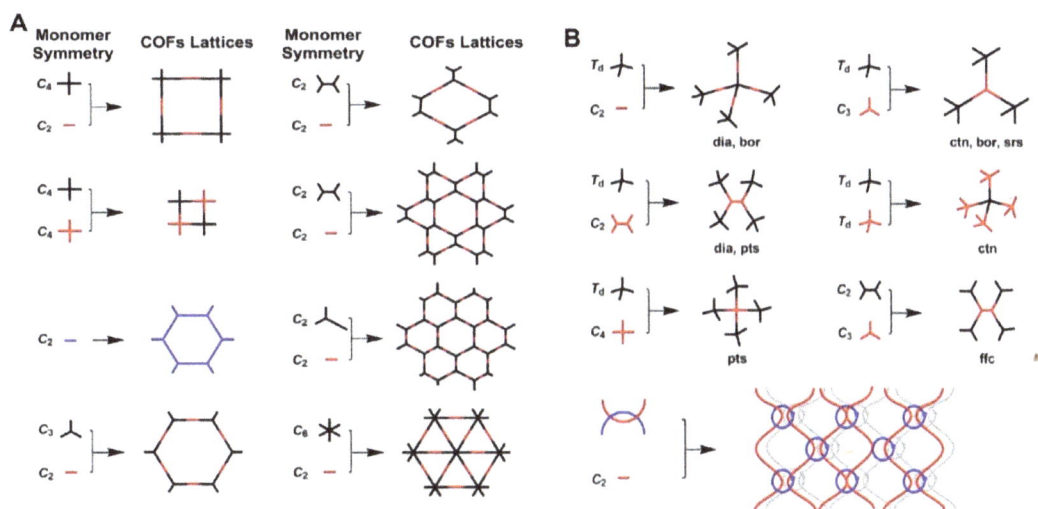

FIGURE 9.1
Schematics for the type of (A) 2D and (B) 3D COF structures based on the type of symmetry. (Adapted with permission [4]. Copyright (2020), American Chemical Society.)

made by selecting monomers with a 2D rigid structure, allowing the formation of a specific morphology with tailored nanopores. In addition, the presence of rigid chemical groups in the monomers prompts an organized structure leading to an increase in crystallinity and defined lattices. Another important feature that one should consider for the design of COFs is the precisely localized π bonds in the monomers, which enable the formation of organized layered architectures since rigid 2D structure prompts self-stacking due to the π–π bonds interlayer interactions. Thus, highly ordered structures can be assembled in the z-axis along with different topologies according to the selected monomers. Based on these concepts, there are several structural designs for 2D COFs based on different building blocks which can lead to atomically precise structures of triangular, tetragonal, and hexagonal lattices, which are presented in Figure 9.2.

For the case of hexagonal COFs, there are a variety of starting materials available which are subdivided into benzene, triphenyl benzene, triazine, and triphenylenes. A few examples of building blocks for COFs include boronic acids, catechol, anhydrides (squaric acids), hydrazine, and benzyl nitriles. The pore size from some of the COF's structures obtained from these monomers can be within the range of around 0.65–5 nm, which enables the adsorption of considerably small atoms, particles, or substances. In this sense, the combination of $C_3 + C_2$ symmetries is favorable for the design of COF hexagonal structures with mesopores. Some of these structures along with their pore size are presented in Figure 9.3.

The tetragonal topology is another possible design for COFs, which can be obtained by using monomers with $C_2 + C_4$ symmetry. Through that several types of backbones can be employed such as triphenyl, biphenyl, phenyl, bipyridine, pyridine, thiophene, thiadiazol, porphyrin, phthalocyanine, among others. This type of monomers allow the construction of COF with micro or mesopores ranging from around 1.8 to 4.4 nm, which can be varied according to the linkers used. One of the variations in the tetragonal topology is that porphyrins and phthalocyanines monomers can form metal complexes with transition-metal such as Ni, Cu, or Zn to compose the COF structure while maintaining the nanopore at

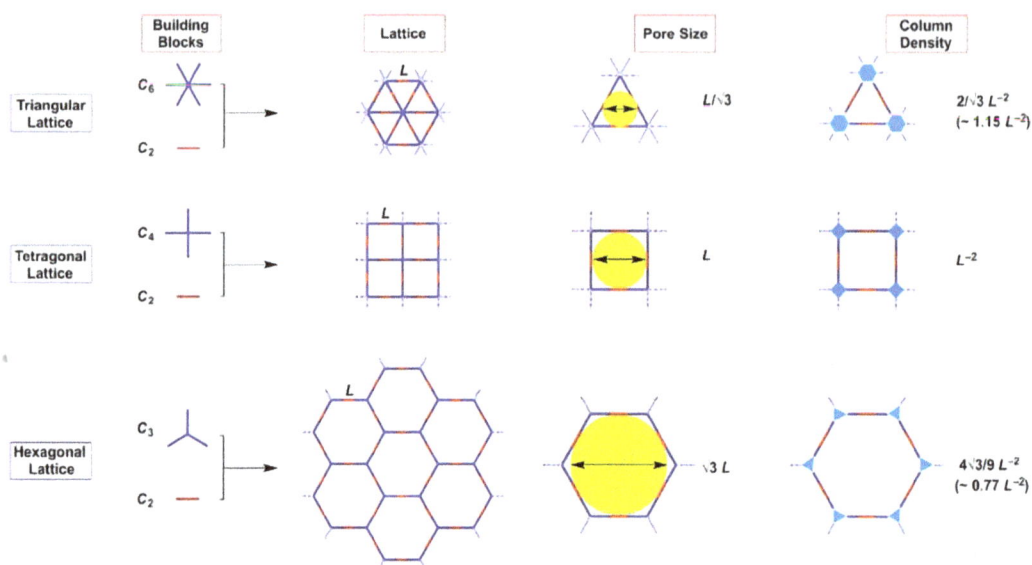

FIGURE 9.2
Variation in pore sizes based on different building blocks to obtain 2D COFs with different topologies. (Adapted with permission [4]. Copyright (2020), American Chemical Society.)

the center. The representation of these tetragonal structures is provided in Figure 9.4. The possible geometries that can be designed go beyond which also include rhombic, kagome, triangular, and heteropored.

For the case of 3D COFs, there is the requirement that at least one of the building blocks must present a tetrahedral (T_d) or orthogonal type of geometries, hence T_d combines with T_d, C_2, C_3, C_4, or $C_2 + C_3$, respectively [5]. Through that, different types of networks can be obtained such as **pts** or **dia**, **cnt** or **bor**, **pts**, **cnt**, and **ffc**. In the sense that, the **ctn** and **bor** type of networks (Figure 9.1b) can be obtained by combining building blocks with T_d and C_3 geometries, which can form non-interpenetrating networks leading to structures with higher surface area [6–8]. The **dia** network presents the relatively larger variety groups of 3D COFs due to the considerable number of linkers available as it can be assembled by combining $T_d + C_2$ building blocks. Despite the myriad of variations possible, there is still a challenge regarding the occurrence of interpenetrating structures which makes it hard to predict the number of folds that a COF would present. Part of this issue is that the parameters that influence this effect are not yet clear. This culminates in COFs with a lower porosity than the one predicted as the multifold interpenetration takes space. Another factor is that there is a relatively smaller number of building blocks suitable for 3D COFs since they must present T_d symmetry or orthogonal nodes. In terms of chemical functions suitable for the synthesis of 3D COFs, there are boronate esters, boroxines, borosilicates, imide, imine, and spiroborates.

The occurrence of chemical reactions is based on the control of kinetic and thermodynamic factors. In that sense, the former relates to irreversible reactions whereas the latter can be related to reversible ones at which bonds can be broken and reformed. This condition enables self-healing properties which leads to a self-arrangement of the structure to find the most stable state, which can be referred to as error checking. Based on that, the type of process that is more influenced by the thermodynamic factors, meaning the

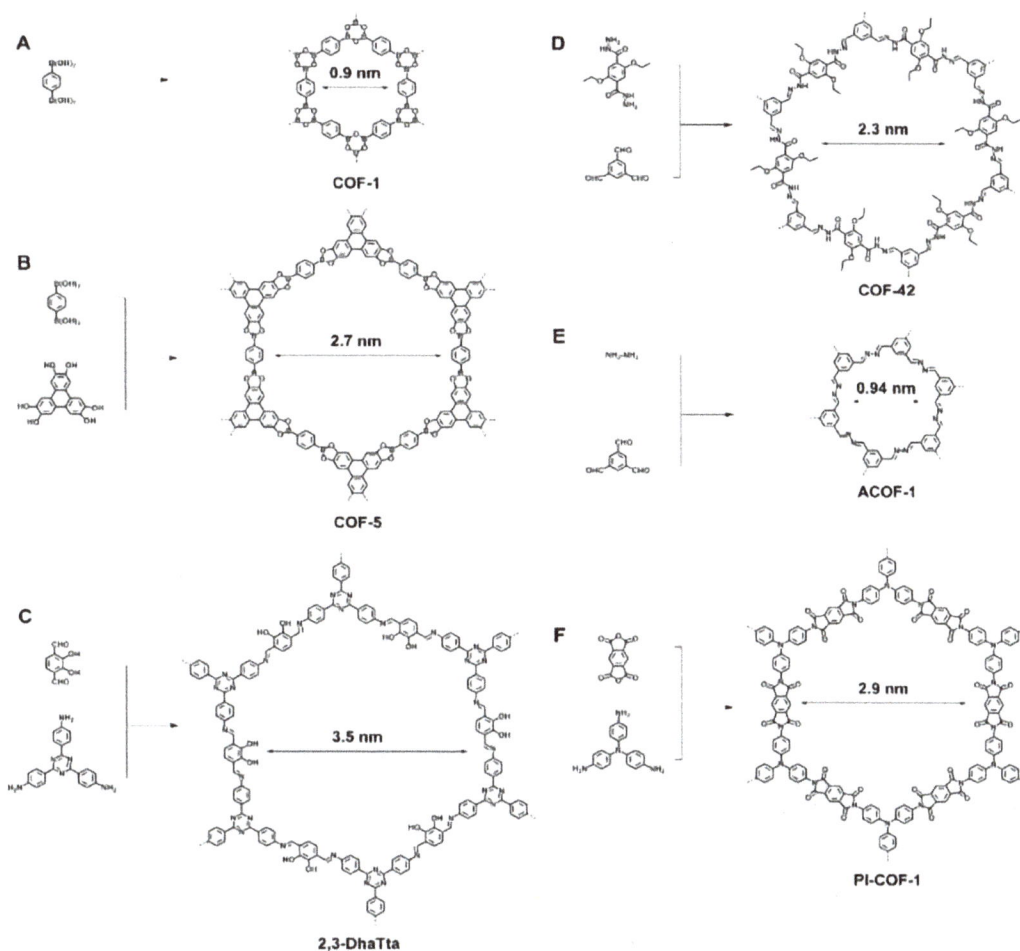

FIGURE 9.3

Hexagonal structures COFs obtained from chemical rigid and aromatic building blocks containing different organic functions with $C_3 + C_2$ type of symmetry. (A) COF-1, linked through the self-reaction of boroxine groups. (B) COF-5, based on boroxine and hydroxyl groups. (C) 2,3-DhaTta COF, based on aldehyde and amine groups. (D) COF-42, based on hydrazine on ketone and aldehyde groups. (E) ACOF-1, based on hydrazine and aldehyde. (F) PI-COF-1, based on anhydride and amine groups. (Adapted with permission [4]. Copyright (2020), American Chemical Society.)

product's stability, rather than surpassing the activation energy barrier, i.e., kinetic factor, is what describes the concept of dynamic covalent chemistry (DCC). Within that line, the chemical reactions such as boronic acid trimerization, condensations of imine or catechol, formation of imide, azine, hydrazone, among others, work as reversible processes for the synthesis of COFs. This self-correction phenomenon functions as a driving force to reach the most thermodynamically stable product, which tends to be more crystalline rather than traditional amorphous polymers. Based on that, it is deemed important to adjust certain parameters to induce higher degrees of crystallinity in COFs. For example, the controlled removal of water that is formed during condensation reactions of boronate esters, imines, or self-trimerization of boronic acids can alter the reaction reversibility which can influence the COF's crystallinity.

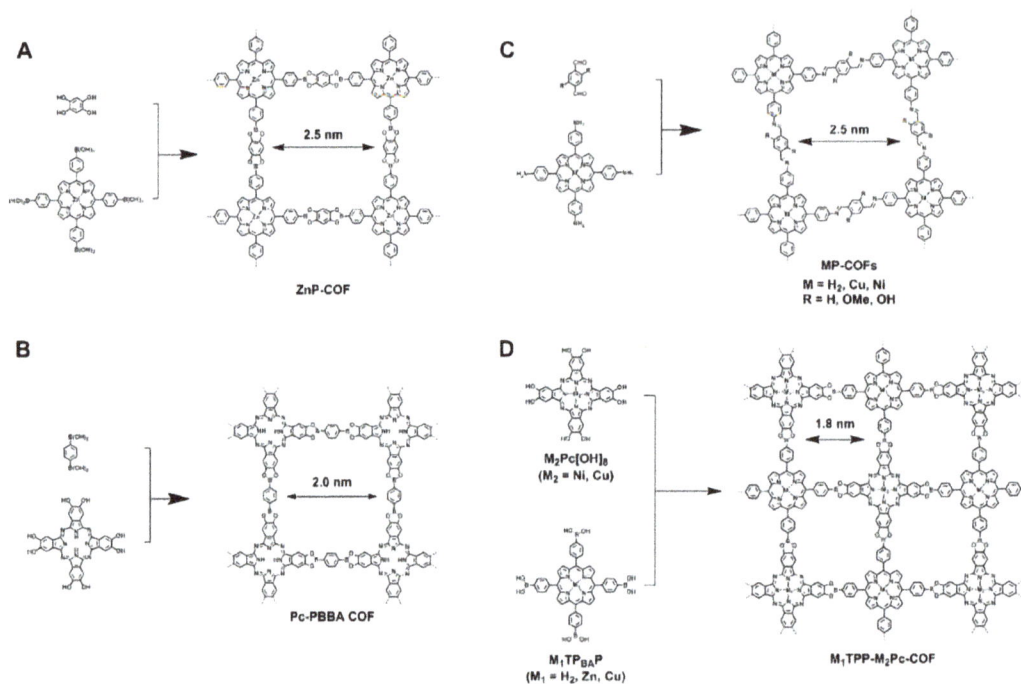

FIGURE 9.4
Structures of tetragonal COFs based on monomers with $C_4 + C_2$ symmetry. (A) ZnP-COF, based on boronate-ester links. (B) Pc-PBBA COF, based on boronate-ester links. (C) MP-COF, based on imine links. (D) M_1TPP-M_2Pc-COF, based on boronate-ester links. (Adapted with permission [4]. Copyright (2020), American Chemical Society.)

However, the long time required for proper crystallization to occur makes it necessary to optimize the process. Some of the strategies for that consists of modulation and seed approach. The modulation approach consists of adding a terminating monofunctionalized ligand that disputes with the monomer to promote a slower but more organized growth. In that sense, it was observed that the rate of monomer addition influences the formation of COF in terms of growth and nucleation, at which the growth has a first-order and the nucleation has a second-order dependence over monomer concentration, which means that growth dominates nucleation over the early stages whereas nucleation dominates over the late stages of reaction [9]. Dichtel et al. [10] performed a growth-dominated process for the synthesis of a boronate COF by slowly adding the monomer, which yielded 2D single crystals of the order of micrometers. The solvent aid in the stabilization of the crystal growth of 2D boronate COF as it was observed that the mixture of 1,4-dioxane and mesitylene prevented aggregation. Yet, the process was optimized by further adding CH_3CN. In another case also based on the modulation approach, Ditchel et al. [11] used a terminal monofunctional monomer, 4-tert-butylcatechol (TCAT), which served as a competitor during the synthesis of COF-5. Even though the formation of COF was slowed there was an increase in crystalline size domain from 23 to 32 nm. The process was further optimized by adding an excess of TCAT which yielded crystals with an average size of 450 nm. That effect was observed because TCAT acted as a nucleation inhibitor whereas inducing the anisotropic growth. On top of that, the authors observed only traces of TCAT into the COF-5, which suggested that modulation accompanied to seed growth can be a feasible procedure to induce crystallinity without incorporating undesired reagents.

The use of different monomers into the same reaction system can yield novel structures which open several possibilities for the synthesis of COFs. Based on this idea, the mixed linker approach can be performed which can be divided into two groups: (i) isostructural mixed linker (IML) or heterostructure mixed linker (HML). The IML approach consists of a linker that yields the same type of structure as if they were being used by themselves. Linkers with similar topologies and different secondary functionalities can be used to synthesize COFs. Oppositely, the HML approach yields a different structure when compared to a linker used by itself, which leads to a COF with a different topology. An example of the IML approach was performed by Nagai et al. [12] at which a condensation reaction among 1,4-benzene-diboronic acid (BDBA), azide-decorated BDBA (N_3-BDBA), and 2,3,6,7,10,11-hexahydroxytriphenylene (HHTP) were performed. Through that, a hexagonal COF skeleton functionalized with azide groups was obtained. Moreover, further chemical modifications in the structure were performed due to the presence of azide functions. An increase of this chemical group around the hexagonal COF walls led to a decrease in pore size and volume. Despite that, the azide groups allowed several types of post-synthetical functionalization through click chemistry to incorporate other organic functions such as aromatics, alkyls, esters, acetyls, or chromophores. The schematics for this process are shown in Figure 9.5. In that sense, a functionalized AcTrz COF-5 was 16 times more selective toward adsorbing CO_2 instead of N_2. This example illustrates one of the possibilities for the applications of COFs as, in this case, its surface was precisely functionalized to adsorb and differentiate CO_2 from N_2. Along that line, several other types of chemical moieties

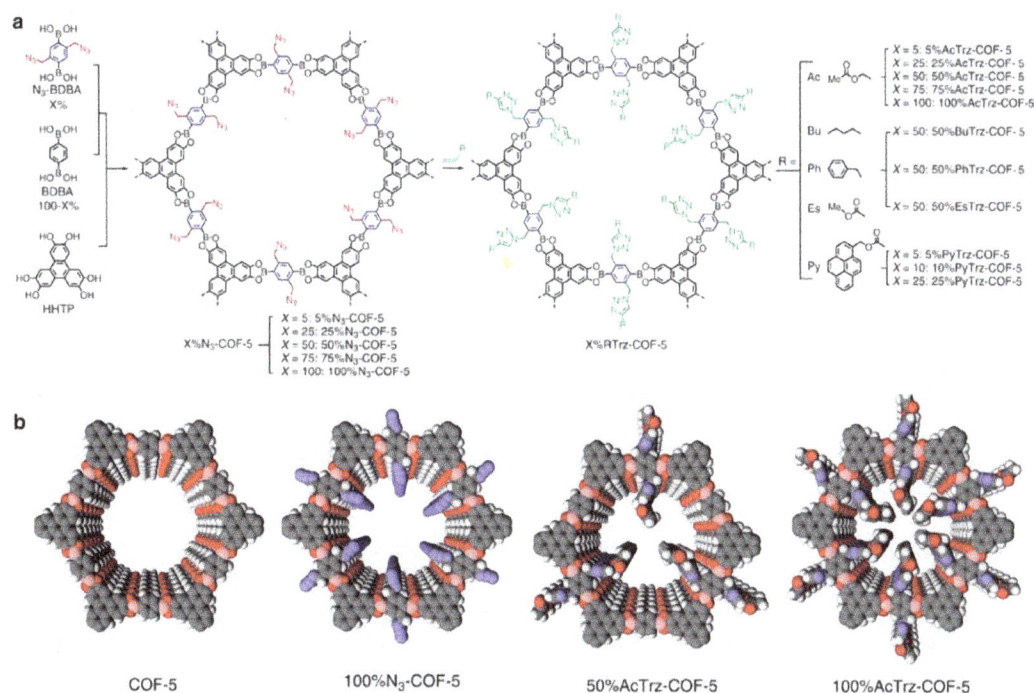

FIGURE 9.5

(a) Scheme for a general synthesis of a hexagonal COF through condensation reaction followed by click chemistry between azide and alkyne anchor groups to introduce other functional groups. (b) Chemical representation for the COF-5 series and the variations in their surfaces. Adapted with permission [12]. Copyright (2011), Springer Nature.

can be introduced into a COF's surface to enable different applications such as functional groups that provide redox, photocatalytic activity, gas adsorption, among others.

In addition to that, the synthesis of COFs can be performed through several methods including solvothermal, microwave-assisted, ionothermal, mechanothermal, interfacial, and under room temperature-controlled synthesis. The solvothermal method is the most explored as the reaction's efficiency is influenced by parameters such as temperature, time, starting materials reactivity, solubility, and catalyst concentration. A usual procedure consists of mixing a solution with components for the edges and vertices of the COF's structure along with solvents and catalysts. Followed by sonication, freeze, and defrost cycles, centrifugation, and purification. The microwave-assisted method offers a quicker approach when compared to solvothermal for means to obtain imine-linked and boronate-ester-linked COFs [13, 14]. One of the advantages of this method is that it can yield COFs with better porosity as the unreacted monomers may be easily extracted during the separation process. The ionothermal synthesis consists of mixing the monomers along with salts under high temperatures and long reaction times to melt the salt and promote its action as solvent as well as catalysts toward the reagents. Despite the high energy demand, it can be considered as a relatively green and efficient method to synthesize 3D COFs. Mechanochemical is a convenient method as it avoids a sealed environment under high temperature, pressure, and inert atmosphere, as the chemical reactions to form the COF structure are obtained through physical grinding. This makes this process more economical and eco-friendly. This approach can also be further optimized by dissolving the monomers into a solvent along with catalysts to enhance reaction rate during the grinding process which aids to improve the COF's crystallinity [15, 16]. Interfacial synthesis can be adapted to synthesize COF films at which the thickness can be controlled as it can vary from 4 to 150 nm [17].

9.3 Electrocatalytic Activities of COFs

The increasing consumption of fossil fuels raises concerns regarding their non-renewability and the harm that it causes to the environment. Also, relying upon one single source of energy is not the desired situation as it is important to have alternatives to supply the demand in case of unexpected events. Hence, depositing efforts to develop efficient and cleaner technologies for the generation of energy is a necessity for mankind. One of the zero-carbon energy generation processes that are receiving considerable attention is the oxygen reduction reaction (ORR) which consists of the electrochemical conversion of O_2 into H_2O with an overall release of energy as it is a more explored technology in fuel cells and metal-air batteries. It was observed that transition metal-based materials that are coordinated with N-doped carbon nanomaterials show appreciable stability and efficiency toward ORR [18, 19]. Accompanied to that it was observed that forming an M–C–N-type of bond, where M is a transition metal can promote a beneficial factor to improve catalytic performance. In addition, it was found that the presence of mesoporous plays an important role in electrolyte diffusion and mass transfer which aids in the ORR efficiency. This discovery increased the interest in the use of COFs for ORR catalysis. Hence, the combination of stable transition metals electrocatalytic sites, mesopore structured nanomaterials, and the presence of carbon-based doped with transition metals and/or heteroatoms can yield a satisfactory performance for ORR.

Based on these concepts, Zhao et al. [20] synthesized a Fe-doped bipyridine-based COF (TpBpy) that was carbonized to introduce Fe–N–C bonds into its structure (TpBpy-Fe). Silica was used as a hard template and *p*-toluenesulfonic acid (PTSA) was used to organize the COF structure for the synthesis. Pyridine moieties functioned as coordinating sites for Fe. The schematics for the synthesis along with the analysis of crystallinity and morphology for the TpBpy-Fe COF are presented in Figure 9.6. The improvement in surface area and pore volume facilitated the interaction with active sites as well as a mass transfer which led to a performance of 0.845 V half-wave potential and 5.92 mA/cm^2 for limiting current density which were competitive values compared to state-of-art Pt/C electrode that displayed 0.852 V and 5.57 mA/cm^2, respectively.

Another widely explored technology in this regard is the electro or photochemical conversion of H_2O into H_2 and O_2. This process consists of the full cell reaction in which at the cathode H_2 is generated through HER whereas at the anode O_2 is generated through OER. These reactions can be efficiently catalyzed by using Pt or Pd for HER and IrO_2 or RuO_2 for OER. However, due to the high cost and scarcity of these noble metals, it is necessary to develop novel electro or photocatalysts that can perform HER and OER reactions under scalable conditions. Hence, the initial challenge lies in employing low-cost and abundant materials for catalysis that can perform water splitting under similar or better conditions when compared to state-of-art electrocatalysts. In energy terms, the water splitting is a non-spontaneous process with Gibbs free energy of 237.1 kJ/mol, which corresponds to the

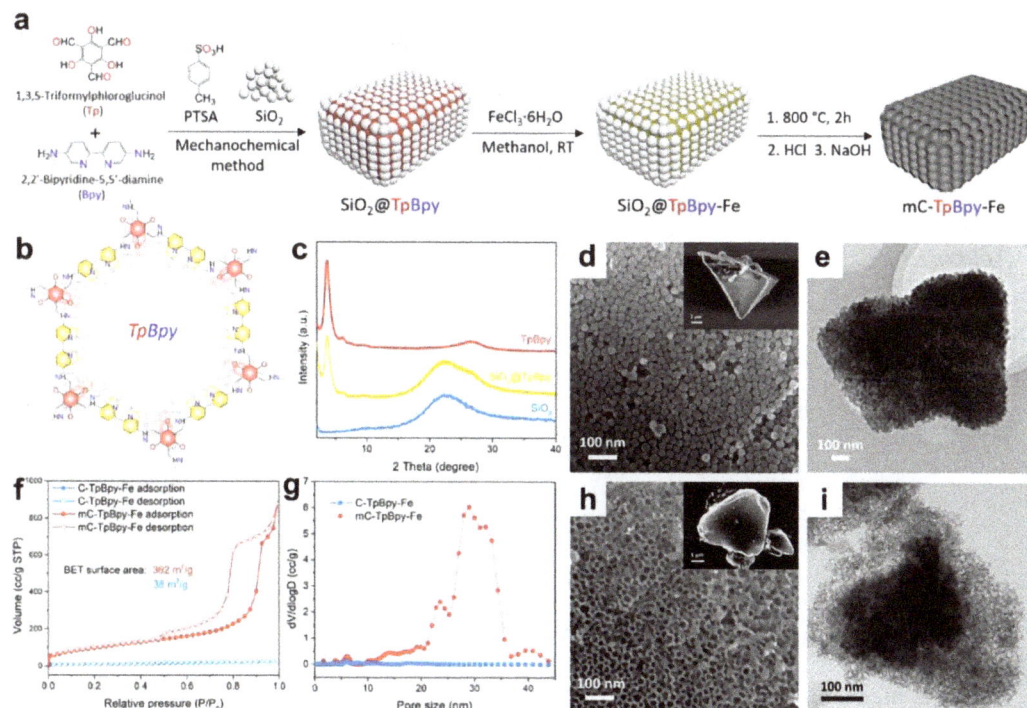

FIGURE 9.6

(a) Synthesis of mC-TpBpy-Fe. (b) TpBpy diagram displaying the N sites available for coordination with Fe. (c) Powder X-ray diffraction (PXRD) patterns. (d) Scanning electron microscopy (SEM). (e) Transmission electron microscopy (TEM) of SiO_2@TpBpy. (f) Surface area analysis. (g) Distribution of pore sizes for various samples. (h) SEM micrograph and (i) TEM images for the mC-TpBpy-Fe COF. (Adapted with permission [20]. Copyright (2019), American Chemical Society.)

electrical potential of 1.23 V ideally. However, due to the sluggishness and thermodynamical barriers the water splitting occurs at a potential higher than 1.23 V. The mechanism of HER has been widely researched as it can occur through the Volmer–Heyrovsky or Volmer–Tafel routes, which are related to the way protons are adsorbed into the substrate's surface along with the requirement of one electron to take place. The proposed chemical reactions for HER in acid media are described in Equations (9.1–9.4) and in alkaline media in Equations (9.5 and 9.6). The OER is the other half cell reaction that complements HER and since it is a multi-step process it displays higher thermodynamic and kinetic barriers making OER usually slower than HER. Likewise HER, OER can also occur in both acid and alkaline media as the reactions are described in Equations (9.7–9.11) and Equations (9.12–9.16), respectively.

HER

Acid media:

$$\text{Volmer: } * + H^+ + e^- \rightarrow * - H \tag{9.1}$$

$$\text{Heyrovsky: } * + H^+ + e^- + * - H \rightarrow H_2 + * \tag{9.2}$$

$$\text{Tafel: } 2H^* \rightarrow H_2 + 2^* \tag{9.3}$$

Alkaline media:

$$\text{Volmer: } * + H_2O + e^- \rightarrow * - H + OH^- \tag{9.4}$$

$$\text{Heyrovsky: } * + H^+ + e^- + * - H \rightarrow H_2 + * \tag{9.5}$$

$$\text{Tafel: } 2H^* \rightarrow H_2 + 2^* \tag{9.6}$$

OER

Acid media:

$$* + H_2O \rightarrow * - OH + H^+ + e^- \tag{9.7}$$

$$* - OH + \rightarrow OH^- \rightarrow * - O + H_2O + e^- \tag{9.8}$$

$$2^* - O \rightarrow 2^* + O_2 \tag{9.9}$$

or

$$2^* - O + H_2O \rightarrow * - OOH + H^+ + e^- \tag{9.10}$$

$$* - OOH + OH^- \rightarrow * + O_2 + H^+ + e^- \tag{9.11}$$

Alkaline media:

$$* + OH^- \rightarrow * - OH \tag{9.12}$$

$$* - OH + OH^- \rightarrow * - O + H_2O \tag{9.13}$$

$$2^* - O \rightarrow 2^* + O_2 \tag{9.14}$$

Or

$$^{*}O + OH^{-} \rightarrow ^{*} - OOH + e^{-} \tag{9.15}$$

$$^{*} - OOH + OH^{-} \rightarrow ^{*} + O_2 + H_2O \tag{9.16}$$

Where * is the active site from the substrate

One of the current challenges in water-splitting technology lies in finding materials capable of performing both HER and OER at a reasonable cost. COFs can be a good candidate for such applications as they can achieve higher cycling as well as chemical stability compared to other nanomaterials. In that sense, Wang et al. [21] developed a 2D COF composed of phthalocyanines and pyrazine groups that acted anchors for Mn and Cr, which enabled it to function as electrocatalysts for ORR, OER, and HER. Such versatility in multi-catalysis might be attributed to the combination of electron-rich segments due to the π–π as well as the well-distributed and exposed metallic centers. Through that, considerably low overpotentials were obtained as Cr-doped COF reached 0.29, 0.35, and -0.239 V whereas Mn-doped COF reached 0.31, 0.44, and -0.014 V for ORR, OER, and HER, respectively. The advantages of incorporating these nitrogenated groups, i.e., phthalocyanine and pyrazine are that they can strongly bond with the transition metals which prevents migration and agglomeration while simultaneously exposing them for catalysis. By performing the mechanistic study, it was observed that the transition metals, C atoms adjacent to N, and phthalocyanines were more prompt to absorb intermediates responsible for both ORR and OER. On the other hand, for the HER process, it was noted that H tended to bind on the pyridinic N. Through that, 2D COF doped with either Mn or Cr displayed triple catalysis properties for HER, OER, and ORR. The scheme of the 2D COF's structure along with the active sites for each catalytic process is presented in Figure 9.7.

9.3.1 Electrocatalytic Activities of COFs for HER

Performing HER under reproducible conditions is a much-desired process as it can provide a viable and sustainable energy source through water-splitting. Even though this process can be efficiently performed by using noble metals, i.e., Pt, Pd, Ir, and Ru this technology becomes applicable if affordable and abundant materials can be used instead. In that regard, COFs can function as metal-free catalysts for HER due to the different structural geometries and arrangement of light elements that grant them catalytic properties like those of transition metals. Sakamoto et al. [22] synthesized a 2D COF containing graphdiyne building blocks and pyrazine groups which yielded 20 nm film thickness with a pore size of 0.8 nm leading to a Brunauer-Emmet-Teller (BET) surface area of 408 m^2/g. The pyrazine-graphdiyne-based COF (PR-GDY) presented considerable thermal stability up to 300°C and provided a considerable improvement in HER performance over a glassy carbon electrode in both acid and alkaline media by displaying -275 mV of onset potential versus a reversible hydrogen electrode (RHE). Also, -475 and -270 mV to reach 10 mA/cm^2 at 0.5 mol/L H_2SO_4 and 0.1 mol/L $NaHCO_3$ and 0.1 mol/L Na_2CO_3 buffer, respectively. Also, the Tafel slope of 75 mV/dec suggested a favorable process for HER. This satisfactory performance is likely attributed to the presence of heteroatoms of N along with the COF's symmetric structure which created electron-rich active sites that can adsorb H as this phenomenon took place through the pyridinic N. Even though the performance for PR-GDY was falling behind those of state-of-art Pt it is worth noting that satisfactory catalytic performance was achieved based on the arrange of C and N atoms that created

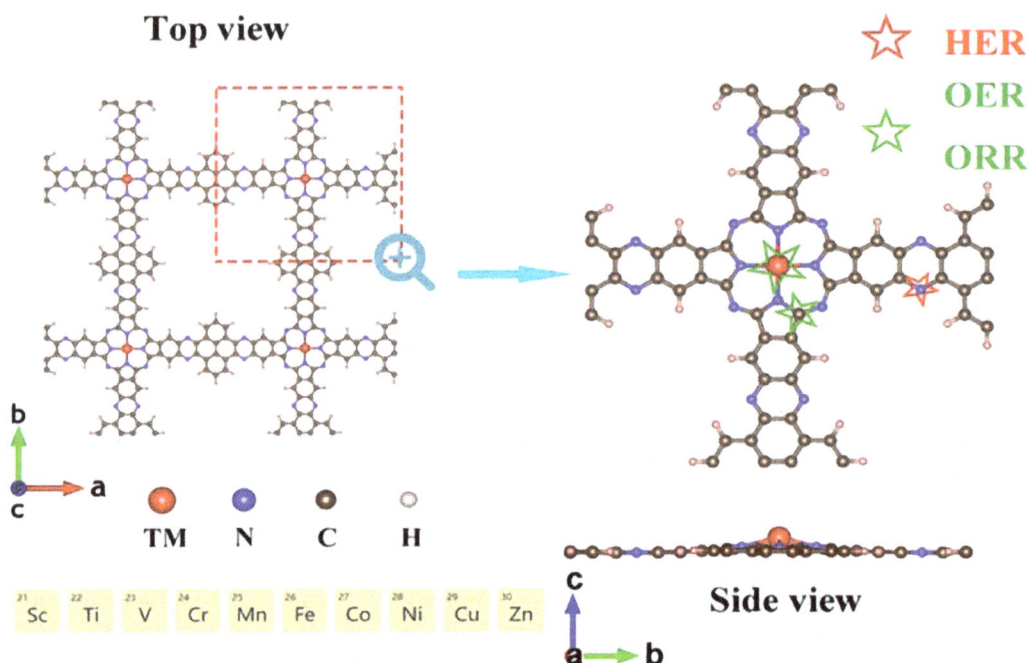

FIGURE 9.7
Diagram for the 2D COF's structure from the top and side perspective with possible active sites for HER, OER, and ORR. (Adapted with permission [21]. Copyright (2020), American Chemical Society.)

evenly distributed active sites that were highly exposed due to the inherently high porosity of COF's. Hence, despite its generally lower efficiency when compared to Pt electrode it shows great potential as a metal-free electrode nanomaterial for the HER process.

What is notable that, the attractive points for the development of COFs lie in their lower cost regarding component composition, accompanied by their high robusticity and electrocatalytic activity. The adjustable parameters such as controllable pore size, and facile tunability creates a broad range of possibilities for electrocatalyst. By focusing on the introduction of other heteroatoms other than N in COF's structures Yang et al. [23] proposed the synthesis of a 2D COF composed of building blocks containing aromatic rings with aldehyde and amine side groups to form imine linkages. The synthesis was performed by solving the monomers in acidic media in the presence of the reaction catalyst. Then, doping with N and P was performed to further improve COF's catalytic properties. In short, the N doping was performed through the pyrolysis process under an inert atmosphere at 1000°C followed by the P doping by using Na_2HPO_2. The summary of this process can be seen in Figure 9.8. The N-P doped 2D COF displayed a 260 mV of overpotential at 10 mA/cm^2 along with high stability during the cycling process with negligible loss in properties. It was noted that a higher concentration of doped P led to an improvement in catalytic activity as the larger presence of P atoms improved the edge effects along with stabilizing the C–N bonds, likely enabling relatively longer retention of intermediates. On top of that, the N-P doped 2D COF presented satisfactory catalytic activity for ORR as well. Hence, the synthesis of COF's holds great potential for a dual metal-free catalytic process with relatively high efficiency.

In another study, Pradhan et al. [24] synthesized metal-free COFs based on the monomers 1,3,5-tris(4-formylphenyl)benzene (TFPB) along with 5,10,15,20-tetrakis

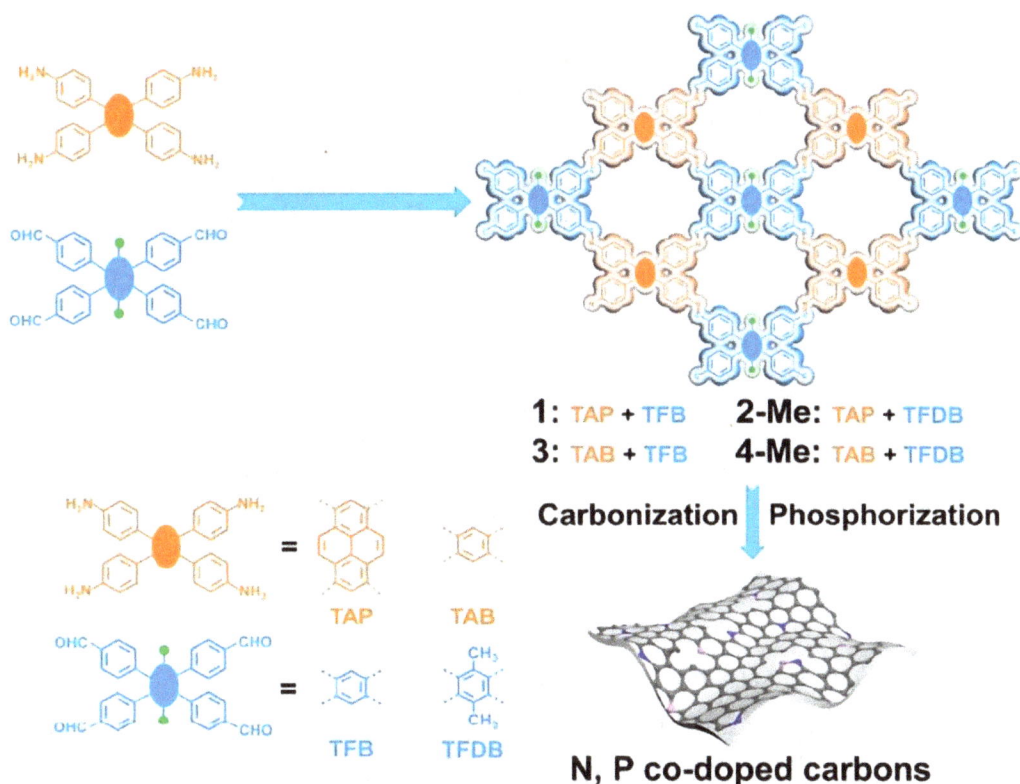

FIGURE 9.8
Synthesis of 2D COF linked through imine bonds and pos doped with N and P heteroatoms. (Adapted with permission [23]. Copyright (2020), American Chemical Society.)

(4-aminophenyl) porphyrin (TPM), which was named TFPB-PAM. The COF was synthesized through the solvothermal method. The electrocatalytic analysis for this material displayed relatively satisfactory performance toward HER by yielding an overpotential of 185 mV and 68.5 mV/decade for the Tafel slope. On top of that, its faradaic efficiency and retention were 98 and 86%, respectively. As for faradaic retention, it was measured at initial current density and after 3,000 cycles. Some of the factors that contributed to the satisfactory HER performance were related to the thin film of TFPB-PAM deposited over the glassy carbon electrode which provided a lower electron transfer resistance. In addition, the highly conjugated structure was likely to improve the electron-transfer step. Another factor that added to the improvement in HER activity was the high surface area that allowed a facile permeation of water into the electrode coated with TFPB-PAM. With the idea to develop an optimized composite for the HER process Zhao et al. [25] fabricated a COF composited with rGO, which served as a support, followed by the incorporation of Ru nanoparticles. The combination of these three nanomaterials to fabricate the electrocatalysts provided satisfactory results for HER as the overpotential as well as Tafel slope were considerably lower, reaching 42 mV at 10 mA/cm^2 along with 46 mV/decade in alkaline media, which suggested a facilitated catalytic effect. On top of that, the COF-based composite also presented good stability and faradaic retention. In that sense, Ru is regarded as a highly efficient metal toward HER catalytic activity as it has been previously reported [26, 27]. One of the approaches lies in promoting the Ru–N where the N source comes from pyridinic groups. Through that, efficient catalytic sites can be introduced into the structure. However, the use of nanoparticles in traditional nanomaterials may lead to agglomeration which can considerably decrease its surface area and therefore exposure to active sites. Hence, the implementation of COFs plays an important role as it allows the even distribution of metallic atoms without causing them to cluster. Additionally, COFs can introduce a synergy effect by facilitating ionic diffusion through their microchannels. COF may display a relatively lower electronic conductivity despite its high degree of conjugation. To address that, rGO can serve as conducting support to facilitate the catalytic process. It is also worth noting that, COF can also function as a spacer between the rGO nanolayers, preventing it from restacking. Based on all these factors it is notable how the overall catalytic performance can be improved through the fabrication of this ternary nanocomposite. The COF/rGO-Ru was also compared to other composites synthesized through the same process that was incorporated with Ni and Co instead of Ru. It was observed that the replacement of Ru with other transition metals caused a decrease in HER performance, which suggested that further optimization may be required to promote the proper catalytic activity with other metals. In this regard it was notable the fabrication of the COF/rGO-Ru showed a type of material that can be a potential competitor against commercial Pt/C electrodes for HER.

9.3.2 Photocatalytic Activities of COFs for HER

Finding efficient ways to use solar radiation as a driving force to the production of fuels such as H_2 is a much-desired technology as it can provide a highly sustainable process for the constant generation of energy. There are a few principles applied to the photocatalytic reactions which can be associated with some of the inherent properties of COFs. For a photochemical reaction to occur first that must be light harvesting. Second, the material should be able to promote an e^-- h^+ charge separation for sufficient time to allow the next reaction steps to take place. Third, promote the migration of the charged species (e^-- h^+) to the surface. Fourth, promote the redox process at the photocatalyst's surface whereas the photoactive species should be integrated into the molecular framework to absorb the photons generated. Then, photons can form excitons which function as an energetical stimulus at the catalytic centers to conclude the reaction, which in this case is the HER process. For that, it deems necessary to use materials that are photochemically stable and possess energy levels that fall within the range to perform the photoreactions. For the case of water-splitting, the conducting band (CB) should be at more negative levels than the H^+/H_2 redox potential (0 V), whereas the valence band (VB) should be at more positive levels than the O_2/H_2O (1.23 V). It's also important that the photocatalysts can maintain the e^-- h^+ separated to improve its efficiency. Within that line, COFs present themselves as viable candidates for photocatalysis due to several factors such as the long length and highly organized π–π conjugation which can broaden the range of light absorption to facilitate the transfer of photoexcited species. Also, the COF's nanochannels allow a fast mass transport along with several tunable processes available to introduce photoactive species that can be coated into the COF's structure. Another possibility lies in using COFs as molecular support to grow photoactive materials over it, which may improve their performance due to the COF's high surface area.

Lotsch et al. [28] proposed the synthesis of COFs able to perform photocatalysis for HER under visible light. The material consisted of the condensation reaction between

1,3,5-tris (4-formylphenyl) triazine (TFPT) and 2,5-diethoxyterephthalohydrazide (DETH). Following that, Pt nanoparticles were also incorporated into the COF's structure. Through that, the Pt embedded COF containing hydrazone groups that were responsible for the photoactivity delivered 1970 μmol/g.h. As a follow-up study at which an azine-based COF was synthesized, it was found that increasing concentration of N in the aryl groups also promoted an increase in H2 production, which at from the absence of N in the COF's structure to the highest concentration of N, in that study, led to the production of 23–1703 μmol/g.h of H_2. Such improvement was likely due to the COF's nanochannels that promoted better migration of photoexcited species along with more exposure to nitrogenated active sites due to its high surface area [29].

In another approach, Pachfule et al. [30] synthesized two β-ketoenamine-based COFs, through an acid solvothermal process, at which one presented acetylene groups (–C≡C–) and the other diacetylene groups (–C≡C–C≡C–). Pt nanoparticles in the form of H_2PtCl_6 were also incorporated into the COF's structure. The sp C containing moieties have been explored due to their specific properties that prompt applications such as optoelectronics, photocatalyst, photochromism, among others. However, the –C≡C– present weaker van der Waals interactions when compared to –C=C, which culminates in a decrease of proper stacking of the COF structure. Hence, to address that it deems important that the COF presents a double bond along with its structure as finding an optimized reaction time is also crucial. For this case, it was observed that performing the reaction for 4 days yielded COFs with higher crystallinity. However, when the reaction time was less than 3 days and around 8 days the authors observed a decrease in crystallinity, suggesting that optimized reaction time is necessary to improve certain properties. Based on that the authors synthesized the two COFs based on 1,3,5-triformylphloroglucinol (TP) that reacted with either 4,4'-(ethyne-1,2-diyl)dianiline (EDDA) to obtain COF-TP-EDDA or 4,4'-(buta-1,3-diyne-1,4-diyl)dianiline (BDDA) to obtain COF-TP-BDDA. During the analysis of H_2 production at visible light irradiation (≥395 nm) yield between the COFs, it was observed that COF-TP-BDDA greatly exceeded the catalytic properties of COF-TP-EDDA as their productions were 324 μmol/h.g and 20 μmol/h.g. The considerable difference in H_2 production performance suggests that the presence of –C≡C–C≡C– played a major role in the catalytic process as it provided a lower bandgap when compared to COF-TP-EDDA. Through that, the photocatalytic process can not only take place under lower energy levels but also through high charge carrier transfer provided by the diacetylene groups.

The development of novel photocatalysts that can deliver satisfactory efficiency while preferably not relying on metallic co-catalysts is a highly desirable process. However, matching the efficiency of state-of-art noble metals and inorganic semiconductors that are commonly explored in this regard remains a challenge. Wang et al. [31] synthesized a COF based on benzo-bis(benzothiophene sulfone) groups that enabled highly photocatalytic activity toward HER while maintaining a steady performance for around 50 h along with a sacrificial electron donor at the range of 420 nm wavelength. Several factors could be attributed to the improved performance. First, high crystallinity and wettability promoted attractive forces between the COF's interface with water. Second, efficient light absorbance. Third, mesopores structures around 3.2 nm facilitated the COF's structure to be dye-sensitized. Through that, there was an HER rate of around 16.3 mmol/g.h. This example displays a promising approach for the synthesis of COF while also showing the main factors that might determine its overall performance toward photocatalysis of HER.

9.4 Conclusion

Energy generation is one of the most discussed matters as it is a core aspect of the development of mankind. Because of this level of importance, it is necessary to develop suitable technologies that can be accessible, sustainable, and efficient. There has been continuous progress in that regard as several technologies are being studied to be further improve. In one perspective, Li-ion batteries (LIB), for instance, are currently the dominant technology. However, the devices fabricated through LIB's technology are likely incapable to supply the demand of a given nation, likely due to a lack of materials available. It is important to develop novel technologies that can rely on abundant resources, which can provide a more stable way of production. On top of that, using alternative sources of energy, preferably renewable ones are a promising way to solve energy-related issues. Based on these two aspects, the generation of energy through renewable resources such as water is highly attractive as it can be replenished while avoiding the emission of carbon for the process. Along with that, COFs which are composed of light and organic elements emerged as a viable option of materials that can be used for the generation of energy through the water-splitting process for the production of H_2 and O_2, as the former was emphasized in this chapter. The interesting properties of COFs which can be summarized in highly ordered domains granting them crystal morphology, high surface area, and tunability that allows the incorporation of catalytic groups for both electro and photo routes, controllable pore size, and synthetical versatility are factors that make these materials highly attractive. However, both water-splitting and COF technologies present drawbacks that must be solved to allow their large-scale application. In terms of water-splitting, there is the need to efficiently perform the HER and OER processes under acceptable input of energy to make the process feasible. Also, it is desirable to use metal-free catalytic routes, due to the non-renewability of those. On the other hand, even though COFs are mostly composed of organic moieties their synthetical approaches are relatively long when compared to inorganic semiconductors or other organic-based materials. Also, the specificity of the monomers required for their synthesis can considerably increase the overall cost. COFs have been introduced around 2005 and already show great potential for applications in several areas along with energy storage, which also includes gas storage and photocatalysis. In addition, satisfactory results have been obtained in regards to overall energy generation. Thus, further exploring the vast possibilities within the potential of COFs is a promising approach.

References

1. Feng X, Ding X, Jiang D (2012) Covalent organic frameworks. Chem Soc Rev 41:6010–6022
2. Diercks CS, Yaghi OM (2017) The atom, the molecule, and the covalent organic framework. Science (80-)355:1585
3. Ding S-Y, Wang W (2013) Covalent organic frameworks (COFs): from design to applications. Chem Soc Rev 42:548–568
4. Geng K, He T, Liu R, Dalapati S, Tan KT, Li Z, Tao S, Gong Y, Jiang Q, Jiang D (2020) Covalent organic frameworks: Design, synthesis, and functions. Chem Rev 120:8814–8933
5. Trewin A, Cooper AI (2009) Predicting microporous crystalline polyimides. CrystEngComm 11:1819–1822

6. Baldwin LA, Crowe JW, Pyles DA, McGrier PL (2016) Metalation of a mesoporous three-dimensional covalent organic framework. J Am Chem Soc 138:15134–15137

7. Fang Q, Zhuang Z, Gu S, Kaspar RB, Zheng J, Wang J, Qiu S, Yan Y (2014) Designed synthesis of large-pore crystalline polyimide covalent organic frameworks. Nat Commun 5:4503

8. El-Kaderi HM, Hunt JR, Mendoza-Cortés JL, Côté AP, Taylor RE, O'Keeffe M, Yaghi OM (2007) Designed synthesis of 3D covalent organic frameworks. Science (80-)316:268–272

9. Li H, Evans AM, Castano I, Strauss MJ, Dichtel WR, Bredas J-L (2020) Nucleation–elongation dynamics of two-dimensional covalent organic frameworks. J Am Chem Soc 142:1367–1374

10. Evans AM, Parent LR, Flanders NC, Bisbey RP, Vitaku E, Kirschner MS, Schaller RD, Chen LX, Gianneschi NC, Dichtel WR (2018) Seeded growth of single-crystal two-dimensional covalent organic frameworks. Science (80-) 361:52–57

11. Smith BJ, Dichtel WR (2014) Mechanistic studies of two-dimensional covalent organic frameworks rapidly polymerized from initially homogenous conditions. J Am Chem Soc 136:8783–8789

12. Nagai A, Guo Z, Feng X, Jin S, Chen X, Ding X, Jiang D (2011) Pore surface engineering in covalent organic frameworks. Nat Commun 2:536

13. Wei H, Chai S, Hu N, Yang Z, Wei L, Wang L (2015) The microwave-assisted solvothermal synthesis of a crystalline two-dimensional covalent organic framework with high CO2 capacity. Chem Commun 51:12178–12181

14. Ritchie LK, Trewin A, Reguera-Galan A, Hasell T, Cooper AI (2010) Synthesis of COF-5 using microwave irradiation and conventional solvothermal routes. Microporous Mesoporous Mater 132:132–136

15. Peng Y, Xu G, Hu Z, Cheng Y, Chi C, Yuan D, Cheng H, Zhao D (2016) Mechanoassisted synthesis of sulfonated covalent organic frameworks with high intrinsic proton conductivity. ACS Appl Mater Interfaces 8:18505–18512

16. Shinde DB, Aiyappa HB, Bhadra M, Biswal BP, Wadge P, Kandambeth S, Garai B, Kundu T, Kurungot S, Banerjee R (2016) A mechanochemically synthesized covalent organic framework as a proton-conducting solid electrolyte. J Mater Chem A 4:2682–2690

17. Zhou D, Tan X, Wu H, Tian L, Li M (2019) Synthesis of C– C bonded two-dimensional conjugated covalent organic framework films by Suzuki polymerization on a liquid–liquid interface. Angew Chemie 131:1390–1395

18. Li S, Cheng C, Zhao X, Schmidt J, Thomas A (2018) Active salt/silica-templated 2D mesoporous FeCo-Nx-carbon as bifunctional oxygen electrodes for zinc–air batteries. Angew Chemie 130:1874–1880

19. Wang J, Huang Z, Liu W, Chang C, Tang H, Li Z, Chen W, Jia C, Yao T, Wei S, Wu Y, Li Y (2017) Design of N-Coordinated Dual-Metal Sites: A Stable and Active Pt-Free Catalyst for Acidic Oxygen Reduction Reaction. J Am Chem Soc 139:17281–17284

20. Zhao X, Pachfule P, Li S, Langenhahn T, Ye M, Tian G, Schmidt J, Thomas A (2019) Silica-templated covalent organic framework-derived Fe-N-doped mesoporous carbon as oxygen reduction electrocatalyst. Chem Mater 31:3274–3280

21. Wang J, Wang J, Qi S, Zhao M (2020) Stable multifunctional single-atom catalysts resulting from the synergistic effect of anchored transition-metal atoms and host covalent-organic frameworks. J Phys Chem C 124:17675–17683

22. Sakamoto R, Shiotsuki R, Wada K, Fukui N, Maeda H, Komeda J, Sekine R, Harano K, Nishihara H (2018) A pyrazine-incorporated graphdiyne nanofilm as a metal-free electrocatalyst for the hydrogen evolution reaction. J Mater Chem A 6:22189–22194

23. Yang C, Tao S, Huang N, Zhang X, Duan J, Makiura R, Maenosono S (2020) Heteroatom-doped carbon electrocatalysts derived from nanoporous two-dimensional covalent organic frameworks for oxygen reduction and hydrogen evolution. ACS Appl Nano Mater 3:5481–5488

24. Pradhan A, Manna RN (2021) Surface-modified covalent organic polymer for metal-free electrocatalytic hydrogen evolution reaction. ACS Appl Polym Mater 3:1376–1384

25. Zhao Q, Chen S, Ren H, Chen C, Yang W (2021) Ruthenium nanoparticles confined in covalent organic framework/reduced graphene oxide as electrocatalyst toward hydrogen evolution reaction in alkaline media. Ind Eng Chem Res 60:11070–11078

26. Peng Y, Lu B, Chen L, Wang N, Lu JE, Ping Y, Chen S (2017) Correction: Hydrogen evolution reaction catalyzed by ruthenium ion-complexed graphitic carbon nitride nanosheets. J Mater Chem A 5:19499

27. Mahmood J, Li F, Jung S-M, Okyay MS, Ahmad I, Kim S-J, Park N, Jeong HY, Baek J-B (2017) An efficient and pH-universal ruthenium-based catalyst for the hydrogen evolution reaction. Nat Nanotechnol 12:441–446

28. Stegbauer L, Schwinghammer K, Lotsch BV (2014) A hydrazone-based covalent organic framework for photocatalytic hydrogen production. Chem Sci 5:2789–2793

29. Vyas VS, Haase F, Stegbauer L, Savasci G, Podjaski F, Ochsenfeld C, Lotsch BV. (2015) A tunable azine covalent organic framework platform for visible light-induced hydrogen generation. Nat Commun 6:1–9

30. Pachfule P, Acharjya A, Roeser J, Langenhahn T, Schwarze M, Schomäcker R, Thomas A, Schmidt J (2018) Diacetylene functionalized covalent organic framework (COF) for photocatalytic hydrogen generation. J Am Chem Soc 140:1423–1427

31. Wang X, Chen L, Chong SY, Little MA, Wu Y, Zhu W-H, Clowes R, Yan Y, Zwijnenburg MA, Sprick RS, Cooper AI (2018) Sulfone-containing covalent organic frameworks for photocatalytic hydrogen evolution from water. Nat Chem 10:1180–1189

10

Covalent Organic Frameworks-Based Nanomaterials for Oxygen Evolution Reactions

Moein Darabi Goudarzi, Negin Khosroshahi, Parisa Miri, and Vahid Safarifard

Department of Chemistry, Iran University of Science and Technology, Tehran, Iran

CONTENTS

10.1 Introduction

Currently, the utilization of fossil fuels by human beings and their unfavorable effect on the surroundings has attracted the attention of countries from all around the world. To escalate energy use, energy storage, conversion, and environmental protection have prompted extensive discoveries as a dependable electrochemical application tool in recent years [1]. Exploring other abundant, renewable, economically, and technologically greener energy sources is crucial. Furthermore, the vast energy resources on earth, including wind, nuclear power, sunlight, hydro/geothermal, and fossil fuels, that can be converted into electricity through appropriate devices, must be stored for further use [2–3]. Electricity storage is essential for portable energy preparation. In practice, high efficiency, pollution less, and stability of energy storage batteries or capacitors for electrochemical water splitting for hydrogen evolution, fuel cells to generate electricity, and for energy storage have been offered as futuristic solutions for energy conservation and transformation [4, 5]. The above-mentioned solutions eliminate our dependence on customary fossil fuels. Presently, thorough research is currently underway on these emerging technologies. In particular, water splitting is a pivotal technology for sustainable and efficient energy production in addition to fuel cells and batteries as auspicious energy storage devices [3, 6]. OER which is an anodic reaction involves several reactions including water splitting and also the process of metal-air batteries charging. H_2O oxidizes to O_2 by 4e⁻ transfer based on the reaction route. But the point is oxygen evolution reaction processes are generally slow reactions due to huge obstacles such as forming O-O bonds, breaking O-H bonds, and removing 4 electrons from water molecules. It is considered beneficial that according to the following reactions, OER can show that RuO_2 and materials based on Ir are great electrocatalysts for OER. Unfortunately, these materials cannot be used widely because of their high cost, limited storage, and their rarity. So, a lot of efforts have been made all around the world to reach more efficient electrocatalysts. Although these technologies have potential uses, they are generally very costly. They require noble electroactive metals like ruthenium oxide (RuO_2) and platinum (Pt) to stimulate key electrochemical reactions including oxygen reduction reaction (ORR), hydrogen evolution reaction (HER), and oxygen evolution reaction (OER) [3, 7]. Unfortunately, the high cost and rarity of these noble metals preclude the commercial use of sustainable and green energy technologies.

Great efforts have been made to synthesize electrocatalysts with abundant materials on earth, consisting of metals and their compounds (e.g., metal alloys, oxides of three-dimensional transition metals and dichalcogenides, and perovskite oxides [8]), and non-metallic carbon nanomaterials (e.g., carbon nanopores, carbon nanotubes, and graphene)) [9, 10]. Two-dimensional COFs are an excellent group of molecular frameworks that are connected via covalent bonds. The unique combination of high crystallinity, great molecular architecture, large surface area, and tunable pore size have introduced COFs as promising candidates for a new generation of selective transport, drug delivery systems, sensors, light-emitting diodes, electrocatalysts, field-effect transistors, and photovoltaics [11]. COFs with specifically controllable structures can be synthesized via selecting suitable linkage motifs and building blocks, which provide an excellent high-performance system for profitable electrocatalysis. Newly, great creativities have been made in design and synthesis COFs to advance efficient technologies in the field of clean energy. The strength of the COFs is functional skeletons, high porosity, well-defined structures, and adjustable and periodic pores. These structures are connected by covalent bonds and through reversible reactions [12]. Stable thermodynamic networks are obtained by reversing the polymerization reaction that corrects the errors [13]. COFs are widely used in optoelectronics [14, 15], sensing [16], catalysis [17, 18], separation [19], storage and sorption [20, 21], and so forth. Yaghi synthesized the first COFs in 2005 [22]. Synthesizing COFs was a success in elimination of "crystallization issue". As an example, the interaction between building blocks which is covalent bonding, usually creates amorphous or weakly crystalline polymers, and the possibility of designing and preparing polymers in primary, secondary, and ternary with highly ordered structures based on the principle of retinal chemistry. In addition, the recent attainment of synthesizing high-quality single crystals makes it possible to decode their structures with atomic precision [23]. Two-dimensional and three-dimensional COFs are expected based on the geometric symmetry of the building blocks. In 2D COFs, a layered structure is made through π-π interaction and organic groups that are bound in 2D sheets. Most of them can be stacked regularly by forming channels. But a few of them accumulate irregularly. Both stacked modes show very specific adjustments. For 3D COFs, their constituents including silane, sp^3 carbon, or boron aids the network expansion into 3D space. In addition, the variety of structural blocks creates multiple combinations, which gives COFs many possibilities in building design [24].

Over the past few years, some excellent literature reviews have summarized the synthesis and application strategies of COF-based materials. In addition to the significant progress of porous organic polymers, a wide variety of COF-based materials have been developed, including strong catalytically active sites with high performance and a specific structural unit with excellent mass transfer [25]. Bulk COF materials can be provided through microwave, mechanical conditions, thermal, or ionothermal, and several researches have explained them in detail.

COF materials have drawn much attention in electrochemical water cracking research. Due to weak conductivity and electrochemical activity, several types of research have been done for developing the composition and structure of COFs after refining to achieve higher OER performance. During the ORR process, H_2O or OH^- generates by O_2 reduction, whereas O_2 generates via H_2O oxidation through OER. OER and ORR processes are the reverse version of each other [26]. In this chapter, we have presented the approaches for COF NSs and thin films that are divided into bottom-up and top-down strategies. The advancement of COFs in OER was highlighted then, and the future vision for the improvement of effective COF-based catalysts for clean energy conversion and storage is also discussed.

10.2 Covalent Organic Frameworks (COFs)

The new type of porous organic materials is named covalent organic frameworks; which are smartly constructed via organic building units and held together by strong covalent bonds. COF materials exhibit excellent potential in different applications, such as adsorption, catalysis, optoelectric, and gas storage. This is due to their well-defined crystalline porous architectures and customized functions. In 2005, Yaghi conducted pioneering research in the field of COFs and attracted other scientists' interest with a variety of expertise from all around the world for developing these structures. COFs can be synthesized using a variety of methods, including solvothermal, ionothermal, microwave, bottom-up, and top-down methods (Figure 10.1) [12]. Here we mostly focus on bottom-up and top-down approaches for preparing COF nanosheets and thin films.

10.2.1 Bottom-Up Strategy

A bottom-up synthesis method is a critical approach for fabricating COF NSs and thin films, the important factor here is organizing precursors well and restricting the condensation reactions among monomers in the substrate's smooth surfaces and also the interface between two phases. In most cases, an on-surface method is used to prepare COF NSs with atomic thickness. Controlling the orientation of COF NSs or thin films during preparation and avoiding chaotic dispersion or accumulation of monomers on surfaces is significant. To direct the polymerization reactions, the chosen substrates, like highly oriented pyrolytic graphite (HOPG) and metals, should have a well-defined single-crystal surface. Surfactants or laminar assembly polymerization (LAP) have lately been used to successfully synthesize few-layered COF NSs [27]. In contrast, fabricating COF thin films using several layers of nanosheet-stacked structures is quite facile, although solvent polarity, monomer concentration, temperature, substrate, and other factors must all be carefully optimized.

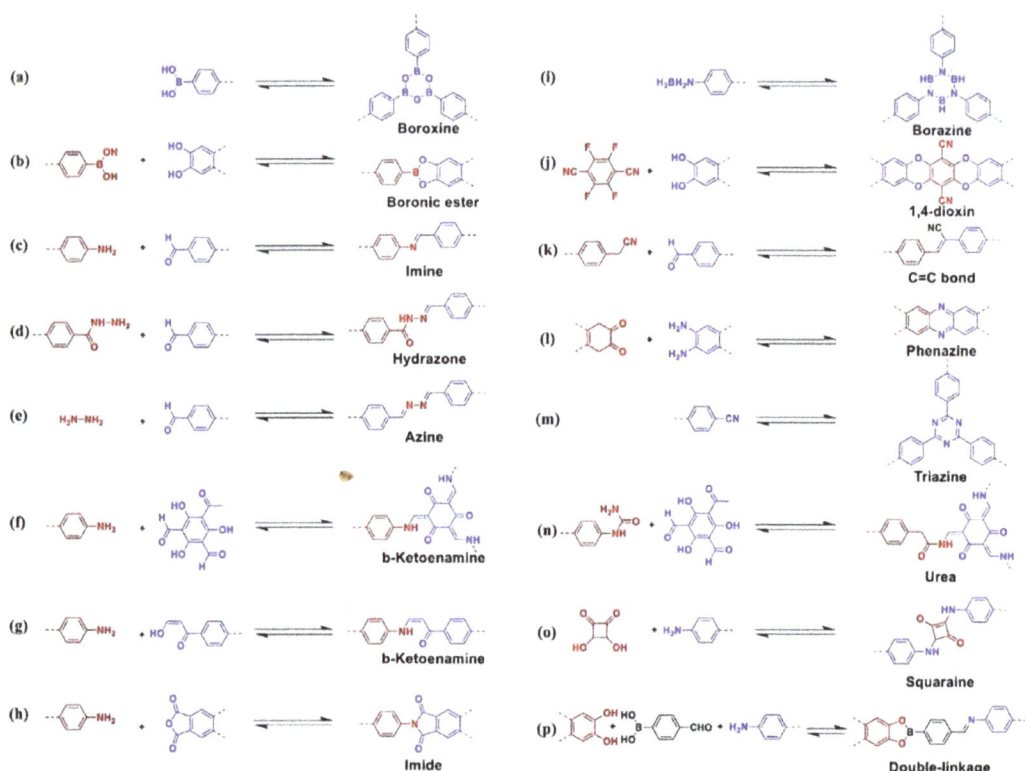

FIGURE 10.1
Various applications and methods for synthesizing COFs [12].

10.2.1.1 Single-Layered Covalent Organic Frameworks (sCOFs)

One of the most efficient methods for producing single-layered COF is on-surface synthesis. Scanning tunneling microscopy (STM) with high resolution can expose the atomic system of sCOFs; sCOFs that contain covalent bands are more chemo- and thermo-stable compared to supramolecular sCOFs which contain non-covalent bonds. Furthermore, the reversibility of the condensation reaction is necessary for error-checking and proof-reading procedures to produce long-range order in COFs. Moreover, due to the reorganization of the monomers driven by the ultra-flat surface, irreversible reactions can be used in synthesizing sCOFs in addition to reversible reactions like imine coupling and boronic anhydridation reactions [28].

10.2.1.2 COF Multilayer Nanosheets and Thin Films

To synthesize ultrathin 2D nanomaterials, the solvothermal synthesis method is frequently used as a common wet-chemical synthesis strategy. Water or an organic solvent act as the reaction medium in a closed system during solvothermal synthesis and the heating temperature needs to be kept above the solvent's boiling point. The use of heated solvents under high pressure can speed up the reaction and improve nanocrystal crystallinity [29]. By using a solvothermal approach, COF nanosheets and thin films can be achieved with or without using a substrate that is immersed in the solvents. As an example, Dichtel

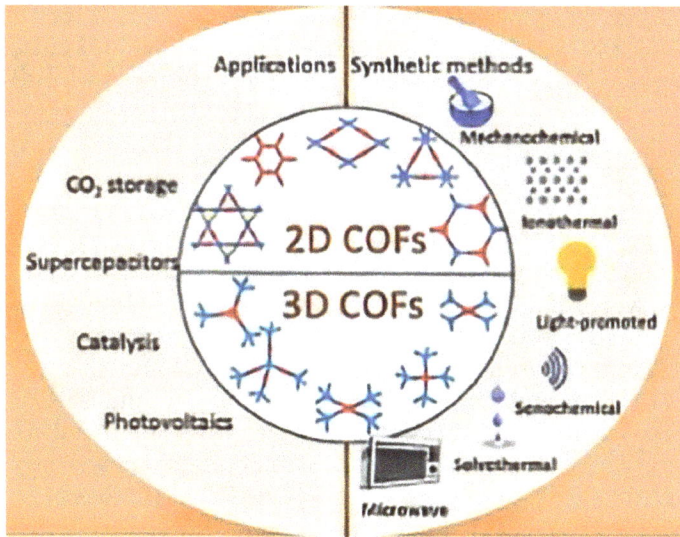

FIGURE 10.2
Schematic of COF-5 films. (Adapted with permission from reference [31]. Copyright (2020) Materials Research Society.)

discovered that at 90°C, a mixture of monomers and solvents can be applied to produce 2D COF-5 thin films on the surface of single-layer graphene (SLG) supported by transparent fused silica (SiO_2), silicon carbide, or copper substrates (Figure 10.2) [30]. The COF film had a considerably higher orientation than the similar COF powder sample, and the polymeric layers were placed normal to the SLG surface.

10.2.1.3 Interface Synthesis

As an example, Zhang's group used the Langmuir–Blodgett method to make Schiff-based 2D COF NSs. The thickness of the structure was approximately about 0.7 nm at the air/water interface [32]. Bao used a shielded Petri dish for placing the reaction solution in it for two days under ambient environmental conditions. After this, another group researched COF films preparation at the solution/air interface. The resulting films had smooth surfaces and their thickness could be tuned from 1.8 nm to 29 nm by changing the monomer concentration and reaction time [33]. Choi discovered that photon irradiation reduced the formation time of COF NSs at the air/water interface to less than 1 h, and it is noteworthy that the width of the resulting COF NSs was 0.75–2.59 nm [34]. Photon energy was found to be important in both speeding imine condensation and simplifying the conversion process from imines which are amorphous to COF crystals, according to the researchers.

10.2.2 Top-Down Method

Layered structures with regularly aligned channels construct bulk 2D COFs. Individual polymeric layers are stacked vertically via quite weak interactions (such as hydrogen bonding and/or Van der Waals force). Via the top-down method, multi-/mono-layered COF NSs can be produced by breaking the interactions between the COFs' layers that are placed near to each other, which is a simple and effective method for synthesizing NSs. Self-exfoliation, mechanical exfoliation, chemical exfoliation, and liquid-assisted exfoliation are the four methods used to exfoliate COFs which have been studied so far [28].

10.2.2.1 Mechanical Exfoliation

Mechanical delamination worked well for a huge number of 2D materials including graphene [29]. Banerjee used this technique to exfoliate COFs for the first time in 2013 [35]. They synthesized eight Schiff-based COFs that were chemically stable. The pore sizes of these materials were ranged from 15 to 24 Å and also, they functionalized them with diamines. A facile and green method was employed for exfoliating these COFs: first, after placing COFs in a mortar, some drops of methanol and soil were added to the contents of the mortar at ambient temperature for thirty minutes; then, via centrifuging the resulting powder after its dispersion into 100 mL of methanol, a clear liquid was achieved and the physical state was altered from dark red powder to the mentioned clear liquid; and finally, the COF NSs in flat nanosheet-like structures were reached by removing the solvent completely via evaporation. The nanosheets were several micrometers long, with a thickness ranging from 3 to 10 nanometers, indicating the presence of only 10–30 COF layers.

10.2.2.2 Liquid-Assisted Exfoliation

External mechanical forces like sonication can also be used to exfoliate layered bulk COFs into ultrathin 2D nanosheets in the liquid phase [28, 36]. A nanosheet suspension was attained after dispersing huge layered materials in a solvent with matched surface energy to COFs and performing sonication then. Sonication generates bubbles in the solution, which causes microjets and shock waves through the layered bulk materials when the bubbles burst. The bulk materials will be introduced into the synthesized strong tensile stress that results in exfoliated sheets containing a few layers.

10.2.2.3 Self-Exfoliation

Despite external exfoliation, the self-exfoliation method tends to deform due to internal pressures produced by the inherent ionic nature. Self-exfoliation is used to exfoliate many ionic COFs (iCOFs) containing charged centers in solvents [28, 37]. As an example, a guanidinium halide-based 2D iCOF (TpTGCl) which had inserted Cl⁻ ions as counter anions in two monomer units (~3.3 Å) was prepared in an eclipsed mode. The TpTGCl layers spontaneously exfoliated into nanosheets after soaking the powder in an aqueous solution because the interlayer distance was raised to ~5.5–6 Å in comparison to the original distance of 3.338 Å [37]. Electrostatic repulsion originating from loosely coupled Cl⁻ ions and positive guanidinium building units greatly reduced the π-π interactions between neighboring layers. TEM and AFM were used to examine micrometer to submicrometer-sized sheets whose height profile was 2–5 nm. The positively charged guanidinium units were substituted with a neutral ligand to establish that the self-exfoliation process was intimately linked to COF's ionic skeleton, and no sheet-like products were seen by TEM. Furthermore, the mechanically grinding technique was predicted to exfoliate as-synthesized TpTGCl, however PXRD patterns did not significantly change.

10.3 Providing Clean Energy by Oxygen Electrocatalysis

ORR arises in a variety of main chemical reaction processes, including bioprocesses (e.g., respiration), corrosion (electrochemical corrosion, etc.), and energy conversion (e.g., fuel cells, metal-air batteries, etc.) [38]. In fuel cells, the role of oxygen reduction on the cathode is critical for electricity production, whereas OER happens in water splitting or metal-air batteries recharging. For stimulating these important but slow chemical reactions, noble metal catalysts are commonly utilized in these devices and their performance is determined by the catalysts. Based on recent discoveries, COFs are great alternatives to noble metals due to their efficiency and being cost-effective for OER and ORR. 2D COFs could be prepared via manipulating functional groups at the edge of pores for gas storage, solar cells, lithium batteries, and fuel cells [39] since they are built of conjugated macromolecular structures that are comparable to graphene [40].

10.3.1 Electrochemical Application of 2D COF Nanosheets

Electrochemical water splitting, which included HER and OER, has been suggested as a viable option for producing green and also renewable sources of energy like hydrogen. Pt, which is a noble metal, has proven to be the best element in HER, whereas noble metal oxides (IrO_x and RuO_x) have been used as benchmark electrocatalysts in OER [41]. There are also some other materials to be used instead of electrocatalysts based on noble metals for water splitting. nonnoble metal catalysts and porous carbon materials that are dopped by metals are two of them [42]. In a water electrolytic system, the electrocatalytic OER is crucial. The OER often requires the assistance of expensive noble metal-based electrocatalysts (RuO_2 and IrO_2) to overcome its slow kinetics, as the OER's energy conversion efficiency is severely challenged by large overpotentials [43]. As a result, much effort has put into designing and also developing alternative electrocatalysts. COF-tailored electrocatalysts are promising materials to replace with noble metal-based catalysts due to their tunable molecular architectures which is an excellent feature in their catalytic activity that can be designed rationally and manipulated perfectly for achieving large surface area and fine electronic characteristics to enhance catalytic activities. COFs are ideal choices for electrocatalytic water splitting applications because of their tunable composition and flexible integration of metal-catalysts [44]. The OER may work in both acidic and alkaline conditions. 4e-transfer steps are involved in the suggested overall reaction pathways for the OER [45]. In acidic and also alkaline solutions, the same species (*OH, *OOH, and *O) are found in the OER processes, just as they are in the ORR. Altogether, the OER can follow one of two paths: (i) direct interaction of two O* to generate O_2, or (ii) creation of the OOH* intermediate. The bonding interactions of intermediates (e.g., *OH, *O, and *OOH) throughout the reaction are considered as a standard metric to assess the overall OER catalytic capacity, even though they may involve both pathways. The OER process, which causes a multi-electron reaction, contains a variety of steps including adsorption, desorption, chemical reaction, and electron transfer [46]. The OER route of an electrocatalyst can be recognized after estimating the Tafel slope. Evaluating this factor reveals the electrons which are involved in the reaction. The small Tafel slope and also determining the rate which is located around the end of a multi-electron OER are two signs of a fine electrocatalyst.

10.3.2 Pristine Metal-Free COFs and Derivatives for the OER

Because good electrical conductivity and the number of active sites are important for electrochemical applications and the COFs that are metal-free lack both of them, only a small amount of electrocatalytic activity toward the OER can be processed by them. Kathiresan et al. recently synthesized a CTF (PD-CTF), which was based on 1,4-phenylenediamine that has a filamentous network structure that formed in a basic environment from reacting cyanuric chloride and 1,4-phenylenediamine [47]. The PD-CTF shows poor catalytic activity when employed as the electrocatalyst for the OER in alkaline conditions, with an onset potential of +1108 mV (vs. Ag/AgCl). The enhanced pyridinic-N sites in the obtained material are the reason for the catalytic activity. Most studies on electrocatalysts without metal use a procedure called post carbonization to boost catalytic activity even more. Finally, A porous COF which was based on ethylene diamine was successfully synthesized and then carbonized to utilize as an electrocatalyst for OER [48]. Materials that are carbonized, exhibited greater OER catalytic activity rather than pristine COFs. This phenomenon occurred under optimized temperature. For reaching a 10 mA.cm^{-2} current density, carbon materials which were based on COFs, coated on carbon paper and showed the initial potential of 1.527 V and overpotential of 297 mV. A great improvement in OER electrocatalysts was obtained when the current density changed to 300 mA cm^{-2} and it showed a low potential of 580 mV.

For comparing the features of different electrocatalysts for OER, scientists introduced some measurements like j_0 (the current density for OER), faradaic efficiency (FE), and η_{10} (the potential needed to produce the catalytic density of 10 mA cm^{-2}). In addition, because of generating oxygen via OER, the challenge of leaching the catalyst which is a reason for bad long-term stability and its gas phase decreases the current density by inhibiting the exposure of the active site. Materials based on transition metals such as Fe/Ni/Co/Mo/Zr/V became promising candidates for being searched in the field of OER catalysis. Among these metals catalysts based on cobalt have appealed to scientists so much [49]. Catalysts based on free-standing transition metals are having their weak points high overpotentials and slow electrochemical activity, also there are water oxidizing complexes(WOCs) that dissociate to a homogenous phase during the catalyzing process. Otherwise, numerous problems like the hard synthesis process hinder the use of carbon materials and metal foam for supporting metal nanoparticles and clusters that are active materials [50]. The COFs with special intrinsic features are a great choice for electrocatalyst preparation. Co-TpBpy was synthesized by coordination of Co species to the bipyridine sites of the TpBpy-COF (Figure 10.3a). It is noteworthy that the above-mentioned COF was produced via Schiff-based condensation of TP and 2-2″–bipyridyl-5,5″-diamine (BPY) [51]. The synthesized Co-TpBpy showed great porosity and a high surface area (450 m^{2}g^{-1}) (Figure 10.3c). In addition, it exposed a TOF of 0.23 s^{-1}, an FE of 95%, an overpotential of 400 mv at a current density of 1 mA cm^{-2} and great cycling stability in an aqueous phosphate buffer (pH = 7, 0.1 M). To prepare TpBpy-COF which its pores are in hierarchical form, Thomas selected polystyrene spheres that are templates and p-toluenesulfonic acid (PTSA) to synthesize the mentioned COF [52]. In comparison with Co-TpBpy and RuO$_2$, Marco-TpBpy-Co reached higher TOF value and it showed enhanced performance in OER (Figure 10.3b). The reason of accelerated O$_2$ gas bubbles transportation that resulted in developed electrocatalytic performance, is the homogeneous and also continuous macroporous structure. It is noteworthy that the Co ions act as active sites via coordinating with pyridine and improve the catalyst's conductivity. The electronic interaction between COFs and metal NPs enhanced the catalytic activity of the catalyst.

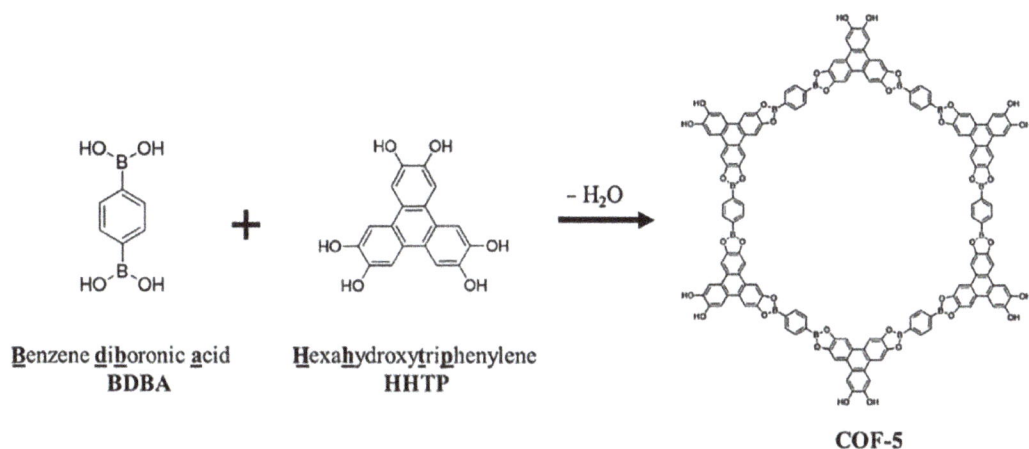

FIGURE 10.3
Schematic illustration of synthesizing COF-5. (Adapted with permission from reference [51]. Copyright (2016) ACS Publications.)

COFs are also good supporters for NPs. IISERP-COF2 is an excellent example that can be synthesized by nonplanar building units and tetrahedral sp^3 nitrogen. The COF flexibility aids it to confine the formation of Co$_x$Ni$_y$(OH)$_2$ NPs about 2 nm size [53]. The wonderful catalytic performance of the synthesized nanocomposite and great kinetics is due to its electronic interaction between the COF and NPs. To produce uniform conduction pathways, 2D benzimidazole-based COF(IISERP-COFs) was loaded by metallic Ni$_3$N NPs. The rate of O$_2$ evolution by the synthesized composite was 230 mmol.h^{-1}.g^{-1} and its overpotential, TOF value, FE, and Tafel slope were 230 mv at 10 mA cm^{-1}, 0.52 s^{-1}, 0.98 at 1 mAm^{-2} and 79 mv dec^{-1}, respectively.

By pyrolyzing COFs into carbon hosts, the conductivity will be increased in OER. Two composites based on RuO$_2$ and carbon derived from COFs were prepared. By controlling the temperature at 400°C and under oxidizing conditions, RuO$_2$@C was synthesized successfully via the pyrolysis method and they exposed great OER performance [54]. For having more accessible active sites, a few of NPs placed on the surface of the pyrolyzed COF.

By adding transition metal into macrocyclic clusters like porphyrin and mixing with a conductive material such as CNTs, another efficient electrocatalyst based on COFs would be synthesized. GDY analog based on metalloporphyrin was synthesized by a Glaser-Hay coupling reaction that occurred on a foam made from copper [55]. The synthesized catalyst had better performance in comparison to RuO$_2$-cobalt centers that are electrocatalytic active sites, expanded pore structure, and fast electron transfer inside the skeleton are the advantages of this catalyst for OER.

10.3.3 Metallized COFs and Nanohybrids for the OER

COFs can be metalized by porphyrin and phthalocyanine. This makes COFs a great choice for designing OER catalysts. Metallophthalocyanine-based COF (CoCMP) was designed successfully and after 1000 cycles, it was still able to continue catalyzing (Figure 10.4) [56]. In another study, for reaching higher efficiency, a multilayer covalent cobalt porphyrin framework was synthesized on MWCNT. The efficiency of the synthesized catalyst was

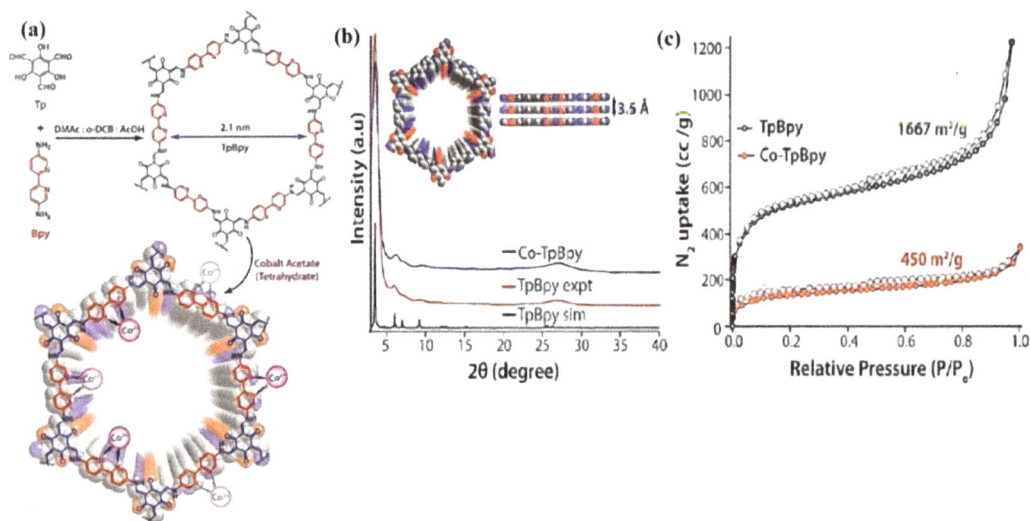

FIGURE 10.4
(a) Schematic representation of the synthesis of TpBpy via proton tautomerized Schiff base condensation and Co-TpBpy via Co(II) impregnation. Comparative (b) powder X-ray diffraction (PXRD), and (c) N2 adsorption isotherms. (Adapted with permission from reference [51]. Copyright (2016) ACS Publications)

more than 86% and was proof of the role of the combination of $(CoP)_X$ and MWCNTs carbon materials in synthesizing OER electrocatalysts.

Another innovative strategy for preparing electrocatalysts is to hybridize other materials like ZIF-67 on benzoic acid-modified COF [57]. By applying calcination, Co_3O_4 on N-doped porous carbon (NPC) was produced. This composite (Co_3O_4/NPs) exhibited a large specific surface area and excellent catalytic efficiency. It is concluded that the COF precursors are having an essential role in catalytic performance [57].

10.3.4 COF-Supported Single-Atom Sites for the OER

Finding a catalyst with high porosity and large surface area is crucial for developing stability and also durability of the catalyst which results in better O_2 evolution. One of the greatest materials that can be selected as platforms for synthesizing single-atoms and ultra-small NPs is COFs. Their stability, porosity, and high surface area are the main advantages of COFs. For enhancing the OER activity, $Co_xNi_y(OH)_2$ and Ni_3N which are ultra-small NPs were incorporated into the pores of IISERP-COF3 under alkaline conditions. Generally, the pore channels and also the framework of IISERP-COF3 that are regularized perfectly, confine Ni_3N active sites which can facilitate mass and charge transfer in the composite [51, 58].

10.4 Conclusion

COFs are constructed by light elements which exhibit significant features including low design, large surface area, and tunable properties compared to other porous materials. Therefore, these advantages have made COFs great candidates for numerous applications like catalysis, gas storage, optoelectric, etc. This chapter focuses on the synthesis of COFs and COF-based nanomaterials. Top-down and bottom-up methods are commonly used in synthesizing COFs. Great advances have been made in developing COFs and as a result, scientists figured out that COFs are wonderful electrocatalysts in OER, ORR, HER, and reduction of CO_2. One of the most famous applications of COFs is OER. OER is an important reaction in water splitting which occurs in an anode. The catalytic activity of COFs is related to their orbital energy and bonding structures. Among all COFs that contain porphyrin, Co and Fe COFs are the best catalysts for OER and ORR rather than Pt and RuO_2. By designing electrocatalysts smartly, newer and more efficient catalysts can be synthesized as green energy converters. This field still needs to be more explored and by innovating strategies, more efficient composites with diverse functions could be achieved. In summary, though existing challenges in this field, we believe that COFs engineered nanomaterials have a great perspective as a versatile electrocatalytic material for energy-related applications.

References

1. Wang, J.; Li, N.; Xu, Y.; Pang, H., Two-dimensional MOF and COF nanosheets: synthesis and applications in electrochemistry. *Chemistry–A European Journal* 2020, *26*(29), 6402–6422.
2. Goodenough, J. B., Electrochemical energy storage in a sustainable modern society. *Energy & Environmental Science* 2014, *7*(1), 14–18.
3. Walter, M. G.; Warren, E. L.; McKone, J. R.; Boettcher, S. W.; Mi, Q.; Santori, E. A.; Lewis, N. S., Solar water splitting cells. *Chemical Reviews* 2010, *110*(11), 6446–6473.
4. Roger, I.; Shipman, M. A.; Symes, M. D., Earth-abundant catalysts for electrochemical and photoelectrochemical water splitting. *Nature Reviews Chemistry* 2017, *1*(1), 1–13.
5. Zou, X.; Zhang, Y., Noble metal-free hydrogen evolution catalysts for water splitting. *Chemical Society Reviews* 2015, *44*(15), 5148–5180.
6. Gray, H. B., Powering the planet with solar fuel. *Nature Chemistry* 2009, *1*(1), 7–7.
7. Li, Y.; Karimi, M.; Gong, Y.-N.; Dai, N.; Safarifard, V.; Jiang, H.-L., Integration of metal-organic frameworks and covalent organic frameworks: Design, synthesis, and applications. *Matter* 2021, *4*(7), 2230–2265.
8. Lin, C. Y.; Zhang, D.; Zhao, Z.; Xia, Z., Covalent organic framework electrocatalysts for clean energy conversion. *Advanced Materials* 2018, *30*(5), 1703646.
9. Qu, L.; Liu, Y.; Baek, J.-B.; Dai, L., Nitrogen-doped graphene as efficient metal-free electrocatalyst for oxygen reduction in fuel cells. *ACS Nano* 2010, *4*(3), 1321–1326.
10. Zhang, J.; Zhao, Z.; Xia, Z.; Dai, L., A metal-free bifunctional electrocatalyst for oxygen reduction and oxygen evolution reactions. *Nature Nanotechnology* 2015, *10*(5), 444–452.
11. Peng, P.; Zhou, Z.; Guo, J.; Xiang, Z., Well-defined 2D covalent organic polymers for energy electrocatalysis. *ACS Energy Letters* 2017, *2*(6), 1308–1314.
12. Abuzeid, H. R.; EL-Mahdy, A. F.; Kuo, S.-W., Covalent organic frameworks: design principles, synthetic strategies, and diverse applications. *Giant* 2021, 100054.
13. Li, W.; Yang, C.-X.; Yan, X.-P., A versatile covalent organic framework-based platform for sensing biomolecules. *Chemical Communications* 2017, *53*(83), 11469–11471.
14. Yu, J.-T.; Chen, Z.; Sun, J.; Huang, Z.-T.; Zheng, Q.-Y., Cyclotricatechylene based porous crystalline material: Synthesis and applications in gas storage. *Journal of Materials Chemistry* 2012, *22*(12), 5369–5373.
15. Zeng, Y.; Zou, R.; Zhao, Y., Covalent organic frameworks for CO2 capture. *Advanced Materials* 2016, *28*(15), 2855–2873.

16. Amini, A.; Kazemi, S.; Safarifard, V., Metal-organic framework-based nanocomposites for sensing applications–A review. *Polyhedron* 2020, *177*, 114260.

17. Wei, P.-F.; Qi, M.-Z.; Wang, Z.-P.; Ding, S.-Y.; Yu, W.; Liu, Q.; Wang, L.-K.; Wang, H.-Z.; An, W.-K.; Wang, W., Benzoxazole-linked ultrastable covalent organic frameworks for photocatalysis. *Journal of the American Chemical Society* 2018, *140*(13), 4623–4631.

18. Zang, Y.; Wang, R.; Shao, P.-P.; Feng, X.; Wang, S.; Zang, S.-Q.; Mak, T. C., Prefabricated covalent organic framework nanosheets with double vacancies: anchoring Cu for highly efficient photocatalytic H 2 evolution. *Journal of Materials Chemistry A* 2020, *8*(47), 25094–25100.

19. Zhang, H.; Li, C.; Piszcz, M.; Coya, E.; Rojo, T.; Rodriguez-Martinez, L. M.; Armand, M.; Zhou, Z., Single lithium-ion conducting solid polymer electrolytes: advances and perspectives. *Chemical Society Reviews* 2017, *46*(3), 797–815.

20. Wang, S.; Ma, L.; Wang, Q.; Shao, P.; Ma, D.; Yuan, S.; Lei, P.; Li, P.; Feng, X.; Wang, B., Covalent organic frameworks: a platform for the experimental establishment of the influence of intermolecular distance on phosphorescence. *Journal of Materials Chemistry C* 2018, *6*(20), 5369–5374.

21. Ali, G. A.; Bakr, Z. H.; Safarifard, V.; Chong, K. F., Recycled Nanomaterials for Energy Storage (Supercapacitor) Applications. *Waste Recycling Technologies for Nanomaterials Manufacturing* 2021, 175–202.

22. Cote, A. P.; Benin, A. I.; Ockwig, N. W.; O'Keeffe, M.; Matzger, A. J.; Yaghi, O. M., Porous, crystalline, covalent organic frameworks. *Science* 2005, *310*(5751), 1166–1170.

23. Evans, A. M.; Parent, L. R.; Flanders, N. C.; Bisbey, R. P.; Vitaku, E.; Kirschner, M. S.; Schaller, R. D.; Chen, L. X.; Gianneschi, N. C.; Dichtel, W. R., Seeded growth of single-crystal two-dimensional covalent organic frameworks. *Science* 2018, *361*(6397), 52–57.

24. Kendall, A. J.; Johnson, S. I.; Bullock, R. M.; Mock, M. T., Catalytic silylation of N2 and synthesis of NH3 and N2H4 by net hydrogen atom transfer reactions using a chromium P4 macrocycle. *Journal of the American Chemical Society* 2018, *140*(7), 2528–2536.

25. Lin, S.; Diercks, C. S.; Zhang, Y.-B.; Kornienko, N.; Nichols, E. M.; Zhao, Y.; Paris, A. R.; Kim, D.; Yang, P.; Yaghi, O. M., Covalent organic frameworks comprising cobalt porphyrins for catalytic CO2 reduction in water. *Science* 2015, *349*(6253), 1208–1213.

26. Suen, N.-T.; Hung, S.-F.; Quan, Q.; Zhang, N.; Xu, Y.-J.; Chen, H. M., Electrocatalysis for the oxygen evolution reaction: recent development and future perspectives. *Chemical Society Reviews* 2017, *46*(2), 337–365.

27. Zhong, Y.; Cheng, B.; Park, C.; Ray, A.; Brown, S.; Mujid, F.; Lee, J.-U.; Zhou, H.; Suh, J.; Lee, K.-H., Wafer-scale synthesis of monolayer two-dimensional porphyrin polymers for hybrid superlattices. *Science* 2019, *366*(6471), 1379–1384.

28. Li, J.; Jing, X.; Li, Q.; Li, S.; Gao, X.; Feng, X.; Wang, B., Bulk COFs and COF nanosheets for electrochemical energy storage and conversion. *Chemical Society Reviews* 2020, *49*(11), 3565–3604.

29. Tan, C.; Cao, X.; Wu, X.-J.; He, Q.; Yang, J.; Zhang, X.; Chen, J.; Zhao, W.; Han, S.; Nam, G.-H., Recent advances in ultrathin two-dimensional nanomaterials. *Chemical reviews* 2017, *117*(9), 6225–6331.

30. Colson, J. W.; Woll, A. R.; Mukherjee, A.; Levendorf, M. P.; Spitler, E. L.; Shields, V. B.; Spencer, M. G.; Park, J.; Dichtel, W. R., Oriented 2D covalent organic framework thin films on single-layer graphene. *Science* 2011, *332*(6026), 228–231.

31. Owen, W. S.; Bible, M. S.; Dohmeier, E. F.; Guthrie, L. R.; Parsons, M. J.; Hendrix, J. W.; Hunt, J. R.; Lowry, M. S., Electronic charge transfer properties of COF-5 solutions and films with intercalated metal ions. *MRS Communications* 2020, *10*(1), 91–97.

32. Dai, W.; Shao, F.; Szczerbiński, J.; McCaffrey, R.; Zenobi, R.; Jin, Y.; Schlüter, A. D.; Zhang, W., Synthesis of a Two-Dimensional Covalent Organic Monolayer through Dynamic Imine Chemistry at the Air/Water Interface. *Angewandte Chemie* 2016, *128*(1), 221–225.

33. Feldblyum, J. I.; McCreery, C. H.; Andrews, S. C.; Kurosawa, T.; Santos, E. J.; Duong, V.; Fang, L.; Ayzner, A. L.; Bao, Z., Few-layer, large-area, 2D covalent organic framework semiconductor thin films. *Chemical Communications* 2015, *51*(73), 13894–13897.

34. Kim, S.; Lim, H.; Lee, J.; Choi, H. C., Synthesis of a scalable two-dimensional covalent organic framework by the photon-assisted imine condensation reaction on the water surface. *Langmuir* 2018, *34*(30), 8731–8738.

35. Chandra, S.; Kandambeth, S.; Biswal, B. P.; Lukose, B.; Kunjir, S. M.; Chaudhary, M.; Babarao, R.; Heine, T.; Banerjee, R., Chemically stable multilayered covalent organic nanosheets from covalent organic frameworks via mechanical delamination. *Journal of the American Chemical Society* 2013, *135*(47), 17853–17861.

36. Peng, Y.; Huang, Y.; Zhu, Y.; Chen, B.; Wang, L.; Lai, Z.; Zhang, Z.; Zhao, M.; Tan, C.; Yang, N., Ultrathin two-dimensional covalent organic framework nanosheets: preparation and application in highly sensitive and selective DNA detection. *Journal of the American Chemical Society* 2017, *139*(25), 8698–8704.

37. Mitra, S.; Kandambeth, S.; Biswal, B. P.; Khayum M, A.; Choudhury, C. K.; Mehta, M.; Kaur, G.; Banerjee, S.; Prabhune, A.; Verma, S., Self-exfoliated guanidinium-based ionic covalent organic nanosheets (iCONs). *Journal of the American Chemical Society* 2016, *138*(8), 2823–2828.

38. Khelashvili, G.; Behrens, S.; Weidenthaler, C.; Vetter, C.; Hinsch, A.; Kern, R.; Skupien, K.; Dinjus, E.; Bönnemann, H., Catalytic platinum layers for dye solar cells: a comparative study. *Thin Solid Films* 2006, *511*, 342–348.

39. Xiang, Z.; Zhou, X.; Zhou, C.; Zhong, S.; He, X.; Qin, C.; Cao, D., Covalent-organic polymers for carbon dioxide capture. *Journal of Materials Chemistry* 2012, *22*(42), 22663–22669.

40. Xu, H.; Gao, J.; Jiang, D., Stable, crystalline, porous, covalent organic frameworks as a platform for chiral organocatalysts. *Nature Chemistry* 2015, *7*(11), 905–912.

41. Hu, C.; Zhang, L.; Gong, J., Recent progress made in the mechanism comprehension and design of electrocatalysts for alkaline water splitting. *Energy & Environmental Science* 2019, *12*(9), 2620–2645.

42. Yan, Y.; He, T.; Zhao, B.; Qi, K.; Liu, H.; Xia, B. Y., Metal/covalent–organic frameworks-based electrocatalysts for water splitting. *Journal of Materials Chemistry A* 2018, *6*(33), 15905–15926.

43. Yang, Y.; Dang, L.; Shearer, M. J.; Sheng, H.; Li, W.; Chen, J.; Xiao, P.; Zhang, Y.; Hamers, R. J.; Jin, S., Highly active trimetallic NiFeCr layered double hydroxide electrocatalysts for oxygen evolution reaction. *Advanced Energy Materials* 2018, *8*(15), 1703189.

44. Hu, H.; Yan, Q.; Ge, R.; Gao, Y., Covalent organic frameworks as heterogeneous catalysts. *Chinese Journal of Catalysis* 2018, *39*(7), 1167–1179.

45. Liao, P.; Keith, J. A.; Carter, E. A., Water oxidation on pure and doped hematite (0001) surfaces: Prediction of Co and Ni as effective dopants for electrocatalysis. *Journal of the American Chemical Society* 2012, *134*(32), 13296–13309.

46. Lu, Z.; Wang, H.; Kong, D.; Yan, K.; Hsu, P.-C.; Zheng, G.; Yao, H.; Liang, Z.; Sun, X.; Cui, Y., Electrochemical tuning of layered lithium transition metal oxides for improvement of oxygen evolution reaction. *Nature Communications* 2014, *5*(1), 1–7.

47. Gopi, S.; Kathiresan, M., 1, 4-Phenylenediamine based covalent triazine framework as an electro catalyst. *Polymer* 2017, *109*, 315–320.

48. Gopi, S.; Giribabu, K.; Kathiresan, M., Porous organic polymer-derived carbon composite as a bimodal catalyst for oxygen evolution reaction and nitrophenol reduction. *ACS Omega* 2018, *3*(6), 6251–6258.

49. Zhao, Q.; Yan, Z.; Chen, C.; Chen, J., Spinels: controlled preparation, oxygen reduction/evolution reaction application, and beyond. *Chemical Reviews* 2017, *117*(15), 10121–10211.

50. Lei, C.; Wang, Y.; Hou, Y.; Liu, P.; Yang, J.; Zhang, T.; Zhuang, X.; Chen, M.; Yang, B.; Lei, L., Efficient alkaline hydrogen evolution on atomically dispersed Ni–N x species anchored porous carbon with embedded Ni nanoparticles by accelerating water dissociation kinetics. *Energy & Environmental Science* 2019, *12*(1), 149–156.

51. Aiyappa, H. B.; Thote, J.; Shinde, D. B.; Banerjee, R.; Kurungot, S., Cobalt-modified covalent organic framework as a robust water oxidation electrocatalyst. *Chemistry of Materials* 2016, *28*(12), 4375–4379.

52. Zhao, X.; Pachfule, P.; Li, S.; Langenhahn, T.; Ye, M.; Schlesiger, C.; Praetz, S.; Schmidt, J.; Thomas, A., Macro/microporous covalent organic frameworks for efficient electrocatalysis. *Journal of the American Chemical Society* 2019, *141*(16), 6623–6630.

53. Mullangi, D.; Dhavale, V.; Shalini, S.; Nandi, S.; Collins, S.; Woo, T.; Kurungot, S.; Vaidhyanathan, R., Low-Overpotential Electrocatalytic Water Splitting with Noble-Metal-Free Nanoparticles Supported in a sp3 N-Rich Flexible COF. *Advanced Energy Materials* 2016, *6*(13), 1600110.

54. Chakraborty, D.; Nandi, S.; Illathvalappil, R.; Mullangi, D.; Maity, R.; Singh, S. K.; Haldar, S.; Vinod, C. P.; Kurungot, S.; Vaidhyanathan, R., Carbon derived from soft pyrolysis of a covalent organic framework as a support for small-sized RuO2 showing exceptionally low overpotential for oxygen evolution reaction. *ACS Omega* 2019, *4*(8), 13465–13473.

55. Huang, H.; Li, F.; Zhang, Y.; Chen, Y., Two-dimensional graphdiyne analogue Co-coordinated porphyrin covalent organic framework nanosheets as a stable electrocatalyst for the oxygen evolution reaction. *Journal of Materials Chemistry A* 2019, *7*(10), 5575–5582.

56. Singh, A.; Roy, S.; Das, C.; Samanta, D.; Maji, T. K., Metallophthalocyanine-based redox active metal–organic conjugated microporous polymers for OER catalysis. *Chemical Communications* 2018, *54*(35), 4465–4468.

57. Zhuang, G.-l.; Gao, Y.-f.; Zhou, X.; Tao, X.-y.; Luo, J.-m.; Gao, Y.-j.; Yan, Y.-l.; Gao, P.-y.; Zhong, X.; Wang, J.-g., ZIF-67/COF-derived highly dispersed Co3O4/N-doped porous carbon with excellent performance for oxygen evolution reaction and Li-ion batteries. *Chemical Engineering Journal* 2017, *330*, 1255–1264.

58. Nandi, S.; Singh, S. K.; Mullangi, D.; Illathvalappil, R.; George, L.; Vinod, C. P.; Kurungot, S.; Vaidhyanathan, R., Low band gap benzimidazole COF supported Ni3N as highly active OER catalyst. *Advanced Energy Materials* 2016, *6*(24), 1601189.

11

Covalent Organic Frameworks as Efficient Electrocatalysts for Oxygen Evolution Reactions

Kayode Adesina Adegoke[1], Thabo Matthews[1], Tebogo Mashola[1], Siyabonga Mbokazi[1],
Thobeka Makhunga[1], Cyril Selepe[1], Memory Zikhali[1], Kudzai Mugadza[1]
Nobanathi Wendy Maxakato[1], and Olugbenga Solomon Bello[2]

[1]Department of Chemical Sciences, University of Johannesburg, Doornfontein, South Africa

[2]Department of Pure and Applied Chemistry, Ladoke Akintola University of Technology,
Ogbomoso, Oyo State, Nigeria

CONTENTS

11.1 Introduction

Energy is fundamental to the continued existence of the humanities. The greater portion of energy feed is from fossil fuels. Worryingly, fossil fuel-based energy production has a negative connotation to the environment and health. These demerits include climatic shifts, ozone layer depletion, and environmental degradation. Above all, the non-renewability nature of fossils impedes their continued use since depletion of the sources is inevitable. Therefore, clean and renewable energy sources are paramount to energy conversion and protection of the earth and all its inhabitants. Renewable energy devices including fuel cells, solar power, wind power, tidal, geothermal, etc. and, clean energy conversion and storage (batteries/supercapacitors). Various energy conversion technologies are under investigation for the improvement of energy production efficacy. These technologies, since its emergent in 2005 [1], include photochemical and electrochemical based oxygen reduction reaction, oxygen evolution reaction, hydrogen evolution reaction, Water splitting, alcohol oxidation reaction, carbon dioxide reduction reaction, and nitrogen reduction reaction.

The energy conversion efficiencies of the alluded technologies are centered on the robustness of electrocatalysts that drives electrocatalytic reactions. Catalyst architectural engineering is the heart of improving the efficacy and efficiency of electrocatalysts, which

DOI: 10.1201/9781003206507-11

are also governed by the engineering protocols guiding electrocatalysis. To date, a range of electrocatalysts has been fabricated and tested for various applications, especially in OER. The applicability of the electrocatalysts is focused on the property–performance relationship, and the properties vary with material size, pore size, distribution, arrangement, porousness, and elemental compositions. Based on these attributes, porous carbonaceous materials have been widely employed as electrode materials. They are cheap, have high chemical stability, are highly porous, and have a high surface area, ensuring great exposure to catalytically active sites, excellent mass transport, and high electrical conductivity for easy electron movement. The major drawback coined to carbons as electrode materials is that they are fabricated using high-temperature pyrolysis and it is very difficult to control their sizes and structures.

Based on the electrocatalyst architectural engineering, engineering protocols, and alterations of property–performance relationship, numerous novel electrochemically active electrode materials for OER have been developed. These novel materials are easily tunable, precisely manipulatable, and have modulated structures implying that they can be designed for different electrocatalytic reaction applications. Of interest have been covalent organic frameworks (COFs) and metal-organic frameworks (MOFs). MOFs have received extensive attention for energy conversion purposes over the years. However, their applicability in the electrochemical reactions has been hindered by their compromised chemical stability in highly alkaline or basic electrolytes and reduced electrical conductance [2].

To that effect, COFs- sub-class porous polymers with molecular frameworks linked by covalent bonds have been developed. The covalent linkage and building motifs geometry direct the COFs to have a 2D or 3D structure [3, 4]. COFs are highly merited to a compounded combination of high surface areas exceeding those of MOFs and zeolites [5], long-range order (facilitating a high-efficiency exciton movement) [6], permanent porosity, highly crystalline, tunable pore sizes, tailorable inimitable molecular architectural structure and electrical conductivity due to π-conjugated structures (enabling the visible light absorption) [7–10] (Figure 11.1). These merits have instigated further development of the COFs – based electrocatalysts for electrocatalysis, sensing, gas storage, separation, purification, etc. Figure 11.2 presents the various design strategies that have been reported for the preparation/fabrication of various efficient COF catalysts for different energy conversion systems. The 2D COFs have intrinsically low bandgaps and high charge mobility, necessitating them for electrocatalysis. Generally, most COFs have tunable bandgaps. Recently, several usage COFs in electrochemical applications have been demonstrated. This chapter focuses on the electrocatalytic behavior of COFs expounding on the principles and mechanism of electrocatalytic OER and the different COF catalysts for OER electrocatalysis.

11.2 OER Electrocatalysis

11.2.1 Principles and Mechanism of Electrocatalytic OER

Electrochemical processes are well-known to be one of the most convenient and efficient ways to convert the energy stored in the chemical bonds of compounds into electrical energy through chemical bond cleavage and creation. This is accomplished through electrochemical conversion of energy via water splitting into hydrogen and oxygen and their subsequent reproduction into the water, thereby resulting in a process that is both

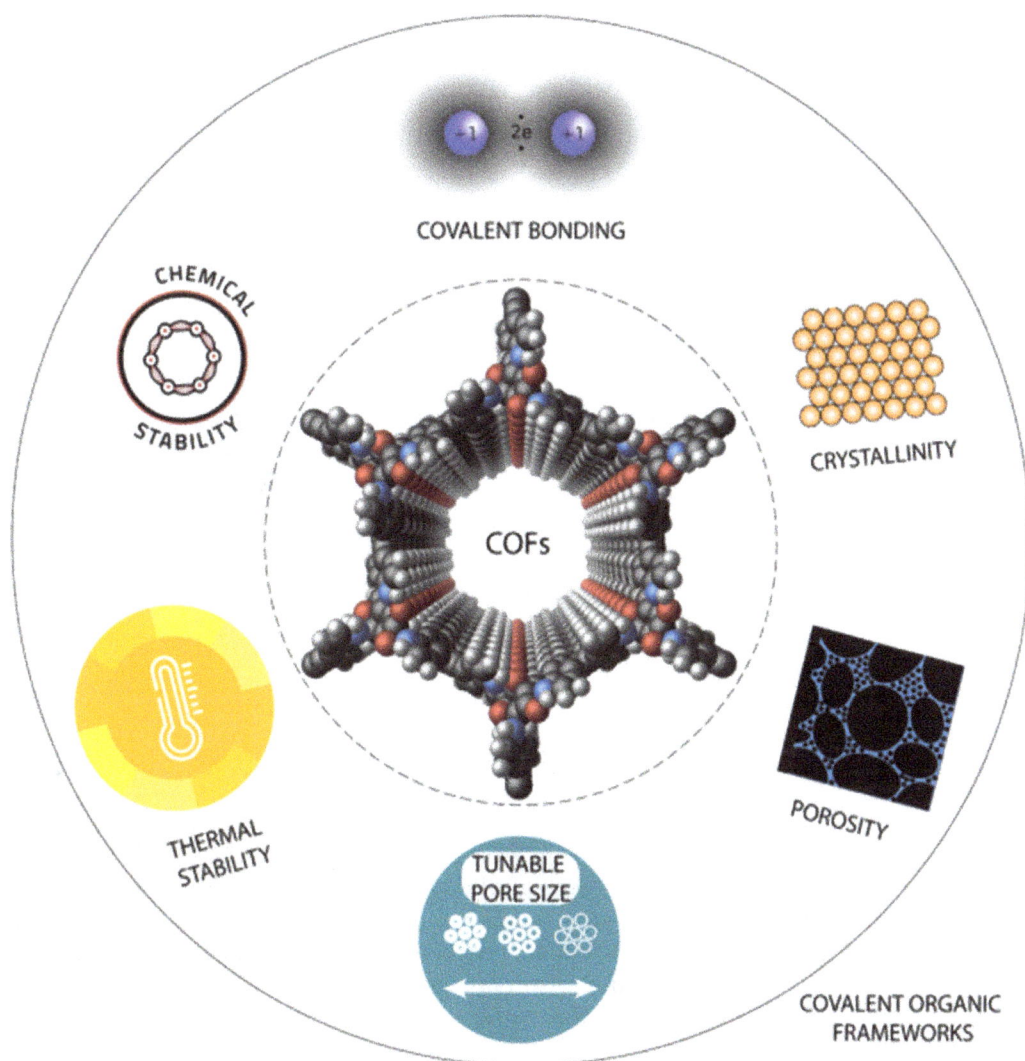

FIGURE 11.1

Vital properties of COFs that are pivotal to the possibility of their usages as electrode materials in electrocatalytic reactions and other applications [11].

imperishable and with slightly no negative impact on the environment. The languid movement of electrons involved during multi-electron transfer pathways such as O-H bond breakage, electron loss from water molecules, and O-O bond creation stifles the dynamics of the OER. The four-electron pathway is more efficient and advantageous in the oxygen evolution reaction than in the oxygen reduction reaction. The OER process is the opposite of the oxygen reduction reaction, in which oxygen is reduced to either water or hydroxide ions, whereas in the OER, water is electrochemically oxidized to oxygen.

In general, OER may function in both low and high pH environments. A four-electron transfer pathway procedure is involved in the proposed net reaction technique for the OER. The same species (*OH, *OOH, and *O) are available in both low and high pH solutions in the established OER processes. The OER typically has two routes: direct reaction of

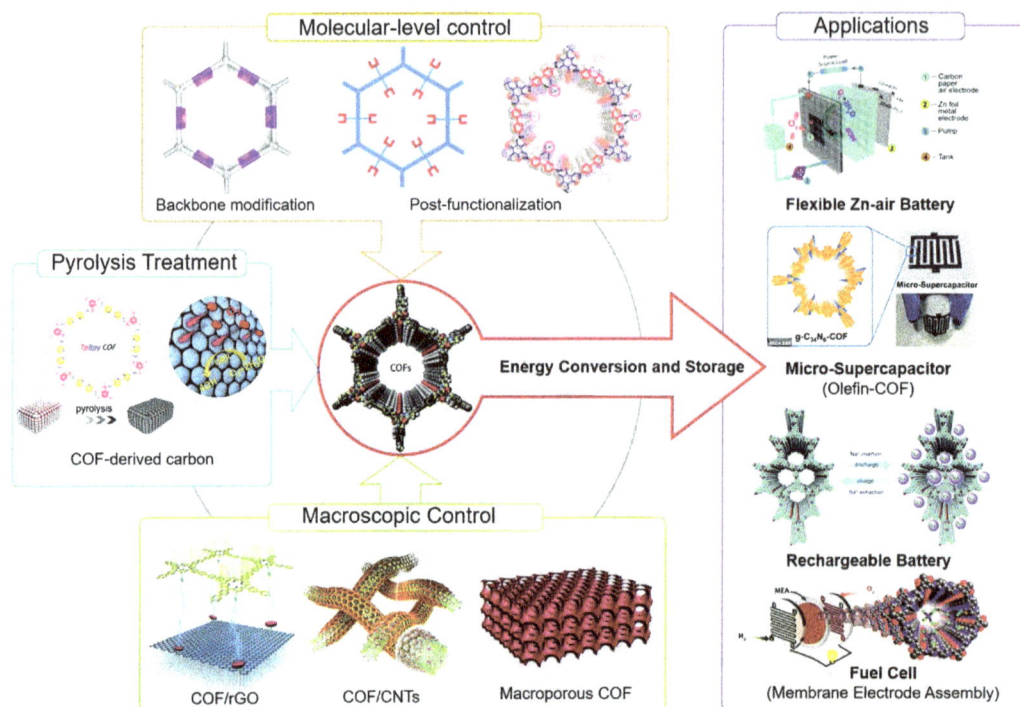

FIGURE 11.2
Various design strategies for the fabrication/synthesis of COF catalysts for energy conversion applications [11].

two *O to yield oxygen and the last one is the formation of an intermediary *OOH. Despite the possibility of both routes being involved, the bonding reaction between intermediates (*O, *OH, and *OOH) during the reaction is a well-known indicator of OER catalytic activity [12]. Surface assimilation (adsorption), desorption, charge transfer, and chemical reaction are all involved in the multi-electron process. The empirically determined Tafel slope can evaluate the electrocatalyst OER pathway, which is associated with the charge transfer involved in the reaction. The smaller Tafel slope, which is generally a sign of excellent electrocatalyst, is implicated by the rate-determining step when it is at the end of the multi-electron OER. The reaction mechanisms of OER can be listed according to the reactions (Equations 11.1–11.12) as follows [13]:

In Acidic media:

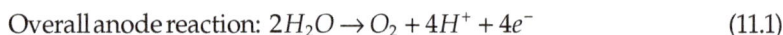

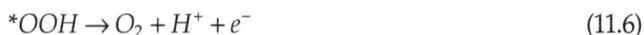

$$\text{Overall anode reaction: } 2H_2O \rightarrow O_2 + 4H^+ + 4e^- \tag{11.1}$$

$$\text{Mechanisms: } * + H_2O \rightarrow *OH + H^+ + e^- \tag{11.2}$$

$$*OH \rightarrow *O + H^+ + e^- \tag{11.3}$$

$$2\overset{*}{O} \rightarrow 2\overset{*}{} + O_2 \tag{11.4}$$

$$*O + H_2O \rightarrow *OOH + H^+ + e^- \tag{11.5}$$

$$*OOH \rightarrow O_2 + H^+ + e^- \tag{11.6}$$

In alkaline media:

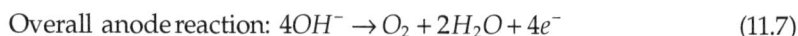

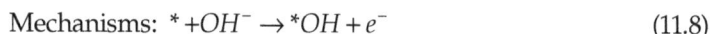

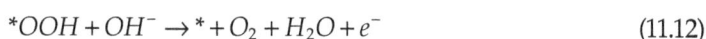

$$\text{Overall anode reaction: } 4OH^- \rightarrow O_2 + 2H_2O + 4e^- \quad (11.7)$$

$$\text{Mechanisms: } * + OH^- \rightarrow *OH + e^- \quad (11.8)$$

$$*OH + OH^- \rightarrow *O + H_2O + e^- \quad (11.9)$$

$$2^*O \rightarrow 2^* + O_2 \quad (11.10)$$

$$*O + OH^- \rightarrow *OOH + e^- \quad (11.11)$$

$$*OOH + OH^- \rightarrow * + O_2 + H_2O + e^- \quad (11.12)$$

11.2.2 COF Electrocatalysts for OER

11.2.2.1 Pristine COF Electrocatalysts for OER

In general, metal-free COFs have poor catalytic activity in the OER processes due to their low electrical conductivity and the small number of bare active sites. Gopi et al. [14] investigated the use of pristine metal-free covalent organic frameworks toward OER. They prepared a filamentous 1,4-phenylenediamine-based covalent triazine framework (PD-CTF) by reacting cyanuric and 1,4-phenylenediamine in alkaline conditions (Figure 11.3a). The PD-CTF showed poor catalytic activity (Figure 11.3b) and stability when used as an electrocatalyst for OER with starting potential of 1.108 V vs. Ag/AgCl. This performance was assigned to the enhanced pyrridic-nitrogen sites found in the final material. Therefore, another study suggested that the post-carbonization process must be applied to COFs to further accelerate the catalytic activity of pristine COFs for OER [14, 15]. Based on this, the authors investigated the effect of the carbonization process on ethylenediamine-based porous COF toward OER. Ethylenediamine-based COFs were prepared, carbonized, and employed as electrocatalysts for the OER. With an optimum temperature, the post-carbonized product demonstrated significantly increased OER catalytic activity compared to pure COF. The covalent organic framework-modified carbon material coated on carbon paper showed the starting potential of 1.527 V and overpotential of 297 mV to obtain a current density of 10 mA [15].

Yang et al. [16] used the solvothermal method to synthesize a new phenazine-linked 2D COF (COF-C$_4$N) by reacting hexaketocyclohexane (HKH) with triphenylenehexamine (TPHA) (Figure 11.3f). Complementary evidence from the DFT calculations showed that the as-synthesized COF-C$_4$N demonstrated superior electrocatalytic activity for OER, N-doped position, stability, and band gap to other materials with similar porosity and structure (Figure 11.3g-j).

11.2.2.2 COF Derivative Electrocatalysts for OER

The pristine metal-free COFs usually possess poor electrocatalytic performance in OER, resulting from less-exposed active centers and poor electrical conductivity. Proper heteroatom-doped position and heteroatom-doped carbon stoichiometry within the metal-free COFs are critical for OER catalysis [17]. Introducing metal ions or complexes and combining COFs with other conductive carbon materials such as graphene, carbon nanotubes,

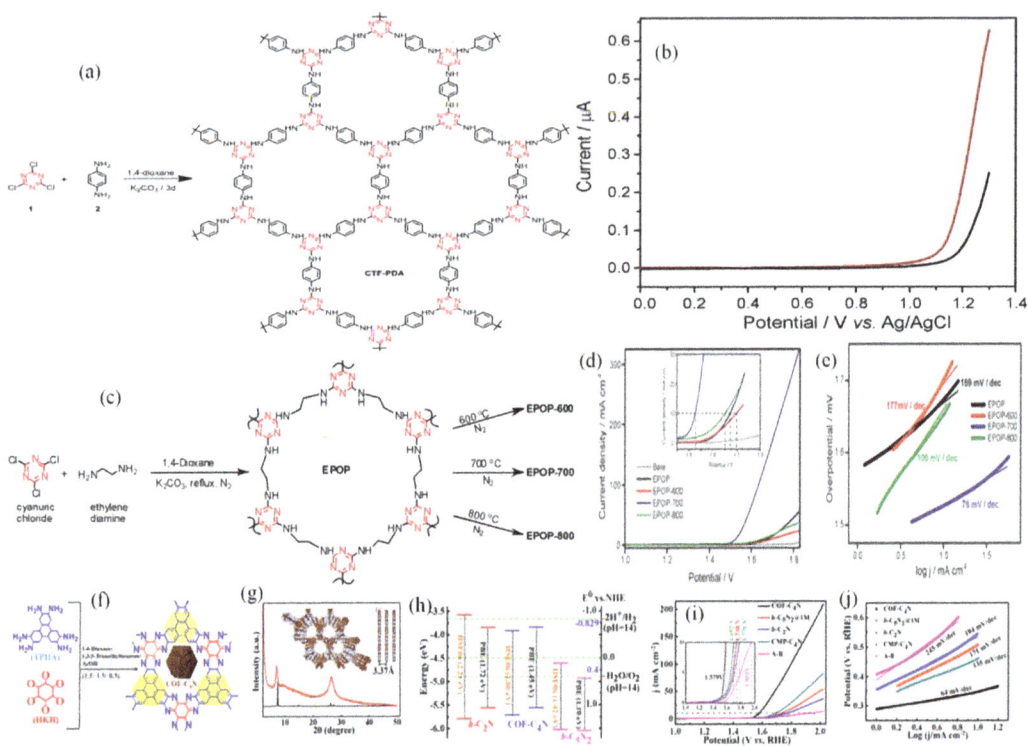

FIGURE 11.3

(a) Synthesis of PD-CTF. (b) Oxygen evolution reaction in 1 M KOH, red –PD-CTF/GC electrode, black- GC electrode. Reproduced with permission [14]. Copyright (2011), Elsevier. (c) Synthesis of ethylene diamine-based porous organic polymer (EPOP) and its carbon composites. Electrochemical performance of EPOP and its carbon composites toward OER. (d) LSV; (e) Tafel plots in 0.1 M KOH solution at 2 mV/s. Reprinted with permission [15]. American Chemical Society. (f) Schematic illustration of the synthesis process of COF-C$_4$N. (g) Observed (red) and simulated (black) powder X-ray diffraction patterns of COF-C$_4$N. The insets show the AA stacking model of COF–C$_4$N. (h) Absolute energy of the conduction band minimum and valence band maximum from PBE and HSE06 calculations and E^θ versus normal hydrogen electrode for h-C$_2$N, COF-C$_4$N, and h-C$_5$N$_2$. (i) LSVs of various OER catalysts in N$_2$-saturated aqueous 1.0 M KOH solution. (j) Tafel plots obtained from OER polarization curves (Reprinted with permission [16]. Copyright (2019), American Chemical Society.)

etc., is another practical way of tackling the conductivity issues of COFs [18]. Gan et al. [19] illustrated the key factors of COF-based electrocatalysts by a series of COF@CNT hybrid electrocatalysts (Figure 11.4). The results revealed that the optimized electrocatalysts catalytic performance relies highly on proper COFs loading on MWCNTs to balance electron and mass transportations. Moreover, the unsaturated coordination sites proved to be favorable for intrinsic activity enhancement [19]. Table 11.1 presents the summary of the electrocatalytic performance of different COF derivative electrocatalysts for OER.

11.2.2.3 COF Composite Electrocatalysts for OER

The COF composites have attracted extensive research as oxygen evolution reaction electrocatalysts due to their intriguing porous structures, tuneable morphologies, and functionalities [26]. When compared to noble metal catalysts, both share low-cost, but pure COFs have poor electrical conductivity and activity since they are unstable; hence post-synthesis

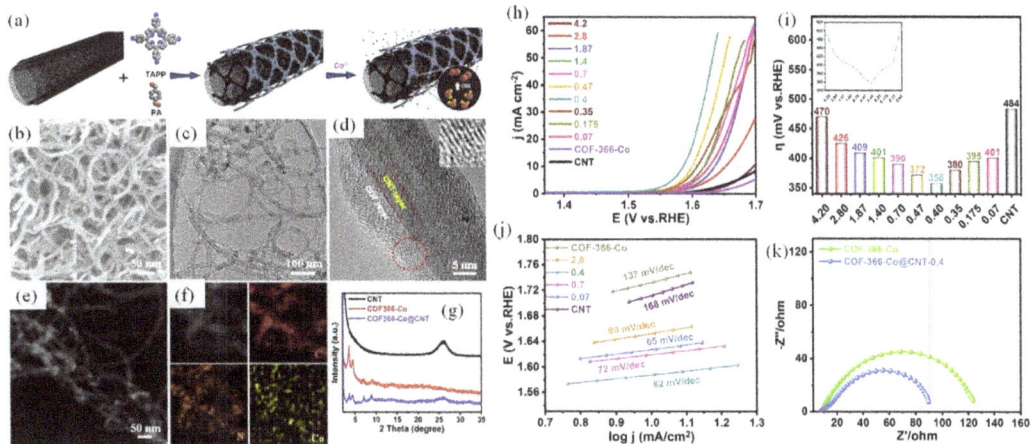

FIGURE 11.4

(a) Schematic representation of the synthesis of COF-366-Co@CNT hybrid catalyst. Morphological and structural information of the COF-366-Co@CNT: (b) SEM image; (c) TEM image; (d) high-resolution TEM image, the inset is the lattice fringes of COF layer; (e) STEM images; (f) STEM image and corresponding C, N, and Co EDS elemental mapping; (g) PXRD patterns of the CNT, COF-366-Co, and COF-366-Co@CNT. The systematic investigation of the electrocatalytic activity of COF-366-Co@CNT-x (x represents the mass ratio of COF monomers input over MWCNTs) for OER in 1 M KOH electrolyte: (h) LSV curves at 5 mV/s with 95% iR-compensation; (i) the overpotential at a current density of 10 mA cm^{-2}; (j) the corresponding Tafel plots; (k) Nyquist plots by EIS at 1.61 V (vs. RHE). (Reproduced with permission [19]. Copyright (2021), Elsevier.)

modifications like incorporating active sites on COFs are essential for enhancing COF electrocatalytic performance. Combining COFs with conductive materials like carbon nanotubes (CNT's), graphene, etc, or metal complexes yields COF-composites with elevated electrocatalytic activity [18].

Benzimidazole-linked IISERP-COF3 composites with a low bandgap were designed to support the even distribution of Ni_3N nanoparticles (NPs) (Figure 11.5a-b) [27]. The IISERP-COF3 organic unit did not only promoted the constraint growth of Ni_3N NPs but also accelerated the charge transfer process, which significantly boosted the OER performance of the COF composite (Figure 11.6a-f). Benzimidazole-linked IISERP-COF3 composite

TABLE 11.1

Summary of Catalytic Performance of COF Derivative Electrocatalysts for OER

COF-derived electrocatalysts	Electrolytes	Onset potential (V vs RHE)	Over-potential (V) at j = 10 mA/cm^2	Tafel slope (mV/dec)	Stability	Cycles	Refs.
NPC	0.1 M KOH	1.50	1.64	97	7 h	-	[20]
Co/CoO@COF	1 M KOH	-	0.278	81.12	5 h	2000	[21]
COF-366-Co@CNT-0.4	1 M KOH	-	0.358	62	-	2000	[19]
FeNi-COP-800	1.0 M KOH	-	0.4	103	175 h	-	[22]
BEAB-Co	1 M KOH	-	0.32	57	10 h	-	[23]
CoCOP	1.0 M KOH	-	0.35	151	7000 s	-	[24]
PSN$_{0.4}$-CoNi/CNT-800	1.0 M KOH	-	0.27	68	10 h	1000	[25]

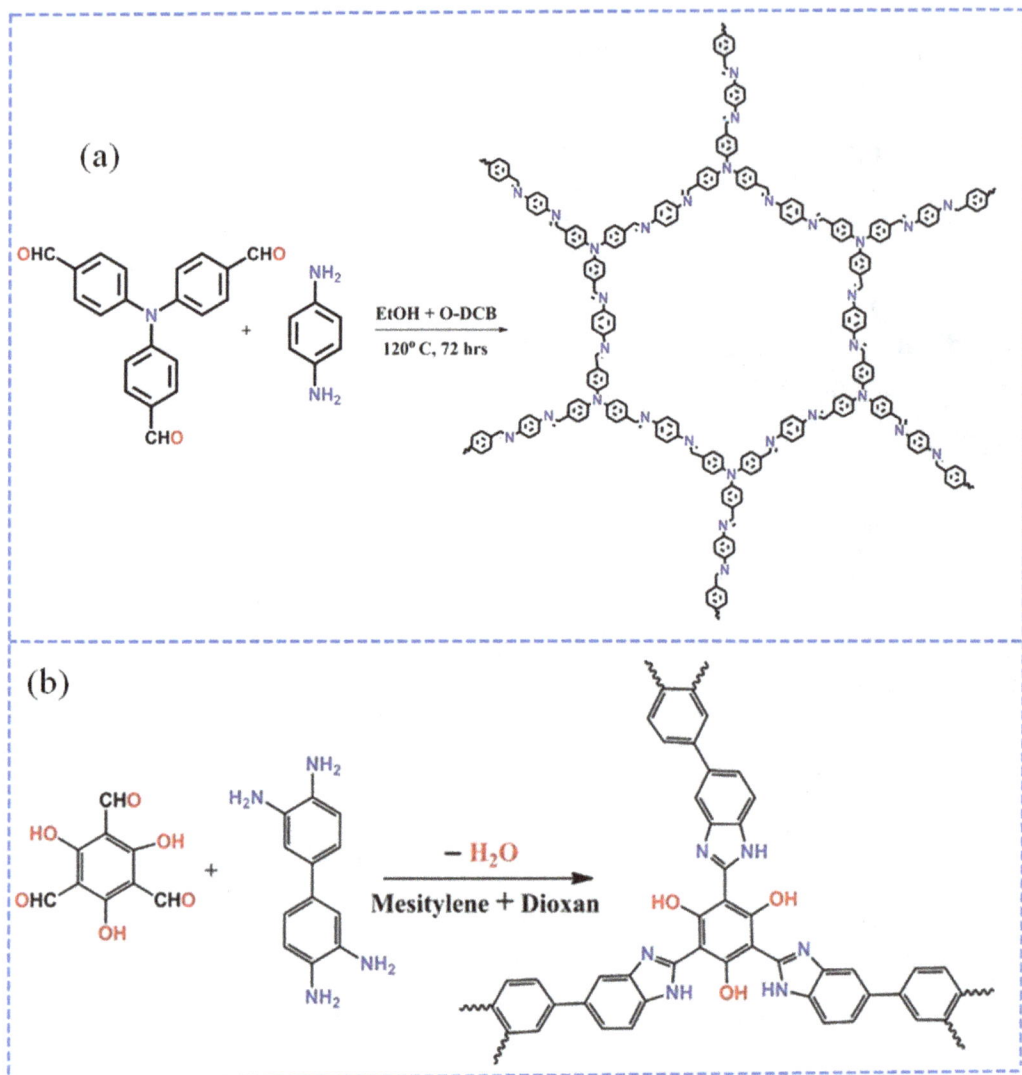

FIGURE 11.5
Preparation reactions of (a) IISERP-COF2 and (b) IISERP-COF3. (Reproduced from [27, 29] with permission from Wiley.)

catalyzes the OER with the overpotential of 230 mV at 10 mA cm^{-2} and oxygen generation rate of 230 mmol g^{-1} h^{-1} [18, 27–29] demonstrated a stable current output over 20 h as presented in Figure 11.6f. A layer structured porphyrin COF (POF, about 4 nm) was integrated into the surface of CNTs and this CNT@POF composite was used as the cathode material to facilitate and speed up the OER and oxygen reduction reaction on the zinc metal-air battery (Figure 11.7a) [30]. This COF composite showed superior electrocatalytic activity to the Pt/C and Ir/C since this composite endowered 0.71 V voltage gap, 237 mW cm^{-2} power density, and over 200 routine cycle endurance (Figure 11.7b) [30, 31].

COF-derived conductive carbons have been developed for oxygen electrocatalysis, and this synthesis strategy is made conducive by pyrolysis. One notable example of a

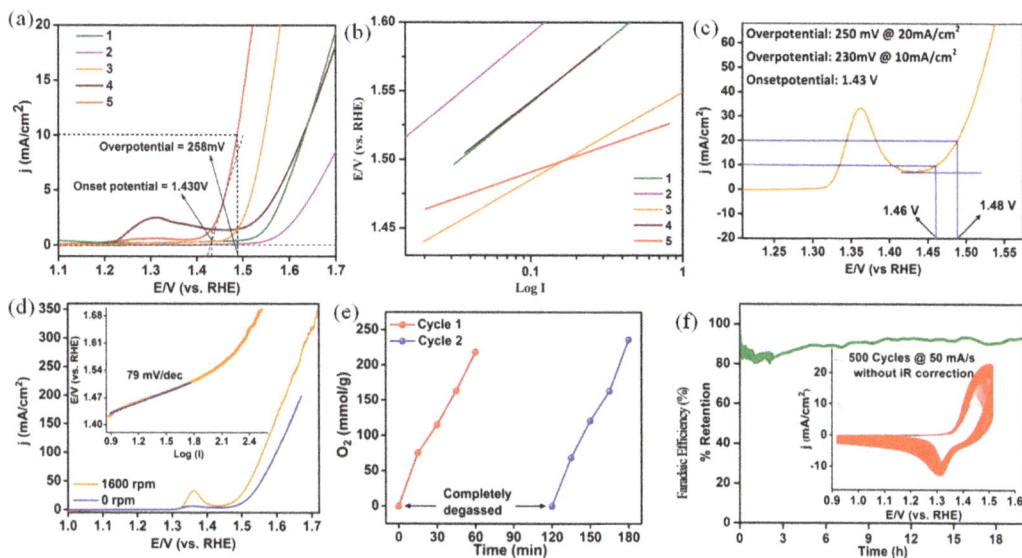

FIGURE 11.6

(a-c) LSV and Tafel plots for the homo and heterometallic composites. (Reproduced from [27] with permission from Wiley). (c-d) LSV plot for the composites and LSV showing the η at high current densities. Inset: Tafel plots (65% IR-compensated). (e) Evolved oxygen amount and the cycling performance of the composites. (f) Chronoamperometry plot showing the stability in the current outputs over 20 h. (Reproduced from [29] with permission from Wiley.)

conductive carbon formed from COFs and used as a precursor material is the Co_3O_4/NPC-2 formed from the ZIF-67/COF composite. Integration of ZIF-67 on the surface of the COF with benzoic acid modification, which was then followed by calcination of the composite, resulted in the formation of the N-doped carbon (NPC) with Co_3O_4 nanospheres that were uniformly distributed within the premises of the COF to result into the formation of the Co_3O_4/NPC-2 [32]. The OER evaluation for this conductive carbon showed an overpotential and Tafel slope superior to the Pt/C. Long cycle stability/durability and high mass activity were comparable to that of the Pt/C.

COFs have attracted surging interest over the past few years due to their wide potential application yet few metal-free electrocatalysts over COFs are reported. A thiadiazole-based COF hybrid (C4-SHz COF) with unique molecular architecture, high porosity, very high surface area, and abundant active sites was developed from a reaction between 3,5-tris(4-formylphenyl)benzene and 2,5-dihydrazinyl-1,3,4-thiadiazole [33]. The synthesized COF composite displayed superior electrocatalytic OER performance and excellent long-term durability in comparison to IrO_2/C. The π-conjugated organic building blocks, porosity, and the high Brunauer–Emmett–Teller surface area are the contributing factors to the electrocatalytic activity of this COF hybrid.

Zhao and team-workers have synthesized β-ketoenamine-based COF with interconnected macro-microporous structures. The produced macroporous COF was more crystalline with a large specific surface area. The Co^{2+} were successfully coordinated within hierarchical pore structure when bipyridine building blocks were inserted into the COF backbone (macro-Tp-Bpy-Co). The final macro-Tp-Bpy-Co exhibited exceptional OER activity, which was significantly superior to that of the pure microporous COFs. A similar observation was reported by Aiyappa et al. that coordinating Co^{2+} with the bipyridine edge resulted in the introduction of TpBpy COF as an OER electrocatalyst under a neutral

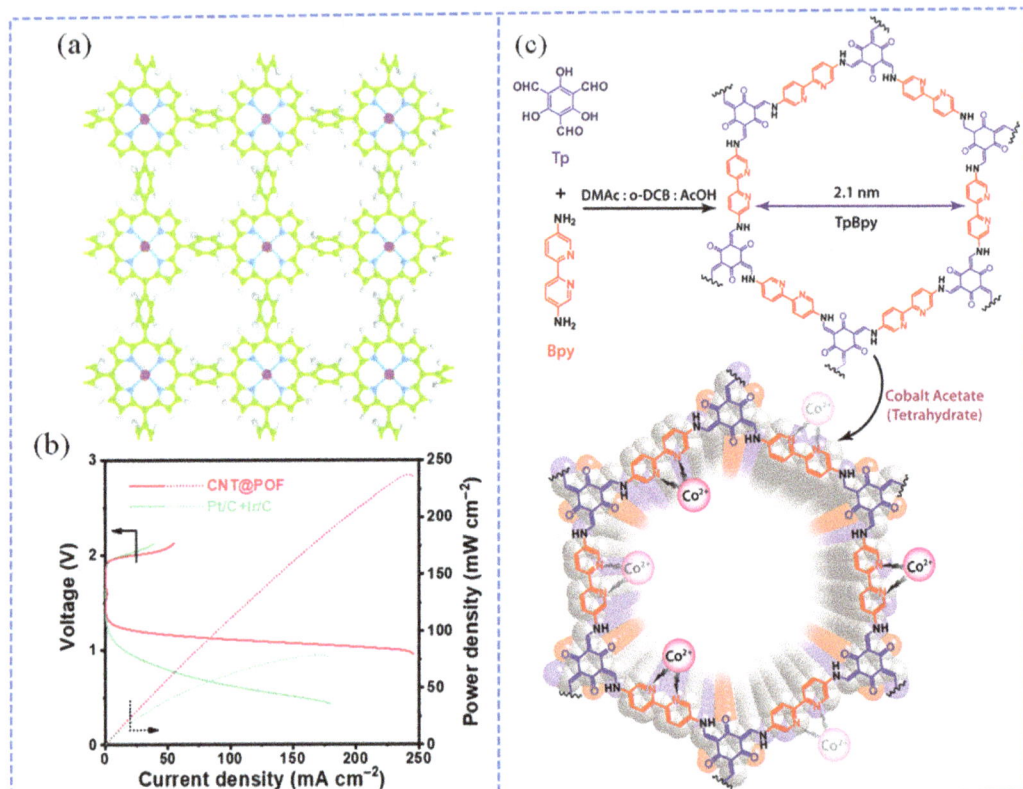

FIGURE 11.7
(a) Structural representation of POF (b) polarization curves of CNT@POF and Pt/C + Ir/C. (Reproduced from [30] with permission from the **Royal Society of Chemistry**). (c) Representation of the Co²⁺ coordinated TpBpy, Co-TpBpy. (Reprinted from [51] with permission from the **American Chemical Society.**)

condition (Figure 11.7c) [31]. The increased mass diffusion qualities in the hierarchically porous COF structures and the readily accessible active Co²⁺-bipyridine sites resulted in increased catalytic activity for the OER [34].

11.3 Concluding Remarks

Covalent organic frameworks can provide promising advantages compared to amorphous porous polymers in regard to catalytic properties such as ion transport and redox kinetics owing to their ordered open pore structures. Designing advanced electrodes with outstanding catalytic properties is possible through structural alteration and functional group introduction into the COF structures. However, issues relating to the synthesis and stability of these materials still prevail. These need proper attention for full utilization of COFs and COF derivatives for OER. To date, the research on the electrocatalytic OER holds great promise for a breakthrough in energy conversion, however, the application in OER is still in its infancy, thereby necessitating more detailed attention to address certain challenges highlighted below.

At present, a limited number of COFs with few linkages have been developed for mass transfer, and to overcome this drawback, the processibility of COFs needs to be improved while maintaining high proton and ionic conductivity to ensure effective design of mass transfer COFs. Research on COFs, especially the use of composite materials, is still in its infancy, and a few challenges still need to be addressed before further practical application of these materials can be implemented:

1. More research on COF composites should focus on increasing the scope of the linkages, functional groups, and the dimensions of COFs to clarify the relationship between the performance and structure of these materials.
2. Research on COF-based conductors only focuses on the ionic conductivity of fuel cells or batteries, so further research needs to be done in this area.
3. The challenges mentioned above, and strategies are only based on 2D COFs with their composite counterparts, so further research also needs to focus on 3D COFs in energy applications.

Acknowledgments

K. A. Adegoke acknowledges the Global Excellence Stature (GES) 4.0 Postdoctoral Fellowships Fourth Industrial Revolution and the University of Johannesburg, South Africa. N. W. Maxakato acknowledges the support received from the National Research Foundation of South Africa: Grant Number 118148 and Centre for Nanomaterials Science Research-University of Johannesburg, and University of Johannesburg, South Africa.

References

1. Yusran Y, Li H, Guan X, Fang Q, Qiu S (2020) Covalent organic frameworks for catalysis. EnergyChem 2:100035
2. Adegoke KA, Maxakato NW (2021) Porous metal–organic framework (MOF)-based and MOF-derived electrocatalytic materials for energy conversion. Mater Today Energy 100816. https://doi.org/1f0.1016/j.mtener.2021.100816.
3. Yahiaoui O, Fitch AN, Hoffmann F, Fröba M, Thomas A, Roeser J (2018) 3D anionic silicate covalent organic framework with srs topology. J Am Chem Soc 140:5330–5333. doi:10.1021/jacs.8b01774.
4. Diercks CS, Yaghi OM (2017) The atom, the molecule, and the covalent organic framework. Science (80-):355. doi:10.1126/science.aal1585.
5. Ma W, Yu P, Ohsaka T, Mao L (2015) An efficient electrocatalyst for oxygen reduction reaction derived from a Co-porphyrin-based covalent organic framework. Electrochem commun 52:53–57
6. Aswani Raj K, Rajeswara Rao M (2021) Crystalline two-dimensional organic porous polymers (covalent organic frameworks) for photocatalysis. Elsevier Inc.
7. Côté AP, Benin AI, Ockwig NW, O'Keeffe M, Matzger AJ, Yaghi OM (2005) Chemistry: Porous, crystalline, covalent organic frameworks. Science (80-) 310:1166–1170
8. Peng P, Zhou Z, Guo J, Xiang Z (2017) Well-defined 2D covalent organic polymers for energy electrocatalysis. ACS Energy Lett 2:1308–1314

9. Hu SY, Sun YN, Feng ZW, Wang FO, Lv Y kai (2022) Design and construction strategies to improve covalent organic frameworks photocatalyst's performance for degradation of organic pollutants. Chemosphere 286:13164. doi:10.1016/j.chemosphere.2021.131646.

10. Jiang T, Jiang W, Li Y, Xu Y, Zhao M, Deng M, Wang Y (2021) Facile regulation of porous N-doped carbon-based catalysts from covalent organic frameworks nanospheres for highly-efficient oxygen reduction reaction. Carbon N Y 180:92–100

11. Zhao X, Pachfule P, Thomas A (2021) Covalent organic frameworks (COFs) for electrochemical applications. Chem Soc Rev 50:6871–6913

12. Cui X, Lei S, Wang AC, Gao L, Zhang Q, Yang Y, Lin Z (2020) Emerging covalent organic frameworks tailored materials for electrocatalysis. Nano Energy 70:104525

13. Zhao X (2021) Chem Soc Rev Covalent organic frameworks (COFs) for electrochemical applications. 6871–6913

14. Gopi S, Kathiresan M (2017) 1,4-Phenylenediamine based covalent triazine framework as an electro catalyst. Polymer (Guildf) 109:315–320

15. Gopi S, Giribabu K, Kathiresan M (2018) Porous organic polymer-derived carbon composite as a bimodal catalyst for oxygen evolution reaction and nitrophenol reduction. ACS Omega 3:6251–6258

16. Yang C, Yang Z-D, Dong H, Sun N, Lu Y, Zhang F-M, Zhang G (2019) Theory-driven design and targeting synthesis of a highly-conjugated basal-plane 2D covalent organic framework for metal-free electrocatalytic OER. ACS Energy Lett 4:2251–2258

17. Zhang H, Zhu M, Schmidt OG, Chen S, Zhang K (2021) Covalent organic frameworks for efficient energy electrocatalysis: Rational design and progress. Adv Energy Sustain Res 2:2000090

18. Wang DG, Qiu T, Guo W, Liang Z, Tabassum H, Xia D, Zou R (2021) Covalent organic framework-based materials for energy applications. Energy Environ Sci 14:688–728

19. Gan Z, Lu S, Qiu L, Zhu H, Gu H, Du M (2021) Fine tuning of supported covalent organic framework with molecular active sites loaded as efficient electrocatalyst for water oxidation. Chem Eng J 415:127850

20. Karajagi I, Ramya K, Ghosh PC, Sarkar A, Rajalakshmi N (2020) Co-doped carbon materials synthesized with polymeric precursors as bifunctional electrocatalysts. RSC Adv 10:35966–35978

21. Ye X, Fan J, Min Y, Shi P, Xu Q (2021) Synergistic effects of Co/CoO nanoparticles on imine-based covalent organic frameworks for enhanced OER performance. Nanoscale 13:14854–14865

22. Liao Z, Wang Y, Wang Q, Cheng Y, Xiang Z (2019) Bimetal-phthalocyanine based covalent organic polymers for highly efficient oxygen electrode. Appl Catal B Environ 243:204–211

23. Li T, Atish C, Silambarasan K, Liu X, O'Mullane AP (2020) Development of an interfacial osmosis diffusion method to prepare imine-based covalent organic polymer electrocatalysts for the oxygen evolution reaction. Electrochim Acta 362:137212

24. Wang A, Cheng L, Zhao W, Shen X, Zhu W (2020) Electrochemical hydrogen and oxygen evolution reactions from a cobalt-porphyrin-based covalent organic polymer. J Colloid Interface Sci 579:598–606

25. Ma DD, Cao C, Li X, Cheng JT, Zhou LL, Wu XT, Zhu QL (2019) Covalent organic polymer assisted synthesis of bimetallic electrocatalysts with multicomponent active dopants for efficient oxygen evolution reaction. Electrochim Acta 321:134679

26. Mullangi D, Shalini S, Nandi S, Choksi B, Vaidhyanathan R (2017) Super-hydrophobic covalent organic frameworks for chemical resistant coatings and hydrophobic paper and textile composites. J Mater Chem A 5:8376–8384

27. Mullangi D, Dhavale V, Shalini S, Nandi S, Collins S, Woo T, Kurungot S, Vaidhyanathan R (2016) Low-overpotential electrocatalytic water splitting with noble-metal-free nanoparticles supported in a sp3 N-rich flexible COF. Adv Energy Mater 6:1600110

28. Chen M, Qi J, Guo D, Lei H, Zhang W, Cao R (2017) Facile synthesis of sponge-like Ni 3 N/NC for electrocatalytic water oxidation. Chem Commun 53:9566–9569

29. Nandi S, Singh SK, Mullangi D, Illathvalappil R, George L, Vinod CP, Kurungot S, Vaidhyanathan R (2016) Low band gap benzimidazole COF supported Ni3N as highly active OER catalyst. Adv Energy Mater 6:1601189. doi: 10.1002/aenm.201601189.

30. Li B-Q, Zhang S-Y, Wang B, Xia Z-J, Tang C, Zhang Q (2018) A porphyrin covalent organic framework cathode for flexible Zn–air batteries. Energy Environ Sci 11:1723–1729

31. Aiyappa HB, Thote J, Shinde DB, Banerjee R, Kurungot S (2016) Cobalt-modified covalent organic framework as a robust water oxidation electrocatalyst. Chem Mater 28:4375–4379

32. Zhuang G, Gao Y, Zhou X, Tao X, Luo J, Gao Y, Yan Y, Gao P, Zhong X, Wang J (2017) ZIF-67/ COF-derived highly dispersed Co3O4/N-doped porous carbon with excellent performance for oxygen evolution reaction and Li-ion batteries. Chem Eng J 330:1255–1264

33. Mondal S, Mohanty B, Nurhuda M, Dalapati S, Jana R, Addicoat M, Datta A, Jena BK, Bhaumik A (2020) A thiadiazole-based covalent organic framework: A metal-free electrocatalyst toward oxygen evolution reaction. ACS Catal 10:5623–5630

34. Zhao X, Pachfule P, Li S, Langenhahn T, Ye M, Schlesiger C, Praetz S, Schmidt J, Thomas A (2019) Macro/microporous covalent organic frameworks for efficient electrocatalysis. J Am Chem Soc 141:6623–6630

12

Emerging Applications of Covalent Organic Frameworks and Their Architectural Aspects for Improved Oxygen Evolution Reactions

Karan Chaudhary and Dhanraj T. Masram

Department of Chemistry, University of Delhi, Delhi, India

CONTENTS

12.1 Introduction

A sustainable and green energy alternative is the prime demand worldwide to solve the issue of environmental pollution and the energy crisis. For resolving this problem, electrochemical water splitting is a green and potential technology [1, 2]. Electrochemical water splitting includes two half-reactions, one is an oxygen evolution reaction (OER) at the anode and the other is a hydrogen evolution reaction (HER) at the cathode [2]. The anodic half-cell reaction of electrochemical water splitting requires high energy because the reaction pathway involves the four-electron transfer, and the reaction has low efficiency. The OER reaction under acidic and basic conditions is described as reactions (i) and (ii), respectively [2].

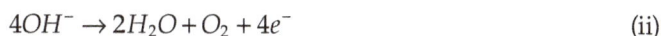

$$2H_2O \rightarrow 4H^+ + O_2 + 4e^- \tag{i}$$

$$4OH^- \rightarrow 2H_2O + O_2 + 4e^- \tag{ii}$$

Therefore, to commercialize the technology of electrochemical water splitting, the need is to develop electrocatalysts that are stable, have low cost, and are highly active. Due to these reasons, over a few years, it has been a topic of main research [1, 2]. At present, RuO_2 and IrO_2 are the known precious-metal-based electrocatalysts that have high OER activity. But their high cost, low stability, and low abundance limit the commercialization of electrochemical water splitting technology [2]. Thus, for the development of efficient electrocatalysts from transition metals having earth-abundance and low cost, a lot of efforts have been made and developed several OER electrocatalysts such as oxides, and nitride, metal-free catalysts, chalcogenides, phosphides, etc. Still, poor operational durability, lower catalytic activity, and lower conductivity are the associated problems with most of these transition metal-based electrocatalysts [3]. Therefore, it is crucial to develop stable and cost-effective OER electrocatalysts that have improved OER efficiency and that can reduce the kinetic barrier.

Covalent organic frameworks (COFs) are organic polymers that belong to a class of porous materials that are crystalline in nature having highly ordered structures and permanent porosity. In comparison to other polymers, COFs have several advantages like being synthetically controllable, structurally predesignable, and functionally manageable [4]. COFs are synthesized using organic units as building blocks that are linked to each other through strong covalent interactions [5]. Yaghi and co-workers in 2005 reported the first example of COF materials as COF-1 and COF-5 [6]. COFs can be synthesized as two-dimensional or three-dimensional frameworks with the presence of several heteroatoms in the structure by judicious selection of organic building blocks [7, 8]. At present, based on present research these organic porous polymer materials have been categorized as polymers of intrinsic micro-porosity (PIMs) [9], covalent organic frameworks [10], covalent triazine framework (CTF) [11], and conjugated micro-porous, and meso-porous polymers (CMPs) [12]. Several methods have been utilized to synthesize these COF materials which include ionothermal [11], mechanochemical [13], solvothermal [6], microwave [14], sonochemical [15], light-promoted [16], interfacial polymerization [17], and vapor-assisted conversion [18]. The merits associated with COFs are pre-designable, low density, porous structure, large accessible surface area, structural regularity, functionable and tunable, and excellent mechanical strength [19, 20]. Since the first report on COFs, a lot of efforts have been done in the development of these porous materials and as result, large numbers of research reports are available now. And this development of COFs is due to its associated merits. Based on these features, COFs have also been applied in various applications including energy storage [21], gas sensor [22, 23], catalysis [24, 25], drug delivery [26], electrocatalysis [27], and many others [4].

To prepare efficient electrochemical electrode material for OER important is proper design and synthesis of material. Thus, basic requirements are high conductivity, porosity, and large surface area for preparing efficient electrode material. In this regard, COF has the advantage of being prepared in desirable chemical structure and tunable porosity by use of proper linkage and building blocks [28]. Along with the associated merits of COFs, they have excellent chemical stability, withstand harsh conditions, and freedom to tune them electronically and structurally which makes them a potential candidate as electrocatalysts for OER [28, 29]. Pristine COFs suffer from low electrocatalytic activity and poor conductivity [28]. Therefore, the need is to incorporate catalysis active metal sites in the COF structure to fabricate a better COF-based material as an efficient OER electrocatalyst. Herein, this chapter involves a discussion on the development of COFs-based nanomaterials as OER electrocatalysts.

12.2 Utilization of COFs and COFs-Based Nanomaterials for Improved Oxygen Evolution Reactions

Li et al. [30] developed new electrocatalysts for the OER reaction in which an interfacial osmosis diffusion method was used that has been developed by them for synthesizing a new imine-based covalent organic polymer to which transition metals were incorporated.

This covalent organic polymer had diyne linkages synthesized at the solvent interface under ambient conditions at room temperature. N chelating sites distribution can be controlled precisely by using this synthesis method that can be used for binding transition metals. BEAB is the novel covalent organic polymer that has been synthesized using bis-(2-ethynylaniline) (BEA) and bis-(2-ethynylbenzaldehyde) (BEB) which had a spherical shape and an average diameter was 1.2μm, and further metal-BEAB complexes were prepared by incorporating transition metals like Co, Fe, and Ni into the structure of BEAB. Further, these BEAB-metal complexes were investigated as electrocatalysts for OER and their OER performance was compared with a commercial RuO_2. After evaluation of all complexes, the best OER performance was demonstrated by the BEAB-Co complex whose results were comparable to commercial RuO_2. At a current density of 10 mA/cm^2, BEAB-Co had a low overpotential which was 320 mV which is 250 mV for commercial RuO_2, while at a current density of 50 mA/cm^2 the overpotential for BEAB-Co was 390 mV which is just 40 mV more than that of commercial RuO_2 (350 mV). BEAB-Co displayed a low Tafel slope which was 57 mV/dec which is comparable to that of commercial RuO_2 (66 mV/dec). While, after 10 h of electrolysis, there was a slight negative shift in the OER polarization curve of BEAB-Co which was 7 mV at 10 mA/cm^2 and 27 mV at 20 mA/cm^2. For the process of OER under alkaline conditions, these results reveal the excellent stability and durability of the BEAB-Co complex. Even the plots for the calculated TOF values of both RuO_2 and BEAB-Co were very close with similar slops, which indicates that the electrocatalytic activity of BEAB-Co complex was comparable to that of RuO_2. Therefore, obtained results and properties exhibited were attributed to the abundant and controlled distributed N chelating sites on BEAB and efficient charge transport on conductive π-conjugated chain in BEAB. This study displayed that method for synthesis is promising and can be used for preparing efficient OER catalysts [30]. Wang et al. [31] showed transition metals (TM) can be firmly anchored on two-dimensional COF of pyrazine and phthalocyanines linkages, and good electrocatalytic activity is exhibited by these TM embedded COFs (TM-COFs) for OER. It was said that the carbon atoms in phthalocyanines and TM atoms serve at OER reaction sites. It was found that Co-, Mn-, and Cr- COFs had the lowest overpotential of 0.28, 0.44, and 0.35 V, respectively for OER and these values were comparable to, or greater than those of noble metal electrocatalysts. Such extremely low potential was due to the synergic effects of COFs and anchored transition metals [31].

Gan et al. [32] reported the construction of an ideal platform using multiwall carbon nanotubes (MWCNTs) and well-known COF to demonstrate the important factors in COF-based electrocatalyst toward OER and refining of catalyst for enhanced performance. In this report, MWCNTs were chosen as conductive substrate and COF was in-situ grown over the substrate (Figure 12.1). COF of p-phthalaldehyde (PA) and 5,10,15,20-tetrakis

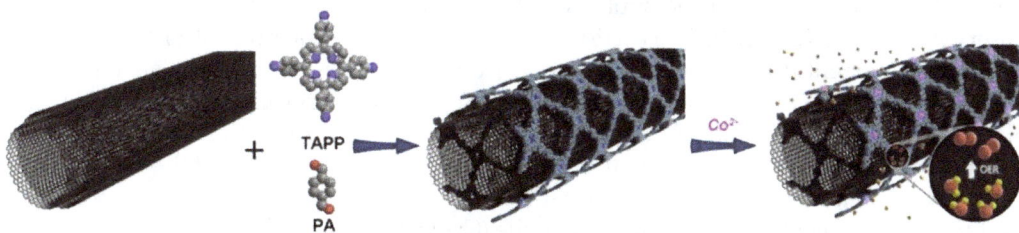

FIGURE 12.1
Illustration showing the synthetic route of COF-366-Co@CNT hybrid catalyst. (Adapted with permission [32]. Copyright (2020) Elsevier.)

(4-aminophenyl) porphyrin (TAPP) was stacked on carbon nanotubes via reversible hydrothermal imine condensation to form an overlapping layered structure. Further, Co ions were added via coordination with porphyrin units which are active centers. Then prepared electrocatalysts were applied for the OER study to study the key factors. Initially, it was observed that the overpotential at 10 mA/cm^2 current density varies with a decrease in mass ratio of COF monomers input over MWCNTs. When the mass ratio was decreased from 4.2 to 0.4, the overpotential decreased from 470 mV to 358 mV, but the further decrease in mass ratio from 0.4 to 0.07 led to an increase in overpotential to 401 mV from 358 mV. Further, the Tafel slope for each hybrid catalyst was below 100 mV/dec and smaller. So, COF-366-Co@CNT-0.4 had the lowest overpotential at the 10 mA/cm^2 current density and the lowest Tafel slope. Also, after 2000 cycles the overpotential at 10 mA/cm^2 current density remained unchanged for COF-366-Co@CNT-0.4 showing greater stability of the catalyst. Then to understand the effect of the active site, the amount of Co^{2+} was varied while keeping the COF@CNT composition constant. It was observed that OER catalytic activity was not the best for the sample with saturated Co^{2+} and overpotential was 398 mV at current density of 10 mA/cm^2. Whereas, the lowest overpotential of 378 mV at current density of 10 mA/cm^2 was of the sample having 2/3 saturated Co^{2+}. Even the lowest Tafel slope was for COF-366-2/3Co@CNT (50 mV/dec). Further, the TOF values were almost unchanged across the potential range when the Co^{2+} amount was increased from 1/3 saturated Co^{2+} to 2/3, but TOF values decreased for increased Co^{2+} amount to saturated. This reduction of TOF values displayed the reduction in the number of conversions normalized to each active site. Therefore, this study showed that the optimized catalytic performance depends upon appropriate COF loading on carbon substrate to balance off mass and electron transportations. Also, for enhanced OER performance the unsaturated Co coordination proved favorable. Moreover, at a 10 mA/cm^2 current density, the optimized catalyst had the overpotential of 350 mV. Based on this study it can be said that fine-tuning of COF-based catalysts can result in enhanced OER catalytic performance [32].

Feng et al. [1] demonstrated the advantages of combining the solid COF material and molecular catalyst that is similar to a homogeneous catalyst for preparing single-site heterogeneous catalyst. In this study, by using 2,5-diaminobenzenesulfonic and 2,4,6-triformylphloroglucinol COF-SO$_3$H was synthesized via solvothermal method having nanofibrous network structure with 1.2 nm pore size and 306 m^2/g surface area, which on immersing in concentrated ammonia water while stirring at room temperature resulted in NH$_4$@ COF-SO$_3$ having same nanofibrous network structure but the surface area was reduced to 110 m^2/g. While using N-sites in cyclen, NiFe complexes were prepared and were immobilized to NH$_4$@ COF-SO$_3$ by an easy cation exchange process to finally achieve single-site catalyst. The prepared catalyst (Cyclen@Ni$_{0.5}$Fe$_{0.5}$)@COF-SO$_3$ maintained nanofibrous network structure but the surface area decreased to 42 m^2/g due to the incorporation of NiFe complex. Then the synthesized catalysts were applied for OER, and it was observed that the (Cyclen@Ni$_{0.5}$Fe$_{0.5}$)@COF-SO$_3$ had an overpotential of only 276 mV at 10 mA/cm^2 current density which in comparison to commercial IrO$_2$ was very low and also lower than other (Cyclen@Ni$_x$Fe$_{1-x}$)@COF-SO$_3$. Also, (Cyclen@Ni$_{0.5}$Fe$_{0.5}$)@COF-SO$_3$ displayed the Tafel slope of 43 mV/dec which is the lowest. Further, at an overpotential of 300mV the TOF was calculated, and (Cyclen@Ni$_{0.5}$Fe$_{0.5}$)@COF-SO$_3$ showed the highest TOF value of 0.69 s^{-1} and was found to be better than efficient OER electrocatalysts reported in the literature. On evaluating the stability of the (Cyclen@Ni$_{0.5}$Fe$_{0.5}$)@COF-SO$_3$, it was found that for 25 h there was a negligible loss of current density proving that it is stable and reusable. The synergy effect of bimetal and high atom-utilization efficiency are the main reasons for this exceptional performance in OER by (Cyclen@Ni$_{0.5}$Fe$_{0.5}$)@COF-SO$_3$. Thus, preparing single-site catalyst using COF paves a new path for preparing more efficient and highly active COF-based electrocatalysts [1].

Gao et al. [5] also used employed the cation exchange method for synthesizing bimetallic COF electrocatalyst. Through solvothermal method, COF-SO$_3$H was synthesized by using 2,5-diaminobenzenesulfonic and 2,4,6-triformylphloroglucinol which on ammoniating produced NH$_4$@COF-SO$_3$. And through the cation exchange process, Co/V metal ions were incorporated into NH$_4$@COF-SO$_3$ for obtaining Co$_x$V$_{1-x}$@COF-SO$_3$. Herein, COF-SO$_3$H, NH$_4$@COF-SO$_3$, and Co$_x$V$_{1-x}$@COF-SO$_3$ had nanofibrous structure. Whereas, surface area of COF-SO$_3$H was 306 m^2/g, surface area of NH$_4$@COF-SO$_3$ was 110 m^2/g, and surface area of Co$_{0.5}$V$_{0.5}$@COF-SO$_3$ was 184 m^2/g. The surface area increased for Co$_{0.5}$V$_{0.5}$@COF-SO$_3$ in comparison to that of NH$_4$@COF-SO$_3$ because ionic radii of Co^{2+} and V^{3+} are smaller than that of NH$_4^+$. Then the synthesized catalysts were applied for OER under alkaline conditions. For Co$_x$V$_{1-x}$@COF-SO$_3$ catalysts, the overpotential was in the range of 345 to 318 mV to reach the current density of 10 mA/cm^2 but Co$_{0.5}$V$_{0.5}$@COF-SO$_3$ having a mass ratio of 1:1 of Co:V had the lowest overpotential (318 mV) to reach the current density of 10 mA/cm^2. The lowest Tafel slope of 62 mV/dec was displayed by Co$_{0.5}$V$_{0.5}$@COF-SO$_3$, and at an overpotential of 300 mV the TOF was calculated and Co$_{0.5}$V$_{0.5}$@COF-SO$_3$ showed the highest TOF value of 0.098 s^{-1} which was found to be superior to efficient Co-based OER electrocatalysts reported in the literature displaying high atom utilization efficiency. Moreover, this catalyst system exhibited reversible switching behavior as repeatedly catalyst-inert phase of COF-SO$_3$H to the catalyst-active phase of Co$_{0.5}$V$_{0.5}$@COF-SO$_3$ can be transformed, which shows catalyst can be regenerated. The Co$_{0.5}$V$_{0.5}$@COF-SO$_3$ showed 99% faradaic efficiency at an overpotential of 366mV during 4h at 50 mA/cm^2 current density. The stability of this Co$_{0.5}$V$_{0.5}$@COF-SO$_3$ was also evaluated, before and after 1000 CV cycles negligible changes were observed in the LSV curves. While during the 40 h chronoamperometric test there was only about an 11.5% decrease in the current density. These results prove that the Co$_{0.5}$V$_{0.5}$@COF-SO$_3$ had excellent stability for OER application [5].

In an another study, Gao et al. [28] synthesized bimetallic COF electrocatalyst in which via solvothermal method COF-SO$_3$H was synthesized by using 2,5-diaminobenzenesulfonic and 2,4,6-triformylphloroglucinol which on ammoniating produced NH4@COF-SO$_3$. And through the cation exchange process, Ni/Fe metal ions were incorporated into NH$_4$@COF-SO$_3$ for obtaining Ni$_x$Fe$_{1-x}$@COF-SO$_3$ (Figure 12.2). Among Ni$_x$Fe$_{1-x}$@COF-SO$_3$ COFs, the lowest overpotential of 308 mV to achieve 10 mA/cm^2 current density was exhibited by the COF having Ni/Fe mass ratio of 0.5:0.5 that is Ni$_{0.5}$Fe$_{0.5}$@COF-SO$_3$. Ni$_{0.5}$Fe$_{0.5}$@COF-SO$_3$ displayed the Tafel slope of 83 mV/dec and the mass activity of Ni$_{0.5}$Fe$_{0.5}$@COF-SO$_3$ was 52.31 A/g at the overpotential of 300 mV. Further, at an overpotential of 300 mV, the TOF value was also calculated for Ni$_{0.5}$Fe$_{0.5}$@COF-SO$_3$ which was found to be 0.1442 s^{-1}. Double-layer capacitance of Ni$_{0.5}$Fe$_{0.5}$@COF-SO$_3$ was 0.237 mF/cm^2 estimated by cyclic voltammogram higher compared to other COFs showing Ni$_{0.5}$Fe$_{0.5}$@COF-SO$_3$ has the largest electrochemical active surface area. Therefore, due to the large electrochemical active surface area, the best OER activity was of Ni$_{0.5}$Fe$_{0.5}$@COF-SO$_3$. The Ni$_{0.5}$Fe$_{0.5}$@COF-SO$_3$ exhibited excellent stability as it was able to maintain 20, 50, and 100 mA/cm^2 current densities at three different applied potentials, respectively. This outstanding performance of Ni$_{0.5}$Fe$_{0.5}$@COF-SO$_3$ was the result of prominent electronic synergetic effects and advantages associated with the COF structure such as favorable charge transfer, enhanced intrinsic activity of single metal sites, and mass transport [28].

Zhao et al. [33] synthesized macroporous TpBpy COF with a surface area of 723 m^2/g via Schiff base reaction by using 1,3,5-triformylphloroglucinol (Tp) and 2,2'-bipyridine-5,5'-diamine (Bpy), while polystyrene spheres were also used as hard templates (Figure 12.3).

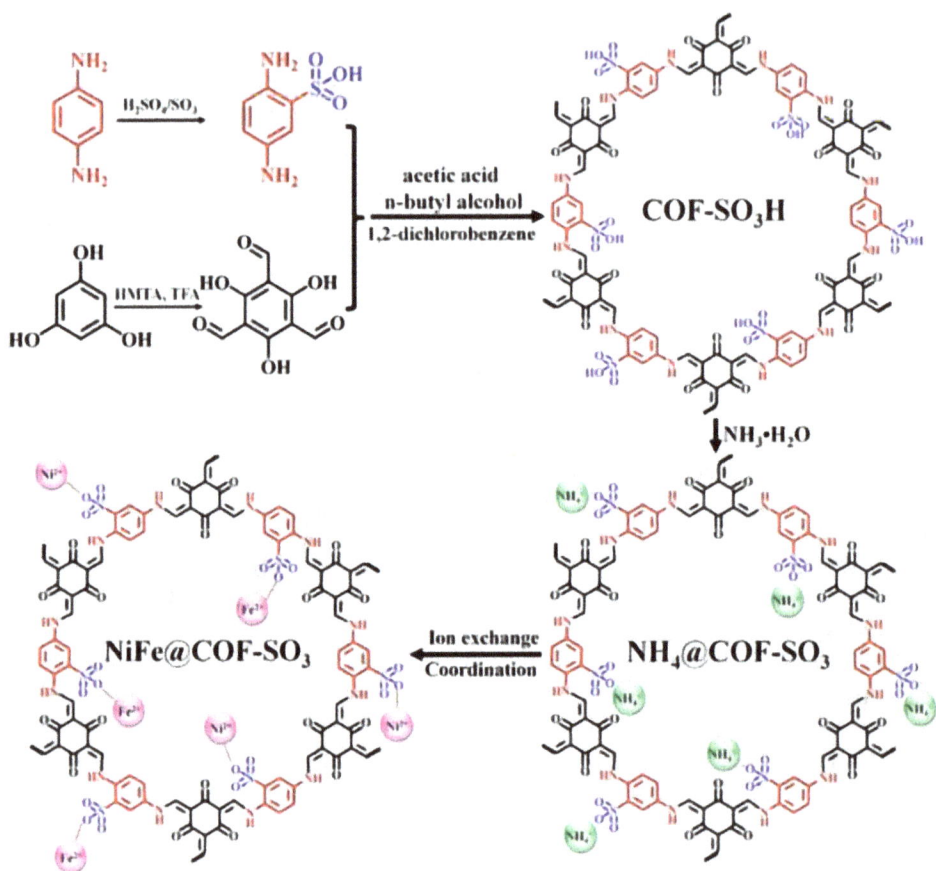

FIGURE 12.2
Illustration showing the synthesis of $Ni_{0.5}Fe_{0.5}$@COF-SO_3 COF. (Adapted with permission [28]. Copyright (2020) American Chemical Society.)

This showed the way to synthesize crystalline COF with inherent microporosity and template-induced macroporosity. As there are bipyridine moieties in the COF backbone, Co^{2+} was coordinated within COF to obtain macro-TpBpy-Co. Then this macro-TpBpy-Co was applied for OER activity in alkaline conditions (0.1 M KOH). It was observed that to achieve 10 mA/cm^2 current density, the overpotential required by macro-TpBpy-Co was 380 mV. And to achieve 50 mA/cm^2 current density, the overpotential required by macro-TpBpy-Co

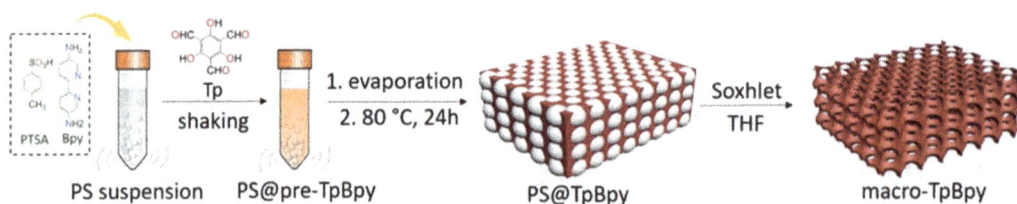

FIGURE 12.3
Illustration showing the synthesis of macro-TpBpy COF. (Adapted with permission [33]. Copyright (2019) American Chemical Society.)

was only 430 mV which was 170 mV lower than that of a standard RuO_2 catalyst (600 mV). Further, macro-TpBpy-Co had the lower Tafel slope of 54 mV/dec whereas RuO_2 had a Tafel slope of 79 mV/dec. Also, higher TOF values were obtained for macro-TpBpy-Co than RuO_2. For OER, Faradaic efficiency was found to be ~98% for this macro-TpBpy-Co revealing the high activity of this electrocatalyst. By chronopotentiometery test at 10 mA/cm^2 current density, stability of macro-TpBpy-Co was studied and after 10 h negligible difference was observed in the required overpotential. Also, only a slight negative shift of 38 mV of the overpotential at 10 mA/cm^2 was observed in the OER polarization after 40000 s. These studies showed that under alkaline conditions, macro-TpBpy-Co had excellent stability in OER [33].

Wang et al. [34] used Schiff-base condensation reaction for the synthesis of a new COF material from 5,10,15,20-Tetrakis(4-aminophenyl)-porphyrin and 2-hydroxy-terephthalaldehyde, to which Co^{2+} was coordinated to finally obtain CoCOP. The prepared CoCOP had a specific surface area of 289 m^2/g and the average pore diameter was 18.63 nm. Then this CoCOP was applied for OER as an electrocatalyst under alkaline conditions. At 1 mA/cm^2 current density, this CoCOP exhibited an overpotential of 166 mV, and the overpotential was only 350 mV at a current density of 10 mA/cm^2. 151 mV/dec was the Tafel slop for CoCOP. At an overpotential of 400 mV, the calculated TON and TOF values of CoCOP were 46.8 and 0.013 s^{-1}. Further, in the stability test, no loss in current density was observed which displayed that CoCOP had good stability [34].

Yang et al. [35] synthesized COF-C$_4$N as a phenazine-linked 2D-COF based on theoretical calculations via solvothermal method by using triphenylenehexamine (TPHA) and hexaketocyclohexane (HKH). The synthesized COF-C$_4$N had an ordered crystalline structure with a highly conjugated basal plane. The pore width was 11.8 Å and the surface area was 105.86 m^2/g for COF-C$_4$N. Then this COF-C$_4$N was applied for OER activity in alkaline conditions and found that it had an overpotential of 349 mV at 10 mA/cm^2 current density which is better than other metal-free COF electrocatalysts. Also, COF-C$_4$N displayed a low Tafel slope (64 mV/dec). By chronoamperometry, the stability of COF-C$_4$N was evaluated and observed that after continuous measurement for 20 h, the current density decreased to 83.61%. An appropriate band gap position of N atoms which promotes the formation of C active sites, COF stability, and crystalline structure are the reasons behind the OER performance of COF-C$_4$N which were revealed from experimental and theoretical studies [35].

Huang et al. [36] prepared Co-PDY a 2D porous COF from Cobalt (II) 5,10,15,20-(tetra-4-ethynylphenyl)porphyrin (Co-TEPP) monomer via Glaser-Hay coupling reaction for which commercial copper foam was used as a catalyst. This Co-PDY COF had a repeating square unit of four Co-coordinated phenylporphyrin rings having butadiyne linkages in between them. Then in alkaline conditions (1M KOH), the catalytic performance of this Co-PDY for OER was evaluated. This Co-PDY had exhibited a low overpotential of 270 mV to achieve the 10 mA/cm^2 current density and displayed a low Tafel slope (99 mV/dec). This excellent OER activity was ascribed to several cobalt active sites, porous framework structure, and butadiyne linkages. While testing the stability, there was a slight decrease observed in the polarization curve as compared to the initial scan after 1000 cycles. The chronopotentiometry measurements displayed that there was no significant decrease in the activity after 10 h of continuous test and even after 20 h the remaining electrocatalytic current was 97.9% of its initial value. This shows that Co-PDY can be an efficient electrocatalyst for developing new devices [36].

Aiyappa et al. [29] prepared a bipyridine-containing TpBpy COF with a high surface area (~1660 m^2/g) via Schiff-base condensation reaction by using 1, 3, 5-triformylphloroglucinol (Tp) and 2,2'-bipyridyl-5,5'-diamine (Bpy). To this TpBpy COF, Co(II) ions were loaded into the framework to obtain Co-TpBpy (Figure 12.4) that had a surface area of 450 m^2/g, which

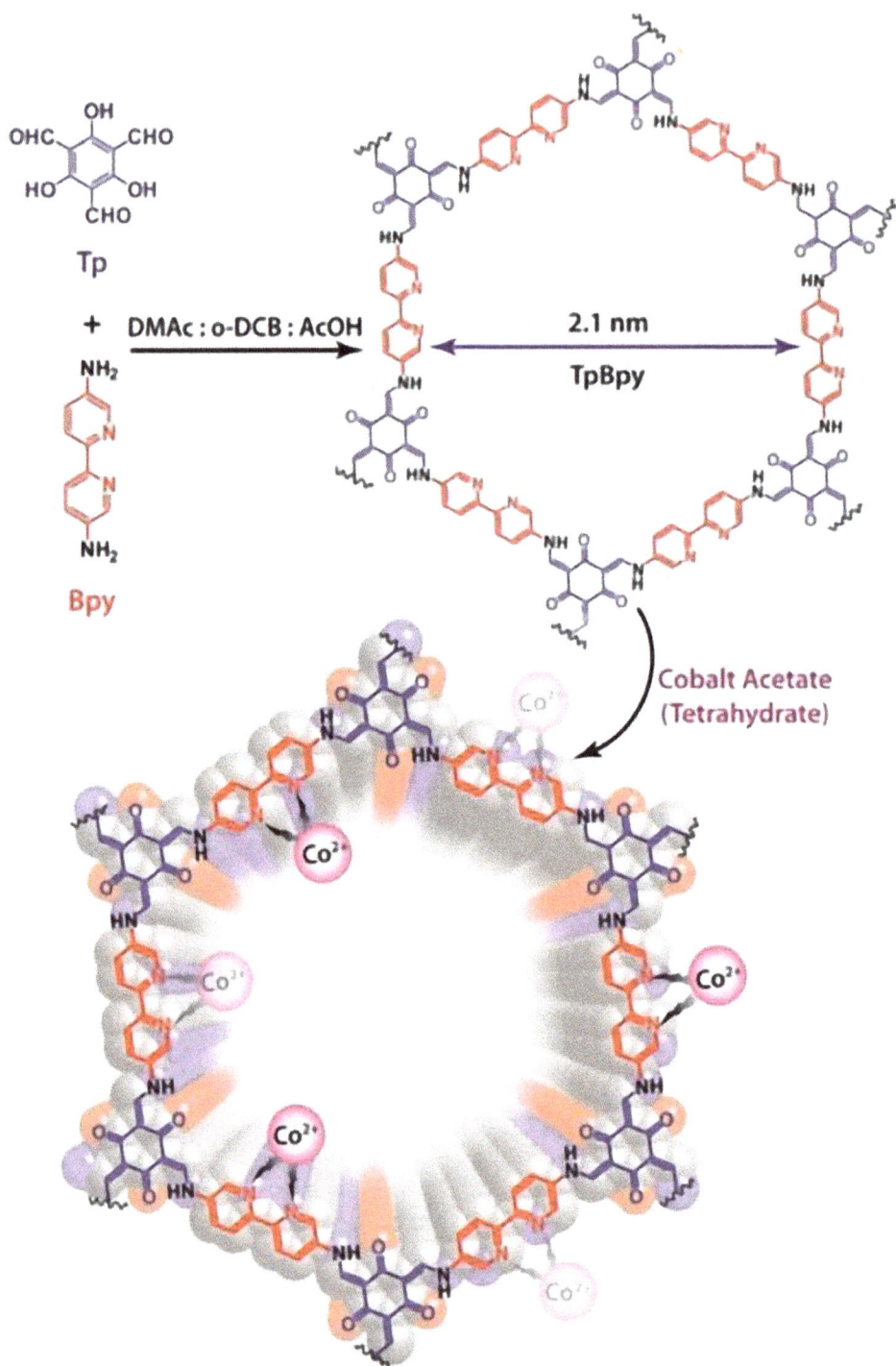

FIGURE 12.4
Illustration showing the Schiff-base condensation method for the synthesis of TpBpy and formation of Co-TyBpy by loading of Co (II) ions. (Adapted with permission [29]. Copyright (2016) American Chemical Society.)

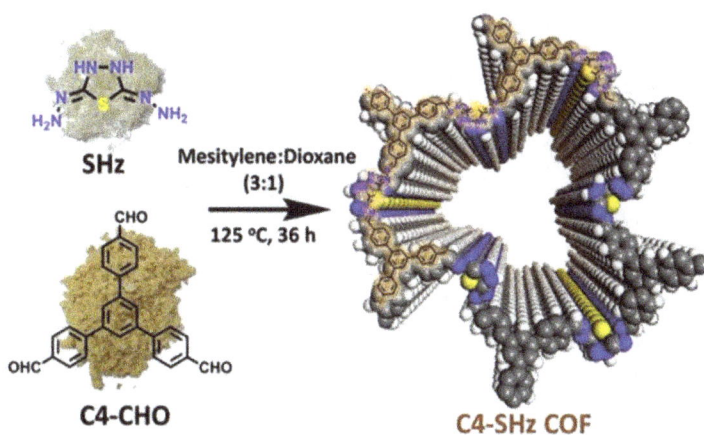

FIGURE 12.5

Illustration showing the proposed structure of C4-SHz COF. (Adapted with permission [3]. Copyright (2020) American Chemical Society.)

TABLE 12.1

COFs and COFs-Based Nanomaterials as Electrocatalysts for OER

S.N.	COF-Based Electrocatalyst	Overpotential (mV)	Tafel Slope (mV/dec)	Electrolyte	Ref.
1	BEAB-Co complex	320 @ 10 mA/cm²	57	1 M KOH	[30]
2	Co-COF	280 @ 10 mA/cm²	-	-	[31]
	Mn- COF	440@ 10 mA/cm²			
	Cr- COF	350 @ 10 mA/cm²			
3	COF-366-Co@CNT-0.4	358 @ 10 mA/cm²	62	1 M KOH	[32]
	COF-366-2/3Co@CNT-0.35	378 @ 10 mA/cm²	50	1 M KOH	
4	(Cyclen@Ni$_{0.5}$Fe$_{0.5}$)@COF-SO$_3$	276 @ 10 mA/cm²	43	1 M KOH	[1]
5	Co$_{0.5}$V$_{0.5}$@COF-SO$_3$	318 @ 10 mA/cm²	62	1 M KOH	[5]
6	Ni$_{0.5}$Fe$_{0.5}$@COF-SO$_3$	308 @ 10 mA/cm²	83	1 M KOH	[28]
7	macro-TpBpy-Co	380 @ 10 mA/cm²	54	0.1 M KOH	[33]
8	CoCOP	350 @ 10 mA/cm²	151	1 M KOH	[34]
9	COF-C$_4$N	349 @ 10 mA/cm²	64	1 M KOH	[35]
10	Co-PDY COF	270 @ 10 mA/cm²	99	1 M KOH	[36]
11	Co-TpBpy	400 @ 1 mA/cm²	59	0.1 M phosphate buffer solution (neutral pH)	[29]
12	C4-SHz COF	320 @ 10 mA/cm²	39	1 M KOH	[3]
13	IISERP_COF3_Ni$_3$N	230 @ 10 mA/cm²	79	1 M KOH	[37]
14	Co/CoO@COF	278 @ 10 mA/cm²	80.11	O$_2$-saturated 1 M KOH	[38]
15	Ni–COF	335 @ 10 mA/cm²	55	O$_2$-saturated 0.1 M KOH	[39]

was finally applied as OER electrocatalyst under neutral pH conditions. The study showed that to generate a 1 mA/cm^2 current density an overpotential of 400 mV was needed. This Co-TpBpy had a low Tafel slope of 59 mV/dec. In LSV, it was observed that Co-TpBpy had remarkable stability as it retained ~94% OER current after 1000 scans from 0.6 V to 1.8 V. From chronoamperomerty measurement, it was shown that for 24 h a steady-state current of 15 ± 3 mA/cm^2 was obtained. The estimated Faradaic efficiency of Co-TpBpy was 0.95 and Co-TpBpy exhibited a TOF value of 0.23 s^{-1} [29].

Mondal et al. [3] synthesized a thiadiazole-based new COF material (Figure 12.5) by reacting 2,5-dihydrazinyl-1,3,4-thiadiazole and 1,3,5-tris(4-formylphenyl)benzene. This new prepared C4-SHz COF had unique molecular architecture, abundant active sites, porosity, and possessed a specific surface area of 1224 m^2/g which is very high. This metal-free electrocatalyst (C4-SHz COF) under alkaline conditions was applied for OER. C4-SHz COF achieved 10 mA/cm^2 current density at a lower overpotential of 320 mV. A lower Tafel slope of 39 mV/dec and 98% Faradaic efficiency was exhibited by C4-SHz COF. The C4-SHz COF showed the mass activity of 286 A/g and specific activity of 0.011 mA/cm^2. For C4-SHz COF, the electrical double layer capacitance was estimated to be 2.75 mF/cm^2, the electrochemically active surface area was estimated to be 68.75 cm^2 and the roughness factor was 968.30. Further, by chronopotentiometry, the durability of C4-SHz COF was studied and found that for more than 11 h the COF can work efficiently at 10 mA/cm^2 current density [3].

The studies discussed above for COFs-based nanomaterials as electrocatalysts for OER have been summarized in Table 12.1.

12.3 Conclusion

The COFs and COFs-based nanomaterials have received great attention as OER electrocatalysts due to extraordinary properties of COFs such as high crystallinity, adjustable pore size, unique structure, large surface area, excellent chemical stability, withstand harsh conditions, and freedom to tune them electronically and structurally. A lot of efforts have been devoted to developing stable and cost-effective COFs and COFs-based nanomaterials as OER electrocatalysts that have improved OER efficiency and that can reduce the kinetic barrier. In the present chapter, recent progress has been summarized that has been made for developing and synthesizing different COFs-based nanomaterials as efficient and improved electrocatalysts for OER. It has been found that different ways that have been involved are adjustment of geometric structures and electronic structures of COFs, modification of the coordination environment, and introduction of active metal centers for improved OER performance of COFs and COFs-based nanomaterials. It is believed that fine-tuning of COFs structure and improved strategies for preparing COFs-based nanomaterials will prove their potential as electrocatalysts for OER and would be beneficial for large-scale application.

References

1. Feng X, Gao Z, Xiao L, Lai Z, Luo F (2020) A Ni/Fe complex incorporated into a covalent organic framework as a single-site heterogeneous catalyst for efficient oxygen evolution reaction. Inorg Chem Front 7:3925–3931

2. Zhou Y, Li J, Gao X, Chu W, Gao G, Wang LW (2021) Recent advances in single-atom electro-catalysts supported on two-dimensional materials for the oxygen evolution reaction. J Mater Chem A 9:9979–9999

3. Mondal S, Mohanty B, Nurhuda M, Dalapati S, Jana R, Addicoat M, Datta A, Jena BK, Bhaumik A (2020) A Thiadiazole-Based Covalent Organic Framework: A Metal-Free Electrocatalyst toward Oxygen Evolution Reaction. ACS Catal 10:5623–5630

4. Geng K, He T, Liu R, Dalapati S, Tan KT, Li Z, Tao S, Gong Y, Jiang Q, Jiang D (2020) Covalent Organic Frameworks: Design, Synthesis, and Functions. Chem Rev 120:8814–8933

5. Gao Z, Yu Z, Huang Y, He X, Su X, Xiao L, Yu Y, Huang X, Luo F (2020) Flexible and robust bimetallic covalent organic frameworks for the reversible switching of electrocatalytic oxygen evolution activity. J Mater Chem A 8:5907–5912

6. Côté AP, Benin AI, Ockwig NW, O'Keeffe M, Matzger AJ, Yaghi OM (2005) Chemistry: Porous, crystalline, covalent organic frameworks. Science (80-) 310:1166–1170

7. Jin Y, Hu Y, Zhang W (2017) Tessellated multiporous two-dimensional covalent organic frameworks. Nat Rev Chem 1:1–11

8. El-Kaderi HM, Hunt JR, Mendoza-Cortés JL, Côté AP, Taylor RE, O'Keeffe M, Yaghi OM (2007) Designed synthesis of 3D covalent organic frameworks. Science (80-) 316:268–272

9. Mc Keown NB, Budd PM (2006) Polymers of intrinsic microporosity (PIMs): Organic materials for membrane separations, heterogeneous catalysis, and hydrogen storage. Chem Soc Rev 35:675–683

10. Zeng Y, Zou R, Zhao Y (2016) Covalent organic frameworks for CO2 capture. Adv Mater 28:2855–2873

11. Kuhn P, Antonietti M, Thomas A (2008) Porous, covalent triazine-based frameworks prepared by ionothermal synthesis. Angew Chemie - Int Ed 47:3450–3453

12. Jiang JX, Su F, Trewin A, Wood CD, Campbell NL, Niu H, Dickinson C, Ganin AY, Rosseinsky MJ, Khimyak YZ, Cooper AI (2007) Conjugated microporous poly(aryleneethynylene) networks. Angew Chemie - Int Ed 46:8574–8578

13. Biswal BP, Chandra S, Kandambeth S, Lukose B, Heine T, Banerjee R (2013) Mechanochemical synthesis of chemically stable isoreticular covalent organic frameworks. J Am Chem Soc 135:5328–5331

14. Campbell NL, Clowes R, Ritchie LK, Cooper AI (2009) Rapid microwave synthesis and purification of porous covalent organic frameworks. Chem Mater 21:204–206

15. Yang ST, Kim J, Cho HY, Kim S, Ahn WS (2012) Facile synthesis of covalent organic frameworks COF-1 and COF-5 by sonochemical method. RSC Adv 2:10179–10181

16. Kim S, Choi HC (2019) Light-promoted synthesis of highly-conjugated crystalline covalent organic framework. Commun Chem 2:1–8

17. Wang C, Li Z, Chen J, Li Z, Yin Y, Cao L, Zhong Y, Wu H (2017) Covalent organic framework modified polyamide nanofiltration membrane with enhanced performance for desalination. J Memb Sci 523:273–281

18. Medina DD, Rotter JM, Hu Y, Dogru M, Werner V, Auras F, Markiewicz JT, Knochel P, Bein T (2015) Room temperature synthesis of covalent-organic framework films through vapor-assisted conversion. J Am Chem Soc 137:1016–1019

19. Ding SY, Wang W (2013) Covalent organic frameworks (COFs): From design to applications. Chem Soc Rev 42:548–568

20. Huang N, Wang P, Jiang D (2016) Covalent organic frameworks: A materials platform for structural and functional designs. Nat Rev Mater 1:1–19

21. Furukawa H, Yaghi OM (2009) Storage of hydrogen, methane, and carbon dioxide in highly porous covalent organic frameworks for clean energy applications. J Am Chem Soc 131:8875–8883

22. Subodh, Prakash K, Masram DT (2020) Chromogenic covalent organic polymer-based microspheres as solid-state gas sensor. J Mater Chem C 8:9201–9204

23. Subodh, Prakash K, Masram DT (2020) A reversible chromogenic covalent organic polymer for gas sensing applications. Dalt Trans 49:1007–1010

24. Subodh, Prakash K, Masram DT (2021) Silver Nanoparticles Immobilized Covalent Organic Microspheres for Hydrogenation of Nitroaromatics with Intriguing Catalytic Activity. ACS Appl Polym Mater 3:310–318
25. Subodh, Prakash K, Chaudhary K, Masram DT (2020) A new triazine-cored covalent organic polymer for catalytic applications. Appl Catal A Gen 593:117411
26. Fang Q, Wang J, Gu S, Kaspar RB, Zhuang Z, Zheng J, Guo H, Qiu S, Yan Y (2015) 3D Porous Crystalline Polyimide Covalent Organic Frameworks for Drug Delivery. J Am Chem Soc 137:8352–8355
27. Patra BC, Khilari S, Manna RN, Mondal S, Pradhan D, Pradhan A, Bhaumik A (2017) A Metal-Free Covalent Organic Polymer for Electrocatalytic Hydrogen Evolution. ACS Catal 7:6120–6127
28. Gao Z, Gong L Le, He XQ, Su XM, Xiao LH, Luo F (2020) General Strategy to Fabricate Metal-Incorporated Pyrolysis-Free Covalent Organic Framework for Efficient Oxygen Evolution Reaction. Inorg Chem 59:4995–5003
29. Aiyappa HB, Thote J, Shinde DB, Banerjee R, Kurungot S (2016) Cobalt-Modified Covalent Organic Framework as a Robust Water Oxidation Electrocatalyst. Chem Mater 28:4375–4379
30. Li T, Atish C, Silambarasan K, Liu X, O'Mullane AP (2020) Development of an interfacial osmosis diffusion method to prepare imine-based covalent organic polymer electrocatalysts for the oxygen evolution reaction. Electrochim Acta 362:137212
31. Wang J, Wang J, Qi S, Zhao M (2020) Stable Multifunctional Single-Atom Catalysts Resulting from the Synergistic Effect of Anchored Transition-Metal Atoms and Host Covalent-Organic Frameworks. J Phys Chem C 124:17675–17683
32. Gan Z, Lu S, Qiu L, Zhu H, Gu H, Du M (2021) Fine tuning of supported covalent organic framework with molecular active sites loaded as efficient electrocatalyst for water oxidation. Chem Eng J 415:127850
33. Zhao X, Pachfule P, Li S, Langenhahn T, Ye M, Schlesiger C, Praetz S, Schmidt J, Thomas A (2019) Macro/Microporous Covalent Organic Frameworks for Efficient Electrocatalysis. J Am Chem Soc 141:6623–6630
34. Wang A, Cheng L, Zhao W, Shen X, Zhu W (2020) Electrochemical hydrogen and oxygen evolution reactions from a cobalt-porphyrin-based covalent organic polymer. J Colloid Interface Sci 579:598–606
35. Yang C, Yang Z-D, Dong H, Sun N, Lu Y, Zhang F-M, Zhang G (2019) Theory-Driven Design and Targeting Synthesis of a Highly-Conjugated Basal-Plane 2D Covalent Organic Framework for Metal-Free Electrocatalytic OER. ACS Energy Lett 4:2251–2258
36. Huang H, Li F, Zhang Y, Chen Y (2019) Two-dimensional graphdiyne analogue Co-coordinated porphyrin covalent organic framework nanosheets as a stable electrocatalyst for the oxygen evolution reaction. J Mater Chem A 7:5575–5582
37. Nandi S, Singh SK, Mullangi D, Illathvalappil R, George L, Vinod CP, Kurungot S, Vaidhyanathan R (2016) Low Band Gap Benzimidazole COF Supported Ni3N as Highly Active OER Catalyst. Adv Energy Mater 6:
38. Ye X, Fan J, Min Y, Shi P, Xu Q (2021) Synergistic effects of Co/CoO nanoparticles on imine-based covalent organic frameworks for enhanced OER performance. Nanoscale 13:14854–14865
39. Zhou W, Yang L, Wang X, Zhao W, Yang J, Zhai D, Sun L, Deng W (2021) In Silico Design of Covalent Organic Framework-Based Electrocatalysts. JACS Au 1:1497–1505

13

Covalent Organic Frameworks-Based Nanomaterials for Oxygen Reduction Reactions

Marcos Martínez-Fernández[1], Emiliano Martínez-Periñán[2], Encarnación Lorenzo[2,3], and José L. Segura[1]

[1]*Departamento de Química Orgánica I, Facultad de CC. Químicas, Universidad Complutense de Madrid, Madrid, Spain*

[2]*Departamento de Química Analítica y Análisis Instrumental & Institute for Advanced Research in Chemical Sciences (IAdChem), Facultad de Ciencias, Universidad Autónoma de Madrid, Madrid, Spain*

[3]*IMDEA-Nanociencia, Ciudad Universitaria de Cantoblanco, Madrid, Spain*

CONTENTS

13.1 Introduction

Oxygen reduction reaction (ORR) is an essential process in different areas such as biological respiration, energy conversion, and hydrogen peroxide generation. The electrocatalysts employed are often classified by attending to the pathway catalyzed in the reaction, which in aqueous media are mainly two:

- The direct reduction of O_2 into H_2O, through a four-electron pathway.
- The reduction of O_2 into H_2O_2 through a two-electron pathway.

DOI: 10.1201/9781003206507-13

The two-electron pathway produces incomplete oxygen reduction, generating hydrogen peroxide as a reaction product, and leading to lower energy conversion. Furthermore, the generation of reaction intermediates and free radicals can have negative effects on the activity of some electrocatalysts [1]. Despite these reasons, the interest in the electrochemical generation of H_2O_2 has recently emerged and new research lines focused on the development of efficient electrocatalysts through the two-electron pathway are under demand.

The four-electron pathway is the preferred one in energy conversion applications such as fuel cells and air batteries due to the higher electric current provided in comparison with the two-electron pathway. In the case of proton exchange membrane fuel cells (PEMFCs), the more promising fuel cell device, hydrogen is oxidized at the anode, and oxygen is reduced at the cathode. Protons are transported from the anode to the cathode through the electrolyte membrane and the electrons are carried over an external circuit load [2]. Both the ORR in the cathode and the hydrogen oxidation reaction in the anode frequently occur on the surfaces of platinum (Pt)-based catalysts. These devices present high energy density, good energy conversion efficiency, zero pollution emissions, and high reliability, which are necessary for obtaining clean energy for residential applications, vehicles, and stationary power systems [3].

Other current energy conversion devices based on the ORR that will play a key role in the energy transformation are the new metal-air batteries (MABs) [4]. The traditional MAB typically consists of a metal anode (usually Li, Al Zn, etc.), a separator soaked in an electrolyte, and an air cathode for the electrocatalytic ORR. The main advantages of MABs are their extremely high energy density, low cost, safety, and abundance of raw materials [5]. Great expectations are placed on the improvements in PEMFCs and MABs to overcome the actual energy crisis and therefore there is a huge interest in the improvement of ORR electrocatalysts design and optimization. At the same time, the desire for improved ORR electrocatalysts is considered a consistent driving force for academic and industrial research all over the world [6].

Platinum group metal (PGM) based electrocatalysts have been traditionally used to achieve ORR through the four-electron pathway. Although platinum reserves and production worldwide are very limited nowadays, the most commonly used material in large-scale is nanostructured Pt, usually supported over carbon conducting materials or platinum alloys (Pt_3Co, Pt_3Ni) [7]. Another important drawback of cathodes used in PEMFCs is that ORR is more than six orders of magnitude slower than the hydrogen oxidation at anodes, so high amounts of electrocatalyst must be included in the fabrication of cathode electrodes thus increasing the cost [3]. In addition, the generation of extremely stable Pt–O and Pt–OH species, and the presence of Pt-based poisons (like chloride) reduce the energy conversion efficiency and slow the ORR kinetics [8].

All the reasons stated above aim to the urgent need for the development of non-Pt and metal-free electrocatalysts for the ORR in the manufacture of energy conversion devices [9]. In this respect, three main groups of novel electrocatalysts have been developed in recent years as an alternative to the traditional electrocatalysts: non-noble metal electrocatalysts, transition metals carbides, and nitrides, and metal-free electrocatalysts. As an alternative to the ORR electrocatalysts outlined above, a new family of compounds named covalent organic frameworks (COFs) are emerging as a powerful alternative. In the following sections, we will review the main approaches developed for the use of COFs as electrocatalysts for the ORR.

13.2 COF-Based Electrocatalysts for the ORR

COFs are porous, crystalline, pre-designable, and hierarchically organized structures that resemble those found in some biomacromolecules. Thus, in analogy to proteins (Figure 13.1A), covalent organic frameworks present a primary structure defined by linkers connected by a specific reversible covalent bond known as linkage. Remarkably, the physicochemical properties should be addressed in the material's design from the linkage perspective since the lower bond reversibility usually enhance the stability of the product [10]. The secondary structure of COFs is related to the pore topology, which is controlled by the relative symmetry of the reactants, as well as their size or their dimensionality [11]. Finally, the tertiary structure of COFs is related to crystal morphology. Usually, COFs are obtained as polycrystalline and insoluble powders and single-crystals COFs are limited to a few examples. However, different COF structures have been reported, such as hollow-spheres, spherulites, core-shell structures, fibers, and even two or three-dimensional entangled structures. These facts limit their implementation in electronic devices due to the inherent insolubility of the COF powders, the polycrystalline nature of the materials, and the uncontrolled crystallite morphology that lower the

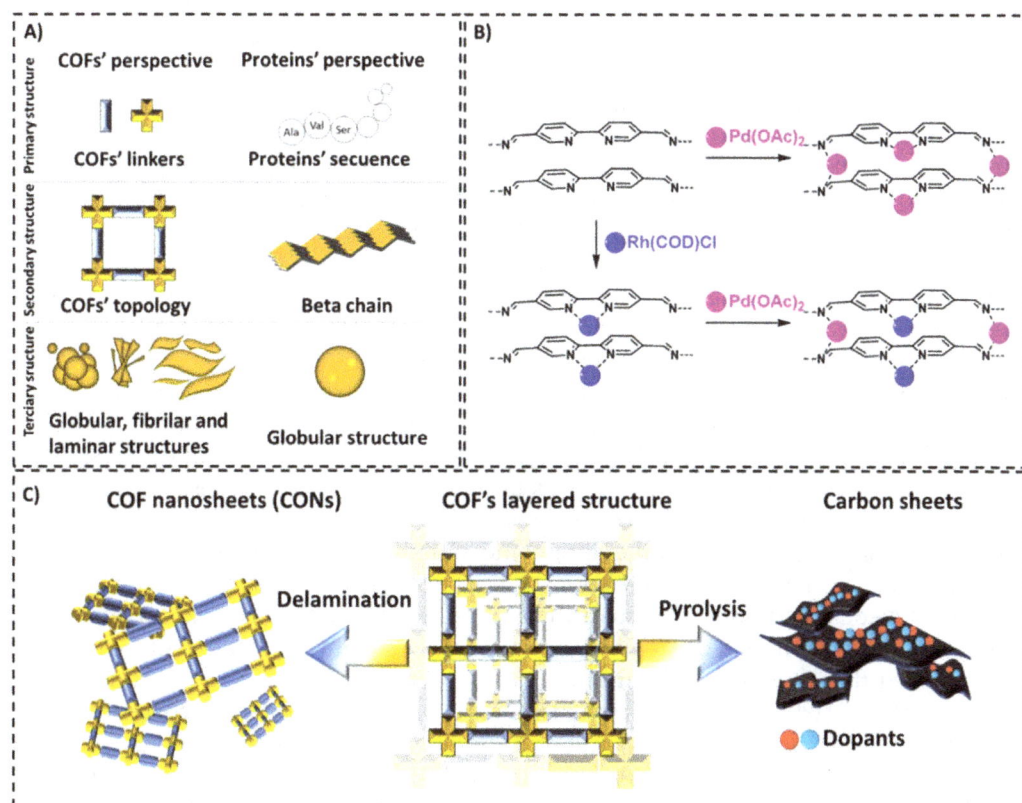

FIGURE 13.1

(A) Comparison between the hierarchical structures of COFs and proteins. (B) Pd & Rh post- synthetic anchoring in an imine/bipyridine-based COF. (C) Common strategies to obtain ORR catalysts from 2D-COFs.

contact area between the COF and an electrode, where an extended laminar structure would be desirable [12].

To address this issue, top-down and bottom-up methods have been developed in the last years to increase the processability of COFs and control the morphology, lateral size, and even the macroscopic shape [13]. On the one hand, top-down methodologies are based on the delamination of already synthesized COFs to obtain covalent organic nanolayers or CONs [14]. In this field, several exfoliation procedures were described, such as the ultrasonic-assisted method, the acid self-exfoliation, the chemical exfoliation, or the mechanical delamination. On the other hand, bottom-up methodologies allow the construction of the desired structures directly from smaller building blocks. Sundry methods were employed in the last few years, highlighting interfacial synthesis, micellar synthesis, or on-surface synthesis.

Following the successful implementation of COFs onto devices, a considerable amount of these polymers has been reported as heterogeneous catalytic systems. The key concept is the incorporation of active centers in an extended network. To illustrate this concept, the Suzuki-Miyaura Pd-mediated C-C cross-coupling reaction between aryl halides and aromatic boronic acids is the best example. Unfortunately, homogeneous catalytic reactions lead to difficult recovery of the Pd complexes. To avoid this problem, palladium can be anchored to the COF scaffold thus leading to a recoverable platform by filtration of the reaction crude. From a synthetic perspective, two complementary strategies can be envisaged for the chelation with palladium. First, a pre-synthetic strategy involves the incorporation of the palladium complex during the formation of the COF. Thus, almost one monomer should include the pre-synthesized complex or its *in situ* incorporation to the polymeric backbone. And secondly, a post-synthetic strategy (Figure 13.1B) involves the modification of the COF skeleton by Pd coordination onto the polymeric matrix chelating points [15].

In addition to the large number of synthetic tools to tune the properties of COFs, efforts have been also devoted to obtaining materials with large surface areas to attain efficient heterogeneous catalytic systems [16, 17]. Catalysts with high surface areas provide sufficient accessible active sites to participate in the chemical reactions, allowing high mass transfer to enhance the catalyst-reactants contact. Covalent organic frameworks usually present high surface areas, to accommodate the catalytic centers. It should be mentioned that eclipsed stacking of two-dimensional COFs (2D-COFs) or non-interpenetrated three-dimensional COFs (3D-COFs) are the most favorable conformations to generate porous networks. Furthermore, pore size is a crucial variable to consider as the reactants must fit into the COFs cavity to generate the desired product and favor its migration outside the pores. Undoubtedly, these advantages influence the kinetics of the process, including the adsorption, the chemical reaction, and the desorption processes [18].

The last variable to consider in the design and synthesis of a heterogeneous COF-based catalyst is the use of a suitable linkage for the framework crystallization [19]. For example, basic and aqueous conditions of Suzuki-Miyaura C-C cross-coupling are incompatible with boronate or boroxine-linked COFs since they are prone to hydrolysis [10], so other linkages must be used that resist electrochemical corrosion over a long time and avoid active site poisoning effects [19]. Electrocatalysis is just the extension of the catalysis concept defined by Berzelius in 1835 (from Greek *kata* (down) and *lyein* (help)), where an energy source is required. Furthermore, electron (negative charges) or hole (positive charges) migration, from the energy source to the active sites is essential to increase de electrocatalytic reaction yield [20]. It should be highlighted that there are two kinds of conductivities present in these frameworks. On the one hand, in-plane conductivity is limited by the linkage polarization or by the existence of conjugation nodes such as imide linkages or *meta*-positions of the trigonal knots. On the other hand, columnar charge migration or transversal conductivity relies mainly on the interactions between neighboring layers, which fundamentally depend on their constitutive units [21, 22]. COF's conductance, at its *bulk* state, is usually poor due to its polycrystalline nature. To improve the conductivity, (i) highly conductive units must be used to build the network, or (ii) additives, such as carbon black or carbon nanotubes, must be added [23]. Finally, exfoliation of 2D-COFs into CONs is the most efficient procedure to assemble electrical devices and enhance the intrinsic conductivity [21].

To sum up, design principles for efficient COF-based electrocatalyst include the obtainment of materials with large surface areas, inherent stability or durability, efficient charge transport, efficient processing, and low cost. To address these issues to obtain efficient ORR catalysts there are mainly two strategies as depicted in Figure 13.1C. The first strategy involves the use of covalent organic frameworks with porous and crystalline structures to obtain homogeneously doped mesoporous carbon sheets [24]. It is well known that graphene and its carbonaceous derivatives catalyze ORR, especially if the polarization of bonds is induced by heteroatomic doping [25]. It should be highlighted that the incorporation of heteroatom dopants is commonly conducted by *in situ* doping or by post-doping of the obtained carbon materials with the use of a heteroatom-containing moiety. However, it is still a challenge to control the distribution of dopants in the carbonaceous matrix due to the uncontrollable pyrolysis process required [26].

The two-dimensional character of 2D-COFs combined with the presence of lightweight elements in these organized structures provides innovative systems which are promising candidates to be used as suitable templates for the pyrolytic synthesis of 2D carbons with a homogenous distribution of heteroatom dopants. Interestingly, in addition to the templating behavior that affords conductive mesoporous 2D sheets [27], edge-exposed active centers are also produced during the pyrolytic treatment, which is intrinsically more active than their basal localized counterparts [28]. Despite some recent strategies to guide the pyrolysis into truly 2D carbon sheets [26], there are still some limitations of the pyrolytic process such as the loss of crystallinity and the high energy costs derived from the pyrolytic treatment (around 900°C are required) [29]. For this reason, the few strategies explored to produce pyrolysis-free ORR catalysts based on COFs are currently receiving a great deal of attention [23, 30].

Besides, regardless of the strategy followed to obtain active catalysts based on COFs, it should be mentioned that it can be combined with the use of metals in the composition. Beyond the metal-based catalysts development (considering precious metals or not), atomically dispersed metallic species are actively studied because they can maximize the use of these metals (noble or non-noble) and exhibit unique selectivity due to the absence of ensembled sites. Usually, single-metal atoms are prone to aggregation because of their thermodynamical instability [31]. Therefore, to provide stability, single-metal atoms are often anchored on supports such as graphene, zeolites, metal alloys, or mono-crystalline metals. Covalent organic frameworks present two types of anchoring points: (i) in the net linkages if they present Lewis basic character to interact with the void metal orbitals in a coordinative bond; (ii) in complexing agents, incorporated in the COF skeleton. These strategies are compatible with the pyrolysis process to synthesize tailored materials with enhanced distribution of metallic species on the carbon matrix rather than the heterogeneous aggregated bulks [32]. In the following sections, we will summarize the different approaches followed in last recent years to develop efficient electrocatalysts based on COFs for the ORR. These novel electrocatalysts will be classified as follows:

- Pyrolyzed COFs: including metalized COFs and metal-free COFs.
- Pyrolysis-free COFs: including metalized COFs and metal-free COFs.

13.3 ORR Electrocatalysts Based on Pyrolyzed-COFs

As stated above, metal-nitrogen-modified carbons (M-N-C) [33] are promising alternative materials to Pt-based electrodes in fuel cells. However, because of their low conductivity, the use of appropriate carbon-based or heteroatom-doped carbon-based [34, 35] supporting material is required to reach higher conductivities and a more efficient mass transfer process for better ORR performance [36]. With this aim, it is important to gain control over the number of metal sites and the disposition of the electrocatalyst over the carbon supporting materials. Thus, a pyrolysis step (high-temperature treatment in an inert atmosphere) is normally essential for the synthesis of M-N-C catalysts and critical to control not only the composition but also the microstructure, which is essential to obtain heterogeneous catalysts with high electronic conductivity and electrocatalytic activity [24].

COFs possess some lightweight atoms (C, N, O, H, etc.) in the structure, which made COFs ideal precursors for heteroatom-doping carbons. At the same time, the heteroatoms in the COF pores can act as binding sites for metal ions, which results in the formation of microstructures, far away from the big aggregates of metallic particles usually obtained after the pyrolysis process in M-N-C catalysts. Thus, because covalent organic frameworks are pre-designable and porous platforms with different heteroatomic species arranged with atomic precision, there is currently a great deal of interest to use the defined structure of COFs to guide the pyrolysis step toward the obtainment of mesoporous and conductive graphitic carbons homogeneously doped with active sites on their skeletons. The pyrolyzed COFs that can be obtained can be classified attending to the following classification based on the specific locations of the dopants in the network.

13.3.1 Surface-Embedded Active Sites

The first COF endowed with metalloporphyrin (M=Fe, Co, or Mn) moieties were synthesized by Dai *and co-workers* [37] by using nickel-catalyzed Yamamoto polymerization reactions. The monomers endowed with the metallic species were prepared by reacting 5,10,15,20-tetrakis(4′-bromobiphenyl)-porphyrin with the corresponding metal-chlorides. The subsequent nickel-catalyzed Yamamoto self-polycondensation yielded crystalline polymers as insoluble powders. The final carbonization step was performed at 950°C to afford highly dispersed metal-nitrogen graphitic structures. In a basic medium (0.1 M KOH), the cyclic voltammograms of the glassy carbon electrodes modified with pyrolyzed COFs in the absence of coordinated metals did not show high ORR electrocatalyst activities. A similar effect was observed when coordinated or non-coordinated 5,10,15,20-tetrakis(4′-bromobiphenyl) porphyrin monomers were pyrolyzed and the obtained product was subsequently used to modify the GC electrode. However, the presence of coordinated metals in the porphyrin centers of the COF materials significantly increases the ORR activity. Thus, the Fe-COF modified electrode showed a similar onset potential (ca. 0.89 V) as that obtained with a Pt/C (ca. 0.96 V) electrode, while the Co-COF modified electrode presented a higher limited current than that of the Pt/C at potentials less than 0.26 V. Regarding the results in acidic medium, although the onset potentials are not as positive as that obtained with Pt/C, two important points must be highlighted: the stability is much better than that of the Pt/C electrode and the electrocatalysis is free from the methanol crossover/CO poisoning effects.

Xu and col. [38] reported the synthesis of a similar structure through the Sonoghasira coupling between 10,15,20-tetrakis(4-trimethylsilyl-ethynylphenyl)porphyrin, 5,10,15, 20-tetrakis-(4 ethynylphenyl) porphyrin, 5,10,15,20-tetrakis (4-bromophenyl) porphyrin

(TBPP-H2) and Co(II) 5,10,15,20-tetrakis (4-ethynylphenyl) porphyrin. Following purification, the polymer obtained was pyrolyzed at 900°C to produce the 10%-Co-N/C catalyst. In their best configuration, ORR appears at a half-wave potential of 0.816 V *vs.* RHE and a limiting current density of 4.74 mA cm^{-2} in alkaline media. This electrocatalyst shows an almost direct four electrons reduction of O_2 to H_2O with relatively low yields of H_2O_2 species. Furthermore, it displays superior stability and great tolerance against methanol compared to Pt/C.

A different strategy to obtain surface-embedded active sites was performed by Mao and co-workers [24]. They managed to synthesize a cobalt/porphyrin-based covalent organic framework (COF-900, Figure 13.2A) by the Schiff-base condensation reaction between terephthalaldehyde and cobalt (II) 5,10,15,20-tetrakis(p-tetraphenylamino)-porphyrin (Co-TAPP) under solvothermal conditions. Finally, the crystalline polymer was pyrolyzed under a nitrogen atmosphere at 900°C, resulting in the formation of the Co–N–C ORR catalyst. Both the potential and current responses obtained with pyrolyzed Co-COF (Co-COF-900) were very close to those obtained with 20% Pt/C electrocatalyst in 0.1 M KOH. The number of electrons involved in ORR was 3.86, suggesting that an almost four-electron reduction process essentially occurs in the ORR process. In this work authors also demonstrated that the pyrolization of the metalized COF produce an ORR electrocatalyst with good activity while if the same process is carried out using the equivalent metal coordinated monomer (Co-TAPP-900), the ORR electrocatalytic activity of the obtained material is far from that obtained with Pt/C (Figure 13.2B).

13.3.2 Edge-Exposed Based Catalyst by Metal Coordination

While in the previous section the active sites were embedded in the carbonaceous wall of the covalent organic framework, in this section we review a complimentary strategy where

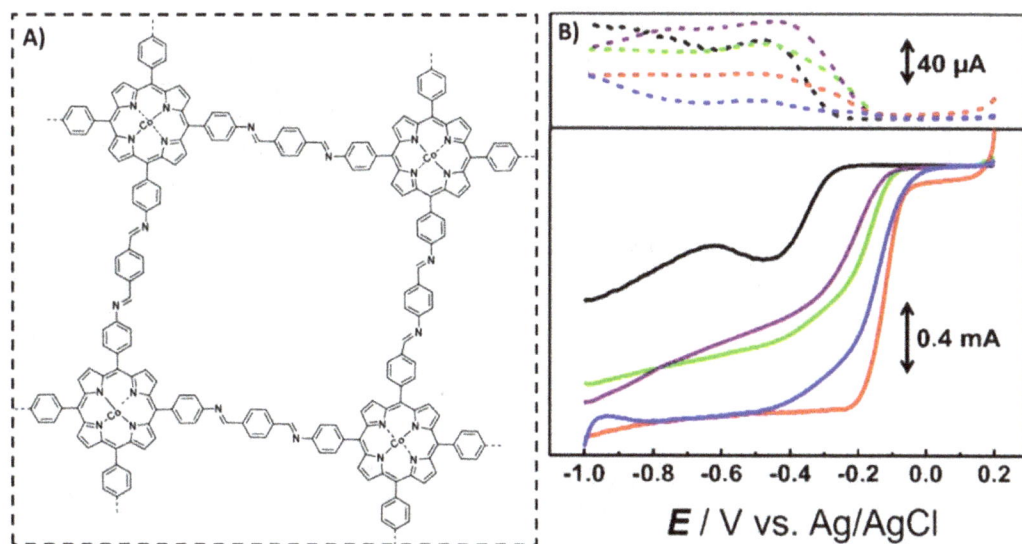

FIGURE 13.2

(A) Structure of COF-900. (B) Rotating ring-disk electrode (RRDE) voltammograms for O_2 reduction at bare GC disk electrode (black line) and GC disk electrodes modified with monomer Co-TAPP-900, COF-900, Co-COF-900, and 20% Pt/C. The upper panel represents the current responses at the Pt ring electrode. (Adapted with permission [24]. Copyright (2015) Elsevier.)

the active sites are not embedded in the network but exposed to the edges. Thus, metallic active sites can be connected to the COF either through coordination with the imine linkages or through coordination with accessible ligands incorporated in the COF structure.

13.3.2.1 Coordination of Metallic Active Sites with Imine Linkages

Edge-exposed Metal-N-C active centers are intrinsically more active than those located at the basal plane [28, 39, 40] because of the enhanced mass transport efficiency. Thus, an effective strategy to incorporate single metal-atom sites to obtain an efficient ORR catalyst is to anchor on the COF matrix the metallic species before the pyrolytic treatment. This was the strategy chosen by Li and co-workers to prepare a COF-derived Fe&N-doped carbon [41]. The first step involved the synthesis of COF-1 by the Schiff-base condensation reaction between 5-tris(4-aminophenyl) benzene (TAPB) and 1,3,5-benzenetricarboxaldehyde (BTC) in an acidic medium at room temperature for 15 min. Then, iron (III) ions were incorporated into COF-1 by coordination between metal ions and nitrogen atoms of the *in situ* synthesized imine linkages-based COF. Finally, pyrolytic treatment under an inert atmosphere yielded the active catalyst as COF-derived Fe&N-doped carbon nanospheres (Figure 13.3, Fe-ISAS/CN). The N-doped carbon nanospheres presented a homogeneous size distribution, with good dispersity in liquid solvents. This facilitates the exposure of isolated single-atom metal sites during catalysis, thus improving the molecular contact between the reactants and the catalytically active sites. Therefore, the Fe-ISAS/CN boosts the catalytic performance in electro-catalysis and in catalyzing organic reactions. A remarkable ORR electrocatalytic activity, close to that reported for commercial 20% Pt/C, is achieved with Fe-ISAS/CN. The ORR mechanism follows a four-electron pathway and the same Tafel slope (78 mV·dec^{-1}) as that reported for 20% Pt/C was observed. In addition, it is worth pointing out that, in comparison with commercial 20% Pt/C, the Fe-ISAS/CN electrode has better stability ($E_{1/2}$ shift of 4 mV for Fe-ISAS/CN *vs.* 25 mV observed for 20% Pt/C) and a total methanol crossover tolerance.

A similar strategy was used by Sun and co-workers [42] to synthesize 3D-COF-300. 3D-COFs are covalent organic frameworks that incorporate three-dimensional monomers to construct expanded skeletons beyond the two-dimensional COFs sheets. Thus, Sun and co-workers synthesized the **3D-COF-300** by a Schiff-base condensation reaction between

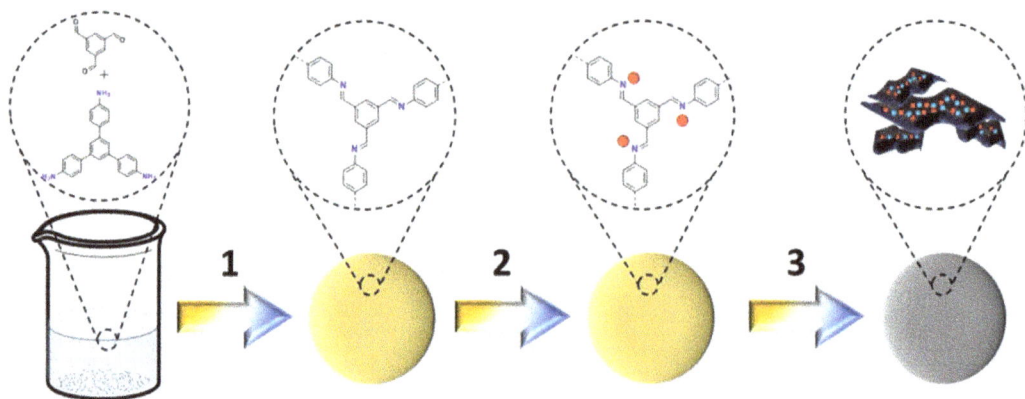

FIGURE 13.3
Schematic illustration for the preparation of electrocatalysts based on COFs through linkage coordination with metals: (1) Polymerization, (2) Imine linkage metalation and (3) Pyrolysis.

4,4′,4″,4‴-methanetetrayltetraaniline and terephthalaldehyde under solvothermal reaction conditions. This three-dimensional COF has chelating points at the nitrogen lone pairs which are used to coordinate Cobalt (II) via impregnation. Finally, the resulting structure was pyrolyzed to obtain edge-exposed Co-Nx-C active centers which are intrinsically more active than those located at the basal plane [28, 39, 40] due to an enhanced mass transport efficiency. The electrocatalytic activity of 3D-COF-300 is better than that shown by Pt/C electrocatalysts in the alkaline medium. It also displays long-term stability and methanol tolerance ability. Even in the challenging acidic medium, the ORR activity is greatly enhanced. Although pyrolyzed Co-COF presents more negative half-wave potential in comparison to that of commercial Pt/C and state-of-the-art single atomically dispersed electrocatalysts in an acidic medium, it showed higher mass activity of 838 A/gCo at 0.9 V *vs.* RHE, which is a consequence of the highly active edge-hosted Co-Nx sites that are fully accessible for ORR.

13.3.2.2 Coordination of Metallic Active Sites with Ligands Incorporated in the COF Structure

A complementary strategy was used by Sun and co-workers [43] that synthesized a bipyridine based-COF (Tp-Bpy-COF, Figure 13.4A) by a Schiff-base condensation reaction between the 2,4,6-trihydroxybenzene-1,3,5-tricarbaldehyde and the [2,2′-bipyridine]-5,5′-diamine. The crystalline polymer had bipyridine units with Lewis-basic nitrogens, which can act as chelating agents to the metal coordination. This COF was then modified by cobalt acetate and iron acetate to obtain a bimetallic COF (Fe,Co-COF) which was subsequently carbonized under N_2 at 900°C. The prepared catalyst has abundant metallic Fe-Co alloy nanoparticles surrounded by standing mesoporous carbon layers and exhibits ultrahigh yields in the ORR as a consequence of the combination of enhanced conductivity, abundant working metallic-active sites with full mass transport toward or away from the catalytic sites. Furthermore, these carbon layers protect the nanoparticles in alkaline conditions, enhancing their durability. The Fe,Co-COF ORR electrocatalytic activity in an alkaline medium is excellent with a 50 mV positive half-wave potential, a higher limited diffusion current density, and a much smaller Tafel slope than a Pt/C electrocatalyst. In addition, Fe,Co-COF also exhibited excellent properties in the OER and HER because of their abundant active sites, high N content, and mesoporous structure, increasing the interest in this kind of electrocatalysts.

To prepare COF-derived carbons with an enhanced accessible surface, Thomas and co-workers synthesized another bypyridine-containing COF, namely Tp-Bpy-COF [29], using a *p*-toluenesulfonic acid-assisted mechanochemical approach. The *p*-toluenesulfonic acid acts as a molecular organizer for the COF growth in the presence of silica nanoparticles (with a diameter of ~24 nm) to enable the formation of mesoporous structures with many bipyridine units. The bipyridine units assist metal ion coordination to afford abundant N- and Fe-doping within the obtained carbon material. The subsequent pyrolysis yields mesoporous carbon mC-Tp-Bpy-COF-Fe material with enhanced porosity (Figure 13.4B). The mesoporous structure and large pore volume efficiently allow the fast diffusion of O_2 and electrolyte to the Fe–Nx electroactive centers, reaching an excellent ORR performance in an alkaline medium. The number of electrons exchanged during the ORR electrocatalysis is 4, furthermore, limited current values and onset potential are quite similar to that obtained for 20% Pt/C.

Another strategy was recently reported by Zeng and co-workers [44] that prepared COF-derived carbons on carbon nanotubes (CNT) to modify CNTs with high N-doping levels and provide more pores for ionic transportation in electrolytes. Thus, they managed to

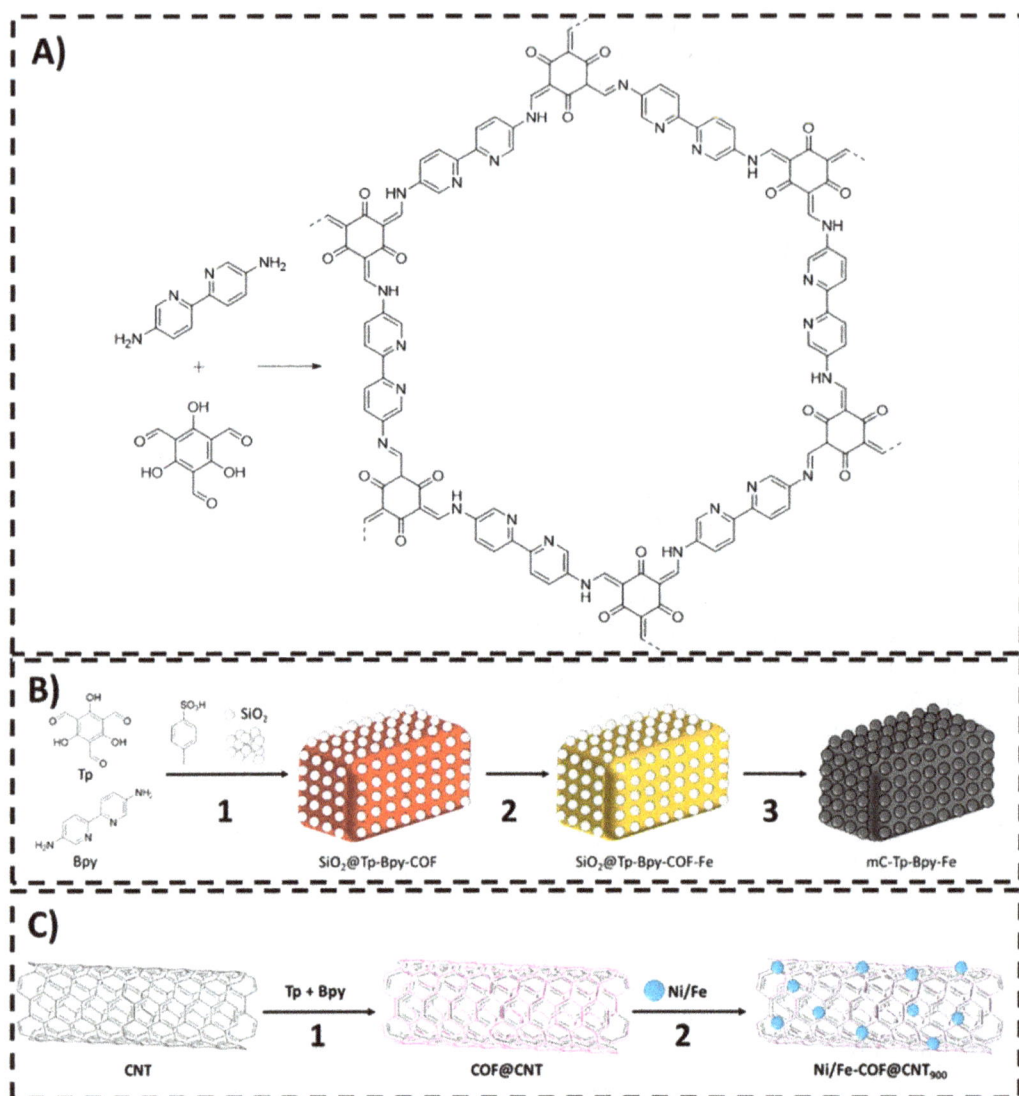

FIGURE 13.4
(A) Schematic structure of TP-Bpy-COF. (B) Synthetic approach to mesoporous carbon mC-Tp-Bpy-Fe: (1) Mechanochemical polymerization, (2) Fe coordination and (3) Pyrolytic treatment. (C) COF@CNTs synthesis and subsequent metalation-pyrolysis process.

grow the Tp-Bpy-COF on the surface of carbon nanotubes by an *in-situ* method to obtain Tp-Bpy-COF@CNT. Finally, iron and nickel nanoparticles were anchored to the bipyridine units and subsequent pyrolysis was carried out to yield Ni/Fe-COF@CNT$_{900}$, a material that is an efficient electrocatalyst for the ORR (Figure 13.4C). The ORR peak potential using a GC electrode modified with CNT is 0.62 V *vs.* RHE. The presence of the COF grown on the CNT improves this value up to 0.66 V *vs.* RHE. The same effect is observed in the presence of Ni and Fe with an increase in the ORR potential of up to 0.67 V *vs.* RHE. A synergetic effect takes place when the COF, Ni, Fe, and the CNTs are present in the pyrolytic process thus improving the ORR peak potential until 0.72 V *vs.* RHE. The Ni/Fe-COF@

CNT onset potential was as high as 1.01 V *vs.* RHE with an $E_{1/2}$ of 0.87 V and a limited current density of 6.0 mA·cm^{-2}, improving the ORR electrocatalytic activity of Pt/C. In addition, the number of electrons involved in the ORR process is 3.9, and only very low yields for H_2O_2 generation (from 0.1% to 5.5%) are observed. It is also worth pointing out that Ni/Fe-COF@CNT has also been successfully applied as an OER electrocatalyst and in the fabrication of a Zn-air battery.

13.3.3 Core-Shell Structures

Despite the good performances exhibited by the COF-based electrocatalysts reviewed above, problems associated with the direct pyrolysis of some COFs are common and include the emergence of undesirable structural changes such as the loss of crystallinity and/or destruction of the defined structures and pore collapse. All these morphology changes usually lead to poor electrochemical performances. To overcome this problem, Yamauchi and co-workers have recently developed an interesting strategy toward hybrid materials to keep the initial porous structure after the pyrolysis step. The strategy is based on the controlled integration of a COF into a graphitic carbon (GC) derived from a metal-organic framework (MOF) to provide a well-defined GC@COF-NC core-shell structure [45]. To obtain this hybrid material, first, micrometer-sized MOF (ZIF-67-MOF) dodecahedrons were synthesized, from the coordination reaction between Co (II) and 2-methylimidazole; The obtained MOF was subsequently calcined at 900°C to yield the nanoparticles of ZIF-67-MOF-derived glassy carbon (GC). A COF endowed with electroactive anthraquinone moieties was assembled to the GC surface by using the Schiff-base condensation reaction between 2,6-diaminoanthraquinone and 1,3,5-triformylphloroglucinol under solvo-thermal reaction conditions. COF nanofibers grow on the surface of the GC particles to form the target core-shell architecture which exhibits both micro and mesopores. Finally, the core-shell heterostructure was calcined to obtain the active catalyst GC@COF-NC. Electrochemical properties of GC@COF-NC worth to be highlighted are a half-wave potential ($E_{1/2}$) of 0.841 V *vs.* RHE which are close to that obtained for 20 wt% Pt/C $E_{1/2}$ = 0.844 V. To highlight the key process of the GC@COF calcination using the **ZIF-MOF** as a template, the electrochemical behavior of COF-derived NC electrocatalyst (without using ZIF-MOF as template) was studied, showing an $E_{1/2}$ of 0.668 *vs.* RHE, far away from the 0.841 V obtained with the template material. Good results regarding the number of electrons exchanged during the ORR (3.90) and electrocatalyst stability are also demonstrated in this work. This result shows the importance of the development of new strategies to keep the initial porous structure after the final pyrolysis step.

13.3.4 Metal-Free Pyrolyzed COFs

Even though much effort has been devoted to obtaining highly electroactive catalysts based on metals, their relatively low stability and the high costs and limited existence of some precious metals hinder their commercial application. Metal-free electrocatalysts derived from organic precursors or renewable sources have been recently used as efficient electrocatalysts not only for the ORR but also in other interesting reactions such as the hydrogen evolution reaction (HER). Therefore, an important issue to be addressed is the development of rational methodologies to create porous heteroatom-doped carbons in which heteroatoms are in specific concentrations. The possibility of using different light-weight heteroatoms (B, N, F, and P) as dopants can effectively tune the electrocatalytic properties of the carbon-nanolayers [46, 47].

In 2019 Zhang and co-workers reported the synthesis of TAPB-DMTA-COF by using the Schiff-base condensation reaction between 1,3,5-tris(4-aminophenil)-benzene and 2,5-dimethoxyterephthaldehyde (DMTA) [48]. The subsequent pyrolysis of this COF followed by phosphorylation produce the P-N-doped derived carbon as the final catalyst. This N, P co-doped carbon catalyst exhibits a remarkable ORR activity in an alkaline medium, being the onset potential 0.87 V *vs*. RHE and showing a limiting current density of 5.6 mA cm^{-2}. The excellent ORR activity might be attributed to the enhanced conductivity and active sites by carbonization and phosphorylation. The number of electrons involved in the ORR is close to four, showing the preferred mechanism for the total reduction of O_2 into H_2O.

A similar approach was followed by Maenosono and co-workers to synthesize an N, P-doped porous carbon by pyrolyzing a metal-free COF precursor [56]. First, a family of intrinsically microporous COF rhombic latticed COFs were grown by Schiff-base condensation of 1,3,6,8-tetrakis(4-aminophenyl)pyrene (TAP) or 1,2,4,5-tetrakis(4-aminophenyl) benzene (TAB) with 1,2,4,5-tetrakis(4-formylphenyl)benzene (TFB) or 1,2,4,5-tetrakis(4-formylphenyl)-3,6-dimethyl-benzene (TFDB). It should be noted that unlike other strategies mentioned above for the syntheses of COFs that involve a reaction between [2 + 3], [3 + 3], or [4 + 2] structural motifs to provide mesoporous hexagonal nets, the use of [4 + 4] tetratopic linkers used in this case provides a microporous tetragonal network. The final step involves the pyrolyzation of the COF to produce N-doped mesoporous carbons which were subsequently phosphorylated with Na_2HPO_2 to yield the active catalyst. The best configuration of this electrocatalyst shows a half-wave potential ($E_{1/2}$) of 0.81 V vs RHE and a limiting current density of 5.5 mA cm^{-2} at 0.40 V vs RHE, with a small Tafel slope of 70 mV dec^{-1}. These values are very close to that of the reference Pt/C in 0.1 M KOH. Great stability during more than seven hours of operating has been proved. This good performance is a consequence of the aromatic cores and functional groups present in the monomers used for the synthesis of these 2D [4 + 4] COF. This electrocatalyst material was also tried as HER electrocatalyst in acid conditions, showing a low overpotential of 260 mV *vs*. RHE and excellent stability of more than 10000 cycles.

As stated above, it has been demonstrated that direct pyrolysis of COFs eventually gives rise to a loss of porous features due to the formation of ill-controlled entangled 3D carbon structures other than conductive 2D graphitic carbons. To overcome this problem, Jiang and co-workers envisaged an innovative strategy by using phytic acid (PA) as a template that prevents the 3D entanglement and the pore collapse after the pyrolysis step [26]. First, a planar COF, TAPT-DHTA-COF, was synthesized by Schiff-base condensation reaction between 4,4′,4″-(1,3,5-triazine-2,4,6-triyl)trianiline (TAPT) and 2,5-dihydroxyterephthalaldehyde (DHTA) under solvothermal conditions. Then, PA loading was achieved by physical adsorption to the COF walls by hydrogen bonding. Finally, the pyrolysis at 1000°C leads to an N-P doped carbon in which the COF porous features remain almost unaltered. Additional features of this precisely designed material includes that (i) PA assists the exfoliation of the stacked layers of TAPT-DHTA-COF into a few layers that help to form truly carbon sheets; (ii) PA guides the conversion to 2D carbons during pyrolysis by hydrogen-bonding to the iminic and alcoholic Lewis-basic positions; PA decomposition leads to a porous structure, and finally, (iv) PA co-dopes the 2D carbon sheets with phosphorus. The pyrolyzed COF material in the absence of the PA pattern (TAPT-DHTA-COF$_{1000}$) exhibits low ORR activity in comparison with Pt/C. When PA is used as a pattern, the ORR electrocatalyst activity of PA@TAPT-DHTA-COF$_{1000}$, increases the onset potential to a value similar to that obtained with Pt/C. The great improvement takes place when NH_3 atmosphere is used instead of argon during the pyrolysis step to yield PA@TAPT-DHTA-COF$_{1000NH3}$. This modification allows the incorporation of nitrogen functional groups on the edges of the porous material, improving the onset potential of ORR and the limited current to values that are better than that found for the Pt/C electrocatalyst. The number of electrons involved in all the cases is close to 3.6, which supports the idea that the ORR mechanism is independent of the porosity of the material unless the diffusion current is affected.

In summary, different strategies have been evaluated to produce conductive, porous, and heteroatom-doped carbon sheets by using COFs as carbonaceous scaffolds. The influence of the position of the active sites in the network has been analyzed showing that the edge-exposed catalytic centers are intrinsically more active than their surface-embedded counterparts. It is worth highlighting that, among the different electrocatalysts based on pyrolyzed-COFs developed for the ORR, the best results have been obtained with COFs modified after their synthesis with metal centers such as Co, Fe, or Ni, generating many edge-exposed active sites after the pyrolytic process. Although platinum-based electrocatalysts are still at the top of best electrocatalytic properties in acidic medium, some examples of COFs-based electrocatalysts have been already reported to get close to these values. Especially remarkable are those based on pyrolyzed metalized COFs with edge-exposed catalytic centers. On the other hand, in basic medium, there are some examples of pyrolyzed-COFs with electrocatalytic properties that compete favorably with Pt-based electrocatalysts. Finally, avoiding the incorporation of metallic species into the catalyst can be envisaged as an efficient strategy for the development of environmentally friendly electrocatalysts. Thus, although the onset potentials obtained with metal-free pyrolyzed COFs are still more negative than that of Pt-based electrocatalysts, some promising results have already been obtained and their still limited performances can be compensated with the environmental benefits.

13.4 ORR Electrocatalysts Based on Non-Pyrolyzed COFs

Despite the benefits obtained in terms of electrocatalytic performance by the pyrolytic treatment of COFs, the high costs derived from pyrolysis and the structural loss produced during the thermal treatment have led to much interest in the development of innovative strategies that allow the use of COFs as electrocatalysts in the absence of pyrolytic treatment. The most promising results have been obtained from the delamination of 2D-COFs into covalent organic nanolayers (CONs) and/or the use of conductive additives to overcome the conductivity issues [49, 50]. The implementation of CONs onto electrodes can be achieved by ultrasonic delamination followed by drop-casting as the easiest and most extended procedure to overcome the processability issues of COFs (Figure 13.5). An important aspect of this strategy is that the pore collapse is avoided which allows to carry out mechanistic studies to improve the ORR yields by rational design of the polymers. As previously stated for the pyrolyzed COFs, the active sites can be embedded in the carbonaceous network of the COF or can be exposed to the pores of the network and can be either metalated or metal-free heteroatom-doped.

13.4.1 Surface-Embedded Active Sites

The first example of non-pyrolyzed metalized COFs (MCOFs) used as ORR electrocatalyst was described by Xiang and co-workers [50]. The active sites in this COF are based on metallomacrocycles, which were first demonstrated to promote ORR in the 1960s. The

FIGURE 13.5
2D-COF ultrasonic delamination into CONs and subsequent electrode modification.

COF, constituted by quasi-phthalocyanine iron (q-FePc) moieties (COF$_{BTC}$, Figure 13.6A), was obtained by using Linstead´s cyclotetramerization of ortho-substituted aromatic dinitriles. The *in-situ* generated COF$_{BTC}$ showed a biphasic crystalline nature in PDRX, coexisting orthorhombic and hexagonal lattices in the 2D conjugated network. Moreover, the abundant positively charged iron centers constrained in the planar direction facilitate the chemical exfoliation of the network in alkaline media. The good processability together with excellent ORR electrocatalytic properties (E$_{1/2}$ = 0.90 V *vs.* RHE, high stability over operating 7 hours) made this iron-phthalocyanine COF material a good candidate to be used in Zn/airflow batteries (ZAFB, Figure 13.6B). An important aspect to highlight of this device is that the electrocatalyst is dissolved in the electrolyte. Therefore, the assembly of electrodes is simplified, avoiding their functionalization. The rechargeability tests by galvanostatic charging and discharging at 5, 10, and 20 mA cm^{-2} show low voltage gaps (0.77 V at 5 mA cm^{-2}, 0.83 V at 10 mA cm^{-2}, 0.96 V at 20 mA cm^{-2}), high energy efficiencies (60.7% at 5 mA cm^{-2}, 59.0% at 10 mA cm$_{-2}$, 52.6% at 20 mA cm^{-2}), and rapid charge–discharge ability. It is worth pointing out that this approach does not require the use of conductive agents to integrate the electrocatalyst over the electrode surface and avoid electrocatalysts' striping and flooding during the flowing process.

A similar strategy was used by Yan and co-workers [51]. They developed a melt polymerization strategy to synthesize a COF constituted by iron-phthalocyanine (FePPC) moieties on top of a carbon-black matrix (FePPC@CB). The FePPc molecules anchor to the carbon black matrix through non-covalent interactions, facilitating the electron transfer process (Figure 13.6D). FePPc@CB displays a half-wave potential (E$_{1/2}$) of 0.908 V *vs.* RHE. This value is 59 and 77 mV higher, respectively, than those reported for single iron-polyphthalocyanine adsorbed on carbon black (FePc@CB) and Pt/C catalysts. Moreover, FePPc@CB reaches a limiting current density of 5.38 mAcm^{-2} which is as high as that obtained with Pt/C. Compared with FePc@CB, the FePPc@CB electrocatalyst possess a high number of active sites, promoting the ORR electrocatalytic activity. Furthermore, in contrast with single iron-polyphthalocyanine adsorbed on carbon black (FePc@CB), free electrons are in the conduction band of FePPc@CB promoting the ORR. For FePPc@CB the number of electrons exchanged during the ORR is 3.93-4.01 and the H$_2$O$_2$ yield is under 3.5%. Moderate stability and good methanol crossover are also reported for this FePPc@CB electrocatalyst.

13.4.2 Edge-Exposed Based Catalyst by Metal Coordination

Non pyrolyzed, metalated COFs with the active sites exposed to the pores have also been developed. Thus, Jiang and co-workers [52] synthesized a metalated COF with edge-exposed platinum nanoparticles (NPs). Typically, to overcome the low Pt utilization

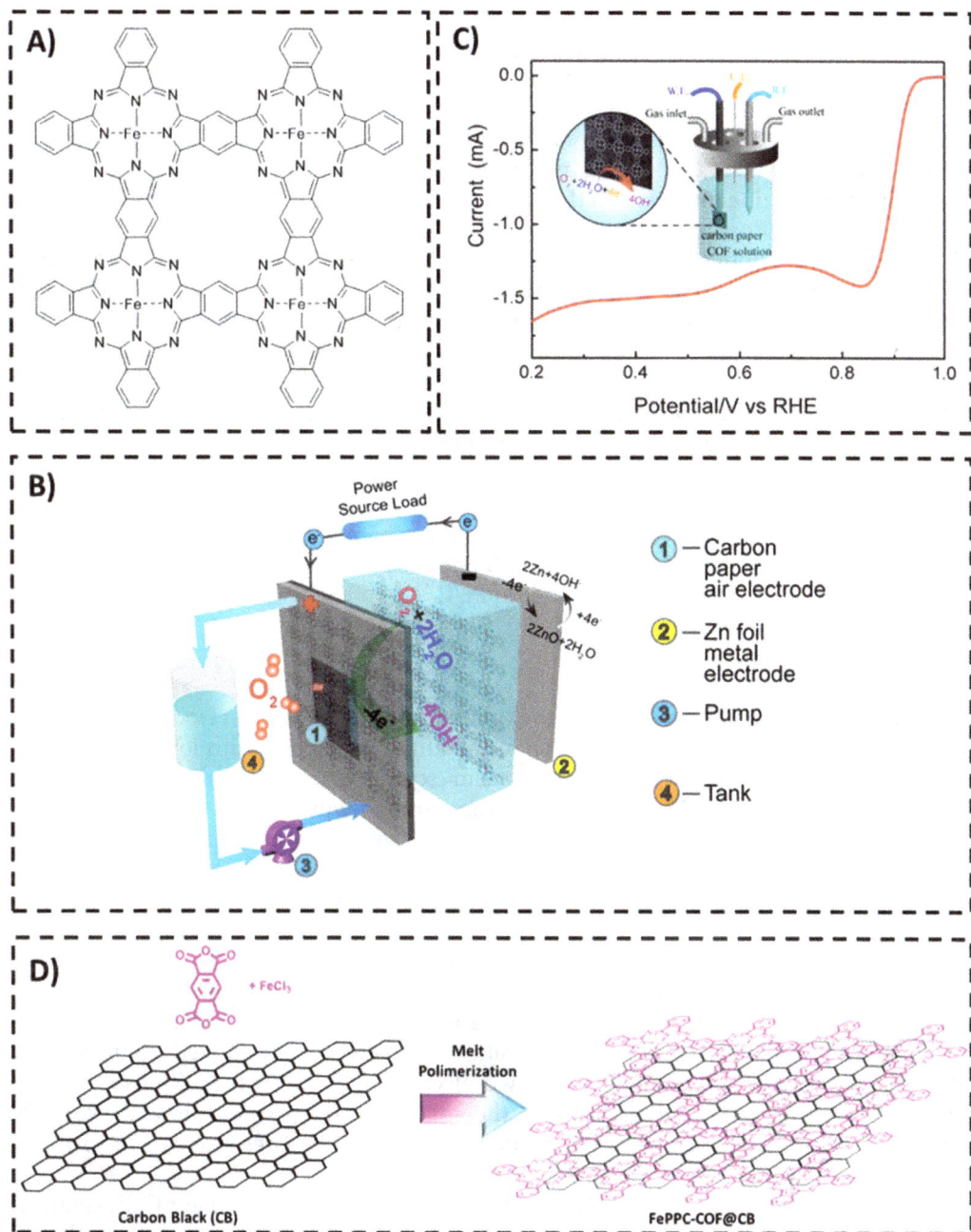

FIGURE 13.6

(A) Structure of COF_{BTC}. (B) Zn/airflow batterie scheme. (C) Linear Sweep Voltammetry (LSV) measurements of the COF_{BTC} solution with carbon paper as electrodes. Inset: Scheme for the detailed operation for the measurements. (B) and (C) adapted with permission [50] Copyright 2019 American Chemical Society. (D) Melt polymerization to FePPC-COF over a carbon black matrix.

efficiency and the sluggish ORR kinetics, the use of porous materials as hosts to downsize Pt NPs has been regarded as a promising method of fabricating alternative catalysts to Pt/C. With this aim, Jiang and co-workers focused on the confinement of Pt NPS on the surface of the COF pores. To address this, an olefin-linked COF with hexagonal topology was synthesized by a Knoevenagel reaction between two monomers with C_3 symmetry, the tris-2,4,6-(4-formylphenyl)-s-triazine and the 2,4,6-trimethyl-s-triazine. The polymer shows multiple triazine nitrogens in the pore surface, which act as nucleation sites for controllable growth of Pt nanoparticles (Pt@COF) resulting in more accessible Pt active sites with a uniform distribution. Pt@COF proved to work as an ORR electrocatalyst in an acid electrolyte with an enhanced performance in comparison with that of commercial Pt/C systems. Pt@COF exhibited an onset potential of 1.04 V and $E_{1/2}$ of 0.89 V *vs.* RHE in 0.1 M $HClO_4$, which were, respectively, 70 and 60 mV more positive than that of Pt/C. The mass activities of Pt@COF obtained at 0.9 and 0.7 V *vs.* RHE were 78 and 174 A gPt^{-1}, while the mass activities of Pt/C at the same potentials were 9.4 and 54 A gPt^{-1} proving its higher electrocatalytic activity. Another relevant property of Pt@COF is its great conductivity because of the nucleation of Pt NPs that promotes charge transport. Thus, electrochemical impedance spectroscopy (EIS) measurements showed resistances of charge transport of 65 and 5800 Ω, respectively, for the Pt@COF and the non-metalated COF. Pt@COF has also been successfully used in a basic medium (0.1 M KOH), showing a more positive onset potential and $E_{1/2}$ than the reference material Pt/C.

13.4.3 Metal-Free COFs

One of the current challenges in the development of COF-based electrocatalysts for the ORR is the obtaining of pyrolysis-free, non-metalized materials with low-lying LUMO energy levels favorable to accept electrons from oxygen. The first example of metal-free COFs was reported in 2017 by Zamora, Herrero, and co-workers that reported the synthesis of a metal-free non-pyrolyzed aza-fused π-conjugated microporous framework that catalyzes the production of hydrogen peroxide [53]. To achieve this material, the polymer synthesis was performed by a condensation reaction between 1,2,4,5-benzene-tetramine and triquinoyl octahydrate to obtain a poly-pyrazine-based material. In an alkaline medium, the onset potential for ORR is ca. 0.65 V *vs.* RHE. This is an example of selectivity toward the production of hydrogen peroxide, as the number of electrons exchanged during ORR is close to two. Due to the well-defined molecular structure of the network, the mechanism of electrocatalysis has been investigated by using DFT calculations. Thus, the preference for the two-electron pathway in the ORR can be rationalized in terms of the lower stability of the adsorbed superoxide state over the COF material in comparison with graphitic materials due to the presence of the nitrogen atoms in the network.

Wu and co-workers reported in 2018 a non-pyrolyzed metal-free COF for ORR catalysis through a 4-electron pathway [30]. Polychlorotriphenylmethyl (PTM) radicals were chosen as the active centers to obtain interacting radicals in an expanded network. Thus, PTM-COFR was obtained in a two-step, one-pot reaction by using the Glaser-Hay acetylenic C_3 homocoupling followed by deprotonation and oxidation to PTM radicals. The polymer was synthesized following the liquid-liquid interfacial method carrying out the reaction in a dichloromethane/water interface, allowing the obtainment of about 71% of radicals anti-ferromagnetically coupled in an extended network (Figure 13.7). This material had a low-lying LUMO energy level favorable to accept electrons from oxygen, making it a potential electrocatalyst for the ORR.

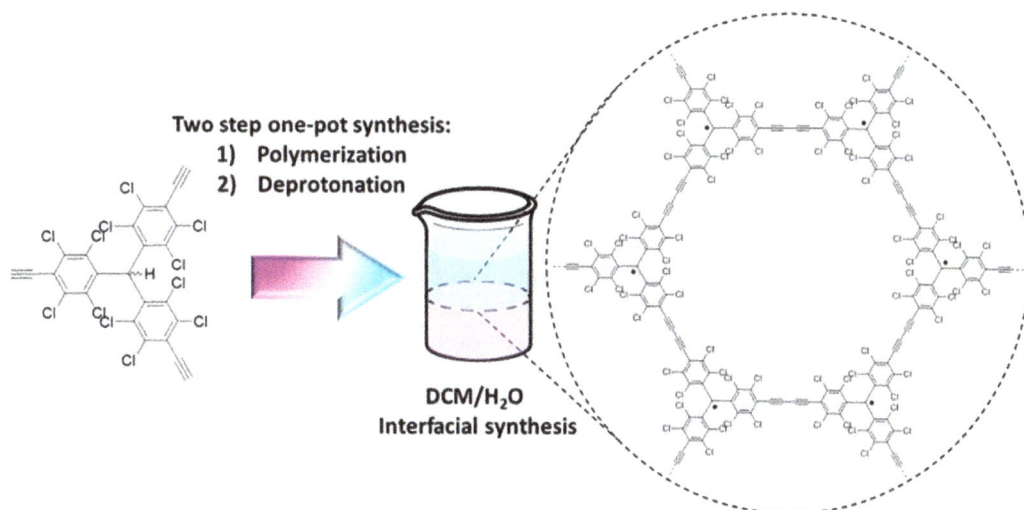

FIGURE 13.7
PTM-COFR interfacial synthesis.

The catalyst was obtained by dispersing PTM-COFR and carbon black (25% w/w) into 1,2-dichloroethane by the previously mentioned ultrasonic-assisted method for 30 min. Then, 10 μL of the dispersion was loaded onto the glassy carbon (GC) disk electrode, then dried and loaded with 0.5 μL of a Nafion solution. The electrochemical behavior of GC electrodes modified with PTM-COFR is dominated by a redox pair at a formal potential of -0.6 V *vs.* Ag/AgCl in a 0.1 M KOH solution saturated with N_2. This redox process can be ascribed to the reduction/oxidation of PTM radicals. When the experiment is carried out in the presence of O_2, an electrocatalytic reduction wave is observed. In this case, the reduction peak potential is observed at -0.25 V *vs.* Ag/AgCl. The electrochemical studies of the ORR process under hydrodynamic conditions showed a reduction process at a half-wave potential of -0.27 V *vs.* Ag/AgCl, which is 0.1 V lower than that observed for the reference material, 20% wt Pt/C. Despite the lower $E_{1/2}$, the achieved current density is higher than that obtained for 20% wt Pt/C. The number of electrons transferred for each oxygen molecule, calculated from the Levich constant, is 3.89, pointing out that the preferred ORR mechanism is through a four electrons pathway. Furthermore, the PTM-COFR stability during ORR operation conditions has been successfully proved until 40000 s with a loss of current density of only 12%.

Inspired by previous studies on ORR electrocatalysts where the active centers are based on the five reconstructed-edge pentagon defects pyrolytically derived carbons [54–58], Yao and co-workers reported in 2020 a thiophene-based covalent organic framework as an ORR electrocatalyst [23]. Three reticular COFs were reported based on the $C_3 + C_2$ solvothermal imine-condensation reactions between tris-(4-aminophenyl)-1,3,5-benzene, and different linear units such as thiophene-2,5-dicarbaldehyde (JUC-527), [2,2′-bitiophene]-5-5′dicarbaldehyde (JUC-528) and terephthaldehyde (PDA-TAPB-COF). Pentagon units with sulfur atoms were introduced into the COF networks for JUC-527 and JUC-528 to test the ability of thiophene units to catalyze the ORR. On the other hand, JUC-528 has a larger pore size and greater thiophene content than JUC-527 to evaluate the mass transport dependence with the increasing content of catalytic centers. Inks of the catalysts were prepared by mixing COF powder with acetylene black and 5 wt% Nafion solution in a (1/1) ethanol/water mixture followed by dispersion, through the ultrasonic-assisted method. Finally, the solution was loaded onto the electrode. The ORR electrocatalytic activity of the COFs was evaluated in 0.1 M KOH. The COFs with thiophene units (JUC-527 and JUC-528) present a better ORR electrocatalyst activity, including higher onset potential and higher current density than the COF without thiophene units (PDA-TAPB-COF). The best performance was determined for JUC-528 with a half-wave potential of 0.70 *vs.* RHE. The electron transfer numbers are 3.81 for JUC-528 and 3.46 for JUC-527, pointing out a preferred four-electron pathway for the ORR. On the other hand, for the COF without thiophene groups (PDA-TAPB-COF) the number of electrons exchanged during the ORR is 2.08, thus suggesting that the two-electron pathway is the preferred one when this COF acts as an electrocatalyst in the ORR. In this case, we can highlight the importance of the presence of thiophene groups regarding the ORR mechanism. This result is also supported by DFT calculations. Finally, the mass activity of the COFs containing thiophene moieties has been compared, showing higher activity in the case of JUC-528, which is ascribed to the higher porosity of this material as determined by BET isotherms.

Segura, Lorenzo, Zamora, and co-workers [49] reported in 2020 a COF with electroactive naphthalenediimide (NDI) moieties as active centers for ORR catalysis. The NDI electroactive units were incorporated by *in-situ* generation during the COF synthesis through the imidation polymerization reaction between tris-(4-aminophenyl)-1,3,5-benzene (C_3 monomer) and 1,4,5,8-naphthalenetetracarboxylic dianhydride (C_2 monomer) under solvothermal reaction conditions. This strategy allows the incorporation of the electron-poor NDI building block directly as the linkage in the network and makes the material energetically favorable to electrons with oxygen. To prepare the suitably modified electrodes, the COF processability interact issues were addressed by using the ultrasonic-assisted method to disrupt the non-covalent interactions between layers and produce COF nanosheets (CONs). The NDI-CONs colloids were then drop-casted onto glassy carbon electrodes (GC) and used to test the ORR electrocatalytic activity (in 0.1 M NaOH) of this metal-free and pyrolysis-free nanomaterial without the use of any additive like Nafion, carbon black, or acetylene black. In the absence of O_2, CV shows a reduction peak at -0.38 V *vs.* SCE ascribed to the reduction of the NDI moiety. In the presence of O_2, an electrocatalytic reduction wave is observed at the same potential as that determined for the NDI moiety, pointing out that this electroactive group is responsible for the ORR electrocatalysis. The number of electrons involved in the ORR electrocatalytic process is 3.6, indicating the prevalence of the four electrons mechanism. To compare the effects of the covalent organic frameworks, the behavior of single NDI monomers was checked under the same conditions, showing a really low onset potential, similar to that of the bare GC electrode, and low current density in comparison with that of the COF-based materials deposited as CONs. Bad ORR electrocatalysis performances were obtained when an amorphous COF material containing the same subunits (NDI and TAPB) was deposited on the GC electrode. Both results reveal the importance of the proper electrocatalysts nanostructure to efficiently electrocatalyst the ORR process.

In summary, different strategies have already been developed to obtain pyrolysis-free COF-based ORR catalysts that avoid structural integrity loss, pore collapse, and the high energetic costs associated with the pyrolysis step. Moreover, the well-defined molecular structure of non-pyrolyzed COFs allows the use of DFT calculations to study the reduction mechanism, shedding light on the design of a more active catalyst for the reduction of oxygen. Concerning the performance, the best results among pyrolysis-free COF electrocatalysts have been obtained with metalized networks. Particularly remarkable are those containing Fe active centers in their structure, especially when the ORR takes place in a basic medium. Reports of electrocatalysis with COFs with coordinated Pt centers in acid and basic medium have been also shown. Nevertheless, the benefit of not using a precious metal will not be overcome until the amount of Pt employed is significantly decreased. The most environmentally friendly strategy with lower associated costs involves the use of metal-free non-pyrolyzed COFs. However, the electrocatalytic performances obtained so far are still below that obtained with Pt-based electrocatalysts, especially in an acidic medium. Finally, it is worth highlighting the good performances obtained in the ORR electrocatalysis in a basic medium with COFs based on electroactive centers that can be used to modulate the electrocatalytic activity.

13.5 Conclusions and Outlook

A selection of the most relevant electrocatalysts based on COFs for the ORR has been presented in this chapter. These innovative electrocatalysts have been revealed in last recent years as a promising alternative to traditional platinum group metal (PGM) based electrocatalysts, showing similar ORR onset potential, current density, and, in some cases, even improving the methanol crossover and stability of PGM electrocatalysts. Some of the most efficient electrocatalysts based on COFs are derived from pyrolyzed materials but drawbacks associated with this methodology have been exposed such as the possibility of porous structure collapse and the difficulty to control the exact composition of the generated material after the pyrolysis treatment. Furthermore, the pyrolysis step requires high temperatures and the use of special equipment, making the big scale production of pyrolyzed electrocatalysts an expensive process. These are the main reasons for the new trend in the field that is attracting the attention of scientists which involves the synthesis of new COF-based materials endowed with electroactive moieties in their structure. This strategy combines the advantages associated with the ordered and porous structure of COFs with the fine control in the composition of the groups responsible for the ORR electrocatalysis. Furthermore, different materials can be designed with the same electroactive moiety as an active center but with fine control on the porous size and distribution to optimize the ORR activity. In addition, these non-metalated pyrolysis-free electrocatalysts based on the presence of electroactive moieties are eco-friendly electrocatalysts that avoid, not only the use of precious metals which present the handicap of their high prices and scarcity, but also the use of other metals that can pollute the environment. From all of the above, it can be envisaged that the ongoing research focused on the syntheses of COFs with different electroactive moieties, pores size, shape, and distribution will pave the way for the development of new families of efficient environmentally friendly electrocatalysts for the oxygen reduction reaction.

Acknowledgments

This work was supported by MICINN (PID2019-106268GB-C33, CTQ2017-84309-C2-1-R; RED2018-102412-T), Comunidad de Madrid (P2018/NMT-4349 TRANSNANOAVANSENS Program, SI3-PJI-2021-00341), and the UCM (INV.GR.00.1819.10759). MMF gratefully acknowledges Comunidad de Madrid for a predoctoral contract.

References

1. S. Bajracharya, A. ElMekawy, S. Srikanth, D. Pant, 6 - Cathodes for microbial fuel cells, in: K. Scott, E.H. Yu (Eds.), Microbial Electrochemical and Fuel Cells, Woodhead Publishing, Boston, 2016: pp. 179–213. https://doi.org/10.1016/B978-1-78242-375-1.00006-X.

2. A. Alaswad, A. Palumbo, M. Dassisti, A.G. Olabi, Fuel Cell Technologies, Applications, and State of the Art. A Reference Guide, in: Reference Module in Materials Science and Materials Engineering, Elsevier, 2016. https://doi.org/10.1016/B978-0-12-803581-8.04009-1.

3. M.K. Debe, Electrocatalyst approaches and challenges for automotive fuel cells, Nature. 486 (2012) 43–51. https://doi.org/10.1038/nature11115.

4. J. Pan, Y.Y. Xu, H. Yang, Z. Dong, H. Liu, B.Y. Xia, Advanced Architectures and Relatives of Air Electrodes in Zn–Air Batteries, Advanced Science. 5 (2018) 1700691. https://doi.org/10.1002/advs.201700691.

5. X. Han, X. Li, J. White, C. Zhong, Y. Deng, W. Hu, T. Ma, Metal–Air Batteries: From Static to Flow System, Advanced Energy Materials. 8 (2018) 1801396. https://doi.org/10.1002/aenm.201801396.

6. A. Kulkarni, S. Siahrostami, A. Patel, J.K. Nørskov, Understanding Catalytic Activity Trends in the Oxygen Reduction Reaction, Chemical Reviews. 118 (2018) 2302–2312. https://doi.org/10.1021/acs.chemrev.7b00488.

7. F. Jaouen, E. Proietti, M. Lefèvre, R. Chenitz, J.-P. Dodelet, G. Wu, H.T. Chung, C.M. Johnston, P. Zelenay, Recent advances in non-precious metal catalysis for oxygen-reduction reaction in polymer electrolyte fuel cells, Energy & Environmental Science. 4 (2011) 114–130. https://doi.org/10.1039/C0EE00011F.

8. A. Morozan, B. Jousselme, S. Palacin, Low-platinum and platinum-free catalysts for the oxygen reduction reaction at fuel cell cathodes, Energy & Environmental Science. 4 (2011) 1238–1254. https://doi.org/10.1039/C0EE00601G.

9. C.R. Raj, A. Samanta, S.H. Noh, S. Mondal, T. Okajima, T. Ohsaka, Emerging new generation electrocatalysts for the oxygen reduction reaction, Journal of Materials Chemistry A. 4 (2016) 11156–11178. https://doi.org/10.1039/C6TA03300H.

10. F. Zhao, H. Liu, S.D.R. Mathe, A. Dong, J. Zhang, Covalent Organic Frameworks: From Materials Design to Biomedical Application, Nanomaterials. 8 (2018). https://doi.org/10.3390/nano8010015.

11. M.S. Lohse, T. Bein, Covalent Organic Frameworks: Structures, Synthesis, and Applications, Advanced Functional Materials. 28 (2018) 1705553. https://doi.org/10.1002/adfm.201705553.

12. J. Wang, N. Li, Y. Xu, H. Pang, Two-Dimensional MOF, and COF Nanosheets: Synthesis and Applications in Electrochemistry, Chemistry – A European Journal. 26 (2020) 6402–6422. https://doi.org/10.1002/chem.202000294.

13. D. Rodríguez-San-Miguel, C. Montoro, F. Zamora, Covalent organic framework nanosheets: preparation, properties and applications, Chemical Society Reviews. 49 (2020) 2291–2302. https://doi.org/10.1039/C9CS00890J.

14. I. Berlanga, M.L. Ruiz-González, J.M. González-Calbet, J.L.G. Fierro, R. Mas-Ballesté, F. Zamora, Delamination of Layered Covalent Organic Frameworks, Small. 7 (2011) 1207–1211. https://doi.org/10.1002/smll.201002264.

15. J.L. Segura, S. Royuela, M. Mar Ramos, Post-synthetic modification of covalent organic frameworks, Chemical Society Reviews. 48 (2019) 3903–3945. https://doi.org/10.1039/C8CS00978C.

16. T. Varga, G. Ballai, L. Vásárhelyi, H. Haspel, Á. Kukovecz, Z. Kónya, Co4N/nitrogen-doped graphene: A non-noble metal oxygen reduction electrocatalyst for alkaline fuel cells, Applied Catalysis B: Environmental. 237 (2018) 826–834. https://doi.org/10.1016/j.apcatb.2018.06.054.

17. J. Liu, Q. Ma, Z. Huang, G. Liu, H. Zhang, Recent Progress in Graphene-Based Noble-Metal Nanocomposites for Electrocatalytic Applications, Advanced Materials. 31 (2019) 1800696. https://doi.org/10.1002/adma.201800696.

18. Y. Zhi, Z. Wang, H.-L. Zhang, Q. Zhang, Recent Progress in Metal-Free Covalent Organic Frameworks as Heterogeneous Catalysts, Small. 16 (2020) 2001070. https://doi.org/10.1002/smll.202001070.

19. H. Zhang, M. Zhu, O.G. Schmidt, S. Chen, K. Zhang, Covalent Organic Frameworks for Efficient Energy Electrocatalysis: Rational Design and Progress, Advanced Energy and Sustainability Research. 2 (2021) 2000090. https://doi.org/10.1002/aesr.202000090.

20. S. Dou, X. Wang, S. Wang, Rational Design of Transition Metal-Based Materials for Highly Efficient Electrocatalysis, Small Methods. 3 (2019) 1800211. https://doi.org/10.1002/smtd.201800211.

21. D.D. Medina, M.L. Petrus, A.N. Jumabekov, J.T. Margraf, S. Weinberger, J.M. Rotter, T. Clark, T. Bein, Directional Charge-Carrier Transport in Oriented Benzodithiophene Covalent Organic Framework Thin Films, ACS Nano. 11 (2017) 2706–2713. https://doi.org/10.1021/acsnano.6b07692.

22. M. Dogru, T. Bein, On the road towards electroactive covalent organic frameworks, Chemical Communications. 50 (2014) 5531–5546. https://doi.org/10.1039/C3CC46767H.

23. D. Li, C. Li, L. Zhang, H. Li, L. Zhu, D. Yang, Q. Fang, S. Qiu, X. Yao, Metal-Free Thiophene-Sulfur Covalent Organic Frameworks: Precise and Controllable Synthesis of Catalytic Active Sites for Oxygen Reduction, Journal of the American Chemical Society. 142 (2020) 8104–8108. https://doi.org/10.1021/jacs.0c02225.

24. W. Ma, P. Yu, T. Ohsaka, L. Mao, An efficient electrocatalyst for oxygen reduction reaction derived from a Co-porphyrin-based covalent organic framework, Electrochemistry Communications. 52 (2015) 53–57. https://doi.org/10.1016/j.elecom.2015.01.021.

25. J.H. Dumont, U. Martinez, K. Artyushkova, G.M. Purdy, A.M. Dattelbaum, P. Zelenay, A. Mohite, P. Atanassov, G. Gupta, Nitrogen-Doped Graphene Oxide Electrocatalysts for the Oxygen Reduction Reaction, ACS Applied Nano Materials. 2 (2019) 1675–1682. https://doi.org/10.1021/acsanm.8b02235.

26. Q. Xu, Y. Tang, X. Zhang, Y. Oshima, Q. Chen, D. Jiang, Template Conversion of Covalent Organic Frameworks into 2D Conducting Nanocarbons for Catalyzing Oxygen Reduction Reaction, Advanced Materials. 30 (2018) 1706330. https://doi.org/10.1002/adma.201706330.

27. C. Yang, S. Tao, N. Huang, X. Zhang, J. Duan, R. Makiura, S. Maenosono, Heteroatom-Doped Carbon Electrocatalysts Derived from Nanoporous Two-Dimensional Covalent Organic Frameworks for Oxygen Reduction and Hydrogen Evolution, ACS Applied Nano Materials. 3 (2020) 5481–5488. https://doi.org/10.1021/acsanm.0c00786.

28. E.F. Holby, G. Wu, P. Zelenay, C.D. Taylor, Structure of Fe–Nx–C Defects in Oxygen Reduction Reaction Catalysts from First-Principles Modeling, The Journal of Physical Chemistry C. 118 (2014) 14388–14393. https://doi.org/10.1021/jp503266h.

29. X. Zhao, P. Pachfule, S. Li, T. Langenhahn, M. Ye, G. Tian, J. Schmidt, A. Thomas, Silica-Templated Covalent Organic Framework-Derived Fe–N-Doped Mesoporous Carbon as Oxygen Reduction Electrocatalyst, Chemistry of Materials. 31 (2019) 3274–3280. https://doi.org/10.1021/acs.chemmater.9b00204.

30. S. Wu, M. Li, H. Phan, D. Wang, T.S. Herng, J. Ding, Z. Lu, J. Wu, Toward Two-Dimensional π-Conjugated Covalent Organic Radical Frameworks, Angewandte Chemie International Edition. 57 (2018) 8007–8011. https://doi.org/10.1002/anie.201801998.

31. B. Qiao, A. Wang, X. Yang, L.F. Allard, Z. Jiang, Y. Cui, J. Liu, J. Li, T. Zhang, Single-atom catalysis of CO oxidation using Pt1/FeOx, Nature Chemistry. 3 (2011) 634–641. https://doi.org/10.1038/nchem.1095.

32. M.M. Sadiq, M.P. Batten, X. Mulet, C. Freeman, K. Konstas, J.I. Mardel, J. Tanner, D. Ng, X. Wang, S. Howard, M.R. Hill, A.W. Thornton, A Pilot-Scale Demonstration of Mobile Direct Air Capture Using Metal-Organic Frameworks, Advanced Sustainable Systems. 4 (2020) 2000101. https://doi.org/10.1002/adsu.202000101.

33. H. Zhao, Z.-Y. Yuan, Design Strategies of Non-Noble Metal-Based Electrocatalysts for Two-Electron Oxygen Reduction to Hydrogen Peroxide, ChemSusChem. 14 (2021) 1616–1633. https://doi.org/10.1002/cssc.202100055.

34. L. Yang, S. Jiang, Y. Zhao, L. Zhu, S. Chen, X. Wang, Q. Wu, J. Ma, Y. Ma, Z. Hu, Boron-Doped Carbon Nanotubes as Metal-Free Electrocatalysts for the Oxygen Reduction Reaction, Angewandte Chemie International Edition. 50 (2011) 7132–7135. https://doi.org/10.1002/anie.201101287.

35. R. Liu, D. Wu, X. Feng, K. Müllen, Nitrogen-Doped Ordered Mesoporous Graphitic Arrays with High Electrocatalytic Activity for Oxygen Reduction, Angewandte Chemie International Edition. 49 (2010) 2565–2569. https://doi.org/10.1002/anie.200907289.

36. A.H.A. Monteverde Videla, L. Zhang, J. Kim, J. Zeng, C. Francia, J. Zhang, S. Specchia, Mesoporous carbons supported non-noble metal Fe–NXelectrocatalysts for PEM fuel cell oxygen reduction reaction, Journal of Applied Electrochemistry. 43 (2013) 159–169. https://doi.org/10.1007/s10800-012-0497-y.

37. Z. Xiang, Y. Xue, D. Cao, L. Huang, J.-F. Chen, L. Dai, Highly Efficient Electrocatalysts for Oxygen Reduction Based on 2D Covalent Organic Polymers Complexed with Non-precious Metals, Angewandte Chemie International Edition. 53 (2014) 2433–2437. https://doi.org/10.1002/anie.201308896.

38. H. Liu, S. Yi, Y. Wu, H. Wu, J. Zhou, W. Liang, J. Cai, H. Xu, An efficient Co-N/C electrocatalyst for oxygen reduction facilely prepared by tuning cobalt species content, International Journal of Hydrogen Energy. 45 (2020) 16105–16113. https://doi.org/10.1016/j.ijhydene.2020.04.024.

39. Z. Zhang, C.-S. Lee, W. Zhang, Vertically Aligned Graphene Nanosheet Arrays: Synthesis, Properties, and Applications in Electrochemical Energy Conversion and Storage, Advanced Energy Materials. 7 (2017) 1700678. https://doi.org/10.1002/aenm.201700678.

40. H.T. Chung, D.A. Cullen, D. Higgins, B.T. Sneed, E.F. Holby, K.L. More, P. Zelenay, Direct atomic-level insight into the active sites of a high-performance PGM-free ORR catalyst, Science. 357 (2017) 479 LP–484. https://doi.org/10.1126/science.aan2255.

41. S. Wei, Y. Wang, W. Chen, Z. Li, W.-C. Cheong, Q. Zhang, Y. Gong, L. Gu, C. Chen, D. Wang, Q. Peng, Y. Li, Atomically dispersed Fe atoms anchored on COF-derived N-doped carbon nanospheres as efficient multi-functional catalysts, Chemical Science. 11 (2020) 786–790. https://doi.org/10.1039/C9SC05005A.

42. Q. Xu, H. Zhang, Y. Guo, J. Qian, S. Yang, D. Luo, P. Gao, D. Wu, X. Li, Z. Jiang, Y. Sun, Standing Carbon-Supported Trace Levels of Metal Derived from Covalent Organic Framework for Electrocatalysis, Small. 15 (2019) 1905363. https://doi.org/10.1002/smll.201905363.

43. D. Wu, Q. Xu, J. Qian, X. Li, Y. Sun, Bimetallic Covalent Organic Frameworks for Constructing Multifunctional Electrocatalyst, Chemistry – A European Journal. 25 (2019) 3105–3111. https://doi.org/10.1002/chem.201805550.

44. Q. Xu, J. Qian, D. Luo, G. Liu, Y. Guo, G. Zeng, Ni/Fe Clusters and Nanoparticles Confined by Covalent Organic Framework Derived Carbon as Highly Active Catalysts toward Oxygen Reduction Reaction and Oxygen Evolution Reaction, Advanced Sustainable Systems. 4 (2020) 2000115. https://doi.org/10.1002/adsu.202000115.

45. S. Zhang, W. Xia, Q. Yang, Y. Valentino Kaneti, X. Xu, S.M. Alshehri, T. Ahamad, Md.S.A. Hossain, J. Na, J. Tang, Y. Yamauchi, Core-shell motif construction: Highly graphitic nitrogen-doped porous carbon electrocatalysts using MOF-derived carbon@COF heterostructures as sacrificial templates, Chemical Engineering Journal. 396 (2020) 125154. https://doi.org/10.1016/j.cej.2020.125154.

46. X. Wang, G. Sun, P. Routh, D.-H. Kim, W. Huang, P. Chen, Heteroatom-doped graphene materials: syntheses, properties, and applications, Chemical Society Reviews. 43 (2014) 7067–7098. https://doi.org/10.1039/C4CS00141A.

47. C. Yang, H.-F. Wang, Q. Xu, Recent Advances in Two-dimensional Materials for Electrochemical Energy Storage and Conversion, Chemical Research in Chinese Universities. 36 (2020) 10–23. https://doi.org/10.1007/s40242-020-9068-7.

48. C. Yang, S. Maenosono, J. Duan, X. Zhang, COF-Derived N, P Co-Doped Carbon as a Metal-Free Catalyst for Highly Efficient Oxygen Reduction Reaction, ChemNanoMat. 5 (2019) 957–963. https://doi.org/10.1002/cnma.201900159.

49. S. Royuela, E. Martínez-Periñán, M.P. Arrieta, J.I. Martínez, M.M. Ramos, F. Zamora, E. Lorenzo, J.L. Segura, Oxygen reduction using a metal-free naphthalene diimide-based covalent organic framework electrocatalyst, Chemical Communications. 56 (2020). https://doi.org/10.1039/c9cc06479f.
50. P. Peng, L. Shi, F. Huo, S. Zhang, C. Mi, Y. Cheng, Z. Xiang, In Situ Charge Exfoliated Soluble Covalent Organic Framework Directly Used for Zn–Air Flow Battery, ACS Nano. 13 (2019) 878–884. https://doi.org/10.1021/acsnano.8b08667.
51. W.-Z. Cheng, J.-L. Liang, H.-B. Yin, Y.-J. Wang, W.-F. Yan, J.-N. Zhang, Bifunctional iron-phtalocyanine metal–organic framework catalyst for ORR, OER and rechargeable zinc–air battery, Rare Metals. 39 (2020) 815–823. https://doi.org/10.1007/s12598-020-01440-2.
52. L. Zhai, S. Yang, X. Yang, W. Ye, J. Wang, W. Chen, Y. Guo, L. Mi, Z. Wu, C. Soutis, Q. Xu, Z. Jiang, Conjugated Covalent Organic Frameworks as Platinum Nanoparticle Supports for Catalyzing the Oxygen Reduction Reaction, Chemistry of Materials. 32 (2020) 9747–9752. https://doi.org/10.1021/acs.chemmater.0c03614.
53. V. Briega-Martos, A. Ferre-Vilaplana, A. de la Peña, J.L. Segura, F. Zamora, J.M. Feliu, E. Herrero, An Aza-Fused π-Conjugated Microporous Framework Catalyzes the Production of Hydrogen Peroxide, ACS Catalysis. 7 (2017) 1015–1024. https://doi.org/10.1021/acscatal.6b03043.
54. Y. Jia, L. Zhang, L. Zhuang, H. Liu, X. Yan, X. Wang, J. Liu, J. Wang, Y. Zheng, Z. Xiao, E. Taran, J. Chen, D. Yang, Z. Zhu, S. Wang, L. Dai, X. Yao, Identification of active sites for acidic oxygen reduction on carbon catalysts with and without nitrogen doping, Nature Catalysis. 2 (2019) 688–695. https://doi.org/10.1038/s41929-019-0297-4.
55. Y. Jia, L. Zhang, A. Du, G. Gao, J. Chen, X. Yan, C.L. Brown, X. Yao, Defect Graphene as a Trifunctional Catalyst for Electrochemical Reactions, Advanced Materials. 28 (2016) 9532–9538. https://doi.org/10.1002/adma.201602912.
56. C. Tang, Q. Zhang, Nanocarbon for Oxygen Reduction Electrocatalysis: Dopants, Edges, and Defects, Advanced Materials. 29 (2017) 1604103. https://doi.org/10.1002/adma.201604103.
57. L. Zhang, Q. Xu, J. Niu, Z. Xia, Role of lattice defects in catalytic activities of graphene clusters for fuel cells, Physical Chemistry Chemical Physics. 17 (2015) 16733–16743. https://doi.org/10.1039/C5CP02014J.
58. Y. Jia, K. Jiang, H. Wang, X. Yao, The Role of Defect Sites in Nanomaterials for Electrocatalytic Energy Conversion, Chem. 5 (2019) 1371–1397. https://doi.org/10.1016/j.chempr.2019.02.008.

14

Recent Development in Covalent Organic Framework Electrocatalysts for Oxygen Reduction Reactions

Kayode Adesina Adegoke[1], Thabo Matthews[1], Tebogo Mashola[1], Siyabonga Mbokazi[1], Thobeka Makhunga[1], Cyril Selepe[1], Memory Zikhali[1], Kudzai Mugadza[1,2], and Nobanathi Wendy Maxakato[1]

[1]*Department of Chemical Sciences, University of Johannesburg, Doornfontein, South Africa*

[2]*Institute of Materials Science, Processing and Engineering Technology, Chinhoyi University of Technology, Chinhoyi, Zimbabwe*

CONTENTS

14.1 Introduction

The scarcity of non-renewable fuels such as oil, coal, and fossil gas, which serve as energy sources, is one of the most pressing issues confronting our century. When fossil fuels are burnt by the power plants, automobiles, and industrial facilities for energy production, greenhouse gases are produced in the surroundings. These are harmful to the living organisms and cause environmental pollution leading to the gradual increase in temperature in the atmosphere (global warming), climate change, and environmental degradation due to the production of acid rains. As a result, new eco-friendly and inexhaustible energy sources must be utilized efficiently to relieve the strain on the energy supply, preserve human productivity and existence, and provide energy with low or no greenhouse gas emissions. Tidal energy, wind energy, fuel cells, and solar energy have all been thoroughly examined in terms of implementation and study. The number of electrochemical redox reactions, such as oxygen reduction at the cathode compartment of regenerative fuel cells or rechargeable metal-air batteries, or even as a green pathway to produce hydrogen

DOI: 10.1201/9781003206507-14

peroxide, and hydrocarbon reactions (electrooxidation of biofuels) are the most important components of these clean energy devices.

Different electrocatalysts have been explored for a long time to reduce the activation energy of the difficult oxygen reduction, hence improving the performance of fuel cells and metal-air batteries. Ever since, platinum has been long studied to be the most effective catalyst for oxygen reduction. However, its low availability and expensiveness have blocked the large-scale application or fuel cell technology commercialization. Furthermore, fuel cell utilization has been hampered by the effect of the fuel crossover, low catalyst durability, and inhibition effects caused by chemisorption of carbon monoxide intermediates onto the catalyst active sites, especially when platinum has been employed as the catalyst [1, 2].

Developing highly efficient and more active catalysts with high durability to catalyze sluggish ORR is still a significant confrontation in fuel cells and metal-air batteries [3]. Nowadays, it is crucial to develop a stable and durable catalyst to facilitate sluggish ORR by lowering the activation energy as much as possible. Metal-free catalysts possess a wide range of unique advantageous properties such as eco-friendly nature, high availability, cost-effectiveness, and resistance to wide pH ranges. Covalent organic frameworks (COFs) are the most common type of porous organic polymer with tunable structural patterns connected by a covalent connection among various metal-free nanomaterials. COFs exhibit outstanding characteristics such as enormous surface area, variable porosity, high degree of crystallinity, and a distinctive chemical structure. Based on these exciting attributes, COFs have been employed in diverse applications due to their exceptional characteristics and are thought to be suitable for electrocatalysis [4]. Figure 14.1 presents various synthetic routes, including ionothermal, mechanochemical, sonochemical, solvothermal, microwave-assisted, and light-promoted processes employed for preparing the COF materials for catalysis and other applications.

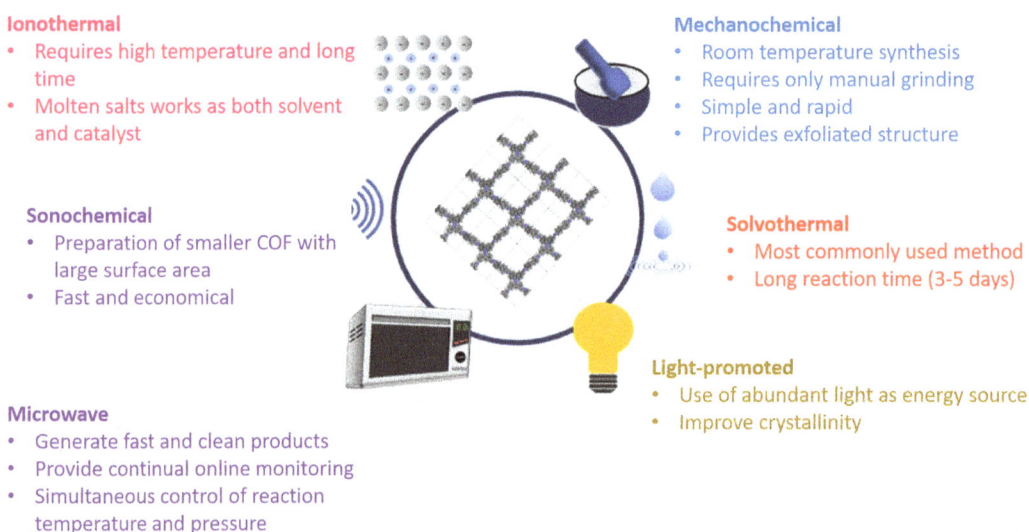

Ionothermal
- Requires high temperature and long time
- Molten salts works as both solvent and catalyst

Sonochemical
- Preparation of smaller COF with large surface area
- Fast and economical

Microwave
- Generate fast and clean products
- Provide continual online monitoring
- Simultaneous control of reaction temperature and pressure

Mechanochemical
- Room temperature synthesis
- Requires only manual grinding
- Simple and rapid
- Provides exfoliated structure

Solvothermal
- Most commonly used method
- Long reaction time (3-5 days)

Light-promoted
- Use of abundant light as energy source
- Improve crystallinity

FIGURE 14.1

Schematic representation of the properties and advantages of various synthetic routes toward COFs. (Reproduced with permission [5]. Copyright (2021), Elsevier.)

14.2 Principles and Mechanism of Electrocatalytic ORR

The ORR process is commonly one of the most important reactions in electrochemistry. It is one of the currently investigated processes due to its significance for developing new eco-friendly and inexhaustible electrochemical conversion of energy and storage devices [6]. The ORR involves a series of multiple-step electrochemical pathways, and it can move either through a two-step mechanism, an undesired or inefficient two-electron route with the formation of H_2O_2 (in an acidic environment (Eqs. 14.1–14.3)) which corrodes fuel cell components or HO_2^- (in high pH environment (Eqs. 14.3–14.6)) as the intermediate or it can move through a more preferable and efficient four-electron route [1]. The performance of ORR catalysts is usually determined using a steady-state polarization technique on a rotating ring-disk electrode (RRDE) or rotating disk electrode (RDE). On an RDE, half-wave potential, onset potential, and limiting current density are frequently used indicators for evaluating the performance of ORR catalysts. The total number of electrons transferred is an indicator usually employed to determine the ORR progress which could be estimated from Koutecky–Levich (K–L) equation on an RDE or measured directly on an RRDE through the equation.

Acidic environment:

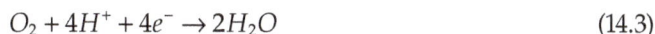

$$O_2 + H^+ + 2e^- \rightarrow H_2O_2 \tag{14.1}$$

$$H_2O_2 + 2H^+ + 2e^- \rightarrow H_2O \tag{14.2}$$

$$O_2 + 4H^+ + 4e^- \rightarrow 2H_2O \tag{14.3}$$

Alkaline environment:

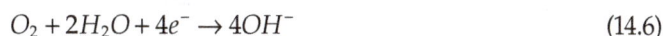

$$O_2 + H_2O + 2e^- \rightarrow 2H_2O^- + OH^- \tag{14.4}$$

$$H_2O + HO_2^- + 2e^- \rightarrow 3OH^- \tag{14.5}$$

$$O_2 + 2H_2O + 4e^- \rightarrow 4OH^- \tag{14.6}$$

14.3 COF Electrocatalysts for ORR

14.3.1 Pristine COF Electrocatalysts for ORR

The exceptional characteristics such as larger surface area, tunable architecture, well-defined pore sizes, etc associated with the COFs have enabled their employment as electrocatalysts for ORR. In general, the heteroatom-containing COFs in their pristine states could be used directly as ORR catalysts. Liu et al. [7] and Kuhn et al. [8] prepared a metal-free covalent triazine-based framework (CTF) having a higher percentage of N atomic pyridine using an ionothermal method with 1,4-dicyanobenzene precursor as shown in Figure 14.2a. The active electrocatalytic N- pyridinic position in the CTF was controlled through the manipulation of molecular structures of a CTF precursor. The electrocatalytic

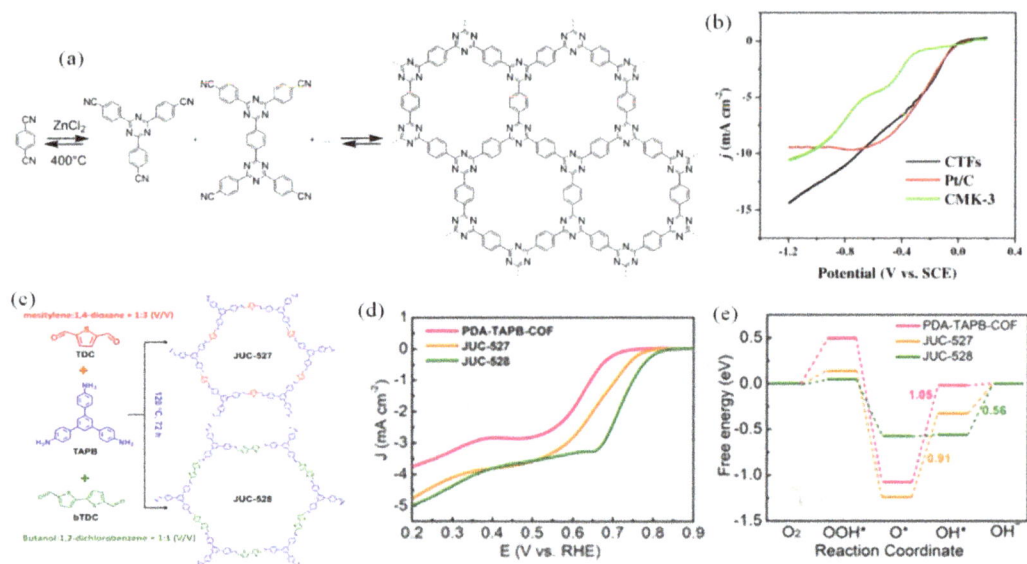

FIGURE 14.2

(a) Synthesis route of metal-free CTFs. Reproduced with permission [8]. Copyright (2008), John Wiley & Sons (b) LSVs of CTFs, Pt/C, and CMK-3. Reproduced with permission [9]. Copyright (2015), Elsevier (c) syntheses and structures of JUC-527 and JUC-528 (d) LSV curves (at 1600 rpm) of PDA-TAPB-COF, JUC-527, and JUC-528. (e) free energy diagrams (at site 5) for PDA-TAPB-COF, JUC-527, and JUC-528. (Reprinted with permission [10]. Copyright (2020), American Chemical Society.)

performance showed that CTF has an onset potential closer to that of commercial (Pt/C) electrocatalyst (Figure 14.2b) and the ~3.6 electrons transferred was determined suggesting four electrons transfer reaction [9]. The preparation of the two metal-free thiophene–S COFs (JUC-527 and JUC-528) electrocatalysts have been reported for their good performance toward ORR (Figure 14.2c) [10]. JUC-527 and JUC-528 were fabricated by employing one and bithiophene–sulfur structures, respectively. Their electrocatalytic performances showed higher ORR activities (Figure 14.2d) than the thiophene-free COF, suggesting that the thiophene–S building block served as an active center for ORR which was also confirmed by the DFT calculation (Figure 14.2e).

14.3.2 COF-Derivative Electrocatalysts for ORR

Tailoring and derivation of the COFs-based electrocatalysts is important in electrocatalyst modulation. Thermodynamically, the adsorption of O*, OH*, and OOH* moieties and associated-dissociated energy of these species are central in manning the ORR kinetics [11]. Based on this, the COFs are often subjected to pyrolysis into heteroatom doped graphene-like carbon matrixes, making the COFs more applicable for ORR [12]. The control of N positions and pore sizes using the Yamamoto polycondensation process has been reported with precursors such as triazine and its derivatives. In their study, there was precise modulation of N moieties and manipulation of the COFs-molecular precursors, which form well-modulated N doped levels and porosity upon carbonization in an inert environment. Zhao et al. utilized p-toluene sulfonic acid-assisted mechanochemical for the fabrication of iron-nitrogen doped mesoporous carbon (mC-TpBpy-Fe). The mC-TpBpy-Fe with 4.50 and 0.58 atomic % N and Fe, respectively, exhibited excellent ORR electrokinetics $E_{\frac{1}{2}}(V) = 0.845$

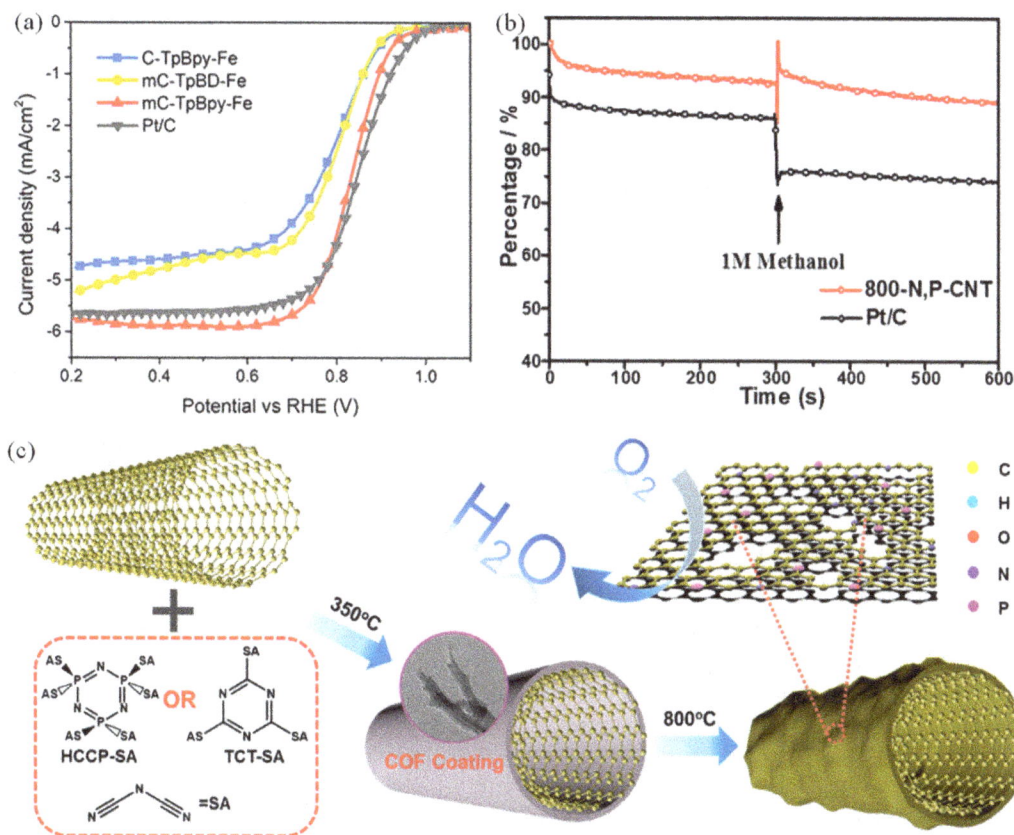

FIGURE 14.3

ORR performance of the obtained carbon catalysts. (a) LSV curves of C-TpBpy-Fe, mC-TpBD-Fe, mC-TpBpy-Fe, and Pt/C. Reproduced with permission [13]. Copyright (2019), American Chemical Society. (b) Chronoamperometric response of 800-N, P–C a constant potential in O_2-saturated 0.1 M KOH solution and electricity change with the addition of methanol, and (c) Illustration of the Preparation of the N, P–C. (Reproduced with permission [15]. Copyright (2017), American Chemical Society.)

and limiting current density of 5.92 mAcm^{-2} comparable to commercial Pt/C $E_{\frac{1}{2}}(V) = 0.852$ and limiting current density of 5.57 mAcm^{-2} (Figure 14.3). Additionally, the number of transferred electron number was approximately 4.0 pointing to a 4e$^-$ ORR pathway [13]. This strategy is extendable to the synthesis of Co-doped carbon species.

Nitrogen, Sulfur Co-doped carbon derived from naphthalene-based COF (CPN-NS) was synthesized and was highly porous and excellent surface area of 1116 m^2/g, 30 mV with more positive $E_{\frac{1}{2}}(V) = 0.868\,V$ Vs Pt/C $E_{\frac{1}{2}}(V) = 0.838\,V$ (Vs RHE), highly stable and following the four-electron pathway, relative to the commercial Pt/C. The CPN-NS had 4.02% N and 4.21% S [14]. The precise control of the heteroatom dopant's positions ensures well-distinct properties. From an N, P COFs coated on CNTs, a nanohybrid codoped with N, P was fabricated [15]. The excellent electrocatalytic activity of the nanohybrid was attributed to synergism between the heteroatom and the carbon matrix. From the synergetic effect, the nanohybrid posed a good $E_{\frac{1}{2}}(V) = 0.162\,V$ vs Ag/AgCl), a high current density of 6.1 mA cm^{-2}, plausible methanol tolerance, and high durability as seen by the retention of ~83% of the current density.

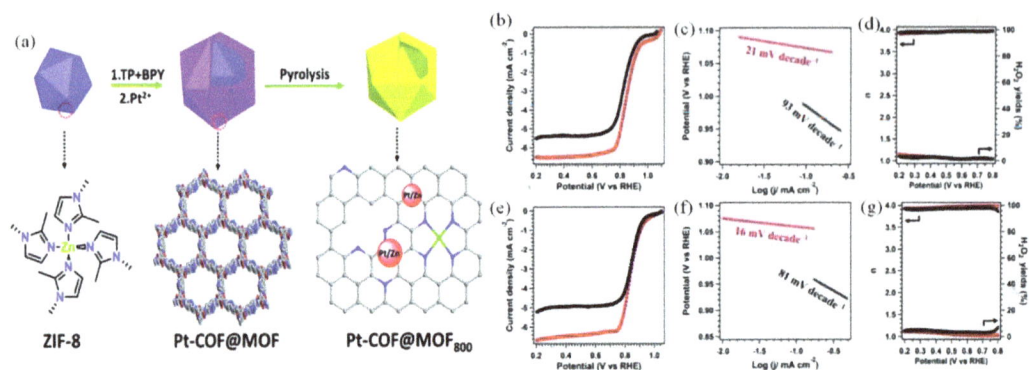

FIGURE 14.4

(a) Schematic of the synthesis procedure of Pt–COF@MOF$_{800}$ catalyst from TP–BPY–COF encapsulated ZIF-8. (b) LSV curves, (c) Tafel slopes, and (d) electron transfer number and H$_2$O$_2$ yield plots for Pt–COF@MOF$_{800}$ and Pt/C. (e) LSV curves, (f) Tafel slopes, and (g) electron transfer number and H$_2$O$_2$ yield plots over Pt–COF@MOF$_{800}$ and Pt/C in oxygen-saturated HClO$_4$ (0.1 M) aqueous solution. (Reproduced with permission [16]. Copyright (2021), Royal Society of Chemistry.)

Guo et al. [16] also prepared the hierarchical confinement of Pt-Zn alloy nanoparticles and single-dispersed Zn atom on COF/MOF-derived carbon as an effective catalyst for ORR (Figure 14.4). They first synthesized an ultra-stable COF/MOF-derived catalyst in which hollow carbon was used as supporting material to confine Pt-Zn intermetallic nanoparticles and Zn atoms (Figure 14.4a). The COF on the surface of a hollow structure of the MOF paved the way for the designing of a hollow structure and the MOF prohibited the breaking down and agglomeration of the COF coating in the pyrolysis step. The as-synthesized catalyst had Pt-Zn intermetallic nanoparticles with an average size of approximately equal to 3 nm decorated on the hollow carbon with uniformly distributed Zinc atoms. Based on the coupling of the active zinc atoms and Pt-Zn nanoparticles or clusters, the catalysts have shown superior catalytic activity (Figure 14.4b–g) toward ORR having a potential of 0.85 V, which was more favorable than those of commercial Pt/C catalysts in alkaline and acidic media. More essentially, the catalyst possessed excellent prolonged durability of 20 hours [16].

A multi-heteroatom porous covalent carbon framework was synthesized through the trimerization of molecularly designed monomers PYPZ by Ronghan Cao et al. High electrocatalytic activity and excellent methanol tolerance were observed when p-PYPZ was used for ORR owing to the abundant N, O heteroatom, hierarchical structure, and large surface area [17]. The summary of the electrocatalytic performance of other selected COF-derivative electrocatalysts for ORR is presented in Table 14.1.

14.3.3 Single Atomic Site COF Electrocatalysts for ORR

Electrocatalysts are required to optimize the performance and understand the reaction mechanisms that occur in various ORR [22]. The high cost, low selectivity, limited durability, and scarcity of Pt catalyst has led to the development of Single-atom catalysts (SACs), which are not only cheap but also are porous crystalline materials that are connected by covalent bonds. They possess superior characteristics which include unique molecular architectures, large surface areas, and tunable pore sizes [23, 24]. Based on this, SACs have been considered as promising candidates for ORR; however, it is difficult to dope several metals such as Ni or Cu into a graphene matrix due to the ease of these metals being

TABLE 14.1

Summary of Electrocatalytic Performance of Selected COF-Derivative Electrocatalysts for ORR

COF-Derived Electrocatalyst	Electrocatalytic Center(s)	Electrolyte	Electrokinetics			Refs
			Diffusion Limiting Current Density (mA/mc²)	Onset (V)	$E_{1/2}$ (V)	
PA@TAPT-DHTACOF$_{1000NH3}$	Pyridinic-N graphiticN	0.1 M KOH	7.2	0 (vs. Ag/AgCl)	0.11 (vs. Ag/AgCl)	[18]
C–COP-4	pyridinic-N graphiticN	0.1 M KOH	5.2	~0.88	0.78	[12]
COP–P–SO₃–Co-rGO	Co-porphyrin	0.1 M KOH	~4.4	0.88	~0.72	[19]
PCN–FeCo/C	FeCo–N–C	0.1 M KOH 0.1 M	~5.4~5.5	1.00 0.90	0.85 0.76	[20]
Ni/Fe-COF@CNT$_{900}$	Ni/Fe clusters	0.1 M KOH	6.0	1.01	0.87	[21]

aggregated at high temperature during heat treatment [25]. COFs can serve as SAC platforms because they possess several heteroatoms with lone pairs of electrons such as O, N, and S to support excellent metal coordination and they are insoluble in common solvents due to their covalently cross-linked structure [22, 26]. Gong et al. [27] reported anchoring metal nanoparticles such as Cu, Co, and Fe into the porous N-doped carbon framework. The resultant single atom COF electrocatalysts showed long-term durability and excellent catalytic activity for ORR in both acidic and alkaline media, thereby demonstrating the potential of single atom COFs as efficient electrocatalysts for ORR [27].

A pyrolysis-free approach was developed by Peng et al. [28] to produce high electrochemical SACs. This was done by fabricating graphene mixed with fully π-conjugated iron phthalocyanine (FePc)-rich COF through intermolecular interaction [28]. The synthesized single atom COF electrocatalyst demonstrated an exceptionally high current density of 25.86 mA cm^{-2} for ORR, which was higher than that of the commercial Pt/C fourfold [28]. Yi et al. [29] reported the use of highly stable SACs on COFs for ORR (Figure 14.5a). In their work, Fe was dispersed on porous porphyrinic triazine frameworks to archive Fe loading of 8.3 wt%. The synthesized electrocatalyst was durable, and it exhibited high activity for ORR (Figure 14.5b–e) in both media studied [29]. Prospects of single atom COF electrocatalysts are promising, and a lot of work remains to improve their performance which is ascribed to their functional versatility, framework tenability, and structural diversity [30, 31].

14.3.4 Metalized-Based COF Electrocatalysts for ORR

The catalytic activity and selectivity of COFs under harsh conditions is a limiting factor due to a shortage of active sites, which restricts their broader practical application. To overcome these disadvantages, the structural properties and stability of COFs can be modified to produce COFs with different functionalities, with metal modification regarded as one of the most efficient for COF functionalization. Metal modification strategy for COFs involves strong coordination reactions between vast heteroatom metal species (atoms, ions, and nanoparticles) with organic building units, which incorporate active metal sites on the COF. One prominent example of skeletal modification of COFs by this strategy involves the

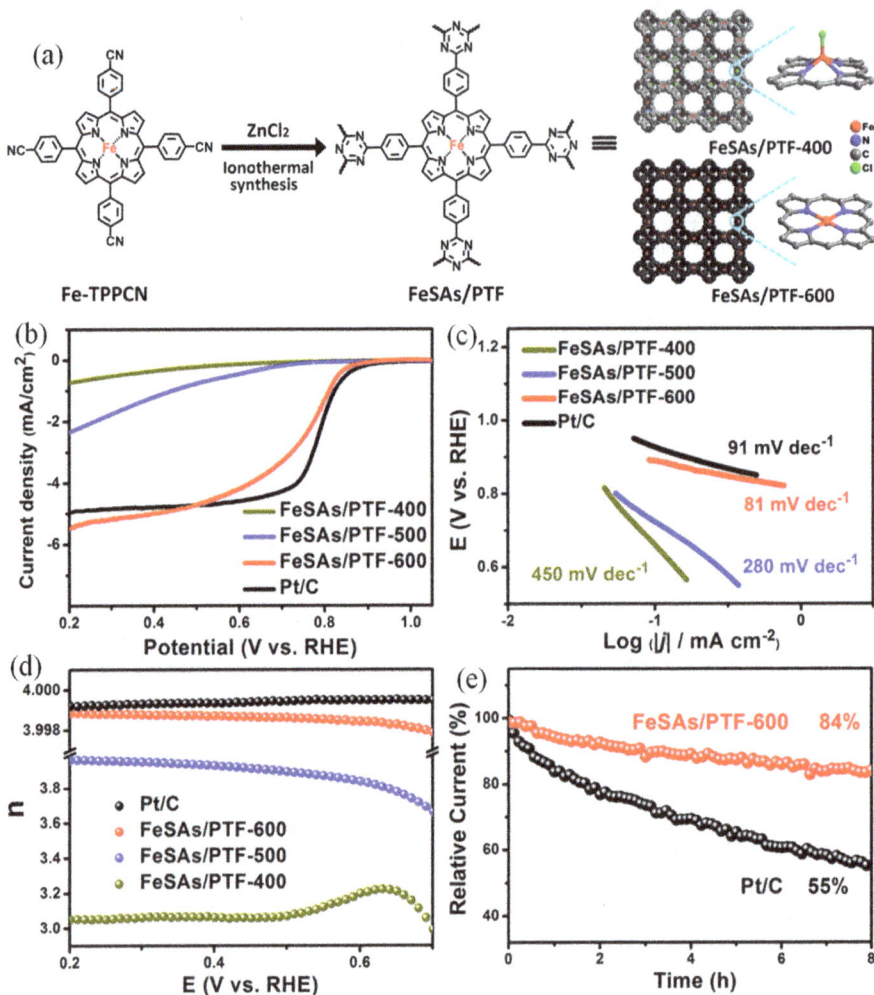

FIGURE 14.5
(a) Schematic illustration of the formation of FeSAs/PTF electrochemical evaluation of catalysts in acidic media. (b) LSVs of FeSAs/PTF-400, -500, and -600 and Pt/C. (c) Corresponding Tafel plots were obtained from the RDE polarization curves. (d) Electron transfer number of different samples obtained by RRDE. (e) Current–time chronoamperometry for FeSAs/PTF-600 and Pt/C. (Reprinted with permission [29]. Copyright (2018), American Chemical Society.)

synthesis of COF-composites by integrating metal-organic frameworks (MOFs) with COFs to endow COFs with destructive functions. Metal-modified COF-composites have a balance of properties between open metal sites of MOFs and strong covalent bonds of COFs (Figure 14.6), which breeds a new set of materials with specific applications [32].

The tunable pore size of COF creates a confined environment for size-selective catalysis. This means that the active sites of metals can be isolated at a molecular level by strong coordination of metals with organic building blocks to attain catalysts with desired electrocatalytic performance. The tunable pore size of COFs means that they can also be separated in catalytic reactions due to their solvent insolubility and these advantages endow metallized based COF with excellent catalytic activity for electrocatalysis. Transition metals with 3d orbitals exhibit great electrocatalytic effects since their energy states can be

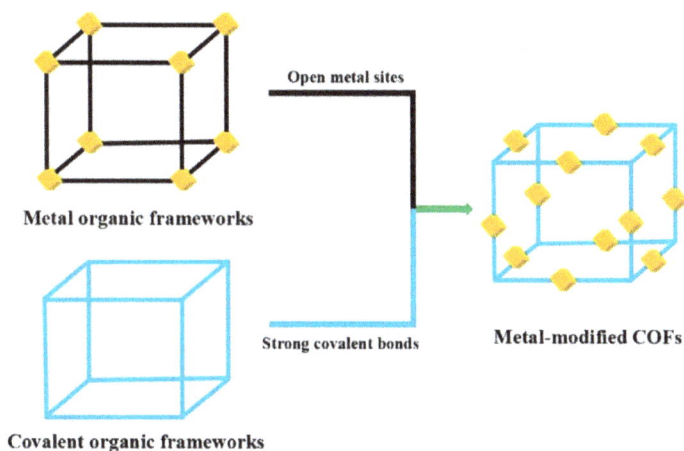

FIGURE 14.6
Schematic representation of a metal-modified COF composite. (Reproduced with permission [32]. Copyright (2021), Elsevier.)

adjusted while COFs act as gas absorbents and catalyst carriers hence hybridization of these metals with COF presents metalized-based COF with potential use in electrocatalysis. One such example of a metal-modified COF is a cobalt nanoparticle modified COF-900 which shows superior electrocatalytic activity for an oxidation-reduction reaction when compared with pristine COF and commercial 20% Pt/C due to the CoO/Co core-shell structures and the effect of CoN_4 sites [33].

The incorporation of metal ions into COFs results in the preparation of conductive carbons with metals residues as participles. This means that metalloporphyrin COFs can be used as precursor materials to produce metal-containing porous conductive carbons where the metal nanoparticles serve as catalytic sites. One notable example is a bifunctional electrocatalyst derived from COF, in which carbon nanotubes (CNT), and COF-derived carbon are used as the support to anchor the bimetallic Ni/Fe clusters and nanoparticles to form $Ni/Fe-COF@CNT_{900}$ (Figure 14.7a–c). The Fe and Ni ions were immobilized in the pore channels of the COF, which effectively hindered the aggregation and migration of these ions during pyrolysis, and the conductive carbon derived from the metalized-based COF featured abundant nitrogen content and high mesoporous volume. The conductive carbon produced ultrahigh ORR performance on the alkaline electrolyte, which is more positive than the one displayed by the commercial Pt/C (Figure 14.7e–g) [21].

Rational design and synthesis of mesoporous electrocatalysts from COF can tackle fundamental challenges to yield practical solutions for efficient energy utilization. A novel approach for developing this COF-derived conductive carbon was initiated from silica templated bipyridine COF, which was used as a precursor material to produce iron-nitrogen doped mesoporous carbon (mC-TpBpy-Fe) from carbonization and template removal. mC-TpBpy-Fe shows large pore volume and surface area, which directly translates to the observed high ORR activity. This large pore volume and surface area massively promoted the mass transfer efficiency and increased the accessibility to the electrocatalytic active sites on this COF-derived conductive carbon, and the ORR performance is superior in comparison to those of commercial Pt/C [13].

Fabricating highly efficient and durable electrocatalysts for ORR remains a challenge. Conjugating Pt nanoparticles onto a nitrogen-rich COF results in a metal-modified COF with

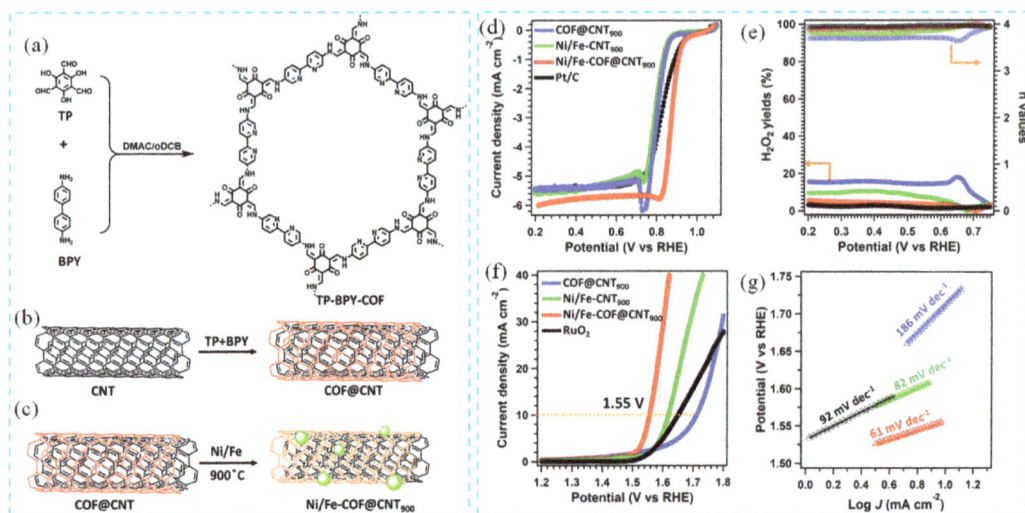

FIGURE 14.7
(a) Synthesis of TP-BPY-COF from TP and BPY monomers. (b) Synthesis of COF@CNT under the same condition as that of TP-BPY-COF. (c) Fabrication of Ni/Fe-COF@CNT$_{900}$. (d) LSV curves of ORR, (e) number (*n*) of electrons transferred, and H$_2$O$_2$ yield plots calculated from the RRDE measurements (f) LSV curves of OER and (g) the Tafel plots for OER. (Reproduced with permission [21]. Copyright (2020), John Wiley & Sons.)

superior ORR compared with conventional Pt/C. The pyridinic nitrogen on the COF acts as the source of nucleation sites for the Pt nanoparticles, ensuring uniform distribution of the Pt on the COF pore channels and surface, which in turn exposes the active sites on the Pt metal to make them more accessible [34]. Zhai et al. also investigated the effect of 2D-COFs being used as support material for platinum toward ORR. The 2D conjugated nitrogen-rich COF with excellent stability toward harsh environments was synthesized and further used to synthesize a Pt decorated COF electrocatalyst via an intermolecular interaction without pyrolysis. The fashionable triazine centers worked as nucleation sites to precisely distribute platinum in a layer over the surface and pore channels of COF, leading to a consistent distribution of Pt and sufficient accessibility of Pt active sites. Notably, Pt/COFs catalyst exhibited much higher electrocatalytic performance than commercial Pt/C electrocatalyst toward ORR. The distinct electron configuration of highly ordered porous structure, and plentiful accessible pores played a considerable role in the excellent ORR performance [34].

14.3.5 Nanohybrid COF Electrocatalysts for ORR

The relatively small pore size that restricts mass transport and low electrical conductivity leads to low charge carrier mobility on COFs are drawbacks to their electrochemical applications. Even though the COF backbone consists of active sites that permit certain charge-transfer pathways within the COF matrix, the electrocatalytic activity of COF toward ORR is still inferior when compared to other typical electrode materials. Based on these associated setbacks, a considerable interest has been devoted by different researchers on COF hybrids that can circumvent/alleviate all the conductivity problems and have controlled morphology. Special interest is given to COF composites that are formed from conducting materials like carbon nanotubes, graphene, and conductive polymers [35].

Hybridization of COF with conductive additives was reported by the Xiang group. It was based on a COF hybrid with high electrocatalytic activity by self-assembling highly ordered porphyrin COP with graphene (COP/rGO). The synergistic effect between the active sites of the COP with the electrical conductivity of rGO resulted in high ORR stability [19]. Xiang et al. developed a fully conjugated Fe-phthalocyanine COF (COF_{BTC}) by in-situ charges exploited with the hydroxyl group inserted in stack layers of the COF_{BTC}. The solution that resulted was referred to as the proper solution since it solved processible issues of COFs toward devices like fuel cells. This soluble COF hybrid (COF_{BTC}) provided superior ORR catalytic performance when compared to the Pt/C and better cycle endurance [36].

COFs have been recently utilized as precursor materials for constructing carbon-based electrocatalysts for ORR. One noticeable example is the Co encapsulated nitrogen-doped graphitic carbon (Co@NGC-600) formed by annealing a COF-hybrid (TZA-COF-rGO-Co) from s-tetrazine based COF on rGO and with reduced Co metal (Figure 14.8a–e). The

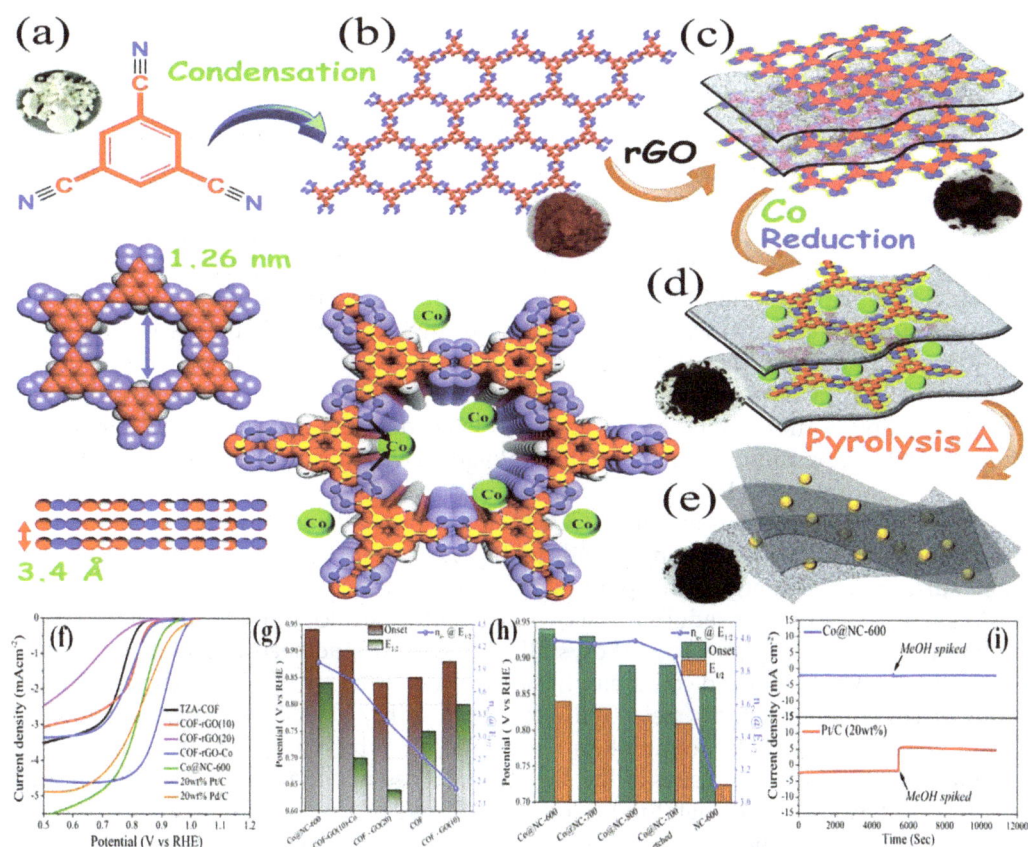

FIGURE 14.8

Schematic fabrication representation of (a and b) TZA-COF, (c) COF-rGO, (d) COF-rGO-Co, (e) pyrolyzed Co@NC-600. The insets in (b–e) show the optical images of the crystalline powders of the corresponding hybrids reflecting the change in color at different stages of catalyst engineering. (f) Electrochemical ORR activity of the catalysts showing the LSV curves of different electrocatalysts. Comparison of the electrocatalytic ORR activity in terms of the onset and half-wave potentials and the number of electrons transferred at $E_{1/2}$ across (g) catalysts at different structural engineering steps and (h) different controlled variants of the CO@NC-600 catalyst. (i) Electrochemical ORR stability showing comparative chronoamperometric curves for Co@NC-600 and Pt/C with MeOH injection. (Reproduced with permission [37]. Copyright (2020), Royal Society of Chemistry.)

carbon-based electrocatalyst (Co@NGC-600) showed superior ORR performance in terms of activity (Figure 14.8f), onset and half-wave potentials (Figure 14.8g–h), and better durability and methanol tolerance (Figure 14.8i) when compared to commercial 20% Pt/C [37].

Similarly, Liu et al. investigated the use of covalent triazine-based frameworks (CTFs) as effective electrocatalysts in alkaline environments for ORR. The ionothermal technique was used to synthesize CTFs. The as-prepared CTF has demonstrated outstanding electrocatalytic activity in alkaline conditions for ORR. Because nitrogen has good electron-donating capabilities, N-doping produced a disorder in the carbon framework and encouraged electron delocalization, increasing active areas for the oxygen reduction process. The onset potential for ORR at CTFs was found to be catalytically comparable to that of marketable Pt/C electrocatalyst. This response was induced by the pyritic sites in the triazine-based framework, which therefore lowered the onset potential of ORR [9].

14.4 Concluding Remarks

Covalent organic framework-based materials have been used as electrocatalysts in many applications, especially in ORR and solid electrolytes in fuel cells and batteries. The electrocatalytic ORR on COFs materials showed promising performances. However, these materials for ORR electrocatalytic ORR are in their infancy and currently, several challenges, such as limited durability, limit the practical use of COFs in energy conversion devices. The collapse of the COFs-based electrocatalysts structure under a strong acidic/basic medium could cause the mass and charge impairment and the inaccessibility of active sites. In addition, the synthesis of COFs by solvothermal treatment of precursors yields insoluble powders with randomly aggregated crystallites which makes it challenging to incorporate these materials in energy-related devices. Recently, COFs composed of cyano-containing monomers and olefin-linkages have improved the recyclability and charge transport of the COFs for electrochemical application.

The construction of metal-modified COFs remains a challenge, and one of the noticeable changes on COFs when doping with metals is crystal damage, morphological variation, surface area variation, and pore blockage of COFs. Metal dopants can also form clusters which can further reduce the active sites on the surface of these materials. Under the premise of retaining the original crystallinity of COFs, the need to maximize surface area and pore volumes need to be emphasized to result in metal COF-composites with higher electrocatalytic activity. Selectivity and recyclability of COFs, e.g., metalized-based COFs remain a major issue. To maintain the excellent catalytic performance of the metal-modified COF, precise metal sites at the skeleton of these materials need to be considered be prioritized. Research on metal-modified COFs has its pros and cons, and to grow more on these materials, further research work should focus on the following aspects:

1. The electrocatalytic efficiency and most functions of metal-modified COFs depend on the incorporation of noble metals, which makes these materials to be cost-effective; hence improving the preparation efficiency, and production of low-cost metalized-based COFs should be an important topic of future research.

2. The type of metal used and the ligand determines the performance of metal-modified COFs under various applications, but the metal-ligand coordination mechanism is unclear since research on these materials is still at the initial stages. To overcome

this setback, more research should be performed to identify the metal sites and reaction pathways in-depth by integrating theoretical simulations with data analysis.

3. The most used organic ligands for metal incorporation are limited, with only imine, porphyrin, and bipyridine as the most used organic ligands for metal-modified COFs. To increase the stability and structural functionality of these materials without adding any functional groups on the most utilized organic ligands, the features and library of organic ligands need to be increased. This will also facilitate the incorporation of multi-metallic loadings on COFs.

4. Advanced characterization techniques need to be developed which ensure more accurate structural information of COFs and metal-modified COFs. Accurate atom calculation methods that will facilitate large-scale measures of bandgap, charge carriers, and electronic absorption bands are needed for accurate analysis of COFs.

Although many challenges still remain, a significant rise in the research of COF-tailored electrocatalysts has paved the way for a new and exciting prospect in a new class of versatile electrocatalytic materials. In general, to fabricate and design efficient COF-based hybrids for electrochemical applications. Different strategies need to be implemented, such as:

I. Controlling the structure of the COFs on a length scale, meaning that the COF-hybrids endowing good mass transfer and high electrical conductivity can be designed by considering that COF at a macroscopic scale length can combine their crystalline structures to produce composites with well-defined morphologies and with desirable electrocatalytic activity.

II. Spatial positions of modified electrocatalytic sites on COF hybrids need to be considered as the electrocatalytic activity on the surface of the COF or the pore channel might change with an introduction of conductive materials. Therefore, special emphasis should be given to catalytic site distribution on COF hybrids to establish proper electrochemical pathways, and further clarify the electrochemical properties of the COF composites.

III. Multiple supporting materials or metal sites on supporting materials should be explored to fabricate COF hybrids with versatile properties. Special emphasis should also focus on the synergistic effect between the COF matrix and the incorporated supporting material to gain extensive insight into the electrocatalytic mechanism of the COF composites.

IV. It is necessary to understand the structure-property relationship for the COF electrocatalysts, and another way to fabricate this will be to utilize computational efforts to reveal the electrocatalytic mechanism between the active sites of COF hybrids and the reactive species. Computational chemistry can then be utilized to accelerate the rational design and construction of high-performance COF composites.

Acknowledgments

K. A. Adegoke acknowledges the Global Excellence Stature (GES) 4.0 Postdoctoral Fellowships Fourth Industrial Revolution and the University of Johannesburg, South Africa. N. W. Maxakato acknowledges the support received from the National Research

Foundation of South Africa: Grant Number 118148, National Research Foundation of South Africa: Grant Number 138083, and Centre for Nanomaterials Science Research-University of Johannesburg, and University of Johannesburg, South Africa.

References

1. Zhang J, Dai L (2015) Heteroatom-doped graphitic carbon catalysts for efficient electrocatalysis of oxygen reduction reaction. ACS Catal 5:7244–7253

2. Jongsomjit S, Prapainainar P, Sombatmankhong K (2016) Synthesis and characterization of Pd – Ni – Sn electrocatalyst for use in direct ethanol fuel cells. Solid State Ionics 288:147–153

3. Wei Q, Tong X, Zhang G, Qiao J, Gong Q, Sun S (2015) Nitrogen-doped carbon nanotube and graphene materials for oxygen reduction reactions. Catalysts 5:1574–1602

4. Mondal S, Mohanty B, Nurhuda M, Dalapati S, Jana R, Addicoat M, Datta A, Jena BK, Bhaumik A (2020) A thiadiazole-based covalent organic framework: A metal-free electrocatalyst toward oxygen evolution reaction. ACS Catal 10:5623–5630

5. Abuzeid HR, EL-Mahdy AFM, Kuo S-W (2021) Covalent organic frameworks: Design principles, synthetic strategies, and diverse applications. Giant 6:100054

6. Gómez-Marín AM, Ticianelli EA (2018) A reviewed vision of the oxygen reduction reaction mechanism on Pt-based catalysts. Curr Opin Electrochem 9:129–136

7. Liu J, Hu Y, Cao J (2015) Covalent triazine-based frameworks as efficient metal-free electrocatalysts for oxygen reduction reaction in alkaline media. Catal Commun 66:91–94

8. Kuhn P, Antonietti M, Thomas A (2008) Porous, covalent triazine-based frameworks prepared by ionothermal synthesis. Angew Chemie - Int Ed 47:3450–3453

9. Liu J, Hu Y, Cao J (2015) Covalent triazine-based frameworks as efficient metal-free electrocatalysts for oxygen reduction reaction in alkaline media. CATCOM 66:91–94

10. Li D, Li C, Zhang L, Li H, Zhu L, Yang D, Fang Q, Qiu S, Yao X (2020) Metal-free thiophene-sulfur covalent organic frameworks: Precise and controllable synthesis of catalytic active sites for oxygen reduction. J Am Chem Soc 142:8104–8108

11. Hansen HA, Viswanathan V, Nørskov JK (2014) Unifying kinetic and thermodynamic analysis of 2 e- and 4 e - reduction of oxygen on metal surfaces. J Phys Chem C 118:6706–6718

12. Xiang Z, Cao D, Huang L, Shui J, Wang M, Dai L (2014) Nitrogen-doped holey graphitic carbon from 2D covalent organic polymers for oxygen reduction. Adv Mater 26:3315–3320

13. Zhao X, Pachfule P, Li S, Langenhahn T, Ye M, Tian G, Schmidt J, Thomas A (2019) Silica-templated covalent organic framework-derived Fe-N-doped mesoporous carbon as oxygen reduction electrocatalyst. Chem Mater 31:3274–3280

14. You C, Jiang X, Wang X, Hua Y, Wang C, Lin Q, Liao S (2018) Nitrogen, sulfur Co-doped carbon derived from naphthalene-based covalent organic framework as an efficient catalyst for oxygen reduction. ACS Appl Energy Mater 1:161–166

15. Li Z, Zhao W, Yin C, Wei L, Wu W, Hu Z, Wu M (2017) Synergistic effects between doped nitrogen and phosphorus in metal-free cathode for zinc-air battery from covalent organic frameworks coated CNT. ACS Appl Mater Interfaces 9:44519–44528

16. Guo Y, Yang S, Xu Q, Wu P, Jiang Z, Zeng G (2021) Hierarchical confinement of PtZn alloy nanoparticles and single-dispersed Zn atoms on COF @ MOF-derived carbon towards efficient oxygen reduction reaction. J Mater Chem A 13625–13630

17. Cao R, Hu F, Zhang T, Shao W, Liu S, Jian X (2021) Bottom-up fabrication of triazine-based frameworks as metal-free materials for supercapacitors and oxygen reduction reaction. RSC Adv 11:8384–8393

18. Xu Q, Tang Y, Zhang X, Oshima Y, Chen Q, Jiang D (2018) Template conversion of covalent organic frameworks into 2D conducting nanocarbons for catalyzing oxygen reduction reaction. Adv Mater 30:1–8

19. Guo J, Lin CY, Xia Z, Xiang Z (2018) A pyrolysis-free covalent organic polymer for oxygen reduction. Angew Chemie - Int Ed 57:12567–12572

20. Lin Q, Bu X, Kong A, Mao C, Bu F, Feng P (2015) Heterometal-embedded organic conjugate frameworks from alternating monomeric iron and cobalt metalloporphyrins and their application in design of porous carbon catalysts. Adv Mater 27:3431–3436

21. Xu Q, Qian J, Luo D, Liu G, Guo Y, Zeng G (2020) Ni/Fe clusters and nanoparticles confined by covalent organic framework derived carbon as highly active catalysts toward oxygen reduction reaction and oxygen evolution reaction. Adv Sustain Syst 4:1–7

22. Zhang H, Zhu M, Schmidt OG, Chen S, Zhang K (2021) Covalent organic frameworks for efficient energy electrocatalysis: Rational design and progress. Adv Energy Sustain Res 2:2000090

23. Altaf A, Baig N, Sohail M, Sher M, Ul-Hamid A, Altaf M (2021) Covalent organic frameworks: Advances in synthesis and applications. Mater Today Commun 28:102612

24. Lin CY, Zhang D, Zhao Z, Xia Z (2018) Covalent organic framework electrocatalysts for clean energy conversion. Adv Mater 30:1–16

25. Ohashi K, Iwase K, Harada T, Nakanishi S, Kamiya K (2021) Rational design of electrocatalysts comprising single-atom-modified covalent organic frameworks for the N2Reduction reaction: A first-principles study. J Phys Chem C 125:10983–10990

26. Díaz U, Corma A (2016) Ordered covalent organic frameworks, COFs and PAFs. From preparation to application. Coord Chem Rev 311:85–124

27. Gong S, Wang C, Jiang P, Hu L, Lei H, Chen Q (2018) Designing highly efficient dual-metal single-atom electrocatalysts for the oxygen reduction reaction inspired by biological enzyme systems. J Mater Chem A 6:13254–13262

28. Peng P, Shi L, Huo F, Mi C, Wu X, Zhang S, Xiang Z (2019) A pyrolysis-free path toward superiorly catalytic nitrogen-coordinated single atom. Sci Adv 5:1–8

29. Yi JD, Xu R, Wu Q, Zhang T, Zang KT, Luo J, Liang YL, Huang YB, Cao R (2018) Atomically dispersed iron-nitrogen active sites within porphyrinic triazine-based frameworks for oxygen reduction reaction in both alkaline and acidic media. ACS Energy Lett 3:883–889

30. Han J, Bian J, Sun C (2020) Recent advances in single-atom electrocatalysts for oxygen reduction reaction. Research 2020:1–51

31. Kim HS, Lee CH, Jang JH, Kang MS, Jin H, Lee KS, Lee SU, Yoo SJ, Yoo WC (2021) Single-atom oxygen reduction reaction electrocatalysts of Fe, Si, and N co-doped carbon with 3D interconnected mesoporosity. J Mater Chem A 9:4297–4309

32. Huang J, Liu X, Zhang W, Liu Z, Zhong H, Shao B, Liang Q, Liu Y, He Q (2021) Functionalization of covalent organic frameworks by metal modification: Construction, properties, and applications. Chem Eng J 404:127136

33. Ma W, Yu P, Ohsaka T, Mao L (2015) An efficient electrocatalyst for oxygen reduction reaction derived from a Co-porphyrin-based covalent organic framework. Electrochem commun 52:53–57

34. Zhai L, Yang S, Yang X, Ye W, Wang J, Chen W, Guo Y, Mi L, Wu Z, Soutis C, Xu Q, Jiang Z (2020) Conjugated Covalent Organic Frameworks as Platinum Nanoparticle Supports for Catalyzing the Oxygen Reduction Reaction. https://doi.org/10.1021/acs.chemmater.0c03614

35. Zhao X, Pachfule P, Thomas A (2021) Covalent organic frameworks (COFs) for electrochemical applications. Chem Soc Rev 50:6871–6913

36. Peng P, Shi L, Huo F, Zhang S, Mi C, Cheng Y, Xiang Z (2019) In situ charge exfoliated soluble covalent organic framework directly used for Zn–air flow battery. ACS Nano 13:878–884

37. Roy S, Mari S, Sai MK, Sarma SC, Sarkar S, Peter SC (2020) Highly efficient bifunctional oxygen reduction/evolution activity of a non-precious nanocomposite derived from a tetrazine-COF. Nanoscale 12:22718–22734

15

Metal-Air Batteries Based on Nanostructured Covalent Organic Frameworks

Maria Mechili[1], Christos Vaitsis[1], Nikolaos Argirusis[2], Pavlos K. Pandis[1], Georgia Sourkouni[3], and Christos Argirusis[1,3]

[1]*Laboratory of Inorganic Materials Technology, School of Chemical Engineering, National Technical University of Athens, Athens, Greece*

[2]*mat4nrg GmbH, Clausthal-Zellerfeld, Germany*

[3]*TU Clausthal, Clausthaler Zentrum für Materialtechnologie, Clausthal-Zellerfeld, Germany*

CONTENTS

15.1 Introduction

Metal-air batteries (MABs) constitute a type of energy conversion technology that incorporates characteristics from both conventional batteries and fuel cells. Briefly, they comprise a metal anode, an aqueous or a non-aqueous electrolyte, and an air cathode (gas diffusion electrode, GDE) allowing ambient air to enter the system and provide the oxidizing material. The first commercial adoption of MABs originates in 1932 [1] when primary zinc-air batteries (ZABs) were developed by Heise and Schumacher which are still utilized as primary power sources. Since then, multiple alternative metals have been reported as candidates for MABs' anode materials such as Fe, Al, Li, Mg, and Na [1, 2].

In the last decade, research regarding MABs has been revived with a specific direction toward enhancing their rechargeability and extending their lifetime. MABs can be mechanically (Mg-air, Al-Air) or electrically recharged (Zn-air, Li-air). The motivation for these efforts is that MABs exhibit an impressive high energy density while keeping a limited environmental footprint and decreased cost. Such characteristics of energy conversion and storage devices are currently of high interest due to continuously expanding energy demands and the emerging need for reducing CO$_2$ emissions [3, 4]. MABs operation can slightly differ depending on the anode and the electrolyte applied, however, relies on an oxidation process occurring on the metal anode which releases free electrons that immediately trigger the reduction process on the cathode. A typical MAB scheme is depicted in Figure 15.1.

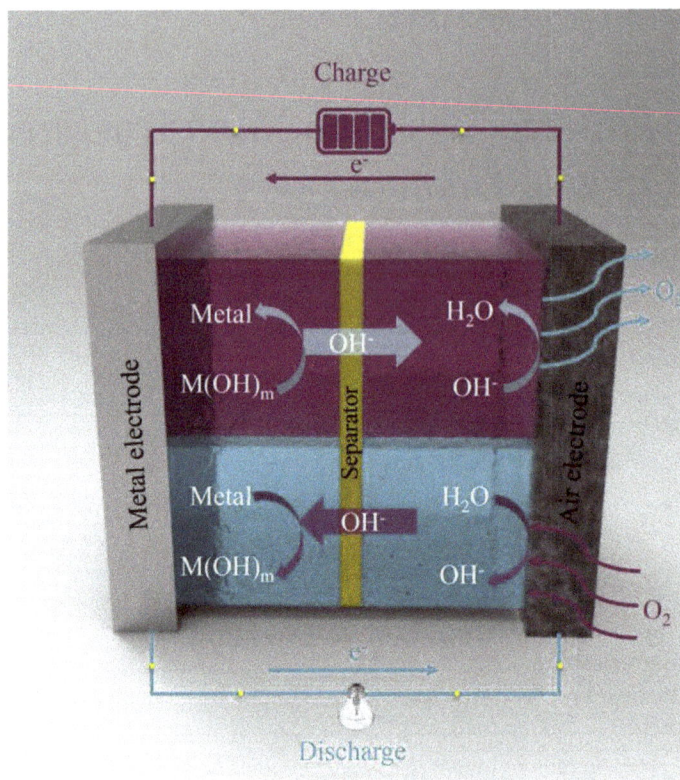

FIGURE 15.1
Working principle of aqueous metal-air batteries. (Adapted with permission [5]. Copyright (2019) Elsevier.)

When referring to aqueous MABs (Zn-Air, Mg-Air, Al-Air) the material that is being reduced in the cathode is usually oxygen and the reaction is typically known as Oxygen Reduction Reaction (ORR).

The reactions are presented below:

Metal Electrode (Anode)

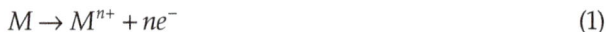

$$M \to M^{n+} + ne^-$$ (1)

Air Electrode (Cathode)

$$O_2 + 4e \to^- 2H_2O + 4OH^- E^0 = 0.4V \text{ vs SHE}$$ (2)

When a conventional cell is charging, the reactions (1)-(2) are reversed, and particularly reaction (3) is called Oxygen Evolution Reaction (OER).

Oxygen Evolution Reaction

$$4OH^- \to O_2 + 2H_2O + 4e^- E^0 = -0.4V \text{ vs SHE}$$ (3)

ORR and OER are reactions of high concern over the last decades, while the research community has invested high effort in exploring electrocatalysts that accelerate them [6]. The materials that can function as both ORR and OER electrocatalysts are denoted as bifunctional electrocatalysts. On the contrary, there are cases like Li-air, Na-air, and K-air batteries, where electrolytes are typically aprotic, and the reactions are depicted as below:

Metal Electrode (Anode)

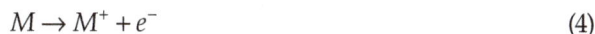

$$M \rightarrow M^+ + e^- \tag{4}$$

Air Electrode (Cathode)

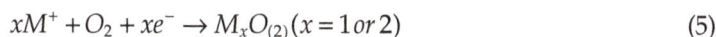

$$xM^+ + O_2 + xe^- \rightarrow M_xO_{(2)}(x = 1\,or\,2) \tag{5}$$

In both cases, the efficient function of rechargeable MABs is still inhibited by early degradation or passivation of the air electrode. The reduction reactions occurring in the air electrode include multiple steps and intermediates, so they are governed by rather slow kinetics and increased overpotentials. Thus, durable electrocatalysts that facilitate cathode reactions are demanded to be adopted [6]. Commonly, the cathode (air electrode) consists of an active catalytic layer that contacts the electrolyte from one side and a macroporous structure that facilitates the entrance of oxidizing molecules from the other side [7].

Of course, some state-of-art electrocatalysts (e.g. Pt, RuO_2, IrO_2) have been demonstrated as effective for both aqueous and aprotic ORR and OER, albeit limited to noble metals. For instance, Pt is effective for ORR facilitation, yet does not give sufficient results during OER, while RuO_2 and IrO_2 are active only regarding OER. A plethora of alternatives have been proposed involving carbonaceous materials [8], single-atom catalysts (SACs) [9], transition metal oxides (TMOs) [10, 11], and metal-organic frameworks (MOFs) [12–14], and multiple of their composites [15]. The electrochemical properties of these materials have been thoroughly investigated and most of them have undergone rational optimization toward better performance, however, the progress has not come to mature results to meet current commercial needs. Very recently, when compared to other conventional materials, covalent organic frameworks (COFs) have appeared in the spotlight as potential oxygen electrocatalysts and consequently applied in MABs. As has been discussed in previous chapters, COFs acquire a highly defined and porous carbon-rich structure that can be rationally pre-designed to adopt different features. Such characteristics favor the specific modulation of intrinsic catalytic activity by tuning the geometry and chemical composition of these networks.

15.2 Zinc-Air Batteries

Among all categories of MABs, zinc-air batteries (ZABs) constitute the most studied battery system, as zinc as an anode has to offer many advantageous properties. Zinc is a material possessing a mature technological background worldwide, is easily accessible, and is a rather environmentally benign option. This anode can also be compatible with alkaline aqueous electrolytes (KOH is mostly used) and exhibits acceptable electrical rechargeability. Recently, many researchers have explored ZAB systems containing non-aqueous electrolytes to overcome destructive phenomena that restrict the long lifetime of Zn anodes [16]. The key to the attractiveness of ZABs is their air-breathing architecture, which attributes them to a prominent energy density reaching a theoretical value of 1353 Wh/kg. ZABs have a theoretical potential of 1.65 V, while they usually work at 1.2 V [17]. Yet, after continuous cycling ZABs exhibit corrosive overpotentials due to ORR/OER high energy demands. Thus, when evaluating catalysts' performance, the Voltage Gap between the charging and discharging voltage, when cycling, is a useful metric. The departure of the voltage while discharging or charging from the open-circuit voltage (OCV) expresses the overpotentials of the corresponding ORR/OER [15].

Besides complete battery tests, the ORR and OER activity of electrocatalysts can be primarily investigated by half-cell tests which include mostly voltage scans of the material in a three-electrode configuration. Typically, the E_{eq} of ORR and OER is 0.4 V at an alkaline pH and when studying new catalysts, it is beneficial to observe the departure from E_{eq} in both reduction and oxidation processes. This departure can be counted in V and is called overpotential (η). An initial metric to evaluate overpotential is E_{onset}, which can be determined by a simple linear scan voltammogram (LSV). However, the metric that is most commonly indicative of the reduction is $E_{1/2}$ (half-wave potential: the potential (or overpotential) at the point where half of the maximum reductive current is reached), $E_{j=10}$ (the potential (or overpotential) needed to reach a current of 10 mA/cm²). Finally, when calculating ΔE ($E_{j=10} - E_{1/2}$), a first assumption of the overall overpotentials needed and the performance of the catalytic material can be made.

Recently, Lin et al. identified a novel activity descriptor, particularly for transition metal (TM) containing COFs, that can be correlated with oxygen catalytic properties and provides fertile ground for further research [18]. According to the Volcano Plots of ORR/OER activities in their study, iron and cobalt are one of the most appropriate TMs for COF enrichment to facilitate the 4e⁻ pathway of ORR while also enhancing OER activity [18]. In this direction, 2,4,6-Trihydroxybenzene-1,3,5-tricarbaldehyde (Tp) - [2,2'-Bipyridine]-5,5'-diamine (Bpy) based COFs enclosing Co or Fe inside their porous structure have been examined regarding their oxygen electrocatalytic properties and have been finally tested as air cathodes in ZAB systems. In 2016, a Tp-Bpy based COF enclosing Co atoms (Figure 15.2) inside its porous structure was tested regarding its OER properties in neutral media for the first time [19]. Similarly, a TP-Bpy-COF was further modified with cobalt acetate and iron acetate and subsequently directly pyrolyzed in the N_2 atmosphere to form a carbon structure enriched with Fe-Co alloy nanoparticles (NPs), which proved to facilitate both ORR and OER in alkaline conditions. After the attachment of metal ions to the walls of the carbonaceous network, the BET surface area was vastly decreased, however, XRD investigation revealed that the structure of the COF was fully preserved. Regarding ORR, the C-Fe, Co-COF-catalyst exhibited a half-wave potential of $E_{1/2} = 0.81$ V, which proved to be higher than the corresponding for a noble Pt-C electrode ($E_{1/2} = 0.76$ V), while furthermore delivering superior durability reducing oxygen for 20h [20]. When explored regarding OER, the composite displayed an encouraging bifunctionality, described by a relatively limited Voltage Gap of $\Delta E = (E_j = 10 - E1/2) = 1.6$ V $- 0.81$ V $= 0.79$ V.

In the same context, Zhao et al. referred to an alternative method to rationally construct the macro-morphology of such electrocatalytically active TM-containing TpBpy based COFs. Primarily, they exploited polystyrene spheres [21] as a hard template, to modulate exactly the porosity of such materials, while in a second publication they refer to SiO_2 being also appropriate for such use [22]. Firstly, a SiO_2@TpBpy-Fe hybrid was created via a Schiff base reaction between 1,3,5-triformylphloroglucinol (Tp) and Bpy in the presence of SiO_2, and afterwards, it underwent a two-step thermal treatment to form the final mesoporous electrocatalyst (Figure 15.3). The catalyst was examined only regarding ORR and delivered a peak power density of around 80 mW/cm² and a specific capacity of 714 mAh/g at 10 mA/cm² in a mechanically rechargeable ZAB. The enhanced catalytic activity was attributed not only to the favorable mesoporous network that permitted fast ion diffusion

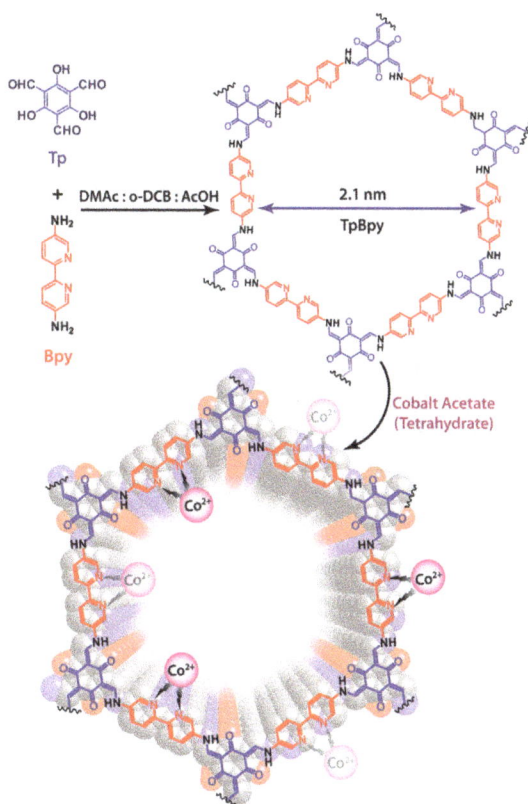

FIGURE 15.2

Schematic representation of the synthesis of TpBpy via proton tautomerized Schiff base condensation and Co-TpBpy via Co(II) impregnation. (Adapted with permission [19]. Copyright (2016) American Chemical Society.)

but also to Fe–N_x sites, which are considered to promote ORR. The contribution of cobalt to the electrochemical properties of pyridine-rich PTCOFs was also studied theoretically by DFT calculations. It was revealed that CoNP-PTCOF delivered decreased ORR/OER overpotentials when compared to pristine PTCOF, thus they were directly implemented in a laboratory ZAB for practical elucidation of the results [23].

In 2018, Peng and his team [24] constructed a soluble 2D COF (denoted as COF_{BTC}) enriched with N-coordinated single Fe atom active sites to implement it in a Zinc-air flow battery. A year later, research was published by the same team, in which the direct mixing of COF_{BTC} with graphene is examined as an efficient pyrolysis-free procedure to construct effective electrocatalysts for liquid ZABs. The *pf* SAC-Fe catalysts had determined Fe-N-C active sites contrary to other conventionally made TM decorated carbons, while furthermore maintaining an exceptional metal-like electrical conductivity due to the intramolecular bonds between the compounds of the composite. When kinetically investigated, the ORR activity of the prepared catalyst outperformed a precious Pt-C electrode, and TEM and HAADF-STEM analyses revealed the existence of Fe atoms still incorporated evenly into the graphene surface. The material had also exceptional cycling stability, exhibiting a negligible voltage gap extension after 300 h of cycling, when applied in an in situ manufactured ZAB. This enhanced behavior is attributed to the superior electrical conductivity of

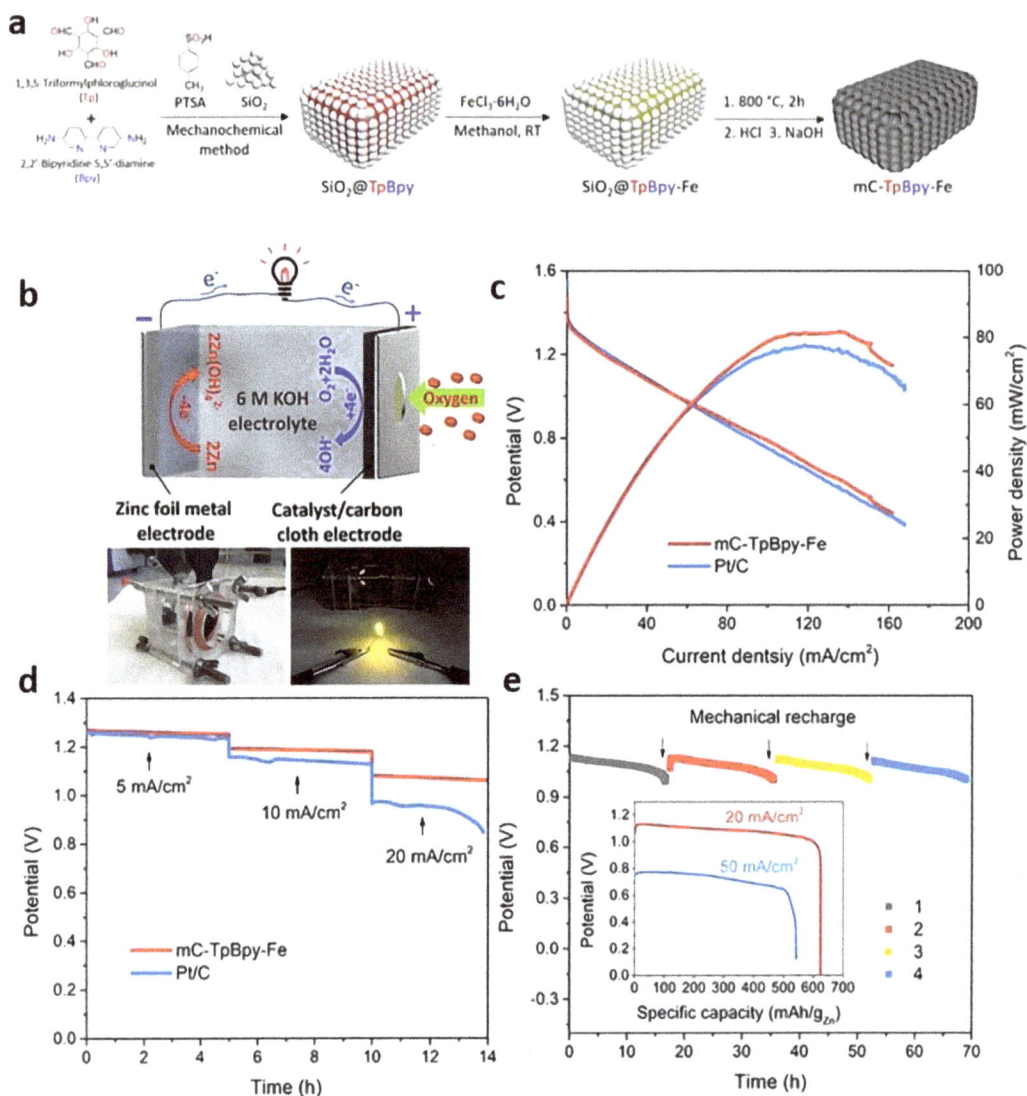

FIGURE 15.3

(a) Synthesis of mC-TpBpy-Fe via PTSA-assisted mechanochemical method, in the presence of silica nanoparticles. (b) Schematic diagram of a Zn–air battery and photographs of a lab-made Zn–air battery with a zinc foil as the anode, mC-TpBpy-Fe loaded on carbon cloth as the cathode (left), and a LED connected to two batteries (right). Comparison of (c) polarization and power density curves of primary Zn–air batteries using mC-TpBpy-Fe and Pt/C as ORR catalysts and (d) discharge curves of the primary Zn–air batteries using mC-TpBpy-Fe and Pt/C as ORR catalysts at various current densities. (e) Long-time durability of a primary Zn–air battery using mC-TpBpy-Fe at 20 mA/cm². Inset: Specific capacities of the Zn–air batteries using mC-TpBpy-Fe normalized to the mass of the consumed Zn at various current densities. (Adapted with permission [22]. Copyright (2021) American Chemical Society.)

the heterostructure and abundantly available oxygen adsorption sites [25]. The combination of graphene with a metal-enriched organic framework to boost fast electron transfer has been also cited by Li et al. who constructed a catalyst denoted as G@POF-Co that exhibited a potential difference between ORR and OER activity of ΔE = 0.85 V in alkaline solution [26]. Later the same team published a work, where the above catalyst was applied in an in situ prepared ZAB system. The contribution of graphene is again highlighted to aid the uniform distribution of small NPs contributing to the larger active surface area, resulting in an air electrode exhibiting promising robustness, as compared to conventional Pt/C and Ir/C cathodes. Particularly, the hybrid displayed a finer Voltage Profile compared to the noble electrodes for both charging and discharging of the battery. When galvanostatically cycling, the ZAB with the Co-G@POF exhibited a critically reduced overpotential regarding OER for 25 short-term cycles [27].

Another carbon architecture that has been explored as a matrix has been carbon nanotubes (CNTs), which are used to enhance conductivity and also regulate the morphology and distribution of catalytic sites. The idea of CNTs composites combined with porphyrin-based heterogeneous catalysts applied as both ORR and OER electrocatalysts can be firstly found in 2017 when a catalyst denoted as PCN-224/MWCNT was synthesized. The ORR/OER electrocatalyst is comprised of cobalt porphyrins attached to MOF cubes, which are further enhanced with the incorporation of multi-walled carbon nanotubes (MWCNTs) and appeared to provide effective facilitation of oxygen reactions [28]. In 2018 Li et al. managed to construct a robust film heterostructure of a porphyrin covalent organic framework grown on CNTs (CNT@POF) and investigated its electrochemical properties in a ZAB. The authors highlighted the superiority of this heterostructure when compared to the performance of mechanically mixed corresponding COFs and CNTs (denoted as CNT + POF). The cathode displayed favorable hydrophilicity and mechanical durability that attributed corrosion resistivity to both a liquid and a flexible rechargeable ZAB. Particularly, the robust construction of this hybrid led to a decisive reduction of the Voltage Gap during the lifetime of the aqueous battery, when galvanostatically cycled at 2 mA/cm² (CNTs@POF: 1.25 V-1.85 V, CNTs + POF: 1.2 V – 2.1 V). Polarization curves also revealed a significant difference in polarization between the prepared catalyst (CNTs@POF catalyst attained 200 mA/cm² with a voltage higher than 1 V) and a noble bifunctional electrode (Pt/C + Ir/C) [29].

Two years later a similar morphology was developed by uniformly coating TP-BPY-COFs on CNTs to subsequently stabilize Ni/Fe clusters on the surface, as seen in Figure 15.4 [30]. The COF serves a critical role in the middle of this heterostructure, as it provides the electrocatalyst with abundant oxygen defects, which promote diffusion mechanisms and works as an immobilization matrix for fine distribution of bimetal Ni/Fe sites. When the Ni/Fe-COF@CNT900 catalyst was evaluated toward ORR and OER it exhibited promising results compared to Pt/C and RuO_2 electrodes ($\Delta E = 0.68$ V) and was indeed successfully durable as an air electrode. When compared again with the above conventional electrodes, the electrocatalyst delivered a far more stable Voltage Gap during galvanostatic cycles at 5 mA cm⁻².

The same year Liu et al. constructed a COF@CNT hybrid (2,4,6-Trihydroxybenzene-1,3,5-tricarbaldehyde-Diaminobenzidine (TPDAB)-Co@CNTs) via an ultrafast coprecipitation procedure and examined its properties as an oxygen bifunctional electrocatalyst [31]. The material possesses inherently a BET surface area of 381 m²/g with favorable mesoporous architecture that promotes accelerated ion kinetics. Moreover, the surface offers enhanced electron transfer kinetics due to the presence of CNTs and plentiful exposed $Co-N_2-O_2$ salophen active sites, which are considered to serve as functional ORR areas. These features contributed so that the electrocatalyst outperformed a noble electrode of Pt/C when evaluated in a secondary ZAB system. The plateau of the charging & discharging Voltage of the cell was not extended after 48 h of continuous cycling at a relatively high current density of 10 mA/cm².

Very recently, Liu et al. constructed a COF-CNT hybrid (denoted as CC-3) by "wrapping" a thienothiophene-containing (TAPTt-COF) around MWCNTs forming one-dimensional

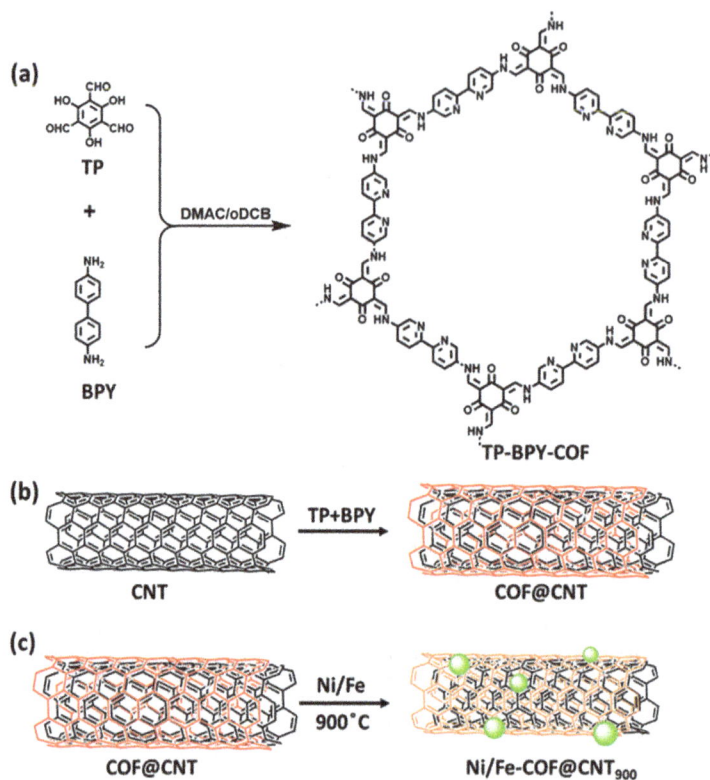

FIGURE 15.4
(a) Synthesis of TP-BPY-COF from TP and BPY monomers. (b) Synthesis of COF@CNT under the same condition as that of TP-BPY-COF. (c) Fabrication of Ni/Fe-COF@CNT900. (Adapted with permission [30]. Copyright (2020) Royal Society of Chemistry.)

van der Waals heterostructures with controllable thickness [32]. The C-S moieties of the COF function as catalytic centers for oxygen reactions, while the existence of CNTs restricts the junction of COF 2D nanosheets caused by π-electron interactions, which inevitably leads to the deactivation of surface catalytic sites. Furthermore, the CNTs may enhance electronic conductivity by creating new *n*-type electronic transfer pathways. Half-cell tests revealed a superior bifunctional performance of the composite ($\Delta E = 0.791$ V), while the material was subsequently applied in an experimental ZAB system. The cell could deliver a specific capacity of 714 mAhg^{-1} and 120 (30-minutes) cycles at a current density of 10 mA/cm. It can be highlighted that even after 120 cycles the Voltage Gap was limited below 1 V (~1.4 V-2.1 V). The overall performance of the ZAB can be found in Figure 15.5.

Except for conductive carbon matrixes, heteroatom doping of COF linkers has proved to enhance donor-acceptor properties and ORR without disturbing the mechanical stability of COFs. N-atoms are commonly known to trigger the neutrality of C atoms leading to a favorable polarization of catalysts. In the case of COFs N-doping can occur in a precise manner aiming at specific location control of N-molecules [33, 34].

A 3D metal-free aerogel catalyst denoted as S-C$_2$NA was recently prepared and implemented in both an aqueous and a solid-state ZAB [35]. The aerogels were prepared via an amination and polymerization process of chloroanilic acid followed by

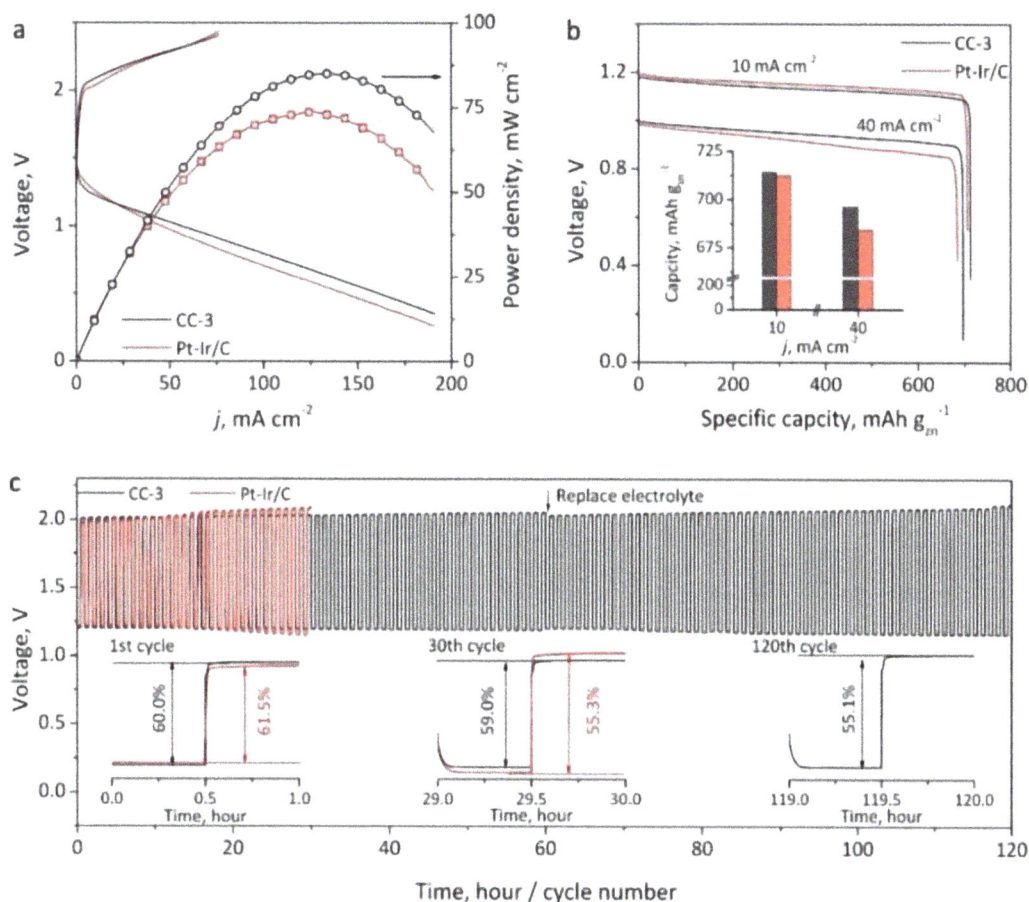

FIGURE 15.5
Performance of rechargeable zinc-air batteries assembled using CC-3 and Pt–Ir/C oxygen catalysts. (a) Galvanodynamic charge/discharge profiles and power density and (b) galvanostatic discharge curves. The inset shows the battery capacity under different discharge current densities. (c) Cycling profiles. The insets show cycling performances at the 1st, 30th, and 120th cycles. (Adapted with permission [32]. Copyright (2021) American Chemical Society.)

freeze-drying and a pyrolysis procedure. The aerogel architecture of these carbon-based and nitrogen-based nanostructured COF nanoribbons forms a highly porous network, abundant with ion transport channels and full of N-C active sites that facilitate O_2 adsorption due to the unsaturated pyridinic and graphitic N species. The electrocatalyst's performance in electrochemical characterization surpassed state-of-art electrodes by exhibiting a ΔE as low as 0.65 V. Furthermore, it was successfully applied in both aqueous and flexible ZAB systems delivering in both cases a Peak Power Density around 200 mW/cm². Particularly referring to the liquid system the electrocatalyst limited the Voltage Gap to lower than 0.8 V for about 375 long-term cycles (2h per cycle) at 10 mA/cm². The same catalyst was also implemented in a Li-Oxygen battery, whose principles and operation will be explained in the next part of the chapter. S-C$_2$NA could attain 200 cycles preserving prominent capacity retention even at the high current density of 50 mAh/g [35]. The performance metrics of COF-based materials applied in ZABs can be found in Table 15.1.

TABLE 15.1

COF-Based Electrocatalysts' Performance in Zinc-Air Batteries

Catalyst (Abbreviation)	Description	Half Cell Test Electrolyte/ Reference	ORR $E_{1/2}$ vs RHE	OER E_{j10} vs RHE	ΔE =($E_{j=10}$ −$E_{1/2}$) vs RHE	Catalyst loading (mg/cm²)	ZAB configuration (cathode support (GDL) / electrolyte / anode)	OCV (V)	Peak Power Density (mW/cm²)	Specific Discharge Capacity & Discharge Current Density (mAh/g⁻¹)	Cycling Life (cycles/min per cycle/ charge & discharge current density mAcm⁻²)	Reference
pfSAC-Fe-COF$_{BTC}$	fully closed conjugated iron phthalocyanine (FePc)–rich covalent organic framework (COF).	0.1 M KOH/ SCE	0.91	-	-	0.2/0.05	GDL/8 M KOH + 0.5 M ZnO/ Zinc foil	1.41	123.43	732/100	300 hr /5	[25]
CNT@POF	Porphyrin covalent organic framework (POF) coated onto a conductive long carbon nanotube (CNT)	-	-	-	-	1.50	Polytetrafluoroethylen (PTFE) treated Carbon layer/6 M KOH + 0.20 M ZnCl₂/Zinc Foil	-	237	772.7/20	200/5/2	[29]
mC-TpBpy-Fe	COF derived iron–nitrogen doped mesoporous carbon	0.1 M KOH Ag/AgCl	0.845	-	-	2	Carbon Cloth (CC)/6 M KOH/Zinc foil	1.5	81	625/20 543/50	-	[22]
TPDAB-Co@CNTs	1,3,5-triformylphloroglucinol and diaminobenzidine derived Co-salophen COF@CNTs	0.1 M KOH Ag/AgCl	0.89	-	-	2	Carbon Paper (CP)/6 M KOH + 0.2 M ZnAc₂/Zinc plate	~1.49	156	-	48 hr /10	[31]
S-C2NA	three-dimensional sulfur-modulated holey C2N aerogels	0.1 M KOH Ag/AgCl	0.88	1.53	0.65	0.5	CP/6 M KOH + 0.2 M ZnAc₂/Zinc plate	~1.5	209	825/25	375/2 hr/10	[35]
COF$_{BTC}$		0.1 M KOH Ag/AgCl	900 mV overpotential	-	-	-	CP/ 8 M KOH + 0.5 M ZnO/ Zinc plate	-	-	-	80 hr/10	[24]
Ni/Fe-COF@CNT900	COF-derived carbon-made catalyst	0.1 M KOH Ag/AgCl	0.87	1.55	0.68	1.0	CP/6 M KOH + 0.2 M ZnCl₂/ Zinc plate	-	105	-	200/-/10	[30]
TAPtt-COF-CNTs	one-dimensional van der Waals heterostructures comprised of 2D-COFs and CNTs	0.1 M KOH Hg/HgO	0.828	389	0.791	0.5	CC/6 M KOH +0.2 M ZnCl₂/Zinc foil	1.477	85	714/10	120/30/10	[32]

15.3 Li-Air Batteries/Li-CO$_2$ Batteries

The concept of Li-air Batteries differs slightly from conventional MABs and was firstly introduced in 1996 by Abraham and Jiang [36]. Contrarily to Zn, a Li cathode is not that stable and by reacting with compounds from the environment can easily undergo cracking, corrosion, and pulverization. As Lithium is explosively reactive with water too, much effort has been devoted to developing compatible and electrochemically stable non-aqueous electrolytes for Li-air batteries and strategies to prevent the contact of Li with other species [37]. The types of batteries that can be distinguished are non-aqueous, aqueous, hybrid, and solid-state batteries [4].

Li-air batteries have attracted commensurate interest, due to their extremely elevated energy density of 5200 Wh/kg [4] and operational voltage of around 2.9 V [36]. These batteries can often also be found as Lithium-Oxygen batteries, when pure oxygen is supplied instead of ambient air, to prevent external interventions [38]. The reason is that air contaminants such as CO_2 and H_2O can restrict the healthy operation of these MABs. For many years, multiple carbon structures have been extensively investigated as air electrodes, however, they proved to be fragile to decomposition due to intermediates (LiO_2, O_2^-) and products of the discharge reactions (Li_2O_2) or parasitic reactions ($LiOH$, Li_2CO_3) [36, 37, 39]. Particularly, the presence of atmospheric CO_2 favors the formation of Li_2CO_3 from Li_2O, which seriously restricts the reversibility and cycle life of the battery.

Toward facing those problems Takechi et al. [40] introduced a Li-O_2/CO_2 battery, which opened new possibilities for the concept of Li-air batteries. The reactions governing the cathode of these cells are the CO_2 reduction reaction (CRR) and CO_2 evolution reaction, instead of ORR and OER. Except for Shinde SS's work [35], there are no other researches detected to report successful COF performance as bifunctional electrocatalysts in conventional Li-air batteries, contrarily to Li-CO$_2$ batteries, where many scientists attempted to construct COF-based materials to facilitate the CRR/CER kinetics [41]. The most critical consideration, when designing catalysts for Li-CO$_2$ battery cathodes, is that the main operation product in the cathode, Li_2CO_3 is found challenging to "break down" in the charging process, which leads to blocked active surface sites from its deposition. In time, this phenomenon restricts fast CO_2 transfer and leads to extreme destructive polarization. Likewise, ZABs, the extent of undesirable polarization can be observed by the departure of discharging and charging voltage from the OCV and has to be restricted by the utilization of multifunctional catalysts.

The majority of electrocatalysts investigated for Li-CO$_2$ batteries have been carbonaceous materials, such as graphene [42] and CNTs [43, 44], and noble metal hybrids containing Ru [45], Au [46], or IrO_2 [47]. However, similar results can be obtained by transition metal-based electrocatalysts, which sometimes are more accessible and cost-effective [48]. Probably the first attempt to introduce a polymeric COF-similar material into a Li-CO$_2$ cell was conducted by Huang et al., who prepared conjugated cobalt polyphthalocyanine (CoPPc) networks via a facile microwave heating procedure. Specific attention was oriented to the catalytic activity of the material concerning the decomposition of Li_2CO_3, which was verified through tests separately from the battery performance. Although the subsequently fabricated battery delivered promising voltage profiles at high current densities (0.05 to 0.25 mA/cm^2) it exhibited limited performance due to extended polarizations probably deriving from severe Li_2CO_3 passivation. In addition, the authors implemented the electrocatalyst in a flexible Li-Co$_2$ battery emphasizing its elasticity and tolerance [49]. One year later, Li et al. constructed a composite combining a hydrazide/hydrazone COF (Tf–DhzOPr) with Ru@CNTs to form a corresponding air electrode electrocatalyst [50]. In this case, the COF part acted as a robust support ambient with reactants' transfer pathways, which prevented early clogging of cathodes' pores, thus limiting the undesirable polarization of the battery, while in reality, Ru molecules performed as active sites for both CO_2 reduction and Li_2CO_3/C decomposition. The hydrazone COF (Tf–DHzOPr) was prepared by condensation of benzene-1,3,5-tricarboxaldehyde (Tf) and 2,5-dipropoxyterephthalohydrazide (DHzOPr) and formed a homogenous hybrid by covering ultimately the Ru@CNTs NPs. Finally, the cell exhibited an initial capacity of 27,348 mAh/g that could be fully recharged at a current density of 200 mA/g and could attain 200 cycles at 1 A/g with a limiting capacity of 1000 mAh/g.

The same year an imine COF was prepared with more effective CO_2 capture properties aiming to be able to incorporate nanosized CO_2 molecules. In this way, when the material would be applied in a Li-CO_2 battery, small CO_2 particles that were captured, when the cell is discharging, would transform into corresponding small $LiCO_3$ particles that could be decomposed more easily. The COF was grown onto graphene leading to a hybrid that preserved the electrical conductivity of graphene while having enhanced porosity and a large active surface area. The corresponding Li-CO_2 cell exhibited superior durability for 56 cycles at 0.5 A/g, which was elucidated by the decreased overpotentials when compared to graphene stand-alone (Figure 15.6) [51]. Nevertheless, the reversible capacities of the battery were preserved only for 26 galvanostatic cycles.

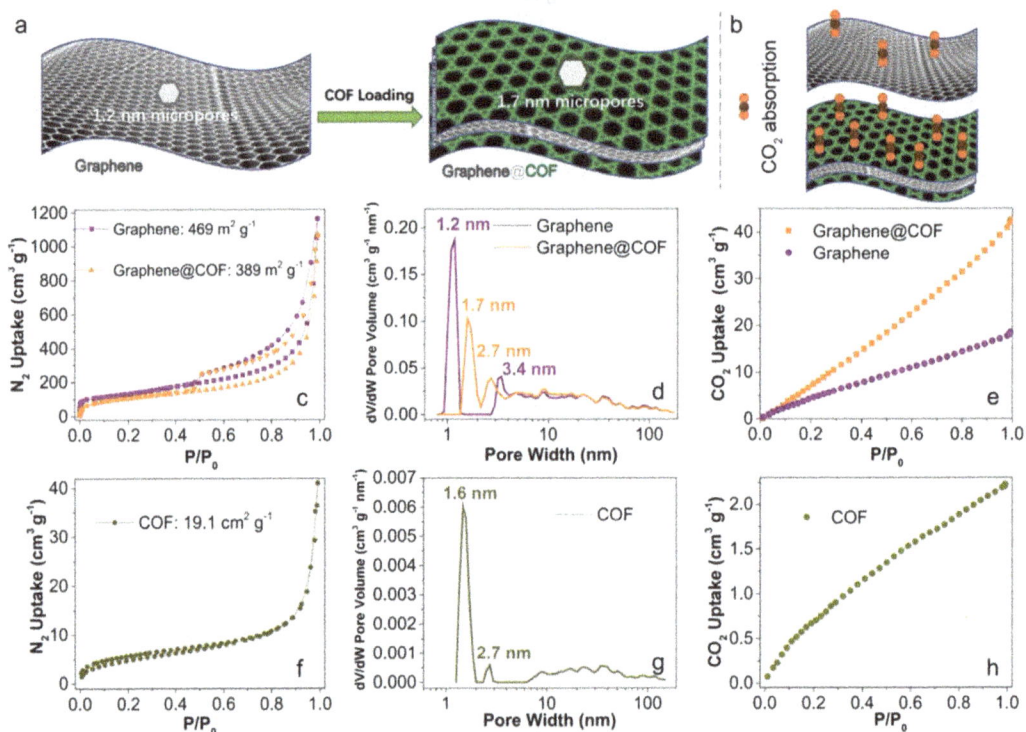

FIGURE 15.6

Top: Scheme of (a) micropore structure and (b) CO_2 absorption ability before and after COF loading onto graphene. Bottom: Electrochemical performance. Electrochemical impedance spectra of (a) graphene battery and (b) graphene@COF battery. (c) OCV changes during 65 h standing. (d) Scheme for CO_2 shuttle effect inhibited by graphene@COF cathode. (e) The scheme for the CO_2 shuttle effect is connived by graphene cathode. (f) Discharge voltage curves of CNTs, graphene, and graphene@COF at 75 mA/g with a cut-off voltage of 1.8 V. (g) The rate capability of the graphene battery and graphene@COF battery at different current densities. (h) Long-term cycling performance of the graphene battery and graphene@COF battery at the current of 0.5 A/g. (Adapted with permission [51]. Copyright (2019) WILEY-VCH.)

TABLE 15.2

COF-Based Electrocatalysts' Performance in Li-CO$_2$ Batteries

Catalyst (Abbreviation)	Description	Catalyst Loading (mg/cm^2)	Li-CO$_2$ Battery Configuration (Cathode Support/Electrolyte/Anode)	Specific Discharge Capacity & Discharge Current Density (mAh/g^1 & mA/g^1)	(Cycles/Current Density(A/g)/Limiting Capacity (mAh/g^1)	Reference
CoPPc		1	CC / 1 M lithium bis(trifluoromethanesulfonyl) imide (LITFSI) in tetraethylene glycol dimethyl ether (TEGDME) /Li foil	13.6 mAh/cm^2	50/0.25 mA/cm^2	[49]
COF@Ru@CNTs	hydrazone COF and Ru nanoparticle-decorated carbon nanotube (Ru@CNT)	0.3	CP / 1 M LiTFSI in TEGDME / Li foil	27'348 / 200	200/ 1 /1000	[50]
graphene@COF		0.15	Ni foam/1 M LiTFSI in TEGDME / Li foil	27' 833/75	56/0.5/-	[51]
TTCOF-Mn	porphyrin-based covalent organic framework single Mn metal sites	0.1/ 0.13	CP/1 M LiTFSI in TEGDME/Li metal tablet	13'018 / 100	180/300/1000	[52]
MnO$_2$/ DQTP-COF-NS	quinone-COF-NSs with MnO2	0.1-0.15	CP/1 M LiTFSI in TEGDME/Li foil	42'802 / 200	120/1/-	[53]

Very recently, Zhang et al. followed a typical strategy to boost the catalytic activity of COF materials, by integrating TM molecules into COF networks [52]. Particularly they exploited a tetrakis(4-aminophenyl)-porphinato manganese(II) (TAPP-Mn) precursor to obtain an Mn-rich porphyrin-based covalent organic framework (TTCOF-Mn) which besides fine porosity and surface area exhibits favorable properties of powerful CO_2 adsorption. The initial capacity of the cell was 13'018 mAh/g while cycle life was examined through 180 cycles at 300 mA/g revealing declined overpotentials. The reaction pathways and mechanisms for CO_2 reduction were carefully examined through DFT calculations which displayed that Mn single metal sites could promote a four-electron accelerated pathway contrarily to Co, Ni, or Cu. Some months later, research was published by Jiang et al. who also combined the advantageous properties of transition metals and COFs. In that case, MnO_2 particles were used to decorate 2,6-diaminoanthraquinone-2,4,6-triformylphloro-glucinol (DQTP)-COF nanosheets. The nanosheets were rationally created by chemical exfoliation in order to reach a thickness as low as 1.87 nm that would support greater specific capacities. Physical examination of the MnO_2/TpPa-COF-NS cathode in between several stages of battery tests reveals the effective decomposition of $LiCO_3$ products which seriously aids the long cycling life of the cell. Particularly, the voltage profile of the cell was relatively stable for even 120 cycles at a surprisingly high current density of 1000 mA/g. Furthermore, charging and discharging overpotentials were carefully observed for multiple current densities (200, 500, and 1000 mA/g) and revealed a superior profile [53]. The performance metrics of COF-based materials applied in Li-Air Batteries can be found in Table 15.2.

15.4 Conclusions

The application of COFs in electrocatalysis has set the ground for their recent implementation in some Metal-Air Battery systems. Such materials are found to offer broad prospects owning to the tunable chemical composition that allows them to obtain different geometries, porosity, and functional moieties, thus electrochemical properties. Moreover, COFs prove to be rather flexible when hybridized with other compounds such as transition metals or nano-sized carbons to form multifunctional heterostructures with more powerful electronic and ionic conductivity. More specifically, COFs can also constitute appropriate porous matrixes for rationally designed TM molecules immobilization instead of just decorating carbonaceous nanoarchitectures.

Yet, to date, pristine COFs have not been widely explored in such applications, as very few research studies can now be found to report COFs located in MABs. Zinc-Air and Li-CO_2 batteries are dominating the reports of COFs applied in MABs. The reported catalysts exhibited promising features and gave encouraging results that were very well elucidated by physical examination and DFT theoretical calculations. A huge advantage of COF materials is that a connection between synthesis procedures and the final product's properties is quite approachable. However, due to the lack of extended reporting of COFs catalysts in this research field, it is not attainable to come to confident assumptions regarding the COFs ability to give a solution to all unresolved ORR/OER or CRR/CER restrictions. Certainly, it is considered unrealistic to depend on a single-phase substrate to tackle all reaction kinetic related challenges in batteries, thus further research is highly suggested to be conducted, directed to ensure firstly repeatability of previous results and afterward explore more possibilities on how to reasonably merge COF materials with other functional substances to achieve widely applicable electrode materials.

References

1. Linden D, Reddy T (2001) Handbook of Batteries: McGraw-Hill Education.
2. Wang H-F, Xu Q (2019) Materials design for rechargeable metal-air batteries. Matter 1 (3):565–595.
3. European Commission (National Energy and Climate Plans: Member State contributions to the EU's 2030 climate ambition, September 2020).
4. Lee J-S, Tai Kim S, Cao R, Choi N-S, Liu M, Lee KT, Cho J (2011) Metal–air batteries with high energy density: Li–Air versus Zn–Air. Adv. Energy Mater. 1 (1):34–50.
5. Liu Q, Pan Z, Wang E, An L, Sun G (2020) Aqueous metal-air batteries: Fundamentals and applications. Energy Storage Mater. 27:478–505.
6. Seh ZW, Kibsgaard J, Dickens CF, Chorkendorff I, Nørskov JK, Jaramillo TF (2017) Combining theory and experiment in electrocatalysis: Insights into materials design. Science 355 (6321):4998.
7. Pan J, Xu YY, Yang H, Dong Z, Liu H, Xia BY (2018) Advanced architectures and relatives of air electrodes in Zn–Air batteries. Adv. Sci. 5 (4):1700691.
8. Wang R, Chen Z, Hu N, Xu C, Shen Z, Liu J (2018) Nanocarbon-based electrocatalysts for rechargeable aqueous Li/Zn-Air batteries. ChemElectroChem 5 (14):1745–1763.
9. Zhang W, Liu Y, Zhang L, Chen J (2019) Recent advances in isolated single-atom catalysts for zinc air batteries: A focus review. Nanomaterials 9:1402.
10. Yi J, Liu X, Liang P, Wu K, Xu J, Liu Y, Zhang J (2018) Non-noble iron group (Fe, Co, Ni)-based oxide electrocatalysts for aqueous Zinc–Air batteries: Recent progress, challenges, and perspectives. Organometallics 38:1186–1199.
11. Vaitsis C, Mechili M, Argirusis N, Kanellou E, Pandis Pavlos K, Sourkouni G, Zorpas A, Argirusis C. 2020. Ultrasound-assisted preparation methods of nanoparticles for energy-related applications. In *Nanotechnology and the Environment*: IntechOpen.
12. Pan Z, Wang X, Yang J, Qiu Y, Xu S, Lu Y, Huang Q, Li W (2019) Hierarchical Co3O4 nano-micro arrays featuring superior activity as cathode in a flexible and rechargeable Zinc–Air battery. Adv. Sci. 6:1802243.
13. Wang X, Ge L, Lu Q, Dai J, Guan D, Ran R, Weng S-C, Hu Z, Zhou W, Shao Z (2020) High-performance metal-organic framework-perovskite hybrid as an important component of the air-electrode for rechargeable Zn-Air battery. J. Power Sources 468:228377.
14. Vayenas M, Vaitsis C, Sourkouni G, Pandis PK, Argirusis C (2019) Investigation of alternative materials as bifunctional catalysts for electrochemical applications. Chimica Techno Acta 6 (4):120–129.
15. Mechili M, Vaitsis C, Argirusis N, Pandis PK, Sourkouni G, Argirusis C (2022) Research progress in transition metal oxide based bifunctional electrocatalysts for aqueous electrically rechargeable zinc-air batteries. Renew. Sustain. Energy Rev. 156:111970.
16. Yi J, Liang P, Liu X, Wu K, Liu Y, Wang Y, Xia Y, Zhang J (2018) Challenges, mitigation strategies, and perspectives in development of zinc-electrode materials and fabrication for rechargeable zinc–air batteries. Energy Environ Sci. 11 (11):3075–3095.
17. Li Y, Dai H (2014) Recent advances in zinc–air batteries. Chem. Soc. Rev. 43 (15):5257–5275.
18. Lin C-Y, Zhang L, Zhao Z, Xia Z (2017) Design principles for covalent organic frameworks as efficient electrocatalysts in clean energy conversion and green oxidizer production. Adv. Mater. 29:1606635.
19. Aiyappa HB, Thote J, Shinde DB, Banerjee R, Kurungot S (2016) Cobalt-modified covalent organic framework as a robust water oxidation electrocatalyst. Chem. Mater. 28 (12):4375–4379.

20. Wu D, Xu Q, Qian J, Li X, Sun Y (2019) Bimetallic covalent organic frameworks for constructing multifunctional electrocatalyst. Chem. Eur. J. 25 (12):3105–3111.
21. Zhao X, Pachfule P, Li S, Langenhahn T, Ye M, Schlesiger C, Praetz S, Schmidt J, Thomas A (2019) Macro/microporous covalent organic frameworks for efficient electrocatalysis. J. Am. Chem. Soc. 141 (16):6623–6630.
22. Zhao X, Pachfule P, Li S, Langenhahn T, Ye M, Tian G, Schmidt J, Thomas A (2019) Silica-templated covalent organic framework-derived Fe–N-doped mesoporous carbon as oxygen reduction electrocatalyst. Chem. Mater. 31 (9):3274–3280.
23. Park JH, Lee CH, Ju J-M, Lee J-H, Seol J, Lee SU, Kim J-H (2021) Bifunctional covalent organic framework-derived electrocatalysts with modulated p-band centers for rechargeable Zn–Air batteries. Adv. Funct. Mater. 31 (25):2101727.
24. Peng P, Shi L, Huo F, Zhang S, Mi C, Cheng Y, Xiang Z (2019) In situ charge exfoliated soluble covalent organic framework directly used for Zn–Air flow battery. ACS Nano 13 (1):878–884.
25. Peng P, Shi L, Huo F, Mi C, Wu X, Zhang S, Xiang Z (2019) A pyrolysis-free path toward superiorly catalytic nitrogen-coordinated single atom. Sci Adv 5 (8):eaaw2322.
26. Li B-Q, Zhang S-Y, Chen X, Chen C-Y, Xia Z-J, Zhang Q (2019) One-pot synthesis of framework porphyrin materials and their applications in bifunctional oxygen electrocatalysis. Adv. Funct. Mater. 29 (29):1901301.
27. Li B-Q, Zhao C-X, Chen S, Liu J-N, Chen X, Song L, Zhang Q (2019) Framework-porphyrin-derived single-atom bifunctional oxygen electrocatalysts and their applications in Zn–Air batteries. Adv. Mater. 31 (19):1900592.
28. Sohrabi S, Dehghanpour S, Ghalkhani M (2018) A cobalt porphyrin-based metal organic framework/multi-walled carbon nanotube composite electrocatalyst for oxygen reduction and evolution reactions. J. Mater. Sci. 53:3624–3639.
29. Li B-Q, Zhang S-Y, Wang B, Xia Z-J, Tang C, Zhang Q (2018) A porphyrin covalent organic framework cathode for flexible Zn–air batteries. Energy Environ Sci. 11 (7):1723–1729.
30. Qing X, Qian J, Luo D, Liu G, Guo Y, Zeng G (2020) Ni/Fe clusters and nanoparticles confined by covalent organic framework derived carbon as highly active catalysts toward oxygen reduction reaction and oxygen evolution reaction. Adv. Sustain. Syst. 4:2000115.
31. Liu J, Cheng T, Jiang L, Zhang H, Shan Y, Kong A (2020) Efficient nitrate and oxygen electroreduction over pyrolysis-free mesoporous covalent Co-salophen coordination frameworks on carbon nanotubes. Electrochim. Acta 363:137280.
32. Liu C, Liu F, Li H, Chen J, Fei J, Yu Z, Yuan Z, Wang C, Zheng H, Liu Z, Xu M, Henkelman G, Wei L, Chen Y (2021) One-dimensional van der Waals heterostructures as efficient metal-free oxygen electrocatalysts. ACS Nano 15 (2):3309–3319.
33. Xiang Z, Cao D, Huang L, Shui J, Wang M, Dai L (2014) Nitrogen-doped holey graphitic carbon from 2D covalent organic polymers for oxygen reduction. Adv. Mater. 26 (20):3315–3320.
34. Jiang T, Jiang W, Li Y, Xu Y, Zhao M, Deng M, Wang Y (2021) Facile regulation of porous N-doped carbon-based catalysts from covalent organic frameworks nanospheres for highly-efficient oxygen reduction reaction. Carbon 180:92–100.
35. Shinde SS, Lee CH, Yu J-Y, Kim D-H, Lee SU, Lee J-H (2018) Hierarchically designed 3D holey C2N aerogels as bifunctional oxygen electrodes for flexible and rechargeable Zn-Air batteries. ACS Nano 12 (1):596–608.
36. Imanishi N, Yamamoto O (2014) Rechargeable lithium–air batteries: characteristics and prospects. Mater. Today 17 (1):24–30.
37. Chen K, Yang D-Y, Huang G, Zhang X-B (2021) Lithium–air batteries: Air-electrochemistry and anode stabilization. Acc. Chem. Res. 54 (3):632–641.
38. Geng D, Ding N, Hor TSA, Chien SW, Liu Z, Wuu D, Sun X, Zong Y (2016) From lithium-oxygen to lithium-air batteries: Challenges and opportunities. Adv. Energy Mater. 6 (9):1502164.
39. Woo H, Kang J, Kim J, Kim C, Nam S, Park B (2016) Development of carbon-based cathodes for Li-air batteries: Present and future. Electron. Mater. Lett. 12 (5):551–567.
40. Takechi K, Shiga T, Asaoka T (2011) A Li–O2/CO2 battery. Chem. Commun. 47 (12):3463–3465.

41. Jiao Y, Qin J, Sari HMK, Li D, Li X, Sun X (2021) Recent progress and prospects of Li-CO2 batteries: Mechanisms, catalysts, and electrolytes. Energy Storage Mater. 34:148–170.

42. Jin Y, Hu C, Dai Q, Xiao Y, Lin Y, Connell JW, Chen F, Dai L (2018) High-performance Li-CO2 batteries based on metal-free carbon quantum dot/holey graphene composite catalysts. Adv. Funct. Mater. 28 (47):1804630.

43. Zhang X, Zhang Q, Zhang Z, Chen Y, Xie Z, Wei J, Zhou Z (2015) Rechargeable Li–CO2 batteries with carbon nanotubes as air cathodes. Chem. Commun. 51 (78):14636–14639.

44. Song L, Hu C, Xiao Y, He J, Lin Y, Connell JW, Dai L (2020) An ultra-long life, high-performance, flexible Li–CO2 battery based on multifunctional carbon electrocatalysts. Nano Energy 71:104595.

45. Qiao Y, Xu S, Liu Y, Dai J, Xie H, Yao Y, Mu X, Chen C, Kline DJ, Hitz EM, Liu B, Song J, He P, Zachariah MR, Hu L (2019) Transient, in situ synthesis of ultrafine ruthenium nanoparticles for a high-rate Li–CO2 battery. Energy Environ Sci. 12 (3):1100–1107.

46. Kong Y, Gong H, Song L, Jiang C, Wang T, He J (2021) Nano-sized au particle-modified carbon nanotubes as an effective and stable cathode for Li–CO2 batteries. Eur. J. Inorg. Chem. 2021 (6):590–596.

47. Wu G, Li X, Zhang Z, Dong P, Xu M, Peng H, Zeng X, Zhang Y, Liao S (2020) Design of ultralong-life Li–CO2 batteries with IrO2 nanoparticles highly dispersed on nitrogen-doped carbon nanotubes. J. Mater. Chem. A 8 (7):3763–3770.

48. Sun X, Hou Z, He P, Zhou H (2021) Recent advances in rechargeable Li–CO2 batteries. Energy & Fuels 35 (11):9165–9186.

49. Chen J, Zou K, Ding P, Deng J, Zha C, Hu Y, Zhao X, Wu J, Fan J, Li Y (2019) Conjugated cobalt polyphthalocyanine as the elastic and reprocessable catalyst for flexible Li–CO2 batteries. Adv. Mater. 31 (2):1805484.

50. Li X, Wang H, Chen Z, Xu H-S, Yu W, Liu C, Wang X, Zhang K, Xie K, Loh KP (2019) Covalent-organic-framework-based Li–CO2 batteries. Adv. Mater. 31 (48):1905879.

51. Huang S, Chen D, Meng C, Wang S, Ren S, Han D, Xiao M, Sun L, Meng Y (2019) CO2 nanoenrichment and nanoconfinement in cage of imine covalent organic frameworks for high-performance CO2 cathodes in Li-CO2 batteries. Small 15 (49):1904830.

52. Zhang Y, Zhong R-L, Lu M, Wang J-H, Jiang C, Gao G-K, Dong L-Z, Chen Y, Li S-L, Lan Y-Q (2021) Single metal site and versatile transfer channel merged into covalent organic frameworks facilitate high-performance Li-CO2 batteries. ACS Central Science 7 (1):175–182.

53. Jiang C, Zhang Y, Zhang M, Ma N-N, Gao G-K, Wang J-H, Zhang M-M, Chen Y, Li S-L, Lan Y-Q (2021) Exfoliation of covalent organic frameworks into MnO2-loaded ultrathin nanosheets as efficient cathode catalysts for Li-CO2 batteries. Cell Reports Physical Science 2 (4):100392.

16

Metal-Sulfur Batteries Based on Nanostructured Covalent Organic Frameworks

Yu Cao, Fusheng Pan, and Jie Sun

Key Laboratory for Green Chemical Technology of Ministry of Education, School of Chemical Engineering and Technology, Tianjin University, Tianjin, China

CONTENTS

16.1 Introduction

Lithium-ion battery (LIB) is the dominant energy storage equipment in recent years, which is more technologically mature and widely used compared with other state-of-art energy storage equipment. At present, the typical anode of LIB is graphite, and the cathode materials mainly include $LiCoO_2$, $LiNi_{1/3}Co_{1/3}Mn_{1/3}O_2$, $LiMn_2O_4$, and $LiFePO_4$, etc. However, the current LIB can only achieve low mass-specific energy (260 Wh kg^{-1}) and volumetric specific energy (780 Wh L^{-1}), which is not enough to meet the demand for commercial energy storage equipment [1–3]. Lithium-sulfur battery (LSB) has high theoretical mass-specific energy (2567 Wh kg^{-1}) and volume-specific energy (2199 Wh L^{-1}) and holds tremendous potential for the next-generation energy storage system [4]. Besides, the advantages of rich reserves, low cost, and no toxicity make LSB more suitable for the resource-conserving and environment-friendly society. The LSB was first reported in 1960, but the development of LSB was stagnant due to the poor cycling life. Until 2009, Nazar et al. have achieved

DOI: 10.1201/9781003206507-16

a breakthrough in the development of LSB by using mesoporous carbon CMK-3 as a host material for sulfur [5]. By melting sulfur into the conductive channel of CMK-3, the S/CMK-3 cathode delivered a high specific capacity and superb cycling stability. In the following ten years, the scientific research and practical applications of LSB have developed very rapidly. However, the low conductivity of sulfur, the polysulfides shuttle, and the dendrite growth at the anode still seriously impeding the progress of commercial LSB. With the development of materials science, porous materials exhibit a wide range of applications prospects, such as adsorption, catalysis, separation, and energy storage. The design and application of porous materials for LSB can effectively solve the above issues, due to the following merits: (1) porous structure can prevent the aggregation of sulfur, reduce the particle size and enhance the contact between the sulfur and conductive matrix, which is conducive to improving the utilization of active sulfur; (2) porous structure can inhibit the diffusion of polysulfides through the physical confinement effect contributing to the superb cycling stability of the LSB; (3) porous structure can provide abundant channels for lithium-ion conduction, which is conducive to reducing the polarization of the battery and improving the reaction kinetics; (4) porous structure can provide space for lithium deposition, alleviate the impact of volume change and induce a homogeneous ion flux distribution for inhibiting the dendrites growth. Typically, the classic porous materials include porous carbon, molecular sieve, polymers of intrinsic microporosity (PIMs), metal-organic frameworks (MOFs), conjugated microporous polymers (CMPs), porous aromatic frameworks (PAFs), hydrogen-bonded organic frameworks (HOFs), and COFs.

Among those materials, COFs, as an emerging type of crystalline organic polymers with the ordered molecular arrangement, are considered to be an advanced functional material for LSB. COFs combine the advantages of polymers and inorganic crystal materials. The polymer characteristics endow COFs with a tunable chemical structure for functionalization and the crystallization makes COFs possess ordered arrangements and structural stability. As shown in Figure 16.1, COFs have the advantages of intrinsically ordered pores, lightweight, abundant functional sites, and high structural stability. When applied to LSB, COFs are mainly used as cathode hosts, separators, lithium metal hosts, and artificial interphase.

1. As hosts of sulfur in the cathode, COFs with uniformly distributed nanopores can accommodate sulfur, avoid its agglomeration, provide the lithium-ion conduction channels and inhibit the polysulfides shuttling. Compared with traditional porous carbon materials, COFs have abundant polar groups and stronger interaction with polysulfides, which is more conducive to confining the diffusion of polysulfides. The π electron conjugate structure of COFs is beneficial for the conduction of electrons perpendicular to the axially stacked pores and promoting charge transfer. However, the electronic conductivity of COF is still limited, which is unfavorable for high-rate performance. This problem is usually solved by compounding COF with other conductive material.

2. As separator materials, COF is usually coated on the surface of the polyolefin separator as a functional layer. The COF layer increases the electrolyte wettability of the separator, which is conducive to the transport of lithium-ion. Besides, the nanopores of COF can block the shuttle of polysulfides, and the large specific surface enables COF to have a large number of active sites for polysulfide adsorption, thus effectively avoiding the side reaction between polysulfides and a lithium anode. However, the simultaneous realization of fast ionic conduction

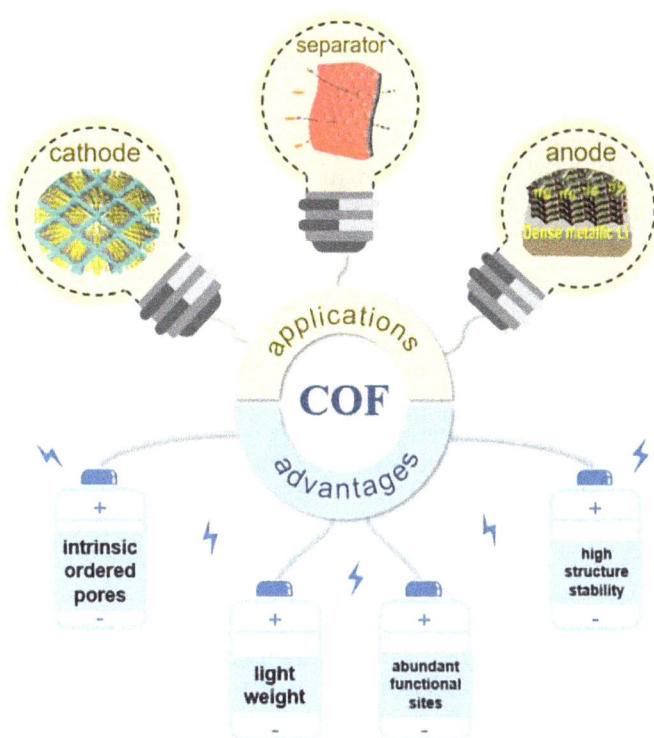

FIGURE 16.1
Advantages of COF and its application in Li-S batteries.

and polysulfide blocking is still extremely difficult. Thus, constructing the ion-selective conducting sites in the nanochannels of COF for fast lithium-ion conduction and polysulfides repulsion can be a promising strategy to improve the electrochemical performance of LSB.

3. As hosts in lithium metal anode, COF can accommodate lithium metal or form artificial solid electrolyte interphase (SEI) to stabilize the interface of lithium metal anode. The uniform pore distribution, abundant lithiophilic sites, and rigid framework of COF favor the development of high-performance lithium metal anode. The pore structure can not only accommodate lithium metal but also alleviate the electrode pulverization caused by the volume expansion of lithium. Besides, the pore structure can indue the homogeneous ion flux and effectively alleviate the dendrite growth, and the rigid structure of COF can further resist the dendrite growth. Furthermore, the COF with abundant lithiophilic sites can reduce the overpotential of lithium-ion deposition. However, the application of COF in lithium metal anode is still limited by the lack of a suitable method for preparing uniform and stable artificial SEI.

This chapter systematically summarized the recent literature on using COFs for LSB and discussed the application prospects. The structure-activity relationship of COF in LSB was also thoroughly analyzed based on the structure and properties of COF.

16.2 Fundamentals of Metal-Sulfur Battery

The LSB is mainly composed of lithium metal anode, sulfur cathode, separator, and electrolyte. The battery structure is shown in Figure 16.2a. A sulfur cathode is generally prepared by mixing sulfur, conductive carbon, and binder on the current collector. Sulfur is served as an active material to provide battery capacity. The theoretical capacity of sulfur is up to 1672 mA h g^{-1}. The conductive carbon functions as an electron conductor. When the sulfur is tightly in contact with conductive carbon, the electrons can be transferred to sulfur for the electrochemical reaction. The role of the binder is to combine the sulfur with conductive carbon and the current collector [6]. The anode of conventional LSB is lithium metal, which has an extremely high theoretical specific capacity (3860 mA h g^{-1}) and the most negative potential (- 3.040 V) [7]. The electrolyte can be generally divided into the liquid electrolyte and solid electrolyte, which is mainly used to transfer ions between the cathode and anode. The typical electrolyte is lithium trifluoromethylsulfonimide (LiTFSI) salt dissolved in ethylene glycol dimethyl ether (DME) and 1,3-dioxolane (DOL) with lithium nitrate additive. The addition of lithium nitrate is beneficial to stabilize the interface of lithium metal anode. Solid electrolytes are generally divided into polymer solid electrolytes and inorganic solid electrolytes, such as PEO, Li_2S-SiS_2, Li_2S-P_2S_5, etc [8]. The separator is generally polyolefin material, which is mainly divided into polypropylene (PP) separator, polyethylene (PE) separator, and their composite separator. The functions of the separator are conducting ions and blocking electrons between the cathode and anode.

16.2.1 Discharge and Charge Process

The charge/discharge process of LSB is based on the multi-electron multiphase conversion reaction (Figure 16.2a). In the discharge process, the lithium metal anode loses electrons and generates lithium ions. The electrons transfer to the sulfur cathode through the external circuit and the lithium ions migrate to the sulfur cathode through the electrolyte. When the sulfur receives the electrons and lithium ions from the anode, the sulfur first

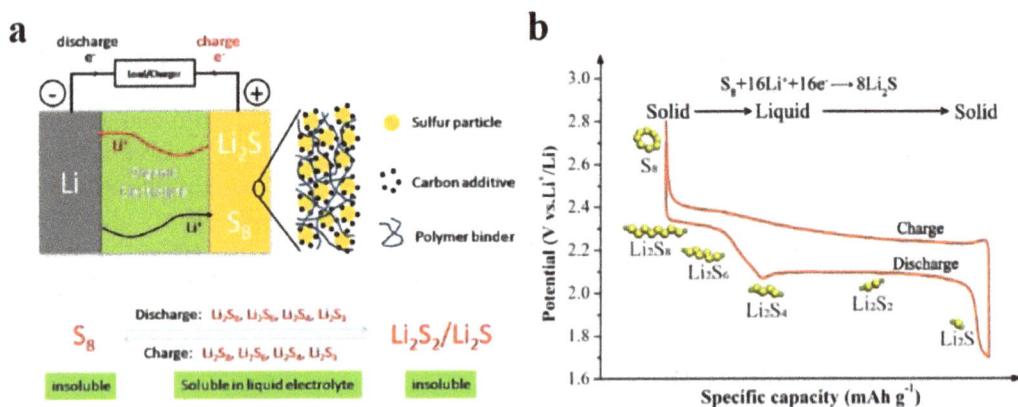

FIGURE 16.2

(a) Illustration of the Li-S cell and charge/discharge process (Adapted with permission from [9]. Copyright (2012) American Chemical Society). (b) Charge and discharge voltage curves of Li-S cell. (Adapted with permission from [10]. Copyright (2017) WILEY-VCH.)

converts to soluble polysulfides, and the polysulfides are further transformed into solid Li_2S_2/Li_2S deposited on the cathode. The charging process is the reverse reaction of the discharge process. A variety of intermediate products will be generated during the discharge process, and the intermediate products correspond to a different voltage. As shown in Figure 16.2b, the discharge process is generally divided into two potential platforms corresponding to the four electrochemical processes. The first process is a solid-liquid two-phase reduction reaction: $S_8 + Li \rightarrow Li_2S_8$, the corresponding potential is 2.2–2.3 V. The Li_2S_8 generated in this process will be dissolved in the electrolyte; the second process is a liquid phase conversion reaction: $Li_2S_8 + Li \rightarrow Li_2S_n$ (2<n<8). This process is corresponding to the long-chain polysulfides converting to the short-chain polysulfides with the fracture of the S-S bond, and the viscosity of the electrolyte increases significantly; the third process is a liquid-solid two-phase reduction reaction: $Li_2S_n + Li \rightarrow Li_2S_2$, $Li_2S_n + Li \rightarrow Li_2S$, and the corresponding potential is 1.9–2.1 V. The solid Li_2S_2 and Li_2S were deposited on the cathode; the fourth process is the solid phase conversion reaction: $Li_2S_2 + Li \rightarrow Li_2S$. In this process, the reaction kinetic rate is relatively slow compared with the other discharge process because both reactants and products are solid phases with poor electronic conductivity [4].

16.2.2 Challenges of Metal-Sulfur Battery

Although the development of LSB has made great successes, some intrinsic physical and chemical properties of sulfur, such as low conductivity and soluble intermediates, lead to the capacity loss and slow reaction kinetics of sulfur cathode, which limited its practical application. The problems of the sulfur cathode can be summarized in the following three aspects [11]:

1. The conductivity of elemental sulfur (5×10^{-30} S cm^{-1} at room temperature) is very low, leading to the poor efficiency of active sulfur and sluggish electrochemical reaction kinetics;
2. The density of elemental sulfur and lithium sulfide is 2.03 g cm^{-3} and 1.67 g cm^{-3}, respectively. The density difference of the active sulfur and discharge product leads to a serious volume change, which results in the electrode structure pulverization and causes safety problems;
3. The polysulfides produced in the discharge process have solubility and diffusivity in the electrolyte, which leads to the shuttle effect and the rapid capacity decay.

The issues affecting the performance of LSB are also related to the lithium metal anode. The inhomogeneous lithium-ion deposition always leads to dendrite growth, which will penetrate the separator and cause the battery to short circuit leading to a safety problem. Besides, the electrode side reaction also leads to the continuous consumption of active lithium and electrolyte [11].

To promote the commercialization of LSB, Manthiram et al. summarized five objectives [12]:

1. The sulfur loading should be greater than 5 mg cm^{-2}. Higher sulfur loading increases the areal capacity of the cathode. Increasing the sulfur loading can maximize the specific energy of the cell because the increased fraction of active material offsets the "dead weight" present from the various inactive components like the current collectors and separator.

2. The carbon content should be less than 5%. To improve the conductivity of sulfur cathode, most works increased their carbon content to 50%, which sharply reduced the energy density.

3. The ratio of electrolyte to sulfur (E/S) should be less than 5 μL mg^{-1}. The electrolyte constitutes the largest weight fraction of a Li-S cell. Therefore, reducing the amount of electrolyte is the key factor to increase the energy density of the battery. In addition, it can also apparently reduce the cost of the battery.

4. The ratio of electrolyte to battery capacity (E/C) should be less than 5 μL (mA h)$^{-1}$. Reducing the amount of electrolyte will significantly reduce the dissolution of polysulfides in the electrolyte, however, decrease the sulfur utilization and lead to poor capacity.

5. The capacity ratio (N/P) of the negative electrode and positive electrode should be less than 5. Typically, the use of excessive lithium can offset the lithium loss caused by electrolyte decomposition and electrode reaction, but the excessive lithium will also reduce the energy density of the battery and cause safety problems.

16.2.3 Strategies Toward the Challenges of Lithium-Sulfur Batteries

Up to now, a large number of works have focused on the strategies for solving the problems of LSB and promoting the process of commercialization. These strategies mainly include the design of cathode, anode, electrolyte, and separator. The cathode design can effectively improve the cycling performance and rate performance of LSB. The addition of host materials with high electronic conductivity in the cathode can solve the problem of sluggish electrode reactions caused by the insulation of sulfur and lithium sulfide. Besides, the design of host material with a porous structure can alleviate the damage of volume expansion and inhibit the polysulfides dissolution. Generally, host materials can be divided into carbon-based materials, polymer materials, and polar inorganic materials. Carbon-based materials, such as carbon nanotube (CNT) [13], carbon nanofiber (CNF) [14], graphene [15], and mesoporous carbon (CMK-3) [5] have the advantages of high conductivity, low density, and good stability. However, due to the non-polarity of carbon materials, they have poor interaction with polar polysulfides and inferior effect on inhibiting shuttle effect. Heteroatom doping can enhance the interaction between carbon-based materials and polysulfides, such as nitrogen-doped graphene [16] and sulfur-doped mesoporous carbon [17]. Conductive polymers with high electronic conductivity have attracted wide interest. The abundant functional groups can adsorb polysulfides, and the flexible structure can inhibit the electrode damage caused by volume expansion. The conductive polymer used for LSB mainly includes polyaniline [18], polyacrylonitrile [19], polypyrrole [20], and polythiophene [21]. However, the limited conductivity of conductive polymer is not conducive to the rapid electrochemical reaction. Polar inorganic materials have strong interaction with polysulfides. Some of them can also catalyze the conversion of polysulfides, such as TiO$_2$ [22], MnO$_2$ [23], and CoS$_2$ [24]. However, the density of metal elements in most polar inorganic materials is generally large, which will reduce the energy density of the battery.

For the problem of lithium metal anode, the common strategy is to add the artificial SEI on the lithium metal surface. The artificial SEI can reduce the side reaction between the lithium metal and electrolyte and inhibit dendrite growth. The preparation methods of artificial SEI on lithium metal surfaces can be divided into three categories. The first method is to prepare artificial SEI based on a solid-gas two-phase interfacial reaction. For example, Li$_3$N thin layer can be synthesized by the reaction of nitrogen and lithium metal surface. Li$_3$N thin layer can not only block the reaction between the polysulfides and lithium metal but also provide high ionic conductivity, which is conducive to improving the electrochemical performance of the LSB [25]. The second method to synthesize artificial SEI is based on the solid-liquid two-phase interfacial reaction. The LiI thin layer is synthesized by using HIO$_3$ solution and lithium metal. The LiI thin layer can realize the rapid conduction of lithium-ion and improve the electrochemical performance of the battery [26]. The third method is to directly cover the solid material on the lithium metal surface to form artificial SEI. This method can be suitable for a wider range of materials, such as alloy material Li-Al alloy [27], inorganic salt LTST [28], hollow carbon ball [29], and Nafion [30].

Electrolyte directly affects the cycling life and safety of the battery. In liquid electrolytes, the serious shuttle effect results in the rapid attenuation of the battery capacity. The most common solvent is DOL and DME. DOL is conducive to the formation of stable SEI on lithium metal anode. DME has good solubility in polysulfides, which is conducive to the conversion reaction of polysulfides. Barchasz et al. [31] explored the effect of the content of oxygen in the solvent on the battery capacity and showed that the initial capacity of the battery increased with the enhancement of oxygen content. For example, the battery using polyethylene glycol dimethyl ether (PEGDME) with high oxygen content exhibits the best electrochemical performance. The common lithium salt for LSB is LiTFSI. LiTFSI has the advantages of high electrochemical stability and ionic conductivity. The viscosity of the electrolyte can be adjusted by changing the type and concentration of lithium salt. For example, adding lithium salt LiFSI to the electrolyte can reduce the viscosity of the electrolyte and improve the ionic conductivity at the same time [32]. The common additive for LSB is LiNO$_3$, which can generate stable SEI and stabilize the lithium metal anode [33]. In addition to a liquid electrolyte, a solid electrolyte is considered to be a more promising electrolyte. Due to the high safety and almost no polysulfide dissolution, the solid electrolyte has attracted many interests. However, its low ionic conductivity and poor compatibility with the electrode surface limit its commercial application.

The basic function of the LSB separator is to prevent direct contact between cathode and anode for avoiding the short circuit. Besides, the separator can also allow the conduction of lithium-ion. The traditional polyolefin separator is generally a polyolefin material composed of a pore with a size of around 100 nm. Polysulfides produced in the reaction process of the sulfur cathode can diffuse across the separator and induce a side reaction with the lithium metal anode, resulting in serious attenuation of battery capacity. The above problems can be effectively solved by modifying the functional layer on the basic polyolefin separator. The modified materials can be divided into carbon-based materials, inorganic materials, and polymer materials [34]. The functions of the modified layer can be roughly divided into: (1) inhibiting the diffusion of polysulfides, (2) regulating the deposition behavior of lithium-ion on lithium metal anode, and avoiding the dendrite growth, (3) enhancing the ionic conductivity and improving the rate-performance of LSB.

16.3 Application of COF in Sulfur Cathode

The design strategy of sulfur cathode mainly focuses on the structure and properties of sulfur host materials. The specific surface area, pore structure, polar groups types, density, conductivity, and stability of the host materials are the main factors affecting the electrochemical properties of the cathode. The large specific surface area and uniformly distributed pore structure provide space for sulfur loading and channels for lithium-ion conduction. Abundant polar groups in the COF skeleton increase the interaction force between the polysulfides and host, contributing to inhibiting the shuttle effect. Besides, the low density of COFs is conducive to improving the energy density of LSB.

16.3.1 COF as Host for Loading Sulfur

In 2014, Wang et al. [35] first introduced the covalent triazine framework (CTF-1) with the triazine group into LSB as the host material for loading active substance sulfur. The CTF has uniformly distributed nanopores with a pore size of 1.23 nm, which is 1.7 times larger than that of the S_8 molecule. After melting solid sulfur into liquid sulfur at 155°C, the sulfur can enter the pores of CTF-1. The XRD characteristic peak of sulfur disappears, indicating that sulfur is completely dispersed into the pores of CTF-1, denoted as S/CTF@155°C. The S/CTF@155°C can provide a high capacity of 1497 mAh g^{-1} at the charge and discharge rate of 0.1 C. In contrast, the S/CTF-1 without melting process can only provide 1015 mAh g^{-1} at 0.1 C. It indicates that the dispersion of sulfur into nanopores of COFs is important for the electrochemical performance of LSB. However, to obtain the long cycling stability of the battery, it is also required that the sulfur host can effectively inhibit the dissolution and diffusion of polysulfides. Tang et al. [36] prepared two kinds of COFs (COF-1 and CTF-1) and compared the interaction force with polysulfides. As shown in Figure 16.3a, the triazine group in CTF-1 has poor adsorption energy of polysulfides (-24 kcal mol^{-1}). In contrast, the adsorption energy of soluble polysulfides in the COF-1 is -34 kcal mol^{-1}, indicating that the positive B element and negative O element of COF-1 have a strong interaction force with S_x^{2-} ion and lithium-ion, respectively. Therefore, S/CTF-1 can only exhibit the capacity of 489 mAh g^{-1} after 100 cycles at 0.2 C, but the S/COF-1 can maintain the capacity of 929 mAh g^{-1}. It is further verified that the interaction force between COF material and polysulfides is very important to improve the cycling performance of LSB. Coskun [37] proposed a strategy to improve the cycling stability of LSB by building a C-S bond between COF and sulfur molecules. As shown in Figure 16.3b, sulfur can catalyze the synthesis of CTF-1, and the C-S bond is formed at the same time.

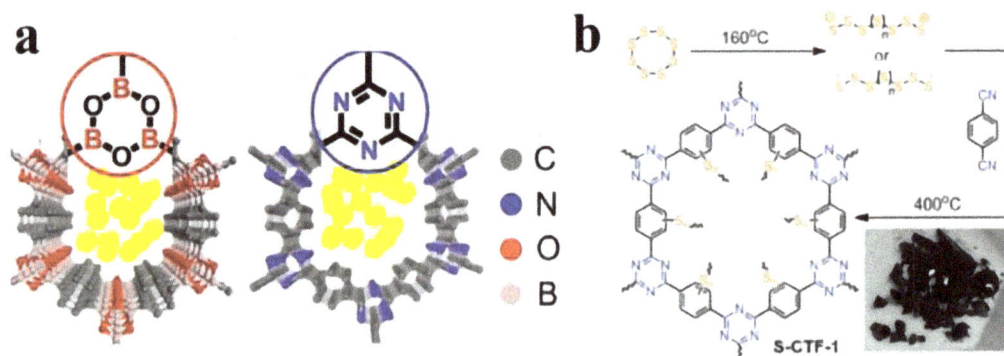

(a) Chemical structure of COF-1/S composite and CTF-1/S composite (Adapted with permission [36]. Copyright (2016) WILEY-VCH). (b) Synthesis of S-CTF-1 from elemental sulfur and an optical image of S-CTF-1. (Adapted with permission [37]. Copyright (2016) WILEY-VCH.)

In this way, the sulfur loading can reach 62 wt%, which is close to the requirements of the commercial sulfur cathode. Besides, the capacity retention rate is up to 85.8% after 300 cycles. Introducing polar groups into COF will be more conducive to the formation of the C-S bond. For example, the fluorine atoms in FCTF-1 benefit from the nucleophilic substitution reaction, to promote the formation of the C-S bond. In addition, the carbon-carbon double bond can also form a C-S bond by reverse vulcanization to realize the combination of a sulfur molecule and COF.

16.3.2 COF Composite as Host for Loading Sulfur

As sulfur host materials, COF enables sulfur cathode with high capacity and good cycle stability. However, COF has intrinsically poor electronic conductivity, which results in rapid capacity fade and poor rate performance. Therefore, to further promote the application of COF in a sulfur host, many works have focused on the problem of poor conductivity of COF. Compounding with conductive materials is considered to be an effective strategy to improve electronic conductivity. Wei et al. [38] synthesized TPPA COF on multi-walled carbon nanotubes (COF@MWCNTs). Due to the π - π interaction between TPPA COF and MWCNTs, TPPA COF can form a continuous coating layer on MWCNTs, which could significantly improve the electronic conductivity. Meng et al. [39] *in-situ* grew P-CTFs on reduced graphene oxide (RGO). As shown in Figure 16.4a, the nanopore structure of P-CTF2 can effectively accommodate the active substance sulfur. Phthalazinone and triazine have strong interactions with polysulfides. Besides, RGO provides a fast electron transfer channel, which is conducive to accelerating the electrochemical reaction. The S/P- CTF@rGO electrode can provide an initial capacity of 1130 mA h g^{-1} at 0.5 C, and the capacity retention rate is up to 81.4% after 500 cycles. In addition to carbon materials, conductive polymers also have high electronic conductivity. As shown in Figure 16.4b, Choi [40] introduced conductive polymer polypyrrole (PPy) into CTF to improve the electronic conductivity of CTF. Therefore, cPpy-S-CTF electrode shows good electrochemical performance, and the capacity retention rate is as high as 86.8% after 500 cycles at 0.5 C. In addition to increasing the electronic conductivity of COF, compounding COF with other polar materials to enhance the chemical adsorption ability of polysulfides is of

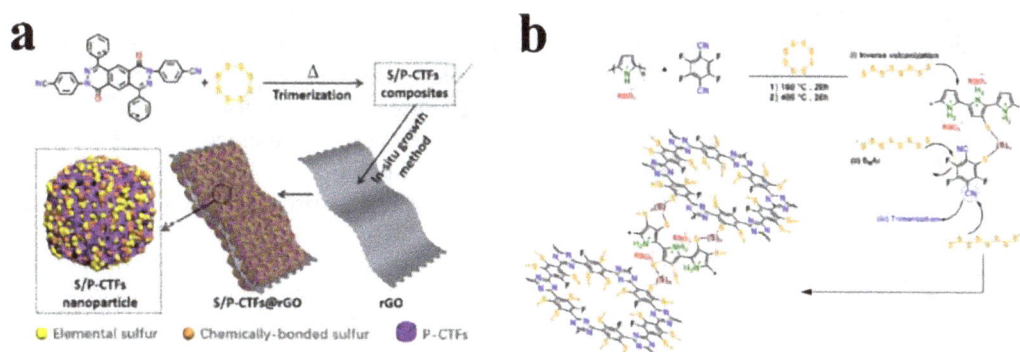

FIGURE 16.4
Schematic representation for the synthesis of (a) S/P-CTF@rGO (Adapted with permission [39]. Copyright (2020) American Chemical Society). (b) cPpy-S-CTF. (Adapted with permission [40]. Copyright (2020) American Chemical Society.)

great significance to increase the cycle life of LSB. Yang [41] synthesized TiO_2/COF composites (HCPT@COF). The TiO_2 nanoparticles with the size of 10 nm are evenly embedded between the layers of COF, and TiO_2 combined with HCPT@COF is a benefit for the adsorption of polysulfides.

16.4 Application of COF in a Separator

The main target of the functional separator of LSB is to block the transport of polysulfide. The pore diameter of the traditional polyolefin separator is about 100 nm, while the dynamic size of polysulfides is less than 1 nm. Driven by the electric field and concentration gradient, polysulfides will cross the separator and cause side reactions with the lithium metal anode, resulting in the continuous attenuation of the battery capacity. Modifying the functional layer on the surface of the polyolefin separator is an effective strategy to inhibit the shuttle effect. COF has been considered to be an ideal functional layer because COF has an intrinsic nanopore structure, which can effectively limit the diffusion of polysulfides. The low density ensures that the introduction of the COF functional layer does not affect the energy density of the battery. Besides, good chemical stability ensures the COF functional layer suitable for the battery system.

16.4.1 Polysulfides-Blocking Separator

The COF pore size affects the obstruct capability for the diffusion of polysulfides. As shown in Figure 16.5a, Cai et al. [42] prepared DMTA-COF with a pore size of only 0.56 nm. The DMTA-COF was mixed super-P and coated on the ceramic membrane. The small pore size of DMTA-COF can effectively block the diffusion of polysulfides. Hu et al. [43] assembled COF-1 with AB stacking mode on graphene film (COF-rGO). As shown in Figure 16.5b, the staggered pore structure in COF-1 can block the diffusion of polysulfides. The capacity retention rate of LSB using COF-rGO separator after 50 cycles is up to 84.3%, while the LSB using traditional Celgard separator can only exhibit the capacity of 59.7%. The excellent cycling stability of LSB using a COF-rGO separator is from the physically blocking effect of COF-1 on polysulfides, which contributes to avoiding side reactions and increasing the cycling life of LSB. However, the problems of polysulfides dissolution and diffusion cannot be completely solved by physical obstruct. The introduction of functional groups into the COF skeleton is benefit for the adsorption of polysulfides and the reactivation of polysulfides dissolved in the electrolyte. As shown in Figure 16.5c, Lee et al. [44] prepared COF-1 on CNT (COF-1/CNT) and reassembled it into a self-supporting interlayer (COF-1 NN). The element B and element O in COF-1 can be served as adsorption sites of polysulfides, while CNT provides electrons for realizing the conversion and reactivation of polysulfides adsorbed in the separator, and greatly reduces the loss of polysulfides in the electrolyte. Li et al. [45] grew ionic COF nanosheets (iCON) on MXene (Ti_3C_2@iCON). As shown in Figure 16.5d, iCON can attract polysulfides through electrostatic interaction, and MXene can chemically adsorb polysulfides. The synergistic effect of iCON and MXene can catalyze the conversion reaction of polysulfides in the separator, which can not only reduce the loss of polysulfides but also improve the reaction kinetics of the battery.

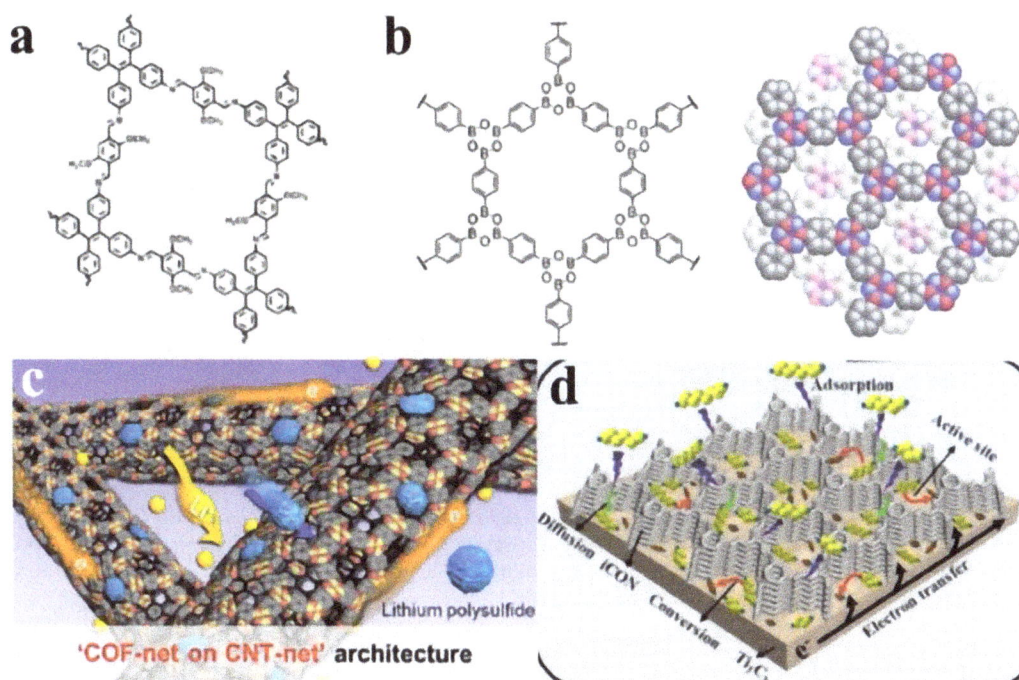

FIGURE 16.5
Chemical structure of (a) CTF-1 (Adapted with permission from [42]. Copyright (2018) American Chemical Society). (b) COF-1 (Adapted with permission [43]. Copyright (2018) WILEY-VCH). (c) Schematic illustrations of ion/electron transport behavior in the COF-1 NN interlayers (Adapted with permission [44]. Copyright (2016) American Chemical Society). (d) Schematic illustrations of polysulfides trapping and conversion process on Ti_3C_2@iCON-PP. (Adapted with permission [45]. Copyright (2021) WILEY-VCH.)

16.4.2 Ion-Selective Separator

The main function of the battery separator is ionic conduction. The electrolyte can infiltrate the separator, while ions can pass through the separator with the electrolyte under the electric field and concentration gradient. In LSB, there are two important ion-conduction behaviors: (1) the lithium-ion conduction is indispensable for electrode reaction, and the conduction rate of lithium-ion directly determines the rate of electrode reaction; (2) The diffusion of intermediate polysulfides will bring side reactions and capacity loss of the battery. Therefore, the ideal LSB separator should have an ion-selective conduction function, which is conducive to the simultaneous conduction of lithium-ion and inhibition of the polysulfides shuttle. Sun et al. [46] introduced the ion-selective COF layer into the LSB separator. As shown in Figure 16.6a, a new type of lithiated COF nanosheet (Li-CON) was assembled with graphene as the separator functional layer (Li-CON@GN). The lithiated sites in Li-CON promote the lithium-ion conduction in the nanopores, and the triazole groups are used as the adsorption sites for polysulfides. The reversible conversion of polysulfides in the separator is realized in cooperation with the electron-conducting graphene. The Li-CON@GN functional layer can not only block the diffusion of polysulfides but also improve the conduction rate of lithium-ion. Besides, Sun et al. [47] further introduced the electrostatic effect into the COF functional layer to realize the ion-selective conduction function. As shown in Figure 16.6b, the TpPa-SO_3Li nanosheets were prepared

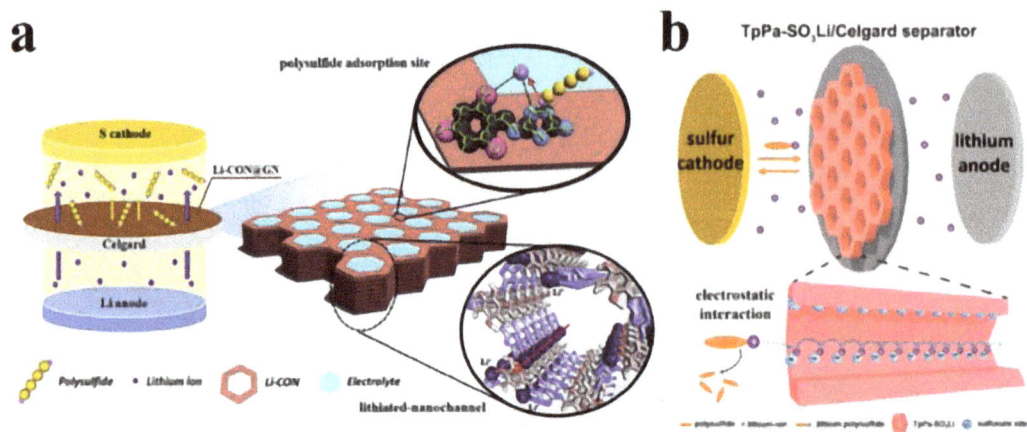

FIGURE 16.6
(a) Schematics of the Li-S cell with the Li-CON@GN/Celgard separator, polysulfide adsorption site, and lithiated-nanochannel (Adapted with permission [46]. Copyright (2021) Elsevier). (b) Schematic of the Li–S cell with the TpPa-SO₃Li/Celgard separator. (Adapted with permission [47]. Copyright (2021) American Chemical Society.)

by interfacial polymerization and chemical lithiation method, assembled on the Celgard separator to prepare the TpPa-SO₃Li functional layer. The sulfonate groups in the nanopores show negative electricity. The sulfonate groups have a moderate attraction to the lithium-ion, which is conductive to the conduction of lithium-ion in the separator. In contrast, the sulfonate groups can inhibit the polysulfides entering the pore of TpPa-SO₃Li, and finally, realize the ion-selective conduction function of the separator. The LSB with TpPa-SO₃Li/Celgard separator can provide a high capacity of 822.9 mAh g^{-1} under the condition of high sulfur loading (5.4 mg cm^{-2}), and the capacity retention rate is up to 78%.

16.5 Application of COF in Lithium Metal Anode

Lithium metal has the advantage of high specific capacity but still suffers from some issues for practical application. (1) The uneven distribution of charge on the surface of lithium metal leads to the local charge accumulation and the lithium dendrite growth, which is easy to puncture the separator, leading to the battery short circuit; (2) The unstable lithium metal interface will continue to consume electrolyte, failing the battery; (3) The side reaction between polysulfides and lithium metal causes the loss of active sulfur. Recently, the study of COF as a host for lithium metal or artificial SEI has made great progress. The advantages of COF can be roughly divided into (1) the nanochannel of COF is conducive to the uniform deposition of lithium-ion. (2) The nanopore structure can accommodate the volume expansion of lithium metal. (3) The rigid structure of COF is beneficial to inhibit the growth of lithium dendrites. (4) The lithiophilic sites of COF are also conducive to reducing the deposition barrier of lithium-ion.

16.5.1 COF as the Host of Lithium Metal

Kang [48] coated COF-LZU1 on the Cu foil through a spin-coating method. The aldehyde groups in COF-LZU1 can interact with TFSI⁻ anion, and confine TFSI⁻ anion in the pores of COF-LZU1, which is a benefit to alleviate local charge accumulation and inhibit the lithium dendrite growth. In addition, the imine bond in COF-LZU1 has good lithium affinity and electrolyte affinity, contributing to the uniform deposition of lithium-ion on the lithium metal. Feng et al. [49] coated lithiophilic COF (TpTt) and acetylene black on the Cu foil. As shown in Figure 16.7a, the triazine ring in TpTt has many lone electrons, which is conducive to attracting lithium-ion, and the aldehyde groups can also coordinate with lithium-ion. Besides, the periodic arrangement of these functional groups in TpTt ensures the uniform deposition of lithium-ion and greatly reduces the growth of lithium dendrites. Coskun et al. [50] used LiTFSI as a catalyst to generate LIF-loaded CTF-1 (CTF-LIF), and used airlaid-paper (AP) to adsorb CTF-LIF to prepare lithium metal host (AP-CTF-LIF). As shown in Figure 16.7b, CTF-1 shows good lithium affinity, and LIF can be served as the interface stabilizer. The AP provides sufficient space for lithium-ion deposition, and the highly stable lithium metal anode was obtained. Lai et al. [51] grew COF-LZU1 on MXene nanosheets to prepare MXene@COF hybrid structure. As shown in Figure 16.7c, the hybrid structure has good crystallinity, hierarchical pore structure, and conductive framework.

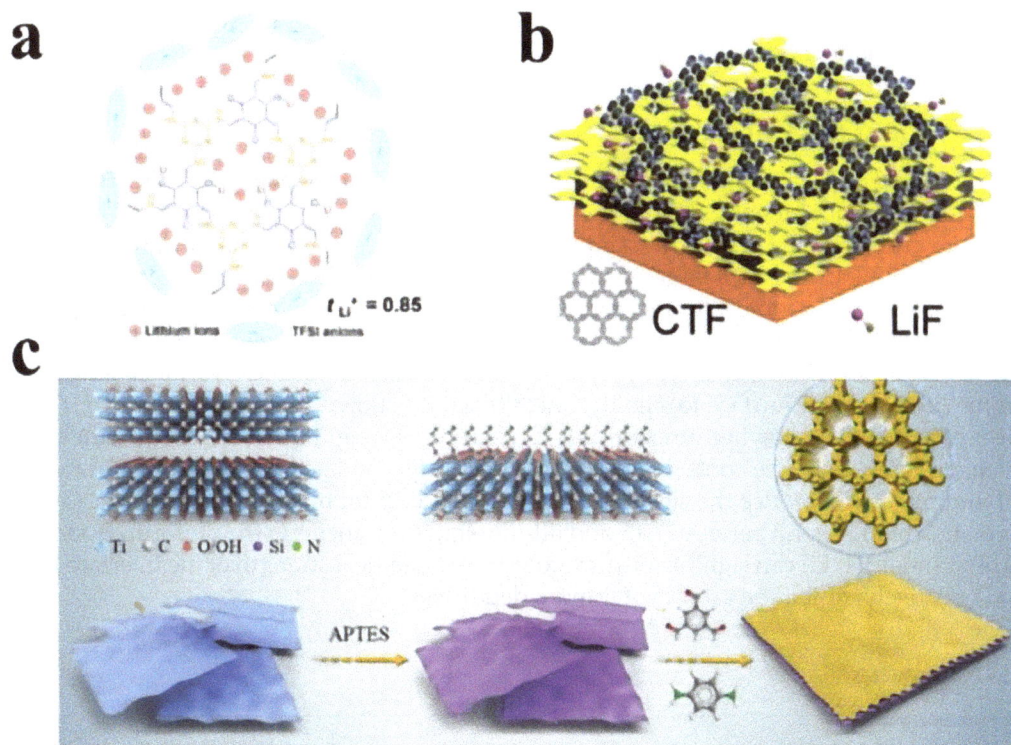

FIGURE 16.7

(a) Schematic illustration of the positive charged TpTt (Adapted with permission [49]. Copyright (2021) American Chemical Society). (b) Schematic illustration of AP-CTF-LiF electrodes upon Li plating (Adapted with permission [50]. Copyright (2019) WILEY-VCH). (c) Schematic illustration of MXene covalent amination, followed by MXene@COF-LZU1 growth. (Adapted with permission [51]. Copyright (2021) WILEY-VCH.)

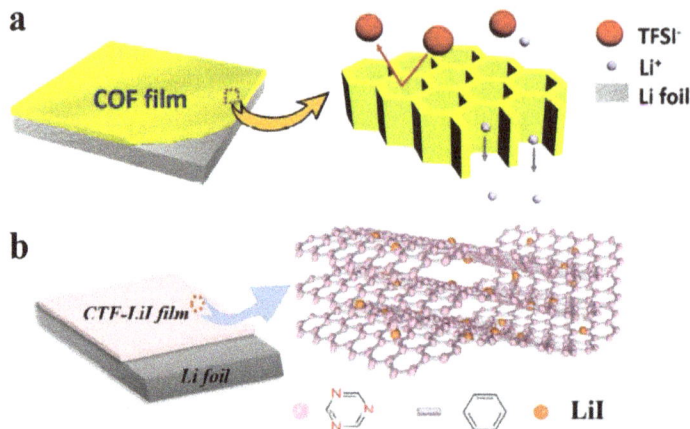

FIGURE 16.8
(a) Schematic diagrams showing the effect of COF film on a Li anode when cycling (Adapted with permission [52]. Copyright (2020) WILEY-VCH). (b) Schematic illustrations of the preparation of CTF-LiI coated Li anode. (Adapted with permission [53]. Copyright (2021) WILEY-VCH.)

When it is used as the host material for lithium metal, the lithium storage performance and charge conduction properties are significantly improved.

16.5.2 COF as the Artificial Solid-Electrolyte Interphase

COFs have made great progress in the artificial SEI for lithium metal anode. The artificial SEI can fulfill the requirement for regulation of lithium-ion deposition and inhibition of lithium dendrite growth. Meng et al. [52] in situ synthesized $COF_{TAPB-PDA}$ layer on the lithium metal. As shown in Figure 16.8a, a large number of uniform distributed nanopores are conducive to the uniform deposition of lithium-ion, and the rigid COF framework can block the growth of lithium dendrite. $COF_{TAPB-PDA}$, as the interface layer of lithium metal, greatly improves the safety of LSB. Zheng et al. [53] prepared a flexible and conformal CTF-LiI coating layer on the lithium metal surface to stabilize metallic Li. As shown in Figure 16.8b, the CTF-LiI layer can play the following functions: (1) CTF-LiI has strong adhesion with lithium metal to ensure good interface contact; (2) CTF-LiI effectively alleviates the volume expansion of lithium metal and reduce the side reactions between lithium metal and electrolyte; (3) The formation of Li-N bond between CTF and lithium-ion is conducive to the uniform deposition of lithium-ion; (4) LiI can significantly improve the mechanical strength of the interface layer and effectively inhibit the growth of lithium dendrites.

16.6 Summary and Perspective

This review systematically summarized the application of COFs in LSB in recent years. COFs play an important role in solving the intrinsic problems of LSB, such as shuttle effect and lithium dendrite growth, through the design of cathode host materials, separator functional layer, and lithium metal host or interphase. Generally speaking, the ordered pore structure and abundant functional sites of COF are very important for its various functions. The nanopores of COF mainly have two functions (accommodating active substances and conducting ions). In the sulfur cathode and lithium metal anode, the nanopores of COF can be used as the "container" of sulfur or lithium metal, which could avoid agglomeration and improve the utilization of active substances. In addition, the nanopores, as the channel of lithium-ion conduction, play an important role in the cathode, anode, and separator, contributing to the fast electrode reaction kinetics and inhibiting lithium dendrite. The interaction of functional sites in COFs with polysulfides is conducive to inhibiting the shuttle effect, and the lithiophilic sites of COFs are beneficial for reducing the overpotential of lithium deposition. However, there are still some bottlenecks in the application of COFs in LSB. Finally, we will discuss the problems and prospects of the application of COFs in the cathode, anode, and separator, respectively.

1. Although COF can be used as the host material of sulfur in the cathode, it cannot directly provide capacity. Besides, the intrinsic electronic conductivity of COFs is poor, thus excessive addition of COFs will also affect the reaction rate of the battery. Therefore, reducing the amount of COF and increasing the content of the active substance sulfur are very important for the practical application of the lithium-sulfur battery. Exploring the novel COF hosts with low density, and high specific surface area is very promising so that sulfur can be well dispersed into the pores without agglomeration, which is conducive to further improving the electrochemical performance of the lithium-sulfur battery.

2. COF can simultaneously inhibit the shuttle effect and conduct lithium-ion in the LSB separator. It is very important to further prepare COF materials with high ionic conductivity, through the design of ordered pore structures and continuous lithiated sites of COFs. Due to the disadvantages of expensive electrolytes, flammability, and side reactions, the COFs separator is promising to be used in a solid electrolyte or little electrolyte condition, which will promote the application and development of LSB. In addition, the preparation of a flexible self-supporting COF membrane is also of great scientific research value for further exploring the ion-conduction behavior of nanopores.

3. COF has a porous structure and abundant lithiophilic sites, which can regulate the lithium-ion deposition behavior. To further reduce the deposition overpotential of lithium-ions and inhibit the growth of lithium dendrites, introduce functional sites into the ordered structure of COF to realize the synergistic effect of ion-conduction and lithium affinity, which is expected to further improve the stability of lithium metal anode.

To summarize, although the applications of COFs in LSB fields are still in the early stage and face many challenges, the improvement in battery performance is very obvious. With the development of preparation technology and the reduction of cost, COFs hold great potential in the development of high-performance batteries.

References

1. Pellow MA, Ambrose H, Mulvaney D, Betita R, Shaw S (2020) Research gaps in environmental life cycle assessments of lithium ion batteries for grid-scale stationary energy storage systems: End-of-life options and other issues. Sustainable Mater. Technol. 23:e00120

2. Lee W, Muhammad S, Sergey C, Lee H, Yoon J, Kang YM, Yoon WS (2020) Advances in the cathode materials for lithium rechargeable batteries. Angew. Chem. Int. Ed. 59:2578–2605

3. Opra DP, Gnedenkov SV, Sinebryukhov SL (2019) Recent efforts in design of $TiO_2(B)$ anodes for high-rate lithium-ion batteries: A review. J. Power Sources 442:227225

4. Zhang SS (2013) Liquid electrolyte lithium/sulfur battery: Fundamental chemistry, problems, and solutions. J. Power Sources 231:153–162

5. Ji X, Lee KT, Nazar LF (2009) A highly ordered nanostructured carbon–sulfur cathode for lithium-sulfur batteries. Nat. Mater. 8(6):500–506

6. Evers S, Nazar LF (2013) New approaches for high energy density lithium-sulfur battery cathodes. Acc. Chem. Res. 46(5):1135–1143

7. Kozen AV, Lin CF, Pearse AJ, Schroeder MA, Han XG, Hu LB, Lee SB, Rubloff GW, Noked M (2015) Next-generation lithium metal anode engineering via atomic layer deposition. ACS Nano 9(6):5884–5892

8. Zhang S, Ueno K, Dokko K, Watanabe M (2015) Recent advances in electrolytes for lithium-sulfur batteries. Adv. Energy Mater. 5(16):1500117

9. Manthiram A, Fu Y, Su YS (2012) Challenges and prospects of lithium-sulfur batteries. Acc. Chem. Res. 46(5):1125–1134

10. Fang R, Zhao S, Sun Z, Wang DW, Cheng HM, Li F (2017) More reliable lithium-sulfur batteries: status, solutions, and prospects. Adv. Mater. 29(48):1606823

11. Yin YX, Xin S, Guo YG, Wan LJ (2013) Lithium-sulfur batteries: Electrochemistry, materials, and prospects. Angew. Chem. Int. Ed. 52(50):13186–13200

12. Bhargav A, He J, Gupta A, Manthiram A (2020) Lithium-sulfur batteries: Attaining the critical metrics. Joule 4(2):285–291

13. Guo J, Xu Y, Wang C (2011) Sulfur-impregnated disordered carbon nanotubes cathode for lithium-sulfur batteries. Nano Lett. 11(10):4288–4294

14. Zheng G, Yang, Y, Cha JJ, Hong SS, Cui Y (2011) Hollow carbon nanofiber-encapsulated sulfur cathodes for high specific capacity rechargeable lithium batteries. Nano Lett. 11(10):4462–4467

15. Wang H, Yang Y, Liang Y, Robinson JT, Li Y, Jackson A, Cui Y, Dai H (2011) Graphene-wrapped sulfur particles as a rechargeable lithium_sulfur battery cathode material with high capacity and cycling stability. Nano Lett. 11(7):2644–2647

16. Qiu Y, Li W, Zhao W, Li G, Hou Y, Liu M, Zhou L. Ye F, Li H, Wei Z, Yang S, Duan W, Ye Y, Guo J (2014) High-rate, ultralong cycle-life lithium/sulfur batteries enabled by nitrogen-doped graphene. Nano Lett. 14(8):4821–4827

17. Wu F, Li J, Tian Y, Su Y, Wang J, Yang W, Li N, Chen S, Bao L (2015) 3D coral-like nitrogen-sulfur codoped carbon-sulfur composite for high-performance lithium-sulfur batteries. Sci. Rep. 5:13340

18. Xiao L, Cao Y, Xiao J, Schwenzer B, Engelhard MH, Saraf LV, Nie Z, Exarhos GJ, Liu J (2012) A soft approach to encapsulate sulfur: Polyaniline nanotubes for lithium-sulfur batteries with long cycle life. Adv. Mater. 24(9):1176–1181

19. Yin L, Wang J, Yang J, Nuli Y (2011) A novel pyrolyzed polyacrylonitrile-sulfur@MWCNT composite cathode material for high-rate rechargeable lithium/sulfur batteries. J. Mater. Chem. 21(19):6807–6810

20. Wang J, Chen J, Konstantinov K, Zhao L, Ng SH, Wang GX, Guo ZP, Liu HK (2006) Sulfur-polypyrrole composite positive electrode materials for rechargeable lithium batteries. Electrochim. Acta 51(22):4634–4638

21. Oschmann B, Park J, Kim C, Char K, Sung YE, Zentel R (2015) Copolymerization of polythiophene and sulfur to improve the electrochemical performance in lithium–sulfur batteries new approaches for high energy density lithium-sulfur battery cathodes. Chem. Mater. 27(20):7011–7017

22. Seh ZW, Li W, Cha JJ, Zheng G, Yang Y, McDowell MT, Hsu PC, Cui Y (2013) Sulfur–TiO_2 yolk-shell nanoarchitecture with internal void space for long-cycle lithium–sulfur batteries. Nat. Commun. 4:1331

23. Li Z, Zhang J, Lou XW (2015) Hollow carbon nanofibers filled with MnO_2 nanosheets as efficient sulfur hosts for lithium–sulfur batteries. Angew. Chem. Int. Ed. 54(44):12886–12890

24. Yuan Z, Peng HJ, Hou TZ, Huang JQ, Chen CM, Wang DW, Cheng XB, Wei F, Zhang Q (2016) Powering lithium–sulfur battery performance by propelling polysulfide redox at sulfiphilic hosts. Nano Lett. 16(1):519–527

25. Ma G, Wen Z, Wu M, Shen C, Wang Q, Jin J, Wu X (2014) A lithium anode protection guided highly-stable lithium–sulfur battery. Chem. Commun. 50(91):14209–14212

26. Jia W, Wang Q, Yang J, Fan C, Wang L, Li J (2017) Pretreatment of lithium surface by using iodic acid (HIO_3) to improve its anode performance in lithium batteries. ACS Appl. Mater. Interfaces 9(8):7068–7074

27. Kim K, Lee JT, Lee DC, Oschatz M, Cho WI, Kaskel S, Yushin G (2013) Enhancing performance of Li–S cells using a Li–Al alloy anode coating. Electrochem. Commun. 36:38–41

28. Kim MS, Kim MS, Do V, Lim YR, Nah IW, Archer LA, Cho WI (2017) Designing solid-electrolyte interphases for lithium sulfur electrodes using ionic shields. Nano Energy 41:573–582

29. Zheng G, Lee SW, Liang Z, Lee HW, Yan K, Yao H, Wang H, Li W, Chu S, Cui Y (2014) Interconnected hollow carbon nanospheres for stable lithium metal anodes. Nat. Nanotechnol. 9:618–623

30. Luo J, Lee RC, Jin JT, Weng YT, Fang CC, Wu NL (2017) A dual-functional polymer coating on a lithium anode for suppressing dendrite growth and olysulfide shuttling in Li–S batteries. Chem. Commun. 53(5): 963–966

31. Barchasz C, Leprêtre JC, Patoux S, Alloin DF (2013) Electrochemical properties of ether-based electrolytes for lithium/sulfur rechargeable batteries. Electrochim. Acta 89:737–743

32. Hu JJ, Long GK, Liu S, Li GR, Gao XP (2014) A LiFSI–LiTFSI binary-salt electrolyte to achieve high capacity and cycle stability for a Li–S battery. Chem. Commun. 50(93):14647–14650

33. Jozwiuk A, Berkes BB, Wei T, Sommer H, Janek J, Brezesinski T (2016) The critical role of lithium nitrate in the gas evolution of lithium-sulfur batteries. Energy Environ. Sci. 9(8):2603–2608

34. Rana M, Li M, Huang X, Luo B, Gentle I, Knibbe R (2019) Recent advances in separators to mitigate technical challenges associated with re-chargeable lithium-sulfur batteries. J. Mater. Chem. A 7(12):6596–6615

35. Liao H, Ding H, Li B, Ai X, Wang C (2014) Covalent-organic frameworks: potential host materials for sulfur impregnation in lithium–sulfur batteries. J. Mater. Chem. A 2:8854–8858

36. Ghazi ZA, Zhu L, Wang H, Naeem A, Khattak AM, Liang B, Khan NA, Wei Z, Li L, Tang Z (2016) Efficient polysulfide chemisorption in covalent organic frameworks for high-performance lithium-sulfur batteries. Adv. Energy Mater. 6: 1601250

37. Talapaneni SN, Hwang TH, Je SH, Buyukcakir O, Choi JW, Coskun A (2016) Elemental-sulfur-mediated facile synthesis of a covalent triazine framework for high-performance lithium-sulfur batteries. Angew. Chem. Int. Ed. 55:3106–3111

38. Zhang X, Wang Z, Lu Y, Mai Y, Liu J, Hua X, Wei H (2018) Synthesis of core-shell covalent organic frameworks/multi-walled carbon nanotubes nanocomposite and application in lithium-sulfur batteries. Mater. Lett. 213: 143–147

39. Guan R, Zhong L, Wang S, Han D, Xiao M, Sun L, Meng Y (2020) Synergetic covalent and spatial confinement of sulfur species by phthalazinone-containing covalent triazine frameworks for ultrahigh performance of Li–S batteries. ACS Appl. Mater. Interfaces 12: 8296–8305

40. Kim J, Elabd A, Chung SY, Coskun A, Choi JW (2020) Covalent triazine frameworks incorporating charged polypyrrole channels for high-performance lithium–sulfur batteries. Chem. Mater. 32:4185–4193

41. Yang Z, Peng C, Meng R, Zu L, Feng Y, Chen B, Mi Y, Zhang C, Yang J (2019) Hybrid anatase/rutile nanodots-embedded covalent organic frameworks with complementary polysulfide adsorption for high-performance lithium–sulfur batteries. ACS Cent. Sci. 5:1876–1883

42. Wang J, Si L, Wei Q, Hong X, Cai S, Cai Y (2018) Covalent organic frameworks as the coating layer of ceramic separator for high-efficiency lithium–sulfur batteries. ACS Appl. Nano Mater. 1:132–138

43. Jiang C, Tang M, Zhu S, Zhang J, Wu Y, Chen Y, Cong X, Wang C, Hu W (2018) Constructing universal ionic sieves via alignment of two-dimensional covalent organic frameworks (COFs). Angew. Chem. Int. Ed. 57:16072–16076

44. Yoo J, Cho SJ, Jung GY, Kim SH, Choi KH, Kim JH, Lee CK, Kwak SK, Lee SY (2016) COF-net on CNT-Net as a molecularly designed, hierarchical porous chemical trap for polysulfides in lithium–sulfur batteries. Nano Lett. 16:3292–3300

45. Li P, Lv H, Li Z, Meng X, Lin Z, Wang R, Li X (2021) The electrostatic attraction and catalytic effect enabled by ionic–covalent organic nanosheets on MXene for separator modification of lithium–sulfur batteries. Adv. Mater. 33:2007803

46. Cao Y, Liu C, Wang M, Yang H, Liu S, Wang H, Yang Z, Pan F, Jiang Z, Sun J (2020) Lithiation of covalent organic framework nanosheets facilitating lithium-ion transport in lithium-sulfur batteries. Energy Storage Mater. 29:207–215

47. Cao Y, Wu H, Li G, Liu C, Cao L, Zhang Y, Bao W, Wang H, Yao Y, Liu S, Pan F, Jiang Z, Sun J (2021) Ion selective covalent organic framework enabling enhanced electrochemical performance of lithium–sulfur batteries. Nano Lett. 21:2997–3006

48. Xu Y, Zhou Y, Li T, Jiang S, Qian X, Yue Q, Kang Y (2020) Multifunctional covalent organic frameworks for high capacity and dendrite-free lithium metal batteries. Energy Storage Mater. 25:334–341

49. Li Z, Ji W, Wang TX, Zhang Y, Li Z, Ding X, Han BH, Feng W (2021) Guiding uniformly distributed Li–Ion flux by lithiophilic covalent organic framework interlayers for high-performance lithium metal anodes. ACS Appl. Mater. Interfaces 13:22586–22596

50. Zhou T, Zhao Y, Choi JW, Coskun A (2019) Lithium-salt mediated synthesis of a covalent triazine framework for highly stable lithium metal batteries. Angew. Chem. Int. Ed. 58:16795–16799

51. Guo D, Ming F, Shinde DB, Cao L, Huang G, Li C, Li Z, Yuan Y, Hedhili MN, Alshareef HN, Lai ZP (2021) Covalent assembly of two-dimensional COF-on-MXene heterostructures enables fast charging lithium hosts. Adv. Funct. Mater. 31:2101194

52. Chen D, Huang S, Zhong L, Wang S, Xiao M, Han D, Meng Y (2020) In situ preparation of thin and rigid COF film on Li anode as artificial solid electrolyte interphase layer resisting Li dendrite puncture. Adv. Funct. Mater. 30:1907717

53. Zheng Y, Xia S, Dong F, Sun H, Pang Y, Yang J, Huang Y, Zheng S (2021) High performance Li metal anode enabled by robust covalent triazine framework-based protective layer. Adv. Funct. Mater. 31:2006159

17

Photocatalysts Based on Covalent Organic Frameworks

Nazanin Mokhtari and Mohammad Dinari

Department of Chemistry, Isfahan University of Technology, Isfahan, Islamic Republic of Iran

CONTENTS

17.1 Introduction

The industrial revolution in human societies and the increase in fossil fuel consumption lead to severe environmental problems. Additionally, the energy shortage and exponential growth of energy demand urged scientists to find clean and renewable energy alternatives. Solar energy is one of the clean and sustainable resources that gained extensive attention during the past decades. Photocatalytic degradation of organic pollutants, CO_2 fixation, water splitting, and solar cells have attracted substantial interest due to the abundance and inexhaustibility of solar energy. There are different approaches for alleviating these issues; however, photocatalysis was one of the most sustainable technologies for making chemical energy from solar energy. The role of semiconductors in photocatalysis is inevitable [1]. The formation of electrons and holes by absorbing photons occurs in semiconductors generating conduction band (CB) and valance band (VB) respectively. The formation of excitons can proceed with organic redox transformations [2].

Traditionally, the leading position of photocatalysis has been occupied by inorganic semiconductors, including silver phosphate (Ag_3PO_4) [3], zinc oxide (ZnO) [4], cadmium sulfide (CdS) [5], and titanium dioxide (TiO_2) [6]. The diversity of inorganic semiconductors has increased over the past several decades; however, the popularity of TiO_2 is maintained due to its low cost, relatively high availability, and durability. Besides TiO_2, other oxides

DOI: 10.1201/9781003206507-17

and sulfides of transition metals such as Ag_3PO_4 and CdS represent the excellent capacity of charge carrier transportation. However, environmental concerns of heavy metal usage and the photo-corrosion effect limited their practical applications [7]. Therefore, finding new semiconductors becomes a long-lasting challenge for investigators. Organic semiconductors have come to the attention of researchers as a new solution to face the challenge of environmental issues. Graphitic carbon nitride (g-C_3N_4) [8], metal-organic frameworks (MOFs) [9], and covalent organic frameworks (COFs) [10] were introduced to be used as promising alternatives for photocatalytic solar energy conversion.

Contrary to traditionally used metal oxides and inorganic materials, COFs' regular structures, high surface areas, and facile modification of functionalities made COFs attract attention in different branches of science. Drug loading and release, energy storage, fluorescence studies, and gas sorption and separation are only some examples of using COFs [11]. They have also displayed attractive advantages to serve as photocatalysts. These include the diverse selections of monomers, precise adjustability of channel size, extended π-conjugated framework, tunable band structure, visible light absorption, good light stability, and, more importantly, the ability to regulate the bandgap of COFs through post-modifications. Therefore, COFs have been used as photocatalysts in organic synthesis, CO_2 fixation, environmental remediation, and water splitting [12–15]. Also, their band structure could be adjusted by preparing composites with other semiconductors [16].

17.2 COF-Based Photocatalyst

Photocatalytic activity is directly related to the efficient transfer and separation of photogenerated charges. COFs having photocatalytic activity usually consist of photoactive functional groups on their building blocks that can act as electron donors or acceptors. Arenes, porphyrins, phthalocyanines, and sulfur and nitrogen heterocyclic rings are the most common functionalities presented in photoactive COFs. Construction of donor-acceptor (D-A) systems allows charge transfer along with the extended conjugated system from the electron donor to the electron-accepting group [17]. Furthermore, to improve the separation efficiency of the photogenerated excitons, using other semiconductors having different band levels could be helpful. Inorganic semiconductors, MOFs, and g-C_3N_4 are the favorite materials used in conjunction with COFs. In the following sections, the photocatalytic activity of the pyrene-based, triazine-based, thiophene-based, phthalocyanine-based, and porphyrin-based COFs and their composites with other semiconductors are reviewed.

17.2.1 Arene-Based COFs

Pyrene and perylene derivatives are always the first options to prepare photocatalytic COFs owing to their excellent properties such as high thermal, chemical, photostability, excellent photoelectric activity, high light-absorption efficiency, and electron-rich nature [18]. In the early stages of using COFs in photocatalytic activities, the main focus was generating photocurrent and photoconductive properties of pyrene-based COFs. PPy-COF (Figure 17.1) synthesized from pyrenediboronic acid is one of the first utilized COFs to generate photocurrent. The ordered well-aligned 2D structure of PPy-COF represents excellent photoconductivity owing to transportation and migration of excitons along with the framework [19].

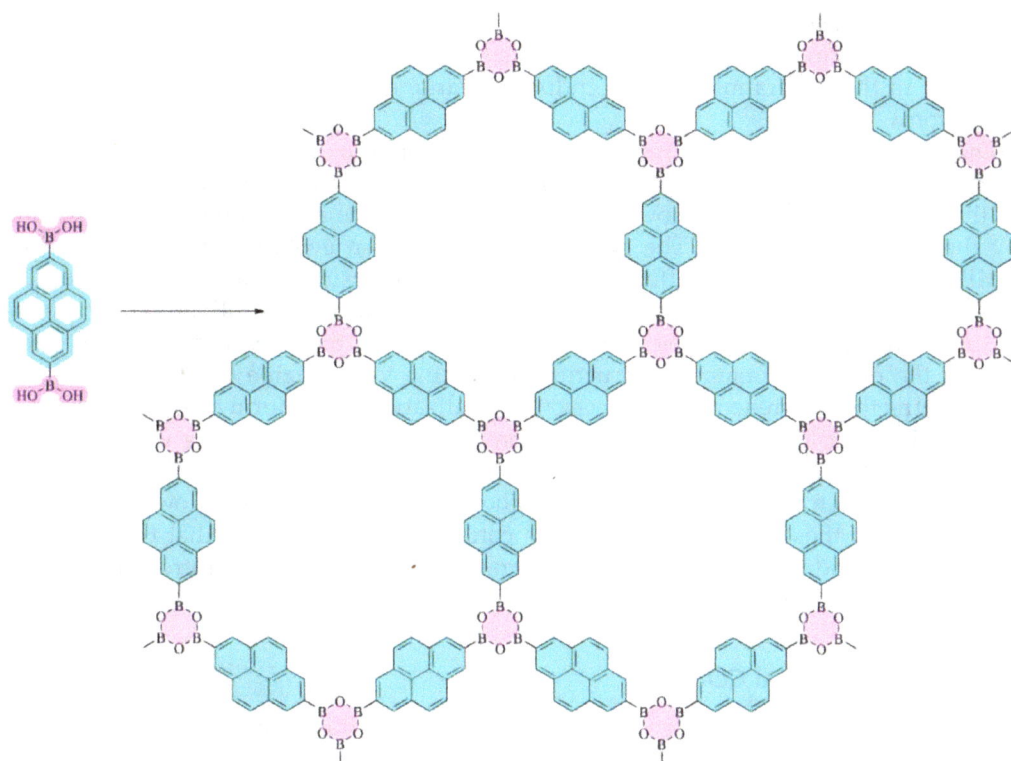

FIGURE 17.1
The schematic view of PPy-COF [19].

The light-harvesting ability of pyrene-based COFs made them good candidates to be used as photocatalysts for organic synthesis or water splitting [15, 20]. Combining pyrene derivatives as electron-donating groups with electron-accepting units could prepare D-A systems in an ordered framework. The ordered structure of COFs provides excellent channels for charge transportation, which reduces the charge recombination; thus, the light absorption shifted toward longer wavelength and lower energies, leading to improved photocatalytic activity. COF-JLU22 is one of the benchmarks of the preparation of D-A COFs. Conjugation of the imine-bond with the 1,3,6,8-tetrakis(4-aminophenyl) pyrene as the electron-donor, and a benzothiazole type dialdehyde as the electron acceptor in COF-JLU-22, which increases the rate of excitons migration along the mesoporous channels [21]. In COF-JLU-22, the light absorption ranges from visible light to the near-infrared region, high efficiency in the photoreduction of phenacyl bromide derivatives, and α-alkylation of aldehydes. In another similar study [22], a D-A type covalent organic polymer (COP) was constructed using 1,3,6,8-tetrabromopyrene (TBP) and 3,8-dibromophenanthroline (DBP) as the electron donor and acceptor respectively. The visible-light absorption around 450 nm led to a $\pi \rightarrow \pi^*$ electron transition in the conjugated system. To obtain the lowest CB, the TPB/DBP ratio was optimized to 3:1. By lowering the CB, the thermodynamic driving force was enhanced, and therefore, the electron migration was facilitated along the COP-TP3:1. The presence of the D-A moieties in the COP-TP structure caused a 14-fold enhancement in the rate of hydrogen evolution compared to the structure without D-A moieties. Upon increasing the number of electron acceptor molecules (DBP) in the structure, the π-π stacking caused an increase in the degree of aggregation, and as a result, the BET surface area dramatically decreased. The viability of the D-A strategy is proved by the increase in the generation of photoinduced charge carriers and the rate of hydrogen evolution.

During the synthesis of a series of conjugated microporous polymers (CMPs) containing pyrene moieties, it was observed that an increase in the pyrene content caused a redshift in the absorption wavelength owing to the decrease in the excitation energy dominated by the pyrene aggregation. The photocatalytic hydrogen evolution using these robust CMPs was resulted in high effectiveness even without using metals as cocatalysts [19]. The number of nitrogen atoms in peripheral aromatic units in the COFs built from pyrene blocks is directly related to the photocatalytic activity of the prepared COF. The extension of the planar π-conjugated structure by linking via an azine bond to tetra-alkyne 1,3,6,8-tetraethinylpyrene renders electron delocalization in the form of excimers (excited dimers). Excimer formation facilitates excitons' transportation, resulting in a redshift in the wavelength to around 670–700 nm. COFs constructed by pyrene blocks with nitrogen-free peripheral aromatic units represent higher photocatalytic activity toward hydrogen evolution than those with higher nitrogen contents. It was asserted that the decreased thermodynamic driving force of exciton migration in COFs with high nitrogen content in peripheral aromatic units caused the higher photocatalytic activity [23].

17.2.2 Perylene-Based COFs

There are few studies on perylene-based covalent organic frameworks and applying them in photocatalytic systems. In 2019, Ascherl [24] and co-workers had prepared a series of perylene-based COFs and investigated their switchable acid chromic properties. It was claimed that the protonation of imine linkages by exposure to the acids vapor caused a remarkable redshift in the absorption spectra. For instance, the COF prepared from tetrakis(4-aminophenyl)perylene and terephthaldehyde represent a redshift from around 400 nm to around 450 nm after exposure to the acid vapor. The application of other types of perylene derivatives was also studied. Yadav et al. [14] had prepared an imide-linked covalent triazine framework (CTF) using 3,4,9,10-perylenetetracarboxylicdiimide and 1,3,5-trichlorotriazine. It was then used as a photocatalyst for the selective production of formic acid from carbon dioxide. The prepared CTF represents an optical bandgap of 2.05 eV, which is suitable for visible-light-driven reduction of rhodium complex ([Cp*Rh(bpy) $H_2O]^{2+}$;Cp* = pentamethylcyclopentadienyl, bpy = 2,2'-bipyridine). After that, NADH was regenerated, and formic acid was produced from CO_2.

17.2.3 Triazine-Based COFs

Triazine is an aromatic six-membered-heterocyclic ring with three nitrogen atoms increasing its electron-accepting ability. Linking triazine building blocks with other electron-releasing units can provide a photoactive electron transfer platform. CTFs, as 2D triazine-based COFs, structures extending organic functional groups outward the triazine ring in a periodically arranged scaffold. The similarities of CTFs with g-C_3N_4 gained the attention of researchers in the photocatalysis field. Moreover, the CTFs band structures can be easily controlled by changing the electron-releasing building blocks linked to the triazine moieties [24, 25]. Based on the theoretical calculations, the number of stacking layers, nitrogen contents, and pore size can affect the electronic properties of CTFs. The latter is due to the quantum confinement effect.

As the involvement of six-membered rings increased, the delocalization of π-electrons increased, and as a result, the bandgap was narrowed. Moreover, the nitrogen content increased the electron delocalization at the p_x and p_y orbitals of the nitrogen atom and caused a contraction in the bandgap. The stacking effect is different from the two previously mentioned factors. AB stacking (staggered form) has a larger bandgap than the AA stacking (eclipsed form), resulting from higher electronic repulsive perturbation in the AB stacking structure [26]. In the case of planarity, triazine building blocks can form more planar structures rather than phenyl units. The transfer of excitons is facilitated in more planar structures. The planarity of the triazine rings was compared to the phenyl rings in recent studies [27]. In a study, a hydrazone-linked CTF was synthesized by a reaction between the 1,3,5-tris-(4-formyl-phenyl)triazine (TFPT) and 2,5-diethoxy-terephthalohydrazide (DETH). The dihedral angle between the phenyl and triazine in TFPT was calculated, and it was small enough to form a π-extended planar structure.

Furthermore, radical anion stabilization is another issue that is of vital importance in photocatalytic processes. A comparison between the electron affinities of COFs shown in Figure 17.2 exhibited that the charge separation in N_3-COF was increased with an increase in the number of nitrogen atoms leading to a higher photocatalytic activity due to its higher radical anion stabilization by the nitrogen atoms [28].

The phenylene spacer length between the two triazine cores is another parameter effective on the photocatalytic activity of CTFs. An increase of phenylene rings between the triazine moieties increases the light absorption and leads to the shrinkage of the bandgap of the CTF (Figure 17.3). It is noteworthy that the bandgap shrinkage is not suitable for the series in Figure 17.3. CTFs with longer phenylene linkers have negligible photocatalytic activity due to the lower thermodynamic driving force for the oxidation of sacrificial hole-scavenger. In Figure 17.3, samples prepared by the Suzuki-Miyaura coupling reaction were

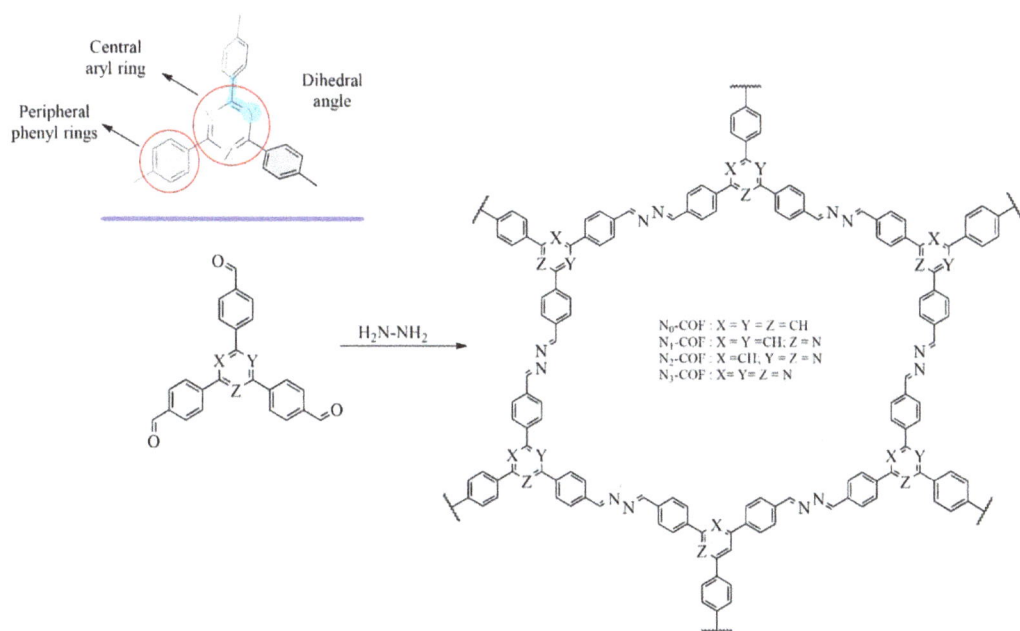

FIGURE 17.2
The highlighted dihedral angle in the picture was decreased by adding nitrogen atoms in X, Y, and Z [28].

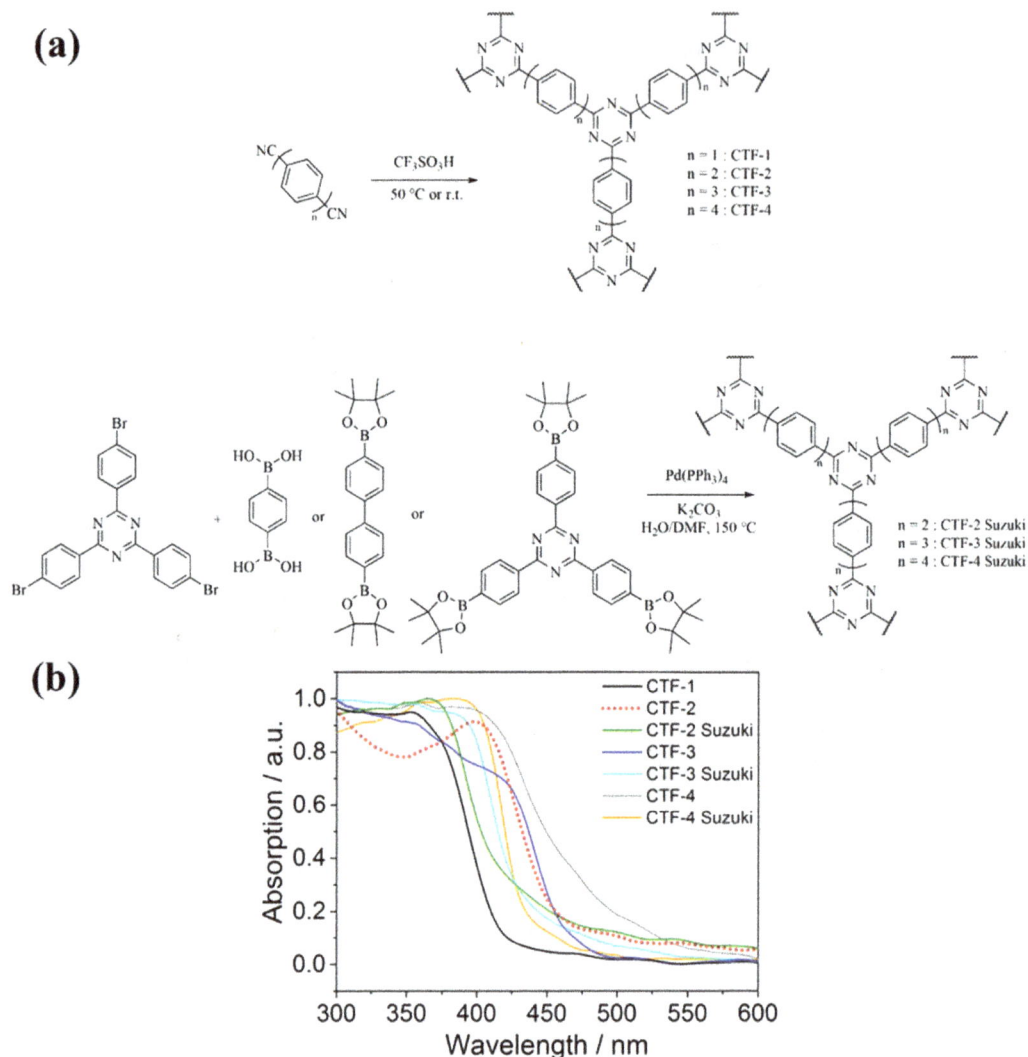

(a)

(b)

FIGURE 17.3

(a) CTFs prepared with two different methods with different phenylene spacer lengths. (b) UV-Vis diffuse reflectance of the CTFs. (Adapted with permission from reference [29], Copyright (2017), Elsevier.)

labeled as *Suzuki*, the other samples were prepared by nitrile trimerization in the presence of trifluoromethanesulfonic acid (TfOH) [29].

There are two different methods for the preparation of CTFs:

1. Self-trimerization of nitrile-containing monomers.
2. Triazine-containing monomers co-condensed with other possible linkages (imines, imides, boronate esters, etc.).

CTF-1 was the first CTF that was prepared from trimerization of 1,4-dicyanobenzene in molten $ZnCl_2$ at high temperatures [30], where $ZnCl_2$ acts as both lewis acid and solvent. The bandgap of CTF-1 found to be 2.42 eV made it suitable for visible-light harvesting in the water oxidation reaction and light-induced hydrogen evolution [26]. The partial carbonization occurred during the preparation of CTF-1 formed residual carbon materials preventing the photon absorption. Therefore, ionothermally synthesized CTF-1 exhibited negligible photocatalytic activity. To enhance the photocatalytic activity of CTF-1, the preparation methods were modified, and different kinds of monomers were used. In 2015, Lotsch and co-workers [31] found a relationship between the incomplete polymerization of oligomers and the yield of amorphous carbon at high temperatures (Figure 17.4a). To avoid carbonization, they propose increasing the reaction time to 150 hours and reducing the temperature to 300°C. In these conditions, the equilibrium presented in Figure 17.4a shifted to the left, leading to the formation of low-molecular-weight phenyl-triazine oligomers (PTOs). PTOs have shorter molecular lengths, causing smaller interlayer distances and stronger π-orbital overlaps, resulting in more efficient charge transfer. Despite the variation of molecular weights, PTOs found to be superb photocatalysts due to their excellent optical properties and carrier dynamics. However, carbonization was inevitable even in longer reaction times. Arne Thomas et al. [32] represent a two-step synthesis for the preparation of CTF-1 without being carbonized. First, an amorphous CTF was prepared in a self-trimerization of 1,4-dicyanobenzene in trifluoromethanesulfonic acid/chloroform at ambient temperatures. Then, a short-term ionothermal process was applied within 30 min to complete the polymerization. In this way, the carbonization was low enough to exhibit a high photocatalytic hydrogen evolution rate. In 2018, the microwave-assisted synthesis of CTF-1 led to a well-ordered and highly crystalline conjugated material with facilitated π-π* electron excitation. The microwave-assisted synthesized CTF-1 represents a superior activity toward g-C_3N_4 regarding its narrower bandgap.

Although the photocatalytic activities of ionothermal and microwave-assisted synthesized CTFs are higher than traditional carbonaceous materials, requiring harsh reaction

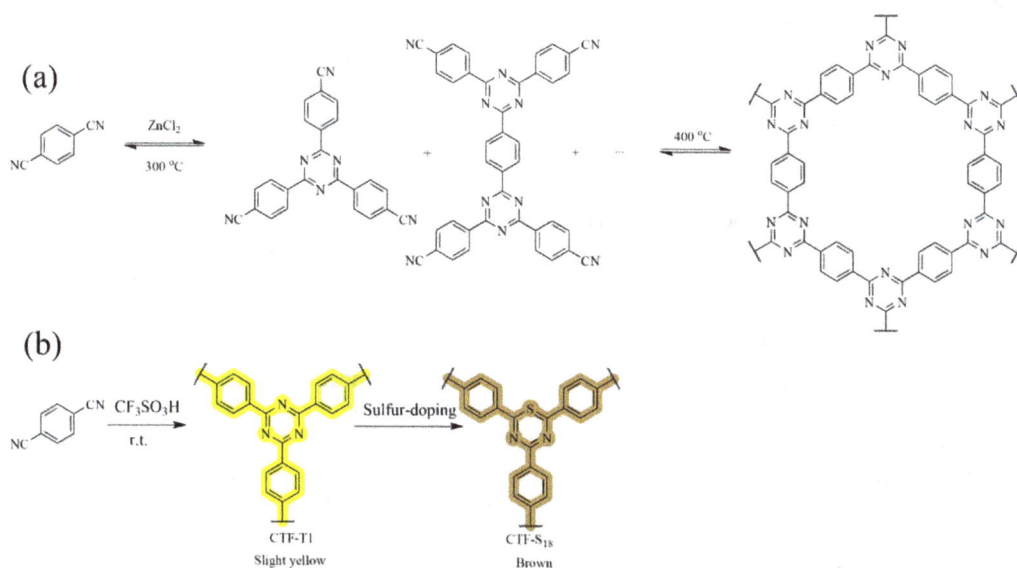

FIGURE 17.4

(a) The equilibrium presented in the formation of CTF-1 started from 1,4-dicyanobenzene. (b) Direct doping of sulfur into CTF-1 [31].

conditions in addition to lower crystallinity and partial carbonization limited their applications. Therefore, Bien Tan and coworkers [33] overcame this problem by developing synthesis methods with milder conditions. This approach used triazine-containing aldehydes and amidines to prepare CTFs with pre-designed structures, high surface area, and tunable functions. Intriguingly, the bandgap can be predicted by changing the building blocks. The crystallinity of the semiconductors significantly affects photocatalytic activity. There are two main methods to increase the crystallinity of the CTFs, (i) slowing the rate of nucleation and (ii) lowering the nuclei concentration. For instance, in situ oxidation of benzyl alcohols to aldehydes or controlling the feeding rate of monomers can increase the crystallinity of the prepared CTF. In a study [34], a highly crystalline CTF was prepared by dropwise addition of terephthalaldehyde solution with a rate of 30 μL min^{-1} to the solution containing terephthalamidine in the presence of Cs_2CO_3 where the (0 0 1) facet was abundant, representing the less number of defects in the unit cells. Benzyl alcohol oxidation was used in another study [35] to slow down the nucleation rate and enhance the crystallinity of a series of CTFs. The lower reaction rate, a higher AA stacking orientation, and superior photocatalytic performance resulted from better charge separation. In a similar study [15], a highly crystalline 2D COF, namely COF JLU5, was prepared in a Schiff-based condensation of 1,3,5-tris-(4-aminophenyl)triazine (TTA) and 2,5-dimethoxyterephthaldehyde. The formation of channels by π-stacking of 2D sheets provides a columnar structure for the diffusion and immigration of excitons. The presence of methoxy groups, as electron-releasing functions, close to imine-linkages, and triazine rings as the electron-accepting moieties constructed a D-A system that improves the photocatalytic efficiency.

Moreover, a fully conjugated melamine-based porous organic polymer was designed and synthesized with a green method. Melamine near the benzene rings led to a facile and fast charge carrier mobility along with the structure [36]. It was asserted that increasing the polymerization temperature caused a redshift in absorption and increased the visible-light harvesting efficiency. β-ketoenamine linked COFs are another class of COFs that possess excellent photocatalytic properties. In a recent study [37], a hybrid triazine-based COF with β-ketoenamine linkages was synthesized. The irreversibility of the β-ketoenamine linkages led to high chemical stability in exposure to light irradiation. The synthesized COF represents an excellent performance in photocatalytic isomerization of *E*-alkenes to *Z*-alkenes. It also became a benchmark for overcoming the stereoselectivity restrictions in organic synthesis.

Different kinds of imine-linked, hydrazone-linked, N-oxide linked, and imide-linked CTFs were also introduced in the photodegradation of organic pollutants. For instance, a series of CTFs were synthesized by Schiff-based condensation of different triazine-containing compounds such as 1,3,5-tris-(4-formyl-phenyl)triazine, 1,3,5-tris(4-aminophenyl)benzene, and 2,4,6-tris(4-hydrazinylphenyl)-1,3,5-triazine for degradation of chemical pollutants [38]. The stereoelectronic properties such as conjugated degree and density of active centers of monomers can control the state of valance and conduction bands. The relationship between the structure and electronic properties of CTFs was well established in different papers. Another approach to tune the electronic properties is doping inorganic semiconductors to COFs. For example, sulfur-doped CTF [39] was synthesized by replacing one of the nitrogen atoms in the triazine ring in the CTF-1 structure with a sulfur atom (Figure 17.4b). Doping sulfur atom caused a shrinkage in the CTF-1 bandgap and reduced it from 2.85 eV to 1.87 eV. It also led to a more negative electrostatic potential in CTF-1. Noteworthy, narrower bandgaps and smaller particle sizes in CTFs are more favorable for visible light-harvesting applications.

17.2.4 Porphyrin-Based COFs

Porphyrins and metal-porphyrins are a well-known chemically-stable class of macrocycles that are widely used in chemical and photochemical transformations. Using porphyrins as the building blocks of COFs synergistically increases the AA stacking adaptation and photosensitizing ability due to the higher rate of excitons transportation through the aligned-pore channels. In 2019, Wang and co-workers [40] used the Knoevenagel condensation to prepare porphyrin-based Sp²-COF. The condensation of 1,4-phenylenediacetonitrile and 5,10,15,20-tetrakis(4-benzaldehyde)porphyrin led to the formation of sp² carbon-conjugated linkages, which provides more stabilization compared to the β-ketoenamines, boronic esters, amide, and imine linkages (Figure 17.5). The sp² carbon-conjugated linkages are not easily affected by the acidic or basic medium, whereas the imine linkages exhibited lower chemical stability due to the higher reversibility in the acidic media. The conjugated double bond enhanced the electron delocalization in the 2D plane of the COF, leading to a redshift (32 nm) in the light absorption edge compared to the imine linkage analogs. Moreover, the photocatalytic activity of the sp² carbon-conjugated linkages toward the aerobic oxidation of amines to imines was superior to the imine linkage analogs.

Using transition metals porphyrins (Mo, Cu, Ni, Co, Fe, Pd, Ir, Rh, Ru, etc.) in the preparation of porphyrin-based COFs increases the photocatalytic activity due to the two more electron excitations of metal to ligand charge transfer (MLCT) and ligand to metal charge transfer (LMCT) besides the π-π* excitations. Pioneered works in preparing metal porphyrin COFs focused on the photoconductivity properties. In 2012, Spitler et al. prepared a series of metal porphyrin COFs and investigated their photophysical properties. Free porphyrin COF (H_2P-COF), copper porphyrin COF (CuP-COF), and zinc porphyrin COF (ZnP-COF) represent different properties. The eclipsed AA stacking of the porphyrin macrocycles in H_2P-COF dominated for the only hole transport channels. Interestingly, the insertion of metals into the structure provides metal-on-metal channels. Opposed to H_2P-COF, the

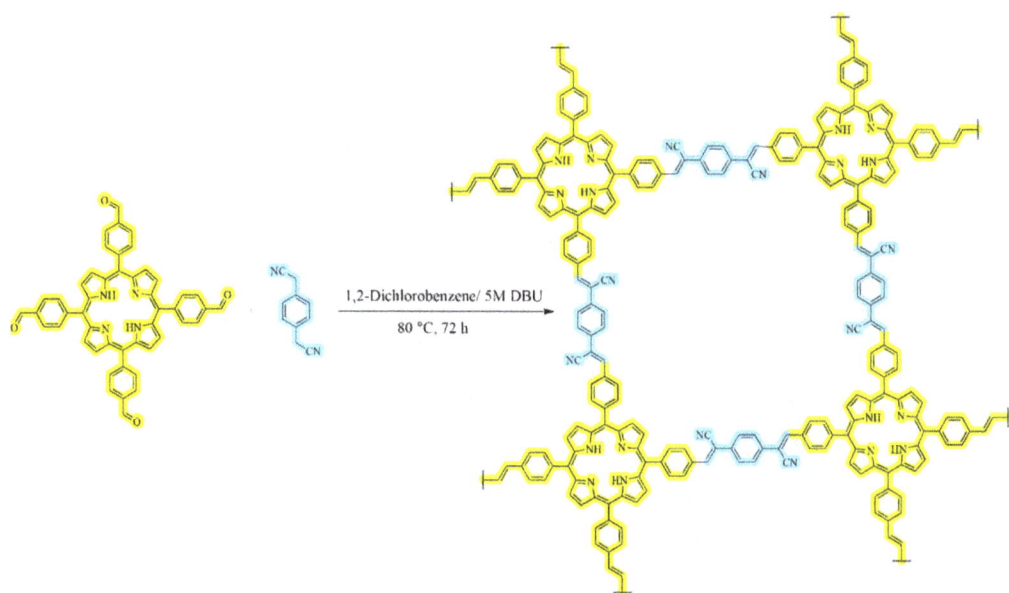

FIGURE 17.5
The schematic view of the porphyrin-based Sp²-COF synthesis procedure [40].

CuP-COF showed MLCT and ZnP-COF has ambipolar conduction allowing a balanced transport. UPC-CMP-1 [41] is another porphyrin-based COF prepared from Sonogashira coupling of iron(III) 5,10,15,20-tetrakis-(4'-bromophenyl) porphyrin and 1,4-diethynylben- zene. Congo red was selectively and efficiently degraded in the presence of UPC-CMP-1 and visible-light irradiation.

17.2.5 Phthalocyanine-Based COFs

As well as porphyrins, phthalocyanines (Pcs) exhibited high chemical and thermal sta- bility and are traditionally used as photocatalysts for organic synthesis. Similarly, metal insertion into phthalocyanine macrocycles created structures with aligned π-systems. The ordered structures of phthalocyanines provide channel-like structure, readily improve the light absorption by facilitating the excitons transfer. For instance, NiPc COF [42] with boronate ester linkages was prepared from the condensation of $[(OH)_8PcNi]$ and 1,4- benzenediboronic acid. Although the monomer $[(OH)_8PcNi]$ did not show photoresponse and photocurrent due to its disordered structure, NiPc COF represents enhanced visible and near-infrared light response and generated photocurrent. Changing the metal ions in Pc, COFs can affect the π-electron density on the macrocycle allowing to control the elec- tronic properties of the COF, especially the excitons transport.

17.2.6 Thiophene-Based COFs

Phthalocyanines, porphyrins, and triazines are nitrogen-rich skeletons used to prepare COFs. However, using other heteroatoms could also be helpful. The electron-donor ability of the sulfur-containing heterocycles was approved in different studies. Thiophene deriva- tives have been put at the center of attention due to their excellent electronic properties. They also used electron-withdrawing materials to prepare D-A systems. SBA-15 is meso- porous silica with hexagonal channels used as a substrate for different materials. In 2017, Zhang and co-workers [43] trimerized 2,5-dicyanothiophene on the SBA-15 to prepare a thiophene-based CTF (CTF-Th@SBA-15). The formation of triazine units, besides the thio- phene moieties, developed a D-A system. The levels of HUMO and LUMO of the prepared material were found to be at +1.75 V and -0.72 V (vs. SCE) respectively. It led to a visible- light absorption at around 520 nm. Comparing the CTF-Th@SBA-15 and pure CTF-Th, rep- resent a marked increase in the surface area and photocatalytic activity of CTF-Th@SBA-15 in benzyl alcohol oxidation to benzaldehyde. The higher photocatalytic activities might be due to the mesoporous structure of the support, making nanoreactors for reactions.

   One of the critical aspects of organic semiconductors is the ability to tune the opto- electronic properties by changing the alignment ordering of donors and acceptors in the structure [44]. It could be explained by defining the symmetry in COFs. Most of the CTFs introduced during the last decade are synthesized via the condensation of symmetrical monomers. Huang and co-workers first introduced the asymmetry in COFs in 2018 [45]. They used 5-(4-cyanophenyl)thiophene-2-carbonitrile (Figure 17.6) in a trimerization reac- tion to prepare an asymmetrical CTF (asy-CTF). In this way, four different D-A domains were formed where thiophene and phenyl ring act as electron donors and triazine rings act as electron acceptors. The four D-A domains enhance the intramolecular electron transfer exhibiting a cascade energy transfer from higher to lower energies. In the asy-CTF there would be a step-by-step excitation rather than the symmetric CTF (sym-CTF). Therefore, asy-CTF represents a higher photocatalytic activity toward photosynthesis of 1,2,3- triphenylphosphindole 1-oxide.

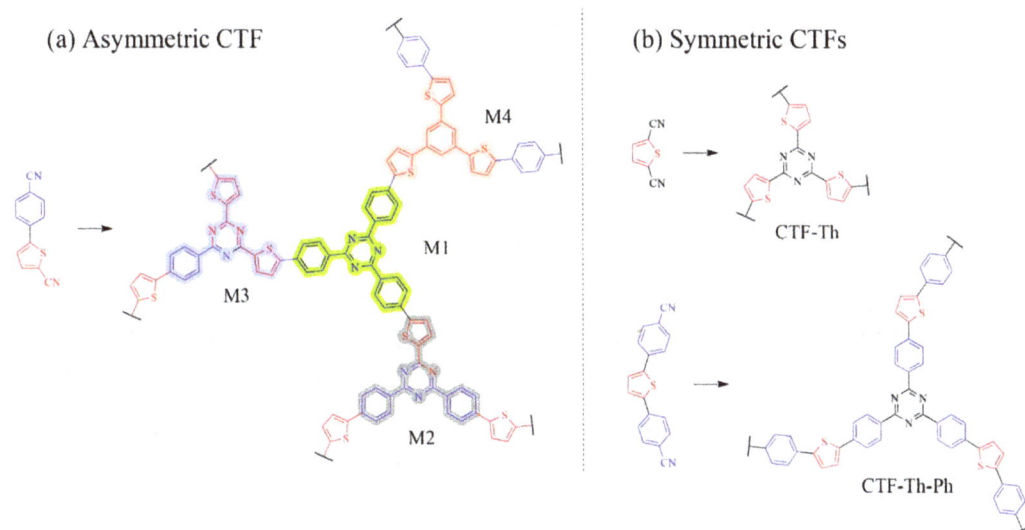

FIGURE 17.6
CTFs with (a) asymmetric structures, (b) symmetric structures [45].

17.2.7 Sulfone-Based COFs

There are some examples of using sulfone-containing COFs as photocatalysts. The impact of the number of sulfone groups on the photocatalytic activity of COFs toward hydrogen evolution was investigated by Wang et al. [46]. Three β-ketoenamine linkages COFs (Figure 17.7) have been prepared using 4,4″-diamino-*p*-terphenyl (TP-COF), 3,7-diaminodibenzo[b,d]thiophenesulfone (S-COF), and 3,9-diamino-benzo[1,2-b:4,5-b′]

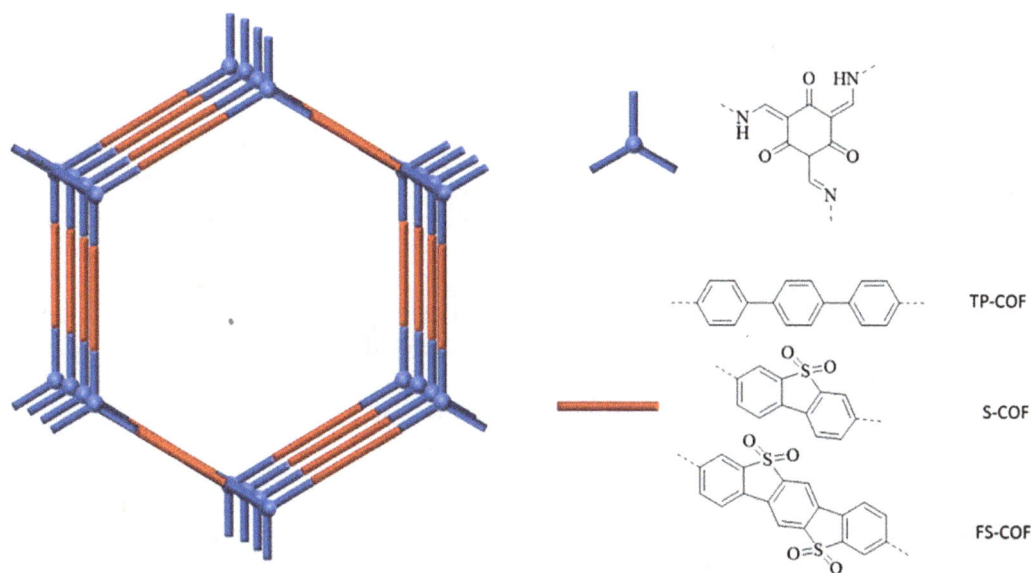

FIGURE 17.7
The schematic view of three different β-ketoenamine sulfone-based COFs [46].

bis[1]benzothiophene-5,5,11,11-tetraoxide (FS-COF). By comparing the hydrogen evolution rate, it was found that the FS-COF with two sulfone groups per unit represents higher photocatalytic activity than S-COF and TP-COF. Moreover, the addition of sulfone groups increased the hydrophilicity by decreasing the water contact angle. Rather than the photocatalytic activity due to the narrower bandgap, FS-COF shows a higher surface area, crystallinity, and AA eclipsed layer stacking. Also, the loading of WS5F as a sensitizer increased the hydrogen evolution rate up to 16.3 mmol h^{-1} g^{-1}.

As well as the number of sulfone groups, another effective parameter on the photocatalytic performance is the length of cross-linker agents. Wang and coworkers [47] explored that the increase in the length of cross-linkers (benzene, biphenyl, and *p*-terphenyl) in a series of sulfone-based CMPs dramatically decreased the rate of hydrogen evolution. It was asserted that the degree of planarity and conjugation decreased in longer cross-linker agents due to the increase in the dihedral angles resulting in poor excitons transportation. The charge recombination is suppressed by transferring light-induced electrons from the HOMO orbitals of electron-donor units (phenyl rings) to the LUMO orbitals of electron-acceptor units (dibenzothiophene dioxide).

17.2.8 Other COFs

Recent studies introduced novel linkages in the preparation of COFs such as triptycene, benzoxazole, and benzothiazole. The mechanochemical grinding method was used to prepare a graphene-like layered triptycene-based COF utilized in photochemical degradation of rhodamine B [48]. Small molecules with benzoxazole moieties are used in the photogeneration of electrons. Benzoxazoles are traditionally prepared in a cascade reaction of *o*-aminophenols with aldehydes. Wang et al. [20] introduced three different benzoxazole-linked COFs (COF LZU-190, LZU-191, LZU-192) and compared their photocatalytic activity toward the transformation of arylboronic acids to phenols. The prepared COFs represent excellent features, including high stability, crystallinity, and high surface area. Among

TABLE 17.1

Photocatalytic Applications of COFs

Photocatalytic Application	COFs		Bandgap (eV)	Reference
Water splitting	Triazine-based	N$_0$-COF	2.6~2.7	[28]
		N$_1$-COF		
		N$_2$-COF		
		N$_4$-COF		
		TFPT-COF	2.8	[27]
		CTF-HUST-A1	2.35	[49]
Degradation of contaminants	B-ketoenamine	TpMa		[13]
	Porphyrin-based	UPC-CMP-1	1.76	[41]
	Triazine-based	CTF-HUST-HC1	-	[34]
Reduction of CO$_2$	Triazine-based	2D-CTF	2.05	[14]
		CT-COF	2.04	[50]
Organic synthesis	Triazine-based	COF-JLU5	2.14	[15]
	Porphyrin-based	Por-sp^{2}c-COF	1.75	[40]
		Por-COF	1.76	[40]
	Pyrene-based	COF-JLU22	2.08	[21]

them, COF LZU-190 represents superior photocatalytic activity in phenol formation from arylboronic acids.

17.3 Photocatalytic Application of COFs

Photoactive COFs have been used in different fields of science. Their potential application as photocatalysts in environmental remediation, organic synthesis, water splitting, and CO_2 reduction was reviewed and summarized in Table 17.1.

17.4 Conclusion

In conclusion, the light-harvesting ability of COFs, besides other magnificent properties, made them potential photocatalysts for several studies, including organic synthesis, CO_2 reduction, environmental remediation, and water splitting. The variety of photoactive building blocks such as pyrene, perylene, sulfone, thiophene, and triazine led to several unique structures. The photocatalytic activity of COFs was modified by suppressing the charge recombination to increase the separation efficiency of photo-induced excitons. The photoactivity of COFs can be predetermined by designing the symmetry of building blocks, linkages of COFs, and using D-A-type materials. Using COFs in preparing photocatalysts can develop clean energy for human societies and create enormous prospects for different applications. Merging COFs with other semiconductors can positively affect the H_2 or O_2 evolution upon light irradiation.

References

1. Wang H., Zhang L., Chen Z., Hu J., Li S., Wang Z., Liu J., Wang X., Semiconductor heterojunction photocatalysts: design, construction, and photocatalytic performances, *Chemical Society Reviews* 2014, 43, 5234–5244.
2. Chen X., Liu L., Peter Y.Y., Mao S.S., Increasing solar absorption for photocatalysis with black hydrogenated titanium dioxide nanocrystals, *Science* 2011, 331, 746–750.
3. Martin D.J., Liu G., Moniz S.J., Bi Y., Beale A.M., Ye J., Tang J., Efficient visible driven photocatalyst, silver phosphate: performance, understanding and perspective, *Chemical Society Reviews* 2015, 44, 7808–7828.
4. Lee K.M., Lai C.W., Ngai K.S., Juan J.C., Recent developments of zinc oxide based photocatalyst in water treatment technology: a review, *Water Research* 2016, 88, 428–448.
5. Cheng L., Xiang Q., Liao Y., Zhang H., CdS-based photocatalysts, *Energy & Environmental Science* 2018, 11, 1362–1391.
6. Li W., Elzatahry A., Aldhayan D., Zhao D., Core–shell structured titanium dioxide nanomaterials for solar energy utilization, *Chemical Society Reviews* 2018, 47, 8203–8237.
7. Li Q., Li X., Wageh S., Al-Ghamdi A.A., Yu J., CdS/graphene nanocomposite photocatalysts, *Advanced Energy Materials* 2015, 5, 1500010.

8. Fu J., Yu J., Jiang C., Cheng B., g-C$_3$N$_4$-Based heterostructured photocatalysts, *Advanced Energy Materials* 2018, 8, 1701503.

9. Dhakshinamoorthy A., Asiri A.M., Garcia H., Metal–organic framework (MOF) compounds: photocatalysts for redox reactions and solar fuel production, *Angewandte Chemie International Edition* 2016, 55, 5414–5445.

10. Wang Y., Vogel A., Sachs M., Sprick R.S., Wilbraham L., Moniz S.J., Godin R., Zwijnenburg M.A., Durrant J.R., Cooper A.I., Current understanding and challenges of solar-driven hydrogen generation using polymeric photocatalysts, *Nature Energy* 2019, 4, 746–760.

11. Mokhtari N., Afshari M., Dinari M., Synthesis and characterization of a novel fluorene-based covalent triazine framework as a chemical adsorbent for highly efficient dye removal, *Polymer* 2020, 195, 122430.

12. Bi J., Fang W., Li L., Wang J., Liang S., He Y., Liu M., Wu L., Covalent triazine-based frameworks as visible light photocatalysts for the splitting of water, *Macromolecular Rapid Communications* 2015, 36, 1799–1805.

13. Lv H., Zhao X., Niu H., He S., Tang Z., Wu F., Giesy J.P., Ball milling synthesis of covalent organic framework as a highly active photocatalyst for degradation of organic contaminants, *Journal of Hazardous Materials* 2019, 369, 494–502.

14. Yadav R.K., Kumar A., Park N.-J., Kong K.-J., Baeg J.-O., A highly efficient covalent organic framework film photocatalyst for selective solar fuel production from CO$_2$, *Journal of Materials Chemistry A* 2016, 4, 9413–9418.

15. Zhi Y., Li Z., Feng X., Xia H., Zhang Y., Shi Z., Mu Y., Liu X., Covalent organic frameworks as metal-free heterogeneous photocatalysts for organic transformations, *Journal of Materials Chemistry A* 2017, 5, 22933–22938.

16. Nguyen H.L., Gándara F., Furukawa H., Doan T.L., Cordova K.E., Yaghi O.M., A titanium–organic framework as an exemplar of combining the chemistry of metal–and covalent–organic frameworks, *Journal of the American Chemical Society* 2016, 138, 4330–4333.

17. Li L., Zhou Z., Li L., Zhuang Z., Bi J., Chen J., Yu Y., Yu J., Thioether-functionalized 2D covalent organic framework featuring specific affinity to Au for photocatalytic hydrogen production from seawater, *ACS Sustainable Chemistry & Engineering* 2019, 7, 18574–18581.

18. Bessinger D., Ascherl L., Auras F., Bein T., Spectrally switchable photodetection with near-infrared-absorbing covalent organic frameworks, *Journal of the American Chemical Society* 2017, 139, 12035–12042.

19. Wan S., Guo J., Kim J., Ihee H., Jiang D., A photoconductive covalent organic framework: self-condensed arene cubes composed of eclipsed 2D polypyrene sheets for photocurrent generation, *Angewandte Chemie International Edition* 2009, 48, 5439–5442.

20. Wei P.-F., Qi M.-Z., Wang Z.-P., Ding S.-Y., Yu W., Liu Q., Wang L.-K., Wang H.-Z., An W.-K., Wang W., Benzoxazole-linked ultrastable covalent organic frameworks for photocatalysis, *Journal of the American Chemical Society* 2018, 140, 4623–4631.

21. Li Z., Zhi Y., Shao P., Xia H., Li G., Feng X., Chen X., Shi Z., Liu X., Covalent organic framework as an efficient, metal-free, heterogeneous photocatalyst for organic transformations under visible light, *Applied Catalysis B: Environmental* 2019, 245, 334–342.

22. Liu Y., Liao Z., Ma X., Xiang Z., Ultrastable and efficient visible-light-driven hydrogen production based on donor–acceptor copolymerized covalent organic polymer, *ACS Applied Materials & Interfaces* 2018, 10, 30698–30705.

23. Stegbauer L., Zech S., Savasci G., Banerjee T., Podjaski F., Schwinghammer K., Ochsenfeld C., Lotsch B.V., Tailor-made photoconductive pyrene-based covalent organic frameworks for visible-light driven hydrogen generation, *Advanced Energy Materials* 2018, 8, 1703278.

24. Ascherl L., Evans E.W., Gorman J., Orsborne S., Bessinger D., Bein T., Friend R.H., Auras F., Perylene-based covalent organic frameworks for acid vapor sensing, *Journal of the American Chemical Society* 2019, 141, 15693–15699.

25. Hug S., Tauchert M.E., Li S., Pachmayr U.E., Lotsch B.V., A functional triazine framework based on N-heterocyclic building blocks, *Journal of Materials Chemistry* 2012, 22, 13956–13964.

26. Jiang X., Wang P., Zhao J., 2D covalent triazine framework: a new class of organic photocatalyst for water splitting, *Journal of Materials Chemistry A* 2015, 3, 7750–7758.
27. Stegbauer L., Schwinghammer K., Lotsch B.V., A hydrazone-based covalent organic framework for photocatalytic hydrogen production, *Chemical Science* 2014, 5, 2789–2793.
28. Vyas V.S., Haase F., Stegbauer L., Savasci G., Podjaski F., Ochsenfeld C., Lotsch B.V., A tunable azine covalent organic framework platform for visible light-induced hydrogen generation, *Nature Communications* 2015, 6, 1–9.
29. Meier C.B., Sprick R.S., Monti A., Guiglion P., Lee J.-S.M., Zwijnenburg M.A., Cooper A.I., Structure-property relationships for covalent triazine-based frameworks: The effect of spacer length on photocatalytic hydrogen evolution from water, *Polymer* 2017, 126, 283–290.
30. Kuhn P., Antonietti M., Thomas A., Porous, covalent triazine-based frameworks prepared by ionothermal synthesis, *Angewandte Chemie International Edition* 2008, 47, 3450–3453.
31. Schwinghammer K., Hug S., Mesch M.B., Senker J., Lotsch B.V., Phenyl-triazine oligomers for light-driven hydrogen evolution, *Energy & Environmental Science* 2015, 8, 3345–3353.
32. Kuecken S., Acharjya A., Zhi L., Schwarze M., Schomäcker R., Thomas A., Fast tuning of covalent triazine frameworks for photocatalytic hydrogen evolution, *Chemical Communications* 2017, 53, 5854–5857.
33. Wang K., Yang L.M., Wang X., Guo L., Cheng G., Zhang C., Jin S., Tan B., Cooper A., Covalent triazine frameworks via a low-temperature polycondensation approach, *Angewandte Chemie International Edition* 2017, 56, 14149–14153.
34. Liu M., Jiang K., Ding X., Wang S., Zhang C., Liu J., Zhan Z., Cheng G., Li B., Chen H., Controlling monomer feeding rate to achieve highly crystalline covalent triazine frameworks, *Advanced Materials* 2019, 31, 1807865.
35. Liu M., Huang Q., Wang S., Li Z., Li B., Jin S., Tan B., Crystalline covalent triazine frameworks by in situ oxidation of alcohols to aldehyde monomers, *Angewandte Chemie International Edition* 2018, 57, 11968–11972.
36. Huang X., Wu Z., Zheng H., Dong W., Wang G., A sustainable method toward melamine-based conjugated polymer semiconductors for efficient photocatalytic hydrogen production under visible light, *Green Chemistry* 2018, 20, 664–670.
37. Bhadra M., Kandambeth S., Sahoo M.K., Addicoat M., Balaraman E., Banerjee R., Triazine functionalized porous covalent organic framework for photo-organocatalytic E–Z isomerization of olefins, *Journal of the American Chemical Society* 2019, 141, 6152–6156.
38. He S., Yin B., Niu H., Cai Y., Targeted synthesis of visible-light-driven covalent organic framework photocatalyst via molecular design and precise construction, *Applied Catalysis B: Environmental* 2018, 239, 147–153.
39. Li L., Fang W., Zhang P., Bi J., He Y., Wang J., Su W., Sulfur-doped covalent triazine-based frameworks for enhanced photocatalytic hydrogen evolution from water under visible light, *Journal of Materials Chemistry A* 2016, 4, 12402–12406.
40. Chen R., Shi J.L., Ma Y., Lin G., Lang X., Wang C., Designed synthesis of a 2D porphyrin-based sp^2 carbon-conjugated covalent organic framework for heterogeneous photocatalysis, *Angewandte Chemie International Edition* 2019, 58, 6430–6434.
41. Xiao Z., Zhou Y., Xin X., Zhang Q., Zhang L., Wang R., Sun D., Iron (III) porphyrin-based porous material as photocatalyst for highly efficient and selective degradation of Congo Red, *Macromolecular Chemistry and Physics* 2016, 217, 599–604.
42. Ding X., Guo J., Feng X., Honsho Y., Guo J., Seki S., Maitarad P., Saeki A., Nagase S., Jiang D., Synthesis of metallophthalocyanine covalent organic frameworks that exhibit high carrier mobility and photoconductivity, *Angewandte Chemie International Edition* 2011, 50, 1289–1293.
43. Huang W., Ma B.C., Lu H., Li R., Wang L., Landfester K., Zhang K.A., Visible-light-promoted selective oxidation of alcohols using a covalent triazine framework, *ACS Catalysis* 2017, 7, 5438–5442.
44. Wu J.-S., Cheng S.-W., Cheng Y.-J., Hsu C.-S., Donor–acceptor conjugated polymers based on multifused ladder-type arenes for organic solar cells, *Chemical Society Reviews* 2015, 44, 1113–1154.

45. Huang W., Byun J., Rörich I., Ramanan C., Blom P.W., Lu H., Wang D., Caire da Silva L., Li R., Wang L., Asymmetric covalent triazine framework for enhanced visible-light photoredox catalysis via energy transfer cascade, *Angewandte Chemie International Edition* 2018, 57, 8316–8320.
46. Wang X., Chen L., Chong S.Y., Little M.A., Wu Y., Zhu W.-H., Clowes R., Yan Y., Zwijnenburg M.A., Sprick R.S., Sulfone-containing covalent organic frameworks for photocatalytic hydrogen evolution from water, *Nature Chemistry* 2018, 10, 1180–1189.
47. Wang Z., Yang X., Yang T., Zhao Y., Wang F., Chen Y., Zeng J. H., Yan C., Huang F., Jiang J.-X., Dibenzothiophene dioxide based conjugated microporous polymers for visible-light-driven hydrogen production, *ACS Catalysis* 2018, 8, 8590–8596.
48. Preet K., Gupta G., Kotal M., Kansal S.K., Salunke D.B., Sharma H.K., Chandra Sahoo S., Van Der Voort P., Roy S., Mechanochemical synthesis of a new triptycene-based imine-linked covalent organic polymer for degradation of organic dye, *Crystal Growth & Design* 2019, 19, 2525–2530.
49. Zhang S., Cheng G., Guo L., Wang N., Tan B., Jin S., Strong-base-assisted synthesis of a crystalline covalent triazine framework with high hydrophilicity via benzylamine monomer for photocatalytic water splitting, *Angewandte Chemie* 2020, 132, 6063–6070.
50. Lei K, Wang D., Ye L., Kou M., Deng Y., Ma Z., Wang L., Kong Y., A metal-free donor–acceptor covalent organic framework photocatalyst for visible-light-driven reduction of CO_2 with H_2O, *ChemSusChem* 2020, 13, 1725–1729.

18

Recent Advancement in Covalent Organic Frameworks for Photocatalytic Activities

El-Sayed M. El-Sayed[1,2,3] and Daqiang Yuan[1,2]

[1]State Key Laboratory of Structural Chemistry, Fujian Institute of Research on the Structure of Matter, Chinese Academy of Sciences, Fujian, Fuzhou, P. R. China

[2]University of the Chinese Academy of Sciences, Beijing, P.R. China

[3]Chemical Refining Laboratory, Refining Department, Egyptian Petroleum Research Institute, Cairo, Egypt

CONTENTS

18.1 Introduction

Due to the increased population worldwide, numerous sustainable and clean energy sources should replace non-renewable ones such as coal, petroleum, and natural gas [1]. These clean fuels, such as hydrogen, can be produced based on a water-splitting-based photocatalytic process [2]. In this regard, several catalytic systems were employed to achieve hydrogen evolution with high efficiency, including zeolites, metal-organic frameworks (MOFs), COFs, and so on. Several photocatalysts exhibited limited applicability due to their instability under harsh conditions, inadequate charge transfer channels,

DOI: 10.1201/9781003206507-18

disordered structures, low surface areas, and narrower band gaps. As a result, employing a photocatalytic system with improved intrinsic physicochemical properties could display enhanced performance. As emerging porous framework materials, COFs have the potential to possess high photocatalytic activity under harsh settings. These materials consist of light elements (C, N, O, B, etc.), resulting in less dense frameworks formed by the covalent linkage of purely organic building blocks to produce dimensional architectures. The prepared COFs are based on various linkages, including boroxines, imines, triazines, esters, olefins, amides, imides, hydrazones, etc. [3, 4]. COFs based on their crystallinity, porosity, high surface area, tunable pore size, and rigid skeletons are amenable to be employed in a variety of applications, including gas adsorption and separation, proton conduction, sensing, drug delivery, energy storage, and catalysis.

Employing COFs as a photocatalytic platform for implementing several environmental and energy applications has been on the rise over the past few decades. This is because COFs exhibit exceptional physicochemical features that empower them to conduct photocatalysis under harsh conditions with high efficiency and stability. In terms of stability, the rigid skeletons of COFs based on diverse covalent linkages are responsible for the stability of prepared materials. This highlights that some linkages are favorable for producing robust thermally stable skeletons and others cannot, implying that some COF types can survive under harsh catalytic settings and others degrade or collapse. In addition to chemical stability, these robust linkages endow the designed COFs with high crystallinity required for COF photocatalytic systems to facilitate charge separation and transport, resulting in high performance and exceptional outcomes over amorphous analogues [5]. After selecting an appropriate linkage type for achieving stability and crystallinity, rational selection of the building units employed in the synthesis is critical. These building blocks produce COFs featuring exact pore diameters and regular channels due to their periodic arrangement in the structure and interaction between layers. In addition, a series of photoactive COF catalysts can be constructed using functional building units containing different heteroatoms (N, O, S, F, and Cl). This results in particular physicochemical characteristics for the resultant COF systems, implying the ability of such systems to be structurally tuned at the atomic level.

Moreover, post-synthetic modification (PSM) is a powerful approach to modulate the physicochemical properties of prepared COFs by introducing other functional groups rationally selected to endow COFs with particular ability to be employed in specific applications with high performance [6]. COFs with high surface areas and pore sizes are obtained when longer building blocks or supercritical carbon dioxide (scCO$_2$) activation are employed. Pore sizes and surface areas have the potential to affect charge diffusion inside COF frameworks, resulting in high performance in their catalytic systems. This hints that anchoring bulky functional groups for modulating the pore size of COFs can alter the physicochemical properties but influence the charge diffusion process, implying that a compromise should be considered between pore size modulation and charge diffusion to afford the optimal COF system. The above-mentioned features, including selecting linkage type and functional building units, are essential for producing COF systems with high crystallinity, porosity, surface area, and stability. Such advantages are high contributors to the working capacity of any COF photocatalytic system under harsh conditions to conduct several catalytic reactions while maintaining the structural integrity for COFs following successive cycles. To synthesize a COF system with compromised features, the synthetic process is the main determinant for expected predesigned structures, emphasizing that several trials should be practiced to rationally select the best synthetic method to accomplish the desired physicochemical characteristics. Although the solvothermal method requires high temperatures and reaction times, it yields highly crystalline COF materials that affect other associated features. This guides us to consider many parameters when synthesizing a COF system with specific properties. For instance, using longer building blocks to obtain high surface area COFs reduces the architectural stability of resultant frameworks.

By screening literature in recent years, COFs were significantly exploited in numerous photocatalytic applications in terms of energy and environmental aspects [7–10]. Typically, photocatalytic COFs were employed for hydrogen evolution, oxygen production, CO_2 reduction, organic transformations, degradation of organic pollutants, etc. Considering the prerequisite inherent properties that a COF system should have, rational selection of COF linkage type and its associated building blocks, followed by optimizing the synthetic conditions, hold great promise in the catalytic activity of COFs in different implementations. These aspects are considered for pure or pristine COFs produced from the covalent linkage of purely organic building units. However, using COF hybrids, COFs based on porous organic cages, or metalated COFs can bring exceptional properties, resulting in high photocatalytic performance outperforming pristine COFs.

In this chapter, we provide a recent update on the employment of COF materials in photocatalysis. First, we introduce the prerequisite features that should be held by a COF system to perform well in a catalytic reaction, highlighting relevant examples. This implies that design principles and synthetic procedures to produce a high-performance COF system will be presented. After that, we discuss the reported applications catalyzed by COFs and their potential implementations that can be addressed by research groups more extensively. We then highlight different novel COF systems that could accomplish unexpected photocatalytic activities in different catalytic reactions, Figure 18.1.

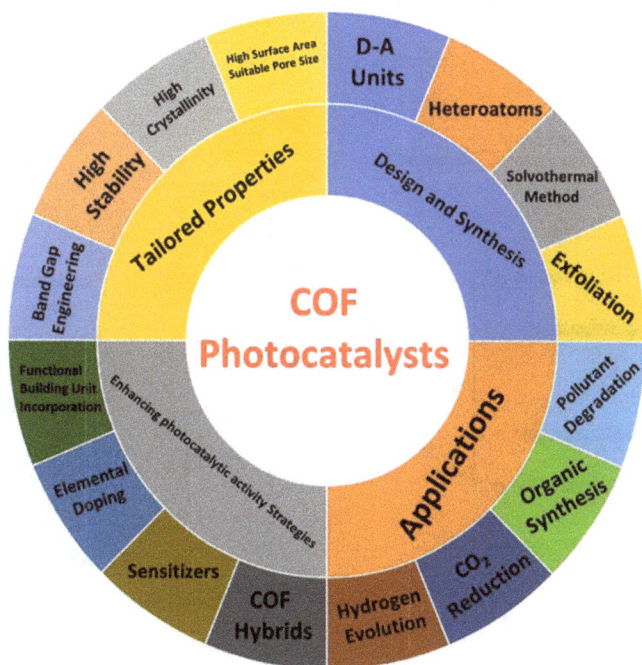

FIGURE 18.1
An overview of topics discussed in this chapter.

18.2 Tailored Properties for Active COF-Based Photocatalysts

Since photocatalytic reactions require particular materials to survive under harsh circumstances, designing robust COF-based photocatalysts is troublesome. In this regard, topology-guided predesign is a major step toward the rational design of COF structures amenable to conducting photocatalytic reactions with excellent outcomes. Designing COFs with a particular focus on their physicochemical properties is the leading pathway that constructs durable crystalline porous photoactive COFs with high surface area, stability, and excellent performance. These features to be found in one COF system require proper selection of covalent linkage-type responsible for constructing COF frameworks, which is highly significant to dictate the inherent chemical and architectural stabilities within COF skeletons. In addition, the diffusion process of photogenerated charges requires adequate pore sizes and surface areas, implying the rational selection of building blocks. Finally, constructing highly crystalline COF-based photocatalysts plays a critical function in facilitating the separation and transport of the generated charges inside the catalytic system.

18.2.1 Band Gap Engineering

Typically, in a given photocatalytic reaction, a series of photochemical processes occur as follows: (a) photocatalytic materials harvest light (UV, visible, or near-infrared) based on their energy band gaps (E_g); (b) when photons of incident light possess higher energy than band gap in a photocatalytic system, holes (h^+) are generated in the valence band (VB) due to excitation of electrons (e^-) from VB to the conduction band (CB); (c) photogenerated electrons and holes are migrated to the surface, and (d) a photoredox cascade reaction occurs on the surface (Figure 18.2) [7]. For a COF system, the energy band gaps rely on the

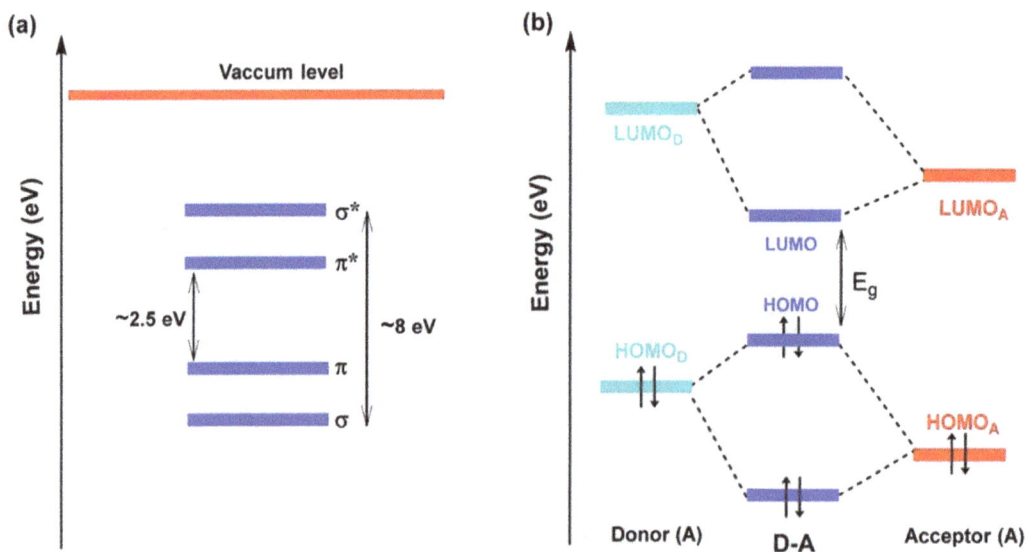

FIGURE 18.2
(a) HOMO-LUMO transitions in a π-conjugated system, (b) Schematic demonstration of hybridizing molecular orbital levels in D-A system via intramolecular interactions. (Adapted with permission [7]. Copyright (2021) Elsevier.)

difference between VB (HOMO, the highest occupied molecular orbital) and CB (LUMO, the lowest unoccupied molecular orbital). When light is absorbed, electronic transitions occur from σ→σ* in UV range and π→π* in visible light, demonstrating the significance of π-conjugated units in COFs for light harvesting. Generally, COF covalent linkages decrease the optical band gap and increase the ability to light harvesting. By introducing heteroatoms or employing selective functional building blocks, COF systems can produce tunable band gaps and increase light absorption. In addition, introducing specific functional groups into COFs can enhance light absorption, such as porphyrin, sulfone, imidazole, etc. By optimizing the band gaps in COFs via rational design, they reach the range of 1.30–2.90 eV featuring a wide optical absorption range up to 800 nm. In summary, tuning band gaps and boosting the light absorption of a given COF yield an ideal photocatalytic system amenable to conducting several implementations for energy and environment aspects.

18.2.2 High Stability

Long-term stability is a prerequisite for a COF system to survive in the presence of harsh photochemical conditions [11]. Such stability originates from the covalent linkage between building blocks that form COF frameworks (Figure 18.3). Among COF types, triazine-linked COFs or CTFs manifested ultrastability due to their robust triazine linkages derived from nitrile trimerization. Due to their nitrogen-rich triazine rings, CTFs are chemically and thermally stable toward photochemical reactions. Since charge migration is improved in photochemical reactions when crystalline COFs are used, crystalline CTFs are required. However, robustness of triazine linkages and their low reversibility limit the number of ordered crystalline CTFs. Although imine-linked COFs are one of the most widely studied, they are unstable under different chemical conditions. Other Schiff-based linkages such as hydrazones and azines are more stable than imine linkages. Several strategies

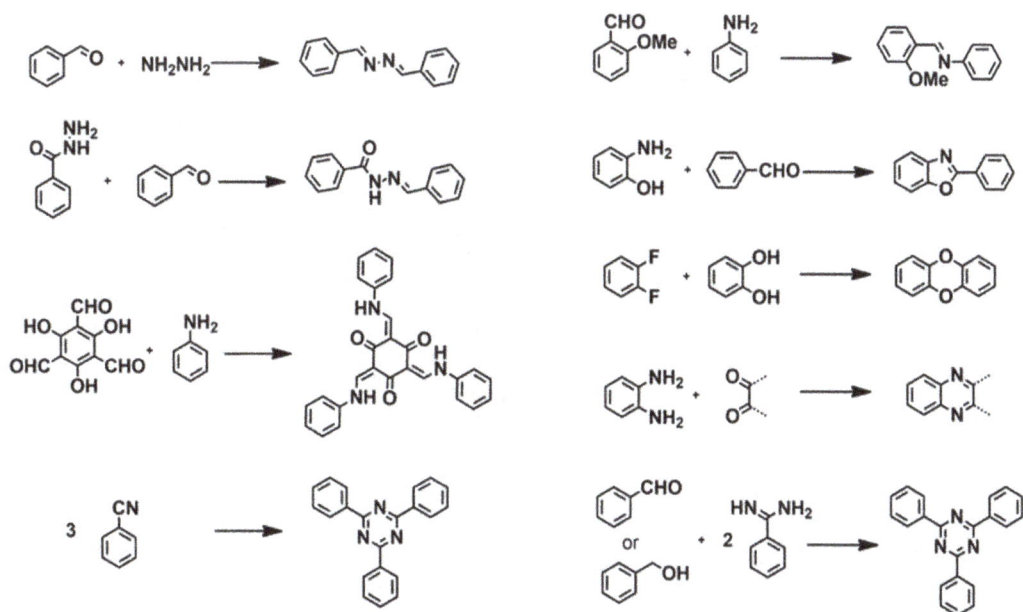

FIGURE 18.3
Synthetic routes to stable COF linkages. (Adapted with permission [11]. Copyright (2019) John Wiley and Sons.)

were employed to chemically covert and lock imine linkages for enhanced stability [12]. For instance, converting imines into enamines produces β-ketoenamine linked-COFs exhibiting chemical stability in acid/base conditions derived from structural stabilization of keto-enol tautomerism. Transforming imine linked-COFs into amide linked-COFs using $NaClO_2$ produces chemically stable frameworks with maintained crystallinity and porosity [13]. Imines can also be transformed into stable azole-linked or thiazole-linked COFs via post-modification reactions [14]. Furthermore, introducing methoxy groups to aldehyde monomers changes the linkage polarity and strengthens the interlayer interactions, resulting in chemically stable COF products. The above-mentioned strategies greatly improved the stability of imine-linked COFs, enriching the COF library with numerous structures due to the facile synthesis of these COFs. Recently, COF photostability has been enhanced by synthesizing sp^2 carbon-linked COFs, offering additional stable materials amenable for practical implementations.

18.2.3 High Crystallinity

Constructing highly crystalline COF materials requires several trials to optimize the reaction conditions for the condensation reaction between different building blocks. The attainment of crystalline materials is highly desirable for enhanced photocatalytic activity in COFs. Unlike amorphous organic polymers, crystalline COFs exhibit several photocatalytic properties including increased long-range order (facilitates exciton migration), π-delocalization (allows for visible absorption), and tunable structures at the atomic and molecular levels. Prior research revealed that COFs with high crystallinity exhibited superior photocatalytic activity to disordered amorphous counterparts due to their ability to decrease defect sites, inhibit potential recombination of charge carriers, and promote exciton migration and charge transport. In this regard, Cooper et al. compared the photocatalytic activity of crystalline COF based on benzo-bis (benzothiophene sulfone) and its amorphous or semicrystalline organic polymer counterparts (Figure 18.4) [5]. Crystalline COF (S-COF) dramatically increased photocatalytic activity for hydrogen evolution reaction over disordered amorphous or semicrystalline analogues (P7 and P10). Typically, S-COF exhibited 10.1 mmol g^{-1} h^{-1} hydrogen evolution, which is ten-fold higher than its amorphous counterparts. In another report, Zhang et al. reported COF based on sp^2 carbon-linked triazine moiety, outperforming its amorphous analog [15]. The crystalline COF achieved high hydrogen evolution with 14.6 μmol h^{-1}, which is nearly two times higher than the amorphous counterpart. This hydrogen evolution rate was dropped when the morphology of COF was changed to cracked rods and particles, implying the influence of morphology and dispersity of material on its light absorption property. Generally, high crystallinity is not only significant to enhance COFs performance in photocatalysis but also advantageous for their structural determination as it dictates atom positions, pore size, bond angles, and lengths. However, less reversibility or irreversibility of COF linkages makes it challenging to obtain crystalline materials. This can be overcome by dynamic covalent chemistry (DCC), which enables covalent bond breaking and reformation, causing error correction that finally makes a crystalline product possible [16]. Several strategies are presented to improve the crystallinity of COF materials, including modulator addition (aniline, benzaldehyde, etc.) to decrease nucleation and increase error-correction, slow addition of monomers to slow down the COF reaction, decrease defects and produce large crystals, and using catalysts (acetic acid for example) to speed up the reaction and increase error-correction [17]. In addition, replacing the linker from an amorphous polymer product with another can result in a crystalline network [18]. Moreover,

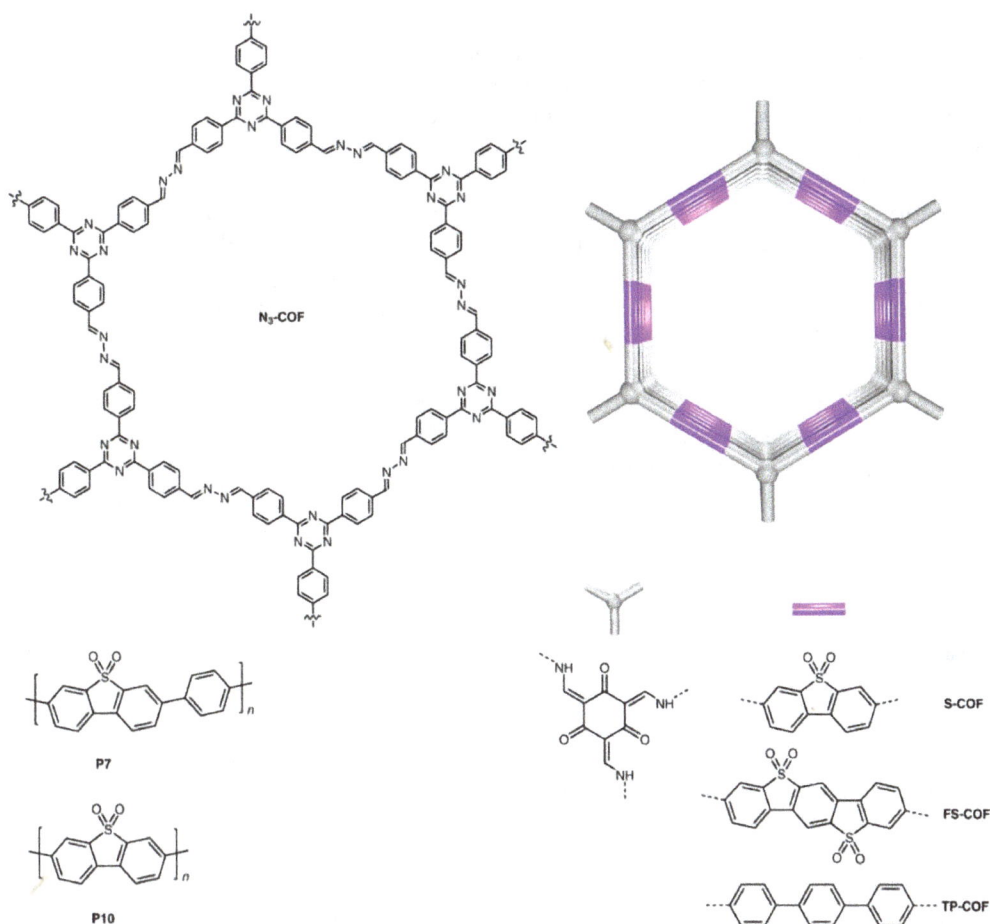

crystallinity is improved using ionic liquids, deep eutectic solvents, and advanced acid catalysts. Different strategies are also highlighted to improve COF crystallinity, including optimizing polycondensation reaction, enhancing planarity, fluorine substitution, engineering of the side chain, etc. Different COF linkages exhibit varied crystallinity behaviors based on their robustness and less reversibility or even different reactivities of building blocks. For instance, imine-linked COFs outperform olefin-linked COFs in crystallinity, which can be enhanced using metal triflates as catalysts. In addition, reaction temperature is a leading factor in determining the crystallinity of COF products. For instance, room temperature mostly produces amorphous kinetically COF products rather than crystalline thermodynamic products, which can be easily produced using the solvothermal method.

18.2.4 High Surface Areas and Adjustable Pore Sizes

Proper selection of linkage types and building blocks for COF formation is highly demanding for obtaining COF materials with high surface area and porosity. Engineering pore chemistry in COFs using a rational selection of their ingredients yields tailored porosity for selective photocatalysis with high performance. Several strategies can be adopted to increase the surface area of COFs. Before COF synthesis, using precise designable longer building blocks by applying the isoreticular expansion approach can afford a specific large surface area with increased tailored porosity, tuned by employing specific functional building units. After COF synthesis, optimal solvent exchange or selecting a proper activation approach can be critical for obtaining optimal surface area with retained porosity. Generally, a trade-off exists between surface area/stability and solubility. This implies that using longer building units with limited solubility can afford high surface area COFs but with limited stability. However, surface area in advanced porous materials is not the only determinant of their targeted applications. For instance, narrower pores in low surface area COFs can exhibit selective adsorption by controlling the diffusion of gases due to their kinetic diameters. In general, photocatalytic active metal nanoparticles (TiO_2, Ni, Pt, Fe, Co, Au, etc.) can be easily deposited on high surface area COFs due to their rich reaction sites and matched pore size between these nanoparticles (1–3 nm) and COFs (0.5–5 nm) [9]. To attain well-desired pore structures in COFs, specific building blocks with varied sizes and shapes are selected and can be subsequently tuned using PSM reactions.

18.3 Design and Synthesis of COF Photocatalysts

As a class of advanced porous materials, COFs are considered purely organic framework materials obtained from the polycondensation reaction between a plethora of building blocks with different geometries and functionalities to produce two- or three-dimensional (2D or 3D) structures with a variety of linkage types. Since their first report in 2005 [19], COFs expanded to provide almost all linkage modes known in chemistry, such as boronate ester, boroxine, borosilicate, borazine, triazine, imide, amide, amine, dioxin, hydrazine, imine, squaraine, spiroborate, azine, thiazole, oxazole, azodioxide, viologen, phenazine, benzobisoxazole, azodioxy, olefin, ester, thiourea, etc. [20, 21]. The diversity of COF linkages results in various properties, implying that COF predesign can dictate the targeted implementations. Photoactive COF-based catalysts can be obtained if a rational design process is considered. This can be accomplished by constructing highly stable structures with excellent crystallinity amenable to enhance light absorption capacity, improve the separation between photogenerated electron-hole pairs, facilitate the transport of charge carriers, and suppress charge recombination. Improved stability, crystallinity, and photocatalytic performance can be attained once proper linkages featuring robust covalent bonds are found in COF skeletons. Other features such as enhanced light absorption or increased photogenerated charge separation are realized by incorporating D-A units in COF structures [22]. The assembly of all ingredients responsible for designing high-performance photocatalytic COFs in addition to optimizing reaction conditions can ultimately produce high-quality COFs (crystallinity, porosity, stability, large surface areas, engineered pore sizes, regular large crystals, etc.) amenable to achieve excellent outcomes in different implementations. However, COF properties can be enhanced by employing particular functional building blocks decorating and engineering the pore system for enriching a required property in COFs to be ready for conducting selective applications. Considering all criteria in the rational design of COF-based photocatalysts, the resultant pristine products are not always efficient. They exhibit limited photocatalytic performance, which can be overcome by constructing COF hybrid systems that integrate multiple properties in a single system. However, their stability under photochemical settings cannot be maintained due to the inherent weak interactions within these hybrids, such as hydrogen bonding, van der Waals forces, and dielectric interactions. This limitation can be circumvented by linking COF hybrid ingredients with strong covalent bonds by PSM approach, which enhances their photostability, maintaining their catalytic performance under harsh reaction conditions. Although high-surface-area COFs are excellent to improve their catalytic performance by integrating photoactive sites such as precious metals, these metals are costly, low recyclable, and potentially toxic, precluding their long-term use in real-world implementations. This highlights the urgent need to construct green, inexpensive, and recyclable COF photocatalysts without noble metals.

When it comes to COF synthesis, the process mostly occurs using a solvothermal method for several days to produce powder or single crystal products. Due to the variety of building units used in COF synthesis, the synthetic method is mainly determined based on their solubility behaviors or stability under a particular synthetic method. For instance, room temperature synthesis is appropriate for highly soluble or sensitive building blocks, whereas high-temperature synthesis using the solvothermal method is highly convenient for thermally stable or sparingly soluble precursors. There are two pathways to consider for COF photocatalysts. The first is the rational selection of synthetic method that can optimize all inherent features to produce COF powders. The second is to make COF powders amenable to being employed in high-performance photocatalysis by producing COF nanofilms or nanosheets using several techniques. Solvothermal, room-temperature, microwave, mechanochemical, ionothermal syntheses, and so on were the methods employed to synthesize COF powders. Among them, the solvothermal method is mainly performed under high temperatures (80–120°C) and autogenous pressure in Pyrex tubes or Teflon-lined stainless autoclaves for 3–7 days. Although high temperatures and long reaction times (slow reaction rates), and the unsuitability of solvothermal method to sensitive precursors or scale-up for practical applications, it is the most widely employed approach to obtain highly crystalline and porous COFs. The solvent as a critical parameter for obtaining crystalline and porous products should be carefully selected due to its high impact on the resultant structures. However, solvent selection relies on trial and error by screening numerous solvent systems that finally achieve large stable crystals with high crystallinity and porosity. For imine-linked COFs, several solvents or solvent systems were employed in the synthetic process, with mesitylene/dioxane as a widely used solvent system catalyzed by acetic acid. To overcome the crystallinity problem in COF synthesis, inducing reversibility during COF reactions is a feasible approach by using reversible condensation reactions such as Knoevenagel, spiro-borane, and imide condensations. For instance, modulated synthesis can enhance the crystallinity of imine-linked COFs using aniline or benzaldehyde. The ionothermal approach is mainly deployed to synthesize high-surface-area covalent triazine frameworks, featuring high electron-hole mobility and possible industrial commercialization. However, this approach requires a high temperature (400°C) with limited monomers or reduced porosity and crystallinity. Microwave synthesis consumes high energy but produces high yield products quickly. Although mechanochemical synthesis is beneficial in producing COF products quickly with solvent-free synthesis or reduced solvent consumption in a simple-to-operate system amenable to be employed in industrial production, the resultant products lack high crystallinity, porosity, or surface area.

To efficiently employ COFs in photochemical applications, their solubility and dispersibility are increased in organic solvents or water by exfoliation, resulting in improved photocatalytic performance. In particular, producing layered organic materials, including nanosheets or 2D films [23], is a general method for layered COF exfoliation. In addition to exfoliation, COF films can be prepared using interfacial synthesis or deposition on substrates. A vapor-phase deposition is a strategy for employing vapor diffusion to deposit atomic layers of COF monomers on specific substrates such as Ag (111), Au (111), copper foil, or pyrolytic graphite, resulting in high-quality COF monolayers. COF films can be accomplished using solvothermal or continuous flow conditions. The thickness and uniformity of the resultant COF films are dictated by the supports employed in synthesis. Besides, interfacial synthesis is a proper strategy to construct COF films at interfaces where COF monomers react. In this strategy, uniform growth of films was accomplished by controlling monomer release at interfaces. Exfoliation is a widely used method to produce COF thin films when COFs are sonicated in delaminating solvents or when mechanical grinding, chemical exfoliation, and self-exfoliation are employed [24]. The underlying mechanism of exfoliation relies on weakening π–π interactions that existed between COF layers when sonicated in particular solvents. For mechanical grinding, several chemically stable bulk COF materials were converted into nanosheets using mechanical forces (sonication, ultrasound, etc.) that disrupt weak interactions between COF layers, resulting in suspension with high dispersibility. When it comes to chemical exfoliation, this strategy relies on incorporating specific molecules into COF frameworks by chemical bonding, allowing COF layers for further disruption via their π–π interactions. In self-exfoliation, COF thin films are produced by inducing exfoliation through internal forces, requiring rationally designed building units for exfoliation. In solution-based exfoliation, reduced crystallinity is identified using water as exfoliation solvent or when exfoliation solution was subjected to mechanical stirring at the boiling point. Accordingly, for mechanical or liquid-phase exfoliation used, it is critical to consider structural features such as surface energy between solvents and exfoliated COFs to develop COF layers with enhanced solubility amenable for several photochemical implementations.

18.4 Strategies to Enhance Photocatalytic Activity in COF-Based Systems

Numerous considerations should be kept in mind when COF photocatalytic systems with enhanced activity are constructed. Typically, the established system should have visible-light absorption with extended capacity, improved electron-hole separation, and inhibited corrosion under long-term reaction times. Several strategies have been practiced to realize such requirements for enhanced outcomes, including functional building block incorporation, elemental doping, sensitizers, and developing COF hybrid systems. As a feasible pathway for modulating electronic or optical properties of COFs, incorporating proper functional building units in COF frameworks is a viable strategy to enhance visible-light absorption and reduce photogenerated electron-hole recombination. Band gap engineering at the molecular level can be accomplished by changing the physicochemical characteristics of COFs by selecting proper building blocks that construct COF photocatalytic systems. At the atomic level, these properties can be tuned by elemental (metal or non-metal) doping. Besides, the band gap can be engineered by introducing anions/cations into COF frameworks to enhance the light-harvesting ability and modulate redox band potentials. In addition, incorporating sensitizers has the potential to extend light absorption toward a higher wavelength range, resulting in enhanced photocatalytic activity. Finally, establishing COF hybrid systems can enhance photocatalytic activity as they can enlarge visible-light absorption, facilitate electron transfer between composites, and efficiently separate photogenerated electron-hole pairs in an enhanced manner outperforming pristine COFs.

18.4.1 Functional Building Unit Incorporation

Photocatalytic active COFs are mainly based on triazine, sulfone, pyrene, benzothiadiazole, thiophene, diacetylene building units, etc., in their skeletons. These systems have reduced optical band gap, extended conjugate structures, efficient visible-light harvesting, accessible active sites, improved charge separation, and accelerated migration of photogenerated excitons. One strategy to improve photocatalytic activity in COFs relies on replacing carbon atoms with nitrogen atoms in the central aryl ring, forming rings with rich nitrogen content that exhibit increased planarity due to reduced dihedral angle between peripheral and central rings. In this case, pyridine-, pyrimidine-, and triazine-based central rings in COF structures exhibit increased crystallinity by increasing nitrogen content from 1–3, resulting in improved exciton migration and increased photocatalytic activity [25]. In addition, improved light-harvesting in the visible region can be greatly achieved in the case of π-conjugated trans-disubstituted-based COFs. This extended π-conjugated system has the potential to exhibit augmented photocatalytic activity due to a small band gap, increased charge separation, and efficient electron-hole transfer.

18.4.2 Elemental Doping

COFs can be photocatalytically active by modulating their optical or electrical properties. This includes tailoring a narrow band gap, promoting visible-light absorption ability, facilitating charge transfer, and increasing the lifetime of charge carriers. Another approach to accomplish this is by doping metals or non-metals such as elemental sulfur or halogens (F, Cl, or Br), particularly in triazine-based COFs. Elemental-doped COFs enhanced the photocatalytic activity of pristine COFs in terms of efficient charge separation, transport, and much narrower band gaps [26].

18.4.3 Sensitizers

To greatly enhance their photocatalytic performance in terms of light-harvesting efficiency, pristine COFs are intriguingly modified using sensitizers that efficiently generate electron-hole pairs. These sensitizers with extended visible light absorption reach the near IR region and can be employed as cocatalyst to improve light absorption and extend electron-hole lifetime, resulting in enhanced photocatalytic activity. Sensitizers examples include Eosin Y (EY). For instance, the photocatalytic activity of Pd-COFs was improved using EY as a sensitizer [27]. The underlying mechanism when EY photosensitizer is integrated into the COF system relies on visible light absorption, electron generation, electron transfer to COFs, and then to Pd catalytic active sites.

18.4.4 COF Hybrids

Although diverse COF systems have been constructed and employed in different photocatalytic reactions, pristine COFs still face limited performance. A viable strategy to circumvent this limitation is constructing integrated COF hybrid structures with synergistic influences on the resultant photocatalytic performance. In this case, the prepared COFs combines the advantages of both hybridized materials that offer remarkable features out-performing their ingredients. For instance, COFs can be loaded with CDS nanoparticles to realize improved photocatalytic activity as COFs serve as π-conjugated support with high photostability, crystallinity, surface area, and porosity. The prepared composites exhibit enhanced visible light absorption and suppressed photogenerated electron-hole recombination. Another COF hybrid (COF grown on MOF) depends on the covalent linkage between aldehyde-based MOF and amine-based building units to construct MOF@COF core-shell structures with high photocatalytic activity [28]. This high performance is originated from the large surface area of the prepared hybrid material with a small optical band gap. When proper band positions are constructed in COF hybrid materials, a photogenerated electron-hole transfer can be easily accomplished from the interface between the two hybridized components to their surface, causing redox and reduction reactions.

18.5 Applications of COF-Based Photocatalysts

18.5.1 Hydrogen Evolution

To overcome the energy crisis, the reliance on sustainable and green fuels is critical, such as hydrogen gas derived from photocatalytic water splitting. The hydrogen combustion process is green as it produces water as a byproduct rather than carbon dioxide in the case of fossil fuels. In water-splitting, light irradiation induces the generation of electron-hole pairs that migrate to the photocatalyst surface, where two half-reactions occur comprising hydrogen evolution reaction (HER) and oxygen evolution reaction (OER) [29]. For a photocatalyst to have high performance in water-splitting reactions, the band gap should be more than 1.23 eV and narrower than 3.10 eV to efficiently utilize visible light ($\lambda > 400$ nm). CB and VB of a photocatalyst should be more negative than 0 eV (vs. NHE) for H^+ reduction to H_2 and more positive than 1.23 eV (vs. NHE) for H_2O oxidation to O_2. This implies the significance of band structures of COFs to actively generate hydrogen gas via photocatalytic water splitting. This can be accomplished by rational design of COF photocatalysts, most notably incorporating D-A structure and π-conjugated system, which facilitate the charge transfer of electron-hole pairs to the catalyst surface where reactions occur.

Selecting proper support for single-metal atom cocatalysts is critical to producing a highly efficient photocatalytic system with high metal loading and sustained durability. In this regard, Zhang et al. successfully employed 2D COF based on β-ketoenamine linkages (TpPa-1-COF) as functional support for highly loaded single-atom platinum (Pt) cocatalyst (Figure 18.5a) [30]. The structural features of this 2D COF comprising peculiar holes and unsaturated coordination nitrogen atoms enable the high atomic dispersion of Pt on the pore walls of COF, forming a coordination microenvironment of a six-coordinated C_3N–Pt–Cl_2 species without agglomeration. Pt_1@TpPa-1 with actual loading of 0.72 wt.% was employed for photocatalytic visible-light-driven hydrogen evolution, reaching a rate of 719 μmol g^{-1} h^{-1} and a large turnover frequency (TOF) of 19.5 h^{-1}. This high activity of Pt_1@TpPa-1 outperformed Pt nanoparticles/TpPa-1 (3.9 times) and bare TpPa-1 (48 times) due to the good dispersion of Pt single atoms on COF support and effective separation and transport of photogenerated charges.

In another report, the influence of different linkages in COF frameworks on photocatalytic hydrogen generation was systematically investigated (Figure 18.5b) [31]. In this

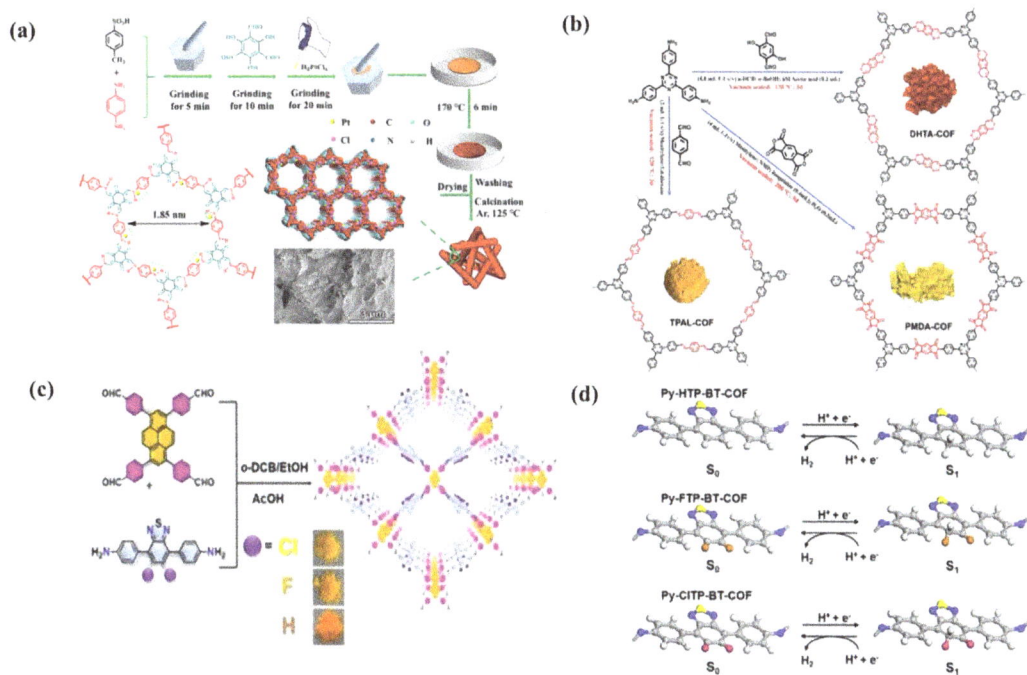

FIGURE 18.5

COF-based photocatalysts for hydrogen evolution (a). Synthesis of single-atom Pt anchored on TpPa-1-COF (Pt1@ TpPa-1). (Adapted with permission [30]. Copyright (2021) American Chemical Society.) (b) Synthesis of PMDA-COF, DHTA-COF, and TPAL-COF. (Adapted with permission [31]. Copyright (2021) Elsevier.) (c) Solvothermal synthesis of Py-XTP-BT-COFs. (Adapted with permission [32]. Copyright (2019) John Wiley and Sons.) (d) A proposed pathway (single-site reaction) of hydrogen evolution reaction on Py-XTP-BT-COFs. C gray, H white, S yellow, N blue, F orange, Cl pink. (Adapted with permission [32]. Copyright (2020) John Wiley and Sons.)

work, 2D triazine-linked COFs were constructed based on 4,4′,4″-(1,3,5-triazine-2,4,6-triyl) trianiline (TAPT), which was condensed with 1,2,4,5-benzenetetracarboxylic anhydride (PMDA), 2,5-dihydroxyterephthalaldehyde (DHTA), and terephthalaldehyde (TPAL), producing PMDA-COF, DHTA-COF, and TPAL-COF respectively. Among them, PMDA-COF with dianhydride substituted imide linkage exhibits the highest hydrogen evolution rate of 435.6 $\mu mol \cdot g^{-1} \cdot h^{-1}$ over DHTA-COF (56.2 $\mu mol \cdot g^{-1} \cdot h^{-1}$) with imine linkage stabilized with hydrogen bonding and TPAL-COF (6.8 $\mu mol \cdot g^{-1} \cdot h^{-1}$) with imine linkage. In addition to light-absorption efficiency, PMDA-COF showed superior separation capability for the photogenerated charge carriers and exhibited energy difference between PMDA and TAPT, resulting in a D-A heterojunction responsible for modifying the electronic band structure and separating reduction and oxidation sites, corroborating the long-term activity of PMDA-COF for 20 h visible-light irradiation.

Chen et al. introduced a halogenation (chlorination or fluorination) modulation approach for a 2D COF containing benzothiadiazole moiety, denoted Py-HTP-BT-COF, to enhance its photocatalytic hydrogen production (Figure 18.5c, d) [32]. In particular, the chlorinated COF (Py-ClTP-BT-COF) revealed a dramatic enhancement in visible-light-induced hydrogen production, with a rate of 177.50 $\mu mol \, h^{-1}$ and excellent apparent quantum efficiency (AQE, 8.45%). According to experimental and computational results, Py-ClTP-BT-COF dramatically improved hydrogen generation efficiency. It can efficiently facilitate charge separation and transport and generate hydrogen at lower activation energy by reducing the energy barrier for forming hydrogen intermediates on COF surface. This engineering approach demonstrates how subtle changes in chemical COF linkages affect their relevant physicochemical characteristics and ultimately influence their photocatalytic performance, particularly solar-to-chemical energy conversion.

18.5.2 CO_2 Reduction

Global warming is caused by the large amounts of CO_2 that evolved from fossil fuels combustion, resulting in the greenhouse effect. This increases the temperature of the atmosphere and adversely affects the ecological system. To circumvent this issue, researchers attempted to convert CO_2 into useful fuels using a photocatalytic reduction approach [33]. The products of CO_2 reduction include formic acid (HCOOH), carbon monoxide (CO), methane (CH_4), etc. These value-added products can be selectively obtained according to the type of cocatalyst employed, the type of photocatalyst used, and matched band gap energy. Unlike MOFs, COFs exhibit high stability under the reaction conditions of CO_2 reduction, and they rapidly and efficiently facilitate the transfer of charge carriers due to their ordered structures with nanochannels. COFs also provide CO_2 adsorption through non-covalent interactions, resulting in high performance with excellent outcomes.

A β-ketoenamine-based COF (TpBb-COF) was employed as a metal-free catalyst for visible-light-induced CO_2 reduction to CO using pure and diluted CO_2 atmosphere [34]. TpBb-COF underwent the photocatalytic reaction in a gas-solid system without the need for photosensitizers or sacrificial agents. The results revealed that, in pure CO_2 atmosphere, TpBb-COF achieved a lower CO production rate (52.8 $\mu mol \cdot g^{-1} \cdot h^{-1}$) than in the case of a 30% CO_2 atmosphere (89.9 $\mu mol \cdot g^{-1} \cdot h^{-1}$) at 80°C, implying that photocatalytic CO_2 reduction exhibits superior performance by using diluted concentrations. Several experiments and calculations found that cyclohexanetrione moiety in the COF framework preferentially adsorbed water via hydrogen bonding, followed by a reaction between adsorbed water with CO_2, which was finally converted into CO.

Symmetry control in developing efficient photocatalyst-based COFs for CO_2 reduction is critical, affecting their physicochemical, electronic, and optical properties. In this regard, two symmetrical building units were employed to construct two COFs (TTzTp and BTzTp) based on tris- or bis-benzothiazole moieties featuring strong affinity to CO_2 reduction due to their uniform distribution of nitrogen and sulfur atoms that allows for improved charge carrier mobility (Figure 18.6a) [35]. Highly crystalline BTzTp with $C_3 + C_2$ symmetry had a large pore size and high surface area, whereas TTzTp exhibited $C_3 + C_3$ symmetry with smaller pore size. Despite the low affinity to CO_2, light absorption with a narrower range, and a broader band gap, BTzTp was highly selective to reduce CO_2 to CO with a higher conversion rate under photocatalytic settings. This unexpected finding was due to the long lifetime of photogenerated charge carriers, which delayed their recombination for the sustained photocatalytic process. This work implies that structural diversity in COF structures in terms of pore dimensions or symmetries has the potential to provide unusual photocatalytic activity in CO_2 reduction.

A fascinating approach to enhance the photocatalytic activity of COFs toward CO_2 reduction can be achieved by decorating their pore walls with metal active sites. Unlike simple coordination of metal centers to COFs, chelating one metal active site between COF and an additional molecular catalyst is a highly active approach to reduce CO_2. In this regard, TFBD-COF decorated with Co^{2+}, followed by coordination to salicylideneaniline (SA), yielding TFBD-COF-Co-SA was compared to SA-based Co^{2+} complex ($Co(SA)_2$) molecular catalyst and TFBD-COF-Co in the photocatalytic reduction of CO_2 to CO (Figure 18.6b) [36].

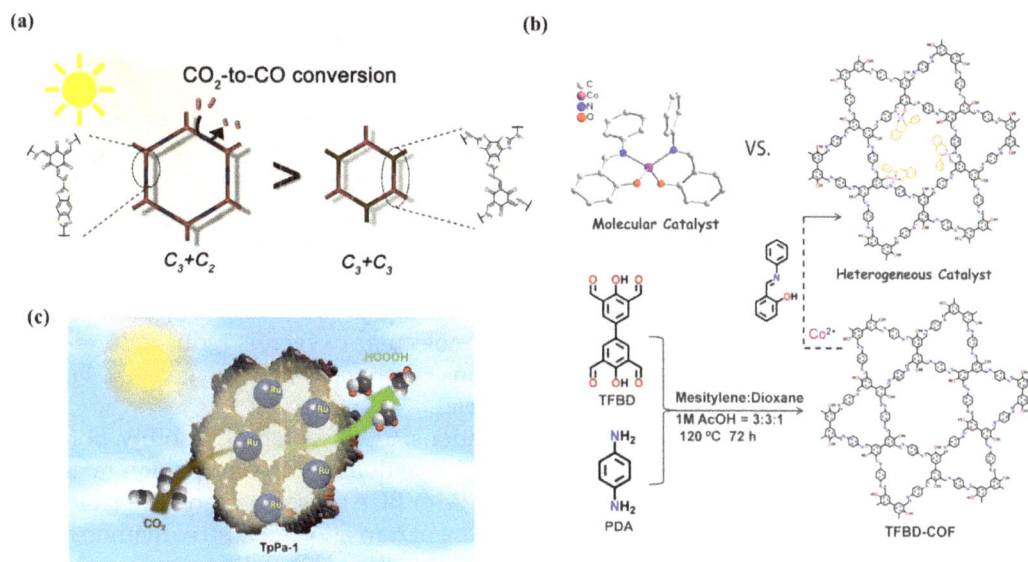

FIGURE 18.6
COF-based photocatalysts for CO_2 reduction. (a) TTzTp (on the right) and BTzTp (on the left) for CO_2 photoreduction to CO. (Adapted with permission [35]. Copyright (2021) American Chemical Society.) (b) Synthesis of TFBD-COF for CO_2 photoreduction to CO. (Adapted with permission [36]. Copyright (2021) American Chemical Society.) (c) TpPa-1 decorated with Ru NPs for CO_2 photoreduction to HCOOH. (Adapted with permission [37]. Copyright (2021) Elsevier.)

$(Co(SA)_2)$ was selected due to its superior activity and unique structure, and TFBD-COF support was chosen due to its porous framework and its resemblance to SA structure. The results revealed that TFBD-COF-Co-SA exhibited a high photocatalytic reduction of CO_2 to CO, with an activity of 7400 μmol·g^{-1} and a selectivity of 90% in 5 h outperforming that of TFBD-COF-Co without SA coordination and comparable to that of $(Co(SA)_2)$.

To guarantee high photocatalytic activity, suppressing electron-hole recombination or separation efficiency is critical in a given photocatalyst material. For photocatalytic CO_2 reduction, the high performance of photocatalyst-based COFs can be accomplished by loading metal nanoparticles on COF supports to produce composite materials with high activity and selectivity. Typically, ruthenium nanoparticles (Ru NPs) were loaded on ketoamine-based COF (TpPa-1) to produce Ru/TpPa-1 and were employed for the first time to reduce CO_2 under visible-light irradiation (Figure 18.6c) [37]. A 3.0 wt.% Ru/TpPa-1 exhibited high photocatalytic activity over TpPa-1 for the production of formic acid with a rate of 108.8 μmol gcat^{-1}h^{-1}. The loading of Ru NPs endowed COF with high photocatalytic performance due to enhanced visible-light harvesting efficiency, the prolonged lifetime of photogenerated charge carriers with a separation efficiency, and facilitated charge transfer. However, such activity is insufficient, requiring rational selection of building units to construct COF materials with modulated composition, structure, and properties.

18.5.3 Organic Transformations

Photocatalytic organic transformations have garnered widespread development since the employment of porous solids as heterogeneous catalysts to induce sustainable organic synthesis, especially when these reactions are driven by visible light [38]. Using photocatalyst-based COFs, in particular, show great promise in organic transformations due to their extended π-conjugated structures, which endow them with high light absorption features. In addition, COFs exhibit high surface area, porosity, and tunable structures, which offer nano-pores with uniform sizes. In this case, COFs serve as molecular flasks accessible to guest molecules to conduct several organic reactions such as coupling, oxidation, hydrogenation, etc. Under this photocatalytic process, COFs possess high chemical stability with retained crystallinity and porosity under harsh reaction conditions, producing quantitative yields of target products in a mild, highly selective, and recyclable process. Typically, the underlying mechanism by which visible-light-driven organic transformations occur includes several steps. First, e⁻/h⁺ pairs are photo-induced, generated, and separated from COFs under irradiation. Second, molecular oxygen is reduced to reactive intermediates, including superoxide radical anion ($\bullet O_2^-$) and single oxygen $^1O_2$. Finally, the substrates are oxidized using h⁺ or active species.

As an excellent platform for organic transformations, COFs with high stability, porosity, surface area, and optoelectronic properties can achieve quantitative conversion rates. For instance, two metal-free COFs (BTZ-X-COF, X = TPA, or BCA) based on benzothiazole moieties were synthesized using a one-pot multicomponent transition metal-free method incorporating elemental sulfur as a key component (Figure 18.7a) [39]. Both porous crystalline COFs showed superior photocatalytic performance regarding visible-light-driven phenylboronic acid transformation to phenol due to their excellent remarkable π-delocalization and optoelectronic characteristics, particularly light-harvesting efficiency. However, BTZ-TPA-COF (< 8 h) outperformed BTZ-BCA-COF (12 h) in the photocatalytic system, reaching 100% conversion with 99% yield. This is because the core framework of BTZ-TPA-COF exhibited electron-rich moiety in addition to its physicochemical and optoelectronic features. Both COFs underwent several photo-oxidation runs with retained crystallinity and

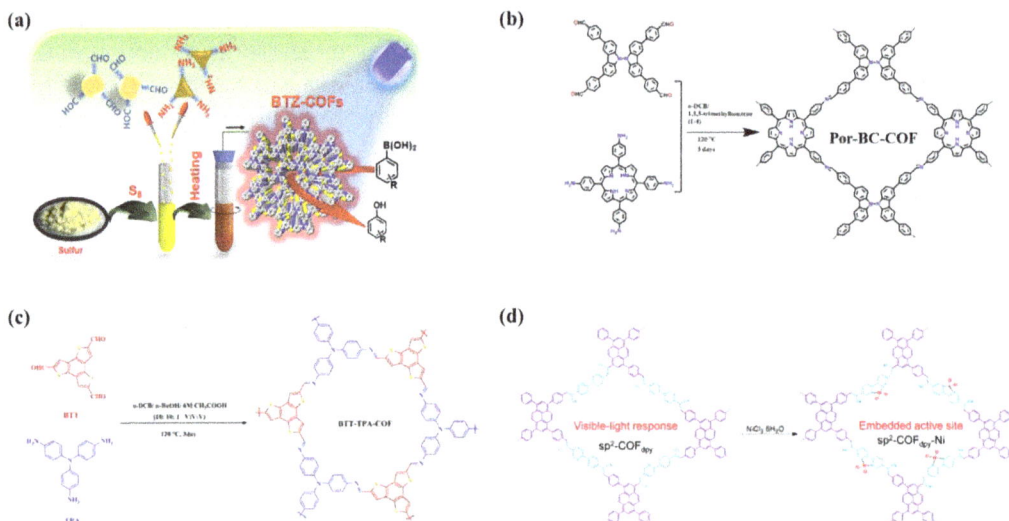

FIGURE 18.7
COF-based catalysts for organic transformations. (a) BTZ-COFs for photo-oxidation of aryl boronic acids to phenols. (Adapted with permission [39]. Copyright (2021) American Chemical Society.) (b) Synthesis of Por-BC-COF for photo-oxidation of amines to imines. (Adapted with permission [40]. Copyright (2021) John Wiley and Sons.) (c) Synthesis of BTT-TPA-COF for benzimidazole synthesis. (Adapted with permission [41]. Copyright (2021) Elsevier.) (d) Synthesis of sp²c-COF_dpy-Ni for C–O cross-coupling. (Adapted with permission [42]. Copyright (2021) Royal Society of Chemistry.)

activity. This work revealed visible-light-induced production of superoxide radicals ($O_2^{•-}$) to investigate the mechanistic reaction pathway.

As essential intermediates used in drug synthetic processes, identifying a low-cost and environmentally friendly approach to produce imines has garnered huge attention. Unlike other conventional materials, 2D Por-BC-COF containing porphyrin (superior visible light absorption features) and bicarbazole (high stability and remarkable charge transfer characteristics) monomers has great potential as metal-free photocatalyst for aerobic oxidative coupling of amines into imines under visible-light irradiation (Figure 18.7b) [40]. The prepared COF exhibited high photocatalytic performance, achieving a high yield of 97%, which is attributed to its high crystallinity, surface area (1200 $m^2 g^{-1}$), proper band structure, and broad light absorption range (200–1300 nm). This work highlights the impact of employing a photocatalyst platform having different building units with diverse functions and properties in the final structure to efficiently catalyze several organic transformations for fine chemical production.

Based on this approach, metal-free photoactive BTT-TPA-COF COF catalyst was solvothermal constructed based on benzotrithiophene (BTT) and triphenylamine (TPA) photosensitive structural units (Figure 18.7c) [41]. In addition to the high surface area, excellent porosity, and high stability, the alternating connection of building units endowed BTT-TPA-COF COF with unique properties, including low band gap, improved light absorption capacity, enhanced separation, and transfer efficiency for photogenerated charge carriers, and distinguished electronic band structure. Under visible-light irradiation, these physicochemical and optoelectronic properties collectively enabled BTT-TPA-COF COF to synthesize 2-arylbenzimidazole compounds with excellent photocatalytic performance, substrate tolerance, and recyclability with retained photocatalytic activity even after eight cycles. This research has the potential to design an efficient photocatalyst with wide applicability in organic synthesis toward fine chemical production.

To broaden the applicability of COFs in organic transformations, new organic syntheses should be investigated using tailored COF photocatalysts. In this aspect, Chen and colleagues explored aryl etherification reactions using a highly stable Ni^{II} embedded vinyl bridged 2D COF (sp^2c-COF_{dpy}-Ni) (Figure 18.7d) [42]. The embedded Ni complexes and the extended delocalized π conjugation system endowed the COF with dual functions as photosensitizer and reactive site. sp^2c-COF_{dpy}-Ni acted as a heterogeneous photocatalyst with high photocatalytic performance in converting aryl bromides into aryl alky ethers as C–O cross-coupling reaction. The photocatalytic reaction was conducted under visible light irradiation (560 nm), and the photocatalytic activity of sp^2c-COF_{dpy}-Ni relied on D-A conjugated structure features, including low band gap and excellent charge separation efficiency. sp^2c-COF_{dpy}-Ni manifested good recyclability with maintained photocatalytic activity, implying the durability of this all-in-one COF photocatalyst and the sustainability of the photocatalytic process, outperforming other traditional homogeneous or semi-heterogeneous photocatalysts and implying the applicability of this approach to other functional group conversions.

18.5.4 Pollutant Degradation

Photocatalytic degradation holds great promise in removing toxic organic materials over other conventional methods, such as adsorption, membrane separation, or biodegradation [43]. Using photocatalysis, complete removal of organic contaminants/pollutants from industrial wastewater, such as organic dyes, antibiotics, phenols, etc., becomes a friendly and efficient process for environmental remediation. In this process, inexpensive, chemically stable, and efficient photocatalyst is preferred to completely degrade organic pollutants into nontoxic CO_2 or H_2O without causing secondary pollution. Although MOFs and other inorganic oxides are conventionally employed in oxidative photodegradation of organic pollutants, their limited stability, inaccessible catalytic active centers, light absorption with narrower range, inefficient charge transfer, and photogenerated electron-hole pairs recombination with a high rate suppressed their implementations. On the other hand, COFs can efficiently perform in the removal process of organic contaminants due to their high stability and superior photoelectronic characteristics. COFs can integrate particular functional centers onto pore walls to efficiently and selectively trap different organic molecules, which facilitate the removal of environmental contaminants. The underlying mechanism by which COFs can photocatalytically degrade organic pollutants relies on producing and separating photo-induced charge carriers or e^-–h^+ pairs when photon energy is higher than the band gap of photocatalyst-based COF. As the active species, superoxide radical anions ($\bullet O_2^-$) and hydroxyl radicals ($\bullet OH$) are generated when molecular oxygen captures photogenerated e^- in CB and H^+ reacts with $\bullet O_2^-$ respectively. Both $\bullet O_2^-$ and $\bullet OH$ exhibit high oxidizing ability, resulting in high efficiency in degrading organic contaminants. As additional active species, holes (h^+) can also capture pollutants or oxidize water after escaping from photocatalyst, which causes effective charge separation. After adsorbing contaminants onto the catalyst surface and conducting redox reactions, the final products are desorbed and transferred to the solution.

As a serious threat to the aquatic environment, several methods have been reported to degrade bisphenol A (BPA), including adsorption, membrane filtration, or advanced oxidation processes (AOPs). In particular, due to their low cost and facile activation techniques, peroxymonosulfate (PMS)-based AOPs exhibit great promise in BPA degradation. Using metal-free COFs toward PMS activation with generated reactive radicals is a promising approach to efficiently degrade BPA when visible light (VL) irradiation is applied. In this regard, COF-PRD (PRD refers to pyridine) was synthesized and well-characterized and was employed for degrading a series of bisphenols, including bisphenol A (BPA), B (BPB), F (BPF), Z (BPZ), and AP (BPAP), indicating excellent performance and diverse degradation pathways (Figure 18.8a) [44]. The degradation kinetics of BPA in the presence of COF-PRD/PMS/VL photocatalytic system outperformed that of COF-PRD/VL by 3.4 times, achieving complete removal of 10 mg L^{-1} of BPA within 150 min. Different active species were collectively contributed to BPA degradation, such as $\bullet O_2^-$, h^+ and $^1O_2$. Under aerobic conditions, PMS was oxidized by h^+, producing $\bullet SO_5^-$ radical and then $^1O_2$, whereas under anaerobic conditions, PMS was reduced with e^-, forming $\bullet SO_4^-$ to efficiently degrade BPA. This work revealed that COF-PRD was an excellent photocatalyst enhanced by PMS activation for degrading BPs, with recyclability and durability for ten runs.

To circumvent the limited recyclability of COFs in antibiotic degradation, constructing COF composite sponge materials rather than conventional COF powders is a promising approach to expand the applicability of COF for environmental remediation. In this aspect, a "reactive seeding" method enabled the successful synthesis of a composite sponge material composed of melamine sponge@COF or MS@COF in a one-pot scenario, allowing small-sized COF-TpTt to uniformly grow on MS fibers (Figure 18.8b) [45]. The exposed active sites and small size, as well as excellent morphological and optical characteristics, endowed MS@COF with enhanced performance for tetracycline antibiotic degradation under visible-light irradiation over pristine COF, reaching 97.3% degradation efficiency in aqueous solution. MS@COF also presented high degradation efficiency (>80%) in real water samples and other tetracycline antibiotics with reasonable recyclability. These results inspire researchers to develop monolithic catalysts based on COFs for environmental protection.

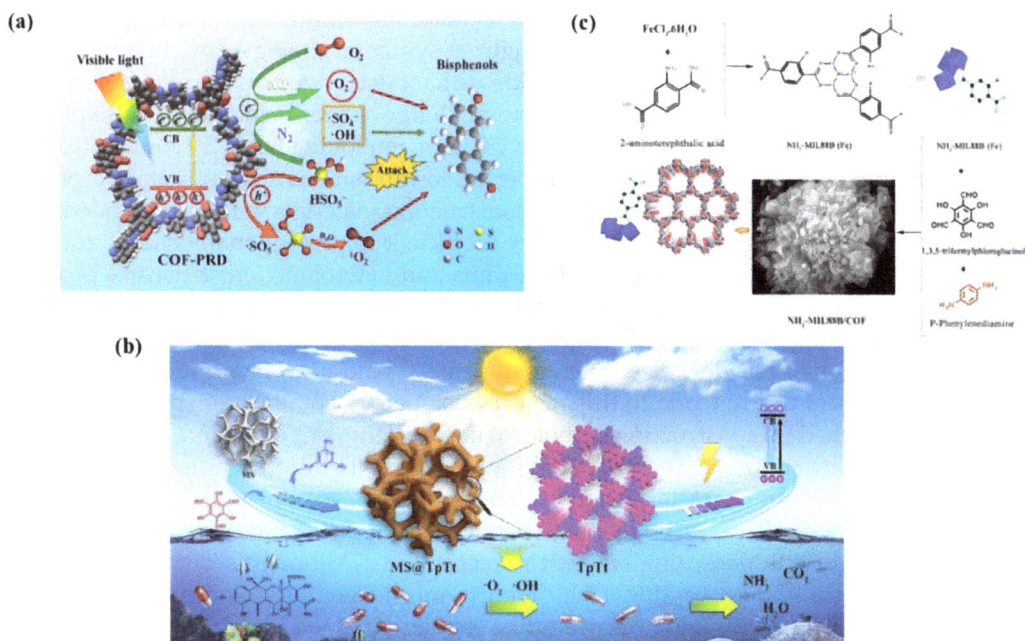

FIGURE 18.8
COF-based catalysts for pollutant degradation. (a) Photodegradation of bisphenols using COF-PRD/PMS/VL system. (Adapted with permission [44]. Copyright (2022) Elsevier.) (b) Photodegradation of tetracycline over melamine sponge@COF. (Adapted with permission [45]. Copyright (2022) Elsevier.) (c) Synthesis of NH$_2$-MIL88B/TpPa-1-COF hybrid for tetracycline and rhodamine B photodegradation. (Adapted with permission [46]. Copyright (2021) American Chemical Society.)

In another report, tetracycline (TC) and rhodamine B (RhB) were efficiently degraded (TC, 86%; Rh, 100%) using a COF hybrid constructed by a covalent linkage between TpPa-1-COF and Fe-based MOF (NH$_2$-MIL88B) (Figure 18.8c) [46]. The photocatalytic reaction activity relied on Fenton-like excitation of H$_2$O$_2$ in the presence of simulated sunlight irradiation, offering excellent degradation efficiency over pristine MOF or COF. The enhanced degradation efficiency was caused by charge transfer induced by charge carrier separation and light absorption efficiency due to a porous hybrid interface between MOF and COF. In addition, more hydroxyl radicals (•OH) were produced due to Fenton-like excitation of H$_2$O$_2$, resulting in high degradation performance. This work indicates the reliable employment of NH$_2$-MIL88B/TpPa-1-COF/H$_2$O$_2$/light system in environmental remediation with high activity and recyclability.

18.6 Conclusion and Future Perspectives

This chapter highlights the design and synthetic strategies of COF-based catalysts and their employment in several photocatalytic applications. However, numerous parameters should be considered when designing new COF-based catalysts to guarantee long-term photoactive materials. At first glance, research should focus on developing a general synthetic method for crystalline COFs as a prerequisite for successful structural elucidation, which empowers the investigation of structure-function relationships to rationally design next-generation materials with high predictability and reproducibility. As optimizing synthetic conditions for COFs is laborious, focusing on high-throughput synthetic strategies might be a reliable approach. In addition, constructing COF hybrids with robust linkages is a promising strategy to enhance photocatalytic activity by suppressing charge carrier recombination. Exploring unusual structures such as organic cage-based COFs or employing 3D COFs could introduce novel structural diversity and intriguing properties toward enhanced photocatalytic performance [47–50]. Another aspect to consider is constructing materials with an expanded light absorption range and high molar absorptivity coefficient to provide many charge carriers. In addition, building COF systems with D-A units and heterojunction structures or fabricating Z-schemes is a reliable strategy to increase light absorption efficiency and improve charge transfer and separation. Although water splitting reaction includes hydrogen and oxygen evolution half reactions, less attention has been paid to oxygen production, which should be a focus of future work. In this regard, concurrent overall water splitting reaction can be accomplished by constructing a photocatalyst system with oxidation/reduction sites or combining semiconductors with varying redox potentials. Another important factor to consider is using in situ characterization methods for COF-based photocatalyst to monitor reaction processes to distinguish reaction intermediates and active centers, which provide a thorough insight into designing efficient photocatalysts. Although several groups have focused on COF-based photocatalysis, their reports are challenging to compare due to varying reaction parameters, implying the need for a standard method for evaluating photocatalytic performance. In the end, it is critical to synthesize highly crystalline and porous COFs or COF hybrids with particular chromophores for enhanced photocatalytic activity using inexpensive, green, and efficient approaches.

References

1. Zhang F, Gallagher KS, Myslikova Z, Narassimhan E, Bhandary RR, and Huang P (2021) From fossil to low carbon: The evolution of global public energy innovation. Wiley Interdisciplinary Reviews: Climate Change 12:e734
2. Villa K, Galán-Mascarós JR, López N, and Palomares E (2021) Photocatalytic water splitting: advantages and challenges. Sustainable Energy Fuels 5:4560–4569
3. Geng K, He T, Liu R, Dalapati S, Tan KT, Li Z, Tao S, Gong Y, Jiang Q, and Jiang D (2020) Covalent organic frameworks: Design, synthesis, and functions. Chem Rev 120:8814–8933
4. Liu R, Tan KT, Gong Y, Chen Y, Li Z, Xie S, He T, Lu Z, Yang H, and Jiang D (2021) Covalent organic frameworks: an ideal platform for designing ordered materials and advanced applications. Chem Soc Rev 50:120–242
5. Wang X, Chen L, Chong SY, Little MA, Wu Y, Zhu W-H, Clowes R, Yan Y, Zwijnenburg MA, Sprick RS, and Cooper AI (2018) Sulfone-containing covalent organic frameworks for photocatalytic hydrogen evolution from water. Nat Chem 10:1180–1189
6. Ding H, Mal A, and Wang C (2020) Tailored covalent organic frameworks by post-synthetic modification. Mater Chem Front 4:113–127
7. You J, Zhao Y, Wang L, and Bao W (2021) Recent developments in the photocatalytic applications of covalent organic frameworks: A review. J Cleaner Prod 291:125822
8. Yang Q, Luo M, Liu K, Cao H, and Yan H (2020) Covalent organic frameworks for photocatalytic applications. Appl Catal, B 276:119174
9. Li H, Wang L, and Yu G (2021) Covalent organic frameworks: Design, synthesis, and performance for photocatalytic applications. Nano Today 40:101247

10. Wang H, Wang H, Wang Z, Tang L, Zeng G, Xu P, Chen M, Xiong T, Zhou C, Li X, Huang D, Zhu Y, Wang Z, and Tang J (2020) Covalent organic framework photocatalysts: structures and applications. Chem Soc Rev 49:4135–4165

11. Guo L and Jin S (2019) Stable covalent organic frameworks for photochemical applications. ChemPhotoChem 3:973–983

12. Cusin L, Peng H, Ciesielski A, and Samorì P (2021) Chemical conversion and locking of the imine linkage: Enhancing the functionality of covalent organic frameworks. Angew Chem, Int Ed 60:14236–14250

13. Waller PJ, Lyle SJ, Osborn Popp TM, Diercks CS, Reimer JA, and Yaghi OM (2016) Chemical Conversion of linkages in covalent organic frameworks. J Am Chem Soc 138:15519–15522

14. Waller PJ, AlFaraj YS, Diercks CS, Jarenwattananon NN, and Yaghi OM (2018) Conversion of imine to oxazole and thiazole linkages in covalent organic frameworks. J Am Chem Soc 140:9099–9103

15. Wei S, Zhang F, Zhang W, Qiang P, Yu K, Fu X, Wu D, Bi S, and Zhang F (2019) Semiconducting 2D triazine-cored covalent organic frameworks with unsubstituted olefin linkages. J Am Chem Soc 141:14272–14279

16. Bourda L, Krishnaraj C, Van Der Voort P, and Van Hecke K (2021) Conquering the crystallinity conundrum: efforts to increase quality of covalent organic frameworks. Mater Adv 2:2811–2845

17. Yang J, Kang F, Wang X, and Zhang Q (2021) Design strategies for improving the crystallinity of covalent organic frameworks and conjugated polymers: a review. Mater Horiz 9:121–146

18. Zhai Y, Liu G, Jin F, Zhang Y, Gong X, Miao Z, Li J, Zhang M, Cui Y, Zhang L, Liu Y, Zhang H, Zhao Y, and Zeng Y (2019) Construction of covalent-organic frameworks (COFs) from amorphous covalent organic polymers via linkage replacement. Angew Chem, Int Ed 58:17679–17683

19. Côté AP, Benin AI, Ockwig NW, O'Keeffe M, Matzger AJ, and Yaghi OM (2005) Porous, crystalline, covalent organic frameworks. Science 310:1166–1170

20. Chen X, Geng K, Liu R, Tan KT, Gong Y, Li Z, Tao S, Jiang Q, and Jiang D (2020) Covalent organic frameworks: Chemical approaches to designer structures and built-in functions. Angew Chem, Int Ed 59:5050–5091

21. Ma B, Li C, Zhang L, Zhai L, Hu F, Xu Y, Qiao H, Wang Z, Ai W, and Mi L (2021) Flexible thiourea linked covalent organic frameworks. CrystEngComm 23:7576–7580

22. Zhao J, Ren J, Zhang G, Zhao Z, Liu S, Zhang W, and Chen L (2021) Donor-acceptor type covalent organic frameworks. Chem Eur J 27:10781–10797

23. Rodríguez-San-Miguel D, Montoro C, and Zamora F (2020) Covalent organic framework nanosheets: preparation, properties and applications. Chem Soc Rev 49:2291–2302

24. Tao Y, Ji W, Ding X, and Han B-H (2021) Exfoliated covalent organic framework nanosheets. J Mater Chem A 9:7336–7365

25. Qian Z, Wang ZJ, and Zhang KAI (2021) Covalent triazine frameworks as emerging heterogeneous photocatalysts. Chem Mater 33:1909–1926

26. Li Y, Gao C, Long R, and Xiong Y (2019) Photocatalyst design based on two-dimensional materials. Mater Today Chem 11:197–216

27. Ding S-Y, Wang P-L, Yin G-L, Zhang X, and Lu G (2019) Energy transfer in covalent organic frameworks for visible-light-induced hydrogen evolution. Int J Hydrogen Energy 44:11872–11876

28. Chen Z, Li X, Yang C, Cheng K, Tan T, Lv Y, and Liu Y (2021) Hybrid porous crystalline materials from metal organic frameworks and covalent organic frameworks. Adv Sci 8:2101883

29. Huang X and Zhang Y-B (2021) Covalent organic frameworks for sunlight-driven hydrogen evolution. Chem Lett 50:676–686

30. Dong P, Wang Y, Zhang A, Cheng T, Xi X, and Zhang J (2021) Platinum single atoms anchored on a covalent organic framework: Boosting active sites for photocatalytic hydrogen evolution. ACS Catal 11(21):13266–13279

31. Lu R, Liu C, Chen Y, Tan L, Yuan G, Wang P, Wang C, and Yan H (2021) Effect of linkages on photocatalytic H2 evolution over covalent organic frameworks. J Photochem Photobiol, A 421:113546

32. Chen W, Wang L, Mo D, He F, Wen Z, Wu X, Xu H, and Chen L (2020) Modulating benzothiadiazole-based covalent organic frameworks via halogenation for enhanced photocatalytic water splitting. Angew Chem, Int Ed 59:16902–16909

33. Nguyen HL and Alzamly A (2021) Covalent organic frameworks as emerging platforms for CO2 photoreduction. ACS Catal 11:9809–9824

34. Cui J-X, Wang L-J, Liu F, Meng B, Zhou Z-Y, Su Z-M, Wang K, and Liu S (2021) A metal-free covalent organic framework as photocatalyst for CO2 reduction at low CO2 concentration in gas-solid system. J Mater Chem A 9:24895–24902

35. Kim YH, Kim N, Seo J-M, Jeon J-P, Noh H-J, Kweon DH, Ryu J, and Baek J-B (2021) Benzothiazole-based covalent organic frameworks with different symmetrical combinations for photocatalytic CO2 conversion. Chem Mater 33(22):8705–8711

36. Yang Y, Lu Y, Zhang H-Y, Wang Y, Tang H-L, Sun X-J, Zhang G, and Zhang F-M (2021) Decoration of active sites in covalent–organic framework: An effective strategy of building efficient photocatalysis for CO2 reduction. ACS Sustainable Chem Eng 9:13376–13384

37. Guo K, Zhu X, Peng L, Fu Y, Ma R, Lu X, Zhang F, Zhu W, and Fan M (2021) Boosting photocatalytic CO2 reduction over a covalent organic framework decorated with ruthenium nanoparticles. Chem Eng J 405:127011

38. Cheng H-Y and Wang T (2021) Covalent organic frameworks in catalytic organic synthesis. Adv Synth Catal 363:144–193

39. Paul R, Chandra Shit S, Mandal H, Rabeah J, Kashyap SS, Nailwal Y, Shinde DB, Lai Z, and Mondal J (2021) Benzothiazole-linked metal-free covalent organic framework nanostructures for visible-light-driven photocatalytic conversion of phenylboronic acids to phenols. ACS Appl Nano Mater 4(11):11732–11742

40. He H, Fang X, Zhai D, Zhou W, Li Y, Zhao W, Liu C, Li Z, and Deng W (2021) A porphyrin-based covalent organic framework for metal-free photocatalytic aerobic oxidative coupling of amines. Chem Eur J 27:14390–14395

41. Luo B, Chen Y, Zhang Y, and Huo J (2021) Benzotrithiophene and triphenylamine based covalent organic frameworks as heterogeneous photocatalysts for benzimidazole synthesis. J Catal 402:52–60

42. Dong W, Yang Y, Xiang Y, Wang S, Wang P, Hu J, Rao L, and Chen H (2021) A highly stable all-in-one photocatalyst for aryl etherification: the NiII embedded covalent organic framework. Green Chem 23:5797–5805

43. Ahmed I and Jhung SH (2021) Covalent organic framework-based materials: Synthesis, modification, and application in environmental remediation. Coord Chem Rev 441:213989

44. Liu F, Dong Q, Nie C, Li Z, Zhang B, Han P, Yang W, and Tong M (2022) Peroxymonosulfate enhanced photocatalytic degradation of serial bisphenols by metal-free covalent organic frameworks under visible light irradiation: mechanisms, degradation pathway and DFT calculation. Chem Eng J 430:132833

45. Lin D, Duan P, Yang W, Huang X, Zhao Y, Wang C, and Pan Q (2022) Facile fabrication of melamine sponge@covalent organic framework composite for enhanced degradation of tetracycline under visible light. Chem Eng J 430:132817

46. Guo X, Yin D, Khaing KK, Wang J, Luo Z, and Zhang Y (2021) Construction of MOF/COF hybrids for boosting sunlight-induced fenton-like photocatalytic removal of organic pollutants. Inorg Chem 60:15557–15568

47. Ma J-X, Li J, Chen Y-F, Ning R, Ao Y-F, Liu J-M, Sun J, Wang D-X, and Wang Q-Q (2019) Cage based crystalline covalent organic frameworks. J Am Chem Soc 141:3843–3848

48. Zhu Q, Wang X, Clowes R, Cui P, Chen L, Little MA, and Cooper AI (2020) 3D cage COFs: A dynamic three-dimensional covalent organic framework with high-connectivity organic cage nodes. J Am Chem Soc 142:16842–16848

49. Li M, Peng Y, Yan F, Li C, He Y, Lou Y, Ma D, Li Y, Shi Z, and Feng S (2021) A cage-based covalent organic framework for drug delivery. New J Chem 45:3343–3348

50. Ji C, Su K, Wang W, Chang J, El-Sayed E-SM, Zhang L, and Yuan D (2021) Tunable cage based three-dimensional covalent organic frameworks. CCS Chem 1–30

19

Covalent Organic Frameworks and Clusters in Storing Hydrogen

Sukanta Mondal[1], Prasenjit Das[2], and Pratim Kumar Chattaraj[2]

[1]*Department of Education, A. M. School of Educational Sciences, Assam University, Silchar, Assam, India*

[2]*Department of Chemistry, Indian Institute of Technology, Kharagpur, India*

CONTENTS

19.1 Introduction

Continuous increment of energy demand with onboard utilization facilities has compelled the scientific community to probe meticulously on the processing, storage, and employment of energy. Amongst all possible energy resources hydrogen is one of the most likely sustainable options to fulfill the climate consistency. The burning of hydrogen produces energy via the release of its sole electron pair and the production of water vapor. On the other hand, biofuels, fossil fuels, and coal produce oxides of carbon (CO_x), nitrogen (NO_x), sulfur (SO_x), etc. Crude fuels, coal, and oil yield ash particles in the combustion process, such by-products contribute to pollution and global warming [1]. The international energy agency (IEA) documented H_2 as a significant constituent to achieve energy secured future [2]. During the last three decades, sufficient research work has been done to achieve potential hydrogen storage material. Considering both the in silico and experimental findings so far instigated H_2 trapping agents include metal-organic frameworks (MOFs), metal hydrides, molecular sheets, various nanostructured materials, covalent organic frameworks (COFs), synthesized hydrocarbons, liquid organic hydrogen carriers (LOHC), ammonia-borane

DOI: 10.1201/9781003206507-19

complexes, etc [3]. These materials can be classified based on constituent elements, morphology, and the type of interaction with hydrogen. Based on morphology, the MOFs and COFs are noted as porous materials. Owing to the high surface area of these frameworks they are considered as potential host for hydrogen storage. A host is considered as potential if reversible storage of H_2 takes place [4]. If the hydrogen binding energy falls in the range 10–50 kJ·mol^{-1}, the host is called reversible H_2 storing material. In addition to most MOFs and COFs, carbon-based materials also possess this characteristic [5]. The guest H_2 molecules are accommodated in the pores of MOFs/COFs and on the surface of carbon-based materials through noncovalent interactions. Such contact is of dominant importance because the interaction energy falls in the reversible range. Moreover, up to 90% onboard efficiency of hydrogen storage materials can facilitate the designing of systems to reach the United States Department of Energy (DOE) target of 0.040 kg/L (ultimatum 0.050 kg/L) by the year 2025 [6]. Omar M. Yaghi and co-workers noted that at NTP the MOF-5, isoreticular metal-organic framework-6 (IRMOF-6), and IRMOF-8 can trap H_2 up to 1.9, 4.2, and 9.1 wt%, respectively [7]. Research on MOFs has been going on globally to achieve a better H_2 storing material. But, due to the high cost of precursor reagents and solvents to prepare MOFs, research on other potential hosts to trap H_2 is still going on [8]. Admonishing tone originates not only from the cost, the cavity size of the different MOFs restricts their particular usage. Hierarchically porous MOFs (HP-MOFs) bring a solution to the problem, particularly, the one which originated due to the pore size of MOFs. The pores in HP-MOFs range from < 2 nm to > 50 nm. Such a vast range can give access to various sized guest species [9] as well as a substantial amount of H_2 gas can also be stored. It is worthy to mention here that the research group of Omar M. Yaghi synthesized COFs [10, 11]. COFs are classified based on their skeleton as two-dimensional (2D) and three-dimensional (3D) [12]. The COFs are composed of the main group elements, H, B, C, N, O, and Si. The stability of COFs originates from the skeletal strong covalent bond between the constituent atoms, C–C, C–O, C–Si, etc. It is reported that via such skeletal linkages composed COF-102, COF-103 possess 3472 m^2/g, 4210 m^2/g porosity, respectively [13]. Han *et al.* noted that the 3D COFs owing to their greater surface area and porosity can store 2.5–3 times more H_2 in comparison to the 2D counterparts [13]. It is reported that due to the increment of pressure from 35 bar to 100 bar the COF-102 reveals 2.79 wt% more H_2 storing ability at 77 K [14]. It was also found that a few 3D COFs can store H_2 up to 26.7 wt% under the pressure of 100 bar at 77 K [15]. But, due to the higher thermal stability and easy synthesis of 2D COFs in comparison to 3D forms, the former is considered as a better H_2 trapping agent [16]. Besides COFs, in this chapter, we would discuss the structure, stability, and hydrogen trapping potential of a few covalent organic clusters and their metal-bound analogues in the subsequent sections. In the next section used computational methods are discussed.

19.2 Theory and Computational Methods

To achieve detailed insight into different chemical and materialistic analyses the practice of density functional theory (DFT) and conceptual density functional theory (CDFT) is becoming indispensable in the present era [17–19]. Computation and calculation of electronegativity (χ), hardness (η), and electrophilicity (ω) reveal further insight into the overall chemical characteristics of a molecule. On the other hand, Fukui functions (f_k) disclose the reactivity of a particular site/center of a molecule [20–22]. Mathematical form to calculate χ of an N electron system is:

$$\chi = -\left(\frac{\partial E}{\partial N}\right)_{v(r)} \tag{19.1}$$

chemical potential (μ):

$$\mu = \left(\frac{\partial E}{\partial N}\right)_{v(r)} \tag{19.2}$$

hardness (η):

$$\eta = \left(\frac{\partial^2 E}{\partial N^2}\right)_{v(r)} \tag{19.3}$$

electrophilicity (ω):

$$\omega = \frac{\mu^2}{2\eta} = \frac{\chi^2}{2\eta} \tag{19.4}$$

$v(r)$ indicates the external potential in the above equations [17–22].
Electronegativity and hardness can also be expressed as (using finite difference method):

$$\chi = (I + A)/2 \tag{19.5}$$

$$\eta = (I - A) \tag{19.6}$$

Here, I and A marks the ionization potential and electron affinity of a system, respectively, and they can be calculated using Koopmans' theorem [23], $I \approx -E_{HOMO}$ and $A \approx -E_{LUMO}$

$$\chi = -\frac{1}{2}(E_{HOMO} + E_{LUMO}) \tag{19.7}$$

$$\eta = -(E_{HOMO} - E_{LUMO}) \tag{19.8}$$

In Equations (19.7) and (19.8), E_{HOMO} and E_{LUMO} specifies the energies of highest occupied and lowest unoccupied molecular orbitals, respectively.
One can use the ΔSCF technique to calculate I and A of a system:

$$I \approx E(N-1) - E(N) \tag{19.9}$$

$$A \approx E(N) - E(N+1) \tag{19.10}$$

Here, E(N – 1), E(N), and E(N + 1) denote the single-point energies obtained using the optimized coordinate of the system considering (N – 1), N, and (N + 1) numbers of electrons, respectively. It is well-known that the minimum electrophilicity principle (MEP) corroborates well with the maximum hardness principle (MHP) for many molecular systems and processes [18–22]. These electronic structure principles help to assess the stability, reactivity, and aromaticity of various molecular systems. It is around one and a half-century; we are still employing the concept of aromaticity to understand many molecular phenomena as well as stability of different molecular systems and intermediates. The simple rules to understand the aromaticity of a molecular moiety are: the system must comprise $(4n + 2)$ numbers of π electrons ($n = 0, 1, 2, 3$, etc), the molecule must be cyclic and planar. The compulsory criterion of aromaticity, the $(4n + 2)$ π electrons count was first proposed by Hückel. Although there are many methods to ascertain the aromaticity of a molecular motif, nucleus independent chemical shift (NICS) [24], harmonic oscillator model of aromaticity (HOMA), multicentre bond index (MCI), and electron localization function (ELF), etc. are used mostly [25]. In this chapter, the NICS values of different molecular systems are discussed. NICS(0) indicates assessment of NICS value at the center of an aromatic ring whereas NICS(1) denotes at 1 Å above the ring. When a closed molecular system (cage-like) reveals (-)Ve NICS value at the cluster center, then such cases are dealt with spherical aromaticity employing the $2(n + 1)^2$ valence electrons rule. This electron count of spherical aromaticity applies to all symmetric conjugated π-networks [26, 27]. Whereas open cage type moieties can also be studied using the "Open-Shell Spherical Aromaticity" introduced by Sola *et al.* [28]. Such molecules must have $(2N^2 + 2N + 1)$ numbers of π-electrons.

The molecular moieties reported in this chapter were studied by using GaussView 3.0 (or GaussView 5.0.8) [29]. Gaussian 03 and Gaussian 09 programming suites were used to accomplish the computations [30, 31]. Optimizations were done to obtain stationary points and thereafter at least local minima were confirmed by harmonic vibrational frequency analysis. Depending on the molecular systems, a series of theoretical levels were employed in conjunction with various basis sets. Used theoretical levels are: MP2, B3LYP, DFT-D-LYP, DFT-D-B3LYP, M052X, M06, M062X, mPW2PLYPD, CAM-B3LYP, LC-wPBE, BP86, LC-BLYP, PBEPBE, and employed basis sets are: 6-31G(d), 6-31G(d,p), 6-311+G(d,p), 6-311+G(d), cc-PVDZ. Vienna ab initio Simulation Package (VASP) [32] was used to do the computations for the modelled periodic systems. Such systems were modelled by XCrySDen [33]. According to the component elements, Projector Augmented Wave (PAW) potentials were used with a kinetic energy cut-off of 550 eV. $1 \times 1 \times 6$ Monkhorst–Pack set of k-points were used to sample the Brillouin zone [34]. Effect of electric field on a limited molecular moiety, particularly, to assess the hydrogen adsorption and desorption processes, were done [35]. The next section includes the hydrogen trapping potential of neutral and ionic clusters.

19.3 Neutral and Ionic Clusters in Storing Hydrogen

Monomeric and polymeric forms of HF and H_2O were studied to understand their underlying bonding (in the polymers) and hydrogen trapping potential. In another study N_6^{4-} and N_4^{2-} rings were modeled to check their stability, aromaticity, and hydrogen trapping potential.

19.3.1 (HF)$_m$ [m = 1 – 8] Clusters

Mondal *et al.* showed that the hydrogen trapping potential of clathrate hydrates increases due to HF doping (substitutional) [36]. It was noted that HF molecules have the potential to form polymeric units. Further, it was found that among the $(HF)_m$ (m = 3–10) units the tetramer, hexamer, and octamer are stable [37]. Interaction energies and the distances of H_2 from the HF monomer and polymeric forms were calculated at the MP2 level of theory and at several other theoretical levels in conjunction with 6-311+G(d,p) basis set [37]. Among the studied theoretical levels, it was noted that at the M052X level, the results are closer to the MP2 level. Using the optimized coordinates of the H_2 adsorbed forms further studies were done to understand the nature of the interaction. Particularly, energy decomposition analysis (EDA) was done. Moreover, the hardness of the $nH_2@(HF)_m$ moieties was calculated to verify the obtained results. Maximum H_2 adsorbed analogues were taken to study the kinetic stability. We did the *ab initio* molecular dynamics simulation using the atom-centered density matrix propagation (ADMP) approach.

It was noted that one HF unit can trap up to 3 H_2 molecules. These three hydrogen molecules are bound via the H and F end of the HF monomer (Figure 19.1). Accordingly, it was found that the dimer, trimer, and pentamer can adsorb up to 6, 9, and 15 H_2 molecules, respectively. On the other hand, the tetramer, hexamer, heptamer, and octamer also can adsorb 12, 18, 21, and 24 H_2 molecules, respectively, but in these cases, the alignment of H_2 molecules is not the same. Calculated interaction energies of adsorption per hydrogen molecule are –3.097, –2.815, –2.873, –2.855, –3.034, –3.218, –3.566 kJ/mol for the dimer,

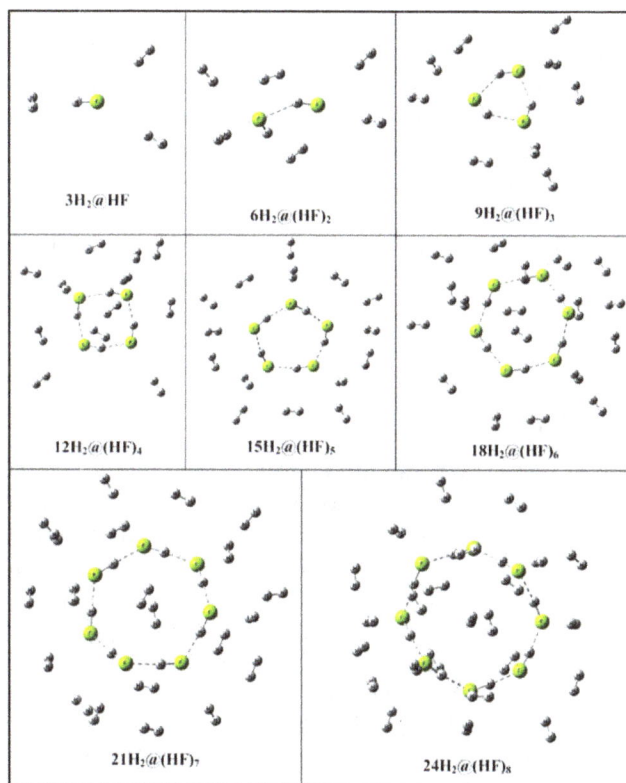

FIGURE 19.1
Minimum energy structures of xH_2 adsorbed $(HF)_m$ clusters at M05-2X/6-311+G(d,p) level of theory. (Adapted with permission [37]. Copyright (2015) Elsevier.)

trimer, tetramer, pentamer, hexamer, heptamer, and octamer, respectively. A decrease in the hardness values in the order of H_2 bound monomer to octamer indicates that the three hydrogen molecule bound HF is the most stable.

From the electron density analysis, it was found that the H····H_2 and F····H_2 contacts are noncovalent. Moreover, the contour plot of total electrostatic potential reveals that the F····H_2 interaction is electrostatic. ADMP simulation for 1 ps was done using the local minima coordinates of $24H_2@(HF)_8$ at 77 K and 298 K. It was noted that up to 1 ps only 10 H_2 molecules remain adsorbed at 77 K whereas only two hydrogen molecules remain in the vicinity at 298 K.

19.3.2 $(H_2O)_n$ [n = 1 – 10] Clusters

Like the HF monomer, the H_2O also can trap up to 3 H_2 molecules (please see Figures 19.1 and 19.2) [37]. Each of the two hydrogens and one oxygen of water molecule can bind one H_2 with the interaction energy of –2.716 kJ/mol. The water dimer can bind 3 hydrogen molecules with the binding energy of –2.057 kJ/mol. Whereas for the $(H_2O)_n$ (n = 3–10) each water molecule can bind only one H_2. It is worthy to mention here that the possible six isomers of water hexamer can bind up to 6 hydrogen molecules. The book isomer of the water hexamer is having greater H_2 binding ability in comparison to other isomers. The water heptamer, octamer, and nonamer bind hydrogens with the interaction energy of –3.757, –3.545, and –3.757 kJ/mol, respectively. Among the water decamers, it was noted that the hydrogen trapping ability of the prism isomer is greater whereas the prism-dot isomer is energetically favorable. The hardness values for hydrogen trapped $(H_2O)_n$ clusters (n = 1–10) show a similar trend to the $(HF)_m$ equivalents. Electron density analysis revealed that the H····H_2 and O····H_2 interactions are noncovalent. And the contour plot of total electrostatic potential revealed that the O····H_2 interaction is electrostatic.

$10H_2@(H_2O)_{10}$ Prism-dot form was taken to assess the kinetic stability. It was noted that 4 H_2 molecules remain bound to the water decamer up to 1 ps at 77 K. The 298 K simulation revealed that up to the 1 ps time scale only two hydrogen molecules remain bound [37].

19.3.3 N_6^{4-} and N_4^{2-} Rings

Polynitrogen molecules are known as high energy density materials [38–41]. Inspired by this idea Duley *et al.* [42] had considered two anionic cyclic planar polynitrogen rings namely, N_6^{4-} and N_4^{2-} to study their aromaticity. Moreover, their objective was to know the hydrogen storing capacity of the cation-bound N_6^{4-} and N_4^{2-} rings. Although the twisted D_2 symmetric geometry of the N_6^{4-} ring is more stable than that of the D_{6h} symmetric structure, but substantial experimental evidence by Volger *et al.* [38] showed that the cyclic N_6 system can exist in planar form. The NICS has been applied for the investigation of the aromaticity of the rings considered for the study. The geometry optimization and the subsequent frequency calculation of all the molecules were performed using B3LYP functional with the help of 6-31G(D) basis set. The NICS(0) and NICS(1) were computed by placing a ghost atom at the center and 1 Å above the cyclic polynitrogen rings, respectively. The NICS scan plots were generated by placing the ghost atom at different distances perpendicular to the planar cyclic rings. The calculated NICS(0) values of N_6^{4-} ring is comparable with that of the benzene, while, the same for N_4^{2-} ring is comparable with cyclobutadiene. Moreover, the NICS scan plots of the N_6^{4-} ring and the benzene nucleus show a similar

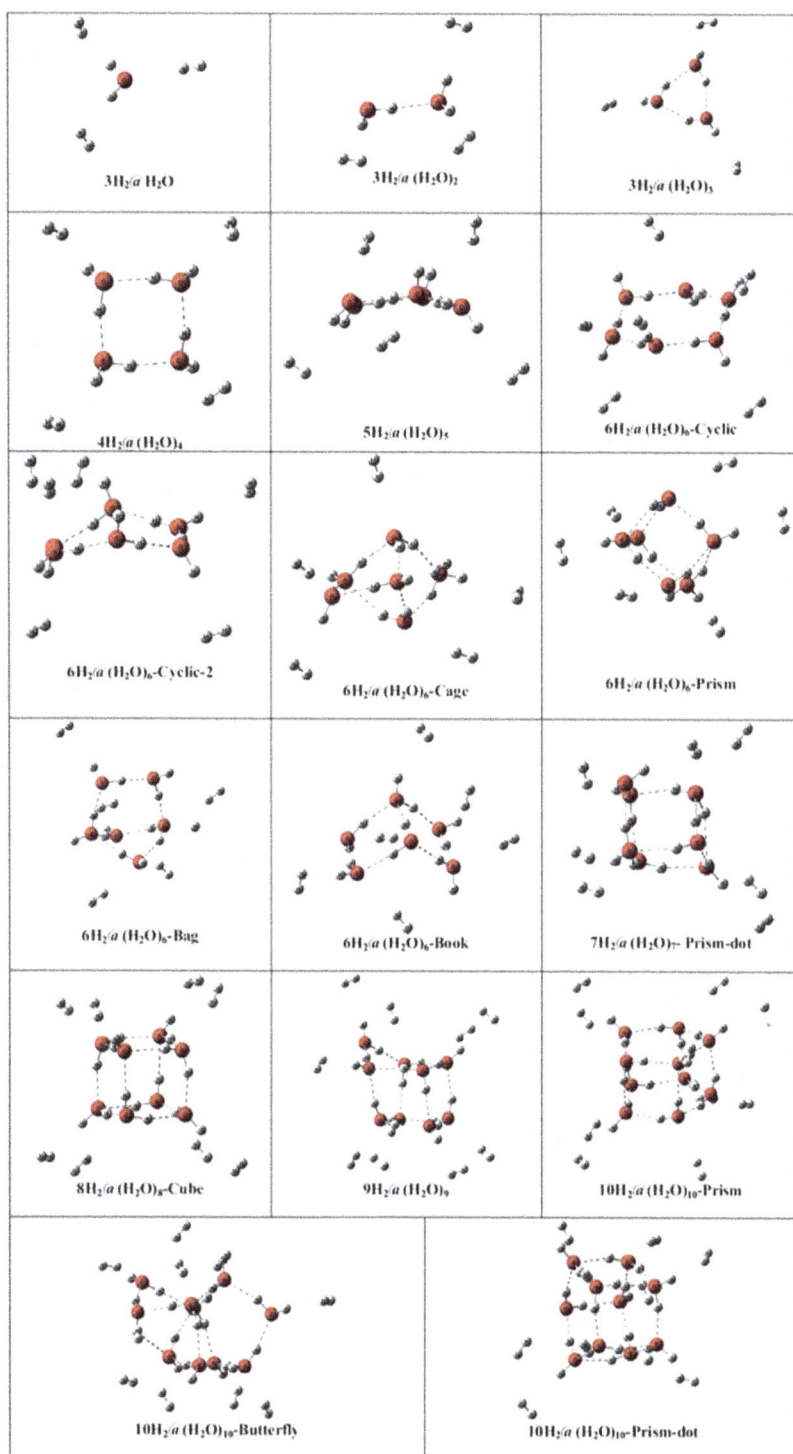

FIGURE 19.2
Minimum energy structures of xH$_2$ adsorbed (H$_2$O)$_n$ clusters at M05-2X/6-311+G(d,p) level of theory. (Adapted with permission from [37]. Copyright (2015) Elsevier.)

pattern but the N_4^{2-} ring shows a different pattern than the cyclobutadiene. The aromaticity is shown by the N_6^{4-} ring having ten π-electrons. But despite of six π-electrons the N_4^{2-} ring shows conflicting aromaticity. The term conflicting aromaticity arises from the simultaneous existence of σ-antiaromaticity and π-aromaticity for a system. The NICS(0) > 0 and NICS(1) < 0 indicates the σ-antiaromaticity and π-aromaticity, respectively. Although the NICS value is positive at the center of the N_4^{2-} ring, the NICS(1) value is negative and the values gradually decrease with increasing the distance above the ring. Both the studied rings were stabilized with the help of suitable counter-ions. The Ca^{2+} and Li^+ ions were used for the stabilization of anionic N_6^{4-} and N_4^{2-} rings, respectively. The cations bound to the rings via cation-π interaction and formed N_6Ca_2 and N_4Li_2 complexes. Figure 19.3 contained the minimum energy structures of N_6Ca_2 and N_4Li_2 complexes and their corresponding maximum number of hydrogen trapped analogues. The aromaticity of the N_6^{4-} ring increases on the binding of Ca^{2+} ion through cation-π interaction, but the conflicting aromaticity is still shown by the N_4^{2-} ring bonded with Li^+. The H_2 molecules binding occurred at the metal centers due to the highly positive charges on them. The negative values of hydrogen adsorption energy (ΔE_{ads}) indicate the hydrogen trapping ability of both N_4Li_2 and N_6Ca_2 clusters. Each Li atom bind with four H_2 molecules with 1.2 kcal/mol adsorption energy. However, six H_2 molecules are attached to each Ca center with 1.3 kcal/mol adsorption energy.

DFT-based computation showed that the N_6^{4-} and N_4^{2-} rings get stabilized through the binding with Ca^{2+} and Li^+ ions, respectively. On binding of Ca^{2+} ion through cation-π interaction, the aromaticity of the N_6^{4-} ring increases. But, both the N_4^{2-} ring and N_4Li_2 system show conflicting aromaticity. The negative values of ΔE_{ads} indicate the hydrogen trapping ability of both N_4Li_2 and N_6Ca_2 clusters. Thus one can conclude that these materials may behave as hydrogen trapping agents.

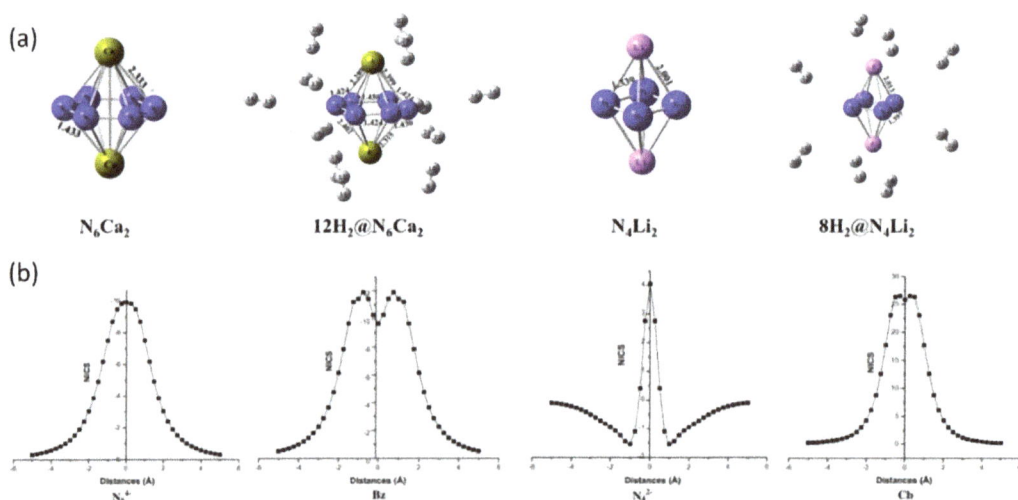

FIGURE 19.3
(a) Optimized geometries of N_6Ca_2 and N_4Li_2 and their corresponding hydrogen-trapped analogs; color code: blue for N, pink for Li, green for Ca, and white for H atoms. (b) NICS-scan plots for N_6^{4-}, N_4^{2-}, benzene (Bz), and cyclobutadiene (Cb). (Adapted with permission [42]. Copyright (2011) Elsevier.)

19.4 Polycyanogen Cages Interacting with H_2

Nitrogen-rich compounds are used as high-energy-density materials that are environmentally acceptable. Mondal *et al.* has taken a nitrogen-rich molecule $C_{12}N_{12}$ and have shown different isomers of this system [43]. The reported possible isomers of $C_{12}N_{12}$ are $C_{12}N_{12}$-A, $C_{12}N_{12}$-B, and $C_{12}N_{12}$-C. The theoretical study predicted the possibility of hydrogen storage by these isomers. Among these three isomers, $C_{12}N_{12}$-A is the most symmetric having D_{6d} point group. While $C_{12}N_{12}$-B and $C_{12}N_{12}$-C isomers have C_3 and C_{2v} symmetry, respectively. Figure 19.4 contained the optimized geometries of the isomers along with the relative energies. The cage wall of the $C_{12}N_{12}$-A isomer is made up of 12 nonplanar five-membered units and there are two planar cyclic C_6 units at the ends of the cage. Altogether, the $C_{12}N_{12}$-A isomer looks like a drum-like structure. The $C_{12}N_{12}$-B structure is formed by the two hexagonal C_3N_3 units having chair conformations, six C_2N_3 units, and six C_3N_2 units. Whereas the isomer $C_{12}N_{12}$-C has a different structure composed of eight C_3N_2 pentagonal units. The calculated relative energies indicated that $C_{12}N_{12}$-C is the most stable isomer. All these isomers show aromatic nature as predicted from the negative values of the NICS(0), computed at the center of the cages. $C_{12}N_{12}$-A and $C_{12}N_{12}$-B have 24 π-electrons and therefore they do not obey the $2(N+1)^2$ π rule of spherical aromaticity. Similarly, the $C_{12}N_{12}$-C isomer has 24 π-electrons and hence does not follow the $(2N^2+2N+1)$ π electron rule of open-shell spherical aromaticity.

The study showed that there are three possible sites to which molecular hydrogen can adsorb exohedrally and they are (i) on N atom; (ii) on C atom; (iii) above the midpoint of the C-N bond. The isomers can adsorb a maximum of 12 H_2 molecules by their N sites. All the isomers show a 7.2 wt% gravimetric hydrogen storage capacity. To support the stability of all the systems, hardness values were calculated. On the gradual increase in the number of molecular hydrogens, the nH_2 loaded $C_{12}N_{12}$-A and $C_{12}N_{12}$-C isomers show an increment of hardness values in a nonlinear fashion. But the variation of hardness for the $C_{12}N_{12}$-B

(a)

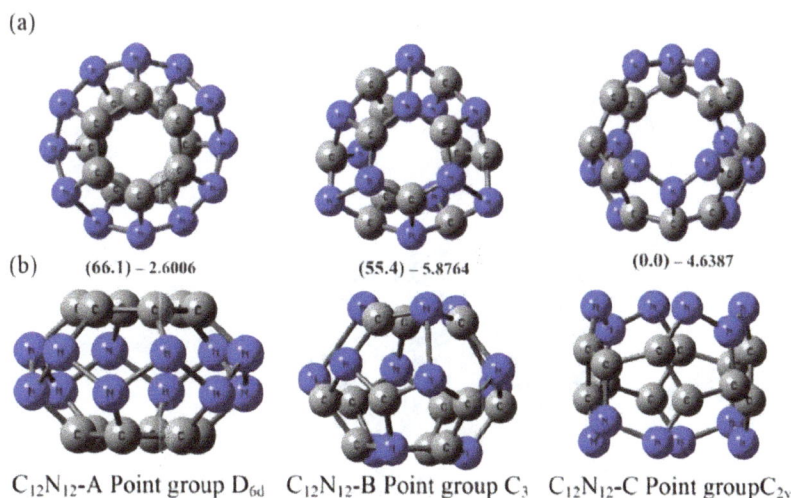

(b) (66.1) – 2.6006   (55.4) – 5.8764   (0.0) – 4.6387

$C_{12}N_{12}$-A Point group D_{6d} $C_{12}N_{12}$-B Point group C_3 $C_{12}N_{12}$-C Point group C_{2v}

FIGURE 19.4
Top view (a) and side view (b) of the modeled structures of $C_{12}N_{12}$-A, $C_{12}N_{12}$-B and $C_{12}N_{12}$-C. (Adapted with permission [43]. Copyright (2013) Royal Society of Chemistry.)

isomer is anomalous. However, the variation of hardness corroborates with the binding energy. In the H_2 loaded analogues of the three isomers, the NICS(0) values are still negative indicating the aromatic nature of the H_2 bound cages. Interestingly, on increasing the number of H_2 molecules the NICS(0) values become more negative for the $C_{12}N_{12}$-A isomer as compared to the other two isomers, indicating the more aromatic character of the former. The NICS(0) trend for nH_2 bound $C_{12}N_{12}$-B isomer is almost similar to the nH_2 bound $C_{12}N_{12}$-A. The trend of NICS(0) values in the nH_2 loaded $C_{12}N_{12}$-A isomer follows the same drift of binding energy and hardness values. Zhou *et al.* showed the effect of the electric field in the hydrogen adsorption and desorption process on BN sheets [35]. In the influence of an electric field with a strength of 0.005 a.u., the binding energy per H_2 molecule increased by 0.46 kcal/mol for the $C_{12}N_{12}$-A analog. The desorption of H_2 molecules is facilitated by the removal of the electric field.

Furthermore, a higher analog of $C_{12}N_{12}$-A isomer, $C_{12}N_{24}$ nanotube was modelled. The most favorable site for H_2 adsorption is above and along the C-C bond, and total 12 H_2 molecules were placed per unit cell. The calculated adsorption energy is 1.73 kcal/mol and the gravimetric H_2 storage capacity is 4.76 wt%. Further, the same host adsorbs 24 H_2 molecules (Figure 19.5) per unit cell with the binding energy of 1.44 kcal/mol and gravimetric storage capacity of 9.10 wt%.

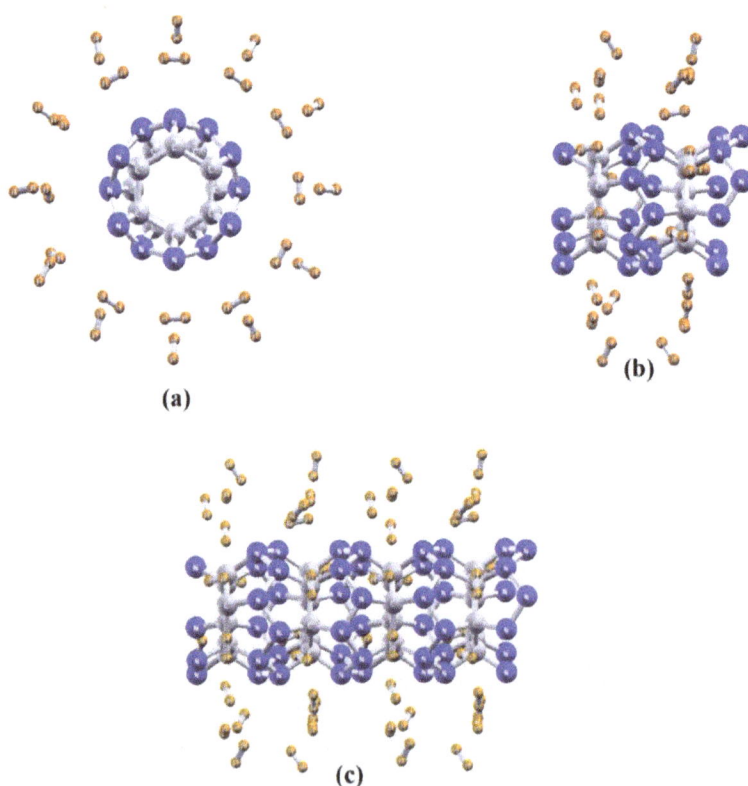

FIGURE 19.5
(a) Top view, (b) side view, and (c) the $1 \times 1 \times 3$ supercell structure of the optimized $C_{12}N_{24}(H_2)_{24}$ unit cell. (Adapted with permission [43]. Copyright (2013) Royal Society of Chemistry.)

19.5 Hydrogen Trapping Potential of ExBox⁴⁺

Barnes *et al.* in 2013 synthesized an interesting molecule ExBox[4+] which is a semi-rigid cyclophane system [44]. The appearance of aromatic six-membered rings and high charge on the atomic sites of the compound inspired Das *et al.* to study the H_2 storage capability of this interesting box typed system [45]. The structure of the compound contains two pyridinium–phenyl–pyridinium chains above and below and two phenyl rings at the side chain (Figure 19.6). The minimum energy structure of this molecule has C_{2v} point group of symmetry.

The geometry optimization and the subsequent frequency calculation were performed using ωB97x-D functional with the help of 6-311G(d,p) basis set. The ADMP simulations and the AIM analysis were also done at the same level of theory.

The theoretical study showed that the compound possesses potential to store H_2 molecules both in endohedral and exohedral fashion. In the endohedral fashion, the most favorable site for adsorption corresponds to the H_2 pointing toward the phenyl ring of the side chain. While for exohedral adsorption the H_2 occurred preferably above the phenyl rings and the least favored sites are the four corners of the molecular box. The interaction of H_2 molecules with the phenyl ring is stronger than that with the pyridine ring. Moreover, the phenyl ring at the above and below chain interacts stronger with the H_2 molecules than that of the side chain phenyl rings. Das *et al.* studied the endohedral adsorption up to three hydrogen molecules. In the case of exohedral adsorption, eight H_2 molecules were placed above the eight six-membered rings and the system $(8H_2)_{exo}$@ExBox[4+] has C_1 symmetry (Figure 19.7). Further, four more H_2 molecules adsorbed at the four corners of the molecule, and the system $(12H_2)_{exo}$@ExBox[4+] is also at the minimum on the potential energy surface (Figure 19.7). For purely endohedral adsorption the interaction energies per H_2 molecule (E_{ads}) are 2.4 kcal/mol and 2.2 kcal/mol for the trapping of two and three H_2 molecules, respectively. For exohedral adsorption, the E_{ads} values are 1.5 kcal/mol for both the 8 and 12 hydrogen trapped analogues. So, these E_{ads} values suggest that the endohedral H_2 molecules interact strongly with the host than the exohedrally adsorbed hydrogens. By the simultaneous endo and exohedral fashion, it is possible to adsorb up to 15 hydrogen

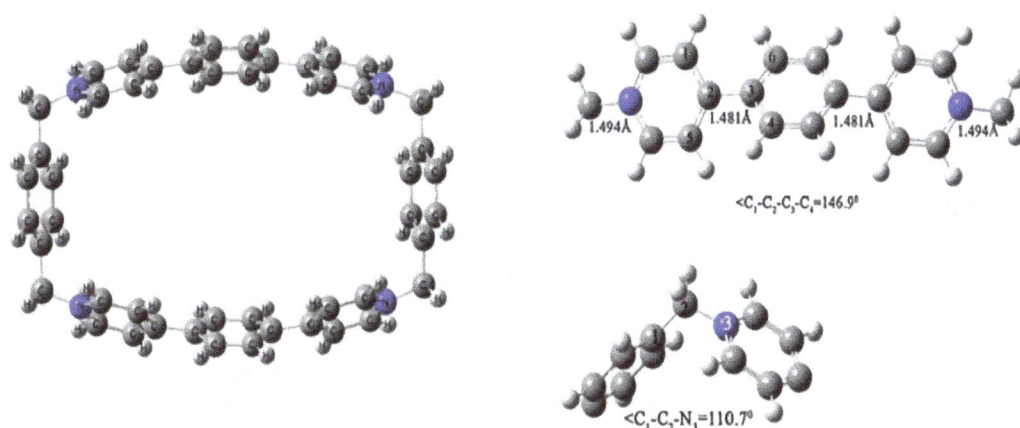

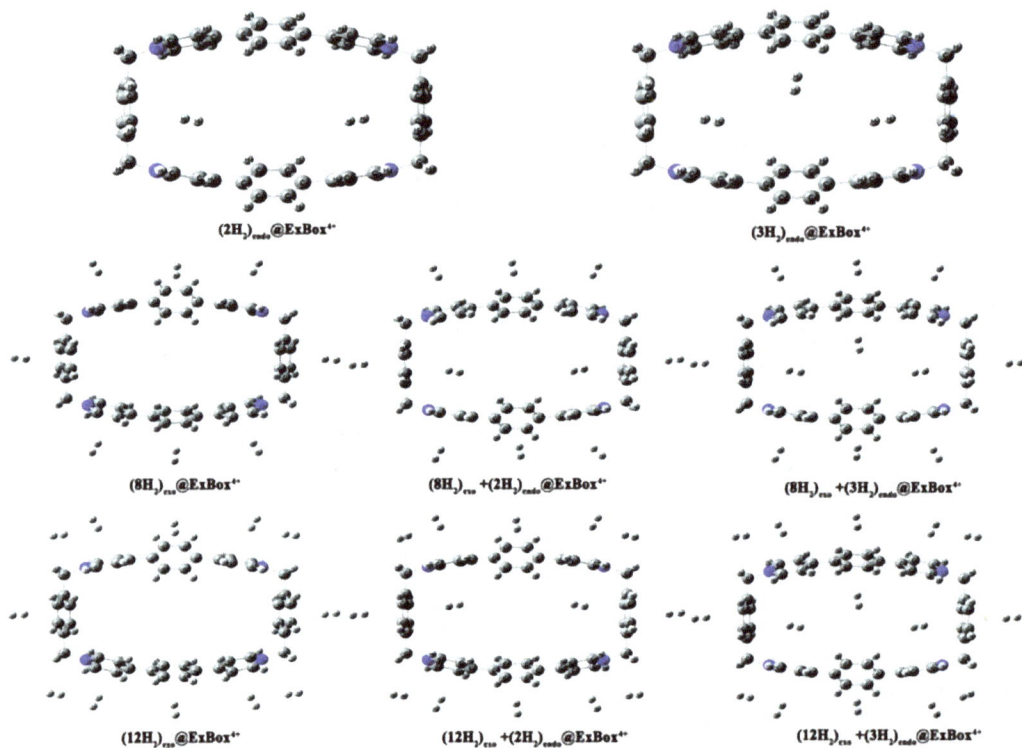

FIGURE 19.7
Optimized geometries of the hydrogen bound ExBox^{4+} complex. (Adapted with permission [45]. Copyright (2014) Royal Society of Chemistry.)

molecules and the system is represented as $(12H_2)_{exo} + (3H_2)_{endo}$@ExBox^{4+} (Figure 19.7) with the E_{ads} 1.7 kcal/mol and 4.3 wt% of storage capacity. The E_{ads} values for simultaneous endo and exohedral adsorption are even higher than that of the purely exohedral adsorption of H_2 molecules. The highest occupied molecular orbital (HOMO) and the lowest unoccupied molecular orbital (LUMO) energy gap ($\Delta E_{H\text{-}L}$) follows the same trend as that of the E_{ads} for the H_2 adsorbed systems. Please note that the $\Delta E_{H\text{-}L}$ indicates the stability of a system. The atoms-in-molecule (AIM) analysis shows the non-covalent interactions among the hydrogen molecules and the parent ExBox^{4+} system.

The ADMP simulation was performed to check the stability of the H_2 adsorbed systems at a finite temperature. For $(8H_2)_{exo} + (3H_2)_{endo}$@ExBox^{4+} system at 77 K temperature within 300 fs simulation time three H_2 molecules desorbed from the system and the other molecules reoriented their positions. At 70 K temperature, four H_2 molecules desorbed from the parent system within 200 fs to 300 fs simulation time. On the other hand, the 77 K simulation of $(12H_2)_{exo} + (3H_2)_{endo}$@ExBox^{4+} system reveals that during the 100 – 200 fs time all three endohedrally adsorbed H_2 molecules dissociated into atomic hydrogens. However, at 200 fs of simulation time, 10 H_2 molecules and three hydrogen atoms adsorbed on the ExBox^{4+} system.

To check the Li doping effect on H_2 adsorption they designed Li$_8$ExBox^{4+} system with eight Li atoms placed above the cyclic aromatic rings (Figure 19.8). The interaction of Li atoms with the ExBox^{4+} system is stronger in an exohedral fashion than in an endohedral

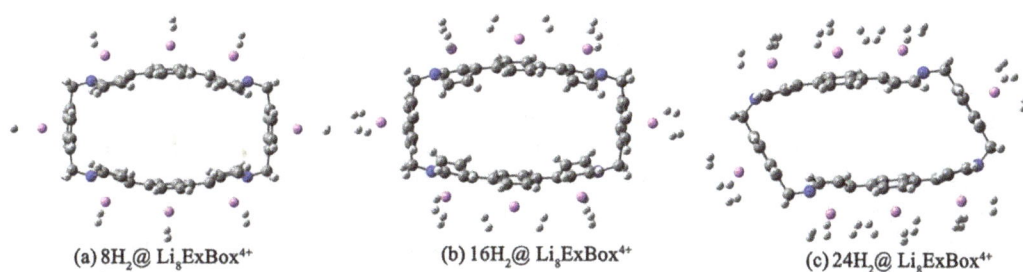

FIGURE 19.8

Optimized geometries of the $nH_2@Li_8ExBox^{4+}$ complex; n = 8 (a), 16 (b), and 24 (c). (Adapted with permission [45]. Copyright (2014) Royal Society of Chemistry.)

manner. The optimized structure of the Li_8ExBox^{4+} system has C_{2h} point group of symmetry. On Li binding, the cage structure is slightly distorted. The metal-bound cage structure remains unaltered on binding with 1 and 2 H_2 molecules per Li center, generating $8H_2@Li_8ExBox^{4+}$ and $16H_2@Li_8ExBox^{4+}$ (Figure 19.8) systems, respectively. Further addition of one more H_2 molecule per Li atom distorts the cage skeleton and the system is represented as $24H_2@Li_8ExBox^{4+}$. The E_{ads} values are 4.8 kcal/mol, 4.7 kcal/mol, and 4.3 kcal/mol for $8H_2@Li_8ExBox^{4+}$, $16H_2@Li_8ExBox^{4+}$, and $24H_2@Li_8ExBox^{4+}$ systems, respectively. On the successive increase in the number of H_2, the E_{ads} values decrease. The Li doped $ExBox^{4+}$ system can adsorb up to 24 H_2 molecules with 6.23 wt% storage capacity, even greater than the bare moiety. The interaction between Li centers and the H_2 molecules is electrostatic and this type of binding is termed quasi-molecular binding. They also plotted the T-P phase diagram to find out the region for the spontaneous hydrogen adsorption.

The ADMP simulation for the $8H_2@Li_8ExBox^{4+}$ system shows that all the H_2 molecules remain bound to the Li centers at 100 K and 77 K temperatures. But at 298 K within 150 fs of simulation time, three H_2 molecules desorbed from the system. In the case of the $16H_2@Li_8ExBox^{4+}$ system, all the adsorbed H_2 molecules remain attached at 100 K and 77 K temperatures up to 300 fs of simulation time. But at 298 K within 50 fs time one H_2 and at 100 fs time, three H_2 molecules desorbed from the system. Finally, for the $24H_2@Li_8ExBox^{4+}$ system, all the H_2 molecules remain bound at 77 K temperature up to 300 fs. But at 298 K within 100 fs of simulation time, 10 H_2 molecules desorbed from the system. So, this Li-doped $ExBox^{4+}$ system would be able to adsorb 24 H_2 at a lower temperature for a certain duration.

19.6 Cucurbit[7]uril as Hydrogen Storing Agent

Chattaraj and co-workers studied cucurbit[7]uril in the search for a suitable hydrogen storage material [46]. Cucurbit[7]uril is a macrocyclic molecule containing seven glycoluril units connected by methylenes. Modeled geometries of repeating glycoluril units and their H_2 adsorbed analogues were studied at ωB97X-D/6-311G(d,p) level of theory. It was noted that the repeating unit of glycoluril $((CH_3)_2C_4H_2N_4O_2(CH_3)_2)$ can bind up to 13 hydrogen molecules. It is worthy to mention here that only the N and O centers of the host moiety interact with the H_2 molecule and yield a stable form. The calculated binding energy of adsorption is 6.6 kJ/mol per H_2. Moreover, the adsorption process is having an enthalpy change of –12.1 kJ/mol, which reflects the exothermic nature. The HOMO-LUMO gap suggests the increase in stability of the system upon adsorption of 13 hydrogen molecules. The binding of hydrogen molecules results from the charge-quadrupole interaction, charge transfer interaction, and a charge-induced dipole type of interaction. The N and O centers show an increase in their charge magnitude on the adsorption of hydrogen molecules. Coming into the macrocyclic cucurbit[7]uril it is worthy to mention that the host possesses an internal cavity, which should facilitate the endohedral adsorption of hydrogen molecules. The binding energy of endohedral H_2 adsorption is 12.5 kJ/mol. On the other hand, the macrocycle can adsorb molecular hydrogen exohedrally also, but with 8.2 kJ/mol. Calculated HOMO-LUMO gaps of CB[7] is 11.428 eV, and for the endohedral and exohedral single H_2 trapped analogues the values are 11.449 and 11.442 eV, respectively. These data indicate that the hydrogen trapped forms are more stable in comparison to the empty cavity and the endohedral trapping of H_2 is more favorable. The CB[7] cavity can keep up to five hydrogen molecules. Further, it was noted that the O and N centers of CB[7] can adsorb up to 52 H_2 molecules (Figure 19.9). The five H_2 molecules inside the cavity are adsorbed with the binding energy values of 12.5, 11.0, 11.8, 12.7, and 11.3 kJ/mol. Five H_2 bound

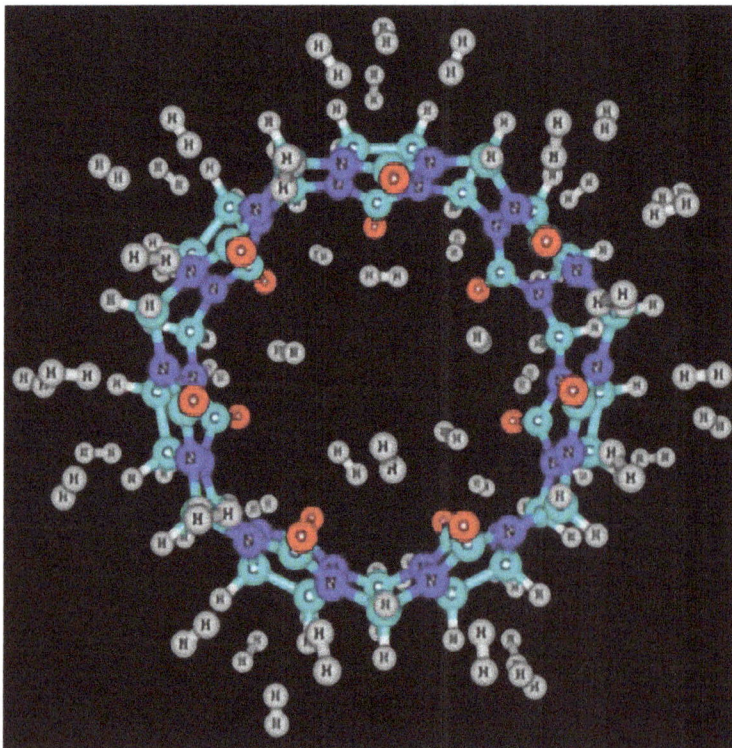

FIGURE 19.9
Optimized geometries of 52H$_2$@CB[7] at ωB97X-D/6-311G(d,p) level of theory. (Adapted with permission [46]. Copyright (2013) Royal Society of Chemistry and the Centre National de la Recherche Scientifique.)

forms further interact with hydrogen molecules via fourteen O centers where each of the oxygen sites binds with one H_2 with the binding energy of 7.7 kJ/mol. Each of these O sites further interacts with one hydrogen molecule with the binding energy of 7.0 kJ/mol. Further, the 33 H_2 bound analog interacts with 19 H_2 molecules via N centers with the binding energy of 7.3 kJ/mol per hydrogen molecule. The HOMO-LUMO gap data were calculated for all the H_2 bound CB[7] analogues and noted that except 19 and 33 H_2 bound forms in all other cases the gap increases indicating higher stability of the H_2 bound analogs. In this present study, CB[7] can bind with molecular hydrogen with up to 8.3 gravimetric wt%. Binding energy of hydrogen storage is comparable to the 3D-COFs, and doped COF-108 [46, 47, 48]. The effect of electric field on hydrogen adsorption was further studied considering the single H_2 adsorbed analog of CB[7]. Electric field polarizes the host cites as well as the adsorbed H_2 molecule. In the electric field, the positive end of the H_2 orients toward the O cite, and attractive interaction takes place. This attraction further increases with the increment of the electric field strength. Calculated binding energy in the presence of an electric field is 28.2 kJ/mol, which is significantly higher in comparison to the field-free case.

19.7 Encapsulation of H_2 by Octa Acid Cavitand

In addition to the carbonaceous molecular frameworks/clusters such as fullerene, carbon nanotubes, cucurbit[n]urils, etc., the basket-shaped octa acid (OA) also possesses a cavity to store several gas molecules. Chakraborty *et al.* studied the gas storage potential of octa acid moiety using DFT-based computations [49]. Although they have studied the guest molecules C_2H_2, C_2H_4, C_2H_6, CO_2, CO, H_2, N_2, NO_2, NO, and rare gasses (Rg), here we will only discuss the hydrogen storage. Optimization and frequency analysis of the OA and its H_2 bound analogues were done at the ωB97X-D/6-311G(d,p) level of theory without any symmetry constraint. The OA does not have a constant cage diameter due to its shape. The inner cavity of OA possesses a diameter of approximately 5.3 Å and the outer diameter is 11.4 Å, whereas at the middle the same is 8.4 Å. In the inner cavity of OA the N site is close to the benzene segment by ~ 3.3 Å, but the outer site is located in the range of 3.7 – 4.4 Å. For the hydrogen molecule bound analog of OA, the H_2 remains well inside the cavity (Figure 19.10). It was noted that the alignment of the hydrogen molecule is nearly perpendicular to the benzene fragments.

Calculation of dissociation energy (D_0) reveals that the magnitude is positive. It indicates that the dissociation would be endothermic. Calculated Gibbs free energy for the same dissociation discloses spontaneity of the process at 298 K. Charge transfer from H_2 to OA was noted and the same was further verified by the natural population analysis (NPA) charge on H_2 interacting site. The direction of charge transfer is OA to hydrogen molecule and the involved orbitals are bonding orbital (BD) of the C-C covalent bond in OA and the anti-bonding orbital (BD*) of the hydrogen molecule. The calculated Wiberg bond index (WBI) is small and indicates noncovalent interaction between OA and H_2. ADMP simulation performed on H_2@OA at 298 K revealed that the hydrogen molecule leaves the OA cavity at 212 fs. Even at low temperature the H_2 escapes the cavity but at 300 fs.

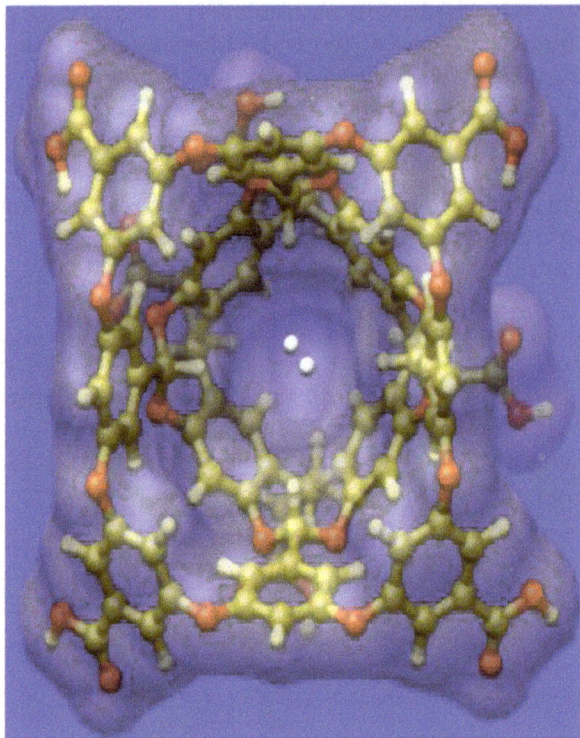

FIGURE 19.10
Surface representation of the minimum energy structure of H₂@OA. (Adapted with permission [49]. Copyright (2016) Springer.)

19.8 Conclusion

The $(HF)_m$ and $(H_2O)_n$ systems can trap molecular hydrogens. Each of the F, H, and O centers in the $(HF)_m$ and $(H_2O)_n$ clusters can interact with H_2. In the cases of HF tetramer, hexamer, and heptamer above and below the molecular plane H_2 molecules remain trapped. HF and H_2O clusters trap molecular hydrogen by the interaction energy ranging from −1.884 to −4.160 kJ/mol. F····H_2 and O····H_2 contacts are of electrostatic type whereas the H····H_2 interactions are having additional noncovalent nature. N_6^{4-} and N_4^{2-} rings show hydrogen trapping potential in their N_6Ca_2 and N_4Li_2 forms. Increment in aromaticity of N_6^{4-} ring due to the binding of Ca^{2+} ion via cation-π interaction was noted. N_4^{2-} ring and N_4Li_2 form reveals conflicting aromaticity. Among the three isomers of $C_{12}N_{12}$ the one that bears D_{6d} point group can store molecular H_2 with 7.2 wt%. And the cluster assembled material $C_{12}N_{24}$ has the potential to store hydrogen with a gravimetric density of 9.1 wt%. The ExBox⁴⁺ host possesses the characteristics of storing H_2 in endohedral and exohedral sites. Li doping generates Li_8ExBox^{4+} species. This doped moiety is thermodynamically stable and can store molecular hydrogen with a gravimetric capacity of 6.23 wt%. ADMP simulation predicts that the lithium-doped moiety can trap 24 H_2 molecules for a certain period at low temperatures. Atoms in molecule analysis disclose that the Li-H interactions are of ionic or van der Waals type. In CB[7] the N and O centers of the host act as hydrogen binding sites. Charge-quadrupole, charge transfer, and ion-induced dipole type of interactions are involved in trapping the molecular hydrogens. CB[7] host can trap up to 52 H_2 molecules. Out of these, five hydrogen molecules are trapped endohedrally and the remaining are present at exohedral sites. HOMO-LUMO gaps of the hydrogen bound CB[7] units disclose the fact that the gradual loading of H_2 molecules increased the stability of the resulting systems. Interaction energy for the adsorption of H_2 molecules increases with the increment of an applied electric field. The guest H_2 molecules interact among themselves as well as with the octa acid surface via a non-covalent type of interactions. The octa acid host can trap molecular hydrogen.

Acknowledgments

SM thanks University Grants Commission, New Delhi for UGC-BSR Research Start-Up-Grant (No. F.30-458/2019(BSR)) and his co-workers whose work is presented in this book chapter. PD thanks to UGC, New Delhi, India for his research fellowship. PKC would like to thank DST, New Delhi, for the J. C. Bose National Fellowship.

References

1. Chu S, Majumdar A (2012) Opportunities and challenges for a sustainable energy future. Nature 488:294–303
2. IEA (2019), The Future of Hydrogen, IEA, Paris, Link: https://www.iea.org/reports/the-future-of-hydrogen (Accessed on 28-11-2021).
3. Mondal S, Chakraborty A, Pan S, Chattaraj PK (2013) Designing of some novel molecular templates suitable for hydrogen storage application: A theoretical approach. In: Nanoscience and Computational Chemistry: Research Progress. Apple Academic Press: CRC Press, Taylor & Francis Group, Florida, USA
4. Jena P (2011) Materials for hydrogen storage: Past, present, and future. J. Phys. Chem. Lett. 2:206–211
5. Srinivasan S, Demirocak DE, Kaushik A, Sharma M, Chaudhary GR, Hickman N, Stefanakos E (2020) Reversible hydrogen storage using nanocomposites. Appl. Sci. 10:4618
6. DOE Technical Targets for Onboard Hydrogen Storage for Light-Duty Vehicles, Hydrogen and Fuel Cell Technologies Office. [Online] Available: https://www.energy.gov/eere/fuelcells/doe-technical-targets-onboard-hydrogenstorage-light-duty-vehicles (Accessed on 18 April 2022).
7. Rosi NL, Eckert J, Eddaoudi M, Vodak DT, Kim J, O'Keeffe M, Yaghi OM (2003) Hydrogen storage in microporous metal-organic frameworks. Science 300:1127–1129
8. DeSantis D, Mason JA, James BD, Houchins C, Long JR, Veenstra M (2017) Techno-economic analysis of metal–organic frameworks for hydrogen and natural gas storage. Energy Fuels 31:2024–2032
9. Cai G, Yan P, Zhang L, Zhou HC, Jiang HL (2021) Metal–organic framework-based hierarchically porous materials: Synthesis and applications. Chem. Rev. 121:12278–12326
10. Zhao H, Guan Y, Guo H, Du R, Yan C (2020) Hydrogen storage capacity on Li-decorated covalent organic framework-1: A first-principles study. Mater. Res. Express 7:035506
11. Côté AP, Benin AI, Ockwig NW, O'Keeffe M, Matzger AJ, Yaghi OM (2005) Porous, crystalline, covalent organic frameworks. Science 310:1166–1170

12. El-Kaderi HM, Hunt JR, Mendoza-Cortés JL, Côté AP, Taylor RE, O'Keeffe M, Yaghi OM (2007) Designed synthesis of 3D covalent organic frameworks. Science 316:268–272

13. Han SS, Furukawa H, Yaghi OM, Goddard WA (2008) Covalent organic frameworks as exceptional hydrogen storage materials. J. Am. Chem. Soc. 130:11580–11581

14. Furukawa H, Yaghi OM (2009) Storage of hydrogen, methane, and carbon dioxide in highly porous covalent organic frameworks for clean energy applications. J. Am. Chem. Soc. 131:8875–8883

15. Klontzas E, Tylianakis E, Froudakis GE (2010) Designing 3D COFs with enhanced hydrogen storage capacity. Nano Lett. 10:452–454

16. Xia L, Liu Q (2016) Lithium doping on covalent organic framework-320 for enhancing hydrogen storage at ambient temperature. J. Solid State Chem. 244:1–5

17. Parr RG, Yang W (1989) Density Functional Theory of Atoms and Molecules. Oxford University Press, New York

18. Chattaraj PK (2009) Chemical Reactivity Theory: A Density Functional View. Taylor and Francis/CRC Press, Florida, USA

19. Chakraborty D, Chattaraj PK (2021) Conceptual density functional theory based electronic structure principles. Chem. Sci. 12:6264–6279

20. Parr RG, Yang W (1984) Density functional approach to the frontier-electron theory of chemical reactivity. J. Am. Chem. Soc. 106:4049–4050

21. Chattaraj PK, Sarkar U, Roy DR (2006) Electrophilicity index. Chem. Rev. 106:2065–2091

22. Mondal S, Chattaraj PK (2014) Stability and structural dynamics of Be_3^{2-} clusters. Chem. Phys. Lett. 593:128–131

23. Koopmans TA (1933) Über die Zuordnung von Wellenfunktionen und Eigenwerten zu den Einzelnen Elektronen Eines Atoms. Physica 1:104–113

24. Schleyer PvR, Maerker C, Dransfeld A, Jiao H, van Eikema Hommes NJR (1996) Nucleus-independent chemical shifts: A simple and efficient aromaticity probe. J. Am. Chem. Soc. 118:6317–6318

25. Krygowski TM, Cyranski M (2001) Structural aspects of aromaticity. Chem. Rev. 101:1385–1420

26. Chen Z, Jiao H, Hirsch A, Thiel W (2001) The $2(N + 1)^2$ rule for spherical aromaticity – further validation. J. Mol. Model. 7:161–163

27. Hirsch A, Chen Z, Jiao H (2000) Spherical aromaticity in I_h symmetrical fullerenes: The $2(N + 1)^2$ rule. Angew. Chem. Int. Ed. 39:3915–3917

28. Poater J, Solà M (2011) Open-shell spherical aromaticity: The $2N^2 + 2N + 1$ (with $S = N + \frac{1}{2}$) rule. Chem. Commun. 47:11647–11649

29. Dennington R, Keith TA, Millam JM (2009) GaussView, Version 3.0, and 5.0.8 Semichem, Inc.

30. Frisch MJ, Trucks GW, Schlegel HB, Scuseria GE, Robb MA, Cheeseman JR, Montgomery Jr. JA, Vreven T, Kudin KN, Burant JC, Millam JM, Iyengar SS, Tomasi J, Barone V, Mennucci B, Cossi M, Scalmani G, Rega N, Petersson GA, Nakatsuji H, Hada M, Ehara M, Toyota K, Fukuda R, Hasegawa J, Ishida M, Nakajima T, Honda Y, Kitao O, Nakai H, Klene M, Li X, Knox JE, Hratchian HP, Cross JB, Bakken V, Adamo C, Jaramillo J, Gomperts R, Stratmann RE, Yazyev O, Austin AJ, Cammi R, Pomelli C, Ochterski JW, Ayala PY, Morokuma K, Voth GA, Salvador P, Dannenberg JJ, Zakrzewski VG, Dapprich S, Daniels AD, Strain MC, Farkas O, Malick DK, Rabuck AD, Raghavachari K, Foresman JB, Ortiz JV, Cui Q, Baboul AG, Clifford S, Cioslowski J, Stefanov BB, Liu G, Liashenko A, Piskorz P, Komaromi I, Martin RL, Fox DJ, Keith T, Al-Laham MA, Peng CY, Nanayakkara A, Challacombe M, Gill PMW, Johnson B, Chen W, Wong MW, Gonzalez C, Pople JA, Gaussian 03, Gaussian, Inc., Wallingford, CT, 2003.

31. Frisch MJ, Trucks GW, Schlegel HB, Scuseria GE, Robb MA, Cheeseman JR, Scalmani G, Barone V, Mennucci B, Petersson GA, Nakatsuji H, Caricato M, Li X, Hratchian HP, Izmaylov AF, Bloino J, Zheng G, Sonnenberg JL, Hada M, Ehara M, Toyota K, Fukuda R, Hasegawa J, Ishida M, Nakajima T, Honda Y, Kitao O, Nakai H, Vreven T, Montgomery Jr. JA, Peralta JE, Ogliaro F, Bearpark M, Heyd JJ, Brothers E, Kudin KN, Staroverov VN, Kobayashi R, Normand J, Raghavachari K, Rendell A, Burant JC, Iyengar SS, Tomasi J, Cossi M, Rega N, Millam JM, Klene M, Knox JE, Cross JB, Bakken V, Adamo C, Jaramillo J, Gomperts R, Stratmann RE,

Yazyev O, Austin AJ, Cammi R, Pomelli C, Ochterski JW, Martin RL, Morokuma K, Zakrzewski VG, Voth GA, Salvador P, Dannenberg JJ, Dapprich S, Daniels AD, Farkas Ö, Foresman JB, Ortiz JV, Cioslowski J, Fox DJ, Gaussian 09, Gaussian, Inc., Wallingford CT, 2009.

32. Kresse G, Furthmüller J (1996) Efficiency of ab-initio total energy calculations for metals and semiconductors using a plane-wave basis set. Comput. Mat. Sci. 6:15–50

33. Kokalj A (2003) Computer graphics and graphical user interfaces as tools in simulations of matter at the atomic scale. Comp. Mater. Sci. 28:155–168

34. Monkhorst HJ, Pack JD (1976) Special points for brillouin-zone integrations. Phys. Rev. B 13:5188

35. Zhou J, Wang Q, Sun Q, Jena P, Chen XS (2010) Electric field enhanced hydrogen storage on polarizable materials substrates. Proc. Natl. Acad. Sci. 107:2801–2806

36. Mondal S, Giri S, Chattaraj PK (2013) Possibility of having HF doped hydrogen hydrates. J. Phys. Chem. C 117:11625–11634

37. Mondal S, Ghara M, Chattaraj PK (2015) Hydrogen trapping potential of $(HF)_m$ (m = 1–8) and $(H_2O)_n$ (n = 1–10) Clusters. Comp. Theo. Chem. 1071:18–26

38. Vogler A, Wright RE, Kunkely H (1980) Photochemical reductive *cis*-elimination in *cis*-diazidobis(triphenylphosphane)platinum(II) evidence of the formation of bis(triphenylphosphane)platinum(0) and hexaazabenzene. Angew. Chem. Ind. Ed. Engl. 19:717–718

39. Chung G, Schmidt MW, Gordon MS (2000) An ab initio study of potential energy surfaces for N_8 isomers. J. Phys. Chem. A 104:5647–5650

40. Engelke R (1992) Ab initio correlated calculations of six nitrogen (N_6) isomers. J. Phys. Chem. 96:10789–10792

41. Ha TK, Nguyen MT (1992) The identity of the six nitrogen atoms (N_6) species. Chem. Phys. Lett. 195:179–183

42. Duley S, Giri S, Sathymurthy N, Islas R, Merino G, Chattaraj PK (2011) Aromaticity and hydrogen storage capability of planar N_6^{4+} and N_4^{2-} rings. Chem. Phys. Lett. 506:315–320

43. Mondal S, Srinivasu K, Ghosh SK, Chattaraj PK (2013) Isomers of $C_{12}N_{12}$ as potential hydrogen storage materials and the effect of the electric field therein. RSC Adv. 3:6991–7000

44. Barnes JC, Juríček M, Strutt NL, Frasconi M, Sampath S, Giesener MA, McGrier PL, Bruns CJ, Stern CL, Sarjeant AA, Stoddart JF (2013) ExBox: A polycyclic aromatic hydrocarbon scavenger. J. Am. Chem. Soc. 135:183–192

45. Das R, Chattaraj PK (2014) Gas storage potential of ExBox^{4+} and its Li-decorated derivative. Phys. Chem. Chem. Phys. 16:21964–21979

46. Pan S, Mondal S, Chattaraj PK (2013) Cucurbiturils as promising hydrogen storage materials: A case study of cucurbit[7]uril. New J. Chem. 37:2492–2499

47. Klontzas E, Tylianakis E, Froudakis GE (2008) Hydrogen storage in 3D covalent organic frameworks. A multiscale theoretical investigation. J. Phys. Chem. C 112:9095–9098

48. Li F, Zhao J, Johansson B, Sun L (2010) Improving hydrogen storage properties of covalent organic frameworks by substitutional doping. Int. J. Hydrogen Energy 35:266–271

49. Chakraborty D, Pan S, Chattaraj PK (2016) Encapsulation of small gas molecules and rare gas atoms inside the octa acid cavitand. Theor. Chem. Acc. 135:119

20

Covalent Organic Frameworks-Based Nanomaterials for Hydrogen Storage

Turkan Kopac

Department of Chemistry, Zonguldak Bülent Ecevit University, Zonguldak, Turkey

CONTENTS

20.1 Introduction

20.1.1 Hydrogen: A Green Energy Carrier

Continuous depletion of nonrenewable fuel sources, such as coal, natural gas, and petroleum; an increase of energy demands, combating greenhouse gases, and global warming have constrained the exploration of less-polluting, sustainable, alternative energy sources renewable in nature [1–4]. In this framework, hydrogen having an appreciable energy content (142 MJ/kg) has drawn considerable attention due to its lightweight, high energy content, energy efficiency, pollution-free, environmentally friendly, abundant, and sustainable characteristics [2–6]. Hydrogen is a nontoxic green energy carrier that does not lead to any particulate, sulfur, or carbon dioxide emissions. The superiorities of hydrogen energy over the other conventional fuels are shown in Figure 20.1. It has a high specific energy content when compared to other conventional fuels on a mass basis. For example, 25 kg of gasoline is identical to 9.5 kg of hydrogen in terms of their energy content [3]. It has the potential to be a predominant energy carrier and promising alternative for conventional fuels [2, 4, 7, 8]. It has been appreciated as an excellent energy carrier specifically for the transportation sector [1, 9]. In recent years, hydrogen energy vehicles have been researched and developed very rapidly [4]. The energy density of hydrogen and specific energy are the major parameters that are highly essential in automotive applications. Specific energy shows the net energy/mass of hydrogen (kWh/kg), while energy density is the net usable energy/volume of hydrogen (kWh/m³) related to the storage system. Hydrogen has an energy density of about ten times lesser than conventional fuels at ambient temperature and pressure, which is a critical handicap for the utilization of hydrogen as a fuel for

DOI: 10.1201/9781003206507-20

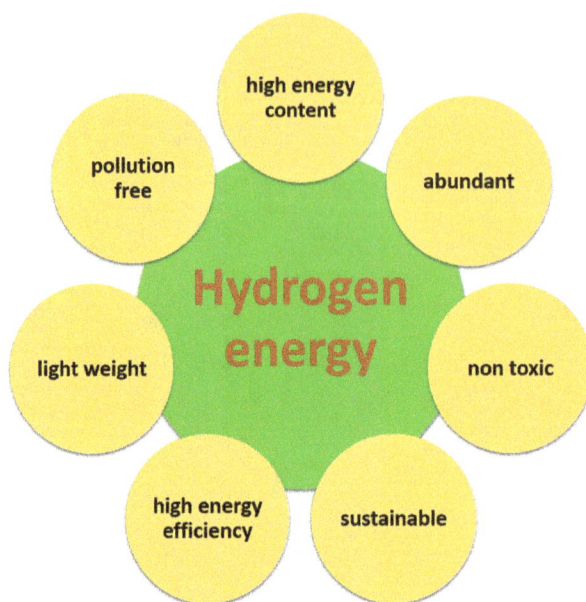

FIGURE 20.1
Hydrogen as a green energy carrier.

transportation applications. Although hydrogen energy density under ambient conditions is lower, it could be possible to increase the volumetric energy density by lower temperature or higher pressure storage of hydrogen [3].

The hydrogen economy approach comprises production, delivery, storage, conversion, and utilization of hydrogen. One of the crucial handicaps for its successful utilization for transportation applications is a safe, cost-effective, and efficient hydrogen storage methodology [2, 3]. Discovering suitable, efficient, low-cost storage materials has been a challenge for the implementation of hydrogen-based technologies [5]. As hydrogen is a gaseous state at ambient conditions, its confined storage is not practical. Hydrogen storage could be accomplished via methods such as compression in tanks, storage in liquefied form, and storage as sorption [10]. However, these storage methods are still far from meeting the onboard vehicular application demands [6].

20.1.2 Hydrogen Storage and Targets

In sorption, hydrogen could either be retained as molecular (physisorption), or atomic hydrogen (chemisorption) [6]. Molecular hydrogen storage depends on physisorption, and leads to lower hydrogen storage capacities under moderate conditions, while chemisorption might take place at ambient conditions, utilizing however costly materials. Under certain conditions, the hydrogen sorption phenomena might be irreversible, even though higher temperatures could facilitate the desorption of hydrogen [6]. The chemisorbed hydrogen might require temperatures higher than 350°C for its release. Physisorption of hydrogen is highly reversible, though cryogenic temperatures are needed for significant gravimetric amounts of hydrogen storage. As a result of weak van der Waals interactions, cryoadsorption of molecular hydrogen progresses with faster kinetics [2].

Storage of hydrogen becomes relatively difficult due to its low density both in liquid ($70.8 \, kg/m^3$, $20.3 \, K$) and gaseous ($0.08 \, kg/m^3$, standard conditions) forms [2]. The usage of hydrogen in vehicular applications relies generally on the presence of a suitable hydrogen storage medium which fulfills many specific limitations, including high hydrogen capacity/mass or volume; low operational energy loss, fast charging kinetics, low stoppage self-discharge, high cycling stability, low charging/recycling costs, and safety. The technical limits related to those criteria are associated with the region of operation and set by various authorities like US DOE (Department of Energy), WENET (World Energy Network Japan), IEA (International Energy Agency). For instance, DOE has put 5.5 wt% H_2 (gravimetric) or $0.040 \, kg/L$ (volumetric) targets for 2025, while 6.5 wt% H_2 and $0.050 \, kg/L$ for ultimate capacity targets for onboard storage systems [11, 12]. Also, the satisfaction of at least 1000 cycles for the durability of the storage material and the refueling time not exceeding 3 min are required [2]. The targets are designed for a 500 km refueling distance with light motor vehicles, and the maximum and minimum acceptable temperature conditions are −40 and 85°C, respectively, for the storage system.

20.2 Hydrogen Storage Materials

Amongst the major challenges in implementing hydrogen-based energy systems is to use a suitable storage medium and the satisfaction of the gravimetric and volumetric targets established [1, 3]. Many materials and sorbents have been investigated as potential candidate medium over other storage techniques, and hydrogen storage at ambient conditions has been a research hot-spot [4, 6]. It has been proposed that the design of new materials, improvement of the specific features of the already existing ones, or process intensification applications could help to further enhance the technologies [1, 13].

Continuing investigations for the identification of new materials for hydrogen storage applications indicate that porous materials, including activated carbons, carbon nanotubes (CNTs), metal-doped CNTs, metal-organic frameworks (MOFs), metal-doped MOFs, COFs, zeolites, clathrates, intermetallic alloys, complex hydrides have attracted attention [3, 4, 14–17]. Activated carbons, zeolites, and MOFs are relatively ideal for gas storage via van der Waals and electrostatic interactions [4, 11, 18–21], whereas complex chemical and metallic hydrides might store appreciable amounts of hydrogen through chemical bonding. Nanoporous materials have advantages, such as the complete reversibility, rapidity, and the lower enthalpy of the adsorption process, in opposition to many metal and complex hydrides. As hydrogen physisorption does not lead to crystallographic phase changes, stability loss of material during successive cycles becomes less. Furthermore, the physical interaction of hydrogen molecules with material surfaces is rather weak as the molecule does not possess any dipole moment and charge, having only a quite low quadrupole moment, along with poor polarizability. Therefore, reasonable hydrogen storage could be attained only at low temperatures. Moreover, as molecular hydrogen physisorption is a surface process, higher adsorbed molecule density within the pores of a porous material contained in the storage tank is required, which would limit the volumetric storage densities for some adsorbents [19].

Most of the existing sorbents might not be stable in humid environments, thus controlling their structures and customizing the functionalities would be hard [4]. The major factors that help to enhance hydrogen storage property in physisorption can be counted as the pore volume, specific surface area, and enthalpy of adsorption, which could be improved by broadened unsaturated metal sites, aromaticity, along with point charges involved in the structure [1]. Pore volume and specific surface area are effective at cryogenic temperatures, while enthalpy of adsorption is significant on H_2 at ambient conditions. It has been shown that H_2 uptake increases with increasing surface area and a linear correlation exist between the hydrogen uptake and specific surface area at 77 K. It was reported that a material NU-100 showed an excess hydrogen uptake value of 99.5 mg/g (77 K, 56 bar) which could be sufficient for practical applications [2].

The challenge with using hydrogen storage materials is to improve the hydrogen storage and reversible characteristics under ambient conditions, at pressures below 10 MPa [5, 19]. Considerable progress has been achieved in the new sorbent synthesis including novel types of porous carbons, amorphous organic polymers, MOFs, and COFs with very high specific surface areas, and consequently sorption capabilities [19]. Carbon-based materials and porous structures such as fullerenes, carbyne networks, and CNTs have been introduced as potential hydrogen storage matrices, due to their good reversible characteristics, larger specific surface areas, and faster kinetics [5, 22, 23]. Nevertheless, a great majority of them failed at ambient conditions owing to improper binding to hydrogen. Studies were conducted by adopting either substitution of elements such as boron/nitrogen to carbon structure, or doping as alkali metals, alkali earth, or transition metals into the porous structure [5].

20.3 COFs and Potential Applications

COFs represent a class of microporous organic crystalline covalent polymeric materials having a periodical topology at which the organic building units are connected covalently. They possess large specific surface areas with stable porosity [2, 16, 24, 25]. Since the work published on the first COF material by Côté et al. [26] in 2005, these materials have attracted much attention and have shown a rapid development [16, 17, 24–28]. The well-defined crystalline porous structures, low density, lightweight, good stability, controllable pore diameter, easier functional modification, large surface area, diversity, and structural designability (Figure 20.2) have offered COF-linked hybrid materials considerable potential and research value in various applications. The synthesis of COFs has shown significant progress with a considerable potential for functionalization. Recently, three-dimensional (3D) COFs have drawn attention as novel COFs, due to their superior porous structures and performances as compared to two-dimensional (2D) analog structures [27]. Design principles, development, synthetic approaches, functionalization, recent advances, and potential applications of COFs have been described in many references [16, 17, 25, 27].

COFs can be considered as organic analogous structures of MOFs, however offer certain advantages over them. These extended frameworks possess high specific surface areas comparable to MOFs and lower densities. As the hydrogen storage capability on a gravimetric basis is directly linked to the material density, COFs not having heavy metal ions indicate the benefit of much higher gravimetric hydrogen storage owing to lower material density. Nevertheless, H_2 storage capacity shows a dramatic decrease with increasing temperature up to ambient temperature, and MOFs or COFs mostly do not exhibit any significant uptake at ambient conditions, while the enthalpy of adsorption becomes the controlling factor. It has been reported that for significant H_2 uptake at ambient conditions, isosteric heat of hydrogen adsorption should be 15–25 kJ/mol, and it could be increased by

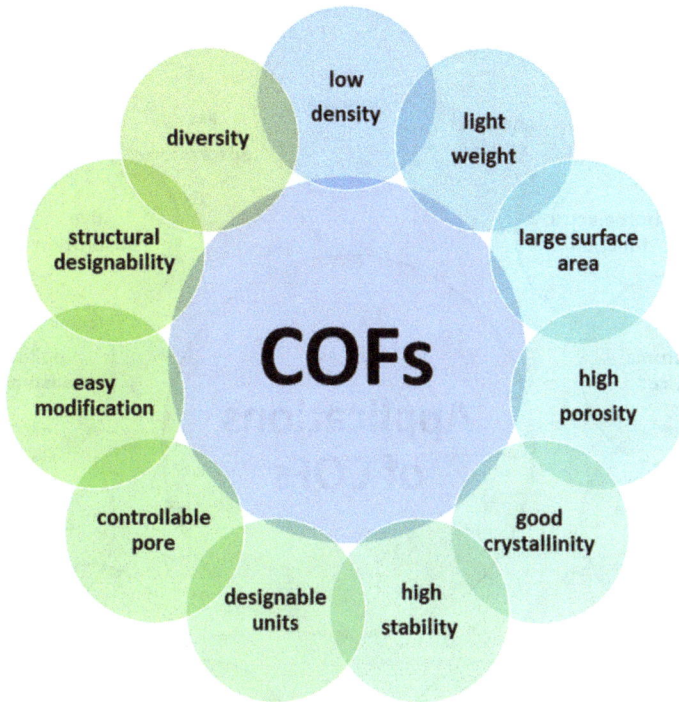

FIGURE 20.2
Multiple advantages of COFs.

the introduction of metal ions and nanoparticles to the framework structure [2]. Figure 20.3 shows the considerable potential and research value of COF-linked materials in various applications, including catalysis, gas adsorption, gas storage, energy storage, separation, optoelectronic device, drug delivery, dye adsorption, photovoltaics, fluorescence sensor, and electronic devices [2, 16, 17, 24, 25, 27, 28].

20.4 Hydrogen Storage Applications of COFs

Development and the advances in solid-state hydrogen storage techniques by the utilization of metallic, intermetallic, complex chemical hydrides, carbon structures, metal-doped CNTs or MOFs, clathrates, and COFs have been presented in some literature [3, 19, 24]. Broom et al. [19] discussed the complete storage system design, density/volume related with the adsorbed phase and interrelation with the entire hydrogen storage capacity of COFs, use of neutron scattering, adsorption isotherms, computational modeling, and H_2 adsorption simulations. Kalidindi and Fischer [2] discussed hydrogen storage with COFs and reviewed the suggested approaches for improved ambient temperature hydrogen storage and the spillover phenomena in metal@MOFs and COFs systems. Konda and Chen [6] highlighted Pd-based nanomaterials and hydrogen spillover effects in MOFs and COFs. It has been pointed out that COFs hold all the benefits of MOFs concerning hydrogen storage. Additionally, as COF frameworks contain light elements, such as C, Si, B, O, their

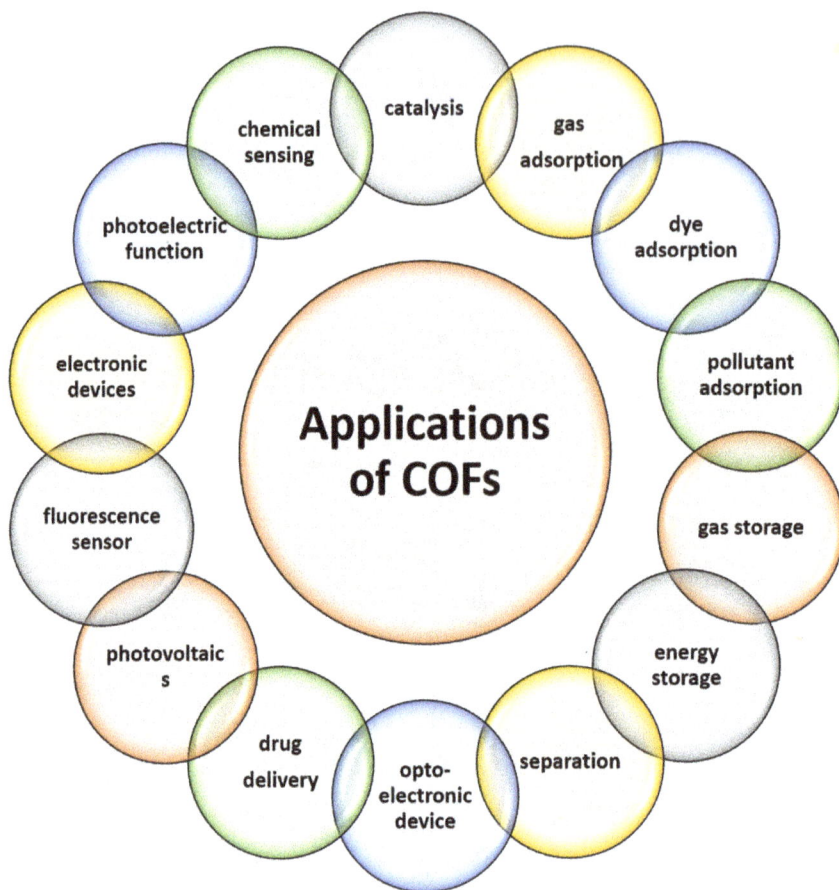

FIGURE 20.3
Applications of COFs.

densities are exceptionally lower. Hence, numerous experimental and theoretical studies have been described on hydrogen storage characteristics of COFs and approaches proposed for enhancing hydrogen storage at cryogenic and ambient temperatures [2]. Table 20.1 presents various hydrogen storage studies on COF structures with the followed methodology, achieved hydrogen uptake results, and comments for different references from the literature.

Garberoglio [29] presented grand canonical Monte Carlo (GCMC) simulation results on hydrogen storage with COF-102, -103, -105, and -108 and reported that a 30% increase for the hydrogen uptake could be attained as compared to MOFs (77, 298 K). All COFs displayed similar qualitative behavior. COF-102 and -103 had the highest adsorption behavior at low pressure due to the more compact microscopic structure of them that led to stronger solid–fluid interactions, but upon saturation indicated lower adsorption than COF-105 and -108 as a result of higher pore volumes of COF-105 and -108. With an increase in loading, lower free volumes of COF-102 and -103 resulted in saturation at a lower pressure than observed with the others. 470 cm³ (STP)/cm³ hydrogen adsorption was obtained at 77 K for COF-102 and -103, whereas COF-105 and -108 had 250 cm³ (STP)/cm³. The considerably lower density of COF-105 and -108 favored their gravimetric adsorption, while the lower

TABLE 20.1

Hydrogen Storage Studies on COFs

COF Structures	Methodology	H₂ Uptake	Comments	Reference
COF-102 COF-103 COF-105 COF-108	GCMC simulations	COF-102, COF-103 470 cm³ (STP)/cm³ COF-105, COF-108 250 cm³ (STP)/cm³ (77 K)	• 30% increased H₂ uptake than MOFs (77 K, 298 K)	[29]
3D-COFs COF-108	Ab initio calculations GCMC simulations	COF-108 21 wt %(77 K, 100 bar) 4.5 wt % (room T, 100 bar)	• Phenyl, B₃O₃/C₂BO rings were the most favorable H₂ adsorption sites (binding energy <1 kcal/mol) • Two times higher gravimetric uptake than MOFs	[30]
COF-5 COF-105 COF-108 COF-102	First-principles-based GCMC simulations	COF-5 3.4 wt % (77 K, 50 bar) COF-105 COF-108 10.0 wt % reversible excess H₂ uptake (77 K) COF-108 18.9 wt % total H₂ uptake (77 K) COF-102 40.4 g/L of H₂ (77 K)	• Good correlation between experimental and theoretical results for COF-5 (3.3 vs 3.4 wt % 77 K, 50 bar) • 3D COFs showing 2–3 times higher H₂ storage capacity than 2D COFs, due to high surface area and free volume • Surface area and free volume should be increased to increase H₂ gravimetric storage capacity • COF systems as the most promising candidates for practical H₂ storage	[31]
COF-1 COF-6 COF-5 COF-8 COF-10 COF-102 COF-103	H₂ isotherm measurements (77–298 K, 1–85 bar)	COF-102 72 mg/g excess H₂ uptake (77 K, 35 bar) Similar to the performance of COF-103 COF-1 15 mg/g COF-5 36 mg/g COF-6 23 mg/g COF-8 35 mg/g COF-10 39 mg/g (77 K)	• Findings placing COFs among the best sorbents for H₂ storage	[32]

(Continued)

TABLE 20.1 *(Continued)*

Hydrogen Storage Studies on COFs

COF Structures	Methodology	H₂ Uptake	Comments	Reference
2D, 3D COFs COF-5 COF-102 COF-103 COF-105 COF-108 COF-202	Molecular dynamic simulations	COF-5 2.53 wt.% COF-102 5.60 wt% COF-103 5.40 wt.% COF-105 2.99 wt.% COF-108 5.94 wt.% COF-202 1.366 wt.%	• H_2 adsorption in seven different adsorption sites at low temperature, all COFs being stable with H_2 molecules adsorbed • Structures need to be tailored where adsorption energy and number of adsorption sites are optimized for H_2 uptake maximization, while keeping the low density and high surface area • Doping COFs with metals (Li, Mg) with high interaction energy with H_2 for increasing the adsorption energy and promoting adsorption capacity at higher temperatures • Benzene rings in organic linkers and near boron–oxygen networks were the most favorable H_2 adsorption sites • Adsorption interaction energy for COFs ~3 kJ/mol	[33]
3D-COFs COF-102-2 COF-102-3 COF-102-4 COF-102-5	Multiscale techniques for optimization of structures GCMC simulations	COF-102-3 26.7 wt% (77, 100bar) 6.5 wt% (300K, 100bar)	• Material design based on ctn network of ultralow density COF-102 by substitution of phenyls rings of TBPM with various aromatic moieties without ctn • COF-102-2, COF-102-3, COF-102-4, COF-102-5 produced by replacing diphenyl, triphenyl, napthalene, pyrene molecules, respectively. • Optimizing structures by multiscale theoretical techniques, storage capacities by GCMC simulations, obtaining total gravimetric and volumetric uptake (77, 300K). • COF-102-3 showing the best gravimetric storage capacity among the proposed 3D COFs (26.7, 6.5 wt% at 77, 300K, 100bar) • Materials showing lower volumetric performance than parent COF-102. COF-102 ~42g/L; COF-102-3 ~35g/L (77 K, 100bar). • Large pores inside the material resulted in empty space dominated by weak fluid–fluid interactions, decreasing volumetric capacity. Pore volume and surface area to be balanced for optimized gravimetric and volumetric storage capacities. • Gravimetric H_2 uptake of materials was superior to their parent structure and other known frameworks. • Gravimetric uptake of COFs could overpass 25 wt % (77 K) reaching DOE's target (6 wt%, room temperature), classifying them among the top H_2 storage materials.	[1]

TABLE 20.1 *(Continued)*

Hydrogen Storage Studies on COFs

COF Structures	Methodology	H_2 Uptake	Comments	Reference
Doping metal atoms on B-doped COFs COF-1	Chemical modification of COFs Ab initio calculations	COF-1 > 6.5 wt %	• H_2 storage capability of COF-1 significantly improved by a two-step doping chemical modification, consisting of B doping of organic units, then, doping of metal atoms forming trapping centers for H_2 molecules on B-doped COFs, ensuring high stability of metal dopants (Ca, Sc, Ti) as active centers for H_2 adsorption. • Benzene ring of COF-1, two-B para substitution energetically most favorable. • Kubas interaction between H_2 molecules and Sc (Ti) atoms and the polarization of H_2 molecules near the Ca atoms responsible, respectively for moderate H_2 binding energies of 0.32 (0.40) and 0.16 eV/H_2 making them suitable for reversible H_2 storage (ambient conditions). • In the macroscopic porous layered Ca-, Sc-, Ti-adsorbed B-COF-1, H_2 storage gravimetric capacity larger than 6.5 wt %. • COFs modified by a two-step doping promising for H_2 storage applications.	[34]
Metal-decorated COF Metal-nanoparticle-loaded porous crystalline 3D hybrid material, Pd@ COF-102 3D		Metal-decorated COF samples (3.5, 9.5 wt% Pd)	• Enhancement of H_2 storage capacities by a factor of 2–3 by Pd impregnation on COF-102 (room temperature, 20 bar) due to Pd hydride formation and hydrogenation of residual organic compounds. • Significantly higher reversible H_2 storage capacity due to the decomposed products of organometallic Pd precursor • Higher uptakes at room temperature than those of similar systems (Pd@ MOFs)	[35]

(Continued)

TABLE 20.1 (Continued)

Hydrogen Storage Studies on COFs

COF Structures	Methodology	H₂ Uptake	Comments	Reference
Metal intercalated 3D COFs (MCOF)	Density functional calculations and thermodynamical analysis	Ca-intercalated COF with diphenylethyne units ~5 wt% (300 K, 20 bar) COF-1 <0.1 wt% (300 K, 20 bar)	• 3D MCOFs with a high density of active metal sites, well-defined crystal structures for H_2 storage • Li, Na, K, Ca intercalation into COF-1, COF-d layers forming thermodynamically stable materials, metal binding stronger in MCOF than in bulk metals • CaCOF-d promising for ambient H_2 storage by optimization of organic building units and linker groups. • Systematic optimization of organic building units and linker groups identified CaCOF-d as most promising for ambient H_2 storage. • Ca the most suitable metal intercalated in the stacking structures for H_2 storage. • Porous 3D CaCOF-d reaching ~5 wt% H_2 storage capacity (ambient conditions) without metal clustering problems, while bare COF-1 <0.1 wt%	[5]
2D COF-1 membrane Li-decorated COF-1	First-principles calculation Ab initio molecular dynamics simulations	Li-decorated COF-1 7.69 wt% COF-1 5.26 wt% (300 K)	• Significant improvement of H_2 storage capacity by Li decoration (7.69 wt%) • Li-COF-1 with 5.26 wt% H_2 storage capacity (300 K), higher than DOE targets (4.5 wt%). • Li-COF-1 as a potential H_2 storage material (ambient condition).	[4]

volumetric capacity of the lower density COF-105 and -108 was due to the lower solid–fluid potential energy. The excess hydrogen adsorption was about 10 wt% at 100 bar for COF-105 and -108, which was nearly 50% higher than the gravimetric capacity of IRMOF-14, the best adsorbing MOF, where the excess amount adsorbed was about 7.5 wt%. Hydrogen adsorption isotherms at 298 K indicated similar behavior to MOFs. The room temperature adsorption was about 0.8 wt% at 100 bar. The result indicated a significant enhancement to the results obtained with MOFs.

Klontzas et al. [30] used a theoretical method to study the hydrogen storage in large surface area 3D-COFs. They performed ab initio calculations and classical GCMC simulations for the investigation of the hydrogen binding sites and uptake under different thermodynamic conditions. The most favorable hydrogen adsorption sites were reported to be the phenyl and B_3O_3 or C_2BO rings (binding energy <1 kcal/mol). The gravimetric uptake was found twice as high as MOFs. COF-108 had 21 wt% gravimetric uptakes at 77 K (100 bar) and 4.5 wt% at ambient temperature.

Han et al. [31] investigated hydrogen storage in 2D and 3D COFs. The predicted H_2 adsorption isotherm based on first-principles-based GCMC simulations was in good compatibility with experimental findings for COF-5 (3.3 vs 3.4 wt% at 50 bar, 77 K). COF-105 and COF-108 showed a 10.0 wt% reversible excess H_2 uptake, which made them the best molecular hydrogen storage media for 77 K. The total H_2 uptake was 18.9 wt% for COF-108 at 77 K. COF-102 showed the best volumetric capacity with a value of 40.4 g/L H_2 at 77 K. 3D COFs showed about two-three times higher hydrogen storage performance than 2D COFs owing to their higher specific surface area and free volumes. Theoretical studies showed that both free volumes and specific surface areas needed to be increased to enhance COFs H_2 gravimetric storage capacity.

Furukawa and Yaghi [32] conducted H_2 isotherm experiments on 2D and 3D COFs at 77–298 K (1–85 bar). The materials were classified according to their pore sizes and structural dimensions, as group 1 (2D; 9Å, COF-1, COF-6), group 2 (2D; COF-5 (27Å), COF-8 (16Å), COF-10 (32Å)), group 3 (3D; 12Å, COF-102, COF-103). COFs of group 3 outperformed the other groups according to their hydrogen storage capacities and were superior to the best MOFs and porous materials. The excess hydrogen uptake of COF-102 and COF-103 were similar with 72 mg/g at 35 bar, 77 K, but higher than COF-1 (15 mg/g), COF-5 (36 mg/g), COF-6 (23 mg/g), COF-8 (35 mg/g), and COF-10 (39 mg/g) at 77 K. It was reported that with these findings, COFs were placed among the best porous hydrogen sorbents.

Assfour and Seifert [33] performed molecular dynamic (MD) simulations for the investigation of the adsorption sites and adsorption energies for hydrogen in 2D and 3D COFs and reported that the molecule adsorption was achieved in seven different adsorption sites at low temperature and all structures were stable. The structures needed to be tailored for the optimization of the adsorption sites and adsorption energies for hydrogen adsorption be maximized, maintaining the low density and the high specific surface area of COFs. The frameworks could also be doped with metal (Li, Mg) atoms ensuring high interaction energy with H_2 for increasing the energy of adsorption to enhance the adsorption at higher temperatures. The most powerful site was the benzene rings of organic linkers and near boron–oxygen networks. Adsorption interaction energy was reported as ~3 kJ/mol. The highest hydrogen uptakes reported were 2.53 wt% COF-5, 5.60 wt% COF-102, 5.40 wt% COF-103, 2.99 wt% COF-105, 5.94 wt.% COF-108, 1.366 wt.% COF-202.

Klontzas et al. [1] carried out a study on the hydrogen storage properties of 3D COF structures which were obtained from the ultralow density COF-102 by the substitution of the phenyls rings with many aromatic moieties without ctn. The structure was replaced with molecules such as diphenyl, triphenyl, napthalene, pyrene to obtain COF-102-2, COF-102-3,

COF-102-4, COF-102-5 structures, respectively. Multiscale theoretical techniques were used for the optimization of the structures and investigation of their storage capacities performing GCMC simulations (77,300 K). The best gravimetric storage capacity was obtained with COF-102-3, as 26.7 wt% at 77 K, 6.5 wt% at 300 K (100 bar), whereas its volumetric capacity was ~35 g/L at 77 K (100 bar). The volumetric performance of the structures was lower than the parent COF-102 (~42 re/L at 77 K, 100 bar). The space formed due to the large pores within the structure was dominated by weak fluid-fluid interactions and decreased volumetric capacity. It was reported that the surface area and pore volume needed to be balanced for the optimization of gravimetric and volumetric hydrogen storage. The gravimetric hydrogen uptake was superior to their parent structure and the other known structures. The GCMC simulations demonstrated that the gravimetric hydrogen uptake of the new materials could attain 25 wt% at 77 K and reached the DOE's target (6 wt%) at ambient temperature, classifying them among the top hydrogen storage materials.

Zou et al. [34] used a 2-step doping approach for the modification of COFs which involved boron (B) doping the structures suppressing the clustering of resulting metal dopants on frameworks in the first step, then metal atoms doping forming trapping centers for hydrogen molecules on B-doped COF structures. They investigated the structural stability during the doping processes and the electronic properties of doped units to the concentration of B, using ab initio calculations. In the benzene ring of COF-1, two-B para substitution was found to be energetically the most favorable as compared to the other B substitutions. On the B-COF-1 layered structure, Ti and Sc preferred the double, whereas Ca favored the single-sided adsorption, due to the different interactions with the substrate and the dopant. Sc and Ti atoms bound 4 hydrogen molecules via Kubas interaction, though Ca atom could adsorb 6 molecules through polarization of hydrogen molecules with the effect of the electric field around Ca atom. The Kubas interaction between Sc (Ti) atoms and the hydrogen molecules along with the polarization of hydrogen molecules in the neighborhood of Ca atoms were responsible for moderate H_2 binding energies, making them suitable for reversible hydrogen storage under ambient conditions. In the macroscopic porous layered Ca-, Sc-, and Ti-adsorbed B-COF-1, hydrogen storage capacity higher than 6.5 wt% was obtained. The work indicated that the two-step doping of COFs were promising for hydrogen storage applications.

Kalidindi et al. [35] studied the hydrogen storage characteristics of Pd@COF-102, a porous crystalline 3D hybrid material, having a very narrow size distribution formed by a gas-phase infiltration method. They used two samples containing 3.5 and 9.5 wt% Pd and reported that hydrogen uptakes at ambient temperature were higher than Pd@MOFs. H_2 storage capacities were improved by 2–3 times at room temperature (20 bar) via Pd impregnation of COF-102, due to the hydrogenation of residual organic products and the formation of Pd hydride. Higher reversible hydrogen storage capacity was obtained.

Gao et al. [5] investigated hydrogen storage on a metal intercalated COF (MCOF), a new form of a 3D network which had well-defined crystal structures. Performing density functional calculations and thermodynamical analysis, they reported that chemically active, stable porous materials could be formed by stacking the COF layers with metals. In MCOFs, metal atoms bound to aromatic organic units rather than B_3O_3 connecting rings. Na, K, Ca, Li were intercalated into COF-1 and COF-d layers and formed stable materials, where the metal binding was stronger in MCOF as compared to the one in bulk metals, and the metal acted as active sites. Metal-metal separation and metal-binding were adjusted by the selection of appropriate linkers and building blocks during the construction of COF layers. CaCOF-d was identified as the most promising hydrogen storage material at ambient conditions. CaCOF-d was reported to reach ~ 5 wt% H_2 storage capacity at 300 K, 20 bar without metal clustering problems, while bare COF-1 < 0.1 wt% MCOF involving a high density of active metal sites was reported to be useful in various applications including hydrogen storage.

Zhao et al. [4] used first-principles calculation to study the hydrogen storage properties of Li-decorated COF-1, and reported that the hydrogen storage capacity (7.69 wt%) was improved considerably by Li decoration. Carrying out ab initio molecular dynamics simulations (300 K), 12 hydrogen molecules were found to be absorbed on double sides of COF-1 unit cell decorated by 6 Li atoms. The hydrogen storage capacity was found as 5.26 wt%, which was higher than the DOE targets (4.5 wt%). It was claimed that Li-COF-1 was promising for hydrogen storage at ambient conditions.

20.5 Conclusions

The review highlighted the recent progress and important aspects of research in the hydrogen storage applications of COFs, which have shown great potential and research value recently. Studies have shown that COFs are promising candidates for H_2 storage owing to their well-defined structures, large specific surface areas, and high pore volumes, particularly for cryogenic H_2 storage and do well in comparison with the most common mesoporous solids, carbon structures, and MOFs, thus have been placed doubtlessly among such powerful porous materials [2, 16, 32]. GCMC simulations showed a 30% enhancement in hydrogen uptake with COF-102, COF-103, COF-105, COF-108 as compared to MOFs with analogous simulations at 77 K and 298 K [29].

Ab initio calculations and classical GCMC simulations revealed phenyl and B_3O_3 or C_2BO rings as the most favorable sites for H_2 adsorption in 3D-COFs. Gravimetric hydrogen uptake of COFs was 2 times higher than MOFs. Gravimetric uptake of COF-108 reached 21 wt% and 4.5 wt%, respectively at 77 K and room temperature (100 bar) [30]. 3D COFs (COF-102, COF-103) outperformed 2D structured COFs (COF-1, COF-5, COF-6, COF-8, COF-10) and showed better H_2 uptake capacities than MOFs and other porous materials. The findings placed COFs among the best sorbents for hydrogen [31].

MD simulations showed that H_2 adsorption was accomplished in seven different adsorption sites in 2D and 3D COFs at low temperatures. For the maximization of the hydrogen uptake, tailoring the structures for the optimization of the adsorption energy and adsorption sites was needed besides the low density and the high surface area of the structures. Doping the structures with metal (Li, Mg) atoms with high interaction energy with H_2 was needed for increasing the adsorption energy and promoting the adsorption capacity at higher temperatures. Benzene rings in the organic linkers and near boron–oxygen networks were the most favored adsorption sites [33].

Attractive H_2 theoretical storage capacities were reported based on GCMC simulations with 3D COFs (COF-105, COF-108), demonstrating that hydrogen uptake of 3D-COFs could overpass 25 wt % at 77 K, reaching the DOE target of 6 wt% at room temperature, classifying them among the top hydrogen storage materials. The hydrogen uptake was superior to the parent structure and other reported frameworks [1]. Hydrogen adsorption properties of COF-1 based on ab initio calculations could be considerably enhanced by a two-step doping chemical modification procedure, consisting of B substitution for C in organic units, and then, providing high stability of metal (Ca, Sc, Ti) dopants as active centers for hydrogen storage. A two-B para substitution was energetically the most favorable site in the benzene ring of COF-1. In the macroscopic porous layered Ca-, Sc-, and Ti-adsorbed

B-COF-1, hydrogen storage higher than 6.5 wt% could be achieved. The two-step doping procedure was effective in the development of materials from COFs by chemical modification for hydrogen storage applications [34].

Hydrogen uptakes on Pd@COF-102, a porous crystalline 3D hybrid material were higher than Pd@MOFs at room temperature. Hydrogen storage capacities were improved by 2–3 times by Pd impregnation on COF-102 at room temperature (20 bar), owing to hydrogenation of residual organic compounds, and Pd hydride formation. A much higher reversible hydrogen storage capacity was obtained through the decomposition products of the organometallic Pd precursor.

3D metal-intercalated COFs based on first-principles calculations had well-defined crystal structures and were chemically active for H_2 adsorption. Ca was the most suitable metal for intercalation in the stacking structures for H_2 storage. 3D CaCOF-d was identified as the most promising material for ambient H_2 storage with 5 wt% H_2 storage capacity without metal clustering problems. MCOF having a high density of active metal sites was considered to be useful in many applications including hydrogen storage [34]. Hydrogen storage properties of Li-COF-1 based on the first-principles calculations and *ab initio* MD simulations showed a 5.26 wt% hydrogen storage capacity which was higher than DOE targets (4.5 wt%), indicating Li-COF-1 as a promising material at ambient conditions [4].

20.6 Perspective Remarks

After all, there still exist many concerns and challenges which require further studies:

Much intensive work is required on mobile hydrogen storage applications of COFs. The development of advanced physical storage systems with high efficiency; low weight, volume, costs; rapid charging/discharging kinetics, long durability would help to direct the technical problems toward hydrogen storage [6]. Experimental difficulties related with pore activation need to be overwhelmed for proper application of COFs in hydrogen storage. More studies are needed on hydrogen spillover effect for enhancement of hydrogen storage in COFs and metal@COF systems at ambient conditions. More studies are needed on alkali metal doping of structures for ambient temperature hydrogen storage. Structural stability of COFs and metal doped COFs for water need to be improved for practical applications [2, 6].

Different experimental and computational approaches to the studies of new materials for hydrogen storage applications need to be addressed. Hydrogen uptake measurements with different materials can be subject to some errors in connection to the reliability and reproducibility of obtained data, as there are not any prescribed directories for hydrogen sorption measurements, which can give rise to inconsistencies in data reporting. There is unclearness in the definitions of the absolute or total capacity of materials that could be evaluated from the measured data using different assumptions. There is a lack of hydrogen adsorption isotherm fitting results at different temperatures, other than 77, 298 K. Both the gravimetric and volumetric capacities of new materials need to be evaluated [19].

It is a challenge to develop a rapid method of synthesis for crystalline, highly stable structured COFs, to synthesize COFs on an industrial scale and reduce the costs for commercialization for prospective practical applications. Further researches are needed to enlighten the interactions between hydrogen with COFs under variable conditions. Further improving the chemical/thermal stability of COFs to be used under harsh conditions requires development [6, 17, 25]. As yet, most of the reported COFs have 2D layered crystalline structures, and it would be of significance to develop more 3D COFs [17, 25]. 3D COFs with easy modifications, highly void frameworks, abundant open channels, and large surface areas are promising for hydrogen storage. Though, the achievements in 3D COFs are not yet sufficient as compared to the 2D COFs. There are still some problems, such as the crystallization and interpenetration in 3D COFs yet to be solved [27].

References

1. E. Klontzas, E. Tylianakis, G.E. Froudakis, Designing 3D COFs with enhanced hydrogen storage capacity, Nano Letters, 10 (2) (2010) 452–454.
2. S.B. Kalidindi, R.A. Fischer, Covalent organic frameworks and their metal nanoparticle composites: Prospects for hydrogen storage. Phys. Status Solidi B, 250 (2013) 1119–1127.
3. R. Zacharia, S.U. Rather, Review of solid state hydrogen storage methods adopting different kinds of novel materials, Journal of Nanomaterials, 2015 (2015) 914845.
4. H. Zhao, Y. Guan, H. Guo, R. Du, C. Yan, Hydrogen storage capacity on Li-decorated covalent organic framework-1: A first-principles study, Materials Research Express, 7(3) (2020) 035506.
5. F. Gao, Z. Ding, S. Meng, Three-dimensional metal-intercalated covalent organic frameworks for near-ambient energy storage. Scientific Reports, 3 (2013) 1882.
6. S.K. Konda, A. Chen, Palladium based nanomaterials for enhanced hydrogen spillover and storage, Materials Today, 19 (2) (2016) 100–108.
7. I. Dincer, Environmental and sustainability aspects of hydrogen and fuel cell systems, International Journal of Energy Research, 31 (1) (2007) 29–55.
8. T. Kopac, A. Toprak, Hydrogen sorption characteristics of Zonguldak region coal activated by physical and chemical methods, Korean Journal of Chemical Engineering, 26 (2009) 1700–1705.
9. I. Dincer, C. Zamfirescu, Sustainable hydrogen production options and the role of IAHE, International Journal of Hydrogen Energy, 37 (21) (2012) 16266–16286.
10. A. Toprak, T. Kopac, Surface and hydrogen sorption characteristics of various activated carbons developed from Rat coal mine (Zonguldak) and anthracite, Chinese Journal of Chemical Engineering, 19 (2011) 931–937.
11. T. Kopac, Hydrogen storage characteristics of bio-based porous carbons of different origin: A comparative Review, International Journal of Energy Research, 45(15) (2021) 20497–20523.
12. DOE Technical Targets for Onboard Hydrogen Storage for Light-Duty Vehicles, https://www.energy.gov/eere/fuelcells/doe-technical-targets-onboard-hydrogen-storage-light-duty-vehicles (accessed July 17, 2021)
13. T. Kopac, Emerging applications of Process Intensification for enhanced separation and energy efficiency, environmentally friendly sustainable adsorptive separations: A Review, International Journal of Energy Research, 45(11) (2021) 15839–15856.
14. S. Kocabas, T. Kopac, G. Dogu, T. Dogu, Effect of thermal treatments and palladium loading on hydrogen sorption characteristics of single-walled carbon nanotubes, International Journal of Hydrogen Energy, 33 (2008) 1693–1699.
15. T. Kopac, F.O. Erdogan, Temperature and alkaline hydroxide treatment effects on hydrogen sorption characteristics of multi-walled carbon nanotube graphite mixture, Journal of Industrial and Engineering Chemistry, 15 (2009) 730–735.
16. H. Guo, L. Zhang, R. Xue, B. Ma, W. Yang, Eyes of covalent organic frameworks: cooperation between analytical chemistry and COFs, Reviews in Analytical Chemistry, 38 (1) (2019) 20170023.
17. L. Deng, J. Zhang, Y. Gao, Synthesis, properties, and their potential application of covalent organic frameworks (COFs), Ch.4, In: M. Krishnappa (ed), Mesoporous Materials, IntechOpen, Rijeka, Croatia (2019).

18. F.O. Erdogan, T. Kopac, Dynamic analysis of sorption of hydrogen in activated carbon, International Journal of Hydrogen Energy, 32 (2007) 3448–3456.

19. D.P. Broom, C.J. Webb, K.E. Hurst, P.A. Parilla, T. Gennett, C.M. Brown, R. Zacharia, E. Tylianakis, E. Klontzas, G.E. Froudakis, Th.A. Steriotis, P.N. Trikalitis, D.L. Anton, B. Hardy, D. Tamburello, C. Corgnale, B.A. van Hassel, D. Cossement, R. Chahine, M. Hirscher, Outlook and challenges for hydrogen storage in nanoporous materials, Applied Physics A, 122 (2016) 151.

20. T. Kopac, Y. Kırca, A. Toprak, Synthesis and characterization of KOH/Boron modified activated carbons from coal and their hydrogen sorption characteristics, International Journal of Hydrogen Energy, 42 (2017) 23606–23616.

21. T. Kopac, Y. Kırca, Effect of ammonia and boron modifications on the surface and hydrogen sorption characteristics of activated carbons from coal, International Journal of Hydrogen Energy, 45 (17) (2020) 10494–10506.

22. T. Kopac, T. Karaaslan, H_2, He and Ar sorption on arc-produced cathode deposit consisting of multiwalled carbon nanotubes-graphitic and diamond-like carbon, International Journal of Hydrogen Energy, 32 (16) (2007) 3990–3997.

23. F.O. Erdogan, T. Kopac, Comparison of activated carbons produced from Zonguldak Kozlu and Zonguldak Karadon hard coals for hydrogen sorption, Energy Sources Part A: Recovery, Utilization and Environmental Effects (2020). https://doi.org/10.1080/15567036.2020.1795310

24. T. Wang, R. Xue, Y. Wei, M. Wang, H. Guo, W. Yang, Development and applications of covalent organic frameworks (COFs) materials: Gas storage, catalysis and chemical sensing, Progress in Chemistry, 30 (6)(2018) 753–764.

25. J. You, Y. Zhao, L. Wang, W. Bao, Recent developments in the photocatalytic applications of covalent organic frameworks: A review, Journal of Cleaner Production, 291 (2021) 125822.

26. A.P. Côté, A.I. Benin, N.W. Ockwig, M. O'keeffe, A. Matzger, O. Yaghi, Porous, crystalline, covalent organic frameworks, Science, 310 (5751) (2005) 1166–1170.

27. X. Guan, F. Chen, Q. Fang, S. Qiu, Design and applications of three dimensional covalent organic frameworks, Chemical Society Reviews, 49 (5) (2020) 1357–1384.

28. O. Yildirim, M. Bonomo, N. Barbero, C. Atzori, B. Civalleri, F. Bonino, G. Viscardi, C. Barolo, Application of metal-organic frameworks and covalent organic frameworks as (photo) active material in hybrid photovoltaic technologies, Energies, 13 (21) (2020) 5602.

29. G. Garberoglio, Computer simulation of the adsorption of light gases in covalent organic frameworks, Langmuir, 23 (2007) 12154–12158.

30. E. Klontzas, E. Tylianakis, G. E. Froudakis, Hydrogen storage in 3D covalent organic frameworks. A multiscale theoretical investigation, The Journal of Physical Chemistry C, 112 (24) (2008) 9095–9098.

31. S.S. Han, H. Furukawa, O.M. Yaghi, W.A. Goddard, Covalent organic frameworks as exceptional hydrogen storage materials, Journal of the American Chemical Society, 130 (35) (2008) 11580–11581.

32. H. Furukawa, O.M. Yaghi, Storage of hydrogen, methane, and carbon dioxide in highly porous covalent organic frameworks for clean energy applications, Journal of the American Chemical Society, 131 (25) (2009) 8875–8883.

33. B. Assfour, G. Seifert, Hydrogen adsorption sites and energies in 2D and 3D covalent organic frameworks, Chemical Physics Letters, 489 (1–3) (2010) 86–91.

34. X. Zou, G. Zhou, W. Duan, K. Choi, J. Ihm, A chemical modification strategy for hydrogen storage in covalent organic frameworks, The Journal of Physical Chemistry C, 114 (31) (2010) 13402–13407.

35. S.B. Kalidindi, H. Oh, M. Hirscher, D. Esken, C. Wiktor, S. Turner, G. V. Tendeloo, R.A. Fischer, Metal@COFs: covalent organic frameworks as templates for Pd nanoparticles and hydrogen storage properties of Pd@COF-102 hybrid material, Chemistry A European Journal, 18 (35) (2012) 10848–10856.

21

Covalent Organic Frameworks-Based Adsorbents for Methane Storage: Experimentation and Simulations

Raghubir Singh[1] and Varinder Kaur[2]

[1]*Department of Chemistry, DAV College, Chandigarh, India*

[2]*Department of Chemistry, Panjab University, Chandigarh, India*

CONTENTS

21.1 Introduction

Methane is the smallest hydrocarbon but it has the power to drive a variety of machines covering a wide range from household equipment to rockets. The natural and anthropogenic origin of methane includes geological processes, biological practices, and industrial activities. Therefore, it exists in all the spheres of the earth; atmosphere, biosphere, lithosphere, and hydrosphere. Predominantly, it is produced in the hydrosphere by the anaerobic decomposition of organic matter and is stored as methane hydrates and clathrates in the cryosphere. Besides, biomass burning, coal mining, wetland methane emissions, ruminant and farm animals, permafrost areas, methanogenesis, rice cultivation, waste treatment, volcanic eruptions, industrial processes, etc. contribute to emerging the methane as the most dominant greenhouse gas in the atmosphere of the earth. Although the

DOI: 10.1201/9781003206507-21

methane emission in the atmosphere is less than the carbon dioxide, however, its impact on global warming is far more than CO_2. The greenhouse effect is responsible for the increase in the levels of carbon dioxide and global warming. It is expected to produce the climate extremes such as floods, droughts, and enhanced temperatures thereby posing a threat to all the ecosystems that are slowly losing their balance [1].

The type of energy sources has been changed over time from wood to natural gas. Initially, coal has been introduced as the energy source to drive the steam engine, and then, it is replaced with liquid fuels with the invention of the internal combustion engine/automobiles. Further development of science technology launched natural gas as a pivotal source for some of the industrial and domestic applications and the game is not over yet. Still, the scientific world is struggling to mitigate the worldwide energy crisis and environmental impacts of the current fuels. Thus, either the modifications in the usage of existing energy sources or searching for sustainable and efficient energy sources are the strategies to meet the current energy demand. In the past few years, the scientific community has managed to utilize the abundantly available methane for day-to-day applications. It is successfully harnessed in cooking cylinders, natural gas dryers, electricity generators, industrial machinery, and rocket fuel. The current research progressions have visualized methane as a replacement for the current fuels and researchers are tirelessly working to achieve something momentous from methane. Undoubtedly, the focus is to use methane as vehicular fuel to reduce the energy demand, costs, and environmental pollution [2, 3]. There are numerous reasons to use methane as fuel and some of the important reasons are listed below.

1. It is highly abundant and available everywhere in nature.
2. It is economical as compared to fossil fuels.
3. It has a more H/C ratio so yields more energy as compared to other fuels.
4. The combustion of gasoline releases highly toxic gases like oxides of N, S, C, and some carcinogens whereas it burns cleanly.
5. The post-burning emission of carbon is comparatively less than fossil fuels.

Methane is lighter than air (density = 0.657 kg m^{-3} and specific gravity = 0.554) and slightly soluble in water (solubility = 22.7 mg L^{-1}). The boiling point of methane is $-161.6°C$ and the melting point is $-182°C$. The critical temperature and critical pressure of methane are 191 K and 46.6 bar, respectively, therefore, its liquefaction at room temperature is abstruse. The liquefaction of methane requires lowering of temperature to $\sim -162°C$ at atmospheric pressure (approx.) to get liquefied natural gas (LNG). Similarly, it is compressed to ~ 200–250 bar when used as compressed natural gas (CNG). Therefore, the use of LNG and CNG as a vehicular fuel requires specific storage cylinders to maintain temperature and pressure conditions, additional space for carrying large fuel tanks, and repeated re-filling due to less driving range, which enhances the cost of transportation and makes their use inconvenient. Moreover, it is not advisable to carry extremely high pressurized fuel tanks during travel and transportation due to safety reasons. Although conversion of methane to other hydrocarbons or oxygenates is suggested for storage however it requires a lot of energy and is an arduous process [4].

The best method to store methane in compact and secure systems for automobile applications is its adsorption in porous sorbents. For this purpose, different inorganic porous materials like zeolites (100 v(STP)/v) [5], coordination polymers (e.g., Cu(BPY)$_2$SiF$_6$ (BPY,4,4-bipyridine: 146 v(STP)/v) [6], activated carbons (100–170 v(STP)/v) [7], and metal-organic frameworks (>170 v(STP)/v) [8] have been investigated. The developed materials have not reached the desired goal in terms of volumetric storage capacity for methane, which is the primary condition for a sorbent. The standards set by the US Department of Energy (DOE) in 2012 include volumetric storage capacity of the adsorbent up to 350 v(STP)/v and gravimetric storage capacity of 0.5 g g^{-1} at room temperature and packing capacity of 263 v(STP)/v. Besides appropriate adsorption capacity, an adsorbent must have high heat capacity, should be hydrophobic, and appropriate adsorption-desorption rate [9].

21.2 Methane Storage by Covalent Organic Frameworks (COFs)

COFs are multidimensional crystalline compounds composed of B, C, O, Si, and H [10]. These compounds can be isolated as discrete units, 1D chains, and 2D or 3D frameworks. The building units in COFs are connected through covalent bonds (i.e., C-C, B-O, C-O, Si-O, Si-C, etc.) to form polymeric networks. These compounds are excellent candidates for methane storage due to large pore volume (range 0.38–5.40 cm^{3}g^{-1}), high surface area (range 1050–6450 m^{2}g^{-1}), and low densities (range 0.17–1.03 gcm^{-3}). Undoubtedly, the density of COF-108 (0.17 g cm^{-3}) is the lowest amongst ever reported crystalline materials. Despite some challenges like complicated synthetic protocols to extend the organic molecules as chains, sheets, or 3D networks, difficult crystallization, and cumbersome characterization of products due to partial or no solubility in common solvents, COFs have been utilized in numerous applications.

21.3 Types of COFs Used for Methane Adsorption

COFs are obtained from various building blocks, which can be bifunctional, trifunctional or tetrafunctional. Depending upon the functionality of precursors, two-dimensional (2-D) and three-dimensional (3-D) frameworks of COFs are constructed. COFs investigated for the adsorption of methane are classified into two classes based on their dimensionality; 2D-COFs and 3D-COFs. The structural detail of each type is discussed below.

21.3.1 Two-Dimensional COFs

Although a variety of 2D-COFs are known few are investigated for methane adsorption by theoretical and experimental means. The COFs used for methane adsorption studies are of two types; 1) Boron-based COFs and 2) Nitrogen-based COFs. The brief synthesis and structural skeletons of these COFs are summarized in Tables 21.1 and 21.2.

21.3.2 Three-Dimensional COFs

Three-dimensional COFs are obtained by using tetrahedral building blocks (Figure 21.1). For instance, self-condensation of tetra(4-dihydroxyborylphenyl)methane (TBPM) generates COF-102. Similarly, tetra(4-dihydroxyborylphenyl)silane (TBPS) is used to obtain COF-103. However, the co-condensation of tetra(4-dihydroxyborylphenyl)methane and

TABLE 21.1

Important Features and Structural Skeleton of Boron-Based COFs

Boron-Based COFs			
COF	Important Features	Structural Skeleton	Ref
COF-1	COF-1 is obtained by the slow dehydration reaction of 1,4-benzenediboronic acid (BDBA) under controlled temperature and pressure conditions in a sealed tube. The reactants disseminate slowly and react to form the white crystalline material. Its structure consists of alternate boroxine and benzene rings. The Cerius models revealed distribution of layers in a staggered fashion to form 2D structure. Therefore, the topology of COF-1 is ABAB-type (similar to graphite). Just like zeolites and porous carbonized materials, the surface area of COF-1 is more than the graphite, clays, and pillared clays. It is microporous with small distribution of mesopores.		[11]
COF-5	COF-5 is obtained from phenylboronic acid and 2,3,6,7,10,11-hexahydroxytriphenylene (HHTP). The later is responsible for the increase in the size of the cavity as it provides a triangular building unit. The structure of COF-5 consists of hexagonal rings in which HHTP units occupies the six corners of the hexagon and are bridged through phenylboronic units. The layers are further joined in eclipsed manner to form a 2D structure. Therefore, the topology of COF-5 is AAAA type similar to boron nitride.		[11]
COF-6	COF-6 is obtained by the co-condensation of HHTP and 1,3,5-benzenetriboronic acid (BTBA). The experimental and simulated PXRD patterns revealed the stacking of layers in eclipsed manner. The layers are stacked via π-π interactions and dative interaction between B and O atoms of the adjacent layers. The COF-6 is a microporous material.		[12]

TABLE 21.1 *(Continued)*

Important Features and Structural Skeleton of Boron-Based COFs

Boron-Based COFs			
COF	**Important Features**	**Structural Skeleton**	**Ref**
COF-8	COF-8 is synthesized by the reaction of HHTP and BTPA under controlled temperature and pressure conditions. The size of one of the building block is enhanced as compared to 1,3,5-benzenetriboronic acid, therefore, cavity size of COF-8 is larger as compared to COF-6. However, the adjacent 1D layers are stacked together by π-π interaction in an eclipsed manner just like COF-6.		[12]
COF-10	The stacking of layers in case COF-10 is also analogous to COF-6 and COF-8 but the skeleton of COF-10 is formed by the precursors HHTP and 4,4′-biphenyldiboronic acid (BPDA). It is a mesoporous material.		[12]
TDCOF-5	The skeleton of TDCOF-5 is constructed by the reaction of 1,4-benzenediboronic acid and hexahydroxytriptycene in a mixture of mesitylene–dioxane. It is microcrystalline solid. The tritopic nature of hexahydroxytriptycene could give eclipsed or staggered conformation to the structure.		[13]

tetra(4-dihydroxyborylphenyl)silane with triangular HHTP results in the formation of COF-105 and COF-108, respectively. This class of COFs is highly porous with high surface area, low densities, and rigid frameworks. The COF-102, COF-103, and COF-105 possess carbon-nitride type topology whereas COF-108 possesses boracite like topology.

21.4 Experimental Studies on Methane Adsorption by COFs

The surface parameters of COFs such as surface area, pore volume, pore size, and density have been investigated by various studies (Table 21.3). One-dimensional COFs, COF-1, and COF-6, possess small pores and show the characteristic adsorption isotherm for

TABLE 21.2

Important Features and Structural Skeleton of Nitrogen-Based COFs

Nitrogen-Based COFs			
COF	Important Features	Structural Skeleton	Ref
ACOF-1	ACOF-1 is azine-linked COF, which can be synthesized by the condensation of 1,3,5-triformylbenzene and hydrazine. It is synthesized in relatively mild reaction conditions than the COFs discussed above. The COF layers are stacked by π-π interactions. The crystalline COF consists of one-dimensional channels, high surface area, and micropores.		[14]
COF-JLU2	COF-JLU2 is azine linked COF and is obtained by the condensation reaction of 1,3,5-triformylphloroglucinol and hydrazine hydrate in the presence of acetic acid. The COF-JLU2 is found thermostable up to 250°C and stable in different solvents. The COF layers are stacked to form 2D structures. It is promising material for the adsorption of gases due to exceptional microporosity, stability, and distinct heteroatomic sites.		[15]
COF-TpAzo	COF-TpAzo is obtained by the Schiff base condensation reaction between triformylphloroglucinol and 4,4′-azodianiline. It is stable in strong mineral acids, bases, and hot water.		[16]

TABLE 21.2 *(Continued)*

Important Features and Structural Skeleton of Nitrogen-Based COFs

		Nitrogen-Based COFs	
COF	Important Features	Structural Skeleton	Ref
IL-COF	IL-COF, a Schiff base obtained by the condensation reaction of 1,3,6,8-tetrakis(p-formylphenyl) pyrene and p-phenylenediamine in 1,4-dioxane. It is robust and is stable at high temperatures upto 400°C.		[14]

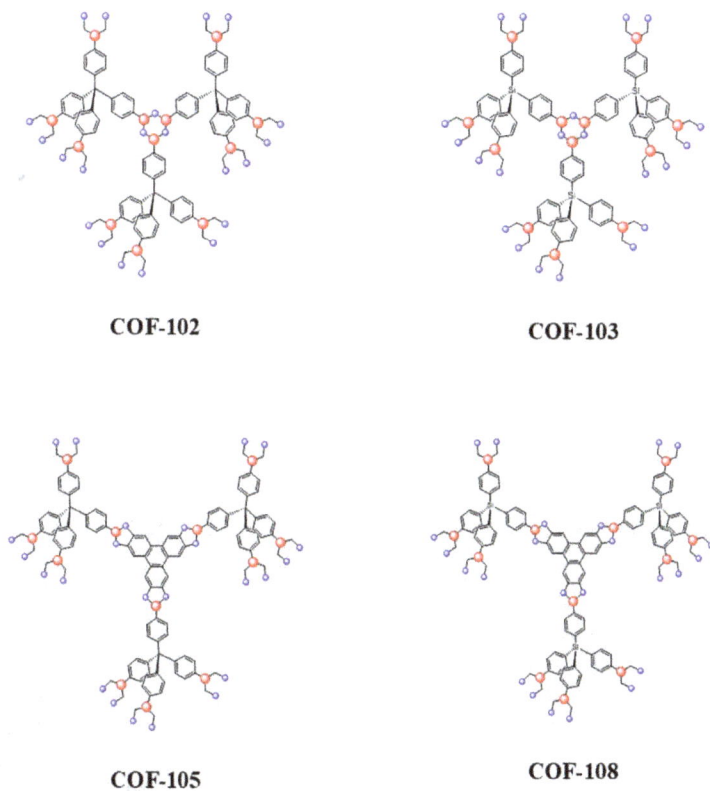

COF-102

COF-103

COF-105

COF-108

FIGURE 21.1
Structures of 3D COFs.

TABLE 21.3

A Summary of Surface Parameters of COFs Such as Surface Area, Pore-Volume, and Pore Size Based on Experimental and Theoretical Studies

		Experimental Data (*Theoretical Data)					
COFs	Type of Porosity	Langmuir Surface Area (S_{Lang}); m^2g^{-1}	Surface Area (S_{BET}); m^2g^{-1}	Total Pore Volume (V_p) cm^3g^{-1} (P/P_o)	Pore Diameter (Å)	*Topology (Space Group)	Ref
COF-1	Microporous material	970	750	0.30	9		[9, 11, 17]
			711	0.32 (0.90)	9		
		*1230		0.38	7	gra (P63/mmc)	
COF-5	Meosporous	1990	1670	1.07	27		[9, 11, 17]
			1590	0.998	27		
		*1530		1.17	27	bnn (P6/mmm)	
COF-6	Microsporous	980	750	0.32	9		[9, 11, 17]
			980	0.32 (0.03–0.05)	6.4		
		*1050		0.55	11	bnn (P6/mmm)	
COF-8	Mesoporous	1400	1350	0.69	16		[9, 11, 17]
			1400	0.69 (0.25–0.30)	18.7		
		*1320		0.87	16	bnn (P6/mmm)	
COF-10	Mesoporous	2080	1760	1.44	32		[9, 11, 17]
			2080	1.44	34.1		
		*1830		1.65	35	bnn (P6/mmm)	
COF-102		4650	3620	1.55	12		[9, 11, 17]
		*4940		1.81	12	$I\bar{4}3d$	
COF-103		4630	3530	1.54	12		[9, 17]
		*5230		2.05	12	$I\bar{4}3d$	
COF-105							[17]
		*6450		4.94	19	$I\bar{4}3d$	
COF-108							[17]
		*6280		5.40	20,11	$P\bar{4}3m$	
IL-COF	Mesoporous	3453	2723				
		*5070		1.21 (0.96)	23	(Cmm2/ Fmm2)	[18]
TDCOF-5	Mesoporous	3832	2497	1.3 (0.95)	26	AB (P63/mmc)	[13]
		*4973			31		

microporous materials. The adsorption saturation of argon in these COFs was observed in the range P/P_0 10^{-4}–10^{-3}. In contrast, the 2D COFs are mesoporous and possess multi-layer adsorption of argon. The adsorption isotherm trends observed in 3D COFs resemble MOFs and are unique due to variable binding sites in the materials.

Few reports are available for the experimental studies on methane adsorption by COFs based on the isotherms obtained on high-pressure gravimetric instruments. In the initial studies, the adsorption isotherms of boron-based COFs were obtained at room temperature and 70 bar for various COFs. The studies revealed that adsorption capacity varies with the porosity of the COFs. The COF-1 and COF-6 show methane uptake up to 10 wt %; COF-5, COF-8, and COF-10 show methane uptake between 10–15 wt %; and COF-102 and COF-103 show methane uptake above 20 wt %. The methane uptake increases with the dimensionality of the COF, i.e., with a trend 1D<2D<3D. Moreover, to saturate the material with a larger pore volume, extra pressure is required. Later on, preliminary studies on B and N COFs such as azine-linked ACOF-1 [14], COF-JLU2 [15], COF-TpAzo [14], TDCOF-5 [13], and IL-COF were done also for the methane adsorption [18].

Isosteric heat of adsorption or the adsorption enthalpy (Q_{st}) is used to evaluate the adsorption efficiency of the material. It can be calculated from the total uptake isotherms using the formula given below.

$$Q_{st} = \frac{\langle U \rangle \langle N \rangle - \langle UN \rangle}{\langle N^2 \rangle - \langle N \rangle \langle N \rangle} + RT$$

Where $\langle \rangle$ refers to the ensemble average; $\langle U \rangle$ is the average energy of the configurations; $\langle N \rangle$ is the number of particles; R is the Boltzmann constant, and T is the temperature [19].

The 2D COFs show similar Q_{st} values due to their structural similarity. The groups of COFs, i.e., COF-5, COF-10, COF-105, and COF-108 show a decrease in the COF-methane interaction or Q_{st} value with the increase in pressure due to occupancy of strong binding sites at the low pressure. However, the 1D COFs, COF-1, and COF-6, have larger Q_{st} values or interactions at low coverage, which indicates that smaller pores are more suitable for methane storage. When the pore size is small, adsorbent-adsorbate interactions are stronger at low pressure because the potential fields of neighboring atoms overlap more strongly. However, these COFs (COF-1, COF-6, COF-8, COF-102, and COF-103) show increase in the interaction or Q_{st} value with the increase in the pressure. It is wise to say that COF-102 and COF-103 are the best adsorbents as their pore diameters and Q_{st} values are appropriate for practical use [9, 17]. The gravimetric uptake of methane by various COFs at 85 bar and 298 K is summarized in Table 21.4 [9].

21.5 Theoretical Studies on Methane Adsorption by COFs

Although more than 300 COFs have been synthesized in the past, their investigation for methane storage is still incomplete. Therefore, computational methods are the best means to rapidly evaluate the methane delivery performance and eliminate the need for laborious experimental procedures. To facilitate the public and private sectors, a computation-ready, experimental (CoRE) COF database is available, which provides the preliminary data on the methane storage calculated by various simulations. The CoRE COF consists of 280 COFs accessible online (https://core-cof.github.io/CoRE-COF-Database/) including 31 3D-COFs with ctn, bor, dia, or pts topology, and2D-COFs (249) with triangular, square, hexagonal, octagonal, heteromorphic, or hybrid pores. A 3D COF, PI-COF-4, exhibits the highest deliverable capacity, i.e., 190 v(STP)/v at 298 K according to these studies which also supported the role of high volumetric surface area in methane delivery. In addition, the database also suggests that 2D COFs with interconnected channels in the three dimensions are promising methane adsorbents [21].

Besides, theoretical studies of some in-silico fabricated COF structures capable of methane storage are also available. This database consists of 69840 covalent organic frameworks based upon 666 distinct organic linkers. Their synthesis is proposed based on Ullmann synthesis and amine condensation. These COFs are supposed to consist of C-C, C=N, -NH- and –C(=O)NH- linkages. The grand-canonical Monte Carlo simulations reveal the

TABLE 21.4

A Comparison of Methane Adsorption Parameters of Various COFs (Gravimetric Uptake and Q_{st} Values)

Type of COF	Gravimetric Uptake at (Pressure and Temperature) (mg g⁻¹)	Qst (KJmol⁻¹) at Zero Coverage	Saturation Pressures	Ref
COF-1	44 mg g⁻¹ (85 bar, 298 K) 40 mg g⁻¹ (35 bar, 298 K)	6.2	66 bar	[9]
COF-5	127 mg g⁻¹ (85 bar, 298 K) 89mg g⁻¹ (35 bar, 298 K)	6.0	NA	[9]
COF-6	68 mg g⁻¹ (85 bar, 298 K) 65 mg g⁻¹ (35 bar, 298 K)	7.0	50 bar	[9]
COF-8	114 mg g⁻¹ (85 bar, 298 K) 87 mg g⁻¹ (35 bar, 298 K)	6.3	NA	[9]
COF-10	124 mg g⁻¹ (85 bar, 298 K) 80 mg g⁻¹ (35 bar, 298 K) 0.58 wt% (1.0 bar, 273 K)	6.6	NA	[9]
COF-102	243 mg g⁻¹ (85 bar, 298 K) 187 mg g⁻¹ (35 bar, 298 K)	3.9	70	[9]
COF-103	229 mg g⁻¹ (85 bar, 298 K) 175 mg g⁻¹ (35 bar, 298 K)	4.4	70	[9]
ACOF-1	1.15 wt% (1.0 bar, 273 K)	16.58		[14]
ILCOF-1	11.2 mmol g⁻¹ (40 bar, 298 K) 0.9 wt% (1.0 bar, 273 K)	13.7		[18]
TDCOF-5	11.5 cm³g⁻¹ (1.0 bar, 273 K) 1.07 wt% (1.0 bar, 273 K)	12.6		[13]
COF-JLU2	3.8 wt% (1.0 bar, 273 K)	20.5		[15]
COF-TpAzo	13.4 mg g⁻¹ (1.0 bar, 273 K)	16.5		[20]

materials as lightweight with high internal surface areas and promising structures for methane storage. The best results are reported for the COF assembled through triazine linkers in the tbd topology. This system shows a deliverable capacity of 216 v STP/v at 65 bar which is similar to some other efficient materials having deliverable capacities greater than 190 v STP/v [22] (available on the Materials Cloud).

Another database of 3D COFs with 4147 structures has been designed based on 620 unique non-interpenetrated structures. These structures are constructed from specified tetrahedral moieties such as tetra(4-anilyl)methane (TAM), tetra(4-dihydroxyborylphenyl) methane (TBPM), and tetra(4-dihydroxyborylphenyl)silane (TBPS) and a possible linker component. The frameworks were extended through two-coordinated (linear) and three-coordinated (trigonal) linkers to obtain imine, boronate ester, or borosilicate linkages [23]. The theoretical studies are mainly accomplished by using the following simulations.

21.5.1 Grand Canonical Monte Carlo Simulations (GCMC)

The grand canonical Monte Carlo method was first developed by Norman and Filinov [24] and then modified by various groups [25]. It is widely used in adsorption studies. In this method, volume, temperature, and chemical potential of adsorbate-adsorbent are kept constant however the number of molecules can vary. It involves the calculation of first principles and force field (FF) parameters which can be executed by Gaussian 03 program package. The force field parameters are used to compute the interaction between methane and the COFs. In this method, potential energies between methane and selected molecules are calculated and the results are fitted into the Morse function (Eq. 1).

$$U_{ij}\left(rU_{ij}\right) = D\left[x^2 - 2x\right], x = exp\left(-\frac{\gamma}{2}\left(\frac{r_{ij}}{r_e} - 1\right)\right)$$

Eq. 1

Where, r_{ij} is the interaction distance in angstroms. D, γ, and r_e are the well depth, the stiffness (force constant), and the equilibrium bond distance, respectively.

The force field parameters are used to envisage the adsorption isotherms by grand canonical Monte Carlo simulations (GCMC). According to these simulations, the gravimetric uptake of COF-105 and COF-108 is more as compared to COF-102 and COF-103 at T = 243K and P = 30 bar. It is obvious because the former set of COFs has a larger pore volume and is less dense than the later set of COFs. The maximum gravimetric uptake for COF-102 and COF-103 can be expected around 40 bar (28.98 and 29.25 wt% respectively), and around 70–80 bar for COF-105 and COF-108 (39.46 and 38.51 wt %, respectively). However, the volumetric uptake of methane in COF-105 and COF-108 is slightly lower than COF-102 and COF-103. Overall, the COFs show high storage capacity toward methane at 243 K under lower pressure (Table 21.5).

The results obtained from the simulations are in good agreement with the experimental results at low pressure (p < 40 bar) [9]. The slight difference in the experimental and theoretical results at high pressure (p > 40 bar) is observed due to the blockage of adsorption pores at the high pressure applied during experimentation. Moreover, the excess gravimetric uptakes of methane are more than other materials like MOFs, IR-MOFs, and PCN. Overall COF-102 and COF-103 possess very high gravimetric storage capacity as well as volumetric storage capacity at room temperature.

Besides, simulations of Li doped COFs (Li-COFs) were executed by using the B3LYP/6-311G(d,p) method for geometry optimization and PW91/6-311g(d,p) method for calculating binding energy. The doping of Li cations in COFs can increase their binding capacity. The theoretical value of binding energy between methane and Li-COFs is about -5.71 kcal/mol (without basis set superposition error BSSE correction). The interaction between methane and the Li$^+$ ion is purely dipole induced as the distance between Li and C (of methane) is 2.36 Å, which suggests no bond formation between the two. Although the methane molecule is non-polar with the symmetrical distribution of electronic charge, the Li$^+$ ion distorts the electronic cloud of the methane and induces polarization. This results in the formation of the induced dipole in the methane and helps in the adsorption of methane on the surface due to induced dipole interactions and London dispersion. The interaction between methane and Li-ion is found in 1:1 stoichiometry. The highest occupied molecular orbital (HOMO) calculated by using the B3LYP/6-311g(d,p) method suggests that 2p orbital

TABLE 21.5

Gravimetric and Volumetric Uptake of Methane Calculated by GCMC at 243 and 298K

	Pressure/ Temperature	COF-102	COF-103	COF-105	COF-108	Ref
Gravimetric uptake (in wt%)	100 bar/243 K	34.63	35.91	54.39	54.68	[25]
Gravimetric uptake (in wt%)	35 bar/298 K	20.39	19.64	23.07	24.13	[25]
Volumetric uptake (in v(STP)/v)	30 bar/243 K	257	238	228	225	[25]
Volumetric uptake (in v(STP)/v)	35 bar/298 K	359	350	412	447	[25]

of Li overlaps with 2p orbital of methane to form a conjugated π bond without any hybridization. The simulations on Li-COFs, i.e., Li-COF-105 and Li-COF-108, executed similarly as discussed above revealed 33% enhancement in the gravimetric uptakes of methane as compared to COF-105 and COF-108(at 35 bar and 243K). In contrast, Li-COF-102 and Li-COF-103 possess 22% increase in the gravimetric uptake as compared to COF-102 and COF-103. The COF-102 has the highest volumetric uptake of 383 v (STP)/v under these conditions. In contrast, at 298 K and 35 bar, the total gravimetric uptake of methane is doubled in Li-COF-105 as compared to COF-105, which further increases with the increase in pressure to 100 bar. This clearly explains that the strong binding capacity of Li-doped materials is responsible for the ultra-high adsorption of methane. In other words, doping of COFs with the Li-ion provides an extra advantage of the material by enhancing the interaction between methane and the adsorbent [25].

21.5.2 Quantum Mechanics (QM) at the MP2 Level

The quantum mechanics method is used to overcome some of the limitations observed in the GCMC method. Actually, the GCMC methods are based on the generic force field parameters which describe the interatomic interactions very well. But, the prediction of adsorption isotherms based on FF is less accurate and the results of theoretical studies do not match with the experimental results. The FF parameters consider only covalent interactions and do not take non-covalent interactions into account, therefore, deviations are observed from the actual adsorption results. It is investigated that methane adsorbs through non-covalent interactions such as London forces (or vander Waals forces) and electrostatic interactions. Therefore to include these interactions in the investigation of methane adsorption on COFs, computations at MP2 level with a resolution of identity (RI-MP2) have been executed. This method is applied to investigate the methane-methane and methane-COFs interactions. The binding energies between methane-methane and methane-COFs can be obtained by basis set superposition error correction using the following Eq. 2.

$$E_{Interactions}^{CP} = E_{Super} - \sum_{(i=1)}^{n} E_{m_{opt}^i} + \sum_{(i=1)}^{n} \left(E_{m_f^i} - E_{m_f^{i*}} \right) \qquad \text{Eq. 2}$$

Here the E_ms represents the energies of the individual monomers. The subscripts opt and f denote the individually optimized monomers and those frozen in their super molecular geometries, respectively. The asterisk (*) denotes monomers calculated with ghost orbitals. The results can be further used to investigate the FF parameters from the Morse potential using Eq. 3.

$$U_{ij}^{Morse}(r_{ij}) = D\left\{ e^{\alpha(1-r_{ij}/r_0)} - 2e^{\alpha/2(1-r_{ij}/r_0)} \right\} \qquad \text{Eq. 3}$$

Where the parameter D is the well depth, r_0 is the equilibrium bond distance, and R determines the stiffness (force constant).

For these studies, various configurations of interaction between COFs and methane can be taken into account: methane-methane (C_3v, C_3v-2, D_3d, and D_3h), benzene-methane (Anti, Syn, Anti-2, Syn-2), boroxine-methane (Anti, Syn, Anti-2, Syn-2), and silane-methane (Anti, Syn, Anti-2, Syn-2). Out of the possible interactions responsible for the uptake of

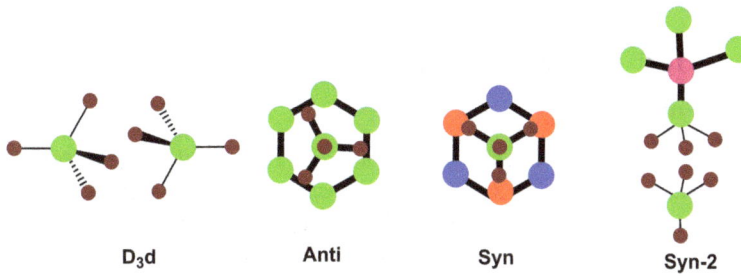

FIGURE 21.2

Possible configurations, D_3d, Anti, Syn, and Anti-2 for methane-methane, benzene-methane, boroxine-methane, and silane-methane based on MP2 studies.

methane by COF surfaces on the basis of MP2 studies include D_3d, Anti, Syn, and Anti-2 in case of methane-methane, benzene-methane, boroxine-methane, and silane-methane, respectively (Figure 21.2).

The experimental and predicted values of methane excess uptake at 298 K and 80 bar were found well in agreement using FF parameters calculated from the MP2 method [17]. The total gravimetric uptake values calculated from this method are summarized in Table 21.6. The difference from the values calculated by GCMC methods Table 21.5 is attributed to the consideration of all possible interactions including methane-methane interactions in this method [25]. The experimental to theoretical methane uptake ratio was found unity at 30 bar for the COF-1 with no excess adsorption above this pressure was observed. Whereas this ratio for other COFs is less than unity. The diameter of pores in COF-1 is small and the stacking of layers is different leading to segregation of pores (i.e., ABAB topology rather than AAAA topology), therefore, the methane saturation occurs at very low pressure. The theoretical methods consider all the pores as equally approachable however ABAB type of stacking in COF-1 makes the pores unapproachable for methane in actual practice. This is the reason for disagreement between experimental and theoretical results.

TABLE 21.6

Total Gravimetric Uptake Values Calculated from MP2 Method [17]

S.No.	Total Gravimetric Uptake (in wt%)	Excess Gravimetric Uptake (Experimental Values at 100 bar; in wt%)	Excess Gravimetric Uptake/Total Gravimetric Uptake Ratio
COF-108	41.5	24.2	0.71
COF-105	40.5	27.6	0.76
COF-103	31.0	26.6	0.90
COF-102	28.4	23.8	0.89
COF-10	19.6	12.2	0.77
COF-8	15.9	10.7	0.81
COF-6	12.3	11.1	0.95
COF-5	16.9	11.7	0.81
COF-1	10.9	10.9	1 (at 30 bar)

21.6 Factors Affecting Methane Adsorption on COFs

21.6.1 Effect of Pore Size

The size of pores affects the adsorption capacity. The COFs with large pores adsorb methane to more extent than the COFs with small pore size. A mesopore can accommodate more guest molecules, so are expected to be appropriate for the adsorption. However, some deviations from the linearity are observed in the case of mesoporous materials. In contrast, microporous materials show a linear response for the gas uptake to the pore volume [9]. For instance, COF-5 and COF-10 deviate from linearity in contrast to COFs with mesopores.

21.6.2 Alkyl Substitution at the Benzene Ring

The computational studies have been done on the COFs with the modified organic skeleton. The meta-hydrogen atoms on the phenyl rings are replaced with some alkyl groups. The substitution of alkyl groups such as methyl, ethyl, n-propyl, i-propyl, t-butyl, and -hexyl to the benzene ring does not show a significant change in the adsorption of methane in COF-102, COF-103, and COF-105. The slight increase in the COFs substituted with bulky alkyl groups is observed due to the increase in the surface area [26]. The substitution of COF-102 (to COF-102-Eth-trans and COF-102-Ant) surpasses the Q_{st} results of COF-102 due to higher surface area and pore volume. The ethyl substituted COF-103 at trans positions (COF-103-Eth-trans) has mild Q_{st}, high surface area, and high pore volume at low pressure [26].

21.6.3 Hetero Groups Substitution (X=–Cl, –Br, –I, –CF$_3$, –NH$_2$, –CN, –OCH$_3$, and –CH$_3$)

The COFs having these groups (COF-X) may be obtained by choosing the substituted building block precursors. The functional groups are substituted in the COF-102, COF-103, and COF5 without disturbing their space group and topology. The COFs are selected based on their binding energy and stable configurations. The adsorption parameters were calculated using GCMC simulations. The FF parameters are obtained by DREIDING force field and van der Waals interactions are simulated by Lennard-Jones potential. The adsorption of methane on the basic COF (COF-102) and its derivatives (COF-102-Xs) increases linearly with the increase in pressure from 0 to 20 bar. The deviation from the linearity at high pressure is due to the less availability of pores for adsorption. Once the methane molecules enter the pores at low temperatures, most of the pores are occupied. So, the entry of additional methane molecules is hindered by the already adsorbed methane molecules. When the adsorption reaches its limit, the curves acquire a constant position. The uptake of methane in the COF-102-Xs (except COF-102-OCH$_3$) is improved as compared to the basic COF-102. The isodensity graphs of methane in the basic and modified COFs show that the density of methane in COF-102 is 5 times higher than the original form. The reason behind the improved adsorption is the availability of more space for the adsorption of methane. The substitution of –Cl groups results in the increase in the distance between phenyl rings due to Cl-Cl repulsions and therefore extended space can accommodate more methane molecules. Secondly, Cl atoms favor the adsorption due to strong interactions with the methane molecules. A similar effect of –CN can be seen on the methane uptake of COF-102-CN.

Similarly, COF-102-NH$_2$ shows enhanced methane adsorption because –NH$_2$ being a small group occupies less space and leaves more adsorption space behind. In contrast, COF-102-OCH$_3$ occupies more space due to the larger size of –OCH$_3$ groups and shows minimum methane uptake. The COF-102-CF$_3$ shows ultra-high adsorption uptake due to rich interaction sites. In contrast, the substitution of methyl does not show a significant effect on the uptake of methane. Out of these alkyl, halo, and –NH$_2$substituents, Cl is found best for improving methane uptake. In COF-102-Xs demonstrate a similar trend followed by COF-103-Xs due to similarity in their structures. The trends in COF-105 and its derivatives are completely different from the COF-102 and -103 derivatives. The methane uptake isotherms show a linear relationship with the pressure without showing any deviation at high pressure. The larger pore size of the COF-105 derivatives allows the saturation of pores at even high pressures. The halogen-substituted COFs show better adsorption than –CF$_3$, –NH$_2$, and –CN substituted COFs. The difference in the uptake values of COFs and their derivatives is not very large because the substitution was done at limited sites only, i.e., 8 atoms out of 1020 atoms in COF-105 and 48 atoms out of 488 atoms in the case of COF-102 and COF-103 [19]. The isosteric heats of adsorption of substituted COFs are more than parent COFs. Especially, the COFs having a substitution of halogens or halogen-based groups (such as Cl. Br, I, and CF$_3$) have higher Q$_{st}$ values. Amine functionalized-COFs, like COF-102-NH$_2$ and COF-103-NH$_2$, possess moderate Qst values which are lower than the –CF$_3$ and –OCH$_3$ substituted COFs (Table 21.7) [19].

21.6.4 Functionalization with Organic Moieties

A combination of density functional theory calculations and grand canonical Monte Carlo (GCMC) simulations was performed on the functionalized COFs such as TpPa1 (Tp: 1,3,4-triformylphloroglucinol; Pa1: p-phenylenediamine), TpBD(BD: benzidine), PI-COF-1, PI-COF-2, and PI-COF-3(PI: polyimide). The functionalization increases the total –Q$_{st}$ (0) values but does not affect the binding location sites for methane adsorption. Moreover, functionalization with methyl and nitro groups increases the methane binding energy as compared to parent COF. For example; methane binding energy of methylated TpPa is –22.67 kJ mol^{-1} and nitrylated TpPa–23.73 kJmol^{-1}. In contrast, fluorinated TpPa has lesser binding energy–15.57 kJ mol^{-1} as compared to its parent COF [27].

21.6.5 Effect of Hydration (or Introducing Ionic Liquids)

Although the hydration of COFs increases the selectivity toward the adsorption of carbon dioxide, it does not enhance the adsorption of methane. A similar effect is observed in the presence of ionic liquids. The main reason behind this is the non-polar nature of methane.

TABLE 21.7

Isosteric Heats of Adsorption of Substituted COFs [19]

COF	Methane Uptake; V(STP)/V	Excess Uptake at 35 bar; V(STP)/V	COF	Methane Uptake; V(STP)/V	Excess Uptake at 35 bar; V(STP)/V
COF-102-Cl	169	148	COF-103-Cl	155	130
COF-102-Br	175	153	COF-103-Br	157	133
COF-102-I	176	156	COF-103-I	163	138
COF-102	149	125	COF-103	137	111
COF-102-NH$_2$		143	COF-103-NH2	151	125

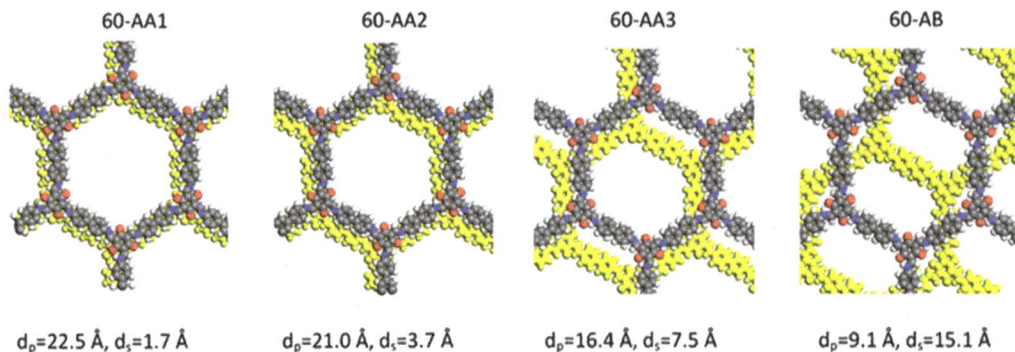

60-AA1	60-AA2	60-AA3	60-AB
d_p=22.5 Å, d_s=1.7 Å	d_p=21.0 Å, d_s=3.7 Å	d_p=16.4 Å, d_s=7.5 Å	d_p=9.1 Å, d_s=15.1 Å

FIGURE 21.3
Structure of TpPa1 and TpBD with 60-AB, 60-AA2, 60-AA1, and 60-AA3 arrangement. (Adapted with permission from [27], Copyright (2021), ACS.)

Therefore, the water or ionic liquid impregnated COFs can be used for the separation of methane and carbon dioxide [28].

21.6.6 Effect of Layer Stacking or Slipping on Methane Adsorption

The stacking of layers in COFs is of two types; 1) Eclipsed stacking (via π–π interactions), 2) Slipped stacking. In eclipsed stacking, layers are stacked in AAAA arrangement and form one-dimensional (1D) pore channels, which are suitable to adsorb the gas molecules. In slipped stacking, the layers are displaced with respect to each other to form ABAB arrangement. The extent of slipping can be varied to tune the porous channels for adsorption. According to grand canonical Monte Carlo (GCMC) simulation (using RASPA code), methane adsorption could be improved in the 2D COFs by slipped stacking of layers. The slipping of layers increases the surface area and as a consequence, methane uptake increases. The isosteric heat adsorption in such systems can be increased by increasing the pressure because due to the rich density of interacting sites. The functionalized slipped systems show better selectivity for carbon dioxide over methane, therefore are promising materials for the separation of these gases [27]. For instance, COFs with slipped structures like TpPa1, TpBD, and PI-COFs (–1, –2, and –3) showed a variation in methane loading due to interlayer slipping (Figure 21.3). The loading order in 60-AB structure of TpPa1 and TpBD is 60-AB > 60-AA2 > 60-AA1 > 60-AA3 > AA however PI-COFs showed maximum uptake for 90-AA1 structure.

21.7 Challenges and Prospects

The major challenge is attaining the appropriate methane delivery from the COFs. Methane delivery is a very crucial parameter to evaluate the practicality of the material as an adsorbent. It is defined as the volume of CH_4 released from the adsorbent at room temperature when the pressure is reduced from 35 to 1 bar. It can be predicted from the total uptake isotherms. The Department of Energy (DOE) fixed the release of methane at standard temperature (298 K) and pressure (1.01 bar) as 180L per liter of the storage vessel. The total uptake and delivery amount of COF-1 are found 195v(STP)/v and 42v(STP)/v, respectively at 30 bar. Therefore, it cannot be used for practical purposes. While, COF-102 and COF-103 show better performance for practical use, i.e., total uptake (255 and 260 v(STP)/v at 100 bar) and delivery amount (229 and 234 v(STP)/v at 100 bar). Therefore, a small pore diameter with accessible aromatic rings and methane-methane interactions are the prerequisites for the efficient adsorption of methane on the COFs. The methane deliveries in the halogenated-COFs and amino-functionalized COFs are found higher than their parent COFs, i.e., 165 V(STP)/V, 169 V(STP)/V, 169 V(STP)/V, and 160 V(STP)/V for COF-102-Cl, COF-102-Br, COF-102-I, and COF-102-NH$_2$ [19]. Therefore, the best solution is to synthesize the COFs with substituted heteroatoms and investigate the materials for their practical use.

Although theoretical studies give a preliminary idea about the expected outcome of the material; however the actual results are generally different. The difference in the theoretical and experimental values is expected due to the following reasons:

1. The van der Waal interactions between methane-methane and methane-COFs are overestimated in the simulations.

2. There is a possibility of deviations in the structures of the COFs at high pressure.

3. The interacting sites in the COFs may not be completely accessible in practice.

The zeolitic-COFs may provide better results for methane adsorption as a computational study of the COFs obtained by the replacement of Si from the zeolite by carbon through various linkers (such as furan, pyrrole, and thiophene, etc) revealed. The computations were done on 300 ZCOFs with 100 topologies of COFs. These materials possess more surface area than zeolites. Out of these, some ZCOFs such as BOZ-S, JSR-N, JSR-O, JSR-S, JST-S, NPT-S, OBW-N, OBW-S, RWY-N, RWY-O, and RWY-S exhibited better methane uptake [29].

References

1. Reay D, Smith P, van Amstel A. Methane and Climate Change: Earthscan; 2010.
2. Karim GA. Natural Gas and Other Gaseous Fuels. Fuels, Energy, and the Environment: CRC Press; 2016, 280–317.
3. Ramanathan A, Dharmalingam B, Thangarasu V. Advances in Clean Energy: Production and Application: CRC Press; 2020.
4. Ramadhas AS. Fuels and trends. Alternative Fuels for Transportation. 2016:1.
5. Menon V, Komarneni S. Porous adsorbents for vehicular natural gas storage: a review. Journal of Porous Materials, 5, 1998, 43–58.
6. Chui SS-Y, Lo SM-F, Charmant JP, Orpen AG, Williams ID. A chemically functionalizable nanoporous material [Cu3 (TMA) 2 (H2O) 3] n. Science, 283, 1999, 1148–50.
7. Himeno S, Komatsu T, Fujita S. High-pressure adsorption equilibria of methane and carbon dioxide on several activated carbons. Journal of Chemical & Engineering Data, 50, 2005, 369–76.
8. Makal TA, Li J-R, Lu W, Zhou H-C. Methane storage in advanced porous materials. Chemical Society Reviews, 41, 2012, 7761–79.
9. Furukawa H, Yaghi OM. Storage of hydrogen, methane, and carbon dioxide in highly porous covalent organic frameworks for clean energy applications. Journal of the American Chemical Society, 131, 2009, 8875–83.
10. Nagai A. Covalent Organic Frameworks: Jenny Stanford Publishing; 2019.

11. Cote AP, Benin AI, Ockwig NW, O'Keeffe M, Matzger AJ, Yaghi OM. Porous, crystalline, covalent organic frameworks. Science, 310, 2005, 1166–70.
12. Cote AP, El-Kaderi HM, Furukawa H, Hunt JR, Yaghi OM. Reticular synthesis of microporous and mesoporous 2D covalent organic frameworks. Journal of the American Chemical Society, 129, 2007, 12914–5.
13. Kahveci Z, Islamoglu T, Shar GA, Ding R, El-Kaderi HM. Targeted synthesis of a mesoporous triptycene-derived covalent organic framework. CrystEngComm, 15, 2013, 1524–7.
14. Li Z, Feng X, Zou Y, Zhang Y, Xia H, Liu X, et al. A 2D azine-linked covalent organic framework for gas storage applications. Chemical Communications, 50, 2014, 13825–8.
15. Li Z, Zhi Y, Feng X, Ding X, Zou Y, Liu X, et al. An azine-linked covalent organic framework: synthesis, characterization and efficient gas storage. Chemistry–A European Journal, 21, 2015, 12079–84.
16. Chandra S, Kundu T, Kandambeth S, BabaRao R, Marathe Y, Kunjir SM, et al. Phosphoric acid loaded azo (– N= N–) based covalent organic framework for proton conduction. Journal of the American Chemical Society, 136, 2014, 6570–3.
17. Mendoza-Cortés JL, Han SS, Furukawa H, Yaghi OM, Goddard III WA. Adsorption mechanism and uptake of methane in covalent organic frameworks: theory and experiment. The Journal of Physical Chemistry A, 114, 2010, 10824–33.
18. Rabbani MG, Sekizkardes AK, Kahveci Z, Reich TE, Ding R, El-Kaderi HM. A 2D mesoporous imine-linked covalent organic framework for high pressure gas storage applications. Chemistry-A European Journal, 19, 2013, 3324–8.
19. Zhao J, Yan T. Effects of substituent groups on methane adsorption in covalent organic frameworks. RSC Advances, 4, 2014, 15542–51.
20. Ge R, Hao D, Shi Q, Dong B, Leng W, Wang C, et al. Target synthesis of an azo (N • N) based covalent organic framework with high CO2-over-N2 selectivity and benign gas storage capability. Journal of Chemical & Engineering Data, 61, 2016, 1904–9.
21. Tong M, Lan Y, Qin Z, Zhong C. Computation-ready, experimental covalent organic framework for methane delivery: screening and material design. The Journal of Physical Chemistry C, 122, 2018, 13009–16.
22. Mercado R, Fu R-S, Yakutovich AV, Talirz L, Haranczyk M, Smit B. In silico design of 2D and 3D covalent organic frameworks for methane storage applications. Chemistry of Materials, 30, 2018, 5069–86.
23. Martin RL, Simon CM, Medasani B, Britt DK, Smit B, Haranczyk M. In silico design of three-dimensional porous covalent organic frameworks via known synthesis routes and commercially available species. The Journal of Physical Chemistry C, 118, 2014, 23790–802.
24. Norman G, Filinov V. Investigations of phase transitions by a Monte-Carlo method. High Temperature, 7, 1969, 216–20
25. Lan J, Cao D, Wang W. High uptakes of methane in Li-doped 3D covalent organic frameworks. Langmuir, 26, 2010, 220–6.
26. Mendoza-Cortes JL, Pascal TA, Goddard III WA. Design of covalent organic frameworks for methane storage. The Journal of Physical Chemistry A, 115, 2011, 13852–7.
27. Sharma A, Babarao R, Medhekar NV, Malani A. Methane adsorption and separation in slipped and functionalized covalent organic frameworks. Industrial & Engineering Chemistry Research, 57, 2018, 4767–78.
28. Vicent-Luna J, Luna-Triguero A, Calero S. Storage and separation of carbon dioxide and methane in hydrated covalent organic frameworks. The Journal of Physical Chemistry C, 120, 2016, 23756–62.
29. Do HH, Kim SY, Le QV, Pham-Tran N-N. Design of zeolite-covalent organic frameworks for methane storage. Materials, 13, 2020, 3322.

22

Covalent Organic Frameworks-Based Nanomaterials for Greenhouse Gases Capture and Storage: CH_4 and CO_2

Heriberto Díaz Velázquez and Rafael Martínez-Palou

Dirección de Investigación en Transformación de Hidrocarburos, Instituto Mexicano del Petróleo, San Bartolo Atepehuacan, CDMX, México

CONTENTS

22.1 Introduction

As is well known, methane is one of the most powerful greenhouse gases (GHGs); it is third just behind water vapor and CO_2. Furthermore, it has been shown that the tropospheric concentration of methane has increased since the period 1000–1750. Its current concentration is about 1.8 ppm [1]. From the two sources of methane, both natural and anthropogenic, human activities have surpassed the natural methane production in current times, being cattle production the most important source of the gas, but also the oil industry plays a significant role on methane production. Natural major reservoirs are wetlands (174 Tg/year), followed by termites and oceans, where the gas is stored as clathrates. Interestingly, most of this gas is lost at the troposphere, where CH_4 is destroyed by $OH\cdot$ radicals, which makes a total imbalance of almost zero, according to the IPCC's fourth assessment report. However, while in the atmosphere, methane can absorb 25 times more heat than CO_2, acting as a very powerful GHG.

Technologies exist to reduce GHG's emissions: reduction of energy consumption, efficient use of energy (both in energy use and conversion), use of fuels with lower carbon content, promotion of natural CO_2 sinks, use of energy sources with low CO_2 emission

DOI: 10.1201/9781003206507-22

levels (such as renewable or nuclear energy) and CO_2 Capture and Geological Storage (CCS). According to the IPCC's report, CO_2 capture, and storage would contribute between 15 and 55% to the global cumulative mitigation effort up to 2100, thus presenting itself as a transition technology that will contribute to the implementation of cleaner, more efficient, and sustainable processes [2]. Recently, ionic liquids (ILs) have been explored, in particular, amino acid-functionalized ionic liquids exhibit absorption values higher than 0.5 mol of CO_2/mol of IL, including anions derived from amino acids [3], as well as the supported IL membranes [4], inorganic structured materials such as zeolites, calcium oxide, active carbon hydrotalcites and lithium zirconate [5], mesoporous materials containing ILs [6], polymers [7], among which the following stand out, MOF [8], and COFs as described in the following section. This chapter covers most of the theoretical work that has been carried out in methane adsorption by COFs and gives a short overview of the advances on CO_2 adsorption by this family of compounds.

22.2 COFs for Methane Storage

22.2.1 Methane in the Energy Production

Despite the negative environmental impact of CH_4, commercially known as natural gas (NG) [9], its utilization as an energy source is increasingly growing, since it can be used as direct fuel, such as vehicular natural gas (VNG), or applied in direct methane fuel cells (DMFC) and electricity generation. In reality, NG's typical composition of methane is between 81 to 95% wt.%. Other applications of NG as fuel include home heating, kilns, turbines, etc. Furthermore, liquid methane is applied as fuel for rockets. The advantage of methane as a power generator lies in that the overall CO_2 released per unit mass is lower than any other hydrocarbon since it holds the highest H/C molar ratio among the rest of hydrocarbons [10]. This high molar ratio produces higher energy per mass of fuel and a cleaner and higher combustion efficiency (considerably reducing the released carbon as CO_2 to about 40% of less GHGs); however, the total energy density is low, compared to that of liquid fuels such as gasoline. One way to increase methane's energy density is by storing it as compressed natural gas (CNG), which is achieved in pressured vessels at about 20.7 MPa. One more way is to cool it up to 112 K at 1 atm to get liquified natural gas (LNG), which is used for long-distance gas transportation. CNG and LNG can reach up to 9.2 and 22.2 MJ/L of energy density, respectively; however, their preparation supposes complex and costly manipulations and, of course, a high degree of technical expertise. A third way to improve the energy density of NG to an acceptable level is to store it as adsorbed gas on porous materials. This last technique, called adsorbed natural gas (ANG), represents the least costly and complex way to elevate NG's energy content at much lower pressures (3.5–6.5 MPa). The US department of energy (DOE) has been modifying the standard acceptable adsorption capacity for a given material for an acceptable application in real conditions [10]. Such threshold is up today 263 cm^3/cm^3-adsorbent at standard conditions (STP: 273 K, 1 atm) at 65 bar and 298 K, which is equivalent to 0.5 g/g-adsorbent at the same conditions. This value is comparable to the amount of CNG obtained at 25 MPa and 298 K (methane density = 11.73 mol/L) [11]. Really few materials have demonstrated to reach the DOE's target for ANG. Among them, they are Ni-MOF-74, HKUST-1 (251 and 267 cm^3/cm^3 respectively), at 6.5 MPa. These materials, however, lack of good mechanical properties.

This is why different materials with better mechanical resistance have been studied, such as activated carbon materials, which can reach the DOE's target value at 10 MPa [11]. The advantages of the ANG technologies, thus, are apparent and the performance of the materials investigated so far, indicate a very high potential for real-life application, especially in natural gas vehicles (NGV).

22.2.2 Materials for Adsorbed Natural Gas Applications

For a given material to be called as a gas adsorbent, a guest material (adsorbate) should be able to enter the pores of the adsorbent, forming numerous layers of adsorbed molecules. There are two modes of adsorption on a porous material. The first, physical adsorption, occurs through weak van der Waals interactions (vdW), meanwhile, the second adsorption mode, the chemical adsorption, is stronger and occurs thanks to covalent interaction (such as coordinative d-π interactions). As expected, the difference between them are the binding energies, which makes chemical adsorption not to be competitive in terms of adsorption-desorption technologies. These technologies, e.g., in VNG applications, must operate at a certain temperature and pressure ranges, such as -40 to 85°C and at pressures ≤35 bar, thus, achieving an energy density >9.2 MJ/L [12]. The adsorbent materials must have good resistance to impurities and a lifetime of about 100 fill/release cycles. It is also critical to consider the amount of working or deliverable methane, which is the amount that is released as a consequence of an external stimulus, such as pressure. For VNG technologies, this is the amount of ANG released between 5 bar and that at the working limit (~35 bar). Finally, the cost of adsorbent production must be competitive to achieve a positive return on investment [13, 14]. In summary, to comply with the DOE target, the following properties are mandatory: good adsorption capacity, high hydrophobicity, moderate adsorption enthalpy, and high heat capacity, along with an efficient charge/discharge rate.

NG adsorption differs from other light gases such as hydrogen or CO_2 in that the whole uptake capacity of a given material depends more on the textural properties of the materials rather than chemical functionalization or post-synthetic modification of the adsorbent, given the intrinsic chemical features of methane, such as non-polarity, high geometrical symmetry (sp^3 tetrahedral around the C-atom), having a high first ionization potential of 13.16 eV and a proton affinity of 130–5 Kcal/mol, granting it with high chemical stability. The physical and chemical properties of methane can be found in the literature [9].

Because of the textural dependence of the materials to act as good methane adsorbents, several porous materials have been studied for ANG technologies. These materials can be classified into inorganic (aluminosilicate zeolites, M-Oxide molecular sieves, activated alumina, silica gel, aluminosilicates), organic (activated carbon, carbon nanotubes, polymeric resins, porous organic polymers, conjugated microporous polymers), and hybrid organic-inorganic materials (MOFs) [12, 13]. Among them, activated carbons, MOFs, and a specific type of porous organic polymers (POPs) called covalent organic frameworks, have gained attention due to their high uptake capacities. Activated carbons have been already tested on NGV applications [12], but still, some drawbacks related to their limitation on pore size distribution, pore-volume, accessible surface area, and surface functionalization are of concern for them to be completely applicable. Metal-organic frameworks (MOFs) are a second class of materials that have been widely studied for their methane adsorption applications [15–18]. This class of materials, however, lack good mechanical resistance under high pressures, and due to their exceptionally high porosity, they have low packing density, thus lowering the volumetric capacity in such systems. Moreover, their production cost is still an issue to take into consideration [13].

22.2.3 Early Investigations on Covalent Organic Frameworks as Methane Adsorbents

Covalent organic frameworks (COFs), as previously described, are a relatively new class of materials belonging to the sub-category of the porous organic polymers (POPs) whose main characteristic is their crystalline nature and structural regularity. The main framework's bond found in COFs comes from boroxine, imine, hydrazone, azine, and amide linkages. Double bond and dioxin linkages can also be found [19, 20]. The earliest report on their syntheses was made by Côté at al. in 2007 [21], for a self-condensation reaction between boronic esters to boroxines. Just one year later, the first application of COFs for methane storage, such as that on COFs-6, 8, and 10, was theoretically studied by Garberoglio et al. [22] using the DREIDING force field method with Lorentz-Berthelot combination rules and their Lennard-Jonnes parameters for the studied adsorbates (He, Ar, CH_4, and H_2). Using this information, the research group found out that both H_2 and methane follow a similar trend on adsorption behavior for all materials, but, regarding the DOE target at that time (167 cm^3 (STP)/cm^3 at 35 bar), only COF-8 could get close to it (155 cm^3(STP)/cm^3 at 75 bar). In general, the expected adsorption capacity was half that of the best performing MOFs (IRMOF-1 and IR-MOF-2). At higher pressures, COF-8 was the most adsorptive material between the studied COFs, due to its intermediate pore size, which is not so small to get quickly saturated, and not so large to allow the appearance of empty voids within the unit cell that lack of good attraction between adsorbent and adsorbate, which happens in COF-10. Moreover, the self-diffusion coefficients calculated for IRMOF-1 are comparable to that COF-6 but lower than that of COF-10.

The theoretical investigations of different 3D COFs led to the work of Lan et al. [23], who in 2009, using Grand Canonical Monte Carlo, (GCMC), found out that COF-102 and COF-103 doped with Li, are capable of significantly enhancing the methane uptake capacity more than two times that of the original non-doped materials at 298K and p = 35 bar, from 127 and 108 v(STP)/v to 290 and 303 v(STP)/v respectively, surpassing the original DOE target for that time. This investigation found that the dispersion forces and induced-dipole interaction that Li cations exerted on the methane molecule, enhances the attraction of this molecule to the adsorbent, improving the binding energy with the pore surface. Once again, the improvement in the uptake capacity is attributed to the large surface areas and free volumes of the COFs. Particularly, the calculated binding energy between methane and Li cations was about -5.71 Kcal/mol. When comparing COF-102 and COF-103 with COF-105 and COF-108, the last two showed better gravimetric uptakes but lower volumetric uptakes because of their lower pore size but larger pore volumes compared to the first ones, coinciding with Garberoglio's analysis for COFs-6, 8, and 10.

The same year, Furukawa and Yaghi [24] finally reported on the experimental adsorption experiments on COF-1, COF-5, COF-6, COF-8, and COF-10 along with those on COF-102 and COF-103 for the gravimetric uptake of the light gases H_2, CH_4, and CO_2. Figure 22.1 shows the synthesis strategy for the studied COFs. COF-1 synthesis was carried out by self-condensation reactions of BDBA; COF-5, -6, -8, and -10 were produced by condensation of HHTP with BDBA, BTBA, TBPA, and BPDA, respectively. COF-102 and -103 were prepared by self-condensation reactions of TBPM and TBPS respectively. These materials were classified into three groups: 1) 2D structures-1D small pores (COF-1 and -6), 2) 2D structures-large 1D pores (COF-5, -8, and -10) and 3) 3D structure-medium sized pores (COF-102, -103). Small pores were defined to be as 9Å, medium-sized pores 12Å and medium-sized pores are within the range 16–32 Å.

Figure 22.2 shows the methane adsorption isotherms of these materials along with that of BPL carbon, where it is possible to see that the uptake capacity of COF-102, -103 (group 3) surpass by far that of the commercial BPL carbon and only COF-1 and -6 (group 1) did

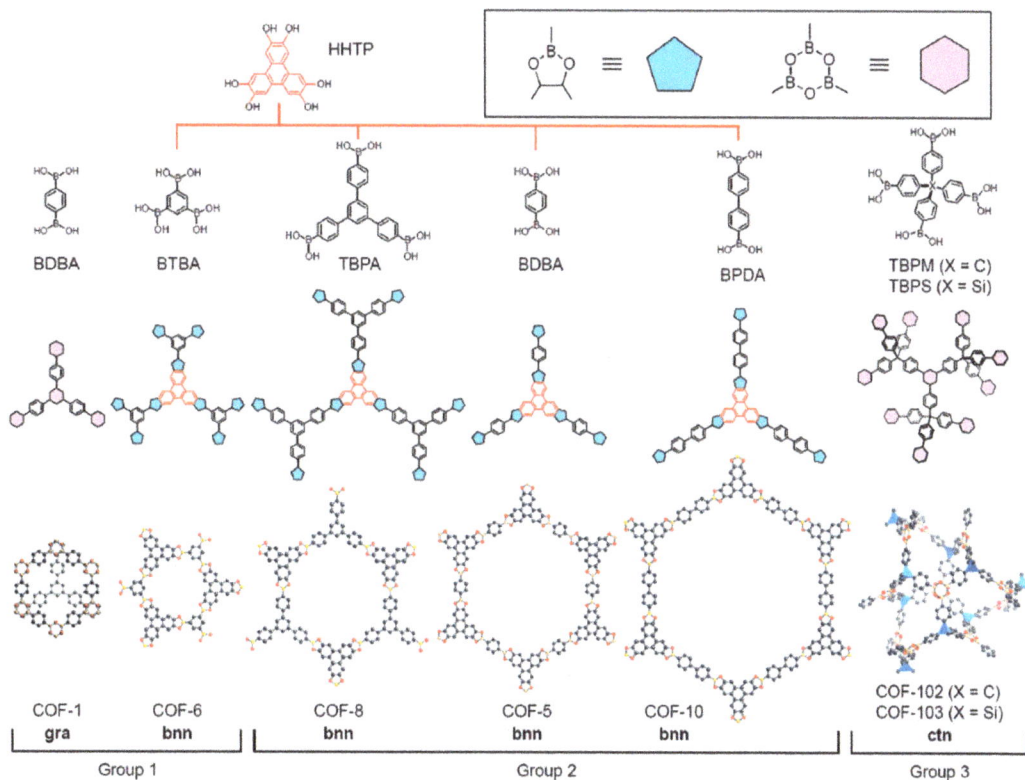

FIGURE 22.1

Synthesis strategy for the COF-1, -5, -6, -8, -10, -102, and -103. (Adapted with permission [24]. Copyright (2009) American Chemical Society.)

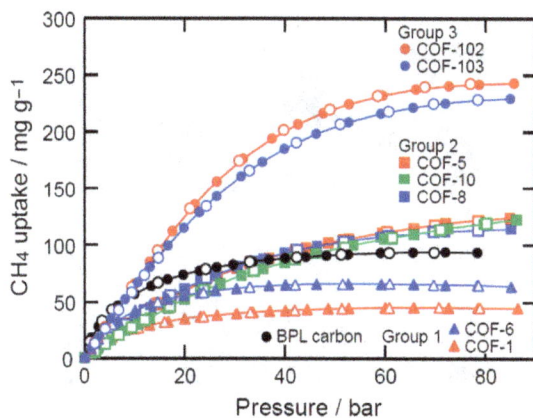

FIGURE 22.2

Methane adsorption isotherms of COF-1, -5, -6, -8, -10, -102, and -103 and comparison with BPL carbon. (Adapted with permission [24]. Copyright (2009) American Chemical Society.)

not show a better performance than this carbon-based material. Here it is possible to note that the pore size is critical for a material to get good adsorption performance. The gravimetric methane capture capacities of COF-102 and -103 at 35 bar and 298 K (187 and 175 mg/g respectively) were also superior to that of zeolites (31–81 mg/g) mesoporous silicas (14–65 mg/g), and even MOFs (160 mg/g), activated carbons, and also competitive to that of high surface carbons and activated anthracite, but lower than the highest performing MOF PCN-14 (253 mg/g). The corresponding methane uptake for COF-102 was calculated to be 136 cm^3 (STP)/cm^3 at 35 bar, which was close to the DOE target for that time. These experimental outcomes demonstrated an almost perfect correlation between the previous theoretical investigation and this experimental exercise.

Mendoza-Cortez et al. [25] further reported on theoretical modeling of Yaghi's COFs, including now COF-8, where GCMC simulations based on force fields (FF) were developed and validated by the experimental results obtained for COF-5 and COF-8 and the methane's equation of state. The ensemble average for the methane adsorption simulation obtained by GCMC of COF-10, -5, -8 and -6, obtained at various pressures, showed the changes in electron density related to the interactions of methane and COF, which reflected the occupancy of the adsorption sites. It was found out that the pores are not filled in case of COF-5, -8, and -10 despite the high pressure applied (100 bar) and that the isotherms show saturation, where the joint between edges gets more methane molecules than the center of the pore. The simulations show that COF-1, having just 7 Å of pore size, can hold up to three molecules per pore, reaching up to 195 v(STP)/v at 30 bar, but with an inefficient delivery amount (42 v(STP)/v) at 30 bar), which makes it not a good candidate for real-life applications. The GCMC calculations found that multilayer formations are in coexistence with the pore filling mechanism. Moreover, pore volumes of about 5 cm^3/g, pore diameters around 12Å, and surface areas >5000 m^2/g, which can generate high volumetric methane uptakes and low binding energies, do not necessarily mean inferior store capacities as originally conceptualized. A more in-depth investigation on the adsorption mechanism was carried out by Yang et al. [26] who observed, by GCMC simulations, that the adsorption of methane follows a stepped mechanism of multilayer formation, where, according to the pressures, in the early stage, the first monolayer of adsorbate is formed, followed by a second layer, from where the local density continues to increase, and the adsorbate starts to deposit on the pore center. Figure 22.3 shows the snapshots taken for

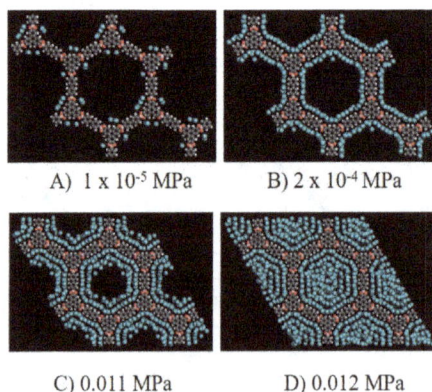

A) 1 x 10^{-5} MPa   B) 2 x 10^{-4} MPa

C) 0.011 MPa   D) 0.012 MPa

FIGURE 22.3

Snapshots of methane molecules adsorbed into COF-10 at different pressures (A) 1 × 10-5 MPa, (B) 2 × 10-5 MPa, (C) 0.011 MPa, and (D) 0.012 MPa. (Adapted with permission [26]. Copyright (2009) American Chemical Society.)

the methane adsorption into COF-10. The results indicate that in the case of methane, intermolecular interactions between adsorbates and between adsorbate-adsorbent, along with pore size and temperature, affect the stepped behavior of methane adsorption. Worth noting that the molecular geometry (nearly spherical) of methane compared to that of CO_2 allowed for a more regular packing of methane rather than that of CO_2.

More works on theoretical investigation and analysis continued to be done rather than experimental real work. In 2011 Mendoza-Cortez et al. [27] theoretically designed 15 COFs by modifying the functional groups in the aromatic rings in-unit linkers and the linking atoms. The condensation types were of boroxine ester-type with Si and/or C. From the results of simulated adsorption, it was shown that inserting alkyl chains into the aromatic rings will not benefit the enhancement of the adsorption, since the interactions between methane and the pore wall become hindered, e.g., when propyl and isopropyl groups are used, the propyl groups provide better uptake, since more molecules of methane are in contact with the pore wall; conversely, the steric hindrance of isopropyl groups prevents methane to accommodate onto the pore wall. Since for energy applications, delivery capacities are a fundamental property of each material, the referred work also included a determination of the delivery capacity of the designed COFs, where COF-103-Eth-trans yielded the best delivery capacity (192 v(STP)/v) at 35 bar. The amount of working methane can be enhanced, however, if the interaction between the gas and framework is reduced at low pressure. As a consequence, the adsorbate-adsorbent interactions should not be as high as not allowing the gas to be delivered for energy applications.

The research on the effect of inserting functional groups on the COF's structure was further investigated by Zhao and Yan [28], who theoretically designed COFs -102, -103, and -105 functionalized in the aromatic ring of the monomers with halogen (Cl, Br and I) as well as $-CF_3$, $-NH_2$, $-CN$, $-OCH_3$, and $-CH_3$ groups and investigated their effect on the methane adsorption behavior using GCMC modeling. To convert the total number of adsorbed molecules to an excess number, the following expression was used:

$$V_{exc} = V_{tot} - \rho Vp \qquad (22.1)$$

Where V_p is the free pore volume that can be used for adsorption and ρ is the bulk density of methane at the given conditions. From the analysis carried out for COF-102, it turned out that the halogen-substituted derivatives are more prone to adsorb a higher amount of methane and among them, COF-102-I was the best performing material (179 v(STP)/v). Iso-density analysis for COF-102-Cl demonstrated that the high repulsion between the Cl atoms produces an enlargement of the spacing between the aromatic moieties of the monomers, making the empty spaces that were not originally filled out with methane, to get the right dimension to host more methane molecules. All of the isotherms of the studied COFs were superior to the original COF-102, except COF-102-OCH_3, since the -OCH_3 group, although enlarges the surface area and provides much more adsorption sites, it is swamped by the volume effect at high pressure. Although methyl groups resemble the CH_4 molecule, its insertion in the COF's structure did not represent a substantial improvement of the storage capacity. Furthermore, all isosteric heat of adsorption was higher than original materials (COFs-102 and 103); however, among the studied materials, there were some discrepancies in the heats of adsorption, where a higher heat not necessarily meant a higher uptake capacity.

The application of new COFs for direct methane storage began with a report by Zhang et al. [29], who reported on the synthesis of the new COF-320 by condensation of tetra-(4-anilyl) methane and 4,4'-biphenyldialdehyde in a 1:1.8 molar ratio using 1,4-dioxane and 3N acetic acid as solvents using a Pyrex tube (120°C/3d). The identity of the material was mainly verified by FT-IR (imine stretching vibrations at 160 and 100 cm^{-1}). The 3D-structure of the new COF was found by single-crystal electron diffraction using the rotation electron diffraction method (RED). COF-320 has an extended framework where the tetrahedral organic building blocks are linked to biphenyl linkers by imine bonds, in a diamond net. The RED data set at 298K found a body-centered orthorhombic unit cell of red parameters $a = 27.93$Å, $b = 31.31$ Å, $c = 7.89$ Å, $V = 6899$ Å^3 in a space group *Imma (No. 74)* with a 9-fold interpenetration of the net and 1-D square-shaped channels running along the c-axis. Regarding textural properties, a pore volume of 0.81 cm^3/g was estimated, and a Langmuir surface area of 2400 m^2/g. Total methane uptake was calculated using Equation (22.1), and this was estimated to be 15 wt. % at 298K, which is lower than COF-102; however volumetric uptake of 176 cm^3/cm^3 can be competitive to that of COF-102 (293 cm^3/cm^3).

22.2.4 Studies on COFs for Methane Separation and Purification

After the work of Lan on Li-doped COFs, Yang and Cao [30] followed the theoretical investigation on the effect of Li-doping on the diffusion and selectivity of methane and hydrogen, since, for specific cases, such as methane reforming reactions, the separation of these gases is an essential process. In this case, the research group worked with COF-105 and -108, since this last material had the lowest density for a crystalline material reported so far. From the self-diffusion coefficients (D_s) of COF-102, -103 and -105, -108, undoped and Li-doped, it was possible to note that at doping, the D_s values increase in all cases, but then they gradually decrease, which is more evident in COF-102 and -103. This was ascribed to the increased interaction between Li and methane in COF-105 and -108, in part due to their large pore sizes leading to a much higher saturation pressure. The diffusion activation, determined by the Arrhenius expression, showed that the activation energy on COF-105, -108 is considerably higher than that of COF-102, -103, about six times higher. The increase on the activation energy accounts for the higher attraction of the Li atoms with the methane molecules, hindering the free flow of gas through the pores. At the same time, doping also can improve the selectivity of CH$_4$/H$_2$ by one order of magnitude, accounting for the much stronger affinity of Li toward methane than that for hydrogen. Note that the undoped COFs had extremely low selectivity.

Chemical modification of the aromatic rings on COF's structure was continued by Hu et al. [31], who modeled COF-102 modified with two halogen atoms inserted instead of hydrogen into the TBPM aromatic rings. The study aimed to analyze the effect of halogen substitution in the adsorption properties of the materials, taking into consideration the previously discussed works. The modeling results showed that the halogen-substituted COFs all yielded better uptake capacity than the original COF-102, and the order of best-performing halogens was -I> -Br> -Cl. The isosteric heats of adsorption, however, showed some inconsistency, since high Q_{st} should be the leading factor at low and medium pressures, and at high pressures, due to saturation, the volumetric factor should dominate the methane storage capacity and Q_{st} is no longer an influencing factor. The theoretical approximation carried out in that work, demonstrated that it would be possible to reach de DOE target (180 v(STP)/v) at 35 bar by the design of six double-I-substituted COF-102, since it was determined to have a delivery uptake of 181.1 ± .7 V(STP)/v at 298 K and 35 bar. Analysis of iso-density distribution showed that the six I atoms on the modified COF-102 are arranged as a triangle that surrounds the B$_3$O$_3$ ring, making one of the I atoms to be nearly close to one ring than the neighboring one, allowing a higher Van der Walls attraction to methane molecules, which results in a higher CH$_4$ density in the vicinity between both groups.

One interesting analysis of the separation selectivity between CO_2/CH_4 mixtures was made by Vicent-Luna et al. [32], who used COF-5, -6, -10, and -102 due to their intrinsic low densities (0.58, 1.08, 0.47, and 0.42 g/cm^3 respectively) and high thermal temperatures, to simulate their behavior on the separation selectivity of such gas mixtures. Their results, using GCMC simulations and Peng-Robinson equation for fugacity, showed excellent accordance of experimental and theoretical isotherms, which was a good starting point for the following analysis. Equimolar mixtures of CO_2/CH_4 were investigated as a function of pressure. From the selectivity results obtained for all studied COFs, a clear preference toward CO_2 was shown for all the materials with pressures up to saturation threshold, with even exclusion of methane in some cases. The highest selectivity values to CO_2 were obtained with COF-102. The largest methane uptake, as previously discussed, is shown by COF-102 (29 molecules/unit cell). Analysis of the selectivity on hydrated COFs (10, 20, and 30 wt.% of H_2O) showed that their selectivity toward CO_2 is even higher than that of dehydrated COFs; however, the uptake capacity gets lowered in both cases. Adding ionic liquids (ILs) to the dehydrated COFs (10 and 30 wt%) also increased the selectivity toward CO_2, but in this case, the uptake capacity was not greatly affected by the IL, e.g., at 10 kPa, the selectivity toward CO_2 for materials containing 10% IL is 8 times larger than that obtained with the dehydrated COF.

The excellent properties of COF-102 were further verified by Gopalsamy et al. [33], who calculated isotherms of CH_4/C_2H_6 mixtures using the Expanded Wang-Landau combined with GCMC simulations. The thermodynamic properties (entropy, enthalpy, and free Gibbs energy) were also determined using statistical mechanics. The high potential to separate CH_4/C_2H_6 mixtures using COF-102 at 300 K was confirmed, as compared with those of COF-105 and COF-108, e.g., at 10 bar and methane molar ratio of 0.4, COF-102 can reach a selectivity of 9.0. An ambitious work was developed by Martín at al. [34], who used a self-developed algorithm to design 620 model structures and relaxed them using semiempirical electronic structure calculations, validated by DFT. The interpenetration of some potential candidates yielded a final set of 4147 structures. The potential methane uptake and delivery were calculated by GCMC simulations and compared with those of COF-102 and -103. The methane deliverable capacity was defined as the difference of methane stored at a pressure of 65 bar (o 35) to that at 5.8 bar (or 1 bar), and it was proved that a delivery capacity up to 181.67 v(STP)/v for the range 65–5.8 bar could be reached. The distribution plots showed that there is a large density of COF structures approaching the empty tank performance and that the materials with the largest porous are close to the ideal behavior of the empty tank. Therefore, large pores, or low-density materials, improve the deliverable capacity, as a higher number of pores are available for vdW interactions to call up more guest molecules.

Investigation on the spatial orientation of the COFs' layers and their effect on their methane adsorption and CO_2/CH_4 separation was carried out by Sharma et al. [35], who introduced the modeling of mesoporous TpPal, TpBD, PI-COF-1, PI-COF-2 and PI-COF-3 along with functionalization of TpBD with NO_2 groups. Figure 22.4 shows the spatial orientation of the slipped and functionalized COFs along with their pore size. Their findings were that slipping improves the methane uptake capacity. On the other hand, the effect of chemical functionalization was not as strong as that of slipping. The beneficial effect of slipping increases with the slipping distance, then, higher uptakes were attained at 10–15 Å slipping distance. In the case of delivery uptake, 60-AB structures (60° slipping at maximum linker distance) of TpBD yielded up to 141 v(STP)/v at 65 bar. In general, 60-AB structures yielded the best results for methane delivery capacity. The total CO_2 binding energy was between 1.2 and 2 times higher than that of methane, which is very good for separation

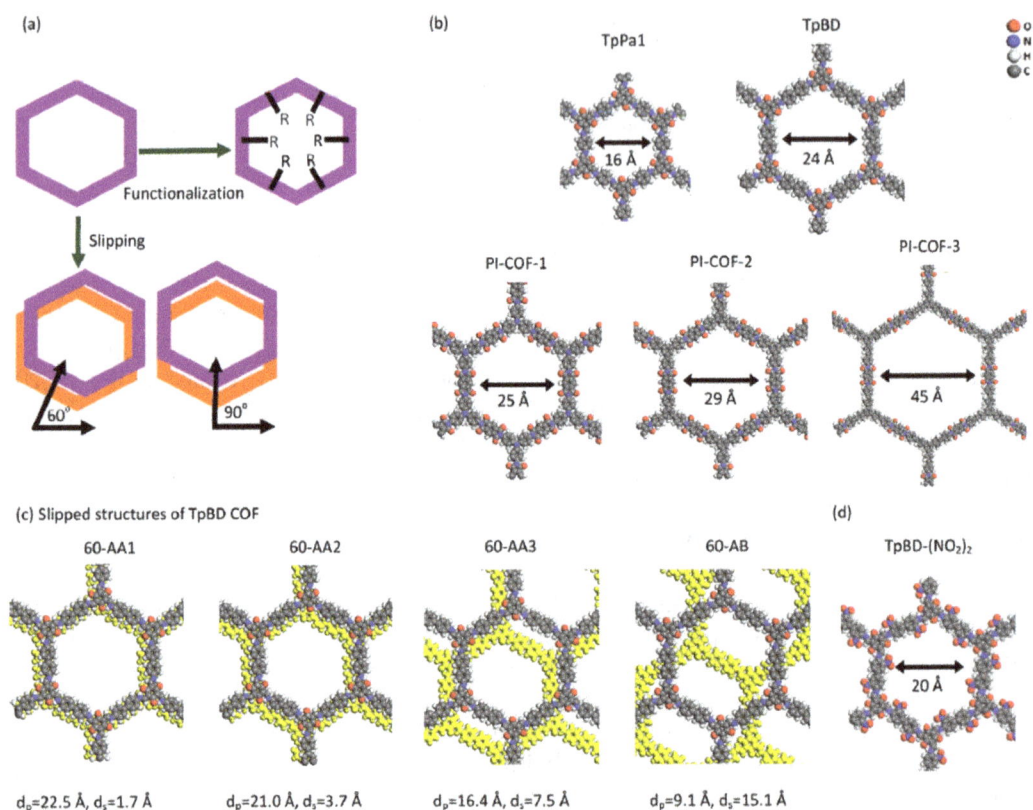

FIGURE 22.4
(a) Spatial orientation of slipped (60° and 90° direction) and functionalized COFs. (b) Eclipsed (AA) structures of TpPa1, TpBD, and PI-COFs with their pore size, (c) slipped structures of TpBD COF (60° direction, d_p and d_s are pore size and slipping distance, respectively), (d) structure of functionalized COF TpBD-$(NO_2)_2$. (Adapted with permission [35]. Copyright (2018) American Chemical Society.)

purposes. Overall, slipping enhances the CO_2/CH_4 selectivity, conversely to functionalization. Despite this good adsorption-desorption outcome, still, much experimental work is needed to realize such results to a real-life level.

Very recently Altundal et al. [36] used a combination of high-throughput computational screening (HTCS), CGMC, and MD simulations to establish a computation-ready experimental COF (CoRE COF) database where a total of 572 COFs were designed to investigate their membrane-based separation capacity of mixtures CH_4/H_2, CH_4/N_2, and C_2H_6/CH_4 mixtures to compare their performance to that of MOFs or other materials and assess the CH_4 separation capacities of the modeled COFs. The separation capacities were evaluated under vacuum swing adsorption (VSA) and pressure swing adsorption (PSA) conditions. Among their findings, it was found that COF-303 gave the best yields in PSA separation of CH_4/N_2 and CH_4/H_2 mixtures. In the case of C_2H_6/CH_4 mixtures, COF-102 gave the best performance. From the results of separation efficiency, due to the high working capacity of the COFs, separation based on the PSA modality gave the best results. Among the materials studied, it was found that COF-300, -303, and PZ-COF2 have outstanding properties for the separation of gas mixtures. The membrane separation efficiency was excellent for CH_4/H_2 separations, which was higher than that of CH_4/N_2 and C_2H_6/CH_4, but all of them outperformed polymers and zeolites in separation efficiency, resulting in promising materials for membrane-based separation applications. Recently, Kessler et al. [37] compared GCMC simulations with classical Density Functional Theory (cDFT) based on perturbed chain statistical associating fluid theory (PC-SAFT) for the adsorption of methane, ethane, butane, and methane/ethane and methane/butane mixtures in the COFs TpPa-1 and 2,3-DhaTph. The results showed that cDFT and GCMC simulations can be used indistinctly for simulating the isotherms, however, the larger the hydrocarbon, the more variation in the results. It was found, interestingly, that ethane showed a higher affinity to the COFs studied at low pressures, and n-butane showed a completely different adsorption behavior than ethane and methane, with an apparent high affinity at low pressures; however, these phenomena were not explained. On an experimental basis, CO_2/CH_4 separations were studied by Fan et al. [38], who synthesized 2D ACOF-1 supported on alumina arranged on a polyethyleneimine matrix. Gas separation was measured from equimolar CO_2/CH_4 methane mixtures. The materials showed high CO_2 selectivity, owing to the high polar groups present in the COF's structure. A separation factor of up to 86.4 was found, with a decrease in permeance for CO_2 from 9.9×10^{-9} to 8.8×10^{-9} mol·cm²/s·Pa, while for CH_4, it goes from 1.1×10^{-10} to 2.0×10^{-10} mol·cm²/s·Pa. This impressive selectivity was not affected by temperature changes from 25 to 120°C.

22.3 COFs for CO_2 Capture

The wide variety of COFs that can be obtained with different compositions, shape and cavity sizes, surface area, and adequate stability have motivated research for application in many areas of chemistry as will be seen throughout this book.

The first COFs synthesized and evaluated in CO_2 capture were boron-based COFs, which can be obtained through different strategies, generating good structural diversity. Despite the high crystallinity, large surface area, low densities, and good thermal stability of boron-based COFs, CO_2 capture capacity is relatively low at low pressures, for example, COF obtained by reaction between 1,3,5-benzenetriboronic acid (BTBA) and 2,3,6,7,10,11– Hexahydroxytriphenylene (HHTP) under solvothermal conditions, can adsorb up to 85 cm³ g⁻¹ at 273 K and 1 bar, which is low compared to other families of COFs based on nitrogen compounds as we will see below. On the other hand, in general, these compounds are very sensitive to humidity, which makes their practical applications very difficult. Different strategies have been described to make post-structural modifications, which could be an alternative to improve the stability of boron-based COFs that needs to be studied in greater depth. [39] One of the most attractive features of COFs is their great structural diversity, which can be obtained through the appropriate selection of the organic monomers used as linkers and nodes. A wide variety of linkages have been described such as imine (the most studied linkage to obtain COF for CO_2 capture application), imide, hydrazone, azine, triazine, boroxine, and others, in addition to post-synthesis conversion and functionalization that allows the structure of these compounds to be properly designed for a specific application [40].

In the particular case of CO_2 capture, research has shown that the performance of these compounds allows obtaining adsorption results competitive with other structured materials [41]. Proper cavity design has enabled the development of prototype COFs that can preferentially adsorb CO_2 when mixed with other gases. One of the COF design strategies

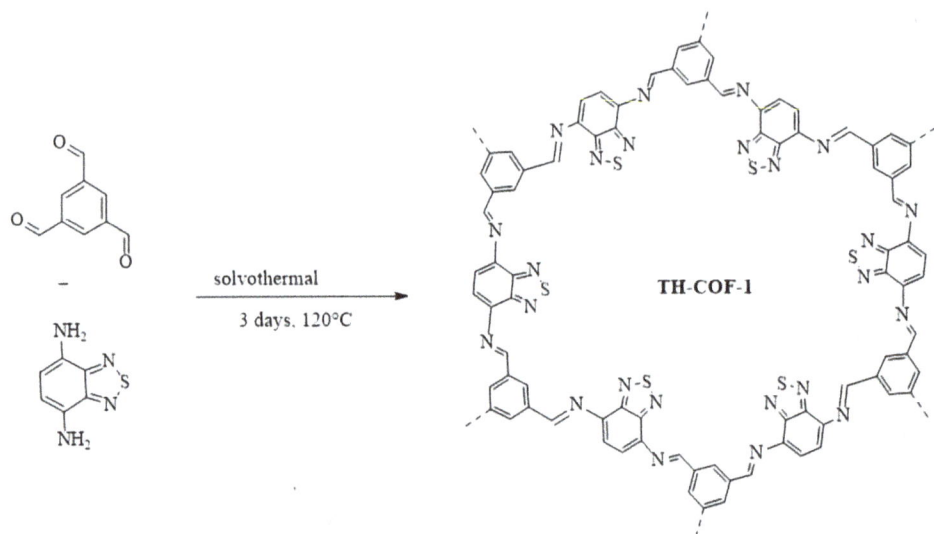

FIGURE 22.5
Bottom-up synthesis of 2D COF incorporating thiadiazole units to increase CO_2 capture.

for this application takes advantage of the property that CO_2 has no dipole moment but exhibits a high quadrupole moment. On the other hand, the introduction of functional groups within the COF's structure that can interact with the CO_2 molecule is an alternative that considerably increases adsorption, e.g., Wang et al. [42] who incorporated dipolar thiadiazol group into a 2D COF by bottom-up synthesis and obtained a material with a selectivity and uptake capacity higher to analogues materials without thiadiazol functionalization. Figure 22.5 shows the synthesis strategy followed

The incorporation of thiadiazole groups into the pore walls resulted in relatively small micropores where more nitrogen-accessible sites were favored, which in turn showed greater selectivity for CO_2 capture. It has also been shown that COFs synthesized from triphenylbenzene (COF-1), triphenyltriazine (COF-2) and triphenylamine (COF-3) have cavities of similar size (~1.8 nm) and groups with similar arrangement of similar basic nitrogen groups that interact relatively strongly with CO_2 favoring an efficient adsorption (Figure 22.6). In this sense, Zhai et al. [43] demonstrated the impact of the weakly interacting units on CO_2 adsorption.

They performed a comprehensive CO_2 adsorption study with COFs without triarylamine units on the backbone showing low CO_2 adsorption (12 and 20 mg/g at 298 and 273 K), suggesting that imine-linked COFs exhibit poor CO_2 sorption. In contrast, COFs with triarylamine units in the hexagonal pore showed a CO_2 adsorption capacity of 33 and 61 mg/g at 298 and 273 K with three triethylamine units in the hexagonal pore and 52 and 105 mg/g at 298 and 273 K with six triethylamine units in the hexagon. Their results revealed that the COF's backbone plays a key role in CO_2 adsorption. Another family of COFs, such as those holding azo groups (N=N), which are obtained by the condensation of aldehydes with diamines, showed that the acid-base N···CO_2 interaction with the nitrogen present in pore walls is crucial to high CO_2 uptake, e.g., Dhankhar & Nagaraja [44] synthesized an Azo COF through the solvothermal reaction between benzene-1,3,5-tricarboxaldehyde and 4,4′-azodianiline, obtaining an Aza 3D-COF (COF-1) of hexagonal structure within the pore channels, with an estimated pore size of 28.4 Å. The CO_2 adsorption isotherms of COF-1

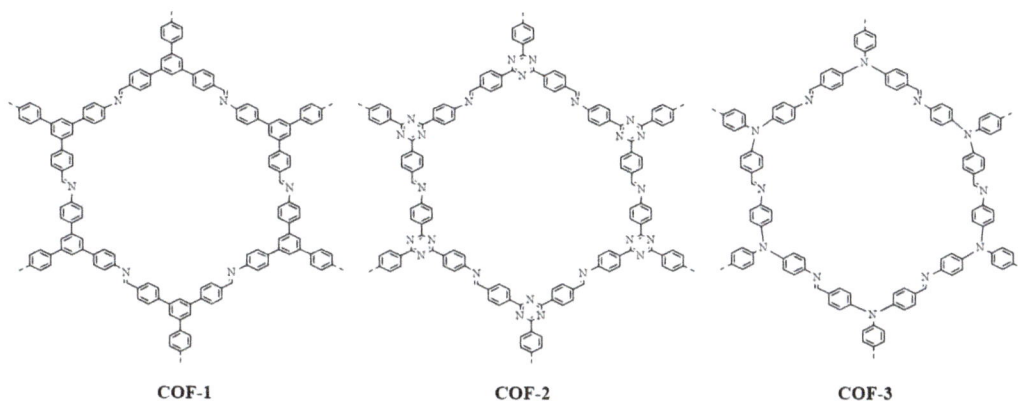

FIGURE 22.6
Channel structure of the COFs synthesized from triphenylbenzene (COF-1), triphenyltriazine (COF-2) and triphenylamine (COF-3).

at 195, 273, and 298 K follow a typical type I profile with volumetric adsorption of 68.5, 26.8, and 15.9 cm^3/g, respectively. In this sense, Ge et al. [45], in 2016, obtained Aza COF through the reaction between 4,4′-Azodianiline (Azo) and 1,3,5-triformyl-phloroglucinol. The stable CO_2-philic and "N_2-phobic" COF obtained showed high selectivity (145 at 298 K) to CO_2 when in mixtures with nitrogen. The CO_2 selectivity in this Aza COF has been explained by N_2-phobicity of azo group. In Figure 22.7 the effect of temperature (273 K and 298 K) from 0.1 to 1 bar of pressure on CO_2 capture for some recently studied COFs can be compared.

22.3.1 COFs Membranes for Selective CO_2 Separation

Recently a novel strategy for the separation of CO_2 from other gases has been based on the use of COF membranes. Due to the low solubility of COFs in volatile solvents, the preparation of COF membranes is not simple; however, COF membranes have adequate properties

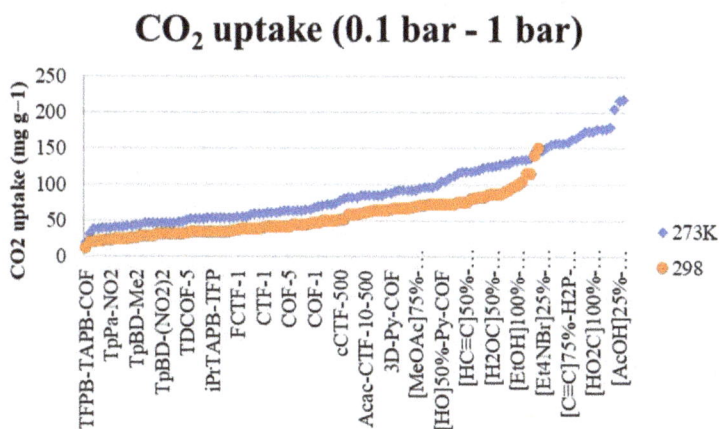

FIGURE 22.7
Effect of temperature on CO_2 capture for some recently studied COFs [41, 46].

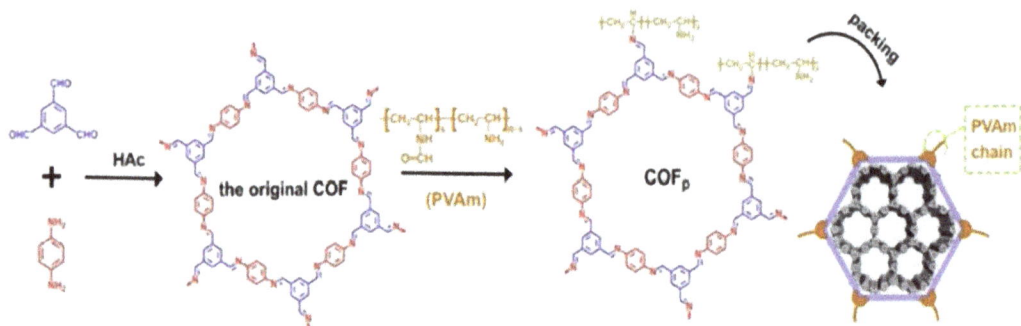

FIGURE 22.8
Synthetic process of mixed matrix membrane of COF with surface modified with Polyvinylamine. (Adapted with permission [48]. Copyright (2019) American Chemical Society.)

to achieve interfacial compatibility with improved polymeric matrices that avoid the formation of non-selective interfacial holes and with well-defined porous structures that show better performance compared to membranes obtained from other inorganic adsorbent materials and even better than that obtained with MOFs. Last year, Xiong et al. [47] reviewed research papers on the subject with special emphasis on COF-based membrane fabrication methods that include mixed matrix membranes, growth *in-situ*, and layer-by-layer stacking. The applications of COF-based membranes in the separation of CO_2 from other gases such as hydrogen, methane and nitrogen are also highlighted.

Membranes formed by continuous layers of COF still require many challenges to overcome, but they show great promise for practical application under adverse separation conditions of temperature, pressure, humidity, and high acidity. Figure 22.8 shows a strategy for the preparation of COF membranes through the modification of their surface with Polyvinyl-amine. The preparation method described by Cao et al. [48] shows an excellent strategy that simultaneously optimizes the polymeric matrix and the packing with which very high permeance and selectivity are achieved for other previously described prototypes. The PVAm-penetrated channels are formed with a pore wall abundant in amino groups and with a pore size suitable for the adsorption and preferential transport of CO_2 for other gases such as nitrogen and methane. Recently, Cao et al. prepared pressure-resistant mixed membranes for CO_2/H_2 separation by modifying the COF surface with segments of the polymer matrix which showed an excellent performance in the separation of these gases with a selectivity that reached up to 17.1 at 2.0 MPa, showing a big potential to be used for hydrogen purification in syngas process [49].

22.3.2 COFs for the Fixation of CO_2

A very attractive alternative to reduce atmospheric CO_2 pollution is through its transformation into other value-added products, however, given the high chemical stability of this compound, its reactivity is quite limited and in the same way, the possibility of having reactions that allow obtaining products from CO_2 with good yields. The most viable alternative in the case of the transformation of CO_2 is its transformation to carbonates through its reaction with epoxides. Carbonates are products of great added value due to their wide application in the polymer industry to obtain polycarbonates, among other uses [50–53].

As described so far, interesting advances have been obtained in the use of COF for CO_2 capture and fixation; however, there are still many research challenges to be solved, such as the diversification of synthesis strategies to obtain different three-dimensional structures with adequate chemical and thermal stability. Optimizing the synthetic strategies to obtain better performance and new COFs that allow obtaining good efficiency at high pressures, depending on the kinetic studies of the CO_2 capture processes from COFs and improving the strategies to obtain COF membranes with high performance in CO_2 capture, will allow COFs to be a competitive alternative to other strategies for CO_2 capture on a large scale soon.

22.4 Conclusions

Most of the research on solid materials for methane adsorption have been focused on activated carbons or MOFs; however, despite the relatively same age of MOFs and COFs, the last are just in their infancy regarding the investigation on their potential as methane adsorbents, although theoretical studies show a very high potential to almost the same level as that of MOFs. The urgency on real laboratory experiments is high and this document can serve as a good basis for those who are interested in probing other COFs as methane adsorbents. In contrast, a lot more experimental work has been carried out on the potential of COFs to adsorb CO_2, from where more available studies and literature is available. This document has a higher emphasis on methane as this gas has direct energy applications, however, CO_2 may also lead to an energy source if it is further reduced to compounds with high energy density.

References

1. Amstel, A. van; van Amstel, A. Methane. A Review. *Journal of Integrative Environmental Sciences*, 2012, *9*(Suppl. 1), 5–30.
2. Olajire, A. A. CO2 Capture and Separation Technologies for End-of-Pipe Applications – A *Review. Energy*, 2010, *35*(6), 2610–2628.
3. Guzmán, J.; Ortega-Guevara, C.; León, R. G. de; Martínez-Palou, R. Absorption of CO2 with Amino Acid-Based Ionic Liquids and Corresponding Amino Acid Precursors. *Chemical Engineering & Technology*, 2017, *40*(12), 2339–2345.
4. Martínez-Palou, R.; Likhanova, N. v.; Olivares-Xometl, O. Supported Ionic Liquid Membranes for Separations of Gases and Liquids: An Overview. *Petroleum Chemistry*, 2015, *54*(8), 595–607.
5. Choi, S.; Drese, J. H.; Jones, C. W. Adsorbent Materials for Carbon Dioxide Capture from Large Anthropogenic Point Sources. *ChemSusChem*, 2009, *2*(9), 796–854.
6. Wan, M. M.; Zhu, H. Y.; Li, Y. Y.; Ma, J.; Liu, S.; Zhu, J. H. Novel CO2-Capture Derived from the Basic Ionic Liquids Orientated on Mesoporous Materials. *ACS Applied Materials and Interfaces*, 2014, *6*(15), 12947–12955.
7. Sun, L.-B.; Kang, Y.-H.; Shi, Y.-Q.; Jiang, Y.; Liu, X.-Q. Highly Selective Capture of the Greenhouse Gas CO2 in Polymers. *ACS Sustainable Chemistry and Engineering*, 2015, *3*(12), 3077–3085.
8. Lin, Y.; Kong, C.; Zhang, Q.; Chen, L. Metal-Organic Frameworks for Carbon Dioxide Capture and Methane Storage. *Advanced Energy Materials*, 2017, *7*(4), 1601296.

9. Crabtree, R. H. Aspects of Methane Chemistry. *Chemical Reviews*, 2002, *95*(4), 987–1007.

10. Hammer, G.; Lübcke, T.; Kettner, R.; Pillarella, M. R.; Recknagel, H.; Commichau, A.; Neumann, H.-J.; Paczynska-Lahme, B. Natural Gas. *Ullmann's Encyclopedia of Industrial Chemistry*, 2006.

11. Casco, M. E.; Martínez-Escandell, M.; Gadea-Ramos, E.; Kaneko, K.; Silvestre-Albero, J.; Rodríguez-Reinoso, F. High-Pressure Methane Storage in Porous Materials: Are Carbon Materials in the Pole Position? *Chemistry of Materials*, 2015, *27* (3), 959–964.

12. Choi, P.-S. J.-M. Y.-K. M.-S. G.-J. S.-J. A Review: Methane Capture by Nanoporous Carbon Materials for Automobiles. *Carbon letters*, 2016, *17*(1), 18–28.

13. Makal, T. A.; Li, J.-R.; Lu, W.; Zhou, H.-C. Methane Storage in Advanced Porous Materials. *Chemical Society Reviews*, 2012, *41*(23), 7761–7779.

14. Alhasan, S.; Carriveau, R.; Ting, D. S.-K. A Review of Adsorbed Natural Gas Storage Technologies. *International Journal of Environmental Studies*, 2016, *73*(3), 343–356.

15. Lin, Y.; Kong, C.; Zhang, Q.; Chen, L. Metal-Organic Frameworks for Carbon Dioxide Capture and Methane Storage. *Advanced Energy Materials*, 2017, *7*(4), 1601296.

16. Li, B.; Wen, H. M.; Zhou, W.; Xu, J. Q.; Chen, B. Porous Metal-Organic Frameworks: Promising Materials for Methane Storage. *Chem*, 2016, *1* (4), 557–580.

17. He, Y.; Zhou, W.; Qian, G.; Chen, B. Methane Storage in Metal–Organic Frameworks. *Chemical Society Reviews*, 2014, *43*(16), 5657–5678.

18. Mahmoud, E. Recent Advances in the Design of Metal–Organic Frameworks for Methane Storage and Delivery. *Journal of Porous Materials 2020 28:1*, 2020, *28*(1), 213–230.

19. Geng, K.; He, T.; Liu, R.; Dalapati, S.; Tan, K. T.; Li, Z.; Tao, S.; Gong, Y.; Jiang, Q.; Jiang, D. Covalent Organic Frameworks: Design, Synthesis, and Functions. *Chemical Reviews*, 2020, *120* (16), 8814–8933.

20. Lohse, M. S.; Bein, T. Covalent Organic Frameworks: Structures, Synthesis, and Applications. *Advanced Functional Materials*, 2018, *28*(33), 1705553.

21. Adrien P. Côté; Hani M. El-Kaderi; Hiroyasu Furukawa; Joseph R. Hunt; Yaghi, O. M. Reticular Synthesis of Microporous and Mesoporous 2D Covalent Organic Frameworks. *Journal of the American Chemical Society*, 2007, *129*(43), 12914–12915.

22. Garberoglio, G.; Vallauri, R. Adsorption and Diffusion of Hydrogen and Methane in 2D Covalent Organic Frameworks. *Microporous and Mesoporous Materials*, 2008, *116*(1–3), 540–547.

23. Lan, J.; Cao, D.; Wang, W. High Uptakes of Methane in Li-Doped 3D Covalent Organic Frameworks. *Langmuir*, 2009, *26* (1), 220–226.

24. Furukawa, H.; Yaghi, O. M. Storage of Hydrogen, Methane, and Carbon Dioxide in Highly Porous Covalent Organic Frameworks for Clean Energy Applications. *Journal of the American Chemical Society*, 2009, *131*(25), 8875–8883.

25. Mendoza-Cortés, J. L.; Han, S. S.; Furukawa, H.; Yaghi, O. M.; William, A.; Goddard, I. Adsorption Mechanism and Uptake of Methane in Covalent Organic Frameworks: Theory and Experiment. *Journal of Physical Chemistry A*, 2010, *114*(40), 10824–10833.

26. Yang, Q.; Zhong, C. Molecular Simulation Study of the Stepped Behaviors of Gas Adsorption in Two-Dimensional Covalent Organic Frameworks. *Langmuir*, 2009, *25* (4), 2302–2308.

27. Mendoza-Cortes, J. L.; Pascal, T. A.; William, A.; Goddard, I. Design of Covalent Organic Frameworks for Methane Storage. *Journal of Physical Chemistry A*, 2011, *115*(47), 13852–13857.

28. Zhao, J.; Yan, T. Effects of Substituent Groups on Methane Adsorption in Covalent Organic Frameworks. *RSC Advances*, 2014, *4*(30), 15542–15551.

29. Zhang, Y.-B.; Su, J.; Furukawa, H.; Yun, Y.; Gándara, F.; Duong, A.; Zou, X.; Yaghi, O. M. Single-Crystal Structure of a Covalent Organic Framework. *Journal of the American Chemical Society*, 2013, *135*(44), 16336–16339.

30. Yang, Z.; Cao, D. Effect of Li Doping on Diffusion and Separation of Hydrogen and Methane in Covalent Organic Frameworks. *Journal of Physical Chemistry C*, 2012, *116*(23), 12591–12598.

31. Hu, J.; Zhao, J.; Yan, T. Methane Uptakes in Covalent Organic Frameworks with Double Halogen Substitution. *Journal of Physical Chemistry C*, 2015, *119*(4), 2010–2014.

32. Vicent-Luna, J. M.; Luna-Triguero, A.; Calero, S. Storage and Separation of Carbon Dioxide and Methane in Hydrated Covalent Organic Frameworks. *Journal of Physical Chemistry C*, 2016, *120*(41), 23756–23762.

33. Gopalsamy, K.; Desgranges, C.; Delhommelle, J. Selectivity and Desorption Free Energies for Methane–Ethane Mixtures in Covalent Organic Frameworks. *Journal of Physical Chemistry C*, 2017, *121*(44), 24692–24700.

34. Martin, R. L.; Simon, C. M.; Medasani, B.; Britt, D. K.; Smit, B.; Haranczyk, M. In Silico Design of Three-Dimensional Porous Covalent Organic Frameworks via Known Synthesis Routes and Commercially Available Species. *Journal of Physical Chemistry C*, 2014, *118*(41), 23790–23802.

35. Sharma, A.; Babarao, R.; Medhekar, N. v.; Malani, A. Methane Adsorption and Separation in Slipped and Functionalized Covalent Organic Frameworks. *Industrial & Engineering Chemistry Research*, 2018, *57*(14), 4767–4778.

36. Altundal, O. F.; Haslak, Z. P.; Keskin, S. Combined GCMC, MD, and DFT Approach for Unlocking the Performances of COFs for Methane Purification. *Industrial & Engineering Chemistry Research*, 2021, *60*(35), 12999–13012.

37. Kessler, C.; Eller, J.; Gross, J.; Hansen, N. Adsorption of Light Gases in Covalent Organic Frameworks: Comparison of Classical Density Functional Theory and Grand Canonical Monte Carlo Simulations. *Microporous and Mesoporous Materials*, 2021, *324*, 111263.

38. Fan, H.; Mundstock, A.; Gu, J.; Meng, H.; Caro, J. An Azine-Linked Covalent Organic Framework ACOF-1 Membrane for Highly Selective CO2/CH4 Separation. *Journal of Materials Chemistry A*, 2018, *6*(35), 16849–16853.

39. Zeng, Y.; Zou, R.; Zhao, Y. Covalent Organic Frameworks for CO2 Capture. *Advanced Materials*, 2016, *28*(15), 2855–2873.

40. Zhuang, Z.; Shi, H.; Kang, J.; Liu, D. An Overview on Covalent Organic Frameworks: Synthetic Reactions and Miscellaneous Applications. *Materials Today Chemistry*, 2021, *22*, 100573.

41. Geng, K.; He, T.; Liu, R.; Dalapati, S.; Tan, K. T.; Li, Z.; Tao, S.; Gong, Y.; Jiang, Q.; Jiang, D. Covalent Organic Frameworks: Design, Synthesis, and Functions. *Chemical Reviews*, 2020, *120*(16), 8814–8933.

42. Wang, L.; Dong, B.; Ge, R.; Jiang, F.; Xiong, J.; Gao, Y.; Xu, J. A Thiadiazole-Functionalized Covalent Organic Framework for Efficient CO2 Capture and Separation. *Microporous and Mesoporous Materials*, 2016, *224*, 95–99.

43. Zhai, L.; Huang, N.; Xu, H.; Chen, Q.; Jiang, D. A Backbone Design Principle for Covalent Organic Frameworks: The Impact of Weakly Interacting Units on CO2 Adsorption. *Chemical Communications*, 2017, *53*(30), 4242–4245.

44. Dhankhar, S. S.; Nagaraja, C. M. Porous Nitrogen-Rich Covalent Organic Framework for Capture and Conversion of CO2 at Atmospheric Pressure Conditions. *Microporous and Mesoporous Materials*, 2020, *308*, 110314.

45. Ge, R.; Hao, D.; Shi, Q.; Dong, B.; Leng, W.; Wang, C.; Gao, Y. Target Synthesis of an Azo (N=N) Based Covalent Organic Framework with High CO2-over-N2 Selectivity and Benign Gas Storage Capability. *Journal of Chemical and Engineering Data*, 2016, *61*(5), 1904–1909.

46. Ozdemir, J.; Mosleh, I.; Abolhassani, M.; Greenlee, L. F.; Beitle, R. R. Jr.; Beyzavi, M. H. Covalent Organic Frameworks for the Capture, Fixation, or Reduction of CO2. *Frontiers in Energy Research*, 2019, *7*(77), 1–32.

47. Xiong, S.; Li, L.; Dong, L.; Tang, J.; Yu, G.; Pan, C. Covalent-Organic Frameworks (COFs)-Based Membranes for CO2 Separation. *Journal of CO2 Utilization*, 2020, *41*, 101224.

48. Cao, X.; Wang, Z.; Qiao, Z.; Zhao, S.; Wang, J. Penetrated COF Channels: Amino Environment and Suitable Size for CO2 Preferential Adsorption and Transport in Mixed Matrix Membranes. *ACS Applied Materials & Interfaces*, 2019, *11*(5), 5306–5315.

49. Cao, X.; Xu, H.; Dong, S.; Xu, J.; Qiao, Z.; Zhao, S.; Wang, J.; Wang, Z. Preparation of High-Performance and Pressure-Resistant Mixed Matrix Membranes for CO2/H2 Separation by Modifying COF Surfaces with the Groups or Segments of the Polymer Matrix. *Journal of Membrane Science*, 2020, *601*, 117882.

50. Wang, J.; Zhang, Y. Facile Synthesis of N-Rich Porous Azo-Linked Frameworks for Selective CO2 Capture and Conversion. *Green Chemistry*, 2016, *18*(19), 5248–5253.
51. Zhi, Y.; Shao, P.; Feng, X.; Xia, H.; Zhang, Y.; Shi, Z.; Mu, Y.; Liu, X. Covalent Organic Frameworks: Efficient, Metal-Free, Heterogeneous Organocatalysts for Chemical Fixation of CO2 under Mild Conditions. *Journal of Materials Chemistry A*, 2018, *6*(2), 374–382.
52. Yu, W.; Gu, S.; Fu, Y.; Xiong, S.; Pan, C.; Liu, Y.; Yu, G. Carbazole-Decorated Covalent Triazine Frameworks: Novel Nonmetal Catalysts for Carbon Dioxide Fixation and Oxygen Reduction Reaction. *Journal of Catalysis*, 2018, *362*, 1–9.
53. Xu, K.; Dai, Y.; Ye, B.; Wang, H. Two Dimensional Covalent Organic Framework Materials for Chemical Fixation of Carbon Dioxide: Excellent Repeatability and High Selectivity. *Dalton Transactions*, 2017, *46*(33), 10780–10785.

23

Covalent Organic Frameworks-Based Membranes and Adsorbents for Water Treatment and Gas Separation

Sana Eid[1,2], **Ahmed Gulzar**[1,3], **Ali Al Najjar**[1,2], **Gabriele Scandura**[1,3], **Twinkle Paul**[1,3],
Dinesh Shetty[4,5], **Hassan Arafat**[1,2], **Georgios N. Karanikolos**[1,2,3,4],
and Ludovic F. Dumée[1,2,3]

[1]*Department of Chemical Engineering, Khalifa University, Abu Dhabi, United Arab Emirates*

[2]*Center for Membranes and Advanced Water Technology (CMAT), Khalifa University,
Abu Dhabi, United Arab Emirates*

[3]*Research and Innovation Center on CO₂ and H₂ (RICH), Khalifa University, Abu Dhabi,
United Arab Emirates*

[4]*Center for Catalysis and Separation (CeCaS), Khalifa University, Abu Dhabi, United Arab Emirates*

[5]*Department of Chemistry, Khalifa University, Abu Dhabi, United Arab Emirates*

CONTENTS

23.1 Introduction

Covalent organic frameworks (COFs) are an emerging class of organic crystalline porous materials that use reticular chemistry to form two- or three-dimensional structures [1, 2]. The building blocks of COFs are light elements, such as C, H, O, N, or B atoms, which are connected through strong covalent bonds [3]. The unique properties of COFs such as high surface area and micropore size distribution have made them promising candidates for separation applications, such as selective extraction of ions or organic molecules by adsorption or membrane processes [4]. The morphological and chemical properties required for water treatment applications include excellent water stability and control over hydrophobic/hydrophilic domains, support charge and size-based exclusion, and capture separation [5]. Furthermore, recent advances in COF synthesis have shown that they could compete with other porous materials like zeolites, activated carbon, and metal-organic

DOI: 10.1201/9781003206507-23

frameworks (MOFs) especially because of their predictable design and structural diversity [6]. However, the use of COFs as separation media is still relatively understudied compared to the use of zeolites, activated carbon, or MOFs, which could be due to their comparatively low crystallinity and the fact that they are newer than those materials [5]. Despite the many challenges to the industrial application of COFs in water treatment, it is worthy to note that they have made significant progress in recent years heralding a promising future for these materials [7].

In this chapter, some recent advances of COFs in water and gas separation treatment processes are discussed. The COF-based separation techniques were classified into two categories including (i) membrane-based separation and (ii) adsorption (packed bed) [5], both of which are discussed critically.

23.2 COF-Based Membranes

COF-based membranes can separate mixtures based on their difference in diffusion rates or molecular sieving effects between the mixture and the COF material [5]. This section discusses the potential application of COF membranes specifically in water desalination and gas separation.

23.2.1 Desalination

Imine-, azine-, and hydrazone- linked COFs are known to have superior structural, thermal, and chemical stability compared to boron or boronate ester-linked COFs making them promising candidates for the fabrication of nanofiltration (NF) membranes in the field of wastewater treatment and valuable resource recovery [8, 9]. A challenge faced when developing a COF-based NF membrane for seawater desalination is that the pore aperture of many COF membranes is often significantly larger than the hydrodynamic diameter of the hydrated ions, allowing for unhindered diffusion of salts across the active layer of the membrane and resulting in low rejections [9]. For example, a COF membrane's pore size could range between 1 and 2 nm [10], whereas the diameter of hydrated ions like Na^+, Mg^{2+}, Fe^{3+}, Cl^-, and SO_4^{2-} are 0.716, 0.856, 0.914, 0.664, and 1 nm respectively [9].

The pore structure of COF materials is known to be highly tunable by both presynthetic and post-synthetic modifications [11]. 2D hydroxyl-functionalized COF membranes, namely IISERP-COF1, were synthesized from a condensation reaction between 1,3,5-tris(4-formylphenyl) benzene (TFP) and 3,5-dia-mino-1,2,4-triazole; and post-synthetic modified into 2D carboxyl-functionalized COF membranes, namely IISERP-COOH-COF1 via ring-opening reactions [9]. The structures of these two COFs are shown in Figure 23.1a.

IISERP-COOH-COF1 membranes offered two advantages over the IISERP-COF1 ones, including (i) a narrower pore aperture thanks to the steric hindrance generated by the added functional groups preventing larger molecules or ions from permeating, and (ii) enhanced-selective transport, thanks to a reduction in the microscale inter-crystalline defects between the COF crystals [9]. Therefore, the ion rejection of the IISERP-COF1 membrane was significantly improved post functionalization [9]. Furthermore, since the carboxyl groups across the IISERP-COOH-COF1 may be deprotonated in aqueous solutions, the IISERP-COOH- COF1 membranes were natively negatively charged at neutral pH,

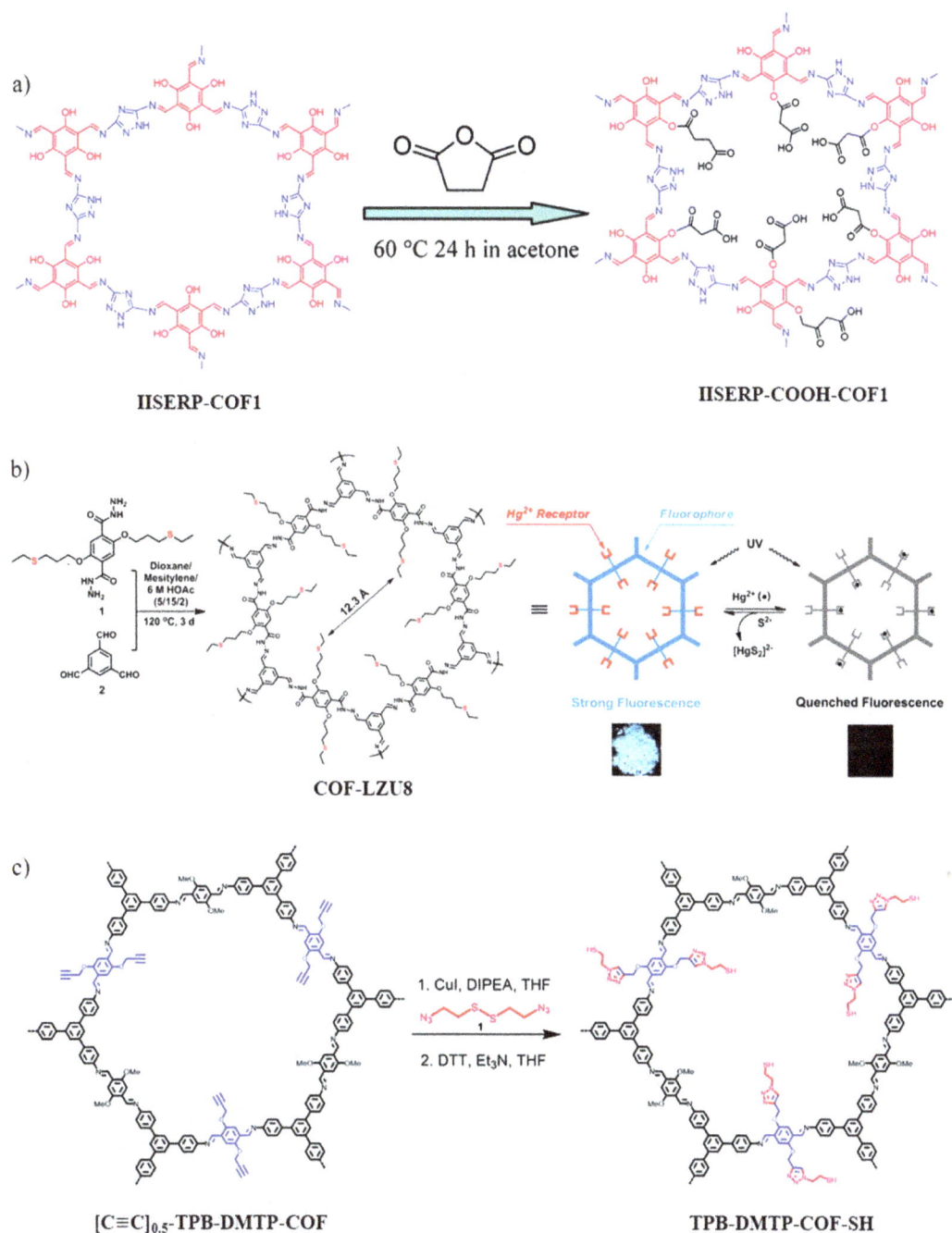

FIGURE 23.1
(a) IISERP-COOH-COF1 structure. (Adapted with permission [1], Copyright (2019), Royal Society of Chemistry);
(b) COF-LZU8 structure and its fluorescence property. (Adapted with permission [2], Copyright (2016),
American Chemical Society); (c) TPB-DMTP-COF-SH structure. (Adapted with permission [3], copyright (2017),
Royal Society of Chemistry.)

aiding toward anion rejection [9]. The BET surface area and pore size of the COF membrane were reduced from about 224 m^2 g^{-1} and 12.7 Å to 190 m^2 g^{-1} and 6.5 Å, respectively after post-modification [9]. Reducing the pore size helped increase the salt rejection (R), whereas the drop in BET surface area resulted in a relatively less potential interaction between the membrane and the ions [9]. Using a secondary growth approach, IISERP-COF1 membranes were grown on 3-amino-propyltriethoxysilane (APTES) modified α-Al$_2$O$_3$ tubes [9]. Crossflow filtration experiments using 2,000 ppm Na$_2$SO$_4$, MgSO$_4$, FeCl$_3$, MgCl$_2$, and NaCl aqueous solutions were conducted at 25°C and a 2 bar trans-membrane pressure to evaluate the nano-filtration performance of the IISERP-COF1 and IISERP-COOH-COF1 membranes. Salts' rejection for Na$_2$SO$_4$, MgSO$_4$, FeCl$_3$, MgCl$_2$, and NaCl in the IISERP-COF1 membrane were 73.2, 76.5, 89.8, 64.6, and 56.4% respectively compared to 96.3, 97.2, 99.6, 90.6, and 82.9% in the IISERP-COOH-COF1 membrane marking a significant increase in salt rejection for all aqueous solutions [9]. The sequence of salt rejection in the membrane: FeCl$_3$ > MgSO$_4$ > Na$_2$SO$_4$ > MgCl$_2$ > NaCl was explained through the salts' different diffusion coefficients [9]. Smaller molecules travel faster than larger ones enabling the smallest molecules to pass quickly through the membrane [9]. Since the pore size distributions of the COF membranes was reduced from 12.7 to 6.5 Å after post-modification [9], even salts with monovalent ions such as NaCl were successfully rejected [9] at higher levels than some commercial NF membranes such as NF270 and NFX (~ 40 - 50% NaCl rejection) [12]. Water pressure normalized flux, on the other hand, only dropped slightly following post-modification, from 0.63 to 0.55 L m^{-2} h^{-1} bar^{-1} [9]. The water flux for both IISERP-COF1 and IISERP-COOH-COF1 membranes was much lower than that for commercialized NF membranes [9, 13]. The operating pressure for commercial NF membranes typically ranges from 5 to 20 bar, and the permeate flux of the composite membranes may reach a permeate flux of 45.9 L m^{-2} h^{-1} at a transmembrane pressure of 9 bar [13].

IISERP-COOH-COF1 is considered an attractive candidate for seawater desalination since there is relatively no difference between the percentages for monovalent and bivalent salt rejection, as mentioned above. In addition, both salt rejection and water flux in the IISERP-COOH-COF1 membrane remained constant for 168 h at 2 bar presenting great stability for nanofiltration separation [9]. IISERP-COOH-COF1 also showed great chemical stability in strongly acidic and basic conditions. Indeed after-treatment of the membrane for 168 h in 1.0 M HCl and 1.0 M NaOH solutions, there was no noticeable change in neither water flux nor salt rejection [9]. However, the challenge of having high water permeance alongside high ion rejection needs to be thoroughly addressed before commercializing this type of membrane for seawater desalination.

COFs' potential application in water desalination was also investigated through computational design and molecular dynamics simulations. TpPa-1, one of the most investigated COFs, was chosen for a study as the parental structure of seven 2D functionalized COF membranes due to its high water stability and tunability [14]. TpPa-1 framework, which consists of the two monomers: 1,3,5-triformylphloroglucinol (Tp) and p-phenylenediamine (Pa), was grafted with the following functional groups: -AM$_2$ (-NHCOCH$_2$CH$_3$), -AMC$_2$NH$_2$ (-NHCOCH$_2$CH$_2$NH$_2$), -OC$_3$OH (-OCH$_2$CH$_2$CH$_2$OH), -OC$_4$H$_9$ (-OCH$_2$CH$_2$CH$_2$CH$_3$), -AMCOOH (-NHCOCH$_2$-COOH), -OBn (-OCH$_2$C$_6$H$_5$), and -AM$_3$ (-NHCOCH$_2$CH$_2$CH$_3$), to produce seven TpPa-X membranes [14]. Not only is the functionality of these membranes different, but so is their aperture size [14]. The aperture diameter (d_a) of the TpPa-1 membrane is 15.8 Å [14], which is considered quite large and would result in poor salt rejection if this membrane was used for water desalination without adding any functional group to its framework. Thus, grafting a functional group into the TpPa-1 framework resulted in a smaller d_a and an enhanced salt rejection [14]. The d_a followed the order of TpPa-AM$_2$ >

TpPa- OC_3OH > TpPa- OC_4H_9 > TpPa- AMC_2NH_2 > TpPa- OBn > TpPa- AMCOOH > TpPa-AM_3 ranging from 7.64 to 5.17 Å [14]. Some of the membranes were hydrophobic like TpPa-AM_2, -AM_3, -OC_4H_9 and -OBn with aliphatic chains and phenyl rings, respectively, whereas others were hydrophilic like TpPa-AMCOOH, -AMC_2NH_2 and -OC_3OH with car-boxyl, amino, and hydroxyl groups, respectively [14].

Molecular simulations demonstrated that water desalination was governed by the mem-brane's functionality and aperture diameter [14]. Water flux and salt rejection were the two main metrics studied and compared with other types of membranes in literature [14]. In particular, the reverse osmosis process was simulated in a system consisting of a feed chamber with NaCl solution comparable to seawater and a permeate chamber, separated by a single layer, 4 Å thick, TpPa-X membrane [14]. Two graphene plates were also placed in the chambers, which self-adjusted their positions during the simulation according to the hydraulic pressures of the chambers. The process was simulated at T = 300 K, under a pressure gradient of 600 bar, while periodic boundary conditions were implemented in the x and y directions as the TpPa-X membrane was considered infinitely large [14]. The net water flow (N_w) versus time (t) curve initially showed a straight line for all the membranes indicating a steady flow of water [14]. All the N_w vs. t curves, except for TpPa-AM_3, eventu-ally reached a plateau due to the depletion of water in the feed chamber [14]. The larger the water flow, the faster the depletion in the feed chamber [14], and since the TpPa-AM_3 membrane had the slowest flow, the plateau was not reached within the first 20 ns [14]. Hydrophilic membranes like TpPa-AMCOOH, -AMC_2NH_2 and -OC_3OH exhibited higher water flux (J_w) than hydrophobic membranes like TpPa-AM_2, -AM_3, -OC_4H_9 and -OBn, especially at small aperture diameters [14].

Water flow over nanoporous graphene membranes showed a similar pattern, in which the J_w was almost double for hydrophilic apertures [15]. The aperture size and wettability of TpPa-X and graphene membranes play a significant role in controlling the water flux. The reason for that is that both TpPa-X and graphene membranes are considered to be two-dimensional single or few layers materials only, where the water density at hydrophilic apertures is higher than that at hydrophobic ones resulting in higher water flux for hydro-philic apertures [14]. ZIF membranes with hydrophobic channels, on the other hand, were found to have a larger water flux than those with hydrophilic ones [16]. That's because water molecules in ZIF membranes experience weak interactions with the hydrophobic channels and hence permeate quickly [17]. Therefore, water flow in a ZIF membrane is controlled by the channel rather than the aperture.

The second important performance metric studied was salt rejection, which was not strongly correlated to the aperture diameter since all the TpPa-X membranes showed R above 95% [14]. In principle, the two membranes with the two smallest aperture diameters should have had the highest R [14]. This was true for TpPa-AM_3 (d_a = 5.17 Å)with an R of 100%, whereas TpPa-AMCOOH (d_a = 5.32 Å) had the lowest R (95.8%) among the seven sim-ulated membranes [14]. The complete salt rejection for TpPa-AM_3 can be attributed to both its small aperture diameter and hydrophobic nature [14]. Even though TpPa-AMCOOH has a close d_a to that of TpPa-AM_3, its -COOH functional groups can form strong interac-tions with the ions, allowing them to pass across the membrane and resulting in a lower R [14]. Therefore, salt rejection is governed by both aperture diameter and wettability. All the seven simulated membranes had relatively high water permeance, ranging from 1216 to 3375 kg m^{-2} h^{-1} bar^{-1} [14], about three orders of magnitude higher compared to com-mercial RO membranes [18], and one or two orders of magnitude higher compared to ZIFs [16]. Overall, this study demonstrated a good potential for TpPa-X membranes in water desalination.

23.2.2 Gas Separation

Carbon capture and energy generation have recently received increasing attention due to their important uses in clean energy and environmental sustainability. Natural gas contains significant concentrations of acidic gases such as CO_2 and H_2S, which shall be removed to boost the heating value and prevent pipeline corrosion [19]. The development of highly selective hydrogen membranes is also sought to support the development and scale-up of global hydrogen roadmaps. Membrane-based technologies may provide advantages in gas separation over other techniques, such as solvent absorption or physical adsorbents, in terms of limited footprint and easy operation, allowing for compact operation and suitability for offshore applications, whereby high efficiency is required. In this section, recent and significant examples of COFs based membranes developed for carbon dioxide and hydrogen separation are discussed.

The structural regularity of imine-, azine-, and hydrazone-linked COFs' makes them promising candidates for CO_2 separation [20] both as neat as well as composite (hybrid) materials. ACOF-1 was synthesized by condensation reaction between, 3,5-triformylbenzene (TFB) and hydrazine hydrate, leading to layered nano-sheets with a well-defined microporous structure [21]. ACOF-1 offers a mean pore size of 0.94 nm [22], which makes it an excellent candidate for gas separation of mixtures containing CO_2, N_2, and/or CH_4. Mixed Matrix Membranes (MMMs) were generated with Matrimid® 5218 as the polymeric matrix using ACOF1 nano-sheets as filler materials [21]. The choices of polymer and COF were based on the high adsorption capacity of ACOF-1 for CO_2 adsorption and good synergy between the polymer and COF. The contents of the filler greatly affected the gas separation performance and increasing the loading of ACOF-1 from 8 to 16 wt % slightly improved the CO_2/CH_4 selectivity, from 31.9 to 32.4, while the gas permeability drastically increased, from 15.3 to 493 Barrer, at 308 K and for an equimolar feed pressure of 4 bar. Such a significant increase can only be due to defect engineering at high nano-sheet loading, leading to enhanced diffusion pathways within the Matrimid polymer. In an experiment designed to evaluate the performance of ACOF1 alone, membranes were synthesized on α-Al_2O_3 substrates [23]. The α-Al_2O_3 substrates were functionalized with the ACOF-1 precursor and later reacted with hydrazine hydrate to form an ACOF-1 layer onto the surface. This layer performed considerably well compared to the ACOF-1 and Matrimid® 5218 MMMs as, at 393K and 1 bar, the CO_2/CH_4 selectivity reached 86.4 with CO_2 permeance of 237.1 Barrer. The feed pressure was critical since, for equimolar mixtures, the CO_2 permeance decreased from 237.1 to 216.4 Barrer between 1 and 2 bar, while the CH_4 permeance increased by ~30% in that same range. This effect was attributed to the blocking of pores by CO_2 for the diffusion of CH_4 at low feed pressure. The dependency of selectivity on the feed temperature was also found to be critical since a decrease of the CO_2/CH_4 selectivity, from 97.1 to 86.4, occurred upon increasing the temperature from 298 to 393 K. The use of ACOF-1 for membrane manufacturing whereby a bilayer membrane was fabricated with COF-LZU1, another imine-linked COF, was also demonstrated onto an amino-functionalized Al_2O_3 substrate. The membranes exhibited small pore size distributions centered at around 0.3–0.5 nm, making them potential candidates for H_2 separation. At 298 K and 1 bar, the H_2 permeance was found to be 732 Barrer, with H_2/CO_2, H_2/CH_4 and H_2/N_2 selectivities at 24.2, 100.2, and 83.9, respectively. These selectivities were considerably higher than those reported for bare ACOF-1 materials (14.1, 24.7, and 21.6) or bare COF-LZU-1 (6, 9.7, and 8.1), which was attributed to the synergistic hybridization of the COF-materials leading to narrower neck formation within the COF matrix.

TpPa-1 and TpBD were also reported for gas separation since the presence of a high density of hydrogen bonds within the structure may be exploited to generate highly flexible and chemically stable structures. PBI-Bul, a polymer exhibiting strong H-bond densities, was used to form self-supported, hybrid membranes with different loadings of TpPa-1 and TpBD [24]. The membranes exhibited high thermal stability of up to 673 K and high mixing of COF particles into the polymer could be achieved for up to 50 wt % loading. The H_2 permeability increased more than three times from 6.2 Barrer for PBI-BUL to 18.8 Barrer, for the TpPa-1@PBI-Bul with 40 wt % filler at 308 K and 20 atm pressure. The selectivity showed no significant changes or trends upon different loadings of the COFs. The H_2/CH_4 was found to be the highest (165.5) for the 40 wt % TpPa-1 membranes while the CO_2/N_2 and CO_2/CH_4 selectivities were low for all the composites formed when compared to bare PBI-Bul, even though the CO_2 permeability increased with the COF loading. This aspect is interesting since most MMMs would be expected to lead to lower selectivity at higher loading due to uncontrolled defects formation [25]. The origin of the results reported in this study may be related to the efficient binding of the COF particles with the polymer matrix.

COF-5 was used for CO_2 separation and, unlike TpPa-1 where the pore structure is partly amorphous, offered a rigid structure and porous network. A mixed matrix membrane was designed by distributing COF-5 in Pebax-1657 [26], a CO_2-phillic and industrially used polymer to capture CO_2 [27, 28]. The feed pressure did not significantly affect the permeability or selectivity of the materials, and within the range of COF loading tested (0.1, 0.4, 0.7, 1, 2, and 3 wt %) an optimal value at around 0.4 was found. For this loading, the CO_2 permeability reached 493 Barrer while the CO_2/N_2 selectivity, at 303 K and 1 bar, was 49.3. The fluctuation in the selectivity and permeability due to the loading may be due to the rigid structure of the COF, namely higher amount of fillers particles made the gas transport more tortuous, and more exposed to a greater density of defects localization at the COF-Pebax interfaces.

COF-300 3D crystals, as opposed to COF 2D nanosheets, were used as filler in two different polymers, namely 6FDA-DAM and Pebax [29]. 6FDA-DAM offers a high permeability to CO_2 due to the bulky pendant -CF_3 groups, preventing efficient packing, while the polyether segments within Pebax allow for the formation of near defect-free MMMs since they can facilitate wrapping around the fillers and offer good performance even in the presence of water [30]. One major advantage of COF-300 is that it is easily chemically tunable, allowing the formation of hybrid filler with poly(ether imine) (PEI), for maximum polymer affinity and CO_2-philicity. A direct relation between the CO_2 permeability and COF or COF-PEI filler loading was established for up to 10 wt % of loading, beyond which point agglomeration of COFs particles started to become detrimental. As expected, when introduced to a relative humidity of ~85%, the 6FDA-DAM showed reduced performance of CO_2 permeability falling by 12%, while for Pebax, the CO_2 permeability increased by around 380%. For the 6FDA-DAM and PEI systems, the hybrid filler of COF-300@PEI allowed a better synergy between the filler and polymeric parts since higher loading of fillers could be achieved without the generation of agglomerates. An optimum hybrid filler content allowed a CO_2/CH_4 and CO_2/N_2 selectivity of 48.3 and 84.2, respectively, with CO_2 permeability of 126 Barrer at 298 K and 1 bar. Although when compared to COF-300 bare membranes, exhibiting selectivities of 30.3 and 56.6 tested for the same gas mixtures, the MMM selectivities values appear much higher, the permeance of CO_2 for the bare COF-300 membrane reached an impressive 1,185 Barrer.

The tunability of COF-300 was exploited for application in H_2 separation using a hybrid COF-MOF material based on [COF-300]-[ZIF-8]. The selectivity for H_2 in an H_2/CO_2

mixture increased by up to 2 times and the hybrid materials exhibited a separation factor of 13.5 while the bare COF and ZIF-8 only had a value of 6 and 9.1, respectively [31]. This is due to the MOF structure serving as a molecular sieve. It closes the gaps between the COF crystals by forming an interlayer of ~200 nm in thickness. COF-MOF synergy was used to prepare a [COF-300]-[UiO-66] composite membrane, which outperformed the respective single-phase membranes in terms of H_2/CO_2 separation [32]. When compared to individual UiO-66 (70,143 Barrer) and COF-300 (109,020 Barrer), the [COF-300]-[UiO-66] composite membrane offered a higher H_2 permeability (118,962 Barrer). In addition, the H_2/CO_2 selectivity of the composite membrane (17.2) was higher than that of the UiO-66 (9.2) and COF-300 (6) bare films. The enhanced separation properties of the composite membrane were explained based on a solution-diffusion mechanism. The model states that permeants dissolve in membrane material and then diffuse down a concentration gradient through the membrane. Different permeants are separated due to differences in the amount of material that dissolves in the membrane and the rate at which the material diffuses through the membrane. This phenomenon was utilized by careful selection of COF and MOF with greater pore sizes than the kinetic diameters of the gas molecules allowing for faster gas diffusion, therefore, higher permeability, while also selecting material with lower isosteric enthalpies of adsorption (Q_{st}) to H_2 than to CO_2 therefore, making the membrane highly selective. The potential of hybrid structures integrating COFs and MOFs appears relevant to balance some compounds' high selectivity and water vapor stability with the surface interactions and tight pore size distributions of other compounds. Further works on co-synthesis of hybrid structures shall however be performed to develop more scalable fabrication routes.

23.3 COF-Based Adsorbents

23.3.1 Adsorbents for Aqueous Mercury Removal

Packed bed-based separation is one of the most important separation processes used in chemical and related industries, where a multi-component mixture is passed through a column packed with adsorbents to remove unwanted components from the mixture based on differences in adsorption capability among components toward the adsorbent [5]. This section discusses the potential application of COFs as adsorbents in the removal of heavy metals, specifically mercury (Hg^{2+}), from water.

Fluorescent COFs with large π-conjugated structures can be used for the dual function of detection and capture of metal ions. COF-LZU8, for instance, with thioether hydrazone bonding was synthesized and used for Hg^{2+} removal delivering high selectivity, sensitivity, and adsorption capacity [33]. The rigid π-conjugation structure served as the fluorophore for signal sensing, whereas the uniformly distributed thioether functionalities served as the receptors for capturing Hg^{2+} [33], as shown in Figure 23.1b. COF-LZU8 displayed a strong UV absorption band at ~390 nm in the solid-state, and the fluorescence was efficiently quenched upon adsorption of Hg^{2+}. COF-LZU8 was also found to maintain its fluorescence capacity in solvents such as acetonitrile, Tetrahydrofuran (THF), Dimethylformamide (DMF), and ethanol, required for detecting Hg^{2+} in solution in real-time [33] displaying good chemical stability. The limit of detection for Hg^{2+} was determined as 25 ppb, which represents a superior sensitivity compared to many thioether-functionalized chemo-sensors [33]. COF-LZU8 also showed excellent selectivity for Hg^{2+} detection in the presence of other metal ions in solution such as Li^+, Na^+, Fe^{2+}, Al^{3+}, and others, as among all the tested metal ions, only Hg^{2+} caused significant fluorescence-quenching of COF-LZU8 [33]. To demonstrate the efficient removal of Hg^{2+} from water, COF-LZU8 (5 mg) was suspended in a dilute aqueous solution of $Hg(ClO_4)_2$ (10 ppm) [33]. The concentration of the remaining Hg^{2+} in water after 3 h of stirring was less than 0.2 ppm [33], showing that over 98% of Hg^{2+} could be successfully eliminated even in extremely dilute solutions. Compared to the thioether-functionalized MOF-5 [34], COF-LZU8 showed a better capability for the Hg^{2+} removal. The thioether-functionalized MOF-5 lowered the concentration of $HgCl_2$ in an ethanol solution from 84 mg L^{-1} to 5 ppm during 6 days indicating ~ 94% removal [34], compared to 98% Hg^{2+} removal during 3 h from a very dilute solution of $Hg(ClO_4)_2$ for COF-LZU8 [33]. This enhanced performance could be attributed to COF-LZU8's 2D eclipsed structure, in which a straight channel with a diameter of 13 Å facilitated the interaction between Hg^{2+} and S atoms [33].

TPB-DMTP-COF-SH (TPB, triphenylbenzene; DMTP, dimethoxyterephthaldehyde) efficiently reacts with Hg^{2+} ions and therefore can be used for the effective and selective sorption of mercury [35]. To synthesize TPB-DMTP-COF-SH, the imine-linked COF ($[HC\equiv C]_{0.5}$-TPB-DMTP-COF) was used as the starting precursor, which was then post-functionalized with triazole and thiol groups through pore-wall surface engineering and click reactions [35], as shown in Figure 23.1c. Triazole groups have a high coordinative capacity toward Hg^{2+}, and thiol groups have a high affinity for mercury because of the sulfur atoms in their ligand making them a promising combination for this application [36]. TPB-DMTP-COF-SH had a high distribution coefficient value ($K_d = 3.23 \times 10^9$), enabling it to adsorb Hg^{2+} ions efficiently from both low and high concentration solutions to maximum acceptable limits within a few minutes [35]. For example, within 10 min, it effectively reduced mercury concentration from 10 ppm to 1.5 ppb, which is lower than the acceptable limit in drinking water standards (2 ppb) [35]. The mercury removal efficiency from the water was 99.98% within 2 min [35]. Hence, TPB-DMTP-COF-SH demonstrated an ultra-high affinity for Hg^{2+} ions with exceptional fast kinetics opening up new possibilities for the application of COFs in mercury removal and environmental remediation.

23.3.2 Adsorbents for Gas Separation

Gas recovery from large stationary point sources, especially power plants, may also achieve with physical or chemical sorbents [37]. Fluidized beds are established and widely used technologies, whereby fluidization is carried out with packed sorbent materials supporting the progressive extraction of gases from mixtures [38]. In this section, the use of COFs for CO_2 and H_2 adsorption is discussed. A range of COF materials were trialed for CO_2 and H_2 capture including, COF-1, -5, -6, -8, -10, -102 and -103 synthesized from polycondensation reactions [39]. The adsorption capacity was measured at both low and high pressure of up to 85 bar, representative of H_2 storage for mobile fueling applications and effluents from power plants respectively. COF-102 showed exceptionally high S_{BET} of 3,620 m^2 g^{-1} and exhibited the highest CO_2 (1,200 mg g^{-1}) and hydrogen (72.4 mg g^{-1}) capacity. COF-102 has one of the lowest bulk densities and the highest total pore volume of all the COFs tested within that series. The pore volume offered almost a linear relation with the gas uptake, while the absolute pore size did not greatly affect the uptake capacity since COF-102 presented a mean pore size of 12 Å compared to 9 Å of COF-1 and 32 Å of COF-10, for example. This systematic study opened the route to the application of COF materials for selective gas capture, by optimizing the pore volume and binding sites within the adsorbent.

TDCOF-5 was derived from 1,4-benzenediboronic acid and hexahydroxytrypticene to 2D mesoporous triptycene by forming a boronate through solvothermal method at 120°C in a mixture of mesitylene and dioxane [40]. COF-5 was used since its solid-state assembly is driven by electron-rich blocks that exhibit favorable p-p stacking interactions that can interfere with access to the boron center. These centers are important for gas capture applications since rendering they render adsorption sites accessible and well distributed across the COF matrix, thanks to the use of triptycene as a modifier. The BET surface area was found to be ~2,497 $m^2 g^{-1}$, which was ~50% higher than that of bare COF-5 (1,670 $m^2 g^{-1}$) [39], supporting a CO_2 uptake capacity up to 92 mg/g at 273 K and 1 bar. Furthermore, the adsorption capacity for H_2 of the TDCOF-5 was found to be almost double that of COF-5, with Q_{st} value at low coverage to be 6.6 kJ mol^{-1} compared to 6 kJ mol^{-1} for COF-5.

Triazine-based COFs designed from adamantine PCTFs (PCTF-3,-4,-5) were formed using 1,3-bis-, 1,3,5-tris-, and 1,3,5,7-tetrakis(4-cyanophenyl) adamantine respectively using trimerization reactions [41]. A Lewis acid, $ZnCl_2$, was used to yield higher S_{BET} rather than Bronsted acid conditions and the PCTF-5 was found to offer the highest S_{BET} of 1183 $m^2 g^{-1}$ of the series with a CO_2 capacity of 114 mg g^{-1} at 273 K and 1 bar. For binary mixtures of CO_2/N_2 and CO_2/CH_4, the selectivities were found to be at 32 and 5, respectively. PCTF-1, -2, -3, and -4, were also used for their controllable surface area, decreasing as the chain length of the branch arm increases, as a means to control effective packing based on the chain length, and therefore the adsorption capacity of the COF [42]. This approach was demonstrated since PCTF-1 offered a surface area of 853 $m^2 g^{-1}$, i.e., 5% higher than PCTF-2 and 50% higher than PCTF-3. Rather than a rigid tetrahedral adamantane core [41], derivatives of triphenylamine were used as the core and the COF was functionalized with benzothiadiazole units to favor CO_2 chemisorption. This approach led to an increase in S_{BET} from 1,090 [41] to 1,404 $m^2 g^{-1}$ [42] and the CO_2/N_2 and CO_2/CH_4 selectivities were found to lay at 56 and 20, respectively. Strong affinity toward CO_2 compared to N_2 could explain the high CO_2/N_2 selectivity and PCTF-4 was also shown to offer the highest H_2 uptake up to 1.3 wt % at 77 K and 1 bar of all the COFs tested.

A series of CTFs with phosphazene core for CO2 adsorption were also synthesized using building blocks of hexakis(oxy)hexabenzonitrile phosphazene (HCPz) under ionothermal conditions and ZnCl2 as catalyst [2]. The effect of the ZnCl2/monomer ratio and the ionothermal reaction temperature and synthesis procedure on the porosity of the Pz-CTFs were systematically explored. The Pz-CTF synthesized using a molar ratio of ZnCl2/monomer of 10 with gradient ionothermal reaction conditions (400°C/25 h; 450°C/13 h; 500°C/1 h; 600°C/1 h) yielded an amorphous, predominantly ultra-microporous material (Pz-CTF6) with high surface area (SBET of 1009 m2 g-1). The primary aspect of the applied gradient ionothermal reaction scheme is that it induces a simultaneous reversible and irreversible trimerization of nitriles, allowing the restructuring of the triazine units, thus resulting in an extended microporous network. Besides, the highly crosslinked, electron-rich phosphazene core provided high structural and thermal stability (up to 500°C) and enhanced CO2-philicity to the synthesized CTFs. As a result, Pz-CTF6 showed a CO2 uptake capacity of 4.19 and 2.47 mmol g-1 at 273 and 298 K, respectively, at 1 bar. In addition, it exhibited a high CO2/N2 selectivity of 147 for a feed containing 85% of CO2 at 50 mbar and 298 K, as determined by the ideal adsorbed solution theory (IAST) method. Furthermore, the developed adsorbents exhibited enhanced hydrophobicity, causing a rather mild reduction in CO2 capacity when humidity conditions were applied.

Imine-linked COFs for CO_2 capture were developed through reversible and irreversible reaction of 1,3,5-triformylphloroglucinol with p-phenylenediamine(PaP-1) and 2,5-dimethyl-p-phenylenediamine(PaP-2) used to form TpPa-1 and TpPa-2, respectively [43]. These COFs showed good stability toward acidic conditions and good thermal stability attributed to irreversible proton tautomerism from the enol–imine (OH) to the keto–enamine form, and TpPa-2 was also found to be stable in basic conditions (9 M NaOH). The CO_2 capacity was found to lay at 153 and 126 mg g^{-1} for TpPa-1 and TpPa-2, respectively. These COFs offer large pore size windows between 1.25 and 1.35 nm and a relatively small BET surface area of 339–535 m^2 g^{-1} when compared to other COFs. TpPa-1 was also synthesized for CO_2 capture using a microwave-assisted solvothermal method, leading to S_{BET} (725 m^2 g^{-1}) [44]. This is significantly higher than that reported for direct solvothermal by nearly 100%, which was attributed to the lower level of amorphization of the framework upon high-intensity energy spiking during the microwave input [44]. This material exhibited a CO_2 capacity of 218 mg g^{-1} and outperformed most COFs utilizing the same class of precursors. It also offered a CO_2/N_2 selectivity of 32 at 273 K and 1 bar, which was attributed to the abundant N-H sites present across the pore walls of the framework, which polarize CO_2 molecules through H-bonds.

Tuning the properties of COF-1 was also explored for CO_2 capture using Schiff base reactions for the construction of a laminar RT-COF-1 from 1,3,5-tris(4-aminophenyl) benzene(TAPB) and 1,3,5-benzenetricarbaldehyde (BTCA) [45]. The CO_2 capacity was found to be 86 mg g^{-1} with a Q_{st} of 16.4 kJ mol^{-1} at 273 K and 1 bar, with the BET surface area being 329 m^2 g^{-1}. The formation of HEX-COF-1 by functionalizing COF-1 with symmetric hexaphenylbenzene (HEX) monomers was also performed [46], which allowed for a major increase in BET surface area up to 1,200 m^2 g^{-1} leading to a CO_2 capacity of 200 mg g^{-1} at 273 K and 1 bar. The azine functionalization of COF-1 (ACOF-1) also offered a relatively high surface area of 1,176 m^2 g^{-1} [22], which enabled a CO_2 capacity of 177 mg g^{-1} with a Q_{st} value of 27.6 kJ mol^{-1}. Tests performed at 0.1 bar and 273 K offered a CO_2/N_2 selectivity of 40.

23.4 Conclusions and Prospects

This chapter dealt with the high potential of COF membranes and adsorbents for capturing salts and gases from model effluents. Recent breakthroughs related to the utilization of COF membranes in water treatment and gas separation were highlighted and innovative approaches to increase COF stability in water (liquid or gas), as well as the crystallinity and porosity of the frameworks, were coupled to effective post-synthetic functionalization to support the practical implementation of these materials. Challenges in the area remain related to the scalability of the materials and the control of defects across the COF membrane structures. Promising hybridizations of COFs with MOFs were also reported, and these findings open new possibilities for employing COFs and their composites as advanced separation systems for industrial and environmental remediation applications.

References

1. H.R. Abuzeid, A.F.M. El-Mahdy, S.-W. Kuo, Covalent organic frameworks: Design principles, synthetic strategies, and diverse applications, Giant, 6 (2021) 100054.
2. V.M. Rangaraj, K.S.K. Reddy, G.N. Karanikolos, Ionothermal synthesis of phosphonitrilic-core covalent triazine frameworks for carbon dioxide capture, Chemical Engineering Journal, 429 (2022) 132160.
3. A. Altaf, N. Baig, M. Sohail, M. Sher, A. Ul-Hamid, M. Altaf, Covalent organic frameworks: Advances in synthesis and applications, Materials Today Communications, 28 (2021) 102612.
4. Z. Xia, Y. Zhao, S. Darling, Covalent Organic Frameworks for Water Treatment, Advanced Materials Interfaces, 8 (2020) 2001507.
5. Z. Wang, S. Zhang, Y. Chen, Z. Zhang, S. Ma, Covalent organic frameworks for separation applications, Chemical Society Reviews, 49 (2020) 708–735.
6. M.-A. Gatou, P. Bika, T. Stergiopoulos, P. Dallas, E.A. Pavlatou, Recent Advances in Covalent Organic Frameworks for Heavy Metal Removal Applications, Energies, 14 (2021).
7. A.K. Mohammed, D. Shetty, Macroscopic covalent organic framework architectures for water remediation, Environmental Science: Water Research & Technology, 7 (2021) 1895–1927.
8. E. Vitaku, W.R. Dichtel, Synthesis of 2D Imine-Linked Covalent Organic Frameworks through Formal Transimination Reactions, Journal of the American Chemical Society, 139 (2017) 12911–12914.
9. C. Liu, Y. Jiang, A. Nalaparaju, J. Jiang, A. Huang, Post-synthesis of a covalent organic framework nanofiltration membrane for highly efficient water treatment, Journal of Materials Chemistry A, 7 (2019) 24205–24210.
10. D.B. Shinde, L. Cao, A.D.D. Wonanke, X. Li, S. Kumar, X. Liu, M.N. Hedhili, A.-H. Emwas, M. Addicoat, K.-W. Huang, Z. Lai, Pore engineering of ultrathin covalent organic framework membranes for organic solvent nanofiltration and molecular sieving, Chemical Science, 11 (2020) 5434–5440.
11. S. Fernandes, V. Romero, B. Espiña, L. Salonen, Tailoring Covalent Organic Frameworks To Capture Water Contaminants, Chemistry - A European Journal, 25 (2019).
12. M. Kammoun, S. Gassara, J. Palmeri, R. Amar, A. Deratani, Nanofiltration performance prediction for brackish water desalination: case study of Tunisian groundwater, DESALINATION AND WATER TREATMENT, 181 (2020) 27–39.
13. J.M. Ochando-Pulido, A. Martínez-Férez, M. Stoller, Analysis of the Flux Performance of Different RO/NF Membranes in the Treatment of Agroindustrial Wastewater by Means of the Boundary Flux Theory, Membranes (Basel), 9 (2018) 2.
14. K. Zhang, Z. He, K.M. Gupta, J. Jiang, Computational design of 2D functional covalent–organic framework membranes for water desalination, Environmental Science: Water Research & Technology, 3 (2017) 735–743.
15. D. Cohen-Tanugi, J.C. Grossman, Water Desalination across Nanoporous Graphene, Nano Letters, 12 (2012) 3602–3608.
16. K.M. Gupta, K. Zhang, J. Jiang, Water Desalination through Zeolitic Imidazolate Framework Membranes: Significant Role of Functional Groups, Langmuir, 31 (2015) 13230–13237.
17. Z. Hu, Y. Chen, J. Jiang, Zeolitic imidazolate framework-8 as a reverse osmosis membrane for water desalination: Insight from molecular simulation, The Journal of Chemical Physics, 134 (2011) 134705.
18. M.M. Pendergast, E.M.V. Hoek, A review of water treatment membrane nanotechnologies, Energy & Environmental Science, 4 (2011) 1946–1971.
19. S. Chaemchuen, N.A. Kabir, K. Zhou, F. Verpoort, Metal–organic frameworks for upgrading biogas via CO2 adsorption to biogas green energy, Chemical Society Reviews, 42 (2013) 9304.

20. H.-C. Zhou, J.R. Long, O.M. Yaghi, Introduction to Metal–Organic Frameworks, Chemical Reviews, 112 (2012) 673–674.
21. M. Shan, B. Seoane, E. Rozhko, A. Dikhtiarenko, G. Clet, F. Kapteijn, J. Gascon, Azine-Linked Covalent Organic Framework (COF)-Based Mixed-Matrix Membranes for CO2/CH4 Separation, Chemistry - A European Journal, 22 (2016) 14467–14470.
22. Z. Li, X. Feng, Y. Zou, Y. Zhang, H. Xia, X. Liu, Y. Mu, A 2D azine-linked covalent organic framework for gas storage applications, Chem. Commun., 50 (2014) 13825–13828.
23. H. Fan, A. Mundstock, J. Gu, H. Meng, J. Caro, An azine-linked covalent organic framework ACOF-1 membrane for highly selective CO2/CH4 separation, Journal of Materials Chemistry A, 6 (2018) 16849–16853.
24. B.P. Biswal, H.D. Chaudhari, R. Banerjee, U.K. Kharul, Chemically Stable Covalent Organic Framework (COF)-Polybenzimidazole Hybrid Membranes: Enhanced Gas Separation through Pore Modulation, Chemistry - A European Journal, 22 (2016) 4695–4699.
25. S. Singh, A.M. Varghese, K.S.K. Reddy, G.E. Romanos, G.N. Karanikolos, Polysulfone Mixed-Matrix Membranes Comprising Poly(ethylene glycol)-Grafted Carbon Nanotubes: Mechanical Properties and CO2 Separation Performance, Ind. Eng. Chem. Res., 60 (2021) 11289–11308.
26. K. Duan, J. Wang, Y. Zhang, J. Liu, Covalent organic frameworks (COFs) functionalized mixed matrix membrane for effective CO2/N2 separation, Journal of Membrane Science, 572 (2019) 588–595.
27. P.D. Sutrisna, J. Hou, H. Li, Y. Zhang, V. Chen, Improved operational stability of Pebax-based gas separation membranes with ZIF-8: A comparative study of flat sheet and composite hollow fiber membranes, Journal of Membrane Science, 524 (2017) 266–279.
28. R.S. Murali, S. Sridhar, T. Sankarshana, Y.V.L. Ravikumar, Gas Permeation Behavior of Pebax-1657 Nanocomposite Membrane Incorporated with Multiwalled Carbon Nanotubes, Ind. Eng. Chem. Res., 49 (2010) 6530–6538.
29. Y. Cheng, L. Zhai, Y. Ying, Y. Wang, G. Liu, J. Dong, L. Ng, Denise Z., S.A. Khan, D. Zhao, Highly efficient CO2 capture by mixed matrix membranes containing three-dimensional covalent organic framework fillers, Journal of Materials Chemistry A, 7 (2019) 4549–4560.
30. H. Wu, X. Li, Y. Li, S. Wang, R. Guo, Z. Jiang, C. Wu, Q. Xin, X. Lu, Facilitated transport mixed matrix membranes incorporated with amine functionalized MCM-41 for enhanced gas separation properties, Journal of Membrane Science, 465 (2014) 78–90.
31. J. Fu, S. Das, G. Xing, T. Ben, V. Valtchev, S. Qiu, Fabrication of COF-MOF Composite Membranes and Their Highly Selective Separation of H2/CO2, Journal of the American Chemical Society, 138 (2016) 7673–7680.
32. S. Das, T. Ben, A [COF-300]-[UiO-66] composite membrane with remarkably high permeability and H2/CO2 separation selectivity, Dalton Transactions, 47 (2018) 7206–7212.
33. S.-Y. Ding, M. Dong, Y.-W. Wang, Y.-T. Chen, H.-Z. Wang, C.-Y. Su, W. Wang, Thioether-Based Fluorescent Covalent Organic Framework for Selective Detection and Facile Removal of Mercury(II), Journal of the American Chemical Society, 138 (2016) 3031–3037.
34. J. He, K.-K. Yee, Z. Xu, M. Zeller, A.D. Hunter, S.S.-Y. Chui, C.-M. Che, Thioether Side Chains Improve the Stability, Fluorescence, and Metal Uptake of a Metal–Organic Framework, Chemistry of Materials, 23 (2011) 2940–2947.
35. L. Merí-Bofí, S. Royuela, F. Zamora, M.L. Ruiz-González, J.L. Segura, R. Muñoz-Olivas, M.J. Mancheño, Thiol grafted imine-based covalent organic frameworks for water remediation through selective removal of Hg(ii), Journal of Materials Chemistry A, 5 (2017) 17973–17981.
36. L. Niu, X. Zhao, F. Wu, Z. Tang, H. Lv, J. Wang, M. Fang, J.P. Giesy, Hotpots and trends of covalent organic frameworks (COFs) in the environmental and energy field: Bibliometric analysis, Science of The Total Environment, 783 (2021) 146838.
37. M.E. Boot-Handford, J.C. Abanades, E.J. Anthony, M.J. Blunt, S. Brandani, N. Mac Dowell, J.R. Fernández, M.-C. Ferrari, R. Gross, J.P. Hallett, R.S. Haszeldine, P. Heptonstall, A. Lyngfelt, Z. Makuch, E. Mangano, R.T.J. Porter, M. Pourkashanian, G.T. Rochelle, N. Shah, J.G. Yao, P.S. Fennell, Carbon capture and storage update, Energy Environ. Sci., 7 (2014) 130–189.

38. S.E. Zanco, M. Mazzotti, M. Gazzani, M.C. Romano, I. Martínez, Modeling of circulating fluidized beds systems for post-combustion CO2 capture via temperature swing adsorption, AIChE Journal, 64 (2018) 1744–1759.
39. H. Furukawa, O.M. Yaghi, Storage of Hydrogen, Methane, and Carbon Dioxide in Highly Porous Covalent Organic Frameworks for Clean Energy Applications, Journal of the American Chemical Society, 131 (2009) 8875–8883.
40. Z. Kahveci, T. Islamoglu, G.A. Shar, R. Ding, H.M. El-Kaderi, Targeted synthesis of a mesoporous triptycene-derived covalent organic framework, CrystEngComm, 15 (2013) 1524–1527.
41. A. Bhunia, I. Boldog, A. Möller, C. Janiak, Highly stable nanoporous covalent triazine-based frameworks with an adamantane core for carbon dioxide sorption and separation, Journal of Materials Chemistry A, 1 (2013) 14990.
42. C. Gu, D. Liu, W. Huang, J. Liu, R. Yang, Synthesis of covalent triazine-based frameworks with high CO2 adsorption and selectivity, Polymer Chemistry, 6 (2015) 7410–7417.
43. S. Kandambeth, A. Mallick, B. Lukose, M.V. Mane, T. Heine, R. Banerjee, Construction of Crystalline 2D Covalent Organic Frameworks with Remarkable Chemical (Acid/Base) Stability via a Combined Reversible and Irreversible Route, Journal of the American Chemical Society, 134 (2012) 19524–19527.
44. H. Wei, S. Chai, N. Hu, Z. Yang, L. Wei, L. Wang, The microwave-assisted solvothermal synthesis of a crystalline two-dimensional covalent organic framework with high CO2 capacity, Chemical Communications, 51 (2015) 12178–12181.
45. A. De La Peña Ruigómez, D. Rodríguez-San-Miguel, K.C. Stylianou, M. Cavallini, D. Gentili, F. Liscio, S. Milita, O.M. Roscioni, M.L. Ruiz-González, C. Carbonell, D. Maspoch, R. Mas-Ballesté, J.L. Segura, F. Zamora, Direct On-Surface Patterning of a Crystalline Laminar Covalent Organic Framework Synthesized at Room Temperature, Chemistry - A European Journal, 21 (2015) 10666–10670.
46. S.B. Alahakoon, C.M. Thompson, A.X. Nguyen, G. Occhialini, G.T. Mccandless, R.A. Smaldone, An azine-linked hexaphenylbenzene based covalent organic framework, Chemical Communications, 52 (2016) 2843–2845.

24

Covalent Organic Framework-Based Nanoparticles for Catalytic Environmental Remediation

Muhammad Sohail Bashir[1,2], Syed Shoaib Ahmad Shah[3], Tayyaba Najam[4], Aqsa Safdar[5,6], and Humaira Bashir[7,8]

[1]*Institutes of Physical Science and Information Technology, Key Laboratory of Structure and Functional Regulation of Hybrid Materials of Ministry of Education, Anhui University, Hefei, Anhui, China*

[2]*Hefei National Laboratory for Physical Sciences at the Microscale, CAS Key Laboratory of Soft Matter Chemistry, Department of Polymer Science and Engineering, University of Science and Technology of China, Hefei, Anhui, China*

[3]*Hefei National Laboratory for Physical Sciences at the Microscale, CAS Key Laboratory of Soft Matter Chemistry, Department of Chemistry, University of Science and Technology of China, Hefei, Anhui, China*

[4]*Institute for Advanced Study and Institute of Microscale Optoelectronics, Shenzhen University, Shenzhen, China*

[5]*School Education Department, Punjab, Pakistan*

[6]*Department of Chemistry, University of the Punjab, Lahore, Pakistan*

[7]*Department of Botany, University of the Punjab, Lahore, Pakistan*

[8]*Stockbridge School of Agriculture, University of Massachusetts Amherst, Amherst, USA*

CONTENTS

DOI: 10.1201/9781003206507-24

24.1 Introduction

With the rapid development of chemical, textile, and manufacturing industries, pollution by industrial wastewaters has become an urgent issue. A great variety of dyes stuff and 4-nitrophenol (4-NP) present in industrial effluents have become a serious health and environmental challenge. Face to this issue, different methods have been developed to remove or degrade the dyes from wastewaters, including chemical precipitation, catalytic reduction, filtration, adsorption, and biological treatment [1–5]. Among these methods, the reduction of organic contaminants such as toxic dyes and 4-NP catalyzed by metallic nanoparticles (M-NPs) in the presence of $NaBH_4$ has gained more attention in recent years [6].

The concept of nanotechnology comes from the lecture delivered by Richard Feynman in 1959 in a symposium. He provided the vision of nanotechnology by predicting that new unique characteristics and properties of a matter will be emerged by controlling the ordering of materials on an ultra-small scale [7]. Nowadays, researchers have confirmed that the compounds exhibited different and unique properties in terms of surface and Coulomb resistance effects and energy level splitting, compared to bulk materials when their size was reduced to a small scale (1–100 nm). Thus, within the past few years, nano-sized M-NPs have gained more research interest due to their unique characteristics regarding optics, catalysis, magnetism, mechanics, and so on. The number of active sites, which generally play a vital task in catalysis, would be enhanced up to 90% as the size of particles is reduced to 1 nm [8]. The small size adds remarkable activity and gives a high specific surface area (SSA) of M-NPs which are highly advantageous in the field of heterogeneous catalysis. However, easy agglomeration/aggregation due to high surface energy, poor stability of surface atoms, and difficult recyclability owing to their extra small size are big challenges for M-NP diverse applications. Scientists have been working to develop supported M-NPs to overcome these grave challenges [9]. It has been revealed that the morphology and properties of the support can strongly influence the activity of M-NPs via electron interaction and spatial captivities. Thus, to get highly stable M-NPs with sustainable activity, scientists have been developing various host substrates, including carbon materials, porous silica, graphene, metal oxide, porous polymers, porous organic polymers (POPs), and metal-organic frameworks (MOFs) [8, 10–12]. However, it is a challenging task to get uniform dispersion of M-NPs with narrow size variation and stability onto the supporting host. In this regard, materials with high porosity and large SSA have been widely employed as effective host materials for M-NP. To date, various porous materials including crystalline MOFs, COFs, and amorphous POPs, have been developed from functional monomers/building-blocks and used as a host for M-NPs to give high active and selective catalysis [13–16].

In 2005, Yaghi et al. [17] reported COFs, new porous and crystalline organic polymers which were prepared via reversible condensations of functional building blocks linked by strong covalent bonds. COFs are well-known to have a highly ordered porous crystalline structure, chemical stability, and can withstand the harsh acidic or basic environment. The higher porosity with uniform pores distribution and large SSA compared to other supporting hosts makes COFs beneficial for their uses in catalysis. Moreover, larger pores density provides more active sites of supported M-NPs, boosts the transportation and diffusion of substrates, and eases the desorption of products, thus accelerating the selectivity and product yield [18, 19]. Recently COFs have been recognized as promising hosts for M-NPs and COF-based M-NPs could have numerous unique advantages as follows:

1. The pre-designable and well-order pores structure of COFs can be advantageous to limit M-NPs growth and hence help to control the size of M-NPs.

2. COFs have well-isolated pore channels, which minimize the aggregation of entrapped M-NPs.

3. Easy functionalization of chemical structures of COFs can enhance the anchoring ability of COFs toward M-NPs.

4. Usually, COFs are resistant to deformation under vigorous reaction conditions which make them stable host materials.

In this chapter, recent progress on synthesis routes and characterizations of COFs-based M-NPs heterostructures and their uses for the sustainable heterogeneous catalytic environmental remediation such as toxic organic dyes and 4-NP in an aqueous system are addressed.

24.2 Synthesis of COF-Based M-NPs Catalysts

To date, numerous COF-based M-NPs catalysts are reported for different applications. However, COF-based M-NPs used for catalytic environmental remediation, that is, catalytic reduction of organic dyes and 4-NP by $NaBH_4$ (a reducing agent) in an aqueous system are in the scope of this chapter.

24.2.1 COF-Based Au-NPs

Gold nanoparticles (Au-NPs) are one of the significant heterogeneous catalysts. However, the high surface/volume leads to high surface energy which usually facilitates Au-NPs aggregation into bigger particles and makes them thermodynamically unstable at longer catalysis, in results reducing their activity and limiting their practical uses. To overcome this grave issue, scientists have implemented several porous organic and inorganic materials as support for M-NPs such as zeolites, porous silica, MOFs, polymers, and porous carbons. However, the weak ligation of M-NPs onto these support materials due to the lack of functional groups is known to be a cause of the decline in their performance [20–22]. COFs have also been practiced as supporting hosts for Au-NPs in recent years and are known to be an efficient supporting hosts owing to their well-ordered structure, high SSA, and good interaction with Au-NPs. However, COF stability in alkaline or acidic medium remains a critical feature for recycling COF-supported M-NPs [23]. Moreover, the anxieties regarding leaching, sintering, recyclability, and stability of supported M-NPs at longer catalysis linger at the forefront. Hence, fabrication of a new supporting host, having excellent ligation with loaded M-NPs and high stability in aqueous alkaline/acidic mediums, is highly desired. Considering this point, Pachfule et al. [24] reported the solution infiltration approach, an easy fabrication route to obtain high stability of COF (TpPa-1: triformylphloroglucinol (Tp); paraphenylenediamine (Pa-1)) anchored with Au-NPs and used for the transformation of toxic 4-NP or 4-NPh to 4-aminophenol (4-AP) or 4-APh in an aqueous medium. In this method, TpPa-1 solution was prepared in methanol, degassed and $HAuCl_4 \cdot 3H_2O$ was added under stirring. After a specified time, the Au ions were reduced by the addition of $NaBH_4$ solution, and the product was filtered, rinsed, and dried under reduced pressure to get solid Au-NPs@TpPa-1 (Figure 24.1). The TpPa-1 was enriched with nitrogen and oxygen, which provided the attraction with Au-NPs. The SSA of Au-NPs@TpPa-1 was found

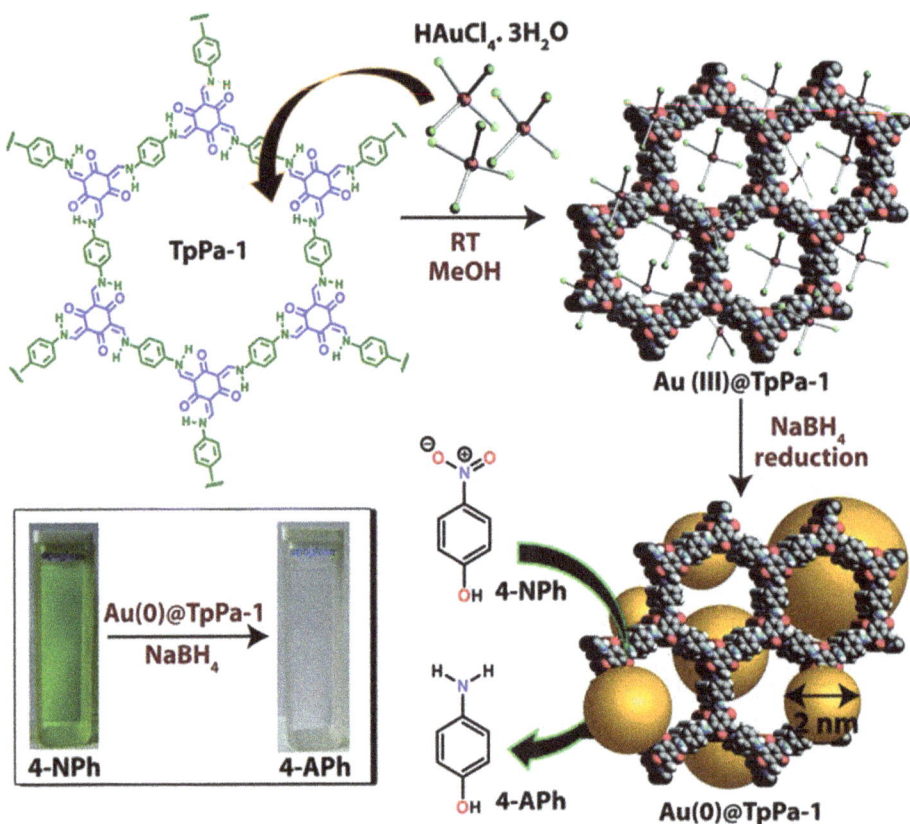

FIGURE 24.1
Synthesis approach of Au-NPs@TpPa-1 catalyst through solution infiltration technique for reduction of 4-NP to 4-AP in an aqueous system. (Adapted with permission [24]. Copyright (2014) Royal Society of Chemistry.)

to be 339 m²/g, lower than the SSA of TpPa-1 of 484 m²/g, which revealed that Au-NPs were stabilized in the pores of TpPa-1. Furthermore, it was found that Au-NPs@TpPa-1 has high stability in ordinary organic solvents, alkaline, and acidic aqueous media. Au-NPs@TpPa-1 was used as a catalyst for 4-NP reduction with NaBH₄ in an aqueous system and good catalytic activity was observed. However, the slight decrease in activity with its recycling used and the large size of Au-NPs (5 ± 3 nm) may create issues of stability of the Au-NPs@TpPa-1 at a longer catalysis time.

The incorporation of functional groups into COF can help to attain the stability of Au-NPs with uniform dispersion and narrow size distribution by increasing the ligation of COF with M-NPs, leading to a decrease in the voyage of the Au-NPs and improving their catalysis activity. Zhang et al. [25] published the propenyl-functionalized COF which was synthesized by the condensation of diethyl 2,5-bis(allyloxy)terephthalohydrazide and 1,3,5-triformylbenzene. Thiol functional groups were introduced from thiolene "click" reaction of 1,2-ethanedithiol by post-modification approach to get a sulfur-grafted COF (*S*-COF). The obtained S-COF was dispersed in methanol and HAuCl₄ was added under stirring to anchor Au ions onto S-COF. Afterward, the solution of NaBH₄ in methanol was added dropwise to reduce the Au ions to get Au-NPs-*S*-COF. The strong coordination of sulfur-containing groups with Au-NPs onto S-COF improved the anchoring of Au-NPs and sustained their uniform dispersion and hence enhanced the activity and stability of

Au-NPs for long-term catalysis. However, the activity of Au-NPs-S-COF was inferior compared with other M-NPs based catalysts, such as Pt, Pd, and Ag, which may be attributed to their different characteristics.

M-NPs with controlled size can be prepared within the pores/cavities of COFs by a confined synthesis approach. These kinds of COF-M-NPs composites can be produced by encapsulating the M-NPs during the process of COFs synthesis. COFs nucleate and grow around the M-NPs to provide the core-shell architecture. In principle, this strategy is enabling to incorporation M-NPs into COFs with the desired chemical characteristics. Keeping this concept in mind, Shi et al. [26] reported a simple technique for the incorporation of Au-NPs with different amounts, shapes, and sizes into 2D COFs. The 2D COFs were synthesized by condensation of 1,3,5-tris(4-aminophenyl)benzene (TAPB) and 2,5-dimethoxyterephthaldehyde. The surface of Au-NPs was functionalized with polyvinylpyrrolidone (PVP) and then functionalized Au-NPs were adsorbed onto the surfaces of a pre-synthesized precursor of COF. It was observed that the encapsulation of Au-NPs did not change the order and porous structure of the COF. The hybrid composite of Au-NPs-COF with stable active sites and large open mesopores provided good catalytic performance with recyclability for 4-NP reduction. However, there is a possibility that the orientation of porous channels or cavities and potential crystal defects in the COFs may be raised by the incorporation of M-NPs. More importantly, hindrance of active sites by incorporation of M-NPs in support can reduce the activity of catalyst.

Although, COF-based Au-NPs have exhibited good catalytic activities, however, their recovery is often troublesome due to their small size. Considering this issue, Xu et al. [27] introduced the magnetic functionality by incorporating Fe_3O_4 in COFs-based Au-NPs catalysts and making their easy recovery after the completion of the reaction by using an external magnet. It was observed that the obtained Fe_3O_4@COF-Au have a core-shell structure. Fe_3O_4@COF-Au was used as a catalyst for the reduction of methylene blue (MB) and 4-NP with $NaBH_4$ in an aqueous system and good activity, easy separation by a magnet, and good stability was observed. Even though, magnetic separation can be used for recycling of M-NPs based catalysts; however, it might be complicated and expensive to use a large number of magnetic apparatus.

Self-sedimentation of catalysts after completion of reaction would be a great functionality for their easy separation and recycling used. In this scenario, Wang et al. [28] reported the *in-situ* reduction of Au-NPs onto sulfhydryl (SH) functionalized COF by using $NaBH_4$, adsorbed on porous montmorillonite (SMt) of spherical shape, and modified with polydopamine (PDA) to get a spherical SMt@COF@Au-NPs hybrid composite. MB reduction with $NaBH_4$ catalyzed by M-NPs is well-known to be a surface phenomenon. It was found that the excellent adsorption capability of SMt for MB enhanced the cluster of MB on the surface of the catalyst which leads to boosting the catalytic reaction. Through optimization of reaction, it was found that the MB reduction rate was dependent on the pH of the reaction mixture, temperature, concentrations of MB, and catalysts amount. More importantly, micro-sized SMt@COF@Au-NPs were easily isolated by self-sedimentation after completion of the reaction, thus recovered quickly from the reaction mixture without any external devices involved and used for subsequent MB reduction. It was found that catalytic activity was remained constant for 20 times reused of SMt@COF@Au-NPs, which indicated the excellent stability of the catalyst.

24.2.2 COF-Based Ag-NPs

Silver nanoparticles (Ag-NPs) have been used for various catalytic reduction reactions owing to their unique properties. COF comprising a unique crystalline and highly porous

structure and Ag-NPs exhibit remarkable performance in catalytic environmental remediation. Thus, the recipe of Ag-NPs and COFs can widen their applications in catalysis.
Even though Ag-NPs exhibited excellent efficiency as catalysts in the degradation of several organic contaminants, the uses of Ag-NPs are limited by their easy agglomeration and
high cost. In general, Ag-NPs are anchored on porous supporting materials not only to
control the cost of Ag-NPs by reducing their quantity and increasing their recycling used
but also to stop their agglomeration without affecting the catalytic efficiency. New materials with high SSA and stability in acidic or alkaline aqueous media and have strong ligation ability with M-NPs are of great importance. In this context, introducing N-rich and
O-rich functional groups onto supporting hosts can not only increase the M-NPs binding
sites but also enhance the interactions of support with M-NPs [29]. Therefore, N and O
groups functionalized COFs are considered to be a good supporting host for M-NPs and
advantageous for boosting the activity and stability of M-NPs catalysts. COFs can be easily modified by N-rich and O-rich functional groups due to the diversification of their
building blocks. The ordered structure of pores not only controls the growth of M-NPs
size but also minimizes their aggregation/agglomeration onto the COFs surface. Wang
et al. [30] reported the fabrication of TPHH-COF via solution infiltration technique to support Ag-NPs to get Ag@TPHH-COF, which exhibited good activity as a catalyst for the
reduction of 4-NP and organic dyes with NaBH$_4$. The combined results of structural characterization of Ag@TPHH-COF revealed that the Ag-NPs with a mean size of 5 ± 3 nm
were evenly distributed into the pores and onto the surface of the TPHH-COF and catalyst have SSA of 143 m^2/g. The activity of Ag@TPHH-COF as a catalyst was studied for
4-NP reduction in water and found that it was a highly active catalyst with a turnover frequency (TOF) value of 4.150 min^{-1}, superior to most Ag-based reported catalysts. Moreover,
the Ag@TPHH-COF was shown good activity as a catalyst for the reduction of several
nitroaromatic compounds including 4-nitroaniline, 2-nitrophenol, nitrobenzene, 1-butyl-
4-nitrobenzene, and 4-nitrotoluene and organic dyes including rhodamine B (RhB), methyl
orange (MO), congo red (CR) and MB at ambient conditions. Through the recycling use of
Ag@TPHH-COF, sustainable activity (97%) after six cycles of reuse was observed, which
revealed the good stability of Ag@TPHH-COF in an aqueous system. However, a slight
reduction in its initial activity was observed due to minor aggregation of Ag-NPs onto
the COFs surface and weight loss during catalyst post-treatment. To solve the recyclability
problem, hybrid composites are generally synthesized by loading M-NPs in nanopores or
nanoshells which stop the coalescence/aggregation of M-NPs, thus enhancing the stability
and recycling performance of M-NPs.

Recent publications revealed that the introduction of the anchoring groups in the pores
of COFs enhanced the ligation of M-NPs, thus making a highly uniform dispersion of
M-NP-based catalysts with good stability. Shen et al. [31] reported 2D N-rich COFs as supporting hosts with opulent anchoring sites to prepare highly dispersed and stable supported Ag-NPs catalysts. The combined results of structural characterization of Ag-NPs@
NCOF revealed the successful anchoring of highly dispersed Ag-NPs (2.93 nm) linked with
N-groups into the pores of NCOF. It was found that AgNPs@NCOF exhibited superior
activity as a catalyst for 4-NP reduction with NaBH$_4$ in an aqueous system. The recycling
use of Ag-NPs@NCOF was shown that it maintained crystallinity without a noticeable
change in activity after five cycles reused, which revealed its good stability. The distinctive
NCOF structure provided the favorable nucleation spots for Ag-NPs, limited the coalescence of Ag-NPs, ligated the Ag-NPs via interface interaction, and hindered their aggregation. However, the recovery and reuse of COFs-based M-NPs is still a vital issue that limits
their practical applications.

The catalyst recovery/isolation and recycling process from the reaction mixture usually decrease their catalytic activity. Therefore, developing new technology for the efficient use of COF-based M-NPs Catalysts for the treatment of organic contaminants is highly desired. The recycling issues of COF-based M-NPs can be resolved by making a fixed-bed continuous reactor for the reduction of organic contaminants. Among various COFs with different structures, spherical COF (SCOF) with high SSA and large porosity can boost the fast transportation of substrates in a continuous process. Moreover, SCOF with high uniformity can facilitate the stability of the M-NPs supported catalyst. Wang et al. [32] reported the loading of Ag-NPs onto pre-synthesized SCOF with an SSA of 500 m^2/g to get Ag-NPs@SCOF which was used for a fixed bed reactor in the continuous reduction of 4-NP, minimizing the recycling issue of catalysts (Figure 24.2). The combined results of structure characterizations of Ag-NPs@SCOF revealed that Ag-NPs were successfully anchored onto SCOF with highly uniform dispersion and a mean diameter of 20 nm. It was found that Ag-NPs@SCOF exhibited better performance as a catalyst toward 4-NP reduction with a rate constant (k) of 1.06 min^{-1} at optimum conditions. Moreover, a good flux (2000 LMH) with catalytic efficiency (> 99%) was observed for fixed-bed flow-through catalysis, which was superior to the reported results with the same reaction. For the 24 h test, a small amount of Ag-NPs was leached out from the catalyst under a continuous fixed bed flow reaction, reflecting the good stability of Ag-NPs@SCOF, which might be ascribed

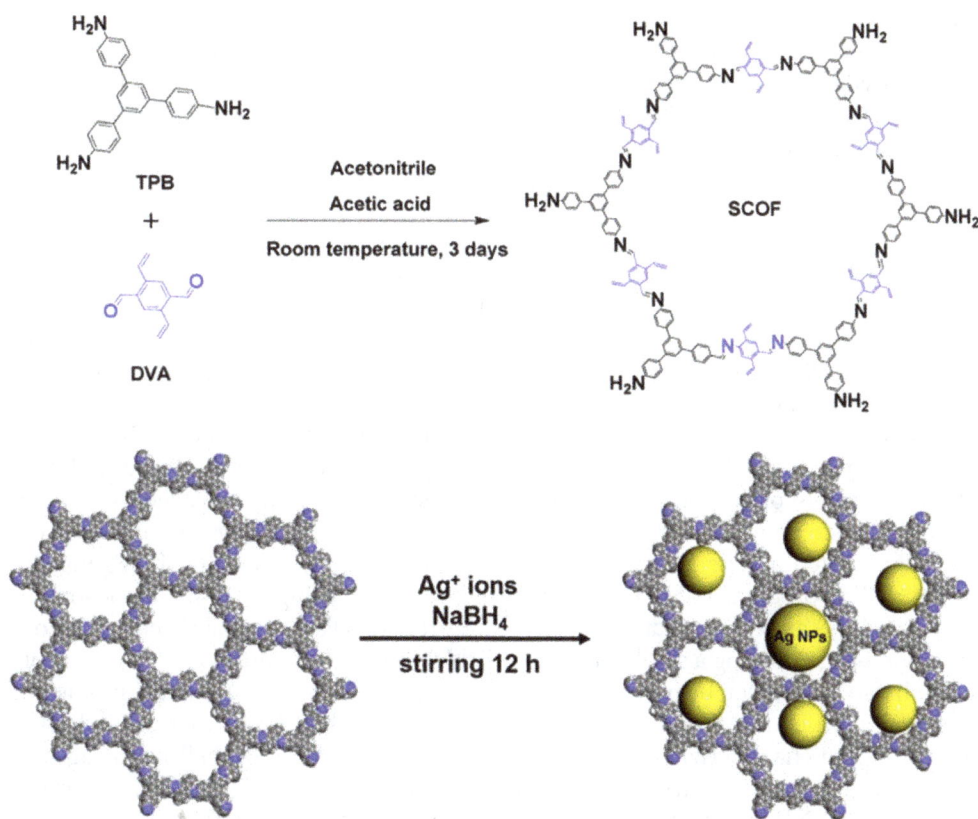

FIGURE 24.2
Synthesis route of Ag-NPs@SCOF through condensation reaction and loading of Ag-NPs on SCOF. (Adapted with permission [32]. Copyright (2021) American Chemical Society.)

to the robust coordination between Ag-NPs and N-functional groups of SCOF and the large size of the SCOF. However, the compression of the catalyst bed by the flow of sub-strate solution can compact the structure of the bed, which leads to the decline of total flux. Moreover, the M-NPs with extra small size can leach out during the flow reaction, which leads to an increase in the processing cost with their difficult recovery. Thus, it is highly important to choose a proper host to load COFs-NPs for continuous fixed bed catalysts.

Sand is well-known to be a low-cost naturally occurring material, has the benefits of open pores, and better hydrophilicity, which is advantageous for fast transportation of sub-strate solution in continuous flow catalytic operation. Furthermore, it is a stable material with a large size and is hard to leach out during flow catalysis. Considering these points, Pan et al. [33] reported the loading of COF-Ag-NPs on the sand to get heterostructure of COF-Ag-NPs@sand and used as a catalyst in fixed bed continuous flow catalytic environmental reme-diation. They prepared COF by the reported process [32] discussed earlier and the COF was used as host materials for Ag-NPs. The characterization of the obtained product revealed that ultrafine Ag-NPs (0.11 wt%) with a mean diameter of 5 nm were well dispersed onto COF-Ag-NPs@sand hybrid, and the catalyst has SSA of 316 m^2/g. The COF-Ag-NPs@sand was used for catalytic reduction of several organic pollutants including MB, 4-NP, and CR in an aqueous system. Both batch and continuous flow operations were performed and good flux (2000 L/m^2/h) with efficiency (> 99%) in reduction of CR, 4-NP, and MB were obtained with fixed bed continuous flow catalysis. More importantly, the reduction efficiency was remained unchanged for 24 h operation and only a minute amount of Ag-NPs was leached out from the catalyst, which revealed the better catalytic performance and stability of the catalyst. Besides the stable structure of the catalyst, the excellent efficiency of the catalyst can be ascribed to the sand properties including its large size which prevented the release of Ag-NPs, and good hydrophilicity for fast transportation of substrate solution.

24.2.3 COF-Based Pd-NPs

Palladium (Pd) NPs are of great importance in catalysis due to their superior activity as a cat-alyst for a wide range of chemical reactions including environmental remediation, coupling, organic synthesis, reforming, redox reaction, hydrogenation, oxidation, and gas storage [34, 35]. The Pd-NPs based hybrid materials have widened use for heterogeneous catalysts in catalytic remediation of organic contaminants such as reduction of dyes and 4-NP in an aqueous system [36]. Yolk–shell nanostructures, which are comprised of M-NPs core sur-rounded by a porous shell, provide a great possibility to achieve the best activity of M-NPs, in which M-NPs core offers the active spot for catalysis, while the porous shell functions as an obstacle to stop aggregation/coalescence of the M-NPs and improve their stability. Cui et al. [37] reported a general approach for the incorporation of several M-NPs inside the pores of COF to produce M-NPs@COF yolk-shell coops by using core-shell nanostructures of M-NPs@ZIF-8 as self-template. Several ligand-free M-NPs were stabilized inside of the hollow nano-cages having a shield of COF shell. Notably, the properties and functions of the yolk-shell nano-cages can be modified by changing the order of the COF shell. In this multi-step strategy, different M-NPs (Au, Ag, Pd) were prepared first and then loaded on ZIF-8. The structure characterizations of Pd@H-TpPa revealed the yolk-shell nano-cages with Pd-NPs core and porous TpPa shell. The obtained Pd@H-TpPa exhibited a good performance as a catalyst in 4-NP reduction with $NaBH_4$ at normal temperature with good stability. The recycling use of Pd@H-TpPa revealed that the activity of the catalyst was remained nearly constant for four times reused without noticeable change. The good catalytic efficiency of Pd@H-TpPa was primarily attributed to its yolk-shell structure which protected the Pd-NPs from aggregation and leaching out by the COF shell, whereas permeability of the COF shell and the pores formed between Pd-NPs cores and COF shell provide a low transport barrier and enhanced the availability of the active sites for catalysis.

Similarly, Miguel et al. [38] reported a facile two-steps technique that facilitates adding the new functionalities to COFs by M-NPs confinement. In the first step, M-NPs were entrapped into amorphous imine-functionalized spheres under normal conditions and in the second step, the spheres were converted to their analogs crystalline COF under acidic conditions at a high temperature (Figure 24.3). The encapsulation of Pd-NPs into COFs (Pd/c-1) spheres was confirmed by structure characterizations of the solids obtained after the completion of the reaction. The same two-steps technique was used to encapsulate Fe_3O_4-NPs and Au-NPs in spherical COFs. These metallic M-NPs (Fe_3O_4, Au, and Pd)/COF were used as catalysis for 4-NP reduction with $NaBH_4$ in an aqueous system and their good performance was observed. It was found that the porous structure of COF provides the transportation of substrates molecules to the entrapped M-NPs. However, their catalyst activity was lower than compared with other COFs based M-NPs hybrid catalysts.

24.2.4 COF-Based Pt-NPs

Platinum (Pt) NPs have also been proved excellent catalysts for the treatment of organic contaminants from wastewaters. Pt-NPs have efficient activity as catalysts but their use is

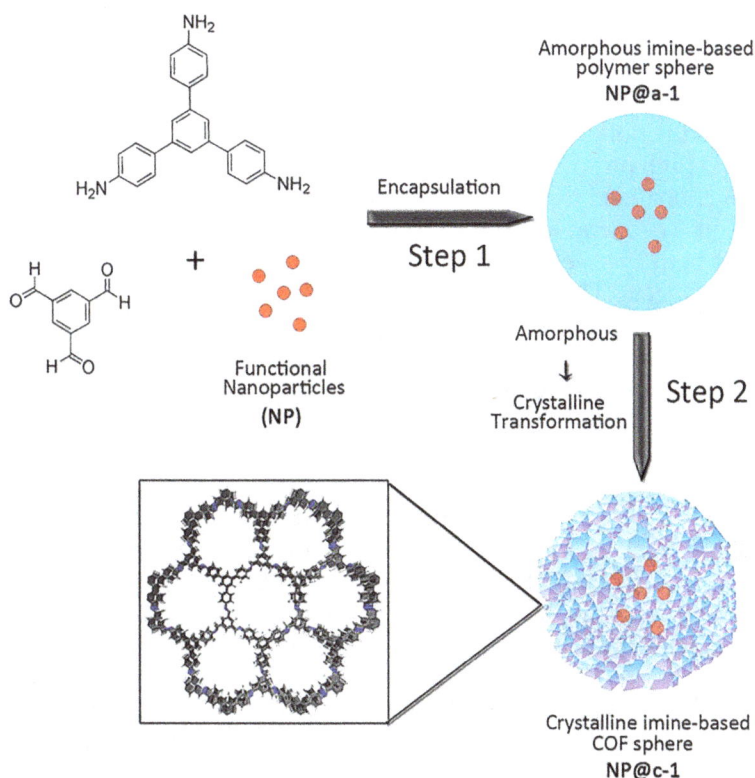

FIGURE 24.3
Schematic two steps encapsulation–crystallization approach for the incorporation M-NPs into COF spheres. (Adapted with permission [38]. Copyright (2017) Wiley Online Library.)

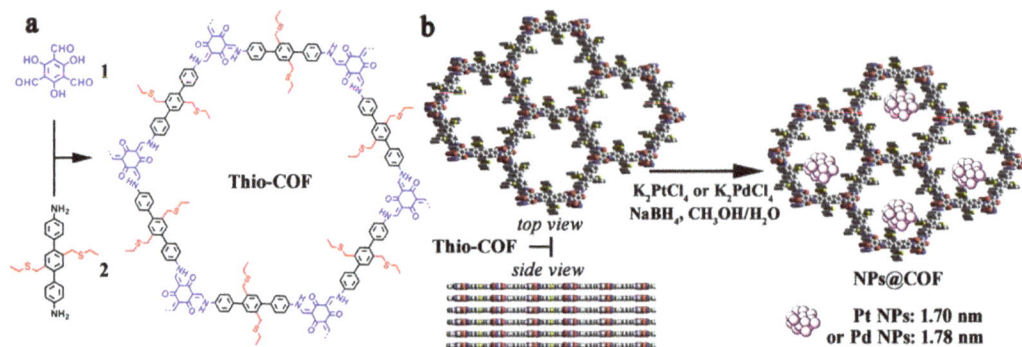

FIGURE 24.4
Schematic illustration of the fabrication of thio-COF (a) and Pt-NPs@COF and Pd-NPs@COF (b), and top and side view of the energy-minimized models (blue, N; yellow, red, O; S; gray, C) of thio-COF. (Adapted with permission [19]. Copyright (2017) American Chemical Society.)

not encouraged due to the high cost of Pt metal. To date, a few publications are available on stabilization of Pt-NPs onto COFs and their use in catalytic reduction of organic contaminants. It has been suggested that the incorporation of M-NPs inside the cavities/pores of COFs are crucial to controlling the size of M-NPs by stopping their coalescence and aggregation. Lu et al. [19] reported the synthesis of highly dispersed Pt-NPs and Pd-NPs with ultrafine size confined in activities of a thioether-functionalized COFs, and their activity as catalysts was observed for the reduction of organic contaminants in water. The scheme for the fabrication of Pd-NPs@COF or Pt-NPs@COF is shown in Figure 24.4. The structure characterization of obtained catalysts revealed that the maximum of Pd-NPs or Pt-NPs were anchored in the pores and some deposit onto the surface of COFs with ultrafine size (1.7 ± 0.2 nm) and high uniformity. It was observed that the well-ordered cavities structure of thioether functionalized COF was vital to get M-NPs (Pt-NPs or Pd-NPs) with high uniform distribution. The metal–sulfur interactions enhanced the stability of M-NPs (Pt-NPs or Pd-NPs) and both the well-order cavities structure and COF shell limited the space for M-NPs to aggregate. The as-prepared Pt-NPs@COF and Pd-NPs@COF were used as catalysts for 4-NP reduction and good activity was observed. More importantly, Pd-NPs@COF and Pt-NPs@COF were easily recovered and reused six times, which revealed their good stability without a significant decline in catalytic activity.

24.2.5 COF-Based Cu-NPs

Copper (Cu) is low-cost metal compare to Pt, Pd, Au, and Ag. It has good electrical conductivity and stability and has been reported as a catalyst for the treatment of organic contaminants such as the reduction of dyes and 4-NP. Usually, conductive materials such as reduced molybdenum disulfide and graphene oxide were used as supporting hosts for Cu-NPs due to their effects as co-catalysts and good Cu-NPs dispersion. Amino group functionalized COFs are alleged to enhance the coordination of Cu-NPs with COFs and improve their uniform dispersion with the assistance of amino groups. However, it is a challenging task to fabricate COFs with abundant amino groups directly because amines can react with the monomers and prevent the fabrication of COFs. Considering this challenge, Yan et al. [39] reported a process of building block interchange strategy to fabricate amino-functionalized COF, but this process needs extra precise control and a longer processing time. Fabricating COFs with abundant amine functional groups by a simple technique is highly significant, and its performance study will widen the uses of COFs as M-NPs supporting hosts. In this context, Jin et al. [40] reported the simple strategy to construct an amino-functionalized COF, TpPa–NH$_2$, through an easy reduction in TpPa–NO$_2$ and used as a supporting host for Cu-NPs to get a hybrid Cu-NPs@COF. The characterizations of Cu-NPs@COF verified that the Cu-NPs were homogenously anchored onto the surface of COF, structure properties of COF were not changed by functionalization and Cu@COF has an SSA of 215 m^2/g. The obtained Cu@COF was used for 4-NP reduction, and its good activity, recyclability, and stability were found as a catalyst. However, ample centrifugation and filtration for recycling COF-based M-NPs not only increase the cost of the process but also affect the activity of catalysts.

To solve the above-mentioned issues, Wang et al. [41] reported the easily recyclable montmorillonite supported Cu-NPs catalyst of spherical shape doping with a COF for a 4-NP reduction in an aqueous system. For the synthesis of COF-SH, a solution of mesitylene and 1, 4-dioxane was prepared, and 2,5-diamino-1,4-benzenedithiol dihydrochloride and 1,3,5-tris (4-formylphenyl)benzene were added to the solution. The aqueous acetic acid was added as a catalyst and the reaction was continued at ambient conditions to get precipitates. The formed precipitates were centrifuged, rinsed with ethanol and THF, and dried under reduced pressure. Afterward, the prepared COF was dispersed in water and CuSO$_4$·5H$_2$O was added under stirring. After the addition of dopamine, the reaction was continued for a fixed time. The formed precipitates were centrifuged and rinsed with water/isopropanol solution and dried under reduced pressure. On the other hand, a dopamine aqueous solution was prepared, SMt was mixed and shacked well to get PDA-coated SMt. Afterward, COF@Cu-NPs and PDA-coated SMt were dispersed in water and mixed for a specific time to get desired product SMt@COF@Cu and rinsed with water and dried under reduced pressure (Figure 24.5). Characterization of an obtained product revealed the catalysts spheres mean size of 16 ± 4 μm with SSA of 74.4 m^2/g. SMt@COF@Cu-NPs was exhibited good performance as a catalyst for 4-NP reduction by NaBH$_4$ with a reduction rate constant (k) of 5.3 min^{-1} and found that COF played the role as a co-catalyst to boost the reaction rate. The excellent performance of the catalyst can be credited to the presence of ample Cu ions at low pH in inclusion to the co-catalysis of COF. Moreover, SMt@COF@Cu-NPs was easily recovered by self-sedimentation without the use of any devices and reused for subsequent reduction and found that the activity of SMt@COF@Cu was remained constant after five cycles of reused, revealing the high stability of the catalyst.

24.2.6 COF-Based Bimetallic Catalysts

Fabrication of bimetallic NPs (BM-NPs) based catalysts have been adopted to control the cost of transition metals NPs and boost the electron transfer operation by increasing the active sites over monometallic counterparts. Magnetic CuFe$_2$O$_4$-NPs are widely used as catalytic environmental remediation due to their better catalytic performance, good stability, and simple separation. Bimetallic composites not only reduce the aggregation of M-NPs but also increase the performance by synergistic effects of both the metals. COF could be a potential scaffold compared to other porous supports for constructing the bi-metallic nanocomposite. Hou et al. [42] reported a simple fabrication technique of core-shell structure to combine CuFe$_2$O$_4$/Ag-NPs and COF for the reduction of 4-NP. The characterizations of the obtained product revealed the core-shell structure (i.e., CuFe$_2$O$_4$/Ag-NPs as core and COF as an outer shell) with a high SSA of 464.21 m^2/g, mesoporous structure, good stability, and quick magnetic response. CuFe$_2$O$_4$/Ag@COF was used as a catalyst in 4-NP reduction and its attraction toward 4-NP was observed via π-π stacking

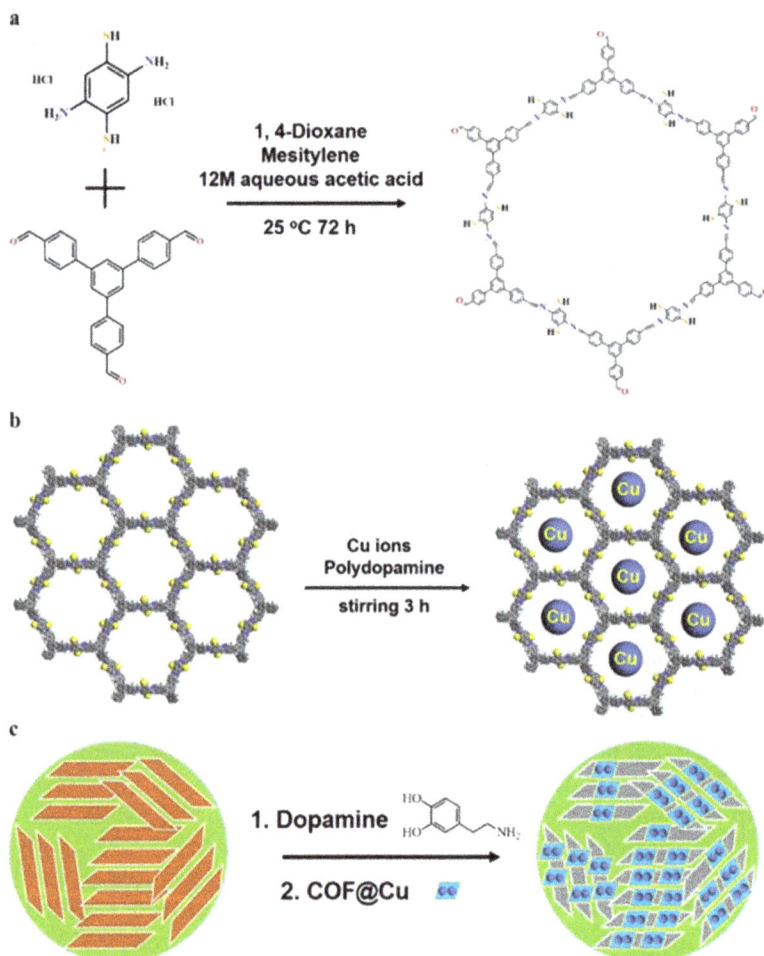

FIGURE 24.5
Fabrication route of SMt@COF@Cu by condensation of 2,5-diamino-1,4-benzenedithiol dihydrochloride and 1,3,5-tris (4-formylphenyl) benzene (a). Chelation of Cu ions via functionalization onto COF and Cu ions reduction via dopamine (b). The synthesis of polydopamine modified SMt as a host to support COF@Cu (c). (Adapted with permission [41]. Copyright (2021) Elsevier.)

which boosted the catalytic activity by three to five times higher compared with $CuFe_2O_4/$ Ag-NPs and $CuFe_2O_4$@COF. It was concluded that the synergetic effect of bi-metals and the highly porous structure of the COF shell make an impact to enhance the performance of $CuFe_2O_4/$Ag@COF as a catalyst for 4-NP reduction.

24.3 Characterizations

The catalytic phenomenon of COF-based M-NPs toward the reduction of organic pollutants depends upon M-NPs size, size dispersion, and morphology of the heterostructure composite. A wide range of methods has been used for the characterization of COF-based

M-NPs. The theoretical details of these techniques are beyond the scope of this chapter. Single-crystal X-ray diffraction (SXRD) has been used to study the structures including the atomic positions, unit-cell parameters, bond angles, and lengths of COFs. Powder X-ray diffraction (XRD) is used to study the crystallinity of COFs which can be further verified by theoretical simulation. Transmission electron microscopy (TEM) is used to find the size, size distribution, and dispersion of supported M-NPs onto COFs. The scanning electron microscope (SEM) technique is used to observe the shape and surface morphology of catalysts. Brunauer–Emmett–Teller (BET) test is useful to measure the SSA of COF and COF-based M-NPs catalysts. Functional groups identification can be done by Fourier transform infrared spectroscopy (FT-IR) and chemical bonding can be studied by solid-state nuclear magnetic resonance (SS-NMR). X-ray photoelectron spectroscopy (XPS) is used for identifying the oxidation state of M-NPs loaded on COFs. Thermogravimetric analysis (TGA) is used to study the thermal stability of COFs and COFs-based M-NPs. The number-average size of particles is measured by the dynamic light scattering (DLS) technique. Notably, the above-mentioned techniques are general to characterize the COFs-based material and are applied for almost all kinds of reported COF-based M-NPs catalysts. For example, Pan et al. [33] characterized the COF-Ag-NPs@sand by SS-NMR, which showed the chemical shift at 140.2 ppm for C = N linkage (Figure 24.6a). The BET test revealed that the SSA of COF was decreased from 1106 m²/g to 316 m²/g by the loading of Ag-NPs, which indicated that Ag-NPs were stabilized in the pores of COFs (Figure 24.6b). XRD not only confirmed the crystal structure of COF but also verified the presence of Ag-NPs on the surface of COF by giving their diffraction peaks (Figure 24.6c). The formation validation of COF-Ag NPs@sand was done by verifying the signals of silicon, N, and Ag in XPS analysis (Figure 24.6d). The amount of Ag-NPs in COF-Ag NPs@sand was obtained by ICP/MS which gave 0.11 wt% of Ag-NPs. The crystallinity of COF was also observed in the TEM photo (Figure 24.6e) and Ag-NPs size (5 nm) and size dispersion onto COF-Ag NPs@sand were observed from the TEM photo (Figure 24.6f). Inter-planar spacing of 0.118 nm, assigned to (222) plane of Ag-NPs crystal was found from HR-TEM photo (Figure 24.6g).

24.4 Monitoring of Catalytic Reduction of Organic Pollutants

4-NP and most of the organic dyes show absorption peaks in an aqueous solution. Thus, their catalytic reduction can be easily examined by UV–vis spectrophotometry by monitoring the intensity of their absorption peaks with a fixed time interval. The decline in the absorption peak of organic pollutants by the addition of COF-based M-NPs and the reducing agent (NaBH$_4$) was observed at a regular time interval until the absorption peak was vanished (Figure 24.7), which indicated the completion of the reaction. The percentage reduction of organic dyes is usually determined from UV-vis spectra using the following Eq. (24.1):

$$\text{Organic pollutant reduction } (\%) = (C_0 - C_t / C_0) \times 100 \tag{24.1}$$

Where, C_0 and C_t are the initial concentration and the concentration of organic pollutant at any time t, respectively.

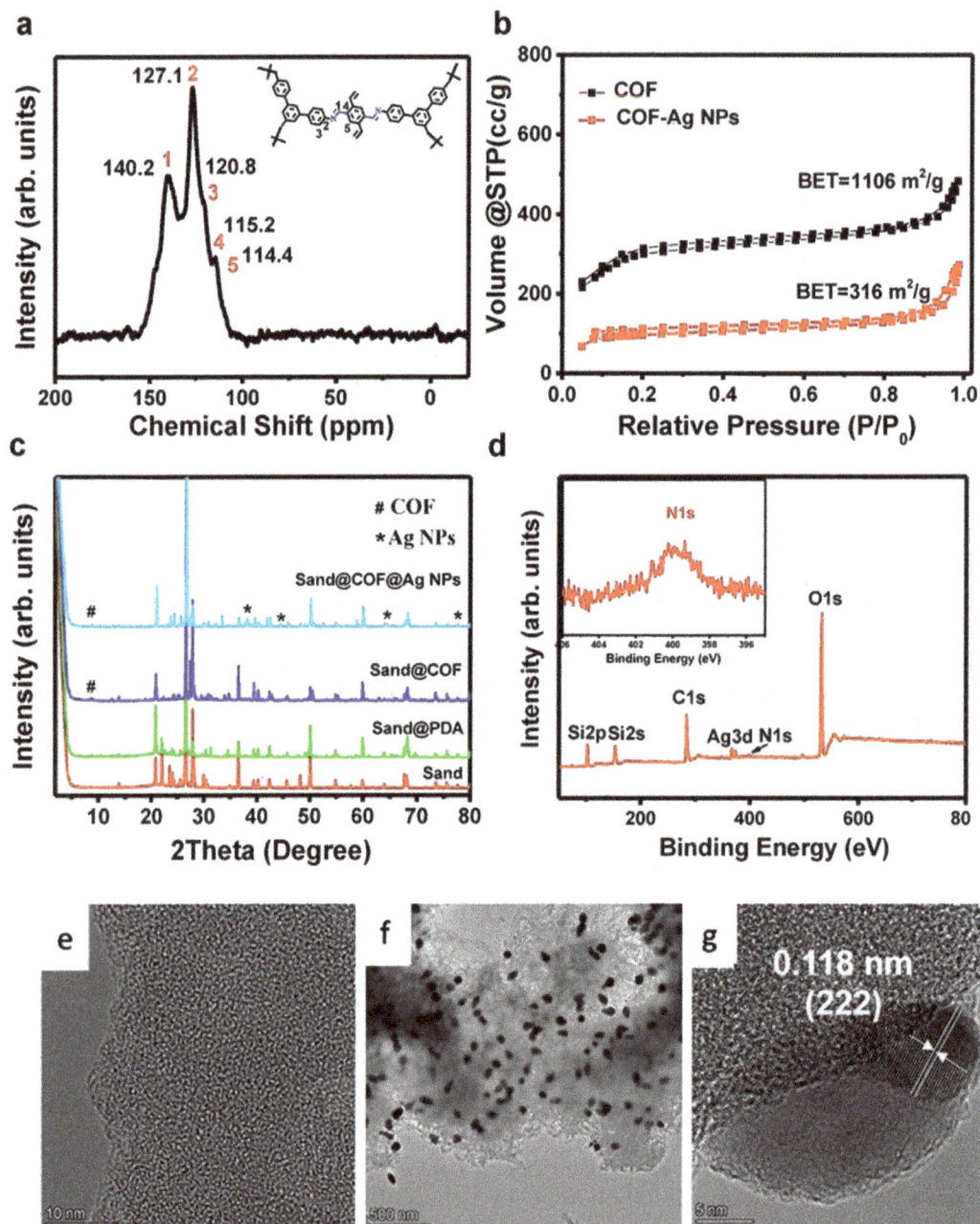

FIGURE 24.6
Structural study of COF-Ag-NPs@sand by SS-NMR (a), BET test N_2 adsorption/desorption curves (b), XRD (c), and XPS (d); TEM photos of COF (e), COF-Ag-NPs (f), and COF-Ag NPs@sand (g) through HR-TEM. (Adapted with permission [33]. Copyright (2021) Elsevier.)

24.5 Kinetics Study of Catalytic Reduction of Organic Pollutants

The activity of COFs-based M-NPs as a catalyst for the reduction of organic pollutants is usually expressed in terms of reaction rate constant (k). In general, the reduction of organic pollutants (4-NP and dyes) catalyzed by COF-based M-NPs followed the Pseudo first-order kinetic model which was formulated in Eq. (24.2)

$$\ln(C_t/C_0) = -kt \text{ or } \ln(A_t/A_0) = -kt \tag{24.2}$$

Where, C_0 and A_0 are the initial concentration and absorption, and C_t and A_t is the concentration and absorption at any time t of organic pollutants.

For example, Pan et al. [33] experimented by adding COF-Ag-NPs@sand as catalysts and NaBH$_4$ as reducer in the aqueous solution of several organic contaminants such as MB, 4-NP, and CR, and their reduction process was observed by UV–vis spectroscopy. It was examined that the absorption peaks at 400 nm (for 4-NP, Figure 24.7a) were disappeared

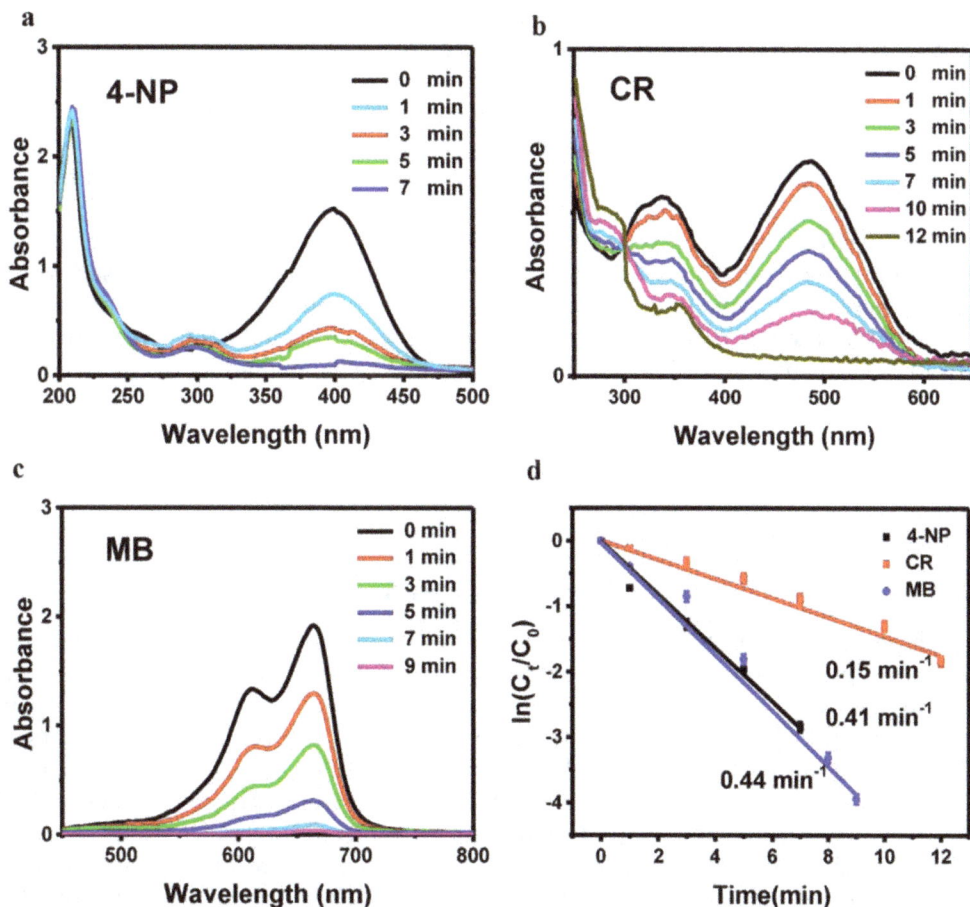

FIGURE 24.7

UV–vis spectra of reduction of 4-NP (a), CR (b), and MB (c) by NaBH$_4$ in the presence of COF-Ag NPs@sand as catalyst and plot of ln (C_t/C_0) versus t for their reduction kinetic study (d). (Adapted with permission [33]. Copyright (2021) Elsevier.)

around 7 min, at 510 nm (for CR, Figure 24.7b) was vanished around 12 min, and at 665 nm (for MB, Figure 24.7c) was extinct around 9 min, which was indicated the completion of their reduction by COF-Ag NPs@sand catalyst. $\ln(C_t/C_0)$ versus t was plotted and found that the reduction of 4-NP, MB, and CR catalyzed by COF-Ag NPs@sand in the presence of $NaBH_4$ followed the Pseudo-first-order reaction kinetics and their k values were found to be 0.15 min^{-1} (for CR), 0.41 min^{-1} (for MB), and 0.44 min^{-1} (for 4-NP) as shown in Figure 24.7d. Moreover, they found that the value of k was the function of several operational factors such as reaction temperature, pH, substrate concentration, reducer concentration, catalyst size and porous properties, and amount of catalyst in the reaction mixture.

24.6 Recyclability of COF-Based M-NPs Catalysts

Separation and reuse of COF-based M-NPs heterogeneous catalysts are the important benchmarks to judge their efficacy to compare with homogeneous catalysts [1, 43]. It is highly desired to separate M-NPs based heterogeneous catalysts from the reaction medium after completion of organic contaminants reduction to reuse them in subsequent reactions. Moreover, transition metals such as Au, Pt, Ag, and Pd are expensive, thus their recycling is necessary to make their maximum reuse to optimize the process. The recovered M-NPs must be capable to do the same task for the next cycle of the same reaction. In general, there are mainly three types of the recycling process, that is, centrifugation, magnetic field, and self-sedimentation for COF-based M-NPs catalysts reported for the degradation of organic pollutants. The stability of catalysts is observed by their recycling performance, retaining of loaded M-NPs structure and morphology during the recovery process which can be confirmed by the characterizations of isolated catalysts.

24.7 Mechanism of 4-NP and Organic Dyes Reduction

The reduction of organic dyes and 4-NP by $NaBH_4$ catalyzed by COF-based M-NPs is a surface phenomenon. It is to note that the presence of both $NaBH_4$ and catalyst are compulsory for the reduction to occur; none of them alone can reduce the organic dyes and 4-NP [36]. The M-NPs anchored on COFs act as a facilitator and offers the surface for the deposition of the substrate molecules and $NaBH_4$ (BH_4^- anions). The role of M-NPs is to facilitate the transfer of an electron from BH_4^- to the dyes or 4-NP molecules and lead to their degradation/reduction through oxidation and reduction reactions (Figure 24.8). Catalysis occurs on the location where M-NPs are accessible on the surface of COF-based catalysts. Hence, the highly porous structure of COFs and uniform dispersion of M-NPs on their surface with ultra-small size are certainly advantageous parameters to the accessibility of active sites (M-NPs), and to the efficiency of catalysis by consequence.

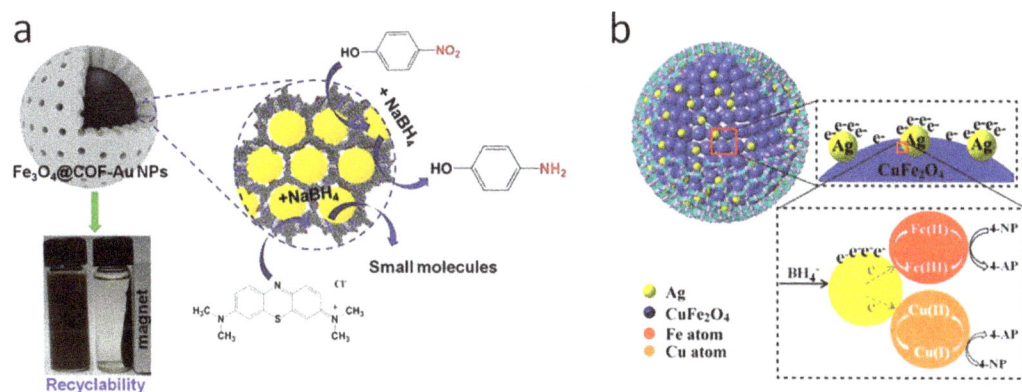

FIGURE 24.8
Illustration of proposed mechanisms for the reduction of organic pollutants by COF-based M-NPs monometallic (a) (Adapted with permission [27], Copyright (2020) Elsevier) and bi-metallic (b). (Adapted with permission [42]. Copyright (2020) Multidisciplinary Digital Publishing Institute.)

24.8 Importance of Organic Pollutants Reduction

The uncontrolled release of organic dyes, frequently used in textile, leather, foodstuff, cosmetics, paper, plastic, and ink industries, is inducing harm to aquatic life. Most of these are oxygen sequestering chemicals and their tenacity in water-bodies has been characterized to decrease light diffusion and thus stop photosynthesis of aquatic flora. Moreover, the presence of 4-NP in wastewaters, originate from several industries including pesticides, pharmaceuticals, petrochemical, dyes and wood, explosives and preservatives is approved priority hazardous by the EPA that have a direct impact on human health and the environment [36, 44–47]. Therefore, it is essential to remediate toxic organic dyes and 4-NP contaminated wastewaters before their discharge into the natural environment and water bodies.

24.9 Conclusion and Perspective

In this chapter, the research progress on the synthesis of COFs-based M-NPs catalysts for the reduction of toxic organic dyes and 4-NP is overviewed and summarized. From the outcomes of several studies, it is concluded that functionalized-COFs-supported M-NPs with ultra-small size and uniform size dispersion of M-NPs not only exhibited excellent activities in catalytic reduction of organic dyes and 4-NP with NaBH$_4$ but also showed remarkable recyclability. Moreover, the strong interaction between M-NPs and functional groups of COFs improved the binding of M-NPs onto the surface of COFs and maintained the uniform dispersion of M-NPs, in results sustaining the catalytic activity of COFs-based-MNPs for long-term use.

Although highly porous and crystalline COFs are considered to be one of the most effective supports for M-NPs catalysts, they suffer from complicated multiple fabrication steps with their low yield. Developing new facile fabrication strategies which gave a high yield of COFs-based M-NPs with their excellent surface properties such as functionalization, high SSA, uniform pore size dispersion with high porosity is still a challenging task to an extent of their practical applications.

References

1. M.S. Bashir, X. Jiang, S. Li, X.Z. Kong, Highly uniform and porous polyurea microspheres: Clean and easy preparation by interface polymerization, palladium incorporation, and high catalytic performance for dye degradation, Frontiers in Chemistry 7(314) (2019).
2. A. Tkaczyk, K. Mitrowska, A. Posyniak, Synthetic organic dyes as contaminants of the aquatic environment and their implications for ecosystems: A review, Science of The Total Environment 717 (2020) 137222.
3. M. Shahid, Z.H. Farooqi, R. Begum, M. Arif, W. Wu, A. Irfan, Hybrid microgels for catalytic and photocatalytic removal of nitroarenes and organic dyes from aqueous medium: A review, Critical Reviews in Analytical Chemistry 50(6) (2020) 513–537.
4. M. Kaykhaii, M. Sasani, S. Marghzari, Removal of dyes from the environment by adsorption process, Chem. Mater. Eng 6(2) (2018) 31–35.
5. M.S. Bashir, S. Zaheer, A.R. Farooq, M. Jamil, A. Safdar, Production of biogas from kitchen waste with inoculum and study the effect of different parameters, Journal of the Chemical Society of Pakistan 41(4) (2019) 563–658.
6. X. Yang, X. Jiang, M.S. Bashir, X.Z. Kong, Preparation of highly uniform polyurethane microspheres by precipitation polymerization and Pd immobilization on their surface and their catalytic activity in 4-nitrophenol reduction and dye degradation, Industrial & Engineering Chemistry Research 59(7) (2020) 2998–3007.
7. C. Toumey, Plenty of room, plenty of history, Nature Nanotechnology 4(12) (2009) 783–784.
8. R. Tao, X. Ma, X. Wei, Y. Jin, L. Qiu, W. Zhang, Porous organic polymer material supported palladium nanoparticles, Journal of Materials Chemistry A 8(34) (2020) 17360–17391.
9. M.S. Bashir, Benign fabrication process of hierarchal porous polyurea microspheres with tunable pores and porosity: Their Pd immobilization and use for hexavalent chromium reduction, Chemical Engineering Research and Design 175 (2021) 102–114.
10. M.S. Bashir, X. Jiang, X. Yang, X.Z. Kong, Porous polyurea supported Pd catalyst: Easy preparation, full characterization, and high activity and reusability in reduction of hexavalent chromium in aqueous system, Industrial & Engineering Chemistry Research 60(22) (2021) 8108–8119.
11. S.S.A. Shah, T. Najam, M. Wen, S.-Q. Zang, A. Waseem, H.-L. Jiang, Metal–Organic Framework-Based Electrocatalysts for CO2 Reduction, Small Structures n/a(n/a) 2100090.
12. S.S.A. Shah, T. Najam, M.K. Aslam, M. Ashfaq, M.M. Rahman, K. Wang, P. Tsiakaras, S. Song, Y. Wang, Recent advances on oxygen reduction electrocatalysis: Correlating the characteristic properties of metal organic frameworks and the derived nanomaterials, Applied Catalysis B: Environmental 268 (2020) 118570.
13. J. Liu, L. Chen, H. Cui, J. Zhang, L. Zhang, C.-Y. Su, Applications of metal–organic frameworks in heterogeneous supramolecular catalysis, Chemical Society Reviews 43(16) (2014) 6011–6061.
14. P. Kaur, J.T. Hupp, S.T. Nguyen, Porous organic polymers in catalysis: Opportunities and challenges, ACS Catalysis 1(7) (2011) 819–835.
15. Y. Song, Q. Sun, B. Aguila, S. Ma, Opportunities of covalent organic frameworks for advanced applications, Advanced Science 6(2) (2019) 1801410.
16. M.B. Hussain, M.S. Khan, H.M. Loussala, M.S. Bashir, The synthesis of a BiOClxBr1–x nanostructure photocatalyst with high surface area for the enhanced visible-light photocatalytic reduction of Cr(vi), RSC Advances 10(8) (2020) 4763–4771.

17. A.P. Cote, A.I. Benin, N.W. Ockwig, M. O'Keeffe, A.J. Matzger, O.M. Yaghi, Porous, crystalline, covalent organic frameworks, Science 310(5751) (2005) 1166–1170.

18. R.K. Sharma, P. Yadav, M. Yadav, R. Gupta, P. Rana, A. Srivastava, R. Zbořil, R.S. Varma, M. Antonietti, M.B. Gawande, Recent development of covalent organic frameworks (COFs): synthesis and catalytic (organic-electro-photo) applications, Materials Horizons 7(2) (2020) 411–454.

19. S. Lu, Y. Hu, S. Wan, R. McCaffrey, Y. Jin, H. Gu, W. Zhang, Synthesis of ultrafine and highly dispersed metal nanoparticles confined in a thioether-containing covalent organic framework and their catalytic applications, Journal of the American Chemical Society 139(47) (2017) 17082–17088.

20. S. Wang, Q. Zhao, H. Wei, J.-Q. Wang, M. Cho, H.S. Cho, O. Terasaki, Y. Wan, Aggregation-free gold nanoparticles in ordered mesoporous carbons: Toward highly active and stable heterogeneous catalysts, Journal of the American Chemical Society 135(32) (2013) 11849–11860.

21. R.J. White, R. Luque, V.L. Budarin, J.H. Clark, D.J. Macquarrie, Supported metal nanoparticles on porous materials. Methods and applications, Chemical Society Reviews 38(2) (2009) 481–494.

22. C.E. Chan-Thaw, A. Villa, P. Katekomol, D. Su, A. Thomas, L. Prati, Covalent Triazine Framework as Catalytic Support for Liquid Phase Reaction, Nano Letters 10(2) (2010) 537–541.

23. S. Proch, J. Herrmannsdörfer, R. Kempe, C. Kern, A. Jess, L. Seyfarth, J. Senker, Pt@MOF-177: Synthesis, room-temperature hydrogen storage and oxidation catalysis, Chemistry – A European Journal 14(27) (2008) 8204–8212.

24. P. Pachfule, S. Kandambeth, D. Díaz Díaz, R. Banerjee, Highly stable covalent organic framework–Au nanoparticles hybrids for enhanced activity for nitrophenol reduction, Chemical Communications 50(24) (2014) 3169–3172.

25. Q.-P. Zhang, Y.-l. Sun, G. Cheng, Z. Wang, H. Ma, S.-Y. Ding, B. Tan, J.-h. Bu, C. Zhang, Highly dispersed gold nanoparticles anchoring on post-modified covalent organic framework for catalytic application, Chemical Engineering Journal 391 (2020) 123471.

26. X. Shi, Y. Yao, Y. Xu, K. Liu, G. Zhu, L. Chi, G. Lu, Imparting catalytic activity to a covalent organic framework material by nanoparticle encapsulation, ACS Applied Materials & Interfaces 9(8) (2017) 7481–7488.

27. Y. Xu, X. Shi, R. Hua, R. Zhang, Y. Yao, B. Zhao, T. Liu, J. Zheng, G. Lu, Remarkably catalytic activity in reduction of 4-nitrophenol and methylene blue by Fe₃O₄@COF supported noble metal nanoparticles, Applied Catalysis B: Environmental 260 (2020) 118142.

28. F. Wang, F. Pan, S. Yu, D. Pan, P. Zhang, N. Wang, Towards mass production of a spherical montmorillonite@covalent organic framework@gold nanoparticles heterostructure as a high-efficiency catalyst for reduction of methylene blue, Applied Clay Science 203 (2021) 106007.

29. M.K. Bhunia, S.K. Das, P. Pachfule, R. Banerjee, A. Bhaumik, Nitrogen-rich porous covalent imine network (CIN) material as an efficient catalytic support for C–C coupling reactions, Dalton Transactions 41(4) (2012) 1304–1311.

30. R.-L. Wang, D.-P. Li, L.-J. Wang, X. Zhang, Z.-Y. Zhou, J.-L. Mu, Z.-M. Su, The preparation of new covalent organic framework embedded with silver nanoparticles and its applications in degradation of organic pollutants from waste water, Dalton Transactions 48(3) (2019) 1051–1059.

31. A. Shen, R. Luo, X. Liao, C. He, Y. Li, Highly dispersed silver nanoparticles confined in a nitrogen-containing covalent organic framework for 4-nitrophenol reduction, Materials Chemistry Frontiers 5 (2021) 6923–6930.

32. N. Wang, F. Wang, F. Pan, S. Yu, D. Pan, Highly efficient silver catalyst supported by a spherical covalent organic framework for the continuous reduction of 4-nitrophenol, ACS Applied Materials & Interfaces 13(2) (2021) 3209–3220.

33. F. Pan, F. Xiao, N. Wang, Towards application of a covalent organic framework-silver nanoparticles@sand heterostructure as a high-efficiency catalyst for flow-through reduction of organic pollutants, Applied Surface Science 565 (2021) 150580.

34. S.H. Ryu, S.J. Choi, J.H. Seon, B. Jo, S.M. Lee, H.J. Kim, Y.-J. Ko, K.C. Ko, T.K. Ahn, S.U. Son, Visible light-driven Suzuki–Miyaura reaction by self-supported Pd nanocatalysts in the

formation of Stille coupling-based photoactive microporous organic polymers, Catalysis Science & Technology 10(16) (2020) 5535–5543.

35. M. Qasim, M.S. Bashir, S. Iqbal, Q. Mahmood, Recent advancements in α-diimine-nickel and -palladium catalysts for ethylene polymerization, European Polymer Journal 160 (2021) 110783.

36. M.S. Bashir, X. Jiang, X.Z. Kong, Porous polyurea microspheres with Pd immobilized on surface and their catalytic activity in 4-nitrophenol reduction and organic dyes degradation, European Polymer Journal 129 (2020) 109652.

37. K. Cui, W. Zhong, L. Li, Z. Zhuang, L. Li, J. Bi, Y. Yu, Well-defined metal nanoparticles@covalent organic framework yolk–shell nanocages by ZIF-8 template as catalytic nanoreactors, Small 15(3) (2019) 1804419.

38. D. Rodríguez-San-Miguel, A. Yazdi, V. Guillerm, J. Pérez-Carvajal, V. Puntes, D. Maspoch, F. Zamora, Confining functional nanoparticles into colloidal imine-based COF spheres by a sequential encapsulation-crystallization method, Chemistry-A European Journal 23(36) (2017) 8623–8627.

39. H.-L. Qian, Y. Li, X.-P. Yan, A building block exchange strategy for the rational fabrication of de novo unreachable amino-functionalized imine-linked covalent organic frameworks, Journal of Materials Chemistry A 6(36) (2018) 17307–17311.

40. D. Jin, B. Wang, X. Wu, J. Li, M. Mu, L. Chen, Construction and catalytic applications of an amino-functionalized covalent organic framework, Transition Metal Chemistry 44(8) (2019) 689–697.

41. F. Wang, F. Pan, G. Li, P. Zhang, N. Wang, Construction of spherical montmorillonite supported Cu-based catalyst doping with a covalent organic framework for 4-nitrophenol removal, Applied Clay Science 214 (2021) 106278.

42. C. Hou, D. Zhao, W. Chen, H. Li, S. Zhang, C. Liang, Covalent organic framework-functionalized magnetic $CuFe_2O_4$/Ag nanoparticles for the reduction of 4-nitrophenol, Nanomaterials 10(3) (2020) 426.

43. Y. Sun, M. Chi, M.S. Bashir, Y. Wang, M. Qasim, Influence of intramolecular π–π and H-bonding interactions on pyrazolylimine nickel-catalyzed ethylene polymerization and co-polymerization, New Journal of Chemistry 45(30) (2021) 13280–13285.

44. K. Shahzad, T. Najam, M.S. Bashir, M.A. Nazir, A.u. Rehman, M.A. Bashir, S.S.A. Shah, Fabrication of periodic mesoporous organo silicate (PMOS) composites of Ag and ZnO: Photo-catalytic degradation of methylene blue and methyl orange, Inorganic Chemistry Communications 123 (2021) 108357.

45. M.A. Nazir, M.S. Bashir, M. Jamshaid, A. Anum, T. Najam, K. Shahzad, M. Imran, S.S.A. Shah, A.u. Rehman, Synthesis of porous secondary metal-doped MOFs for removal of rhodamine B from water: Role of secondary metal on efficiency and kinetics, Surfaces and Interfaces 25 (2021) 101261.

46. M. Sultan, A. Javeed, M. Uroos, M. Imran, F. Jubeen, S. Nouren, N. Saleem, I. Bibi, R. Masood, W. Ahmed, Linear, and crosslinked polyurethanes based catalysts for reduction of methylene blue, Journal of Hazardous Materials 344 (2018) 210–219.

47. M.S. Bashir, N. Ramzan, T. Najam, G. Abbas, X. Gu, M. Arif, M. Qasim, H. Bashir, S.S.A. Shah, M. Sillanpää, Metallic nanoparticles for catalytic reduction of toxic hexavalent chromium from aqueous medium: A state-of-the-art review, Science of the Total Environment 829 (2022) 154475.

25

Covalent Organic Frameworks in Polymer Nanocomposites with Superior Thermo-Mechanical Properties and Electrochemical Applications

Parisa Najmi, Navid Keshmiri, Bahram Ramezanzadeh, and Mohammad Ramezanzadeh

Surface Coating and Corrosion Department, Institute for Color Science and Technology, Tehran, Iran

CONTENTS

25.1 Introduction

In the past decade, porous materials such as metal-organic frameworks (MOFs), silica (organosilica, micro/mesoporous silica), zeolites, and covalent organic frameworks (COFs) became hotspots in various applications. COFs have received outstanding attention among other organic materials in recent decades. This explosion of interest is attributed to their unique structure with nanoscale pores (ranges from 7 Å to 4.7 nm) as well as their possession of remarkable properties [1], including tunable pore size and geometry, high surface

DOI: 10.1201/9781003206507-25

area (4210 m^2/g), low density (0.17 g/cm^3), superior crystallinity, intrinsic adaptability, and flexibility, leads to a broad range of applications such as sorption [2], separation [3], drug delivery [4], energy [5], catalyst [6], and optoelectronics [7]. COFs are classified as porous materials that predominantly contain C, H, O, N, and B atoms and, for the first time, were synthesized by Yaghi in 2005 [2, 8, 9]. Based on the different building blocks and units, they can be categorized into three main classes: B-O (boronate, boroxine, and borosilicate), C=N (imine, hydrazine, and squaraine), and C–N (triazine and imide) [2]. One of the main hot spots that fascinated the researchers is the covalent chemistry/bonding of organic and inorganic molecules, resulting in many vital advancements in science and technology. The fabrication and modification of organic-based molecules with covalent bonding have changed the way of our life to a large extent in every aspect, including pharmaceuticals, chemicals, and polymers [10, 11]. MOFs and COFs are the two most popular classes of porous materials.

In contrast to COFs, MOFs comprise coordination bonds, meaning that their structures are weaker than covalent bonds, making them frangible. Therefore, covalent bonding through inorganic complexes/molecules has led to valuable catalytic efficiency in high activity and selectivity. However, this pinpoint preciseness and diversity with which covalent chemistry on such molecules is examined have not been translated to either the building units of extended structures or their modification. This is the next-generating idea of MOFs and COFs.

This chapter concentrates on the design and synthesis of COFs, their novel applications in metal corrosion protection, and polymer composites, and their thermo-mechanical properties improvement.

25.2 Design of COFs

There are two types of reaction control procedures, making the covalent bonds of COFs reversible or irreversible. Typically, the crystallization of COFs is in kinetic control, which results in irreversible covalent bonds. In comparison, higher reversibility can be achieved through dynamic covalent chemistry to correct errors efficiently. The design and synthesis of COFs can be divided into two groups, including kinetic products and thermodynamic products. Indeed, dynamic covalent chemistry is a concept in which the reaction is most thermodynamically controlled and thus offers reversible reactions. In other words, the small activation barrier (ΔG^*) leads to the formation of the kinetic intermediates very quickly, consequently forming kinetic products, which results in amorphous products. However, the thermodynamic products have lower Gibbs free energy, resulting in the re-equilibration of the reaction and hence more stable and crystalline product formation. In the first stage, the cross-linking polymers or hyper branching polymers are designed via building blocks. In the second stage, pores are formed. In the last stage, the correction of pores will be held via the dynamic covalent chemistry (DCC), and the crystalline COF will be formed in the reaction solution.

One of the vital issues in the fabrication and synthesis process of COFs is designing the building blocks since it may contribute to the correctness of the pores, enhancing the crystallinity of the final product. Furthermore, the formation of 2D COFs is much more straightforward and accessible than 3D COFs, since 2D COFs have more structural diversity than 3D ones [2].

25.3 Synthesis of COFs

Developing a productive and easy processing method for various materials production is pivotal since it leads to commercialization and industrialization. Therefore, various methods have been discovered and examined to fabricate COFs to meet the demand for several applications. This section will discuss the most important COF synthesis methods, including solvo/hydrothermal, ionothermal, microwave, mechanochemical, and room-temperature techniques. Figure 25.1 illustrates different types of COF synthesis methods.

25.3.1 Solvothermal

This synthesis condition is the most widely used procedure to fabricate COFs. The most straightforward technique is that organic monomers (precursors) and suitable solvents are placed in a tube, degassing with different cycles. Then, the tube is set to the specified temperature and reaction time, taking about two to nine days. Ultimately, the collected precipitate is washed with proper solvents to eliminate the unreacted monomers and other extra ingredients. Figure 25.2 demonstrates a typical solvothermal synthesis technique. The critical points that should be considered in synthesizing COFs with the solvothermal method are the solubility of the precursors, reaction rate, crystal nucleation and growth rate, and self-healing structure when selecting the raw materials and solvents. In addition, the type of solvents and the ratio of raw materials play an essential role in framework formation and crystallization. For instance, the combination of either dioxane–mesitylene, or DMAc–o-dichlorobenzene, and or HF–methanol solvents are mostly used for Boronate

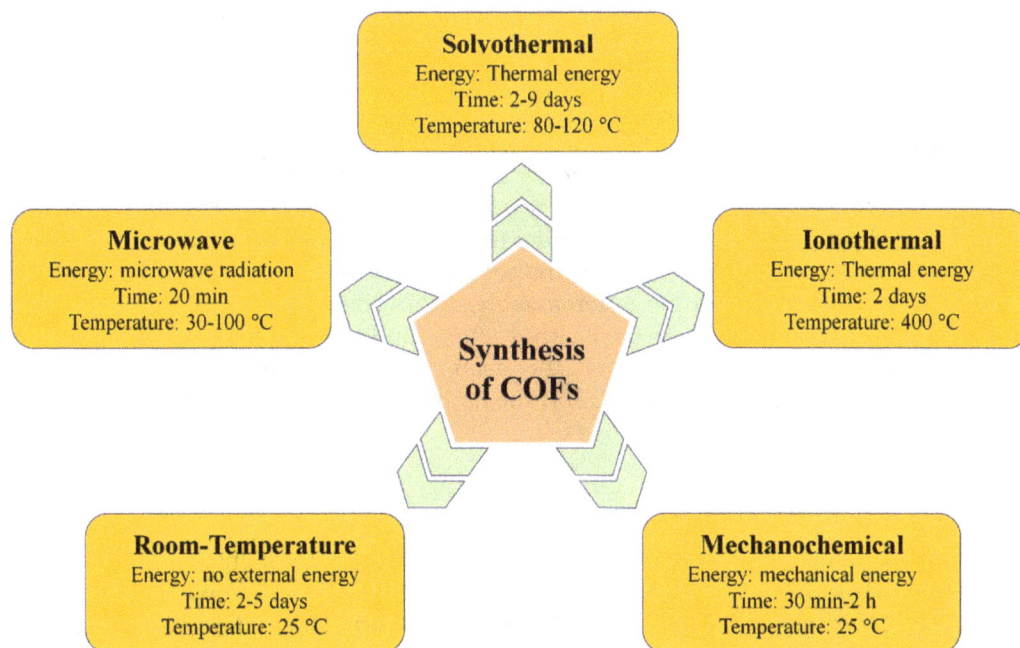

FIGURE 25.1
Various synthesis methods.

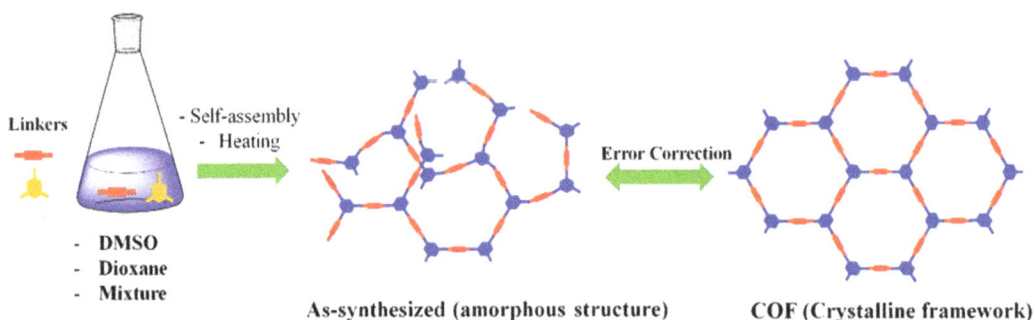

FIGURE 25.2
Schematic diagram of typical solvothermal synthesis method.

ester COFs. The other example is that for imine-linked COFs, the mixture of dioxane–aqueous acetic acid is used [12].

Moreover, temperature accounts for another vital factor in ascertaining the reversibility of the reaction. Overall, the temperature in the synthesis of COFs is adjusted between 80 and 120°C depending on the reactivity of the employed building blocks. Despite some advantages such as easy processing and a wide variety of building blocks, this method of COFs synthesis is not eco-friendly due to the required solvents and produced wastes; besides, this method needs soluble precursors.

25.3.2 Ionothermal

Thomas and coworkers first employed this synthesis method to synthesize triazine-based COFs (CTFs) [13]. The $ZnCl_2$ is utilized as a solvent and catalyst to fabricate crystalline COFs. Typically, aromatic nitrile, including 1,4-dicyanobenzene, 2,6-dicyanobenzene, and 1,3,5-tri(4-cyanophenyl)benzene, are solvated fused (zinc chloride) $ZnCl_2$ and kept for almost 40 h at 400°C. Finally, the mixture is cooled, and the precipitate is purified, resulting in the crystalline COF. CTFs are the class of COFs that possess high thermal, mechanical, and chemical stabilities; however, they suffer from low crystallinity due to the poor reversibility of trimerization reaction. In other studies, Maschita et al. [14] manufactured polyimide-linked COFs under the Ionothermal condition in zinc chloride. It is interesting to note that monomers are not required to be soluble in this synthesis condition, and the reaction time is remarkably diminished. The two vital issues that limit the ionothermal method are the harsh reaction condition and the amorphous structure of the final product. As mentioned, the reaction condition makes the availability of the organic building blocks confined, and the amorphous architecture does not satisfy the broad molecular range.

25.3.3 Microwave Synthesis

Although this synthesis technique has a great record in synthesizing a wide range of organic materials, it was not used to produce COFs until 2009 [15]. Cooper and coworkers employed this method to fabricate COFs for the first time. The rudimentary process is similar to the solvothermal method but quicker due to its microwave heating ability. This procedure is used to synthesize 2D and 3D COFs, including COF-5 and COF-102, in only 20 minutes, while 72 h is spent via the solvothermal method [9]. Furthermore, in some

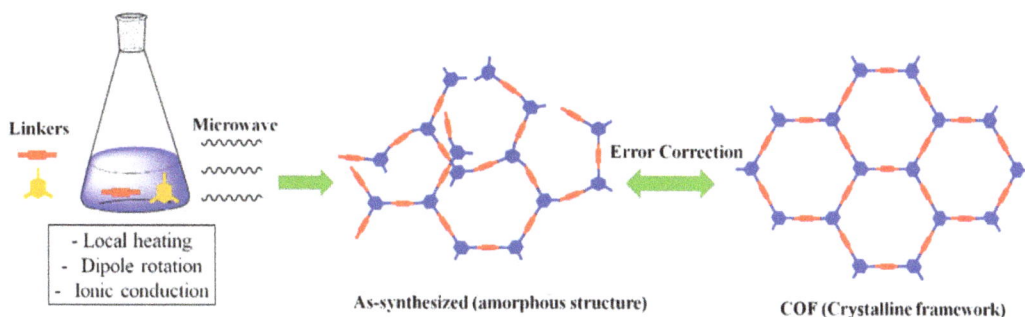

FIGURE 25.3
Schematic diagram of typical microwave synthesis method.

cases, the BET results of the synthesized COF-5 with microwave and solvothermal were reported to be 2019 and 1590 m^2/g, respectively. Figure 25.3 depicts the condition of the microwave to synthesize COFs. Indeed, the reactions can be performed in a shorter time and quicker via the microwave synthesis method based on its microwave heating capability, obtaining large-scale production and industrialization. In other words, the employment of this synthesis method outperforms the solvothermal technique via (1) Availability of industrial-scale synthesis, (2) A sealed tube is not required, and (3) The solvent extraction in microwave synthesis will eliminate the unreacted reagents more efficiently than solvothermal ones.

25.3.4 Mechanochemical

In generally used techniques such as the solvothermal method, the used solvents and materials solubility plays a key role. That is to say, the solubility not only may limit the proper selection of building blocks but also the produced solvent waste may have detrimental effects on human life [16]. However, mechanochemical synthesis, the traditional large-scale method to fabricate porous organic polymers and COFs, tackles these cons with grinding [17–19]. For instance, COFs based on β-ketoenamine were first synthesized via manually grinding in a mortar binder [20]. Intriguingly, the appearance of the mixture changed to dark red, illustrating polycondensation and COF formation. The comparison of β-ketoenamine based COF synthesized with mechanochemical and solvothermal processes demonstrated that the samples had similar chemical stability. However, the mechanochemically manufactured samples were less crystalline than solvothermal ones. To overcome this drawback, liquid cooperation grinding was applied to construct crystalline COFs [21]. The mentioned mechanical triturating might be further beneficial in fabricating covalent-organic nanosheets with superior stability against water, acids, and bases [22]. The classic model of COF synthesis with the help of mechanochemical technique is displayed in Figure 25.4. Moreover, a state-of-the-art approach was introduced by Banerjee et al. to synthesize COFs with twin-screw extruders [23]. These COFs possess a high crystalline structure, excellent porosity, and high surface area besides the potency to be molded in any shape. In addition, the imine-based COFs are the type of materials suitable for use in mechanochemical synthesis. Eventually, mechanochemical synthesis is an environmentally friendly and time-saving procedure to manufacture COFs on an industrial scale compared to other synthesis methods.

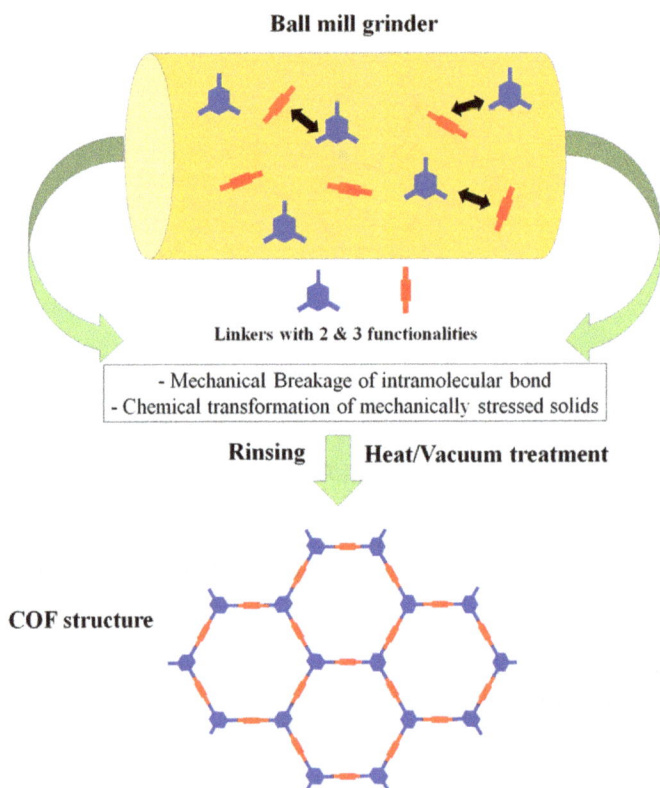

Ball mill grinder

Linkers with 2 & 3 functionalities

- Mechanical Breakage of intramolecular bond
- Chemical transformation of mechanically stressed solids

Rinsing   **Heat/Vacuum treatment**

COF structure

FIGURE 25.4
Schematic diagram of typical mechanochemical synthesis method.

25.3.5 Room-Temperature (RT)

This method has not been thoroughly investigated yet. However, Wang et al. discovered that the imine-linked COFs could be rationally produced at room temperature [24, 25]. The advantages of this method to the traditional ones are avoiding sealed tubes and controlling various parameters. For instance, m-cresol and DNSO were employed as solvents to obtain RT-COF-1 by Zamora et al. [26]. One of the significant advances of their work was that they compounded two methodologies, including room-temperature synthesis and microfluid technology, to fabricate fibrous microstructure COFs [27]. Moreover, as room-temperature synthesis has a moderate condition, it can be used to synthesize COFs with fragile precursors and on sensitive substrates [28].

25.4 Dynamic Linkages and Building Blocks

One of the significant characteristics of COFs is their dynamic linkage, resulting in various kinds of COFs and various performances. Until now, five primary dynamic linkages based on boron and nitrogen have been discovered. Each group of dynamic linkages has different kinds of building blocks. For instance, the B-O linkages consider boroxine, boronic ester, spiroborate, and borazine. According to their crystallization, topology, high surface area, and gas-storage applications, the boron-linked COFs have been extensively investigated. The cyclotrimeric anhydrides of boronic acid are called boroxine. The evaluation of the structure of these derivatives was performed using theoretically and experimentally analyses [29]. One example is that the X-ray patterns of triphenylboroxine demonstrated that this boroxine derivative is flat [30]. The primary way to produce boroxines is the dehydration of boronic acids [31]. Yaghi et al. were pioneers in synthesizing boroxine-based COFs, conducting via self-condensation of polyboronic acid [9]. The two-dimensional COF-1 was synthesized via the self-condensation of 1,4-phenylenediboronic acid. While the three-dimensional COF-102 and COF-103 were synthesized using self-condensation of tetra(4-dihydroxyborylphenyl) methane and its silane analogous, respectively [32]. The BET results demonstrated that all the three COFs had a surface area of 3472, 4120, and 711 m²/g for COF-102, COF-103, and COF-1, respectively. Moreover, microwave condition was employed by Cooper and coworkers to synthesize COF-5 and COF-105, resulting in an accelerated reaction compared to the typical solvothermal method and the same physical properties [15].

One good example of boronic ester-based linkages is 2,3,6,7,10,11-hexahydroxytriphenylene (HHTP), that Yaghi and coworkers produced COF-5, COF-6, COF-8, and COF-10 under solvothermal condition for approximately 72–120 h at 85–100°C [9, 33]. The two-dimensional COFs can be produced via incorporating tetrahydroxybenzene and diboronic acid benzene derivatives, using the mixture of tetrahydrofuran and methanol solvents in the inert atmosphere [34]. In Table 25.1, a summary of various linkages with their temperature formation and chemical stability has been given.

TABLE 25.1

Different Linkages and Their Formation Temperature and Chemical Stability for the Construction of COFs

Bond	Linkage	Formation Temperature (°C)	Water Stability	Chemical Stability (Acid)	Chemical Stability (Base)	Ref
B-O	Boroxine	120	Sensitive	Sensitive		[9]
	Boronate ester	120	Sensitive	Sensitive		[9]
	Spiroborate	120	2 days	Sensitive	2 days	[35]
	Borosilicate	120	Sensitive	Sensitive	Sensitive	[36]
C-N	Imine	120	Sensitive (better than B-O)	Sensitive (better than B-O)	Sensitive (better than B-O)	[37]
	Hydrazone	120	Sensitive (better than imine)	Sensitive (better than imine)	Sensitive (better than imine)	[38]
	Imide	200	Excellent	Excellent	Excellent	[39]
	Amide	120	Sensitive	1 day in 12 M HCl	1 day in 1 M NaOH	[40]
	β-kentoenamine	120	More than one week in boiling water	More than one week in 9 M HCl	More than one week in 9 M NaOH	[41]
	Azine	120	1 day in water	1 day in 1 M HCl	1 day in 1 M NaOH	[2]
	Phenazine	120	1 day in water	1 day in 1 M HCl	1 day in 1 M NaOH	[42]
	Squaraine	85	1 day in water	1 day in 1 M HCl	1 day in 1 M NaOH	[2]
	Triazine	150	Stable	Stable	Stable	[13]
	Melamine	Microwave	Stable	Stable	Stable	[2]
N-N	Azodioxide	37	Not reported	Not reported	Not reported	[43]
B-N	Borazine	120	Not reported	Not reported	Not reported	[44]

25.5 Stability of Covalent Organic Frameworks

"Stability" is not defined as a settled property, and researchers should concern various parameters that have been defined in advance. This is attributed to the application, concentration, and duration of materials used in a harsh environment, including corrosive media, water, and organic solvents. According to the mentioned properties, stability can be divided into four groups: thermal, mechanical, chemical, and water stability. Consequently, the stability illustrates the structural tolerance and resilience when the COFs are exposed to operational circumstances. It should be noticed that thermal stability and chemical stability are related to each other as heating and elevating temperature may interfere with chemical structure, activating the crystalline lattice breakdown. These organic frameworks must have different kinds of stabilities due to their broad range of applications. For instance, chemical stability is pivotal when the materials are used in aqueous solutions with pH gradients, such as absorption and wastewater treatment applications. In fuel production and catalytic applications, both thermal and chemical stabilities play a significant role. Mechanical stability is another critical factor when it comes to shaping these materials. Take the imine-linked COFs, produced under mechanochemical synthesis, as examples, which can be molded into any desired shape.

25.5.1 Thermal Stability

The most attractive materials used in the industry are those with superior thermal stability since they may have a huge range of applications. Thermal stability in the lattices of COFs means their structural ability to tolerate irreversible changes concerning physical and chemical points whenever they are exposed to high temperatures. As a result of exposure to the high temperature, the structural decomposition would occur in the form of linkers dehydrogenation, amorphization, melting, and graphitization, taking place gradually at different heating levels.

Thermogravimetric analysis (TGA) and variable temperature X-ray diffraction (VT-XRD) are the most used techniques to evaluate the thermal stability of the synthesized COFs. The latter is a detailed method for investigating thermal stability, which is conducted under an inert atmosphere. Moreover, COFs are considered eminently thermally stable porous materials, and thermogravimetric analysis confirmed this view. However, the measurement condition may result in the misbehavior of the materials as in the actual process; they are simultaneously subjected to heat and oxygen/moisture.

The organic strut is one of the primary parameters in the determination of the thermal stability of COFs. According to the structure of the linker mentioned above, they are classified into different models, which are called multifunctional spacers. For instance, boronate ester COFs have shown better thermal stability in comparison with imine ones. This phenomenon might be attributed to either the more van der Waals surfaces in boronate linkers or the inherent rigidity of boronate esters compared to imine linkers.

The other vital factors consist of lattice and pore structure (size and functionality). COFs with smaller lattice sizes are more stable than larger ones due to their smaller free volumes, fewer deformation points, and lower available phonon modes. On the other hand, the functionalization of pores leads the COF to lower thermal stability, which could be assigned to the considerable number of available entropic states in these frameworks. This phenomenon is much the same as the case in linear polymeric systems when entropic structural distortions occur. Therefore, all the mentioned factors may substantially disrupt the material when extra thermal energy is applied to the system [45].

25.5.2 Mechanical Stability

Some specific conditions such as vacuum or pressure would significantly affect the mechanical stability of COFs and hence their applications. From the engineering point of view, the consequence of the instability of these porous lattices under vacuum or pressure would be the failure of the structure or phase changes. Although COFs are zeolitic-like structures, they possess lower mechanical stability than zeolites. As the insoluble powder materials (COFs) are pressed into pellets, this low difference between zeolites and COFs became an obstacle. Because in COFs, it results in damage to the structure to a great extent. Subsequently, it is demanded to control and monitor the amount of the applied pressure.

Moreover, finding a method to balance the brittleness of these porous frameworks and the compression required to mold them obtains crystalline pellets with well-aligned structures [46]. In situ synthesis of COFs is one of the methods, performing the mechanically stable membranes. The synthesis of ketoenamine COFs similar to terracotta is the other strategy that has been employed to manufacture COFs with good mechanical properties without the need for a matrix. In this procedure, the amine is blended with p-toluenesulfonic acid. The obtained blend is combined with the aldehyde and water to produce a paste that provides highly porous and crystalline material [23]. Banerjee et al. demonstrated that the different shapes in centimeter-sized pieces could be molded with this methodology, and the continuous industrial-scale two-screw extruding is possible. The utilization of microfluid technology in the synthesis of COFs is another elaborate technique that can modify the mechanical stability of COFs. For instance, the formation of imine-based COF under microfluidic conditions leads to supremely crystalline porous structures with superior mechanical stability. Moreover, the COFs can be directly drawn on any subject [27].

25.5.3 Chemical and Water Stability

Even though many COFs have been synthesized until now, their chemical and hydrolytic stability is still an issue to be solved. That is to say; these porous materials are inclined to decomposition in connection with chemical environments such as organic solvents. Indeed, during the formation of COFs, water acts as a byproduct, making the backward reaction possible, and hence the COFs decomposition takes place [47]. The B-O-linked COFs, specially boroxine and boronates, are prone to degrade with humidity and water molecules. Therefore, the evaluation of the chemical stability of COFs is a pivotal aspect that needs to be considered before practical applications like gas storage or separation. Since the gas storage applications in industry are conducted in aqueous environments, the utilized adsorbents containing COFs are subjected to various kinds of solvents, including water and acidic oxides [48]. Hence, while the COFs have substantial gas storage applications, they cannot be used industrially due to their low chemical stability. Thus, there is a crucial demand to ameliorate the stability of these useful materials against chemical solvents for potential large-scale applications. Lavigne and coworkers developed one way to improve chemical/hydrolytic stability. They used long-chain alkyl groups to reduce the hydrophilicity of the COF backbone, delaying its hydrolysis. Although modification of the COFs with organic materials may affect their physical properties and applications (porosity and gas storage properties), researchers used pyridine to eliminate the hydrolysis of boronate ester in COF-5 [49]. Due to the high chemical/hydrolytic instability of

B-O linkages, the focus of researchers is inclined to the more stable imine-based linkages, which are less prone to decompose because they are produced through pH-induced reactions. The research done by Banerjee and coworkers [39] showed that via intramolecular hydrogen bonding in the synthesis process of imine-based COFs, satisfactory and remarkable stability in 3M HCl was achieved. Indeed, the enhancement of chemical stability was carried out via some strategies, including the decrement of the nucleophilicity of the imine bonds, reinforcing the interlayer stacking, and post-synthetic modifications. However, the stability improvement of these COFs is limited to 2D materials, and active research is still being performed to stabilize 3D ones.

25.6 Applications

The perfect engineered and crystalline COFs make their usage possible in a huge range of applications. It means the covalent bonds between layers result in substantial thermally stable COFs, making them good candidates for many applications. Several covalent organic frameworks are stable up to 800°C and even more suitable for most demands. Although they are good candidates for heat resistance, they bear other restrictions such as chemical stability. According to the captivating structural architecture of the COFs, they can be employed in many fields (Figure 25.5). There are plenty of design techniques, which can be employed to have the anticipated performance of the synthesized material for any application. In this part, the advanced applications of COFs in anti-corrosion coatings will be discussed.

25.6.1 Anti-Corrosion Properties of COFs

Corrosion is an undesirable phenomenon that has detrimental effects on metal-based compounds. Many attempts have been made to diminish these effects or at least slow the corrosion reactions. The foremost strategies have comprised the utilization of corrosion inhibitors, application of organic/inorganic coatings, and employment of anodic/cathodic protection. Based on the properties mentioned earlier of COFs, these highly porous materials can be employed to mitigate corrosion reactions. Moreover, they can act as nanocontainers due to their tunable pore structures/sizes to carry organic/inorganic inhibitors and release them when needed. In this regard, stability in every aspect, such as thermal, mechanical, chemical, and hydrolytic, plays an important role. That is to say; in the inhibitors, the synthesized materials must possess good water solubility to eliminate sediment production. However, the low water solubility of COFs in the polymeric matrix would positively impact the barrier anti-corrosion performance of the coatings. The utilized COFs in the polymeric matrix can dissociate on-demand and mitigate the corrosion reactions. Furthermore, the COFs should have reasonable chemical stability when it comes to their applications in acidic, alkaline, and saline solutions with corrosive cations and ions. Besides, thermal stability is another moot point since anti-corrosion additives are employed in media with elevated temperatures.

As mentioned before, due to the nanostructure of COFs, they can also act as nanocontainers or hosts, meaning that corrosion protective materials like organic and inorganic inhibitors can be loaded into their structure and enhance their dispersion in the polymeric matrix as well as anti-corrosion properties.

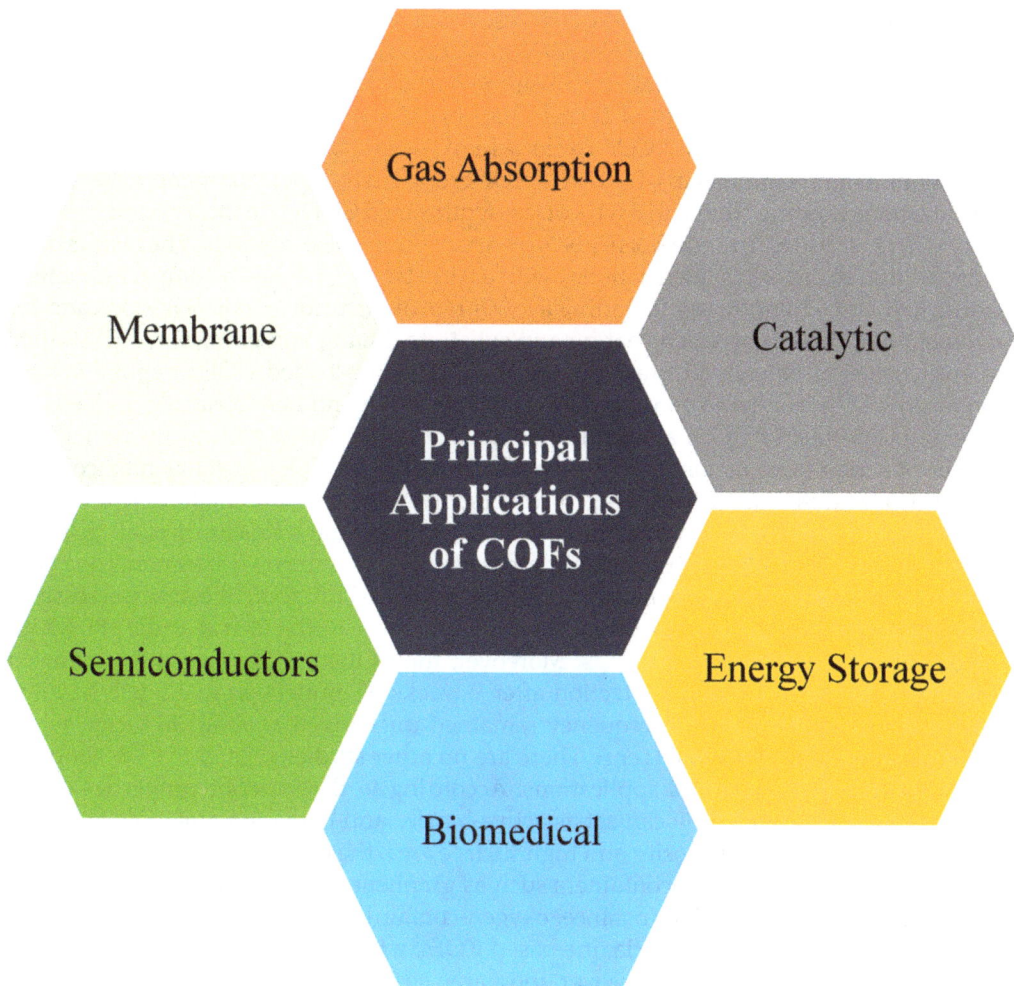

FIGURE 25.5
Principal applications of COFs.

25.6.1.1 Prospective COFs as Corrosion-Resistant Materials

Although Yaghi and coworkers synthesized COFs in 2005 for the first time, their applications in many areas have not been investigated yet. Corrosion-resistant COFs and their applications in anti-corrosion coatings and composites are the hotspots that a few pieces of research have been conducted in the last two years. Wen et al. [50] synthesized triazine-based COFs under solvothermal conditions. The prepared COF was combined with two different types of surfactants (CDEA and T-85) to augment the abrasion and anti-corrosion properties of the samples. They found that the combination of CDEA/COF reduced the friction coefficient and wear rate by 72.2% and 98%, respectively. At the same time, the T-85/COF samples diminished the friction coefficient and wear rate by almost 74.4% and 95.8%.

Moreover, the anti-corrosion properties of the samples illustrated that the CDEA/COF and T-85/COF had good protection ability. This phenomenon might be assigned to the van der Waals and hydrogen bonding interactions between CDEA, T-85, and COF. Indeed, at first, COF adsorbs on the surface of the immersed disk, and then either CDEA or T-85 interacted with formed COF, resulting in barrier protection and mitigating corrosion reactions.

In 2021, Wen Li et al. [51] synthesized an amine-based COF under mechanochemical conditions. One of the shortcomings of 2D materials in polymeric matrices is their dispersion. In this study, the amine-based COF could achieve self-dispersion, corresponding to the interactions of amine groups in COF with amine groups in the epoxy hardener. The diverse concentrations ranging from 0 to 6 wt% of the synthesized COF were incorporated into the epoxy resin to evaluate the anti-corrosion potency of the applied coatings. The results demonstrated that the epoxy-loaded samples with amine-based COF had a compact structure, leading to the two hypotheses, including a lower rate of diffusion of corrosive ions and the increment of adhesion between coating and substrate. Zhang and coworkers [52] studied the anti-corrosion potency of graphene and benzotriazole-loaded COF in epoxy matrix. They used COF-1 as a nano container for graphene oxide and benzotriazole. Indeed, the supreme properties of COFs alongside the tunable surface area and an overwhelming number of active sites demonstrated the potential ability of COFs to act as nanocontainers. In this research work, Zhang and coworkers illustrated that coating loaded with COF depicted better anti-corrosion performance. More interestingly, Keshmiri et al. [53] synthesized a novel and ultrastable melamine-based COF assembled carbon nanotube, and studied its anti-corrosion performance. This COF with 1163 m^2/g surface area possessed a high capacity for zinc cations absorption and 418 ppm release after 24 h at acidic pH, which is promising for self-healing coatings. Moreover, the COF-modified nanotubes showed outstanding barrier corrosion protection after 9 weeks of immersion in 3.5 wt. % saline solution the impedance at low frequency remained unchanged at about 10 $\Omega.cm^2$, while the blank sample dropped to 6 $\Omega.cm^2$. There are no other studies related to COF cooperation in anti-corrosion coatings applications. According to the superior properties of the COFs and their potential, they can act as active-barrier anti-corrosion materials in future works. According to their porosity and high surface area, these newly developed materials can be linked with other nanocontainers such as graphene oxide, carbon nanotubes, and metal-organic frameworks. For instance, oxygen-containing groups of carbon nanotubes can react with amine groups of triazine-based COFs, which can be a good nanoinhibitor in anti-corrosion applications. Moreover, inorganic corrosion inhibitors such as zinc and cerium cations can make chelation with these nanocontainers released in corrosive media when demanded.

25.6.2 Thermomechanical Properties

The modification of architecture-property functions of polymeric coatings is conducted either by chemical cross-linking or physical blending. The main purpose of these amendments is the augmentation of the toughness of the coatings. Although most polymeric coatings are well-known with substantial mechanical properties with remarkable modulus and strength, they are struggling with low thermal stability besides their brittleness when exposed to impact. Several methods have been implemented to overcome such shortcomings of the coatings. Among all the employed methods, incorporating fillers into the matrix is the most practical one concerning the experimental view. However, the compatibility of the organic and inorganic ingredients in these composite systems is the most vital challenge, which leads the nanoparticles to agglomerate, causing considerable interference in the performance of the coatings. Owing to the higher surface area and lower percentage consumption of nanomaterials compared to the macro/micro-sized particles, the nanoparticles filled polymer-based coatings have attracted overwhelming attention in recent years. These nanoparticles could enhance the physical, mechanical, and thermal properties of the coatings to a great extent.

Incorporating micro/nano-COFs into the polymer matrix is one of the most widely used methods to elevate the thermomechanical properties of the coatings. COFs, due to their organic bondings, are well-matched with organic polymeric systems compared to other additives such as carbon black, silica, and alumina. Although these materials are fragile and their processability is difficult, some methods are used to tackle these issues, including interfacial polymerization of small monomers, vacuum filtration, and utilization of molecular organizers (such as p-toluene sulfonic acid). For example, Wang et al. [3] studied the enhancement of the mechanical properties of COF-42 by introducing linear polymers into it. The outstanding results were achieved as Young's modulus increment from 9 MPa to 914 MPa for the neat COF-42 and Poly-COF-42, respectively. Moreover, the organic ligands in COFs can interact with the polymeric system and improve the curing mechanism, enhancing the thermal-mechanical properties.

One of the most growing applications for COFs is to utilize them as nanofillers to improve the flame resistance of the coatings. Qiu et al. [54] developed an NH_2-functionalized triazine-based COF integrated with black phosphorus (BP) into the epoxy resin to enhance its flame retardancy as well as mechanical properties. They found that using only 2 wt% of the modified COF in the epoxy resin resulted in the superior decrement of peak heat release rate of 62% and total heat release of 44.3%. Meanwhile, the mechanical properties such as storage modulus increased dramatically by approximately 67%, incorporating 1 wt% of the nanofiller. Mu et al. [55] studied the influence of Co_3O_4-COF loaded polypropylene on flame retardancy and carbon monoxide (CO) suppression. They found that only 0.2 wt% of the nanoparticles decreased the total smoke amount by 22%. In another study, Xiao et al. [56] investigated the flame resistance and mechanical properties of novel graphene conjugated COF incorporated into the epoxy matrix. The surface modification of graphene was conducted via amine-based COF under solvothermal conditions. The results demonstrated that the storage modulus of the loaded epoxy was about 34.66 GPa, while for the neat epoxy, it was 28 GPa, and it increased by almost 24%.

It is undeniable that the usage of polymeric materials has led to a dramatic expansion in the risk of fire incidents available in our daily life and industry. One of those widely used polymeric materials is thermoplastic polyurethane (TPU). Although it possesses outstanding properties, including anti-corrosion, anti-erosion, high chemical stability, and substantial mechanical properties, it is easily flammable. Its applications range from medical equipment to aerospace. TPU may have harmful effects on human life as it can release an overwhelming number of toxic gases. Thus, there is an urgent need to reduce its fire risk. Afshari et al. [57] studied the thermal/mechanical efficiency of the triazine-based COF-loaded TPU. The results depicted that by incorporating 3 wt% COF, the peak heat release and total heat release decreased remarkably by approximately 51.6% and 47.3%, respectively. In addition, the tensile test outcomes revealed that both tensile properties like strength and elongation at break raised by 69% and 12%, respectively. In another study, Mu et al. [58] used exfoliated COF with ultrasound and sodium lignosulfonate based on the condensation reaction between 1,3,5-triformylphloroglucinol (TFP) and p-phenylenediamine (PDA). Chitosan biopolymer was filled with different amounts of the modified COF (0, 0.2, 0.4, 0.8, 1.6, and 3.2 wt%). The incorporation of the nanofillers into the chitosan not only enhanced the mechanical performance and acted as physical barriers but also contributed to the graphitization degree. Young's modulus and fracture strength increased by 94.3% and 27.9%, respectively. This is attributed to the good dispersion and compatibility of the modified COF with chitosan, confirming the hydrogen bonding interactions between the COF and chitosan. Meanwhile, the total heat release decreased by 33.4%. Based on the graphite-like structure of COFs and their covalent bonding through van der Waals forces, they have the privilege to get exfoliated in the presence of physical or chemical techniques. Moreover, Schiff-based COFs, due to their high content of nitrogen elements with high thermal stability, can be good candidates for enhancing the thermomechanical performance of the coatings. Mu et al. [59], in another work developed boron and nitrogen-doped COFs to ameliorate thermomechanical properties of PVA nanocomposites.

As the COFs are covalently bonded, their organic ligands may overreact with the functional group of the polymeric matrix. For instance, the epoxide groups of epoxy resin can have a strong interaction with amine-based COFs, resulting in the ring-opening of the epoxide groups and acting as a secondary hardener. Subsequently, the applied external stresses/pressures would be delivered from the polymeric matrix to the COF. Therefore, developing the linear cross-linked lattice at the COF/polymer interface would be preferred based on these phenomena. Moreover, this method would help in the decline of fragileness and improvement in the toughness specifications.

25.7 Outlook and Future Perspective

Recent years have testified to a boom in COF chemistry and applications. The various perspectives of new COFs have been summarized in this chapter. According to the unique structure of COFs, including flexible structure design, low density, permanent porosity, tunable pore size, low to high crystallinity, and more importantly, the versatility of available organic building blocks makes them good candidates for functional design and synthesis as well as potential materials for many applications. As summarized, these unique structures can be fabricated under diverse conditions such as solvothermal, ionothermal, microwave, mechanochemical, and room temperature; each one has its pros and cons. Generally, these micro/nanomaterials suffer from some limitations, including industrial-scale production (production method point of view), thermochemical stability, and processability.

Moreover, their production procedure has developed dramatically; however, the most widely used method is still solvothermal, which needs to be broadened. Importantly enough, the two prominent families produced (boron-based and nitrogen-based) each demonstrated similar thermal stability. However, the nitrogen-based showed higher chemical stability against various solvents in acidic and basic conditions. The more chemical stability of the nitrogen-based COFs motivated researchers to be inclined toward processing these materials compared to their boron-based analogous. We believe that in the near future, there will be other alternative solutions and reactions to manufacture more COFs.

Furthermore, their potential as a corrosion inhibitor has been investigated. The intrinsic supramolecular structure, covalent bonding, and porous architecture made these micro/nanomaterials fascinating to incorporate into the polymeric matrix as corrosion inhibitors. In addition, COFs are porous and can act as nanocontainers for other organic-inorganic nano-based inhibitors. That is to say, modification and growth of COFs with/on other nanomaterials like GO or CNT would be promising measures to develop high-performance polymer composites with supreme anti-corrosion and thermomechanical properties.

References

1. Diercks, C. S.; Yaghi, O. M. The Atom, the Molecule, and the Covalent Organic Framework. *Science (80-.)*, 2017, *355* (6328), eaal1585. https://doi.org/10.1126/science.aal1585.
2. Nagai, A. *Covalent Organic Frameworks*; Nagai, A., Ed.; Jenny Stanford Publishing, 2019. https://doi.org/10.1201/9781003004691.
3. Wang, Z.; Yu, Q.; Huang, Y.; An, H.; Zhao, Y.; Feng, Y.; Li, X.; Shi, X.; Liang, J.; Pan, F.; et al. PolyCOFs: A New Class of Freestanding Responsive Covalent Organic Framework Membranes with High Mechanical Performance. *ACS Cent. Sci.*, 2019, *5* (8), 1352–1359. https://doi.org/10.1021/acscentsci.9b00212.
4. Bai, L.; Phua, S. Z. F.; Lim, W. Q.; Jana, A.; Luo, Z.; Tham, H. P.; Zhao, L.; Gao, Q.; Zhao, Y. Nanoscale Covalent Organic Frameworks as Smart Carriers for Drug Delivery. *Chem. Commun.*, 2016, *52* (22), 4128–4131. https://doi.org/10.1039/C6CC00853D.
5. Li, J.; Jing, X.; Li, Q.; Li, S.; Gao, X.; Feng, X.; Wang, B. Bulk COFs and COF Nanosheets for Electrochemical Energy Storage and Conversion. *Chem. Soc. Rev.*, 2020, *49* (11), 3565–3604. https://doi.org/10.1039/D0CS00017E.
6. Xu, Y.; Shi, X.; Hua, R.; Zhang, R.; Yao, Y.; Zhao, B.; Liu, T.; Zheng, J.; Lu, G. Remarkably Catalytic Activity in Reduction of 4-Nitrophenol and Methylene Blue by Fe3O4@COF Supported Noble Metal Nanoparticles. *Appl. Catal. B Environ.*, 2020, *260*, 118142. https://doi.org/10.1016/j.apcatb.2019.118142.
7. Ren, X.; Liao, G.; Li, Z.; Qiao, H.; Zhang, Y.; Yu, X.; Wang, B.; Tan, H.; Shi, L.; Qi, X.; et al. Two-Dimensional MOF and COF Nanosheets for next-Generation Optoelectronic Applications. *Coord. Chem. Rev.*, 2021, *435*, 213781. https://doi.org/10.1016/j.ccr.2021.213781.
8. Ramezanzadeh, M.; Ramezanzadeh, B. *Thermomechanical and Anticorrosion Characteristics of Metal-Organic Frameworks*; Khan, A., Verpoort, F., Asiri, A. M., Hoque, M. E., Bilgrami, A. L., Azam, M., Naidu, K. C. B. B. T., Eds.; Elsevier, 2021. https://doi.org/10.1016/B978-0-12-822099-3.00012-5.
9. Cote, A. P. Porous, Crystalline, Covalent Organic Frameworks. *Science (80-.)*, 2005, *310* (5751), 1166–1170. https://doi.org/10.1126/science.1120411.
10. Najmi, P.; Keshmiri, N.; Ramezanzadeh, M.; Ramezanzadeh, B. Synthesis and Application of Zn-Doped Polyaniline Modified Multi-Walled Carbon Nanotubes as Stimuli-Responsive Nanocarrier in the Epoxy Matrix for Achieving Excellent Barrier-Self-Healing Corrosion Protection Potency. *Chem. Eng. J.*, 2021, *412*, 128637. https://doi.org/10.1016/j.cej.2021.128637.
11. Keshmiri, N.; Najmi, P.; Ramezanzadeh, M.; Ramezanzadeh, B. Designing an Eco-Friendly Lanthanide-Based Metal Organic Framework (MOF) Assembled Graphene-Oxide with Superior Active Anti-Corrosion Performance in Epoxy Composite. *J. Clean. Prod.*, 2021, *319*, 128732. https://doi.org/10.1016/j.jclepro.2021.128732.
12. Feng, X.; Ding, X.; Jiang, D. Covalent Organic Frameworks. *Chem. Soc. Rev.*, 2012, *41* (18), 6010–6022. https://doi.org/10.1039/c2cs35157a.
13. Kuhn, P.; Antonietti, M.; Thomas, A. Porous, Covalent Triazine-Based Frameworks Prepared by Ionothermal Synthesis. *Angew. Chemie Int. Ed.*, 2008, *47* (18), 3450–3453. https://doi.org/10.1002/anie.200705710.
14. Maschita, J.; Banerjee, T.; Savasci, G.; Haase, F.; Ochsenfeld, C.; Lotsch, B. V. Ionothermal Synthesis of Imide-Linked Covalent Organic Frameworks. *Angew. Chemie Int. Ed.*, 2020, *59* (36), 15750–15758. https://doi.org/10.1002/anie.202007372.
15. Campbell, N. L.; Clowes, R.; Ritchie, L. K.; Cooper, A. I. Rapid Microwave Synthesis and Purification of Porous Covalent Organic Frameworks. *Chem. Mater.*, 2009, *21* (2), 204–206. https://doi.org/10.1021/cm802981m.
16. Friščić, T. Supramolecular Concepts and New Techniques in Mechanochemistry: Cocrystals, Cages, Rotaxanes, Open Metal–Organic Frameworks. *Chem. Soc. Rev.*, 2012, *41* (9), 3493. https://doi.org/10.1039/c2cs15332g.

17. Friščić, T.; James, S. L.; Boldyreva, E. V.; Bolm, C.; Jones, W.; Mack, J.; Steed, J. W.; Suslick, K. S. Highlights from Faraday Discussion 170: Challenges and Opportunities of Modern Mechanochemistry, Montreal, Canada, 2014. *Chem. Commun.*, 2015, *51* (29), 6248–6256. https://doi.org/10.1039/C5CC90113H.

18. James, S. L.; Adams, C. J.; Bolm, C.; Braga, D.; Collier, P.; Friščić, T.; Grepioni, F.; Harris, K. D. M.; Hyett, G.; Jones, W.; et al. Mechanochemistry: Opportunities for New and Cleaner Synthesis. *Chem. Soc. Rev.*, 2012, *41* (1), 413–447. https://doi.org/10.1039/C1CS15171A.

19. Zhang, P.; Dai, S. Mechanochemical Synthesis of Porous Organic *Materials. J. Mater. Chem. A*, 2017, *5* (31), 16118–16127. https://doi.org/10.1039/C7TA04829G.

20. Biswal, B. P.; Chandra, S.; Kandambeth, S.; Lukose, B.; Heine, T.; Banerjee, R. Mechanochemical Synthesis of Chemically Stable Isoreticular Covalent Organic Frameworks. *J. Am. Chem. Soc.*, 2013, *135* (14), 5328–5331. https://doi.org/10.1021/ja4017842.

21. Das, G.; Balaji Shinde, D.; Kandambeth, S.; Biswal, B. P.; Banerjee, R. Mechanosynthesis of Imine, β-Ketoenamine, and Hydrogen-Bonded Imine-Linked Covalent Organic Frameworks Using Liquid-Assisted Grinding. *Chem. Commun.*, 2014, *50* (84), 12615–12618. https://doi.org/10.1039/C4CC03389B.

22. Chandra, S.; Kandambeth, S.; Biswal, B. P.; Lukose, B.; Kunjir, S. M.; Chaudhary, M.; Babarao, R.; Heine, T.; Banerjee, R. Chemically Stable Multilayered Covalent Organic Nanosheets from Covalent Organic Frameworks via Mechanical Delamination. *J. Am. Chem. Soc.*, 2013, *135* (47), 17853–17861. https://doi.org/10.1021/ja408121p.

23. Karak, S.; Kandambeth, S.; Biswal, B. P.; Sasmal, H. S.; Kumar, S.; Pachfule, P.; Banerjee, R. Constructing Ultraporous Covalent Organic Frameworks in Seconds via an Organic Terracotta Process. *J. Am. Chem. Soc.*, 2017, *139* (5), 1856–1862. https://doi.org/10.1021/jacs.6b08815.

24. Ding, S.-Y.; Cui, X.-H.; Feng, J.; Lu, G.; Wang, W. Facile Synthesis of –C · N– Linked Covalent Organic Frameworks under Ambient Conditions. *Chem. Commun.*, 2017, *53* (87), 11956–11959. https://doi.org/10.1039/C7CC05779B.

25. Ding, S.-Y.; Wang, W. Covalent Organic Frameworks (COFs): From Design to Applications. *Chem. Soc. Rev.*, 2013, *42* (2), 548–568. https://doi.org/10.1039/C2CS35072F.

26. Montoro, C.; Rodríguez-San-Miguel, D.; Polo, E.; Escudero-Cid, R.; Ruiz-González, M. L.; Navarro, J. A. R.; Ocón, P.; Zamora, F. Ionic Conductivity and Potential Application for Fuel Cell of a Modified Imine-Based Covalent Organic Framework. *J. Am. Chem. Soc.*, 2017, *139* (29), 10079–10086. https://doi.org/10.1021/jacs.7b05182.

27. Rodríguez-San-Miguel, D.; Abrishamkar, A.; Navarro, J. A. R.; Rodriguez-Trujillo, R.; Amabilino, D. B.; Mas-Ballesté, R.; Zamora, F.; Puigmartí-Luis, J. Crystalline Fibres of a Covalent Organic Framework through Bottom-up Microfluidic Synthesis. *Chem. Commun.*, 2016, *52* (59), 9212–9215. https://doi.org/10.1039/C6CC04013F.

28. Medina, D. D.; Rotter, J. M.; Hu, Y.; Dogru, M.; Werner, V.; Auras, F.; Markiewicz, J. T.; Knochel, P.; Bein, T. Room Temperature Synthesis of Covalent–Organic Framework Films through Vapor-Assisted Conversion. *J. Am. Chem. Soc.*, 2015, *137* (3), 1016–1019. https://doi.org/10.1021/ja510895m.

29. Korich, A. L.; Iovine, P. M. Boroxine Chemistry and Applications: A Perspective. *Dalt. Trans.*, 2010, *39* (6), 1423–1431. https://doi.org/10.1039/B917043J.

30. Thilagar, P.; Chen, J.; Lalancette, R. A.; Jäkle, F. Reversible Formation of a Planar Chiral Ferrocenylboroxine and Its Supramolecular Structure. *Organometallics*, 2011, *30* (24), 6734–6741. https://doi.org/10.1021/om200947v.

31. Snyder, H. R.; Kuck, J. A.; Johnson, J. R. Organoboron Compounds, and the Study of Reaction Mechanisms. Primary Aliphatic Boronic Acids 1. *J. Am. Chem. Soc.*, 1938, *60* (1), 105–111. https://doi.org/10.1021/ja01268a033.

32. El-Kaderi, H. M.; Hunt, J. R.; Mendoza-Cortes, J. L.; Cote, A. P.; Taylor, R. E.; O'Keeffe, M.; Yaghi, O. M. Designed Synthesis of 3D Covalent Organic Frameworks. *Science (80-.)*, 2007, *316* (5822), 268–272. https://doi.org/10.1126/science.1139915.

33. Côté, A. P.; El-Kaderi, H. M.; Furukawa, H.; Hunt, J. R.; Yaghi, O. M. Reticular Synthesis of Microporous and Mesoporous 2D Covalent Organic Frameworks. *J. Am. Chem. Soc.*, 2007, *129* (43), 12914–12915. https://doi.org/10.1021/ja0751781.

34. Tilford, R. W.; Mugavero, S. J.; Pellechia, P. J.; Lavigne, J. J. Tailoring Microporosity in Covalent Organic Frameworks. *Adv. Mater.*, 2008, *20* (14), 2741–2746. https://doi.org/10.1002/adma.200800030.

35. Du, Y.; Yang, H.; Whiteley, J. M.; Wan, S.; Jin, Y.; Lee, S.; Zhang, W. Ionic Covalent Organic Frameworks with Spiroborate Linkage. *Angew. Chemie Int. Ed.*, 2016, *55* (5), 1737–1741. https://doi.org/10.1002/anie.201509014.

36. Hunt, J. R.; Doonan, C. J.; LeVangie, J. D.; Côté, A. P.; Yaghi, O. M. Reticular Synthesis of Covalent Organic Borosilicate Frameworks. *J. Am. Chem. Soc.*, 2008, *130* (36), 11872–11873. https://doi.org/10.1021/ja805064f.

37. Uribe-Romo, F. J.; Hunt, J. R.; Furukawa, H.; Klöck, C.; O'Keeffe, M.; Yaghi, O. M. A Crystalline Imine-Linked 3-D Porous Covalent Organic Framework. *J. Am. Chem. Soc.*, 2009, *131* (13), 4570–4571. https://doi.org/10.1021/ja8096256.

38. Uribe-Romo, F. J.; Doonan, C. J.; Furukawa, H.; Oisaki, K.; Yaghi, O. M. Crystalline Covalent Organic Frameworks with Hydrazone Linkages. *J. Am. Chem. Soc.*, 2011, *133* (30), 11478–11481. https://doi.org/10.1021/ja204728y.

39. Fang, Q.; Zhuang, Z.; Gu, S.; Kaspar, R. B.; Zheng, J.; Wang, J.; Qiu, S.; Yan, Y. Designed Synthesis of Large-Pore Crystalline Polyimide Covalent Organic Frameworks. *Nat. Commun.*, 2014, *5* (1), 4503. https://doi.org/10.1038/ncomms5503.

40. Waller, P. J.; Lyle, S. J.; Osborn Popp, T. M.; Diercks, C. S.; Reimer, J. A.; Yaghi, O. M. Chemical Conversion of Linkages in Covalent Organic Frameworks. *J. Am. Chem. Soc.*, 2016, *138* (48), 15519–15522. https://doi.org/10.1021/jacs.6b08377.

41. Kandambeth, S.; Mallick, A.; Lukose, B.; Mane, M. V.; Heine, T.; Banerjee, R. Construction of Crystalline 2D Covalent Organic Frameworks with Remarkable Chemical (Acid/Base) Stability via a Combined Reversible and Irreversible Route. *J. Am. Chem. Soc.*, 2012, *134* (48), 19524–19527. https://doi.org/10.1021/ja308278w.

42. Guo, J.; Xu, Y.; Jin, S.; Chen, L.; Kaji, T.; Honsho, Y.; Addicoat, M. A.; Kim, J.; Saeki, A.; Ihee, H.; et al. Conjugated Organic Framework with Three-Dimensionally Ordered Stable Structure and Delocalized π Clouds. *Nat. Commun.*, 2013, *4* (1), 2736. https://doi.org/10.1038/ncomms3736.

43. Beaudoin, D.; Maris, T.; Wuest, J. D. Constructing Monocrystalline Covalent Organic Networks by Polymerization. *Nat. Chem.*, 2013, *5* (10), 830–834. https://doi.org/10.1038/nchem.1730.

44. Jackson, K. T.; Reich, T. E.; El-Kaderi, H. M. Targeted Synthesis of a Porous Borazine-Linked Covalent Organic Framework. *Chem. Commun.*, 2012, *48* (70), 8823. https://doi.org/10.1039/c2cc33583b.

45. Evans, A. M.; Ryder, M. R.; Ji, W.; Strauss, M. J.; Corcos, A. R.; Vitaku, E.; Flanders, N. C.; Bisbey, R. P.; Dichtel, W. R. Trends in the Thermal Stability of Two-Dimensional Covalent Organic Frameworks. *Faraday Discuss.*, 2021, *225*, 226–240. https://doi.org/10.1039/D0FD00054J.

46. Rodríguez-San-Miguel, D.; Zamora, F. Processing of Covalent Organic Frameworks: An Ingredient for a Material to Succeed. *Chem. Soc. Rev.*, 2019, *48* (16), 4375–4386. https://doi.org/10.1039/C9CS00258H.

47. Lanni, L. M.; Tilford, R. W.; Bharathy, M.; Lavigne, J. J. Enhanced Hydrolytic Stability of Self-Assembling Alkylated Two-Dimensional Covalent Organic Frameworks. *J. Am. Chem. Soc.*, 2011, *133* (35), 13975–13983. https://doi.org/10.1021/ja203807h.

48. Kandambeth, S.; Dey, K.; Banerjee, R. Covalent Organic Frameworks: Chemistry beyond the Structure. *J. Am. Chem. Soc.*, 2019, *141* (5), 1807–1822. https://doi.org/10.1021/jacs.8b10334.

49. Du, Y.; Mao, K.; Kamakoti, P.; Ravikovitch, P.; Paur, C.; Cundy, S.; Li, Q.; Calabro, D. Experimental and Computational Studies of Pyridine-Assisted Post-Synthesis Modified Air Stable Covalent–Organic Frameworks. *Chem. Commun.*, 2012, *48* (38), 4606. https://doi.org/10.1039/c2cc30781b.

50. Wen, P.; Lei, Y.; Li, W.; Fan, M. Synergy between Covalent Organic Frameworks and Surfactants to Promote Water-Based Lubrication and Corrosion Resistance. *ACS Appl. Nano Mater.*, 2020, *3* (2), 1400–1411. https://doi.org/10.1021/acsanm.9b02198.

51. Li, W.; Zhang, X.; Zhang, C.; Yu, M.; Ren, J.; Wang, W.; Chen, S. Exploring the Corrosion Resistance of Epoxy Coated Steel by Integrating Mechanochemical Synthesized 2D Covalent Organic Framework. *Prog. Org. Coatings*, 2021, *157*, 106299. https://doi.org/10.1016/j.porgcoat.2021.106299.

52. Zhang, M.; Yu, X.; Lin, Y.; Liu, J.; Wang, J. Anti-Corrosion Coatings with Active and Passive Protective Performances Based on v-COF/GO Nanocontainers. *Prog. Org. Coatings*, 2021, *159*, 106415. https://doi.org/10.1016/j.porgcoat.2021.106415.

53. Keshmiri, N.; Najmi, P.; Ramezanzadeh, M.; Ramezanzadeh, B.; Bahlakeh, G. Ultrastable Porous Covalent Organic Framework Assembled Carbon Nanotube as a Novel Nanocontainer for Anti-Corrosion Coatings: Experimental and Computational Studies. *ACS Appl. Mater. Interfaces*, 2022. https://doi.org/10.1021/acsami.1c24185

54. Qiu, S.; Zou, B.; Zhang, T.; Ren, X.; Yu, B.; Zhou, Y.; Kan, Y.; Hu, Y. Integrated Effect of NH2-Functionalized/Triazine Based Covalent Organic Framework Black Phosphorus on Reducing Fire Hazards of Epoxy Nanocomposites. *Chem. Eng. J.*, 2020, *401*, 126058. https://doi.org/10.1016/j.cej.2020.126058.

55. Mu, X.; Pan, Y.; Ma, C.; Zhan, J.; Song, L. Novel Co3O4/Covalent Organic Frameworks Nanohybrids for Conferring Enhanced Flame Retardancy, Smoke and CO Suppression and Thermal Stability to Polypropylene. *Mater. Chem. Phys.*, 2018, *215*, 20–30. https://doi.org/10.1016/j.matchemphys.2018.04.005.

56. Xiao, Y.; Jin, Z.; He, L.; Ma, S.; Wang, C.; Mu, X.; Song, L. Synthesis of a Novel Graphene Conjugated Covalent Organic Framework Nanohybrid for Enhancing the Flame Retardancy and Mechanical Properties of Epoxy Resins through Synergistic Effect. *Compos. Part B Eng.*, 2020, *182* (November 2019), 107616. https://doi.org/10.1016/j.compositesb.2019.107616.

57. Afshari, M.; Dinari, M. A Novel Triazine-Based Covalent Organic Framework: Enhancement Fire Resistance and Mechanical Performances of Thermoplastic Polyurethanes. *Compos. Part A Appl. Sci. Manuf.*, 2021, *147*, 106453. https://doi.org/10.1016/j.compositesa.2021.106453.

58. Mu, X.; Zhan, J.; Feng, X.; Cai, W.; Song, L.; Hu, Y. Exfoliation and Modification of Covalent Organic Frameworks by a Green One-Step Strategy: Enhanced Thermal, Mechanical and Flame Retardant Performances of Biopolymer Nanocomposite Film. *Compos. Part A Appl. Sci. Manuf.*, 2018, *110*, 162–171. https://doi.org/10.1016/j.compositesa.2018.04.030.

59. Mu, X.; Cai, W.; Xiao, Y.; He, L.; Zhou, X.; Wang, H.-J.; Guo, W.; Xing, W.; Song, L. A Novel Strategy to Prepare COFs Based BN Co-Doped Carbon Nanosheet for Enhancing Mechanical Performance and Fire Safety to PVA Nanocomposite. *Compos. Part B Eng.*, 2020, *198*, 108218. https://doi.org/10.1016/j.compositesb.2020.108218.

26

Recent Development in Covalent Organic Frameworks-Based Materials for Supercapacitors

**Christos Vaitsis[1], Maria Mechili[1], Nikolaos Argirusis[2],
Pavlos K. Pandis[1], Georgia Sourkouni[3], and Christos Argirusis[1,3]**

[1]*Laboratory of Inorganic Materials Technology, School of Chemical Engineering,
 National Technical University of Athens, Zografou/Athens, Greece*

[2]*mat4nrg GmbH, Clausthal-Zellerfeld, Germany*

[3]*TU Clausthal, Clausthaler Zentrum für Materialtechnologie, Clausthal-Zellerfeld, Germany*

CONTENTS

26.1 Introduction

Over the years, energy and environmental issues have been considered crucial global concerns. Traditional fossil fuel consumption has been sharply increasing, resulting in global warming and elevated gas emissions. Energy demands keep rising due to industrialization, thus environmentally benign power solutions are necessary for a brighter future. Energy storage devices, such as batteries and supercapacitors have been extensively studied to improve their characteristics, regarding energy/power density, lifetime/self-discharge energy loss, cost, operating conditions, etc.

Compared to batteries, supercapacitors (SCs) possess lower energy density but higher power density, albeit both are effectively used for different applications. SCs could be categorized into electric double-layer capacitors (EDLCs) and pseudocapacitors, based on their charge storage mechanism. Specifically, in EDLCs the charge is stored at the electrode and electrolyte interfaces, while in pseudocapacitors, the energy is stored through reversible and fast redox reactions (faradaic reactions). The electrodes are a critical component for the SC performance control, with high conductivity being a determining property for its efficiency. Additionally, large specific area, pore size distribution, temperature stability, and cost-effectiveness have to be considered during the selection and optimization of electrode design [1]. As electrical conductivity is a requirement for high-performance SCs, carbon is one of the most popular electrode materials, especially in EDLCs, thanks to its many different forms, such as activated carbon, graphite, graphene (and their oxides), single- and multi-walled nanotubes (SWCNTs, MWCNTs), etc.

Since their first appearance in 2005, covalent organic frameworks (COFs) have been used in various electrochemical applications, including Oxygen Evolution Reaction (OER), Oxygen Reduction Reaction (ORR), CO_2 Reduction Reaction (CO_2RR), different types of batteries and supercapacitors [2]. Distinguished for their high porosity, surface area, crystallinity, chemical stability, and topology diversity, researchers have been exploring synthetic strategies and new ligands for intriguing molecular design, which could further improve their performance and open up new possibilities. In the field of SCs, studies are mainly focused on morphology control, the introduction of redox-active groups, and composite synthesis. This chapter aims to bring out the most recent development in COF materials, both pristine and hybrids, underlining inherent COF weaknesses and ways to enhance their traits for such applications.

26.2 Pristine COFs

26.2.1 Functionalization

COFs can be functionalized by integrating groups such as hydroxyl, carboxyl, amine, keto, etc. either during the initial synthesis by selecting the preferred monomers or by post-synthetic modifications, to enhance their properties for the desired applications [3]. Li et al. prepared a benzobisthiazole-linked COF based on DABT monomer [4]. As the second monomer BTA, TFP, or both were used to investigate the effect degree of phenolic hydroxyl groups. More specifically, along with the pure BTA and TFP, partially functionalized COFs were prepared by fine-tuning the molar ratios of TFP/BTA to 0.25, 0.5, and 0.75. As the presence of hydroxyl groups was increased by adding TFP, the specific capacitances raised from 166 (non-functionalized) to 197, 248, 366, and finally 724 F/g for the fully functionalized COF at 1 A/g respectively. The electrochemical performance could be attributed to the benzobisthiazole unit and intramolecular hydrogen bonds facilitating the electron transfer, leading to excellent cycling stability by retaining 96% over 10′000 cycles.

Similarly, a study about the presence of keto groups showed a twofold increase in capacitance from 392 to 752 F/g at 1 A/g [5]. The COF construction was based on the imine condensation of TFTTF as electron donor and either DAA or DAQ as electron acceptor. Additionally, the assembled ASC exhibited 183 F/g at 1 A/g delivering an energy density of 57 Wh/kg at a power density of 858 W/kg. Khayum et al. have used the same diamine linkers in various ratios along with Tp for the development of thin films [6]. The π-electron rich anthracene linkers could improve the mechanical properties of the film by enhancing the crystallite interaction without the need for a binder, while still keeping it flexible (Figure 26.1a). As shown in Figure 26.1b–c increasing the DAQ:DAA ratio boosts the specific capacitance up to 154 F/g at 1.56 mA/cm². Keto groups were inserted in increasing steps in the benzidine (BD) ditopic diamine to construct Tp-BD COFs with different amounts of oxygens [7]. More precisely, diamines with one, two or four keto moieties were utilized

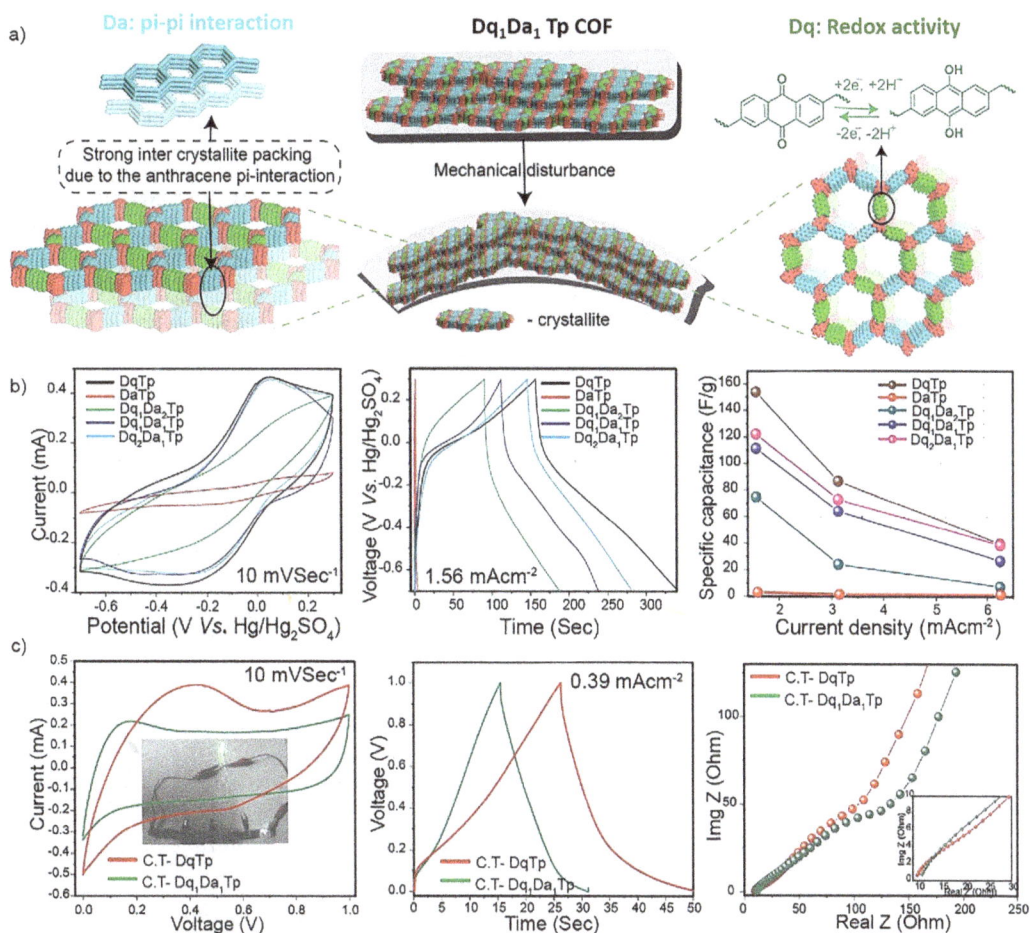

FIGURE 26.1

(a) Hybrid property of Dq_1Da_1Tp COF thin sheet [the Da (cyan linker) moiety is responsible for the strong interlocked crystallite interaction, whereas the Dq (green linker) could serve as an active redox unit in the framework]. (b) Three-electrode characterization using CV and GCD, and the plot represents current density vs specific capacitance. (c) Device characterization: CV (inset: 3.5 V LED lighted up by the series connection of four flexible devices), charge-discharge, and impedance analysis of CT-DqTp and CT-Dq_1Da_1Tp COF SC devices. (Adapted with permission from [6]. Copyright (2018) ACS.)

(denoted as 1KT-BD, 2KT-BD, and 4KT-BD) leading to specific capacitances of 61, 256, and 583 F/g at 0.2 A respectively. 2KT-Tp and 4KT-Tp COFs displayed retentions greater than 92% over 20'000 cycles. Additionally, the same monomers were used without the amine group, which exhibited lower values of 222 and 507 F/g (in the case of 2KT and 4KT). The assembled ASCs (2KT-Tp//AC and 4KT-Tp//AC) exhibited capacitances of 86 and 251 F/g at 0.8 A/g with practically no deterioration (99%) after 10'000 cycles at 5 A/g.

Another example was reported by Kandambeth et al. with the Hex-aza-COFs based on an aza-fused π-conjugated system [8]. The COFs were prepared via the condensation reaction of HKCH and aromatic tetramines (either 1,2,4,5-benzenetetramine or 1,2,4,5-tetramino-benzo-quinone). As expected, the quinone variant showed higher specific capacitance (585 vs 220 F/g at 1A/g). Additionally, a phenazine COF (based on TAP) exhibited 663 F/g, whereas as the negative electrode in RuO_2 ASC, it delivered an energy density of 23.3 Wh/kg at 661.2 W/kg.

26.2.2 Covalent Triazine Frameworks (CTFs)

Covalent Triazine Frameworks (CTFs) have also been used in SCs exhibiting good results [9–11]. They are mainly synthesized via trimerization of aromatic nitriles, possessing strong covalent bonds, along with N-rich content, which can be beneficial in energy storage applications [12]. A CTF based on PEBN was constructed at different temperatures (400, 600, 800°C), although the lower temperature variants were further treated at 800°C as well [10]. Nevertheless, CTF-800 had superior properties, possessing more micropores (a ratio of 42:58 micro:meso&macro) and the highest surface area (BET 1954 m^2/g). As an electrode, it exhibited a specific capacitance of 625 F/g at 0.5 A/g and exceptional cycling stability with 96% retention over 20'000 cycles at 30 A/g. Moreover, the assembled coin cell SC showed good results in an ionic liquid electrolyte ([EMIM][BF$_4$]) at a potential window of 0–3 V and could maintain a capacitance of 116 and 136 F/g at sub-ambient temperatures of -20 and 0°C respectively.

Gao et al. examined the effect of halogen atoms by using TN, TFTN, or TCTN monomers, possessing simple H, F, or Cl respectively [11]. The theoretical calculations had indicated that the halogen presence in the CTF could lower the energy gap, facilitate the electron transfer and possibly enlarge the interlayer space of two adjacent CTF layers due to atom repelling effects, especially for F atoms, which have high electronegativity. The FCTF displayed capacitance of 379 F/g at 1 A/g and could retain 96.8% after 10'000 cycles at 5 A/g, compared to 55 and 176 F/g at 1 A/g of HCTF and ClCTF respectively.

26.2.3 Melamine-Based COFs

Various efforts have been made with melamine-based COFs coupled with aldehyde [13], thiophene [14], or squaraine [15] moieties. Li et al. utilized PDC for the Schiff base aldehyde-amine condensation, leading to a COF with π-π stacking interactions, favoring fast charge transfer and reducing ionic diffusion resistance [13]. The interlayer C–H···N H-bonding provided the COF with a performance of 335 F/g at 1 A/g, accessing around 20% of the redox-active triazine units. The ASC managed to retain 88% after 20'000 cycles, exhibiting 29.2 Wh/kg at 750 W/kg. The combination of melamine with thiophene-based monomer (specifically DBT) generated a high heteroatom content that could facilitate ion transfer from the electrolyte to the COF channels (average pore diameter of 1.3 nm) [14]. The COF reached a capacitance of 407 F/g at 1 A/g in a simple setup, whereas in the ASC with N-doped graphitized chitosan as a negative electrode, it achieved 90 F/g at 1 A/g and 92.1 Wh/kg at 800 W/kg, with a retention of 83% over 30'000 cycles. Another interesting monomer choice was squaric acid, where the synthesized COF, coated on Ni foam, displayed 177 F/g at 0.3 A/g and preserved 94% after 10'000 cycles, albeit with some fluctuations in the window of 75–125% [15].

26.2.4 Various COFs

Zhang and co-workers have studied olefin-linked COFs based on the Knoevenagel condensation of TMTA with TFPT (g-$C_{30}N_6$-COF) or the longer TFBT (g-$C_{48}N_6$-COF) [16], as well as DCTMP with TFPT (g-$C_{34}N_6$-COF) [17]. The N-rich triazine moieties offered strong π-π stacking for the growth along one direction forming long fibers, along with active sites in the open nano-channels. The COFs were interwoven with SWCNTs for the construction of hybrid films to enhance conductivity as well as electron transfer, due to π-electron cloud stacking (Figure 26.2). The assembled planar micro SCs delivered volumetric energy

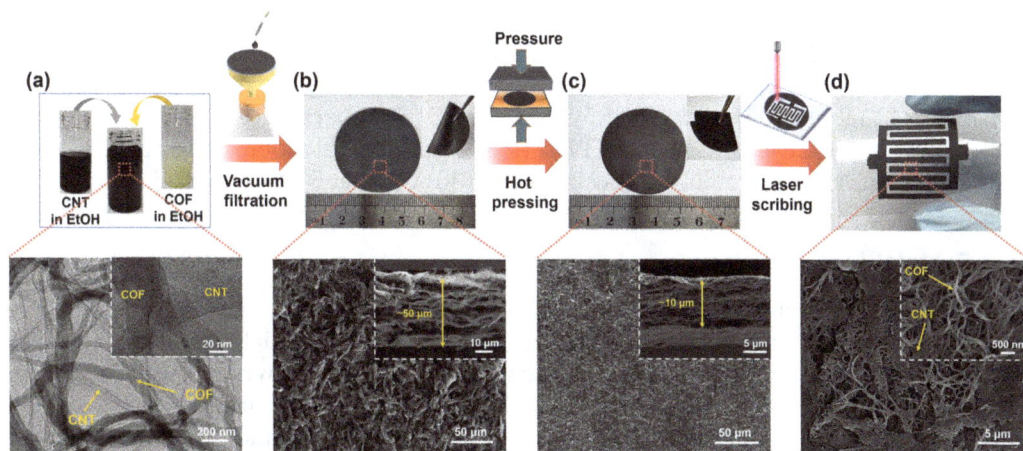

FIGURE 26.2
The fabrication process of a COF for micro-SC (MSC). (a) COF/CNT dispersion in ethanol (EtOH). (b) Photograph of as-prepared COF/CNT film. (c) Photograph of COF/CNT film after hot-pressing treatment. (d) Photograph of the as-fabricated COF-MSC electrodes on a flexible PET substrate. Below: top-view of SEM/TEM images of a respective process step. (Adapted with permission from [16]. Copyright (2020), Elsevier.)

densities of 8.9 (g-$C_{30}N_6$-COF) and 8.7 mWh/cm³ (g-$C_{48}N_6$-COF) at 1.3 W/cm³ and 4.9 mWh/cm³ (g-$C_{34}N_6$-COF) at 1.04 W/cm³.

Halder et al. have prepared a TpOMe-DAQ COF with high stability in concentrated acidic (18 M H_2SO_4, 12 M HCl) and alkaline solutions (9 M NaOH) [18]. The chemical stability was gained due to steric and hydrophobic imine protection bonds by the –OCH_3 functionality and interlayer C–H⋯N H–bonding. The areal capacitance of the electrode reached 1600 mF/cm² (or 169 F/g) with good cycling stability over 100'000 cycles. Apart from the more frequently used DAA/DAQ, NDA has also been employed as a monomer for SC applications. Its combination with TFP afforded a COF electrode which demonstrated a pseudocapacitance of 348 F/g at 0.5 A/g [19]. Iqbal et al. synthesized a 2D COF via the condensation of PTA and PMDA [20]. The nanopores and micron-grade sheets and clusters of the COF structure assisted with the diffusion of the LITFSI electrolyte and the symmetric SC delivered a specific capacitance of 163 F/g at 0.5 A/g with an energy density of 35.7 Wh/kg at 250 W/kg and retention of 84.1% after 30'000 cycles at 1 A/g. Sajjad et al. have prepared a phosphine-based COF (Pa and TFPP building blocks), which have been more scarce in literature for such applications [21]. The COF electrode displayed a capacitance of 100 F/g at 1 A/g in a 3M Na_2SO_4 electrolyte. Very recently, Xiong et al. reported the use of TAPA-TFPA COF taking advantage of interlayer π-π stacking promoting electron conduct in 2D and 3D directions [22]. In a three-electrode setup, it displayed around 185 F/g at 1 A/g, with a retention of 111% over 5000 cycles at 5 A/g; the increased percentage could be explained due to the pore expansion effect generated by the ion shuttle during charge-discharge.

El-Mahdy and co-workers prepared carbazole-based COFs via the [3+3] condensation of Car-3NH₂ and triformyl linkers of different planarity [23]. The COFs presented hollow morphology, either micro-spherical or micro-tubular through the Ostwald ripening effect, but were proven inefficient as electrode materials with capacitances lower than 20 F/g even at 0.2 A/g. A similar tubular morphology could be attained with the [3+2] condensation of TFPT and DAHQ, forming a sponge-like surface due to the lack of interaction of crystallites on the outer part [24]. This COF managed to reach 256 F/g at 0.5 A/g with good stability over

1850 cycles (98.8% retention) and an energy density of 43 Wh/kg. Good capacitances (at a current density of 2 A/g) were achieved with β-ketoenamine-linked COFs via [3+3] Schiff base polycondensations of TFP with tris(aminophenyl) moieties, featuring amino (291.1 F/g), carbazole (149.3 F/g) and pyridine (185.5 F/g) units [25]. Comparable results were displayed with TPPDA-based COFs consisted of TPPyr (188.7 F/g, 85.6% retention) or TPTPE (237.1 F/g, 86.2% retention); cycling stability was calculated at 10 A/g over 5000 cycles [26].

26.2.5 Metal Incorporation

Li et al. have incorporated Ni^{2+} into a COF based on 1,2,4,5-benzenetetraamine and 2,5-dihydroxy-1,4-benzenedicarboxaldehyde [27]. The Schiff base aldehyde-amino reaction competes with the formation of Ni(II)-salphen units, and slows down the nucleation step. The Ni-COF exhibited a high capacitance of 1257 F/g at 1 A/g, due to high-density redox centers and conjugated planar structure, with 94% capacitance retention over 10'000 cycles. The assembled ASC delivered 417 F/g at 1 A/g and an energy density of 130 Wh/kg at 839 W/kg. The effect of Ni^{2+} presence was further studied by comparing the Ni-COF with the same COF that did not include the metal. The electrical conductivity of the simple COF was almost 1500 times lower and it only achieved a capacitance of 184 F/g, with higher charge transfer resistance as proved by electrochemical impedance spectroscopy (EIS; Figure 26.3)

26.2.6 COF Derived Materials

Xue et al. prepared a COF linked by –NH– bonds via the condensation of Pa and TCT [28]. The thermal treatment at 700°C under N_2 afforded the derived N-doped porous carbon

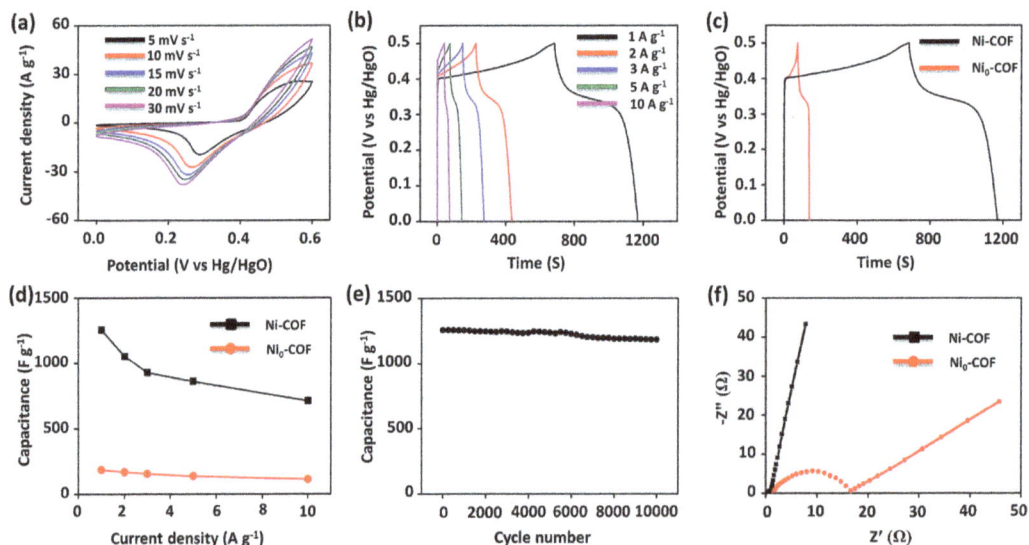

FIGURE 26.3
Electrochemical properties of Ni-COF in a three-electrode system. (a) CV curves at scan rates from 5 to 30 mV/s. (b) GCD curves at different current densities (1–10 A/g). (c) Comparison of GCD curves at 1 A/g for Ni-COF and Ni_0-COF. (d) Capacitance vs. current density for Ni-COF and Ni_0-COF. (e) Cycling stability measurement at a current density of 1 A/g. (f) Comparison of Nyquist plots of Ni-COF and Ni_0-COF. (Adapted with permission from [27]. Copyright (2019), RSC.)

structure, which presented a 9-fold increase to BET surface area from 44 to 398 m^2/g. The enhanced properties of the derived product "expanded" onto the electrochemical performance, increasing the specific capacitance from 283 to 450 F/g at 1 A/g. Regarding its stability, the pristine COF retained around 88% over 20'000 cycles, whereas the porous carbon displayed a retention of 99.2% after 5000 cycles at 1 A/g and 77% after 3000 cycles at 5 A/g.

Zhou et al. synthesized a melamine-boroxine COF (with 4-FPBA), along with the use of copper nitrate as a catalyst [29]. The calcination at 1000°C led to the B-N-doped carbon capsules; the N and B atoms can generally enhance the carbon hydrophilicity and lower the valence band. The charge-discharge curves at 20 A/g kept the retention above 90% over 1400 cycles. COF-5 was carbonized at 1000°C under N_2, receiving a B-doped carbon with a specific capacitance of 15.3 mF/cm^2 at 40 mA/g [30]. The electron deficiency in the boron atom can attract anions toward the surface and the performance could be further improved by tuning the boron content and optimizing the surface area. The carbonization of a TPB-based COF afforded microporous materials with narrow pore size [31]. Specifically, a distribution of 0.5–1.5 nm was observed when the materials were further activated at 800 or 900°C under CO_2, whereas at 950°C a mesoporous honeycomb-like structure was received with 2–5 nm pore size. The 900°C variant electrode displayed superior specific capacitance (278 F/g) at 1 A/g, most probably due to the compatibility of micropores with the effective ion sizes of the SC system and the facilitation of the electrical double-layer formation.

The pyrolysis process is naturally a critical factor to the properties of the derived material. Tp-DAQ COF was carbonized either on its own or after its hybridization with potassium carbonate at various temperatures [32]. The presence of K_2CO_3 affected the electrochemical performance to a high degree, as for the temperature of 700°C, the specific capacitance increased from 279 to 768 F/g at 1 A/g, whereas at 850°C, the values spiked from 38 to 1711 F/g. Even at extreme current densities of 500 A/g they were maintained at 439 (700°C) and 856 F/g (850°C), where the superior performance of 850°C samples could be ascribed to the higher O/N content. Additionally, a symmetric SC was assembled by mixing the derived carbon with CNT and constructing hybrid thin-film electrodes, which exhibited values of 480 and 700 F/g at 1 A/g.

26.3 COF Composites

In the case of developing highly electrochemically active materials, significant efforts have been carried out for the preparation of COF composites in order to implement additional functionalities. Carbon composites have been very popular among electrochemical applications since carbon offers high conductivity, mechanical strength and chemical robustness. TpPa-COF was nanocoated onto SWCNTs via in-situ polymerization, showing a specific capacitance of 153 F/g at 0.5 A/g [33]. Wang et al. synthesized 2D COF nanosheets and combined it with reduced graphene oxide (rGO) [34]. The ILCOF-1 (based on TFPPy and PPD) prevented the (re)stacking of graphene by interrupting the distance-dependent van der Waals interaction. The research group produced composites with different GO mass fractions (10, 20, 30%) presenting that 20% ratio delivered the optimal specific capacitance of 321 F/g (and volumetric of 237 F/cm^3) at 1 A/g compared to 270 (10%) and 283 F/g (30%). The 1D fiber fabricated SC showed a high energy density of 10.3 Wh/kg and 7.9 mWh/cm^3 in aqueous and gel electrolyte respectively, whereas an ionic liquid electrolyte (EMIMBF$_4$) achieved a stack energy density of 87 Wh/L at a power density of 638 W/L. Another

COF-rGO composite, based on BTA and DPPD, was prepared via in-situ Schiff base coupling at room temperature [35]. The hybrid exhibited a specific capacitance of 239.1 F/g at 0.5 A/g (192.7 F/g at 1 A/g and 158.1 F/g at 2 A/g); the electrochemical performance could be explained due to ion conduction, π-conjugation, and pyridine redox sites.

A promising approach was followed by Li et al., who grew the COF on 2D graphene sheets, stacked to form an aerogel with a hierarchical porous structure [36]. The composite was obtained via in-situ Schiff base condensation between the aldehyde of Tp and amine groups of DAQ in the presence of GO. The aerogel yielded an increased specific capacitance of 269 F/g (an equivalent of 404 C/g) at 0.5 A/g and 222 F/g at 10 A/g with a retention of 83%. For comparative purposes, a simple mix of COF-carbon black only delivered 22.4 F/g at 0.5 A/g. Very recently, a similar route was reported, with a surfactant modification, by firstly preparing the COF and then mixing a positively charged CTAB-COF dispersion with the negatively charged GO to achieve electrostatic self-assembly (Figure 26.4) [37]. The aerogel composite (DAQ-COF/GA) was received after 12 h at 180°C and it exhibited 378 F/g at 1 A/g (or 308 F/g at 70 A/g with retention of 81.5% over 20'000 cycles). A good energy density of 30.5 Wh/kg at 700 W/kg was displayed by the ASC (GA//DAQ-COF/GA) with 88.9% stability after the same number of cycles in 1M H_2SO_4 electrolyte.

Xu et al. prepared the same COF as a DAAQ-Tp-COF membrane with the addition of a hydroxyl-ended hyperbranched polymer (OHP) template and was then impregnated in a CNT film [38]. The COF@OHP complex implementation prevented the agglomeration and possible collapse of COF and CNT film. The composite membrane electrode exhibited a high capacitance of 249 F/g and a retention of 80% after 10'000 cycles, which could be attributed to the strong π-π interaction between the COF and CNT film. Kong et al.

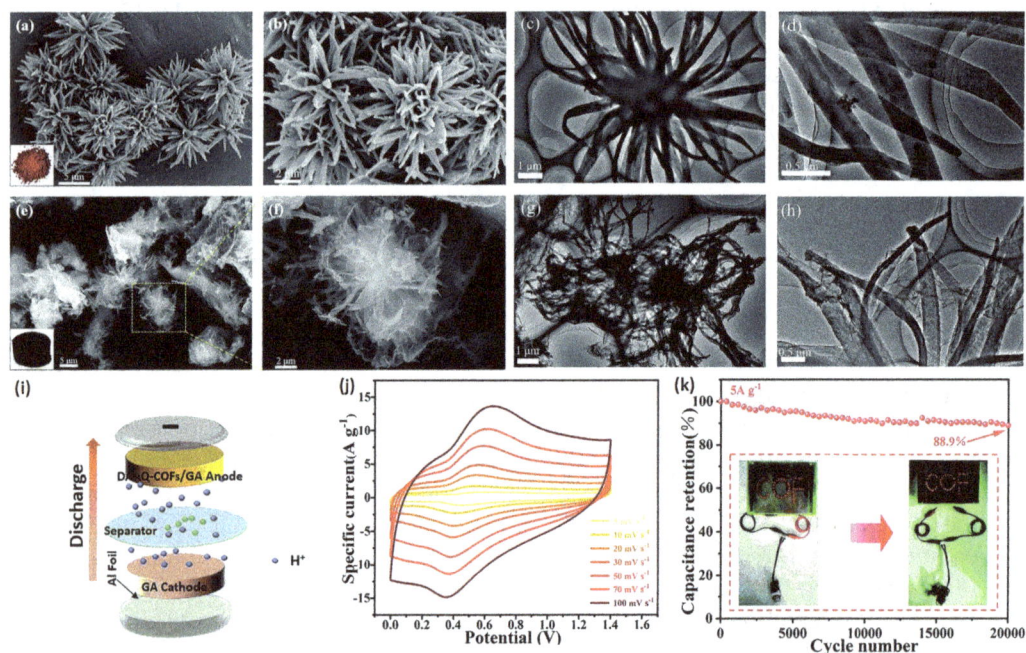

FIGURE 26.4
SEM images of (a, b) DAAQ-COFs and (e, f) DAAQ-COFs/GA. TEM images of (c, d) DAAQ-COFs and (g, h) DAAQ-COFs/GA. (i) Schematic illustration of GA//DAAQ-COF/GA ASC. (j) CV curves with different scan rates. (k) Cycling stability test at 5 A/g for 20'000 cycles. (Adapted with permission from [37]. Copyright (2021) RSC.)

reported a tube-type core-shell structure by growing DAQ-Tp-COF onto carboxylated MWCNTS (c-CNTS) [39]. Simple mixing of the COF and c-CNTs only slightly increased the capacitance value from pure c-CNTs to 52.3 F/g at 0.5 A/g, with negligible Faradaic process in the COF. Instead, the controlled in-situ growth raised the capacitance of c-CNT@ COF-1,-2,-3 (depending on initial c-CNTs mass loading of 100, 200, 300 mg) to 261.3, 364.7 and 376.2 F/g at the same current density, whereas the calculations showed that 38.4, 83, and 99.1% of the COF redox active sites were utilized. EIS analysis revealed that the charge transfer resistances of c-CNT-COF-3 were much lower, while from the diffusion responses it was observed that the electrolyte penetrates the thin COF nanolayer much faster.

Yang et al. prepared NKCOF-8, based on Azo-NHBoc and TFPA monomers, displaying the proton conduction Grotthuss mechanism due to phenolic hydroxyl and azo groups [40]. The COF was in-situ hybridized with CNTs, leading to a capacitance of 440 F/g at 0.5 A/g. The presence of enriched redox-active N=N groups along with the introduction of CNTs electronic conductivity facilitated the reaction kinetics resulting in an energy density of 71 Wh/kg at 42 kW/kg for the constructed NKCOF-8//AC ASC. Besides commonly used carbon materials, conductive polymers, such as Polyaniline (PANI) can also effectively enhance COF electrochemical properties. Liu et al. have prepared TpPa-COF on PANI via in-situ polycondensation process [41]. The composite showed capacitance of 95 F/g at 0.2 A/g (and 50 F/g at 50 A/g) and a retention of 83% after 30'000 cycles, indicating good stability. Peng and co-workers added a 3D porous carbon along with PANI to their composite based on COF-LZU1 [42]. More specifically, the spherical COF was formed via dehydration condensation between BTA and Pa and bonded simultaneously to the 3D carbon surface (derived from kenaf stem), while the PANI array was grown via chemical oxidation polymerization to finally receive the 3D-KSC/COF-LZU1/PANI composite. The specific capacitance of the hybrid reached 583 mF/cm² at 0.1 mA/cm² and retained 81.5% of the initial value after 2000 cycles. On the other hand, composites prepared without PANI or without the COF were limited to 115.7 and 107.5 mF/cm² respectively. Similarly, Dutta et al. modified the COFs (triazine-based) with polymerization of aniline [43]. The two pristine COFs, TCOF-1 (based on TFPOT and Pa) and TCOF-2 (based on TFPOT and TAPT) exhibited very low specific capacitances of less than 20 F/g, whereas their composites with PANI reached 154 and 275 F/g at 0.5 A/g respectively (pure PANI had a capacitance of 220 F/g). The increased performance of TCOF-2 could be attributed to the synergistic effect of faster ion diffusion within the framework (whisker-like morphology) and reversible redox reactions.

Another viable electrode material choice can be polypyrrole (PPy), endowed with rich porosity and interconnected hollow structure. Wang et al. synthesized hollow COF-316 micro-flowers of 5–7 μm particle size through a process of self-assembly, Ostwald-ripening and epitaxial growth (Figure 26.5) [44]. PPy was electrochemically polymerized onto COF-316 (HHTP-TFTN monomer blocks) through hydrogen bonds, improving the composite stability and accelerating the charge transfer, achieving an areal capacitance of 783 μF/cm² at 3 μA/cm² (fully stable for 3400 cycles at 20 μA/cm²).

Another route for attractive hybrids is the combination of COFs with Metal-Organic Frameworks (MOFs). Falling in the same category, both MOFs and COFs share most of their intriguing properties, but also some of their problems. With their many synthesis methods [45], MOFs have been expanding their applications over the years, but still, they are not ideal for electrochemical applications as pristine materials [46]. A recent example of such a combination was the core-shell structure prepared by a post-synthetic modification (PSM) via an aza-Diels-Alder cycloaddition [47]. Specifically, as shown in Figure 26.6, the research group started with the surface condensation of the UiO-66-NH₂ MOF core with

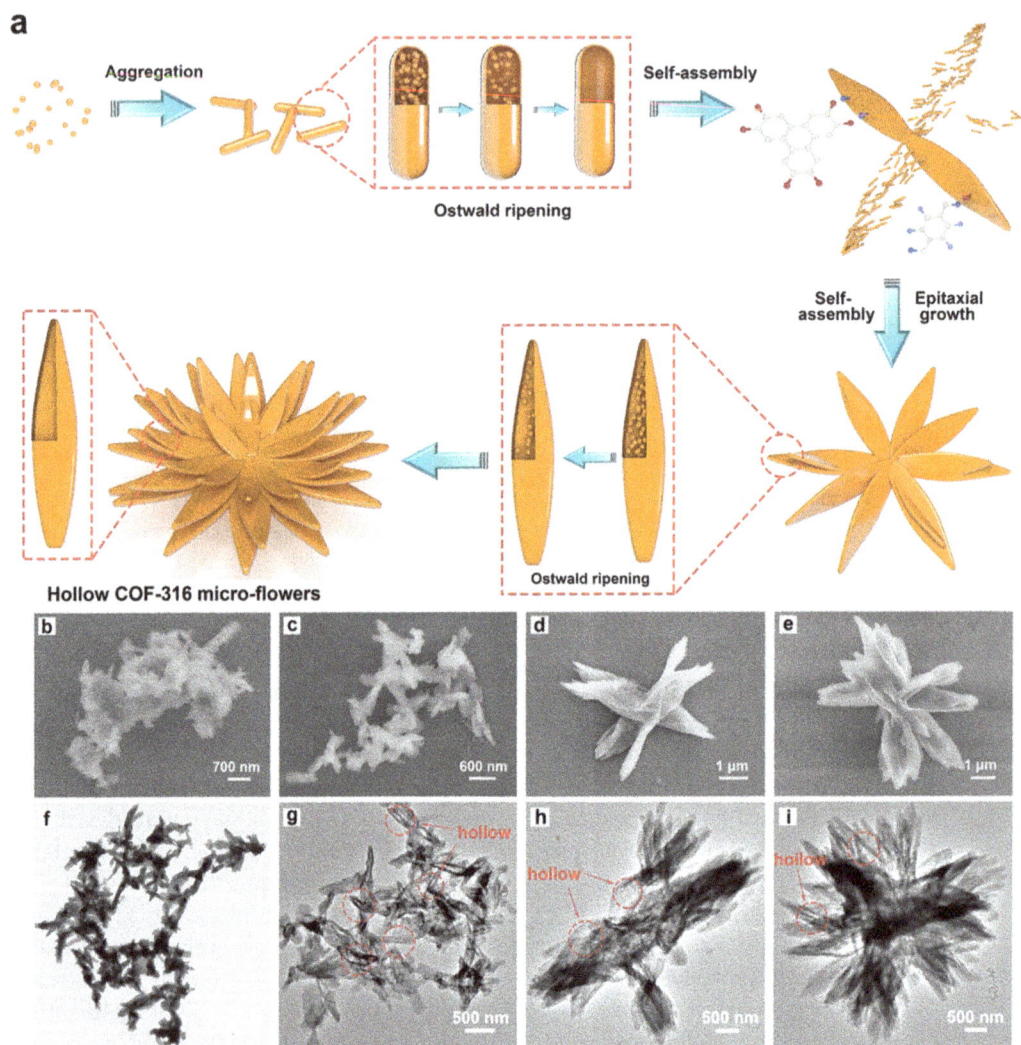

FIGURE 26.5
Formation mechanism and time-dependent structure characterization of hollow COF-316 micro-flowers. (a) The schematic diagram for the formation process of hollow COF-316 micro-flowers, following the process of Ostwald ripening, self-assembly, and epitaxial growth. (b–e) SEM images and (f–i) TEM images of COF-316 intermediates synthesized at 6, 12, 24, and 48 h respectively. (Adapted with permission from [44]. Copyright (2021), Wiley.)

BTA to receive UiO-66(CHO) and then with Pa followed by the PSM reaction between the aryl imine of MOF@COF-LZU1 and phenylacetylene to convert the imine bridges to the corresponding quinoline-linked aza-MOF@COF final structure (Figure 26.6). The specific capacitance reached 20.35 μF/cm^2 and the use of molecular spacers provided a volumetric energy density of 1.16 F/cm^3. Regarding the cycling stability, the composite had a retention of 89.3% at 0.2 A/cm^2 over 2000 cycles, compared to 69.5% for the pristine UiO-66-NH$_2$ and 73.9% for MOF@COF-LZU1.

Metal oxides have been widely studied in electrochemical storage [48], while pure metals can be utilized as well in various morphologies. For instance, Han et al. nanocoated

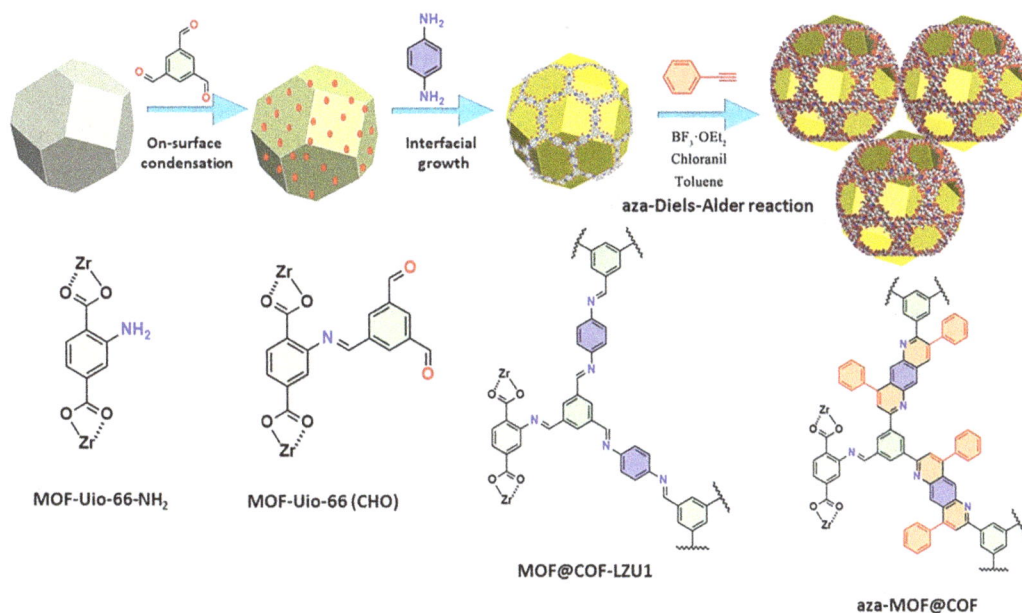

FIGURE 26.6

Schematic illustration of the aza-MOF@COF synthetic route. (Adapted with permission from [47]. Copyright (2020), Wiley.)

Ni nanowires (NWs) with a COF to combine the NiNWs for their conductivity and COF nanocoating for the high porosity and efficient penetration of electrolyte [49]. SEM/TEM analysis revealed that the thickness of TpPa-COF coating was around 180 nm. The specific capacitance of NiNWs@TpPa-COF reached 426 F/g at 2 A/g and 314 F/g at 50 A/g.

26.3.1 Derived COF Composites

Sun et al. attempted a molecular pillar approach in order to grow COF nanosheets on graphene [50]. With controllable synthesis the orientation of COF-1 growth was kept perpendicular to the graphene surface via the edge-on anchoring of BDBA monomers; the thickness was controlled from 3 (BDBA:GO ratio of 0.5:1) to 15 nm (ratio of 5:1) based on the loading of BDBA (Figure 26.7). The composite was then carbonized at 600°C under Ar and the final material, denoted as CNS-RGO had kept the initial orientation. The study showed that the ratio of 3:1 had the best electrochemical performance with a specific capacitance of ~162 F/g at 2 A/g. Additionally, in order to investigate the orientation effect, they prepared a horizontal variant, which only delivered ~97 F/g. Similarly, Zhang et al. followed the same initial procedure and then attached MnO_2 nanosheets onto the porous carbon skeleton [51]. The combination of rGO/C and MnO_2 slightly increased the surface area and also facilitated the electrolyte ion access into the electrodes. The composite exhibited a specific capacitance of 215.2 F/g at 0.15 A/g and a retention of 72% after 2500 cycles. The assembled SC (with AC as the negative electrode) provided an energy density of 21.2 Wh/kg at a power density of 190.4 W/kg.

Vargheese and co-workers prepared a COF based on biphenyl and cyanuric chloride via Friedel-Crafts reaction, which was then carbonized and subsequently added to $KMnO_4$/H_2SO_4 solution to receive the flower-like MnO_2@N-PC [52]. The aqueous symmetric SC

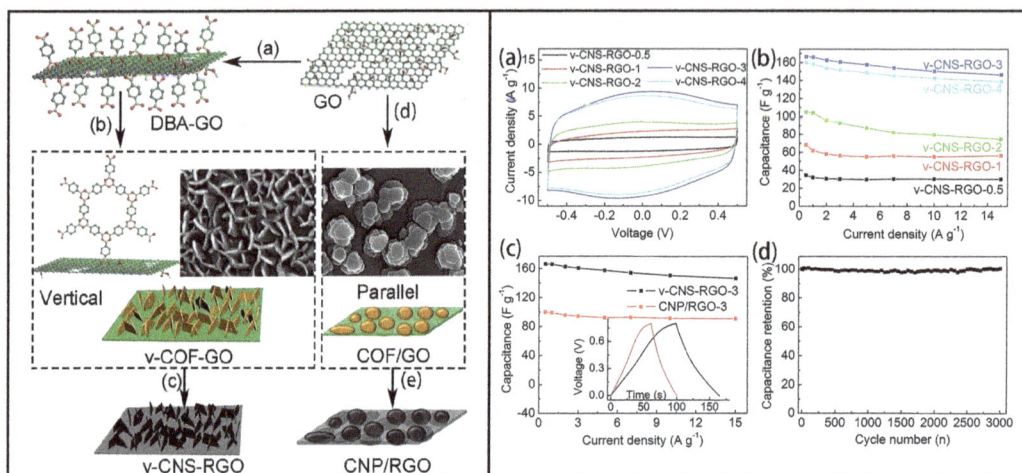

FIGURE 26.7

Left: (a) Covalent functionalization of GO with DBA (b) Growth of vertical COF-1 nanosheets using DBA as molecular nucleation sites grafted on GO in mesitylene/dioxane solution (v-COF-GO). (d) COF-1 platelets parallel to GO are formed at the same reaction conditions in absence of DBA functionalization (COF/GO). Carbonization of (c) v-COF-GO and (e) COF/GO to produce carbon nanosheets oriented vertical or parallel to RGO surface. Right: (a) CV curves of v-CNS-RGO electrodes prepared with different ratios of DBA to DBA-GO (100 mV/s), (b) Capacitance for v-CNSRGO electrodes as a function of loading ratio, (c) Capacitance for carbonized samples with vertical (v-CNS-RGO-3) and parallel (CNP/RGO-3) orientation of nanosheets. Inset: GCD curves of v-CNS-RGO-3 and CNP/RGO-3 electrodes at 1 A/g, (d) Cycling performance of the v-CNS-RGO-3 electrode. (Adapted with permission from [50]. Copyright (2018), Wiley.)

delivered a capacitance of 134.5 F/g with a specific energy of 42.1 Wh/kg and specific power of 750 W/kg at 0.5 A/g. The cycling stability was examined over 40'000 charge/discharge cycles, with 76% retention, where the fading was observed during the first 10'000 cycles and could be explained due to the irreversible reduction/oxidation of Mn^{4+} to Mn^{2+} and Mn^{7+}. Moreover, a 9 V symmetric SC was constructed by connecting a series of six cells, delivering 53.5 Wh/kg and 5.6 kW/kg (19% capacitance loss after 10'000 cycles).

26.4 Conclusions and Outlook

The rapid development of COF materials over the last decade has brought new exciting possibilities in energy applications. A variety of different COFs have been already explored, but still, novel COF materials are being created to enrich the library with abundant functionalities, as the presence of redox-active moieties can increase the pseudocapacitance. In order to achieve superior electrochemical performance, electron exchange and ion transfer should be optimized by designing either new COFs or synthetic approaches that are focused on facilitating mass transport. The introduction of other materials, such as conducting polymers, carbon, or metal complexes cannot only enhance the electrocatalytic efficiency but also the composite stability. Nevertheless, the COF conductivity is always one of the core issues that need to be addressed, by improving the framework conjugation degree or hybridizing with the aforementioned materials.

References

1. Poonam, Sharma K, Arora A, Tripathi SK (2019) Review of supercapacitors: Materials and devices. J. Energy Storage 21:801–825.
2. Zhao X, Pachfule P, Thomas A (2021) Covalent organic frameworks (COFs) for electrochemical applications. Chem. Soc. Rev. 50 (12):6871–6913.
3. Segura JL, Royuela S, Mar Ramos M (2019) Post-synthetic modification of covalent organic frameworks. Chem. Soc. Rev. 48 (14):3903–3945.
4. Li T, Yan X, Liu Y, Zhang W-D, Fu Q-T, Zhu H, Li Z, Gu Z-G (2020) A 2D covalent organic framework involving strong intramolecular hydrogen bonds for advanced supercapacitors. Polym. Chem. 11 (1):47–52.
5. Li T, Yan X, Zhang WD, Han WK, Liu Y, Li Y, Zhu H, Li Z, Gu ZG (2020) A 2D donor-acceptor covalent organic framework with charge transfer for supercapacitors. Chem. Commun. 56 (91):14187–14190.
6. Khayum MA, Vijayakumar V, Karak S, Kandambeth S, Bhadra M, Suresh K, Acharambath N, Kurungot S, Banerjee R (2018) Convergent covalent organic framework thin sheets as flexible supercapacitor electrodes. ACS Appl. Mater. Interfaces 10 (33):28139–28146.
7. Li M, Liu J, Li Y, Xing G, Yu X, Peng C, Chen L (2021) Skeleton engineering of isostructural 2D covalent organic frameworks: Orthoquinone redox-active sites enhanced energy storage. CCS Chemistry 3 (2):696–706.
8. Kandambeth S, Jia J, Wu H, Kale VS, Parvatkar PT, Czaban-Jóźwiak J, Zhou S, Xu X, Ameur ZO, Abou-Hamad E, Emwas AH, Shekhah O, Alshareef HN, Eddaoudi M (2020) Covalent organic frameworks as negative electrodes for high-performance asymmetric supercapacitors. Adv. Energy Mater. 10 (38):2001673.
9. Li Y, Zheng S, Liu X, Li P, Sun L, Yang R, Wang S, Wu ZS, Bao X, Deng WQ (2018) Conductive microporous covalent triazine-based framework for high-performance electrochemical capacitive energy storage. Angew Chem Int Ed Engl 57 (27):7992–7996.
10. Vadiyar MM, Liu X, Ye Z (2019) Macromolecular polyethynylbenzonitrile precursor-based porous covalent triazine frameworks for superior high-rate high-energy supercapacitors. ACS Appl. Mater. Interfaces 11 (49):45805–45817.
11. Gao Y, Zhi C, Cui P, Zhang KAI, Lv L-P, Wang Y (2020) Halogen-functionalized triazine-based organic frameworks towards high performance supercapacitors. Chem. Eng. J. 400:125967.
12. Liu M, Guo L, Jin S, Tan B (2019) Covalent triazine frameworks: synthesis and applications. J. Mater. Chem. A 7 (10):5153–5172.
13. Li L, Lu F, Xue R, Ma B, Li Q, Wu N, Liu H, Yao W, Guo H, Yang W (2019) Ultrastable triazine-based covalent organic framework with an interlayer hydrogen bonding for supercapacitor applications. ACS Appl. Mater. Interfaces 11 (29):26355–26363.
14. Li L, Lu F, Guo H, Yang W (2021) A new two-dimensional covalent organic framework with intralayer hydrogen bonding as supercapacitor electrode material. Microporous Mesoporous Mater. 312:110766.
15. Xue R, Gou H, Zhang L, Liu Y, Rao H, Zhao G (2021) A new squaraine-triazine based covalent organic polymer as an electrode material with long life and high performance for supercapacitors. New J. Chem. 45 (2):679–684.
16. Zhang F, Wei S, Wei W, Zou J, Gu G, Wu D, Bi S, Zhang F (2020) Trimethyltriazine-derived olefin-linked covalent organic framework with ultralong nanofibers. Sci. Bull. 65 (19):1659–1666.
17. Xu J, He Y, Bi S, Wang M, Yang P, Wu D, Wang J, Zhang F (2019) An olefin-linked covalent organic framework as a flexible thin-film electrode for a high-performance micro-supercapacitor. Angew Chem Int Ed Engl 58 (35):12065–12069.
18. Halder A, Ghosh M, Khayum MA, Bera S, Addicoat M, Sasmal HS, Karak S, Kurungot S, Banerjee R (2018) Interlayer hydrogen-bonded covalent organic frameworks as high-performance supercapacitors. J. Am. Chem. Soc. 140 (35):10941–10945.

19. Das SK, Bhunia K, Mallick A, Pradhan A, Pradhan D, Bhaumik A (2018) A new electrochemically responsive 2D π-conjugated covalent organic framework as a high performance supercapacitor. Microporous Mesoporous Mater. 266:109–116.
20. Iqbal R, Badshah A, Ma Y-J, Zhi L-J (2020) An electrochemically stable 2D covalent organic framework for high-performance organic supercapacitors. Chin. J. Polym. Sci. 38 (5):558–564.
21. Sajjad M, Tao R, Qiu L (2021) Phosphine based covalent organic framework as an advanced electrode material for electrochemical energy storage. J. Mater. Sci. Mater. Electron. 32 (2):1602–1615.
22. Xiong S, Liu J, Wang Y, Wang X, Chu J, Zhang R, Gong M, Wu B (2022) Solvothermal synthesis of triphenylamine-based covalent organic framework nanofibers with excellent cycle stability for supercapacitor electrodes. J. Appl. Polym. Sci. 139 (3):51510.
23. El-Mahdy AFM, Young C, Kim J, You J, Yamauchi Y, Kuo SW (2019) Hollow microspherical and microtubular [3+3] carbazole-based covalent organic frameworks and their gas and energy storage applications. ACS Appl. Mater. Interfaces 11 (9):9343–9354.
24. El-Mahdy AFM, Hung YH, Mansoure TH, Yu HH, Chen T, Kuo SW (2019) A hollow microtubular triazine- and benzobisoxazole-based covalent organic framework presenting sponge-like shells that functions as a high-performance supercapacitor. Chem. Asian J. 14 (9):1429–1435.
25. El-Mahdy AFM, Hung Y-H, Mansoure TH, Yu H-H, Hsu Y-S, Wu KCW, Kuo S-W (2019) Synthesis of [3+3] β-ketoenamine-tethered covalent organic frameworks (COFs) for high-performance supercapacitance and CO2 storage. J. Taiwan Inst. Chem. Eng. 103:199–208.
26. El-Mahdy AFM, Mohamed MG, Mansoure TH, Yu HH, Chen T, Kuo SW (2019) Ultrastable tetraphenyl-p-phenylenediamine-based covalent organic frameworks as platforms for high-performance electrochemical supercapacitors. Chem. Commun. 55 (99):14890–14893.
27. Li T, Zhang W-D, Liu Y, Li Y, Cheng C, Zhu H, Yan X, Li Z, Gu Z-G (2019) A two-dimensional semiconducting covalent organic framework with nickel(ii) coordination for high capacitive performance. J. Mater. Chem. A 7 (34):19676–19681.
28. Xue R, Gou H, Liu Y, Rao H (2021) A layered triazinyl-COF linked by –NH– linkage and resulting N-doped microporous carbons: preparation, characterization and application for supercapacitance. J. Porous Mater. 28 (3):895–903.
29. Zhou Z, Zhang X, Xing L, Liu J, Kong A, Shan Y (2019) Copper-assisted thermal conversion of microporous covalent melamine-boroxine frameworks to hollow B, N-codoped carbon capsules as bifunctional metal-free electrode materials. Electrochim. Acta 298:210–218.
30. Umezawa S, Douura T, Yoshikawa K, Takashima Y, Yoneda M, Gotoh K, Stolojan V, Silva SRP, Hayashi Y, Tanaka D (2021) Supercapacitor electrode with high charge density based on boron-doped porous carbon derived from covalent organic frameworks. Carbon 184:418–425.
31. Kim M, Puthiaraj P, Qian Y, Kim Y, Jang S, Hwang S, Na E, Ahn W-S, Shim SE (2018) High performance carbon supercapacitor electrodes derived from a triazine-based covalent organic polymer with regular porosity. Electrochim. Acta 284:98–107.
32. Yan D, Wu Y, Kitaura R, Awaga K (2019) Salt-assisted pyrolysis of covalent organic frameworks to porous heteroatom-doped carbons for supercapacitive energy storage. J. Mater. Chem. A 7 (47):26829–26837.
33. Han Y, Zhang Q, Hu N, Zhang X, Mai Y, Liu J, Hua X, Wei H (2017) Core-shell nanostructure of single-wall carbon nanotubes and covalent organic frameworks for supercapacitors. Chin. Chem. Lett. 28 (12):2269–2273.
34. Wang C, Liu F, Chen J, Yuan Z, Liu C, Zhang X, Xu M, Wei L, Chen Y (2020) A graphene-covalent organic framework hybrid for high-performance supercapacitors. Energy Storage Mater. 32:448–457.
35. Xu L, Wang F, Ge X, Liu R, Xu M, Yang J (2019) Covalent organic frameworks on reduced graphene oxide with enhanced electrochemical performance. Microporous Mesoporous Mater. 287:65–70.
36. Li C, Yang J, Pachfule P, Li S, Ye MY, Schmidt J, Thomas A (2020) Ultralight covalent organic framework/graphene aerogels with hierarchical porosity. Nat Commun 11 (1):4712.
37. An N, Guo Z, Xin J, He Y, Xie K, Sun D, Dong X, Hu Z (2021) Hierarchical porous covalent organic framework/graphene aerogel electrode for high-performance supercapacitors. J. Mater. Chem. A 9 (31):16824–16833.

38. Xu Z, Liu Y, Wu Z, Wang R, Wang Q, Li T, Zhang J, Cheng J, Yang Z, Chen S, Miao M, Zhang D (2020) Construction of extensible and flexible supercapacitors from covalent organic framework composite membrane electrode. Chem. Eng. J. 387:124071.
39. Kong X, Zhou S, Strømme M, Xu C (2021) Redox active covalent organic framework-based conductive nanofibers for flexible energy storage device. Carbon 171:248–256.
40. Yang Y, Zhang P, Hao L, Cheng P, Chen Y, Zhang Z (2021) Grotthuss proton-conductive covalent organic frameworks for efficient proton pseudocapacitors. Angew Chem Int Ed Engl 60 (40):21838–21845.
41. Liu S, Yao L, Lu Y, Hua X, Liu J, Yang Z, Wei H, Mai Y (2019) All-organic covalent organic framework/polyaniline composites as stable electrode for high-performance supercapacitors. Mater. Lett. 236:354–357.
42. Peng C, Yang H, Chen S, Wang L (2020) Supercapacitors based on three-dimensional porous carbon/covalent-organic framework/polyaniline array composites. J. Energy Storage 32:101786.
43. Dutta TK, Patra A (2021) Post-synthetic modification of covalent organic frameworks through in situ polymerization of aniline for enhanced capacitive energy storage. Chem. Asian J. 16 (2):158–164.
44. Wang W, Zhao W, Chen T, Bai Y, Xu H, Jiang M, Liu S, Huang W, Zhao Q (2021) All-in-one hollow flower-like covalent organic frameworks for flexible transparent devices. Adv. Funct. Mater. 31 (29):2010306.
45. Vaitsis C, Sourkouni G, Argirusis C (2019) Metal organic frameworks (MOFs) and ultrasound: A review. Ultrason. Sonochem. 52:106–119.
46. Vayenas M, Vaitsis C, Sourkouni G, Pandis PK, Argirusis C (2019) Investigation of alternative materials as bifunctional catalysts for electrochemical applications. Chimica Techno Acta 6 (4):120–129.
47. Peng H, Raya J, Richard F, Baaziz W, Ersen O, Ciesielski A, Samori P (2020) Synthesis of robust MOFs@COFs porous hybrid materials via an Aza-Diels-Alder reaction: Towards high-performance supercapacitor materials. Angew Chem Int Ed Engl 59 (44):19602–19609.
48. Mechili M, Vaitsis C, Argirusis N, Pandis PK, Sourkouni G, Argirusis C (2022) Research progress in transition metal oxide based bifunctional electrocatalysts for aqueous electrically rechargeable zinc-air batteries. Renew. Sust. Energ. Rev. 156:111970.
49. Han Y, Hu N, Liu S, Hou Z, Liu J, Hua X, Yang Z, Wei L, Wang L, Wei H (2017) Nanocoating covalent organic frameworks on nickel nanowires for greatly enhanced-performance supercapacitors. Nanotechnology 28 (33):33LT01.
50. Sun J, Klechikov A, Moise C, Prodana M, Enachescu M, Talyzin AV (2018) A molecular pillar approach to grow vertical covalent organic framework nanosheets on graphene: Hybrid materials for energy storage. Angew Chem Int Ed Engl 57 (4):1034–1038.
51. Zhang H, Lin L, Wu B, Hu N (2020) Vertical carbon skeleton introduced three-dimensional MnO2 nanostructured composite electrodes for high-performance asymmetric supercapacitors. J. Power Sources 476:228527.
52. Vargheese S, Muthu D, Pattappan D, Kavya KV, Kumar RTR, Haldorai Y (2020) Hierarchical flower-like MnO2@nitrogen-doped porous carbon composite for symmetric supercapacitor: Constructing a 9.0 V symmetric supercapacitor cell. Electrochim. Acta 364:137291.

27

Covalent Organic Frameworks-Based Nanomaterials as Electrode Materials for Supercapacitors

Hani Nasser Abdelhamid[1,2,3]

[1]*Advanced Multifunctional Materials Laboratory, Department of Chemistry, Faculty of Science, Assiut University, Assiut, Egypt*

[2]*Proteomics Laboratory for Clinical Research and Materials Science, Department of Chemistry, Assiut University, Assiut, Egypt*

[3]*Nanotechnology Research Centre (NTRC), The British University in Egypt, El-Shorouk City, Suez Desert Road, Cairo, Egypt*

CONTENTS

27.1 Introduction

Supercapacitors (SCs) are energy-storage devices that offer advantages such as high power density, low cost, and high cycle stability [1]. General Electric® patented the first supercapacitor in 1957 using activated charcoal. The market for supercapacitors is estimated to be $150–$200 million annually. Supercapacitors combine the properties of the traditional capacitor (i.e., high power) and batteries (i.e., high energy storage). They can be used for electric vehicles, portable electronic devices, and light rail trains. The charging mechanism of supercapacitors can be divided into three categories: (i) the electrical double-layer capacitors (EDLCs) that implies they are formed by ion adsorption/desorption of ions; however, they have low energy density; (ii) pseudocapacitors (battery-type electrode) that implies reversible faradaic reactions; and (iii) hybrid supercapacitors (HSCs) that combine both mechanisms in the same system. HSCs offer higher energy density compared to SCs. HSCs can be operated using a wide operating voltage window, consisting of EDLCs and battery-type charging mechanisms offering both advantages of EDLCs and pseudocapacitors electrode [2]. Advanced materials may overcome the energy storage limitations offering high electrochemical performance.

The first covalent organic framework (COFs), consisting of boronate ester with boroxine linkages, was reported by Cote *et al.* in 2005 [3]. The synthesis procedure for COFs-based composite has been reviewed [4]. They exhibit good properties, including high surface area, chirality, high photophysical stability, and good conductivity. COFs were applied for energy, environmental and analytical-based technologies, including adsorption, sensing, heterogeneous catalysis, organocatalysis, and water treatment [5, 6]. They have been used as electrode materials. COFs advanced SCs applications such as flexible, micro, and self-powered SCs [7]. COFs have advanced supercapacitors applications [8, 9].

This book chapter offered an introduction to COFs, COFs-based nanocomposites, and their applications as electrode-based materials for supercapacitors. It includes a brief introduction to COFs and supercapacitors. The electrochemical methods used to evaluate the electrochemical performance of COFs, such as cyclic voltammetry (CV) and galvanostatic charge/discharge cycles (GCDC), are involved. The synthesis of COFs is briefly discussed. The types of COFs such as Redox-COFs, two-dimensional COFs (2D COFs), and ionic COFs (iCOFs) are discussed. Nanocomposites of COFs containing materials such as carbon nanomaterials, metal oxides, metal-organic frameworks (MOFs), and polymers are promising materials for supercapacitors. Electroactive materials charged via EDLCs suffer from a limited specific capacitance leading to low energy density. On the other side, the electroactive materials charged via the mechanism of pseudo-capacitors exhibit higher specific capacitance and energy density compared to supercapacitors materials charged via the EDLCs mechanism. However, they exhibit low electrical conductivity leading to low power density. Inorganic electroactive materials such as transition metals, oxides, and hydroxides lack high structural instability. At the same time, conductive polymers (CPs) show weak cyclic life. Thus, COF-based nanocomposites for those nanomaterials, i.e., metal oxides/hydroxides or CPs, could address the challenges of both mechanism types offering high electrochemical performance as hybrid supercapacitors.

27.2 Electrochemical Energy Storage (EES): Supercapacitors

Electrochemical energy storage (EES) systems include batteries, fuel cells, and capacitors. ESS devices such as batteries and fuel cells store electrical energy as chemical energy such as redox reactions (reduction-oxidation) at the anode and cathode. The reactions at the anode (i.e., oxidation process) occur at lower electrode potentials than at the cathode (reduction process); thus, terms such as negative (- ve) and positive electrode (+ ve) are used, for describing anode and cathode, respectively. On the other side, a supercapacitor can store electrical energy mainly as a charge with and without chemical reaction, as explained below. In general, the EES system consists of anode, cathode, and electrolyte. The components of EES systems with short definitions are summarized in Table 27.1.

Supercapacitors are promising energy storage devices [1]. They can be applied for applications in mobile electronic equipment and electric vehicles. They exhibit several advantages, such as high power density. The supercapacitor electrodes can be charged via electric double-layer capacitors or pseudocapacitance. EDLCs mechanism depends on electrostatic charge separation at the interface between electrolyte–electrode. On the other hand, the pseudocapacitance mechanism is due to either superficial or multi-electron transfer via Faradic reactions. EDLCs offered high capacitance, while pseudocapacitance offered fast charge-discharge properties. The electroactive materials have been considered the most important component to improve the electrochemical performance compared to other components of a supercapacitor. They are responsible for the energy density,

TABLE 27.1

Summary for Components and Process for Supercapacitors

Term	Definition
Cell	An electrochemical device contains anode and cathode electrodes.
Anode (Negative Pole)	The electrode involves the release of electrons into the external circuit, i.e., oxidation reaction.
Cathode (Positive Pole)	The electrode involves gaining electrons from the external circuit, i.e., reduction reactions.
Active mass	The material generates an electrical current via a chemical reaction.
Electrolyte	The material provides high ionic conductivity between the positive and negative electrodes of a cell.
Current density	It represents the current response to the applied voltage for the specific mass of an electroactive material.
Separator	A physical barrier locates between the two electrodes. It can be such as gel electrolyte, cellulose membrane, microporous plastic film, or other porous material filled with an electrolyte.
Open-circuit voltage (OCP)	The voltage for an electrochemical cell with no external current flows.
Discharge	An operation of a cell delivers electrical energy to an external source.
Charge	Electrochemical operation in which electrical charges are stored.
Internal impedance	The resistance for the current flow of a cell.

i.e., the amount of energy that can be stored. Several electroactive materials can be used for supercapacitors, including (i) carbon nanomaterials, e.g., carbon nanotubes (CNTs), graphite, graphene, carbide-derived carbon (CDC), and activated carbons (ACs); (ii) transition metal oxides (TMOs); (iii) transition metal sulfides (TMSs); (iv) metal hydroxides; (v) polymers, e.g., CPs; and (vi) porous coordination polymers (PCPs) or MOFs. Carbon nanomaterials store charge via EDLCs. At the same time, TMOs and conductive polymers store charge via pseudocapacitance mechanism. Combining the two different type, i.e., EDLC and pseudocapacitive offer the advantages of both modes and may advance supercapacitors for high electrochemical performance.

The mechanism principle of the supercapacitor is different than for other EES such as batteries and fuel cells. The energy produced from the fuel cell is due to an electrochemical conversion of the reaction between a fuel source (e.g., hydrogen, natural gas (CH_4), or methanol and an oxidant (e.g., oxygen (O_2), air, or hydrogen peroxide). In batteries, the charge storage mechanism for an electroactive material takes place via a chemical reaction or ion diffusion. Thus, batteries suffer from low charge and discharge rates leading to low power density. In contrast, supercapacitors store the charge on the surface of an electroactive material offering high power density compared to batteries. A supercapacitor device stores electrical energy at the interface between the electrolytes and the conductor. Supercapacitors offer several advantages such as long lifetime, low cost, and high reversibility for charge and discharge processes. Thus, they are promising for energy storage devices.

Supercapacitors consist of two electrodes separated by a porous separator. The charges move and accumulate on the electrodes via the application of a voltage to the electrodes. The electrode materials with a large surface area offer high capacitance (Table 27.2). There is high demand for supercapacitors that offer high energy densities and power densities. They ensure energy for a long time. The energy density is directly proportional to the capacitance. A supercapacitor with high capacitance ensures high energy density. The speed of charging and discharging is determined by power density. The power density and energy density can be plotted using the Ragone plot (Figure 27.1). Ragone plot shows

TABLE 27.2

Summary for Equations Used for the Calculation of Supercapacitors

Items	Definition	Equation	Equation terms
Capacitance (C)	Ability to charge storage	$C = \varepsilon_0 \varepsilon_{r_A}/d$	ε_0: Dielectric constants of free space ε_r: Dielectric constants insulating material A: Electrode surface area d: Distance between the electrodes
Voltammetry specific capacitance (C_V, F/g)	Capacitance using CV curve	$C_V = \dfrac{\int I(V)dV}{v \times m \times \Delta V}$	$\int I(V)dV$: the integrated area of the CV curve I: Current density (A/g) m: Mass of electroactive material (g) v: Potential scan rate (mV/s) ΔV: Potential window (V)
Gravimetric specific capacitance (C_D, F/g)	Galvanostatic charge-discharge curve	$C_D = \dfrac{I \times \Delta t}{m \times \Delta V}$	I: Applied current (A) Δt: Discharging time (s) ΔV: Potential window (V) m: Mass of electroactive material (g)
Specific capacitance (C_{SC}, F/g)	Capacitance for full symmetric cells	$C(F) = \dfrac{\left(\int I(V)dV/2\right)}{v \times \Delta V}$ $C_{Sc}(F/g) = \dfrac{4C}{m}$	C (F): Capacitance calculated from the CV $\int IdV$: Calculated integrated area of the CV curve v: Potential scan rate (mV/s) ΔV: Potential window (V) m: Overall mass for both electrodes (g)
Energy density (E)	Energy storage capacity	$E = \dfrac{CV^2}{2}$	C: Capacitance V: Voltage
Power density (P_{max})	Determine charge or discharge speed	$P_{max} = \dfrac{V^2}{4ESR}$	ESR: Equivalent series resistance
The specific energy density (E_{sp}, Wh kg^{-1})		$E_{sp} = \dfrac{1}{2}C_{sp}(\Delta V_{max})^2 \times \dfrac{1000}{3600}$	C_{SP}: Specific capacitance from CV or GCDC (F/g) ΔV_{max}: Maximum potential range (V)
Specific power density (P_{sp}, W kg^{-1})		$P_{sp} = \dfrac{E}{t} \times 3600$	t: Discharging time (s)
Electrochemical active surface area (E_{CSA})		$E_{CSA} = C_{dl}/C_s$	C_{dl} (mF): EDL capacitance C_s (μF/cm): Specific capacitance of the electrode

that the fuel cells and SCs exhibit high energy and high power, respectively. Batteries show intermediary power and energy capabilities between fuel cells and SCs. So far, no single electrochemical device with high performance competes with a combustion engine or gas turbine [10]. Thus, further efforts should be paid to improve the current technologies.

The electrochemical performance of electroactive materials can be determined using several electrochemical measurements, including cyclic voltammetry (CV), galvanostatic charge/discharge curve, and electrochemical impedance spectroscopy (EIS). The CV curve determines the capacitive feature and the mechanism of the capacitance of electroactive materials (Figure 27.2). CV curves of an electroactive material are usually collected at different scan rates. The CV curve of a typical capacitor is rectangular for materials charged with EDLC. The typical curve for CV shows a mirror image in the zero-current line, and the potential is independent of the capacitance. These features can be observed for electrode materials such as carbon that exhibit EDLC capacitance. While, electrode

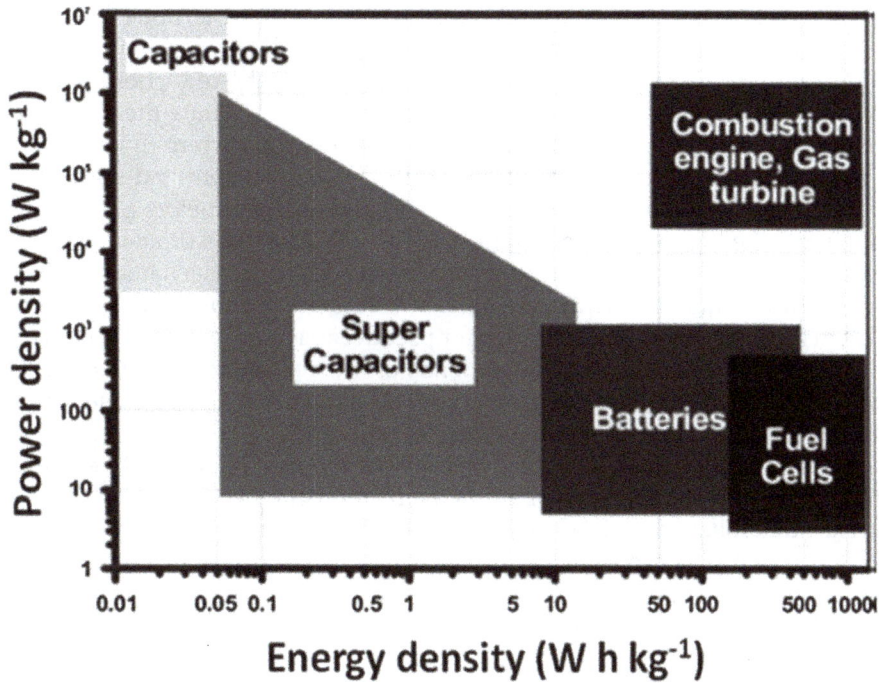

FIGURE 27.1

Ragone plot for electrochemical energy conversion systems. (Adapted with permission [10]. Copyright (2004), American Chemical Society.)

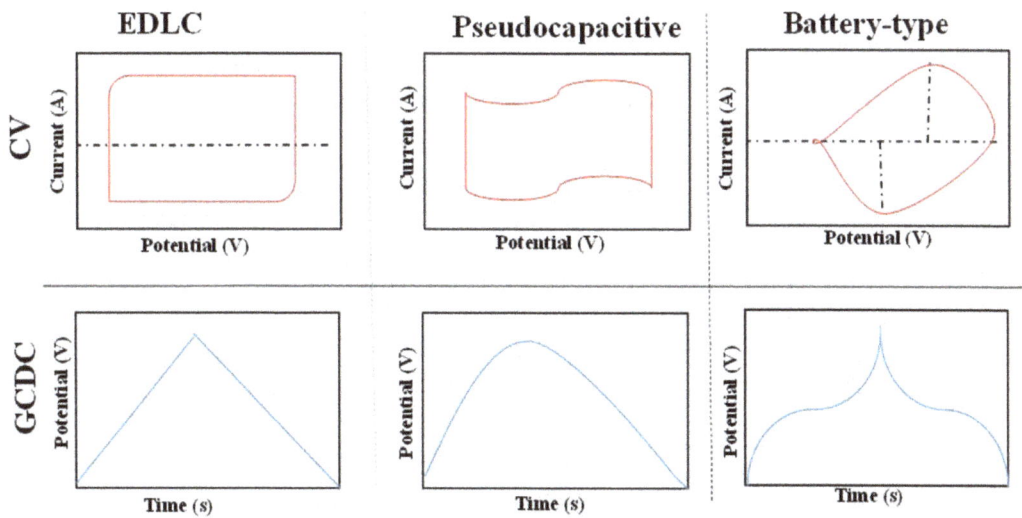

FIGURE 27.2

Shape of CV and GCDC for electroactive material charged via EDLC, pseudocapacitive, and battery-type mechanisms.

materials such as metal oxides with pseudocapacitive exhibit the same features, except that the voltammogram misses the rectangular shape.

The CV curves for battery-like materials show peaks for the reduction-oxidation process. CV curves are collected using different scan rates to evaluate the electrochemical performance of the investigated material. They represent the current (i) with the applied voltage range (voltage window). The relationship between current and scan rate can be determined using Equation $i(V) = a v^b$, where "a" and "b" parameters are constants. "a" value represents the proportional constant. The "b" value can be estimated from the slope of the plot of logI versus logV. For the materials that follow the mechanism of EDLC, the "b" value is always equal to 1. Thus, the current values are proportional linearly with the scan rate. On the other hand, the "b" value for pseudocapacitive materials is close to 1. The "b" value for the battery-type material depends on the nature of the materials. It can be 0.5 for conventional bulk battery-type materials and higher than 0.5 for those with specific structural design electrodes or nanomaterial electrodes [11].

GCD curves (GCDC) can also be used to evaluate the electrochemical performance of material (Figure 27.2). The GCDC represents the relationship between potential and time. The materials that follow the EDLC mechanism show a linear dependency of potential on time. Thus, the GCDC curve shows a triangle shape. Pseudocapacitive materials exhibit the same shape, i.e., triangle, but with a small deviation of the linear dependence of potential and time. The GCDC curve for battery-type materials shows obvious plateaus during the charge and discharge process. The capacitance of electroactive materials can be calculated from CV curves or GCDC (Table 27.2). The charge storage capability for the materials followed EDLC and pseudocapacitance is represented using the Farad (F) unit. At the same time, battery-type materials should be described in coulomb (C) or mAh.

Electrochemical impedance spectroscopy is used to evaluate the reaction kinetics of the charge transfer process [12]. The electrical conductivity of an electroactive material determines the charge-transfer process at the interface between electrode and electrolyte. EIS spectra can be evaluated using Nyquist plots (Figure 27.3). Nyquist impedance plot usually shows main three segments:

    I. In the region of high to middle-frequency: A semicircle segment is due to the existence of a charge transfer resistance (R_{ct}). The diameter of this semicircle is equal to R_{ct}. Therefore, the semicircle with a larger diameter indicates the high R_{ct} values

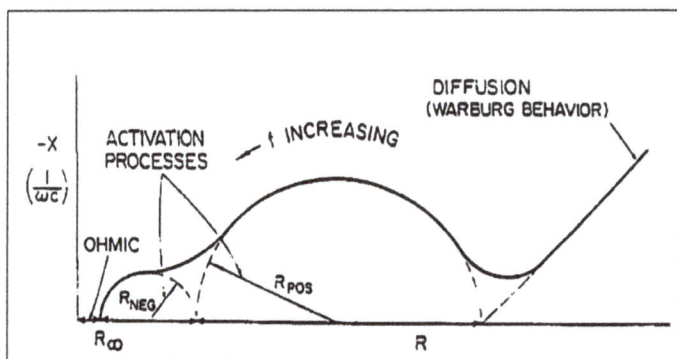

FIGURE 27.3
EIS spectrum and their processes. (Adapted with permission [10]. Copyright (2004), American Chemical Society.)

meaning the low electrical conductivity of the electrode material. Consequently, a weak electrochemical performance for the tested materials.

II. At the high frequencies, the real axis intercept of the semicircle represents the internal resistance (R_s). This resistance generates due to the electrolyte, the electrode material, and the interface between the electroactive material and a current collector (e.g., carbon cloths, nickle foam, copper foam).

III. A linear part from middle to low-frequency regions, a straight line with a slope of 45°, is known as the Warburg resistance (W). This part represents ion diffusion.

27.3 Covalent Organic Frameworks

COFs are porous organic polymers of small organic molecules containing functional groups such as carboxylic, carbonyl, cyano, amine, hydroxyls, and boric acid (Figure 27.4). The frameworks of COFs are formed via linkages such as C=C linkages, boroxine, boronate ester/boric acid ester, borosilicate, imine, squaraine, imide, azodioxide, borazine, azine, triazine, hydrazine, hydrazine, phenazine, and dioxin [13]. A general overview of the synthesis of COFs can be found in Ref. [13]. COFs materials with pore channels such as hexagonal channels can be obtained through self-condensation using building units with the symmetry of C_2 (linear, $C_2 + C_2 + C_2$), C_2 and C_3 (triangles), and two C_3 (i.e., $C_3 + C_3$). The length of the building blocks determines the pore size of the prepared COFs. The other channels, such as square channels of COFs, can be achieved using building units with the symmetry of C_4 (square) and C_2 (i.e., $C_4 + C_2$) or via the reaction between the two molecules with C_4 symmetry (i.e., $C_4 + C_4$). COFs materials with the triangular channel are synthesized via the reaction of C_6 symmetric (hexagon) building units with C_2. COFs

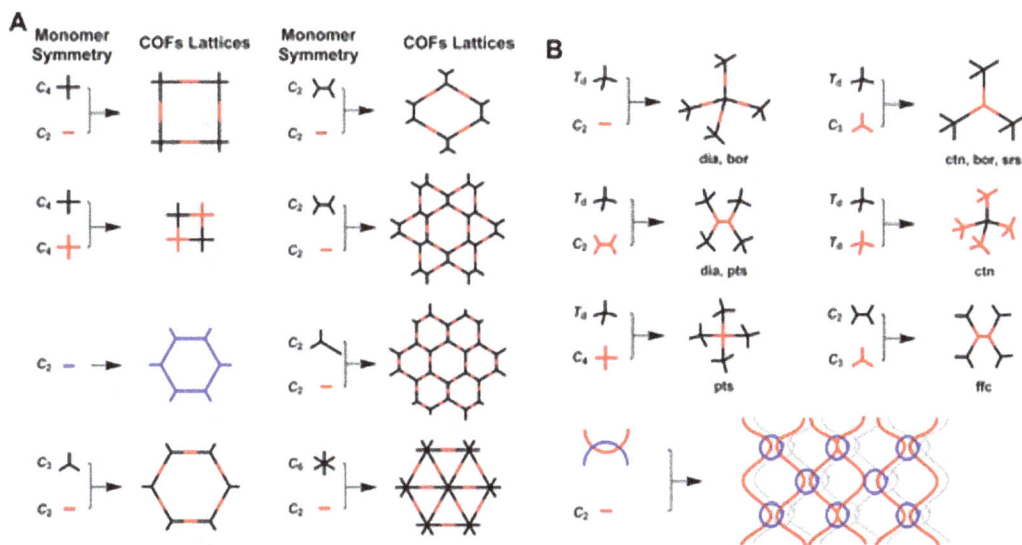

FIGURE 27.4
Basic topological diagrams used for the design of COFs with the morphology of A) 2D and B) 3D. (Adapted with permission [14]. Copyright (2020), American Chemical Society.)

with rhombic pore structures are obtained using rectangular molecules with C_2 symmetry and linear building blocks with C_2 symmetry. Most of these building units produce COFs with the morphology of 2D. For the synthesis of 3D COFs, building units with tetrahedron symmetric are usually used. 3D COFs can be synthesized via self-condensation of T_d symmetric (tetrahedron), the reaction of C_2 or C_3 monomers with T_d. 3D COFs suffer from interpenetrating making. Therefore, it is challenging to design or predict the topology and structure of 3D COFs.

The length and the symmetry of the monomers determine the size and shape of pores for COFs crystals. For instance, micropore and mesopore structures can be obtained via square channels and hexagon COFs, respectively. COFs materials with triangular channels provide COFs with micropore structures offering the highest density and lowest aperture. The shape, as well as the size of the aperture affect the material's performance. COFs materials with large pore sizes and specific surface areas enable high electrochemical performance.

COFs can be synthesized using various methods such as solvothermal methods, ionothermal synthesis, microwave-assisted solvothermal synthesis, mechanochemical synthesis, and stirring at room temperature (Table 27.3). Each method exhibits its advantages and disadvantages. No single method can fit all the requirements. Thus, the judicious selection of the synthesis method is highly important. Synthesis COFs materials at room temperature are promising for low energy demand. The synthesis methods offer several advantages compared to other methods. The solvothermal reaction involves the heating of the reactants in a closed ampoule/autoclave or open system such as reflux. The synthesis procedure takes place at a high temperature for several days. The method produces highly crystalline COFs with high porosity, good yields and holds the potential for large-scale production. However, it requires a high reaction temperature and a long reaction time. Solvothermal methods can be suitable for the synthesis of Schiff based (-C=N- bonds) COFs.

The microwave-assisted solvothermal method has been reported for the synthesis of COFs materials using a polar solvent for a better heating effect. Microwave synthesis of COF exhibits a material with a specific surface area of > 2019 cm^2/g, high crystallinity without requirement of high temperature and high pressure. The ionothermal synthesis method belongs to solvothermal methods using melting salts, including ionic liquids (ILs), as a green solvent. It was used to synthesize porous covalent triazine frameworks (CTFs) [15]. The materials were synthesized via trimerization of benzene with two cyano groups in the melting state of $ZnCl_2$ at 400 °C. The method can be used for the synthesis of triazine-based COFs with high surface area and excellent chemical stability. However, the high temperature of the synthesis procedure requires monomers with high thermal stability. It also requires intensive purification to remove the incorporated metal ions used during the synthesis procedures. The materials usually exhibit low crystallinity. Low melting points salts such as ILs were also reported for ionothermal synthesis [16].

Mechanochemical methods involve the synthesis via mechanical energy via shear, friction, impact, and extrusion. It can be mainly used for solid precursors. The mechanochemical method, such as mechanical grinding, can be assisted using a few drops of liquid such as water, organic solvent, or ionic liquids. A solvent-free mechanochemical method was used to synthesize three chemically stable COFs called TpPa-1, TpPa-2, and TpBD (Tp, Pa-1, Pa-2, and BD refer to 1, 3, 5-triformyl phloroglucinol, with *p*-phenylenediamine, 2,5-dimethyl-*p*-phenylenediamine, and benzidine, respectively) at room temperature [17]. The properties of COFs, such as surface area depend on the milling frequency or time. The mechanochemical method offers several advantages such as low cost, simplicity, requires short reaction time, and environmental friendliness operation.

TABLE 27.3

Summary of the Procedures Used for the Synthesis of COFs

Synthesis Method	Temperature	Solvent	Time	Advantages	Disadvantages
Solvothermal strategy	85–120 °C	Solvent Organic or non-aqueous solvent	2–9 days	• High crystallinity • Uniform particle size • Low defects	• Long reaction time • High temperature • Large consumption of solvents • High pressure
Ionothermal synthesis	400 °C	Molten zinc chloride	40 h	• Large specific surface area • High mobility • Large-scale production	• High reaction temperature • Precursors should have high thermal stability • High temperature (400 °C) • Low crystallinity
Microwave-assisted solvothermal synthesis	100 °C	Organic solvent	30–60 min	• Fast heating rate • Fast energy transfer • Easy control • Short time • High yield	• Low crystallinity • Require polar solvent • High energy consumption
Mechanochemicalsynthesis	RT	No solvent	10–60 min	• Fast process • Environmentally friendly • Small or no organic solvents • Simple operation	• Impure materials • High energy consumption • Low crystallinity • Low porosity • Low specific surface area
Room temperature (RT) solution synthesis	RT	Organic solvent	30 min	• Simple • Environment friendly • Potential for large-scale production	

27.4 COFs Nanocomposites

COF-based nanocomposites are attractive compared to pristine COFs [18]. The synthesis procedure for COFs composite has been reviewed in Ref. [4]. COFs-based nanocomposites exhibit the properties of COFs and nanomaterials, endowing the applications such as energy-storage devices. COFs can be combined with other organic, inorganic, and hybrid materials, including 1) Organic polymers, 2) Small molecules, 3) Metal nanoparticles, 4) Metal oxides, 5) Quantum dots, 6) Carbon nanomaterials, 7) Mxene, 8) Hybrid materials such as metal-organic polyhedral (MOP), and MOFs, and 9) COF@COF for two different materials were also reported. COF-based composites exhibit the advantages of both components, i.e., COFs and the other nanomaterials. They show higher electron conductivity and better electrochemical performance compared to pristine COFs.

27.5 COFs as Material for Supercapacitors

COFs are electroactive materials for electrode fabrication of supercapacitors [19, 20]. They exhibit several advantages such as large specific surface area, tunable pore size/structure, lightweight, and active functional groups (Figure 27.5). Materials with large surface areas are widely used as electrode materials.

A Schiff-based COF of 1,3,5-tris-(4-aminophenyl)triazine and 2,6-diformyl-4-methyl phenol was prepared through a solvothermal method via the condensation reaction. They show high capacitance with a specific capacitance of 354 F/g at a scan rate of 2 mV/s. The materials showed good cyclic stability (95% retention of its specific capacitance after

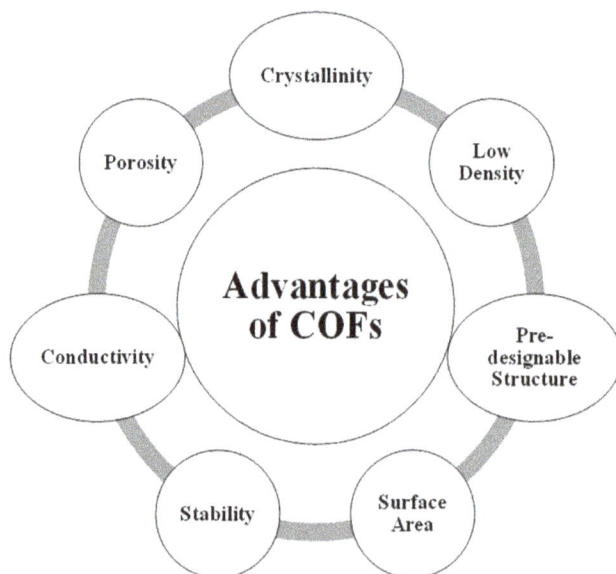

FIGURE 27.5
Advantages of COFs.

1000 cycles) [21]. The high electrochemical performance of the synthesized COFs is due to several reasons, including the high specific surface area (651 m^2/g), extended π-π conjugation, and inherent microporosity [21]. The high surface area of COF materials ensures a high specific capacitance of 546 F/g at 500 mA/g in an acidic solution [22]. It showed also capacitance of 92 mF/cm^2 at 0.5 mA/cm^2 in the solid-state device with a P$_d$ of 98 μW/cm^2 at 0.5 mA/cm^2 [22]. The materials can be recycled for 10 000 cycles. It has pyridyl-hydroxyl functionalized COFs, which enable keto-enol tautomerism and the hydrogen bonding ability of the hydroxyl units offering chemical stability with high electrochemical performance [22].

NWNU-COF-1 was synthesized via the condensation reaction of melamine and 2,4,6-trichloro-1,3,5-triazine forming –NH– bonds [23]. The synthesis procedure is simple and requires cheap reagents. NWNU-COF-1 showed a specific surface area of 301 m^2/g with a pore size of 1.41 nm. It exhibited pseudocapacitance with a specific capacitance value of 155.38 F/g at 0.25 A/g in the characteristic of a 6 M KOH in a three-electrode configuration. It retain 100% capacitance retention after 20 000 GCDC cycles [23]. NWNU-COF-4 was synthesized via the condensation between 2,4,6-trihydroxypyrimidine and trinitrophenol [24]. NWNU-COF-4 exhibited a surface area and pore size of 21.33 m^2/g and 1.351 nm, respectively. It showed pseudocapacitance of 133.44 F/g and 114.12 F/g at 0.3 A/g and 3 A/g, respectively. It exhibited retention rate of 94% after 10000 GCD cycles [24].

Conjugated COFs exhibit high electrochemical performance due to the conjugation. The conjugated polymer was synthesized via piperazine-1,4-dicarboxaldehyde and pyrrole in the presence of FeCl$_3$ and acetic acid [25]. The material was reported as an electrode for a supercapacitor with the capacitance of 571 F/g at 1.0 A/g, offering energy of 15.4 Wh/kg at a specific power of 292 W/kg. It showed high stability with 80% capacitance retention after 9000 cycles at 6.0 A/g [25]. Low dimension COFs such as 2D COFs enable high charge storage offering high electrochemical capacitance. The preparation of 2D COFs can be achieved via several methods such as self-assembly, surfactant-directed method, solid-supported growth, template-directed method, and decomposition. 2D COFs can be effective electrode-based materials for supercapacitors. 2D COF of PI-COF was synthesized via the solvothermal method [26]. It was investigated as electrode materials exhibiting a specific capacitance of 163 F/g and 96 F/g at 0.5 A/g and 40 A/g, respectively [26]. PI-COF can be used over a wide potential window of 0–2.5 V. It showed a superior energy density of 35.7 Wh/kg at a power density of 250 W/kg. It can be recycled for 30 000 charge/discharge cycles at 1 A/g with an efficiency of 84.1% of their initial capacitance (137 F/g) [26]. The solvothermal condensation of 1,3,5-triformyl phloroglucinol (TFP) with 1,5-diaminonaphthalene (NDA) offered the synthesized 2D Schiff-based COF [27]. The synthesized materials showed electrochemical capacitance of 379 F/g at 2 mV/s. It retained 75% after 8000 GCDC cycles. The material exhibited high capacitance due to π-conjugation, ion conduction, and high porosity. The high electrochemical performance of the prepared material could be attributed to the π-electronic conjugation and high ion conduction inside the framework.

COFs with redox groups enable high pseudocapacitance. COFs with redox groups such as carbonyl [28], N-containing heterocycles [29], S-containing compounds [30], or radical species [31] offer high electrochemical capacitance. COFs with high redox-active groups such as azo functional groups undergo a proton/electron transfer reaction enabling high capacitance via the pseudocapacitance mechanism [32]. iCOF are COF materials with charged and redox properties [33]. They are a subclass of COFs that contain a repeating unit bearing ionic or ionizable groups. iCOFs with various groups were reported, including anionic functional groups such as sulfonate, silicate, and imidazolate COFs, and cationic functional groups. iCOFs are similar to polyelectrolytes (biological (e.g., DNA, polypeptides) or synthetic polymers (e.g., polystyrene sulfonate)); however, the former materials are crystalline and exhibit porosity. The synthesis of iCOFs can be achieved via the polymerization of monomers bearing ionic groups or via post-synthetic procedures [34]. iCOF-based materials can be synthesized via acid-catalyzed condensation reactions between amines and aldehydes. In 2015, the first iCOFs were synthesized via the condensation of polyols with trimethyl borates to form spiroborate-linked iCOFs [35]. The spiroborate-linked anionic COFs contained dimethylammonium ions were treated with Li^+ ions to obtain Li^+ countercations. iCOFs containing Li^+ exhibit a conductivity of 3.05×10^{-5} S/cm. They exhibit high potential to be used as electrode materials for supercapacitors.

COFs are important precursors for the synthesis of carbon nanomaterials via the carbonization method. The method depends on the synthesis of COFs with a designed structure that was heated at high temperatures for the carbonization process [36]. F/N co-doped porous carbons (FNCs) were synthesized using F- and N-rich CTFs as precursors [37]. The aqueous-based SC using the synthesized materials offered a specific capacitance of 326 F/g at 1 A/g with a maximum energy density of 31.4 Wh/kg in the two-electrode system [37]. It exhibited excellent recyclability with insignificant decay after 10 000 GCDC cycles in three-electrode systems [37].

27.6 COFs-Based Nanocomposites as Supercapacitor Electrodes

Supercapacitors as electrochemical storage devices are different from other devices such as batteries. They are different from batteries in the charge/discharge mechanisms. They store the charge physically, while batteries store the charge as electrochemical reactions. Electroactive materials used for supercapacitors can provide storage charge via 1) EDLCs, 2) pseudocapacitors, and 3) hybrid capacitors (Table 27.4). Each mechanism exhibits advantages and disadvantages. Among these mechanisms, hybrid supercapacitors meet both advantages of EDLCs and pseudocapacitors.

Supercapacitors follow the EDLCs mechanism to store charge electrostatically (i.e., non-Faradaic process) process. EDLCs mechanism implies no shifting of the charge between electrodes and electrolyte. This mechanism exhibits several advantages, such as high reversibility and high cycling stability. Electrode materials following EDLCs mechanism should exhibit large specific surface area, good mechanical, high chemical stability, and high electrical conductivity to ensure high electrochemical capacitance. Several materials can be used for supercapacitors based on the EDLCs mechanism, including materials such as carbon-based nanomaterials, e.g., ACs, CNTs, G, CDC, etc.

Pseudocapacitors (battery-type capacitors) store charge through a Faradaic process such as electrosorption, oxidation-reduction (Redox) reactions, and intercalation mechanism. They imply only charge transfer between electrolyte and electrode materials. The process is reversible, and the material can be recyclable. The Faradaic processes ensure higher capacitance and energy density than non-Faradaic supercapacitors based on EDLCs. The electrode materials should exhibit chemical affinity toward the ions adsorbed on the surface of the electrode. The charge storage via pseudocapacitors increases linearly with the applied voltage. Several materials such as TMOs, TMSs, and CPs can store the charge via redox behavior.

Hybrid supercapacitors combine both mechanisms of EDLCs and pseudocapacitors. In such capacitors, electrode materials can follow both mechanisms, i.e., EDLCs and pseudocapacitors, to increase the capacitance or energy density and decrease the anode potential.

TABLE 27.4

Comparison of EDLC, Pseudocapacitors, and Hybrid Capacitors

	EDLCs	Pseudocapacitors	Hybrid Capacitor
Principles	Charge storage via electrostatic formation of double-layer, i.e., non-Faradaic	Charge storage via Faradic process such as redox reaction, electrosorption	Combine both non-Faradaic and Faradaic process
Capacitance	Low	High	High
Energy density	Low	High	High
Power density	Low	High	High
Factors affecting performance	• Specific surface area • Chemical stability • Electrical conductivity • Mechanical performance	• The oxidation state of electroactive materials • Charge of both electrolyte and electrode materials • Electrolyte composition Applied voltage	• Specific surface area • Chemical stability • Electrical conductivity • The oxidation state of electrode materials • Mechanical properties
Reversibility	Highly reversibility	Low compared to EDLCs	High
Cycling	High cycling stability	Low compared to EDLCs	High
Examples	Carbon nanomaterials such as carbide-derived carbon, ACs, CNTs, G	Metal oxides, conductive polymers	Metal oxides/polymers, Polymer/carbon, Li/Carbon

Hybrid supercapacitors offer several advantages, such as high capacitance, high energy density, and high power density. Hybrid materials such as carbon materials/CPs, carbon nanomaterials/TMOs, CPs/TMOs were reported for HSCs. Supercapacitors can be assembled via two classes, i.e., asymmetric (ASCs) and symmetric (SCs). An asymmetric supercapacitor involves the use of two different electrodes. At the same time, symmetric supercapacitors use similar electrodes. Asymmetric supercapacitors exhibit high capacitance over symmetrical SCs because of the combination of pseudocapacitance and EDLC mechanisms. ASCs of two different materials such as activated carbon (negative electrode (-ve), anode, oxidation) and metal oxide (positive electrode (+ve), cathode, reduction electrode).

27.7 Applications of COFs Nanocomposites for Supercapacitors

COFs are porous materials with the potential use as electroactive materials for supercapacitors (Figure 27.6). They can be used as pristine materials or after conjugation with other materials such as nanoparticles. Pristine COFs as electrode materials offer several advantages such as high surface area, high porosity, and good chemical and physical stability. They are electroactive materials with good capacitance. However, the electrochemical capacitance is still below the high demand due to the low conductivity and charge transfer. On the other side, COFs-based nanocomposites containing high conductive materials such as carbon nanomaterials offer high capacitance. Nanocomposites of COFs exhibit the advantages of both constituents, i.e., COFs and the other nanomaterials leading to higher electrochemical performance. COFs can be conjugated with other nanomaterials such as polymers, carbon nanomaterials, metallic nanoparticles, metal oxides, quantum dots, MOFs, and COFs (Figure 27.7). The synthesis of COFs composites can be achieved via in-situ and ex-situ procedures.

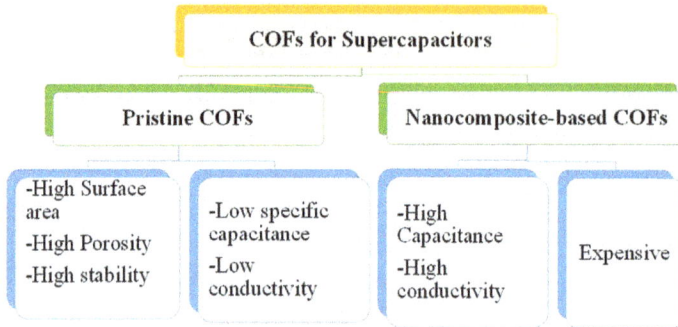

FIGURE 27.6
COFs for supercapacitors, advantages, and disadvantages.

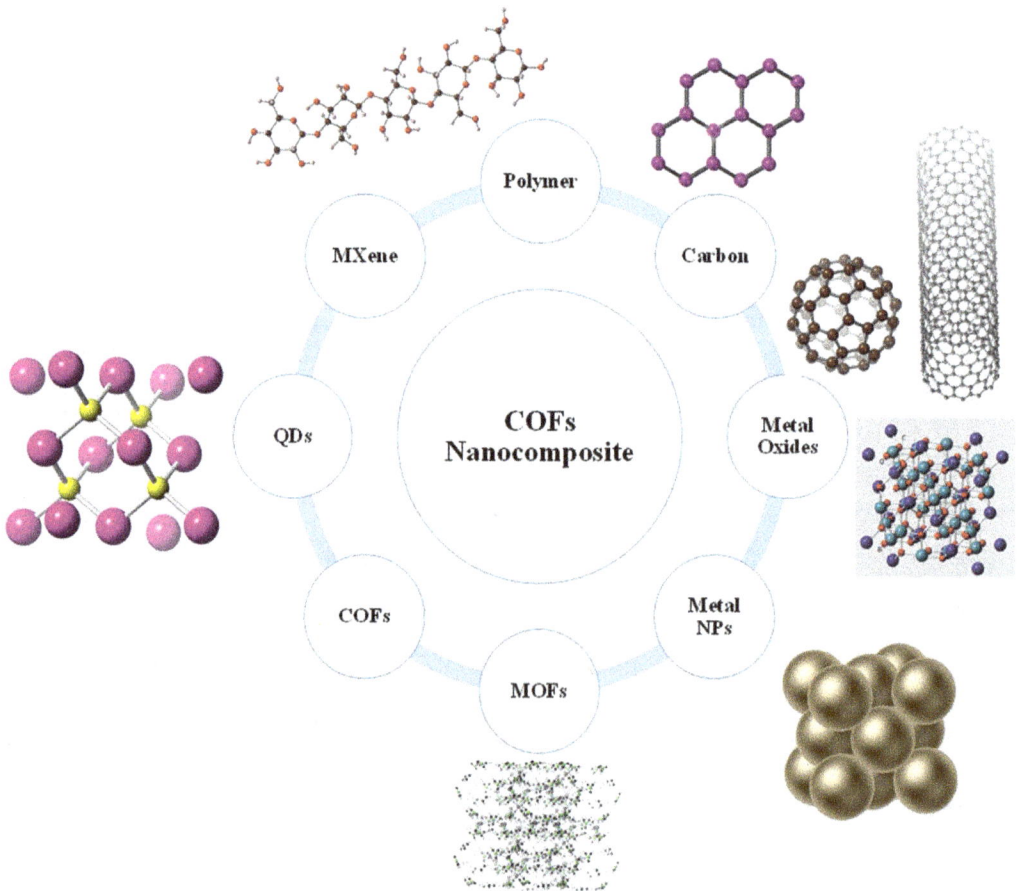

FIGURE 27.7
COFs-based nanocomposites.

COFs are promising sources for the synthesis of porous carbons with nitrogen heteroatom [38]. N-doped porous carbon was derived using triazine-based COF as a precursor [36]. N-doped porous carbon has a high surface area of 801 m²/g and a specific capacitance of 505 F/g at 0.5 A/g. It exhibited excellent cyclic stability with 89% capacitance retention even after 10000 cycles at 3.0 A/g [36]. COFs are porous materials that enable the synthesis of porous carbon nanomaterials. The synthesized carbon nanomaterials can be doped with a heteroatom such as nitrogen, oxygen, or fluoride. COFs such as iCOFs can be easily conjugated with other materials such as polymers bearing ionic functional groups. The ionic (anionic or cationic) functional groups of COFs have strongly interacted with the opposite charge of ionic species via interactions such as electrostatic forces. The strong interactions enable a composite with robust mechanical and electrochemical properties.

The functional groups of COFs can be modified with metal ions to improve the conductivity. Conductive COF of Meso-tetra(p-hydroxyphenyl) porphyrin (THPP), and phthalic anhydride (PA) was modified with manganese (Mn) to synthesize THPP-PA-Mn [39]. THPP-PA-Mn showed a capacitance of 90.9 F/g at 2.5 A/g [39]. It exhibited an energy density of 12.6 Wh/kg and 10.8 Wh kg⁻¹ at 1.0 kW/kg and 19.3 kW/kg, respectively [39]. Nanocomposites of COFs and carbon-based nanomaterials offer high electrochemical performance. Carbon-based materials such as G, AC, and multi-wall carbon nanotube (MWCNTs) exhibit several advantages such as high electrical conductivity, large surface area, high chemical and thermal stability, wide functionality, and low cost. COF/MWCNTs composite was synthesized via the solvothermal method. 2D $COF_{TTA-DHTA}$ (consisting of 4,4′,4″-(1,3,5-triazine-2,4,6-triyl)trianiline (TTA) and 2,5-dihydroxy terepthaldehyde (DHTA)) was in-situ grown on amino-functionalized MWCNTs [40]. NH_2-f-MWCNT@$COF_{TTA-DHTA}$ showed capacitance of 127.5 F/g and 98.7 F/g at a current rate of 0.4 A/g and 2 A/g, respectively. It exhibited higher capacitance compared to $COF_{TTA-DHTA}$ (10 F/g at 0.1 A/g) and NH_2-f-MWCNT (65 F/g at 0.5 A/g). The amino group of MWCNT offered a 30% increment of capacitance compared to that of MWCNT@$COF_{TTA-DHTA}$ (92.4 F/g at 0.4 A/g) [40]. One-pot synthesis of a composite consisting of imine-based COF (denoted as COF-300) and oxidized MWCNTs (ox-MWCNTs) was reported [41]. ox-MWCNTs@COF improved coating and mechanical properties of COF materials [41]. COF was impregnated on a microporous CNT film (CNTF) for the preparation of a composite membrane (CHCM) [42]. CHCM-based electrodes showed capacitance of 249 F/g and 425 F/g using phosphoric acid (H_3PO_4) and sulfuric acid (H_2SO_4, 2 M) as electrolytes, respectively. A few-layered 2D redox-active COFs (DAAQ-TFP COF) was conjugated on the surface of carboxylated carbon nanotubes (c-CNTs) to obtain core-shell c-CNT@COF nanofibers [43]. The prepared c-CNT@COF nanofibers offered a flexible energy-storage device [43].

COFs/G composite was synthesized via a one-step method via the condensation between the formyl group and the amine group of diamine or NH_2-modified reduced GO [44]. The polymerization via condensation is simple. The synthesized material, i.e., COFs/NH_2-rGO, showed high electrochemical performance as a supercapacitor offering a specific capacitance of 533 F/g at a current density of 0.2 A/g using Na_2SO_4 (1.0 M) as an electrolyte with excellent stability [44]. It offered higher capacitance than COFs and NH_2-rGO that showed capacitance of 226 and 190 F/g, respectively. A phosphine-based porous organic polymers/rGO aerogel composites were prepared via the growth of polymer on rGO aerogel [45]. The material was used as a cathode for an ASC device using AC materials as an anode. The ASC device showed energy density and power density of 33.3 Wh/kg and 3996 W/kg, respectively. It exhibited high recycling stability with a retention value of 88% after 12000 cycles [45]. An anthraquinone-based COFs/G aerogel (DAAQ-COFs/GA) electrode was fabricated via the electrostatic self-assembly between negatively charged GO nanosheets and modified positively charged DAAQ-COFs nanoflower [46]. The synthesized electrode exhibited a high specific capacitance of 378 F/g at 1 A/g and fast kinetics [46]. Furthermore, the binder-free DAAQ-COFs/GA and pure graphene aerogel (GA) electrodes were assembled into an asymmetric supercapacitor (ASC), offering an energy density of 30.5 Wh/kg at a power density of 700 W/kg [46].

Metal oxides-based COFs materials were reported as electroactive materials for supercapacitors. They exhibit large capacities and high energy densities. The capacitance of metal oxides-based materials is due to the reversible Faradic redox reaction with the electrolytes. Metal oxide such as Fe_3O_4 can be used as a template for the synthesis of hollow COFs materials. A template-directed synthesis method was reported for the formation of hollow COF materials via the etching of the Fe_3O_4 template using HCl [47]. The presence of metal oxide in COFs nanocomposite offers the potential to use as electroactive materials for electrode fabrication for supercapacitors.

27.8 Conclusions

COFs are promising materials for supercapacitors. They are porous, have a high surface area, are lightweight, and are pre-designable materials. They exhibit several advantages such as high electronic conjugation, good conductivity, tunable functional groups that enable charging and discharge via EDCL, and pseudocapacitance mechanisms. COFs with high microporosity, π-conjugated skeleton, 2D morphology, and redox-active enable high energy-storage capacitance. Further efforts are highly required for further investigation of key parameters affecting the material's performance as well as commercialization. The application of COF nanocomposites is promising for electrode fabrication. Although the method of conjugating nanomaterials with COFs materials improves the electrochemical performance as electrode materials, the specific surface areas of the composite are sharply decreased. Thus, the rationale design of the composite should be adjusted to enable a high surface area with superior electrochemical performance. Several points should be further investigated, including the development of a cost-effective, efficient, large-scale synthesis method, understanding the structure-performance relationship, developing synthetic methods to afford desirable composite materials using computational simulations, exploring other nanomaterials to be integrated with COFs for high electrochemical performance, and addressing key factors that govern the material's performance.

References

1. Wang F, Wu X, Yuan X, Liu Z, Zhang Y, Fu L, Zhu Y, Zhou Q, Wu Y, Huang W (2017) Latest advances in supercapacitors: from new electrode materials to novel device designs. Chem Soc Rev 46:6816–6854
2. Jayakumar A, Antony RP, Wang R, Lee J-M (2017) MOF-derived hollow cage Ni x Co 3– x O 4 and their synergy with graphene for outstanding supercapacitors. Small 13:1603102
3. Côté AP, Benin AI, Ockwig NW, O'Keeffe M, Matzger AJ, Yaghi OM (2005) Chemistry: Porous, crystalline, covalent organic frameworks. Science (80-) 310:1166–1170

4. Kumar S, Kulkarni VV, Jangir R (2021) Covalent-organic framework composites: A review report on synthesis methods. ChemistrySelect 6:11201–11223
5. Tan W, Wu X, Liu W, Ye F, Zhao S (2021) Synchronous construction of hierarchical porosity and thiol functionalization in COFs for selective extraction of cationic dyes in water samples. ACS Appl Mater Interfaces 13:4352–4363
6. Abdellah AR, Abdelhamid HN, El-Adasy A-BAAM, Atalla AA, Aly KI (2020) One-pot synthesis of hierarchical porous covalent organic frameworks and two-dimensional nanomaterials for selective removal of anionic dyes. J Environ Chem Eng 8:104054
7. Pei C, Choi MS, Yu X, Xue H, Xia BY, Park HS (2021) Recent progress in emerging metal and covalent organic frameworks for electrochemical and functional capacitors. J Mater Chem A 9:8832–8869
8. Singh V, Byon HR (2021) Advances in electrochemical energy storage with covalent organic frameworks. Mater Adv 2:3188–3212
9. Zhang B, Wang W, Liang L, Xu Z, Li X, Qiao S (2021) Prevailing conjugated porous polymers for electrochemical energy storage and conversion: Lithium-ion batteries, supercapacitors and water-splitting. Coord Chem Rev 436:213782
10. Winter M, Brodd RJ (2004) What are batteries, fuel cells, and supercapacitors? Chem Rev 104:4245–4269
11. Jiang Y, Liu J (2019) Definitions of pseudocapacitive materials : A brief review. Energy Environ Mater 2:30–37
12. Wang S, Zhang J, Gharbi O, Vivier V, Gao M, Orazem ME (2021) Electrochemical impedance spectroscopy. Nat Rev Methods Prim 1:41
13. Lin J, Zhong Y, Tang L, Wang L, Yang M, Xia H (2021) Covalent organic frameworks: From materials design to electrochemical energy storage applications. Nano Sel nano.202100153
14. Geng K, He T, Liu R, Dalapati S, Tan KT, Li Z, Tao S, Gong Y, Jiang Q, Jiang D (2020) Covalent organic frameworks: Design, synthesis, and functions. Chem Rev 120:8814–8933
15. Kuhn P, Antonietti M, Thomas A (2008) Porous, covalent triazine-based frameworks prepared by ionothermal synthesis. Angew Chemie Int Ed 47:3450–3453
16. Guan X, Ma Y, Li H, Yusran Y, Xue M, Fang Q, Yan Y, Valtchev V, Qiu S (2018) Fast, ambient temperature and pressure ionothermal synthesis of three-dimensional covalent organic frameworks. J Am Chem Soc 140:4494–4498
17. Biswal BP, Chandra S, Kandambeth S, Lukose B, Heine T, Banerjee R (2013) Mechanochemical synthesis of chemically stable isoreticular covalent organic frameworks. J Am Chem Soc 135:5328–5331
18. Liu Y, Zhou W, Teo WL, Wang K, Zhang L, Zeng Y, Zhao Y (2020) Covalent-organic-framework-based composite materials. Chem 6:3172–3202
19. Zhang K, Kirlikovali KO, Varma RS, Jin Z, Jang HW, Farha OK, Shokouhimehr M (2020) Covalent organic frameworks: Emerging organic solid materials for energy and electrochemical applications. ACS Appl Mater Interfaces 12:27821–27852
20. Kumar R, Naz Ansari S, Deka R, Kumar P, Saraf M, Mobin SM (2021) Progress and perspectives on covalent-organic frameworks (COFs) and composites for various energy applications. Chem Eur J 27:13669–13698
21. Bhanja P, Bhunia K, Das SK, Pradhan D, Kimura R, Hijikata Y, Irle S, Bhaumik A (2017) A new triazine-based covalent organic framework for high-performance capacitive energy storage. ChemSusChem 10:921–929
22. Haldar S, Kushwaha R, Maity R, Vaidhyanathan R (2019) Pyridine-rich covalent organic frameworks as high-performance solid-state supercapacitors. ACS Mater Lett 1:490–497
23. Xue R, Guo H, Yue L, Wang T, Wang M, Li Q, Liu H, Yang W (2018) Preparation and energy storage application of a long-life and high rate performance pseudocapacitive COF material linked with –NH– bonds. New J Chem 42:13726–13731
24. Guo H, Wang M, Xue R, Yao J, Wang X, Zhang L, Liu J, Yang W (2019) A new COF linked by an ether linkage (–O–): synthesis, characterization and application in supercapacitance. RSC Adv 9:13458–13464

25. Zhang Y, Cheng L, Zhang L, Yang D, Du C, Wan L, Chen J, Xie M (2021) Effect of conjugation level on the performance of porphyrin polymer based supercapacitors. J Energy Storage 34:102018

26. Iqbal R, Badshah A, Ma Y-J, Zhi L-J (2020) An electrochemically stable 2D covalent organic framework for high-performance organic supercapacitors. Chinese J Polym Sci 38:558–564

27. Das SK, Bhunia K, Mallick A, Pradhan A, Pradhan D, Bhaumik A (2018) A new electrochemically responsive 2D Π-conjugated covalent organic framework as a high performance supercapacitor. Microporous Mesoporous Mater 266:109–116

28. Han C, Li H, Shi R, Zhang T, Tong J, Li J, Li B (2019) Organic quinones towards advanced electrochemical energy storage: recent advances and challenges. J Mater Chem A 7:23378–23415

29. Song Z, Zhou H (2013) Towards sustainable and versatile energy storage devices: an overview of organic electrode materials. Energy Environ Sci 6:2280

30. Oyama N, Tatsuma T, Sato T, Sotomura T (1995) Dimercaptan–polyaniline composite electrodes for lithium batteries with high energy density. Nature 373:598–600

31. Jähnert T, Häupler B, Janoschka T, Hager MD, Schubert US (2014) Polymers based on stable phenoxyl radicals for the use in organic radical batteries. Macromol Rapid Commun 35:882–887

32. Yang Y, Zhang P, Hao L, Cheng P, Chen Y, Zhang Z (2021) Grotthuss proton-conductive covalent organic frameworks for efficient proton pseudocapacitors. Angew Chemie 133:22009–22016

33. Liang X, Tian Y, Yuan Y, Kim Y (2021) Ionic covalent organic frameworks for energy devices. Adv Mater 33:2105647

34. Ying Y, Tong M, Ning S, Ravi SK, Peh SB, Tan SC, Pennycook SJ, Zhao D (2020) Ultrathin two-dimensional membranes assembled by ionic covalent organic nanosheets with reduced apertures for gas separation. J Am Chem Soc 142:4472–4480

35. Du Y, Yang H, Whiteley JM, Wan S, Jin Y, Lee S, Zhang W (2016) Ionic covalent organic frameworks with spiroborate linkage. Angew Chemie Int Ed 55:1737–1741

36. Vargheese S, Kumar RTR, Haldorai Y (2019) Synthesis of triazine-based porous organic polymer: A new material for double layer capacitor. Mater Lett 249:53–56

37. Gao Y, Cui P, Liu J, Sun W, Chen S, Chou S, Lv L-P, Wang Y (2021) Fluorine/nitrogen co-doped porous carbons derived from covalent triazine frameworks for high-performance supercapacitors. ACS Appl Energy Mater 4:4519–4529

38. Kim G, Yang J, Nakashima N, Shiraki T (2017) Highly microporous nitrogen-doped carbon synthesized from azine-linked covalent organic framework and its supercapacitor function. Chem Eur J 23:17504–17510

39. Cheng Z, Qiu Y, Tan G, Chang X, Luo Q, Cui L (2019) Synthesis of a Novel Mn(II)-porphyrins polycondensation polymer and its application as pseudo-capacitor electrode material. J Organomet Chem 900:120940

40. Sun B, Liu J, Cao A, Song W, Wang D (2017) Interfacial synthesis of ordered and stable covalent organic frameworks on amino-functionalized carbon nanotubes with enhanced electrochemical performance. Chem Commun 53:6303–6306

41. Moya A, Hernando-Pérez M, Pérez-Illana M, San Martín C, Gómez-Herrero J, Alemán J, Mas-Ballesté R, de Pablo PJ (2020) Multifunctional carbon nanotubes covalently coated with imine-based covalent organic frameworks: exploring structure–property relationships through nanomechanics. Nanoscale 12:1128–1137

42. Xu Z, Liu Y, Wu Z, Wang R, Wang Q, Li T, Zhang J, Cheng J, Yang Z, Chen S, Miao M, Zhang D (2020) Construction of extensible and flexible supercapacitors from covalent organic framework composite membrane electrode. Chem Eng J 387:124071

43. Kong X, Zhou S, Strømme M, Xu C (2021) Redox active covalent organic framework-based conductive nanofibers for flexible energy storage device. Carbon N Y 171:248–256

44. Wang P, Wu Q, Han L, Wang S, Fang S, Zhang Z, Sun S (2015) Synthesis of conjugated covalent organic frameworks/graphene composite for supercapacitor electrodes. RSC Adv 5:27290–27294

45. Sajjad M, Tao R, Kang K, Luo S, Qiu L (2021) Phosphine-based porous organic polymer/rGO aerogel composites for high-performance asymmetric supercapacitor. ACS Appl Energy Mater 4:828–838
46. An N, Guo Z, Xin J, He Y, Xie K, Sun D, Dong X, Hu Z (2021) Hierarchical porous covalent organic framework/graphene aerogel electrode for high-performance supercapacitors. J Mater Chem A 9:16824–16833
47. Yang H, Cheng X, Cheng X, Pan F, Wu H, Liu G, Song Y, Cao X, Jiang Z (2018) Highly water-selective membranes based on hollow covalent organic frameworks with fast transport pathways. J Memb Sci 565:331–341

Index

For Product Safety Concerns and Information please contact our EU
representative GPSR@taylorandfrancis.com
Taylor & Francis Verlag GmbH, Kaufingerstraße 24, 80331 München, Germany

www.ingramcontent.com/pod-product-compliance
Lightning Source LLC
Chambersburg PA
CBHW080120220326
41598CB00032B/4902